T0301972

PROBLEMS IN
LINEAR ALGEBRA
AND
MATRIX THEORY

PROBLEMS IN
LINEAR ALGEBRA
AND
MATRIX THEORY

Fuzhen Zhang
Nova Southeastern University, USA

NEW JERSEY · LONDON · SINGAPORE · BEIJING · SHANGHAI · HONG KONG · TAIPEI · CHENNAI · TOKYO

Published by

World Scientific Publishing Co. Pte. Ltd.

5 Toh Tuck Link, Singapore 596224

USA office: 27 Warren Street, Suite 401-402, Hackensack, NJ 07601

UK office: 57 Shelton Street, Covent Garden, London WC2H 9HE

Library of Congress Cataloging-in-Publication Data

Names: Zhang, Fuzhen, 1961– author.

Title: Problems in linear algebra and matrix theory / Fuzhen Zhang,
    Nova Southeastern University, USA.

Other titles: Linear algebra

Description: Third edition. | New Jersey : World Scientific Publishing Co., [2022] |
    Revised edition of: Linear algebra : challenging problems for students. 2nd ed.
    Johns Hopkins University Press, c2009. | Includes bibliographical references and index.

Identifiers: LCCN 2021049419 | ISBN 9789811239793 (hardcover) |
    ISBN 9789811239083 (paperback) | ISBN 9789811239090 (ebook for institutions) |
    ISBN 9789811239106 (ebook for individuals)

Subjects: LCSH: Algebras, Linear--Problems, exercises, etc.

Classification: LCC QA184.5 .Z48 2022 | DDC 512/.5--dc23/eng/20211007

LC record available at https://lccn.loc.gov/2021049419

British Library Cataloguing-in-Publication Data

A catalogue record for this book is available from the British Library.

For any available supplementary material, please visit
https://www.worldscientific.com/worldscibooks/10.1142/12338#t=suppl

Printed in Singapore

To my family

Contents

Preface to the Third Edition

This is the revised and expanded edition of the problem book *Linear Algebra: Challenging Problems for Students*, now entitled *Problems in Linear Algebra and Matrix Theory*. The first two editions of the book were published by the Johns Hopkins University Press (in 1996 and 2009, respectively). This new edition contains about fifty-five examples and many new problems, based on my lecture notes of *Advanced Linear Algebra* classes at Nova Southeastern University (NSU-Florida) and short lectures *Matrix Gems* at Shanghai University and Beijing Normal University.

The book is intended for upper division undergraduate and beginning graduate students. The students are assumed to have completed a calculus course and a linear algebra course, both at introductory level.

The Advanced Linear Algebra course I taught at NSU was a 3-hour credit course for 16 weeks; the class met twice a week for 14 weeks, plus the midterm and the final. So the total meeting time was 48 hours. My goal of the course was to present basic, important, and elegant results in linear algebra and matrix theory to students. The class was run in a seminar style; the materials were not taught in the way of Definition-Theorem-Proof. In each class, I went over a few examples with students engaged in discussions, then randomly assigned some problems as homework.

"Math is fun!", as a student exclaimed in a class. Yes, indeed, I had a lot of fun presenting the materials to my students; I also had a lot of fun collecting, working on, and compiling these problems.

I am thankful to many colleagues and friends who helped with the revision, including Elaine Gan, Yongjian Hu, Zejun Huang, Rajat Khandelwal, Pan Shun Lau, Yongtao Li, Zhongshan Li, Xin Liang, Ricky Ng, Aysha Nuhuman, Edward Poon, Meiyue Shao, Wasin So, Tin-Yau Tam, Qingwen Wang, Shengqian Wang, Xiangxiang Wang, and Xiaodong Zhang. In particular, I want to thank Roger Horn for lifelong advising as my most trustable and authoritative source, to Yue Liu for carefully reading the draft of this edition with many corrections and suggestions, and to Fupeng Sun who worked out all the problems in the second edition with corrections.

Fuzhen Zhang

Preface to the Second Edition

This is the second, revised, and expanded edition of the linear algebra problem book *Linear Algebra: Challenging Problems for Students*. The first edition of the book, containing 200 problems, was published in 1996. In addition to about 200 new problems in this edition, each chapter starts with definitions and facts that lay out the foundations and groundwork for the chapter, followed by carefully selected problems. Some of the new problems are straightforward; some are pretty hard. The main theorems frequently needed for solving these problems are listed on page xiii.

My goal has remained the same as in the first edition: to provide a book of interesting and challenging problems on linear algebra and matrix theory for upper-division undergraduates and graduate students in mathematics, statistics, engineering, and related fields. Through working and practicing on the problems in the book, students can learn and master the basic concepts, skills, and techniques in linear algebra and matrix theory.

During the past ten years or so, I served as a collaborating editor for *American Mathematical Monthly* problem section, associate editor for the *International Linear Algebra Society Bulletin* IMAGE Problem Corner, and editor for several other mathematical journals, from which some problems in the new edition have originated. I have also benefited from the math conferences I regularly attend; they are the International Linear Algebra Society (ILAS) Conferences, Workshops on Numerical Ranges and Numerical Radii, R. C. Thompson (formerly Southern California) Matrix Meetings, and the International Workshops on Matrix Analysis and Applications. For example, I learned Problem 4.25 from M.-D. Choi at the ILAS Shanghai Meeting in 2007; Problem 4.107 was a recent submission to IMAGE by G. Goodson and R. Horn; some problems were collected during tea breaks.

I am indebted to many colleagues and friends who helped with the revision; in particular, I thank Jane Day for her numerous comments and suggestions on this version. I also thank Nova Southeastern University (NSU) and the Farquhar College of Arts and Sciences (FCAS) of the university for their support through various funds, including the President's Faculty Research and Development Grants (Awards), FCAS Minigrants, and FCAS Faculty Development Funds.

Readers are welcome to communicate with me at zhang@nova.edu.

Preface to the First Edition

This book is written as a supplement for undergraduate and first-year graduate students majoring in mathematics, statistics, or related areas. I hope that the book will be helpful for instructors teaching linear algebra and matrix theory as well.

Working problems is a crucial part of learning mathematics. The purpose of this book is to provide a suitable number of problems of appropriate difficulty. The readers should find the collection of two hundred problems in this book diverse, interesting, and challenging.

This book is based on my ten years of teaching and doing research in linear algebra. Although the problems have not been systematically arranged, I have tried to follow the order and level of some commonly used linear algebra textbooks. The theorems that are well known and found in most books are excluded and are supposed to be used freely. The problems vary in difficulty; some of them may even baffle professional experts. Only a few problems need the Jordan canonical forms in their solutions. If you have a little elementary linear algebra background, or are taking a linear algebra course, you may just choose a problem from the book and try to solve it by any method. It is expected that readers will refer to the solutions as little as possible.

I wish to dedicate the book to the memory of my Ph.D. advisor, R. C. Thompson, a great mathematician and a founder of the International Linear Algebra Society (ILAS). I am grateful to C. A. Akemann, R. A. Horn, G. P. H. Styan, B.-Y. Wang, and X.-R. Yin for guiding me toward the road of a mathematician. I would also like to thank my colleagues J. Bartolomeo, M. He, and D. Simon for their encouragement. Finally, I want to thank Dr. R. M. Harington, of the Johns Hopkins University Press, for his enthusiastic cooperation.

Frequently Used Notation

$\mathbb{R}$	real number field
$\mathbb{C}$	complex number field
$\mathbb{F}$	scalar field ($\mathbb{F} = \mathbb{C}$ or $\mathbb{R}$)
$\mathbb{R}^n$	vectors of n real components
$\mathbb{C}^n$	vectors of n complex components
$\mathbb{P}_n[x]$	polynomials of real coefficients with degrees less than n
$M_{m \times n}(\mathbb{F})$	$m \times n$ matrices with entries from $\mathbb{F}$
$M_n(\mathbb{F})$	$n \times n$ matrices with entries from $\mathbb{F}$
$\dim V$	dimension of vector space V
I_n, I	identity matrix of size $n \times n$ or of inferred size from context
$A = (a_{ij})$	matrix A with (i, j)-entry a_{ij}
$r(A)$	rank of matrix A
$\operatorname{tr} A$	trace of matrix A
$\det A$	determinant of matrix A
$|A|$	determinant of matrix A (mostly for block matrices)
A^{-1}	inverse of matrix A
v^t, A^t	transpose of vector v, transpose of matrix A
$\bar{A}$	conjugate of matrix A
A^*	conjugate transpose of matrix A, i.e., $A^* = \bar{A}^t$
$\operatorname{adj}(A)$	cofactor matrix or adjugate of matrix A
$\operatorname{Im} A$	image or column space of matrix A, i.e., $\operatorname{Im} A = \{Ax\}$
$\operatorname{Ker} A$	kernel or null space of matrix A, i.e., $\operatorname{Ker} A = \{x \mid Ax = 0\}$
$\operatorname{Im} \mathcal{A}$	image or range of transformation $\mathcal{A}$, i.e., $\operatorname{Im} \mathcal{A} = \{\mathcal{A}(x)\}$
$\operatorname{Ker} \mathcal{A}$	kernel of transformation $\mathcal{A}$, i.e., $\operatorname{Ker} \mathcal{A} = \{x \mid \mathcal{A}(x) = 0\}$
$\operatorname{Span} S$	vector space spanned by the elements in set S
$A \geq 0$	A is positive semidefinite (Hermitian)
$A \geq B$	$A - B$ is positive semidefinite for Hermitian A, B
$\operatorname{diag}(d_1, \ldots, d_n)$	diagonal matrix with $d_1, \ldots, d_n$ on the main diagonal
$A \circ B$	Hadamard product of A and B, i.e., $A \circ B = (a_{ij}b_{ij})$
$A \otimes B$	Kronecker product of A and B, i.e., $A \otimes B = (a_{ij}B)$
$\langle u, v \rangle$	inner product of vectors u and v
$\|v\|$	norm or length of vector v
$\|A\|_F$	Frobenius norm of matrix A
$\|A\|_{\mathrm{sp}}, \|A\|_{\mathrm{op}}, \|A\|_2$	spectral norm, operator norm, or 2-norm of matrix A
$\lambda_i(A)$	eigenvalues of A (usually decreasingly ordered if all real)
$\sigma_i(A)$	singular values of A (usually decreasingly ordered)
$\lambda_{\max}(A), \lambda_{\min}(A)$	largest, smallest eigenvalues of A
$\sigma_{\max}(A), \sigma_{\min}(A)$	largest, smallest singular values of A
$W \oplus V$	direct sum of vector spaces W and V
$W^\perp$	space of the vectors that are orthogonal to all elements in W

Frequently Used Theorems

Throughout this book, unless otherwise stated, the underlying scalar field for vector spaces and matrices is the complex numbers $\mathbb{C}$ (or the real numbers $\mathbb{R}$).

- **Cauchy–Schwartz inequality:** Let V be an inner product space equipped with an inner product $\langle \cdot, \cdot \rangle$. Then for all vectors x and y in V

$$|\langle x, y \rangle|^2 \leq \langle x, x \rangle \langle y, y \rangle.$$

Equality holds if and only if x and y are linearly dependent.

- **Theorem on the eigenvalues of AB and BA:** Let A and B be $m \times n$ and $n \times m$ matrices, respectively. Then AB and BA have the same nonzero eigenvalues, counting algebraic multiplicities. Thus (cyclic invariance)

$$\mathrm{tr}(AB) = \mathrm{tr}(BA).$$

- **Schur triangularization theorem:** Let A be an $n \times n$ matrix. Then there exists an $n \times n$ unitary matrix U such that

$$A = UTU^*,$$

where T is upper-triangular; the eigenvalues of A are on the diagonal of T.

- **Jordan decomposition theorem:** Let A be an $n \times n$ matrix. Then there exists an $n \times n$ invertible (complex) matrix P such that

$$A = P^{-1}JP,$$

where $J = J_1 \oplus \cdots \oplus J_k$ is a Jordan canonical form of A, i.e., J is a block-diagonal matrix with the Jordan blocks $J_1, \ldots, J_k$ of A on the diagonal.

- **Spectral decomposition theorem:** Let A be an $n \times n$ normal matrix with eigenvalues $\lambda_1, \lambda_2, \ldots, \lambda_n$. Then there exists an $n \times n$ unitary matrix U (whose columns are corresponding orthonormal eigenvectors) such that

$$A = U \, \mathrm{diag}(\lambda_1, \lambda_2, \ldots, \lambda_n) U^*.$$

In particular, if A is positive semidefinite, then all $\lambda_i \geq 0$; if A is Hermitian, then all λ_i are real; and if A is unitary, then all $|\lambda_i| = 1$.

- **Singular value decomposition (SVD) theorem:** Let A be an $m \times n$ matrix with rank r. Then there exist an $m \times m$ unitary matrix U and an $n \times n$ unitary matrix V such that

$$A = UDV,$$

where D is the $m \times n$ matrix with the singular values of A as the (i, i)-entries, $i = 1, 2, \ldots, r$, and other entries 0. If $m = n$, then D is diagonal.

Chapter 1

Vector Spaces

Definitions, Facts, and Examples _____

Vector Space. A vector space involves four things – two nonempty sets V and $\mathbb{F}$ and two algebraic operations $+$ and $\cdot$ on the elements of V and $\mathbb{F}$:

V set of *vectors* (the elements of V are called vectors),

$\mathbb{F}$ set of *scalars*, ($\mathbb{F}$ is a number field, mostly $\mathbb{C}$ or $\mathbb{R}$),

$+$ operation between vectors, and

$\cdot$ operation between scalars and vectors.

Vector addition is an operation on the vectors in V; it assigns any two vectors u and v in V to a vector in V. The resulting vector is called the *sum* of u and v and denoted by $u + v$. So, $u + v \in V$ for all $u, v \in V$.

Scalar multiplication is an operation between scalar λ in $\mathbb{F}$ and vector v in V; it assigns the pair λ and v to a vector in V, called the *product* of λ and v and written as $\lambda \cdot v$, or simply λv. So, $\lambda v \in V$ for all $\lambda \in \mathbb{F}$, $v \in V$.

We say that V is a *vector space* over $\mathbb{F}$ if the following are satisfied:

(1) $u + v = v + u$ for all $u, v \in V$.

(2) $(u + v) + w = u + (v + w)$ for all $u, v, w \in V$.

(3) There is a *zero* element $0 \in V$ such that $v + 0 = v$ for all $v \in V$.

(4) For each $v \in V$, there is an element $-v \in V$ such that $v + (-v) = 0$.

(5) $\lambda(u + v) = \lambda u + \lambda v$ for all $\lambda \in \mathbb{F}$ and $u, v \in V$.

(6) $(\lambda + \mu)v = \lambda v + \mu v$ for all $\lambda, \mu \in \mathbb{F}$ and $v \in V$.

(7) $(\lambda \mu)v = \lambda(\mu v)$ for all $\lambda, \mu \in \mathbb{F}$ and $v \in V$.

(8) $1v = v$ for all $v \in V$.

1

Some Important Vector Spaces.

- The *xy-plane* (also called the *Cartesian plane*) is a vector space over
  $\mathbb{R}$. Here we view the *xy*-plane as the set of arrows (directed line
  segments) in the plane, all with initial point O, the origin. Define the
  addition by the parallelogram law, which states that for two vectors
  u and v, the sum $u + v$ is the vector defined by the diagonal of the
  parallelogram with u and v as adjacent sides. Define λv to be the
  vector whose length is $|\lambda|$ times the length of v, pointing in the same
  direction as v if $\lambda \geq 0$ and otherwise pointing in the opposite direction.
  Note that the extreme case where the terminal point of the arrow
  coincides with O gives the zero vector for which the length of the
  arrow is 0 and any direction may be regarded as its direction. The
  xy-plane can be identified with the space $\mathbb{R}^2$ (see $\mathbb{F}^n$ below) or $\mathbb{C}$.

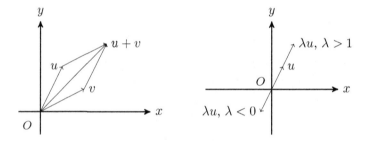

Figure 1.1: Vector addition and scalar multiplication

- The three-dimensional Euclidean space (or the ordinary *xyz*-space)
  is a vector space over $\mathbb{R}$. It consists of all arrows starting from the
  origin in the *xyz*-space with vector addition and scalar multiplication
  similarly defined (by the parallelogram rule) as above for the *xy*-plane.
  This space can be identified with the space $\mathbb{R}^3$ (see $\mathbb{F}^n$ below).

  The spaces $\mathbb{R}^2$ and $\mathbb{R}^3$ will help the reader understand and visualize
  many concepts of vector spaces.

- $\mathbb{F}^n$ is a vector space over a field $\mathbb{F}$, where n is a positive integer and

$$
\mathbb{F}^n = \left\{ \begin{pmatrix} x_1 \\ x_2 \\ \vdots \\ x_n \end{pmatrix} \,\middle|\, x_1, x_2, \ldots, x_n \in \mathbb{F} \right\}.
$$

Here addition and scalar multiplication are defined, respectively, by

$$
\begin{pmatrix} x_1 \\ x_2 \\ \vdots \\ x_n \end{pmatrix} + \begin{pmatrix} y_1 \\ y_2 \\ \vdots \\ y_n \end{pmatrix} = \begin{pmatrix} x_1 + y_1 \\ x_2 + y_2 \\ \vdots \\ x_n + y_n \end{pmatrix}, \quad \lambda \begin{pmatrix} x_1 \\ x_2 \\ \vdots \\ x_n \end{pmatrix} = \begin{pmatrix} \lambda x_1 \\ \lambda x_2 \\ \vdots \\ \lambda x_n \end{pmatrix}.
$$

In particular, $\mathbb{R}^n$ is a vector space over $\mathbb{R}$; $\mathbb{C}^n$ is a vector space over $\mathbb{C}$.

Note: In the context (of isomorphism) of vector spaces, it usually makes no difference whether to write a vector in $\mathbb{F}^n$ as a row or a column. Sometimes we may write the vectors in $\mathbb{F}^n$ as rows for convenience. However, when a matrix-vector product, say Ax, is involved, it is clear from the context that x is a column vector.

The *length* or *Euclidean norm* of vector $x \in \mathbb{C}^n$, denoted by $\|x\|$, is

$$
\|x\| = \left(|x_1|^2 + |x_2|^2 + \cdots + |x_n|^2 \right)^{\frac{1}{2}}, \quad \text{where } x = (x_1, x_2, \ldots, x_n).
$$

We say that x is a *unit vector* if $\|x\| = 1$ and that $y = (y_1, y_2, \ldots, y_n)$ and $z = (z_1, z_2, \ldots, z_n)$ are *orthogonal* if $y_1 \overline{z_1} + y_2 \overline{z_2} + \cdots + y_n \overline{z_n} = 0$.

- $M_{m \times n}(\mathbb{F})$, the set of all $m \times n$ (size or order) matrices whose entries are from a scalar field $\mathbb{F}$, is a vector space over $\mathbb{F}$. An $m \times n$ *matrix* A over $\mathbb{F}$ is an array of m rows and n columns in the form

$$
A = \begin{pmatrix} a_{11} & a_{12} & \cdots & a_{1n} \\ a_{21} & a_{22} & \cdots & a_{2n} \\ \cdots & \cdots & \cdots & \cdots \\ a_{m1} & a_{m2} & \cdots & a_{mn} \end{pmatrix}.
$$

For simplicity, we write $A = (a_{ij})_{m \times n}$ or $A = (a_{ij})$. If $m = n$, we use $M_n(\mathbb{F})$ for $M_{m \times n}(\mathbb{F})$. Two matrices $A = (a_{ij})$ and $B = (b_{ij})$ are *equal* if they have the same size and $a_{ij} = b_{ij}$ for all i and j.

The *addition* of two $m \times n$ matrices is defined by adding the corresponding entries, and the *scalar multiplication* of a matrix by a scalar is obtained by multiplying every entry of the matrix by the scalar. In symbols, if $A = (a_{ij})$, $B = (b_{ij}) \in M_{m \times n}(\mathbb{F})$, and $\lambda \in \mathbb{F}$, then

$$
A + B = (a_{ij} + b_{ij}), \quad \lambda A = (\lambda a_{ij}).
$$

Note: If $m = 1$ or $n = 1$, $M_{m \times n}(\mathbb{F})$ can be identified with $\mathbb{F}^n$ or $\mathbb{F}^m$.

Matrices can also be multiplied when they have appropriate sizes. Let A be a $p \times n$ matrix and let B be an $n \times q$ matrix. The *matrix product* AB of A and B is a $p \times q$ matrix whose (i, j)-entry is given by

$$a_{i1}b_{1j} + a_{i2}b_{2j} + \cdots + a_{in}b_{nj}, \quad i = 1, 2, \ldots, p, \ j = 1, 2, \ldots, q.$$

So, to add two matrices, the matrices must have the same size, while to multiply two matrices, the number of columns of the first matrix must equal the number of rows of the second matrix. Note that even though AB is well defined, BA may not be. Moreover, $AB \neq BA$ in general for square matrices A and B of the same size.

The *zero matrix* of size $m \times n$, written as $0_{m \times n}$ or 0 when the size is immaterial, is the $m \times n$ matrix whose entries are all 0. The $n \times n$ *identity matrix*, symbolized by I_n (or simply I when the size is clear from context), is the n-square matrix whose diagonal entries are equal to 1 and off-diagonal entries are all 0. A square matrix A is said to be *invertible,* or *nonsingular,* if there exists a matrix B such that $AB = BA = I$. Such a matrix B is called the *inverse* of A, denoted by A^{-1}. It is immediate that $(A^{-1})^{-1} = A$. If X and Y are $n \times n$ nonsingular matrices, then XY is nonsingular and $(XY)^{-1} = Y^{-1}X^{-1}$.

Besides the properties on addition and scalar multiplication of a vector space, matrices satisfy the following rules for multiplication:

(a) $0A = A0 = 0$. (The three 0's may have different sizes.)

(b) $AI = IA = A$. (The two I's may have different sizes.)

(c) $(AB)C = A(BC)$.

(d) $A(B + C) = AB + AC$.

(e) $(A + B)C = AC + BC$.

(f) $k(AB) = (kA)B = A(kB)$, where k is a scalar.

The product of matrices is associative (the above property (c)), and

$$ABC = \left(\sum_{j,\,k} a_{ij}b_{jk}c_{kl} \right).$$

For a square matrix A and a positive integer k, A^k is the product of k copies of A. Conventionally, $A^0 = I$. For an $m \times n$ matrix $A = (a_{ij})$, we convert the rows of A to columns to get an $n \times m$ matrix whose (i, j)-entry is a_{ji}. Such a matrix is called the *transpose* of A and is denoted by A^t. We take the conjugate of each entry of A to obtain the *conjugate* $\bar{A} = (\overline{a_{ij}})$ of A which has the same size as A. We write A^* for the *conjugate transpose* of A, i.e., $A^* = (\bar{A})^t$. $A^* = A^t$ if A is real. The following properties will be used throughout the book:

(i) $(A^t)^t = A$; $(A^*)^* = A$; $(A^{-1})^* = (A^*)^{-1}$ if A is invertible.

(ii) $(A + B)^t = A^t + B^t$; $(A + B)^* = A^* + B^*$.

(iii) $(AB)^t = B^t A^t$; $(AB)^* = B^* A^*$. In particular, for vectors, $(x^t y)^t = y^t x$ and $(x^* y)^* = y^* x$, where $x, y \in \mathbb{C}^n$.

(iv) $(kA)^t = kA^t$; $(kA)^* = \bar{k}A^*$, where k is a complex number.

An $n \times n$ complex matrix $A = (a_{ij})$ is said to be *Hermitian* if $A^* = A$; *symmetric* if $A^t = A$; *real symmetric* if A is real and symmetric; *skew-Hermitian* if $A^* = -A$; *normal* if $A^* A = AA^*$; *upper-triangular* if $a_{ij} = 0$ whenever $i > j$; *lower-triangular* if $a_{ij} = 0$ whenever $i < j$; *diagonal* if $a_{ij} = 0$ whenever $i \neq j$, written as $A = \text{diag}(a_{11}, a_{22}, \ldots, a_{nn})$; *unitary* if $A^* A = AA^* = I$; and *orthogonal* if $A^t A = AA^t = I$. A is *real orthogonal* if A is real and orthogonal.

Two column vectors x and y of $\mathbb{C}^n$ are *orthogonal* if and only if $y^* x = 0$.

Let A be a matrix. A *submatrix* of A is a matrix that consists of the entries of A lying in certain rows and columns of A. For example, let

$$A = \begin{pmatrix} 1 & 2 & 3 \\ 4 & 5 & 6 \\ 7 & 8 & 9 \end{pmatrix}, \quad B = \begin{pmatrix} 2 & 3 \\ 5 & 6 \end{pmatrix}, \quad C = \begin{pmatrix} 3 \\ 6 \\ 9 \end{pmatrix}.$$

B is the submatrix of A lying in rows 1 and 2 and columns 2 and 3 of A. C is the 3rd column of A; it is the submatrix of A obtained by deleting the first two columns. Sometimes it is useful and convenient to *partition* a matrix into submatrices. For instance, we may write

$$A = \begin{pmatrix} 1 & 2 & 3 \\ 4 & 5 & 6 \\ 7 & 8 & 9 \end{pmatrix} = \begin{pmatrix} X & B \\ U & V \end{pmatrix} = (P_1, P_2, P_3),$$

where P_1, P_2, P_3 are columns of A, and

$$X = \begin{pmatrix} 1 \\ 4 \end{pmatrix}, \quad B = \begin{pmatrix} 2 & 3 \\ 5 & 6 \end{pmatrix}, \quad U = (7), \quad V = (8, \ 9).$$

If $A = (A_{ij})$ is a partitioned matrix, then $A^t = (A_{ji}^t)$ and $A^* = (A_{ji}^*)$. For 2×2 block matrices, we have explicitly: if $M = \begin{pmatrix} A & B \\ C & D \end{pmatrix}$, then

$$M^t = \begin{pmatrix} A^t & C^t \\ B^t & D^t \end{pmatrix}, \quad M^* = \begin{pmatrix} A^* & C^* \\ B^* & D^* \end{pmatrix}.$$

- $\mathbb{P}_n[x]$ is a vector space over $\mathbb{R}$, where n is a positive integer and $\mathbb{P}_n[x]$ is the set of all polynomials in x of degree (strictly) less than n with real coefficients. A constant polynomial $p(x) = c$ is said to have degree 0 if $c \neq 0$, or degree $-\infty$ if $c = 0$. The *addition* and *scalar multiplication* are defined for $p, q \in \mathbb{P}_n[x]$, and $\lambda \in \mathbb{R}$ by

$$(p+q)(x) = p(x) + q(x), \quad (\lambda p)(x) = \lambda(p(x)).$$

Denote by $\mathbb{P}[x]$ the set of all real polynomials of any finite degree. Then $\mathbb{P}[x]$ is a vector space over $\mathbb{R}$ with respect to the above operations for polynomials. (Note: $\mathbb{R}$ may be replaced with a field $\mathbb{F}$.)

- $\mathcal{C}[a, b]$ is a vector space over $\mathbb{R}$, where $\mathcal{C}[a, b]$ is the set of all real-valued continuous functions on the interval $[a, b]$. Functions are added and multiplied in the usual way, i.e., if f and g are functions, then $(f+g)(x) = f(x) + g(x)$ and $(\lambda f)(x) = \lambda(f(x))$, where $\lambda \in \mathbb{R}$. Let $\mathcal{C}(\mathbb{R})$ be the vector space of real-valued continuous functions on $\mathbb{R}$.

Example 1.1 Let V be the set of 2×2 matrices of the form $\left(\begin{smallmatrix} 1 & x \\ 0 & 1 \end{smallmatrix}\right)$, $x \in \mathbb{F}$. For $A = \left(\begin{smallmatrix} 1 & a \\ 0 & 1 \end{smallmatrix}\right)$, $B = \left(\begin{smallmatrix} 1 & b \\ 0 & 1 \end{smallmatrix}\right)$ in V and λ in $\mathbb{F}$, define vector addition $\dot{+}$ (by the ordinary matrix multiplication) and scalar multiplication $\dot{\times}$ to be

$$A \dot{+} B = \begin{pmatrix} 1 & a+b \\ 0 & 1 \end{pmatrix} = AB, \quad \lambda \dot{\times} A = \begin{pmatrix} 1 & \lambda a \\ 0 & 1 \end{pmatrix}.$$

Show that V is a vector space over $\mathbb{F}$ with respect to the operations.

Solution. We need to verify that all conditions in the definition of a vector space are satisfied. For instance, the addition is commutative, i.e.,

$$A \dot{+} B = \begin{pmatrix} 1 & a+b \\ 0 & 1 \end{pmatrix} = \begin{pmatrix} 1 & b+a \\ 0 & 1 \end{pmatrix} = B \dot{+} A$$

and the scalar multiplication is distributive, that is,

$$
\begin{aligned}
\lambda \dot{\times} (A \dot{+} B) &= \lambda \dot{\times} \begin{pmatrix} 1 & a+b \\ 0 & 1 \end{pmatrix} = \begin{pmatrix} 1 & \lambda(a+b) \\ 0 & 1 \end{pmatrix} \\
&= \begin{pmatrix} 1 & \lambda a + \lambda b \\ 0 & 1 \end{pmatrix} = \begin{pmatrix} 1 & \lambda a \\ 0 & 1 \end{pmatrix} \dot{+} \begin{pmatrix} 1 & \lambda b \\ 0 & 1 \end{pmatrix} \\
&= (\lambda \dot{\times} A) \dot{+} (\lambda \dot{\times} B).
\end{aligned}
$$

The zero vector is $0 = \left(\begin{smallmatrix} 1 & 0 \\ 0 & 1 \end{smallmatrix}\right)$ because $A \dot{+} \left(\begin{smallmatrix} 1 & 0 \\ 0 & 1 \end{smallmatrix}\right) = A$ for all $A \in V$. Moreover,

$$\begin{pmatrix} 1 & a \\ 0 & 1 \end{pmatrix} \dot{+} \begin{pmatrix} 1 & -a \\ 0 & 1 \end{pmatrix} = 0, \quad \text{thus} \ -A = \begin{pmatrix} 1 & -a \\ 0 & 1 \end{pmatrix};$$

$$(-1) \dot{\times} A = \begin{pmatrix} 1 & -a \\ 0 & 1 \end{pmatrix} = -A \quad \text{and} \quad 0 \dot{\times} A = \begin{pmatrix} 1 & 0 \\ 0 & 1 \end{pmatrix} = 0. \quad \square$$

(Throughout the book, a $\square$ is placed at the end of each example.)

Question: In Example 1.1, if the addition is defined by $A \uplus B = \begin{pmatrix} 1 & ab \\ 0 & 1 \end{pmatrix}$, is V a vector space with respect to the operations $\uplus$ and $\dot{\times}$?

Note: In the above, we used zeros (0's) for both numbers and matrices for convenience. One can tell which is which in the context. This is often seen in math; and it causes no confusion. Additionally, we use 0's for 0s (zeros) in plural. Likewise, we use u's (or u_i's) for us, v's for vs, etc.

Linear Independence. The definition of a vector space V ensures that the sum of two vectors, say v_1 and v_2, is a vector, i.e., $v_1 + v_2 \in V$. For another vector v_3, then $(v_1 + v_2) + v_3$ makes sense. Since $(v_1 + v_2) + v_3 = v_1 + (v_2 + v_3)$, we simply write the sum as $v_1 + v_2 + v_3$. In this way, we can add any finitely many vectors. (Adding infinitely many vectors could be meaningless.)

Let $v_1, v_2, \ldots, v_n$ be n vectors of a vector space V over a field $\mathbb{F}$ and let $\lambda_1, \lambda_2, \ldots, \lambda_n$ be scalars from $\mathbb{F}$. Then the vector

$$v = \lambda_1 v_1 + \lambda_2 v_2 + \cdots + \lambda_n v_n$$

is called a *linear combination* of the vectors $v_1, v_2, \ldots, v_n$, and the scalars $\lambda_1, \lambda_2, \ldots, \lambda_n$ are called the *coefficients* of the linear combination. If all the coefficients are zero, then $v = 0$. There may exist a linear combination of the vectors $v_1, v_2, \ldots, v_n$ that equals zero even though the coefficients $\lambda_1, \lambda_2, \ldots, \lambda_n$ are not all zero. In this case, we say that the vectors $v_1, v_2, \ldots, v_n$ are *linearly dependent*, or $\{v_1, v_2, \ldots, v_n\}$ is a linearly dependent set. In other words, the vectors $v_1, v_2, \ldots, v_n$ are linearly dependent if and only if there exist scalars $\lambda_1, \lambda_2, \ldots, \lambda_n$, not all zero, such that

$$\lambda_1 v_1 + \lambda_2 v_2 + \cdots + \lambda_n v_n = 0. \tag{1.1}$$

The zero vector 0 itself is linearly dependent as $\lambda 0 = 0$ for $\lambda \neq 0$.

Vectors $v_1, v_2, \ldots, v_n$ are *linearly independent* (or $\{v_1, v_2, \ldots, v_n\}$ is a linearly independent set) if they are not linearly dependent, i.e., $v_1, v_2, \ldots, v_n$ are linearly independent if (1.1) holds only when all the coefficients $\lambda_1, \lambda_2, \ldots, \lambda_n$ are zero. If $\{v_1, v_2, \ldots, v_n\}$ is a linearly independent set, then every nonempty subset of $\{v_1, v_2, \ldots, v_n\}$ is linearly independent.

We generalize linear independence for an infinite set S of vectors: the set S is said to be *linearly independent* if any finitely many vectors from S are linearly independent (again, we avoid infinite sums of vectors in this book). For example, $\{1, x, x^2, \ldots\}$ is a linearly independent set in $\mathbb{P}[x]$.

Here is a useful fact: $\alpha_1, \ldots, \alpha_n \in \mathbb{C}^m$ are linearly dependent if and only if $Ax = 0$ for a nonzero $x = (x_1, \ldots, x_n)^t \in \mathbb{C}^n$, where $A = (\alpha_1, \ldots, \alpha_n)$, as

$$Ax = x_1\alpha_1 + \cdots + x_n\alpha_n = (\alpha_1, \ldots, \alpha_n) \begin{pmatrix} x_1 \\ \vdots \\ x_n \end{pmatrix} = 0.$$

Example 1.2 Let A be an $n \times n$ real matrix such that for some positive integer k, $A^k = 0$ but $A^{k-1} \neq 0$. Show that there is a vector $v \in \mathbb{R}^n$ such that $v, Av, A^2v, \ldots, A^{k-1}v$ are linearly independent (vectors in $\mathbb{R}^n$).

Solution. Since $A^{k-1} \neq 0$, there is a nonzero $v \in \mathbb{R}^n$ such that $A^{k-1}v \neq 0$. We claim that the vectors $v, Av, A^2v, \ldots, A^{k-1}v$ are linearly independent. Let $x_0v + x_1Av + x_2A^2v + \cdots + x_{k-1}A^{k-1}v = 0$. Multiplying both sides by A^{k-1} from the left and noticing that $A^mv = 0$ for $m \geq k$, we get $x_0A^{k-1}v = 0$. Thus, $x_0 = 0$. In a similar way, applying A^{k-2} yields $x_1 = 0$. Inductively, $x_2 = \cdots = x_{k-1} = 0$. Therefore, the statement is true. $\square$

Dimension and Bases. The largest "number" of linearly independent vectors in a vector space V is called the *dimension* of V, denoted by $\dim V$. If that is a finite number n, we say that V is *finite* or *n-dimensional* and write $\dim V = n$. If there are arbitrarily large (countable or uncountable) linearly independent sets in V, we say V is *infinite dimensional* and denote $\dim V = \infty$. A set S is said to be a *basis* of the vector space V if

(1) S is a linearly independent set, and

(2) Each vector in V is a (finite) linear combination of vectors in S.

It is known that every vector space has a basis (which can be a uncountable set). The proof of this requires other machinery; and it lies outside our scope. Unless otherwise stated, our vector spaces are finite dimensional.

If there exist n vectors in the vector space V that are linearly independent and any $n+1$ vectors in V are linearly dependent, then $\dim V = n$. In this case, any set of n linearly independent vectors is a basis for the vector space V. The vector space of one element, zero, is said to have dimension 0 with empty set as its basis. Note that the dimension of a vector space also depends on the underlying number field of the vector space.

For a scalar field $\mathbb{F}$, the dimension of the vector space $\mathbb{F}^n$ is n, and the vectors $e_1 = (1, 0, 0, \ldots, 0), e_2 = (0, 1, 0, \ldots, 0), \ldots, e_n = (0, 0, \ldots, 0, 1)$ (or written in columns) comprise a basis for $\mathbb{F}^n$. We refer to $\{e_1, e_2, \ldots, e_n\}$ as the *standard basis* of $\mathbb{F}^n$. (Note: $\mathbb{F}^n$ has different bases.)

For $M_{m \times n}(\mathbb{F})$, let E_{ij} be the $m \times n$ matrix with (i, j) entry 1 and 0 elsewhere, $i = 1, 2, \ldots, m$, $j = 1, 2, \ldots, n$. Then these matrices E_{ij} are linearly independent and form a basis of $M_{m \times n}(\mathbb{F})$ as $A = (a_{ij}) = \sum_{i,j} a_{ij} E_{ij}$. The dimension of $M_{m \times n}(\mathbb{F})$ is mn. If $m = n$, the dimension of $M_n(\mathbb{F})$ is n^2, and

$$E_{ip} E_{qj} = \begin{cases} E_{ij} & \text{if } p = q \\ 0 & \text{if } p \neq q. \end{cases} \tag{1.2}$$

Let $\{\alpha_1, \alpha_2, \ldots, \alpha_n\}$ be a basis of the vector space V, and let v be any vector in V. Since $v, \alpha_1, \alpha_2, \ldots, \alpha_n$ are linearly dependent $(n + 1$ vectors), there are scalars $\lambda, \lambda_1, \lambda_2, \ldots, \lambda_n$, not all equal to zero, such that

$$\lambda v + \lambda_1 \alpha_1 + \lambda_2 \alpha_2 + \cdots + \lambda_n \alpha_n = 0.$$

Since $\alpha_1, \alpha_2, \ldots, \alpha_n$ are linearly independent, we see $\lambda \neq 0$. It follows that

$$v = x_1 \alpha_1 + x_2 \alpha_2 + \cdots + x_n \alpha_n,$$

where $x_i = -\lambda_i / \lambda$, $i = 1, 2, \ldots, n$. Again, due to the linear independence of $\alpha_1, \alpha_2, \ldots, \alpha_n$, such an expression of v as a linear combination of $\alpha_1, \alpha_2, \ldots, \alpha_n$ is unique. We call the n-tuple $(x_1, x_2, \ldots, x_n)$ the *coordinate* of v under the ordered basis $\{\alpha_1, \alpha_2, \ldots, \alpha_n\}$; we also say that $x_1, x_2, \ldots, x_n$ are the *coordinates* of v under the basis. When coordinates are under discussion, the basis $\{\alpha_1, \alpha_2, \ldots, \alpha_n\}$ is an ordered set by convention.

Example 1.3 If $\{u, v\}$ is a basis for a vector space V, show that $\{w_1, w_2\}$ is also a basis of V, where $w_1 = u + v, w_2 = u - v$. If $w \in V$ has coordinate $(-2, 3)$ under the basis $\{u, v\}$, what is the coordinate of w under $\{w_1, w_2\}$?

Solution. Let $x_1 w_1 + x_2 w_2 = 0$. Then $(x_1 + x_2)u + (x_1 - x_2)v = 0$. Since u, v are linearly independent, we have $x_1 + x_2 = 0$ and $x_1 - x_2 = 0$. It is immediate that $x_1 = x_2 = 0$. Thus, w_1 and w_2 are linearly independent.

Note that $u = \frac{1}{2}(w_1 + w_2)$ and $v = \frac{1}{2}(w_1 - w_2)$. If $z \in V$ has coordinate (a, b) with respect to $\{u, v\}$, i.e., $z = au + bv$, then

$$z = au + bv = \frac{a + b}{2} w_1 + \frac{a - b}{2} w_2.$$

It follows that $w = -2u + 3v$ has coordinate $(\frac{1}{2}, -\frac{5}{2})$ under $\{w_1, w_2\}$. $\square$

Subspaces. Let V be a vector space over a field $\mathbb{F}$ and let W be a nonempty subset of V. If W is also a vector space over $\mathbb{F}$ with respect to the vector addition and scalar multiplication of V, then W is said to be a *subspace* of V. One may check that W is a subspace of V if and only if W is closed under the operations of V, that is, (i) if $u, v \in W$ then $u + v \in W$ and (ii) if $v \in W$ and $\lambda \in \mathbb{F}$ then $\lambda v \in W$. It follows that, to be a subspace, W must contain the zero vector 0 of V. $\{0\}$ and V are trivial subspaces of V. A subspace W of V is called a *proper subspace* of V if $W \neq V$.

Example 1.4 Show that the set of real matrices that commute with $\left(\begin{smallmatrix} 1 & 0 \\ 2 & 3 \end{smallmatrix}\right)$ is a subspace of $M_2(\mathbb{R})$. Find the dimension and a basis of the subspace.

Solution. Let $A = \left(\begin{smallmatrix} 1 & 0 \\ 2 & 3 \end{smallmatrix}\right)$ and $V = \{X \in M_2(\mathbb{R}) \mid AX = XA\}$. Clearly, $0 \in V$. If X and Y commute with A, namely, $AX = XA$ and $AY = YA$, then

$$A(X + Y) = AX + AY = XA + YA = (X + Y)A.$$

Thus, $X + Y \in V$. For any $k \in \mathbb{R}$, $A(kX) = k(AX) = k(XA) = (kX)A$, so $kX \in V$. V is closed under the operations and is a subspace of $M_2(\mathbb{R})$.

To find the dimension and a basis of V, we first find the elements in V. Let $\left(\begin{smallmatrix} a & b \\ c & d \end{smallmatrix}\right)\left(\begin{smallmatrix} 1 & 0 \\ 2 & 3 \end{smallmatrix}\right) = \left(\begin{smallmatrix} 1 & 0 \\ 2 & 3 \end{smallmatrix}\right)\left(\begin{smallmatrix} a & b \\ c & d \end{smallmatrix}\right)$. Multiplying out both sides and equating the corresponding entries, we obtain $b = 0$ and $d = a + c$. So V consists of all matrices $\left(\begin{smallmatrix} a & 0 \\ c & a+c \end{smallmatrix}\right)$. Since $\left(\begin{smallmatrix} a & 0 \\ c & a+c \end{smallmatrix}\right) = \left(\begin{smallmatrix} a & 0 \\ 0 & a \end{smallmatrix}\right) + \left(\begin{smallmatrix} 0 & 0 \\ c & c \end{smallmatrix}\right) = a\left(\begin{smallmatrix} 1 & 0 \\ 0 & 1 \end{smallmatrix}\right) + c\left(\begin{smallmatrix} 0 & 0 \\ 1 & 1 \end{smallmatrix}\right)$, we see that every element in V is a linear combination of $\left(\begin{smallmatrix} 1 & 0 \\ 0 & 1 \end{smallmatrix}\right)$ and $\left(\begin{smallmatrix} 0 & 0 \\ 1 & 1 \end{smallmatrix}\right)$. On the other hand, $\left(\begin{smallmatrix} 1 & 0 \\ 0 & 1 \end{smallmatrix}\right)$ and $\left(\begin{smallmatrix} 0 & 0 \\ 1 & 1 \end{smallmatrix}\right)$ are linearly independent. It follows that the dimension of V is 2 and $\left\{\left(\begin{smallmatrix} 1 & 0 \\ 0 & 1 \end{smallmatrix}\right), \left(\begin{smallmatrix} 0 & 0 \\ 1 & 1 \end{smallmatrix}\right)\right\}$ serves as a basis of V. $\square$

Let W_1 and W_2 be subspaces of a vector space V. The *intersection*

$$W_1 \cap W_2 = \{\, v \mid v \in W_1 \text{ and } v \in W_2 \,\}$$

is a subspace of V, and so is the *sum*

$$W_1 + W_2 = \{\, w_1 + w_2 \mid w_1 \in W_1 \text{ and } w_2 \in W_2 \,\}.$$

$W_1 \cap W_2$ is the largest subspace contained in both W_1 and W_2, while $W_1 + W_2$ is the smallest subspace containing both W_1 and W_2.

Sum $W_1 + W_2$ is called a *direct sum*, denoted by $W_1 \oplus W_2$, if every element v in $W_1 + W_2$ can be uniquely written as $v = w_1 + w_2$, where $w_1 \in W_1$, $w_2 \in W_2$, that is, if $v = v_1 + v_2$, where $v_1 \in W_1$, $v_2 \in W_2$, then $w_1 = v_1$ and $w_2 = v_2$. In particular, if $0 = w_1 + w_2$, then $w_1 = w_2 = 0$.

Let S be a nonempty subset of a vector space V. We define the *span* of S, denoted by $\operatorname{Span} S$, to be the set of all (finite) linear combinations of the vectors in S. In particular, if S is a finite set, say, $S = \{v_1, v_2, \ldots, v_k\}$, then

$$\operatorname{Span} S = \{\, \lambda_1 v_1 + \lambda_2 v_2 + \cdots + \lambda_k v_k \mid \lambda_1, \lambda_2, \ldots, \lambda_k \in \mathbb{F} \,\}.$$

$\operatorname{Span} S$ is a subspace of V and it is the smallest subspace containing S. We say that the subspace $\operatorname{Span} S$ is *spanned* (or *generated*) by S. It follows that S is a basis of a vector space V if and only if S is a linearly independent set and S spans V. Thus, if S is a linearly independent set, then S is a

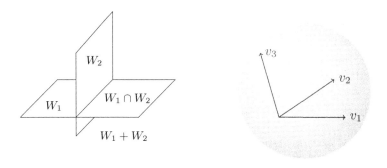

Figure 1.2: Sum, intersection, and span

basis of Span S. Note that Span S is *finitely generated* by S; Span S need not be finite dimensional. For example, $\mathbb{P}[x]$, the space of all polynomials of finite degrees, is spanned by the infinite set $\{1, x, x^2, \dots\}$.

Let V be a vector space of dimension n and let $v_1, v_2, \dots, v_k \in V$ be linearly independent vectors, $k < n$. We can extend $\{v_1, v_2, \dots, v_k\}$ to a basis of V as follows. Let $v_{k+1} \notin \text{Span}\{v_1, v_2, \dots, v_k\}$. Then $v_1, v_2, \dots, v_k, v_{k+1}$ are linearly independent. Repeat this process with $\text{Span}\{v_1, v_2, \dots, v_{k+1}\}$ and so on, until we obtain n linearly independent vectors $v_1, v_2, \dots, v_k, v_{k+1}$, $\dots, v_n$. Thus, $\text{Span}\{v_1, v_2, \dots, v_n\} = V$ and $\{v_1, v_2, \dots, v_n\}$ is a basis of V.

Given a nonzero column vector $u_1 \in \mathbb{C}^n$, we can extend u_1 to a special basis of $\mathbb{C}^n$. If $n = 1$, $\{u_1\}$ is a basis already. Let $n \geq 2$. Solve the linear equation $u_1^t x = 0$ to get a nonzero solution $x = u_2 \in \mathbb{C}^n$. Then u_1 and u_2 are orthogonal, i.e., $u_1^t u_2 = 0$. We claim that u_1 and u_2 are linearly independent. If $a u_1 + b u_2 = 0$, then $a u_1^t u_1 + b u_1^t u_2 = 0$. Thus $a u_1^t u_1 = 0$, implying $a = 0$, consequently $b u_2 = 0$, yielding $b = 0$.

If $n = 2$, then $\{u_1, u_2\}$ is a basis for $\mathbb{C}^2$. If $n > 2$, we can find a nonzero vector u_3 that is orthogonal to u_1 and u_2 by considering the linear equation system $u_1^t x = 0$ and $u_2^t x = 0$. Repeating this process produces n nonzero vectors $u_1, u_2, \dots, u_n$ that are mutually orthogonal, i.e., $u_i^* u_j = 0$ if $i \neq j$. One proves that $u_1, u_2, \dots, u_n$ are linearly independent. We call a basis $\{u_1, u_2, \dots, u_n\}$ in which basis vectors are mutually orthogonal an *orthogonal basis* of $\mathbb{C}^n$. If, in addition, every u_i has length 1, i.e., $u_i^* u_i = 1$, $i = 1, 2, \dots, n$, we say that $\{u_1, u_2, \dots, u_n\}$ is an *orthonormal basis* of $\mathbb{C}^n$. The standard basis $\{e_1, e_2, \dots, e_n\}$ is an orthonormal basis of $\mathbb{C}^n$.

Thus, given a nonzero vector u_1, there exist $u_2, \dots, u_n \in \mathbb{C}^n$ such that $\{u_1, u_2, \dots, u_n\}$ is an orthogonal basis for $\mathbb{C}^n$. By scaling, let $v_i = \frac{1}{\|u_i\|} u_i$, $i = 1, 2, \dots, n$. Then $\{v_1, v_2, \dots, v_n\}$ is an orthonormal basis of $\mathbb{C}^n$.

Example 1.5 Let V be a vector space. If W_1 and W_2 are two subspaces of V such that $W_1 \cup W_2 = V$, show that $W_1 = V$ or $W_2 = V$.

Solution. Suppose $W_1 \neq V$ and $W_2 \neq V$. Since $W_1 \cup W_2 = V$, we see neither $W_1 \subseteq W_2$ nor $W_2 \subseteq W_1$. Let $w_1 \in W_1 \setminus W_2$ and $w_2 \in W_2 \setminus W_1$. Set $w = w_1 + w_2 \in V$. If $w \in W_1$, then $w_2 = w - w_1 \in W_1$, a contradiction. Similarly, if $w \in W_2$, then $w_1 = w - w_2 \in W_2$, a contradiction. □

Given an $m \times n$ matrix A over a scalar field $\mathbb{F}$, there are three important spaces associated to A. The space spanned by the rows of A is a subspace of $\mathbb{F}^n$, called the *row space* of A. The space spanned by the columns of A is a subspace of $\mathbb{F}^m$, called the *column space* of A. The column space of a matrix A is also known as the *image* or *range* of A, denoted by $\operatorname{Im} A$; this origins from A being viewed as the mapping from $\mathbb{F}^n$ to $\mathbb{F}^m$ defined by $x \mapsto Ax$. Both terms and notations are in practical use. Thus

$$\operatorname{Im} A = \{\, Ax \mid x \in \mathbb{F}^n \,\} \subseteq \mathbb{F}^m.$$

If we denote the columns of A by $\alpha_1, \ldots, \alpha_n$, then

$$\operatorname{Im} A = \{x_1\alpha_1 + \cdots + x_n\alpha_n \mid x_1,, \ldots, x_n \in \mathbb{F}\} = \operatorname{Span}\{\alpha_1, \ldots, \alpha_n\}.$$

All solutions to the equation system $Ax = 0$ form a subspace of $\mathbb{F}^n$. This space is called the *null space* or *kernel* of A, symbolized by $\operatorname{Ker} A$. So

$$\operatorname{Ker} A = \{\, x \in \mathbb{F}^n \mid Ax = 0 \} \subseteq \mathbb{F}^n.$$

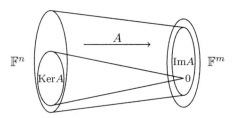

Figure 1.3: Image and kernel

Let $\dim \operatorname{Ker} A = s$ and let $u_1, \ldots, u_s$ be a basis of $\operatorname{Ker} A$. We can extend the basis of $\operatorname{Ker} A$ to a basis of $\mathbb{F}^n$: $u_1, \ldots, u_s, v_1, \ldots, v_r$, where $s + r = n$. We claim that $Av_1, \ldots, Av_r$ are linearly independent. Let $l_1 Av_1 + \cdots + l_r Av_r = 0$. Then $A(l_1 v_1 + \cdots + l_r v_r) = 0$, so $l_1 v_1 + \cdots + l_r v_r \in \operatorname{Ker} A$. Since $u_1, \ldots, u_s, v_1, \ldots, v_r$ are linearly independent, all l_i's are zero.

Now for every $v \in \mathbb{F}^n$, let $v = x_1 u_1 + \cdots + x_s u_s + y_1 v_1 + \cdots + y_r v_r$. Then $Av = y_1 Av_1 + \cdots + y_r Av_r$. So $\{Av_1, \ldots, Av_r\}$ is a basis of Im A. Thus

$$\dim \text{Im } A + \dim \text{Ker } A = n. \tag{1.3}$$

Example 1.6 Determine the null space Ker A and the range Im A for

$$A = \begin{pmatrix} 1 & 1 & -1 \\ 1 & 0 & 1 \\ 3 & 2 & -1 \end{pmatrix}.$$

Solution. Denote the column vectors of A by $c_1 = (1, 1, 3)^t$, $c_2 = (1, 0, 2)^t$, and $c_3 = (-1, 1, -1)^t$. So $A = (c_1, c_2, c_3)$ and $Ax = x_1 c_1 + x_2 c_2 + x_3 c_3$, where $x = (x_1, x_2, x_3)^t$. To find the null space of A, consider the system of linear equations $Ax = 0$, i.e., $x_1 c_1 + x_2 c_2 + x_3 c_3 = 0$. By reduction of variables, we can express x_1 and x_2 in terms of x_3, and obtain solutions: $x_1 = -x_3$, $x_2 = 2x_3$, $x_3 = x_3$ (which is free to choose). It follows that the dimension of the null space of A is 1, $(-1, 2, 1)^t$ is a basis, and Ker $A = \{a(-1, 2, 1)^t \mid a \in \mathbb{F}\} = \text{Span}\{(-1, 2, 1)^t\}$ (where $\mathbb{F} = \mathbb{C}, \mathbb{R}$, or $\mathbb{Q}$). Im A consists of all linear combinations in the form $x_1 c_1 + x_2 c_2 + x_3 c_3$, that is,

$$\text{Im } A = \{ x_1 c_1 + x_2 c_2 + x_3 c_3 \mid x_1, x_2, x_3 \in \mathbb{F} \}.$$

Observing that $c_3 = c_1 - 2c_2$, we obtain Im $A = \text{Span}\{c_1, c_2\}$.   □

Example 1.7 Let A be an $m \times n$ real matrix. Show that

$$\dim \text{Im } A^t = \dim \text{Im } A. \tag{1.4}$$

Solution. Since for any real column vector x, $x^t x = 0$ if and only if $x = 0$, we see that $Ax = 0 \Leftrightarrow x^t A^t Ax = 0 \Leftrightarrow A^t Ax = 0$. Thus, for $x_1, \ldots, x_k \in \mathbb{R}^n$, $Ax_1, \ldots, Ax_k$ are linearly independent if and only if $A^t Ax_1, \ldots, A^t Ax_k$ are linearly independent. If $\{Ax_1, \ldots, Ax_k\}$ is a basis of Im A, then $A^t(Ax_1), \ldots, A^t(Ax_k)$ are linearly independent vectors in Im A^t. Hence, $\dim \text{Im } A \leq \dim \text{Im } A^t$. Replacing A with A^t, we get $\dim \text{Im } A^t \leq \dim \text{Im}(A^t)^t = \dim \text{Im } A$. It follows that $\dim \text{Im } A^t = \dim \text{Im } A$.   □

Slightly modifying the above proof extends the statement for complex matrices. Let $A \in M_{m \times n}(\mathbb{C})$. Then $\dim \text{Im } A^* = \dim \text{Im } A$. On the other hand, $\dim \text{Im } \bar{A} = \dim \text{Im } A$. So $\dim \text{Im } A^* = \dim \text{Im } \bar{A}$. We conclude that

$$\dim \text{Im } A = \dim \text{Im } \bar{A} = \dim \text{Im } A^* = \dim \text{Im } A^t. \tag{1.5}$$

Dimension Identity. Let W_1 and W_2 be subspaces of vector space V. Then

$$\dim W_1 + \dim W_2 = \dim(W_1 + W_2) + \dim(W_1 \cap W_2). \qquad (1.6)$$

A proof goes as follows. Assume that W_1 and W_2 are finite dimensional. Let $\dim W_1 = s$, $\dim W_2 = t$, $\dim(W_1 \cap W_2) = r$, and let $\{\alpha_1, \ldots, \alpha_r\}$ be a basis of $W_1 \cap W_2$. Extend $\{\alpha_1, \ldots, \alpha_r\}$ to a basis for W_1: $\{\alpha_1, \ldots, \alpha_r, \beta_{r+1}, \ldots, \beta_s\}$, and to a basis for W_2: $\{\alpha_1, \ldots, \alpha_r, \gamma_{r+1}, \ldots, \gamma_t\}$. We show that

$$\{\beta_{r+1}, \ldots, \beta_s, \ \alpha_1, \ldots, \alpha_r, \ \gamma_{r+1}, \ldots, \gamma_t\}$$

is a basis for $W_1 + W_2$. For any $v \in W_1 + W_2$, by definition, $v = w_1 + w_2$ for some $w_1 \in W_1$ and $w_2 \in W_2$. So, w_1 is a linear combination of α's and β's, and w_2 is a linear combination of α's and γ's. Thus $v = w_1 + w_2$ is a linear combination of α's, β's, and γ's. Therefore, every element in $W_1 + W_2$ is a linear combination of $\beta_{r+1}, \ldots, \beta_s, \alpha_1, \ldots, \alpha_r, \gamma_{r+1}, \ldots, \gamma_t$. Next we show that these vectors are linearly independent. Consider

$$x_{r+1}\beta_{r+1} + \cdots + x_s\beta_s + y_1\alpha_1 + \cdots + y_r\alpha_r + z_{r+1}\gamma_{r+1} + \cdots + z_t\gamma_t = 0,$$

or

$$x_{r+1}\beta_{r+1} + \cdots + x_s\beta_s = -(y_1\alpha_1 + \cdots + y_r\alpha_r + z_{r+1}\gamma_{r+1} + \cdots + z_t\gamma_t).$$

The left side is an element in W_1 and the right side lies in W_2. We get

$$x_{r+1}\beta_{r+1} + \cdots + x_s\beta_s \in W_1 \cap W_2,$$

implying that $x_{r+1}\beta_{r+1} + \cdots + x_s\beta_s$ is a linear combination of $\alpha_1, \ldots, \alpha_r$. As α's and β's are linearly independent, we see that $x_{r+1} = \cdots = x_s = 0$. It follows that $y_1, \ldots, y_r$ and $z_{r+1}, \ldots, z_t$ are all equal to 0. Consequently, $\beta_{r+1}, \ldots, \beta_s, \alpha_1, \ldots, \alpha_r, \gamma_{r+1}, \ldots, \gamma_t$ are linearly independent.

Note that $\dim(W_1 + W_2) = s + t - r$. The dimension identity follows.

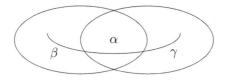

Figure 1.4: Dimension identity

Here is a typical use of the dimension identity, and the fact will be repeatedly used. Let W_1 and W_2 be subspaces of a vector space V of dimension n. If $\dim W_1 + \dim W_2 > n$, then $W_1 \cap W_2$ contains a nonzero vector.

Chapter 1 Problems

1.1 Determine whether each of the following sets is a vector space (with respect to the usual addition and multiplication of numbers). Find the dimension and a basis for each vector space.

    (a) $\mathbb{C}$ over $\mathbb{C}$.

    (b) $\mathbb{C}$ over $\mathbb{R}$.

    (c) $\mathbb{R}$ over $\mathbb{C}$.

    (d) $\mathbb{R}$ over $\mathbb{Q}$, where $\mathbb{Q}$ is the field of rational numbers.

    (e) $\mathbb{Q}$ over $\mathbb{R}$.

    (f) $\mathbb{Q}$ over $\mathbb{Z}$, where $\mathbb{Z}$ is the set of all integers.

    (g) $\mathbb{S} = \{\, a + b\sqrt{2} + c\sqrt{5} \mid a, b, c \in \mathbb{Q} \,\}$ over $\mathbb{Q}$, $\mathbb{R}$, or $\mathbb{C}$.

1.2 Referring to the definition of a vector space V over a field $\mathbb{F}$, show that

    (a) The zero vector 0 for addition is unique.

    (b) For every vector v, $-v \in V$ is unique.

    (c) For any $\lambda \in \mathbb{F}$ and $v \in V$, $\lambda v = 0$ if and only if $\lambda = 0$ or $v = 0$.

    (d) For every vector $v \in V$, $(-1)v = -v$.

    (e) For any $\lambda \in \mathbb{F}$ and $v \in V$, $-(\lambda v) = (-\lambda)v = \lambda(-v)$.

    (f) For any $\lambda \in \mathbb{F}$ and $u, v \in V$, $\lambda(u - v) = \lambda u - \lambda v$.

    (g) For any $u, v, w \in V$, $u + v = w$ if and only if $u = w - v$.

1.3 Show that the condition $u + v = v + u$ (addition commutativity) in the definition of a vector space can be derived from other conditions.

1.4 Consider $\mathbb{R}^2$ over $\mathbb{R}$. Give an example of a subset of $\mathbb{R}^2$ that is

    (a) closed under addition but not under scalar multiplication;

    (b) closed under scalar multiplication but not under addition.

1.5 Let $V = \{\, (x, y) \mid x, y \in \mathbb{C} \,\}$. Under the standard addition and scalar multiplication for ordered pairs of complex numbers, is V a vector space over $\mathbb{C}$? over $\mathbb{R}$? over $\mathbb{Q}$? If so, find the dimension of V.

1.6   Why does a vector space V over $\mathbb{F}$ ($\mathbb{F} = \mathbb{C}$, $\mathbb{R}$, or $\mathbb{Q}$) have either one element or infinitely many elements? Given $v \in V$, is it possible to have two distinct vectors u, w in V such that $u+v = 0$ and $w+v = 0$?

1.7   Let V be the collection of all real ordered pairs in which the second number is twice the first one, that is, $V = \{(x, y) \mid y = 2x, \ x \in \mathbb{R}\}$. If the addition and multiplication are defined, respectively, to be

$$(x_1, y_1) + (x_2, y_2) = (x_1 + x_2, y_1 + y_2), \quad \lambda \cdot (x, y) = (\lambda x, \lambda y),$$

show that V is a vector space over $\mathbb{R}$ with respect to the operations. Is V also a vector space with respect to the above addition and the scalar multiplication defined instead by $\lambda \odot (x, y) = (\lambda x, 0)$? (Note: The reason for the use of the symbol $\odot$ instead of $\cdot$ is to avoid confusion when the two operations are discussed in the same problem.)

1.8   Let $\mathbb{H}$ be the collection of all 2×2 complex matrices of the form

$$\begin{pmatrix} a & b \\ -\bar{b} & \bar{a} \end{pmatrix}, \quad a, b \in \mathbb{C}.$$

Show that $\mathbb{H}$ is a vector space (under the usual matrix addition and scalar multiplication) over $\mathbb{R}$. Is $\mathbb{H}$ also a vector space over $\mathbb{C}$?

1.9   Let $\mathbb{R}^+$ be the set of all positive real numbers. Show that $\mathbb{R}^+$ is a vector space over $\mathbb{R}$ with respect to the addition

$$x \boxplus y = xy, \quad x, \ y \in \mathbb{R}^+$$

and the scalar multiplication

$$a \boxdot x = x^a, \quad x \in \mathbb{R}^+, \ a \in \mathbb{R}.$$

Find the dimension of the vector space. Is $\mathbb{R}^+$ also a vector space over $\mathbb{R}$ if the scalar multiplication is instead defined as

$$a \boxtimes x = a^x, \quad x \in \mathbb{R}^+, \ a \in \mathbb{R}?$$

1.10   Let $\{\alpha_1, \alpha_2, \ldots, \alpha_n\}$ be a basis of an n-dimensional vector space V. Show that $\{\lambda_1 \alpha_1, \lambda_2 \alpha_2, \ldots, \lambda_n \alpha_n\}$ is also a basis of V for any nonzero scalars $\lambda_1, \lambda_2, \ldots, \lambda_n$. If the coordinate of a vector v under the basis $\{\alpha_1, \alpha_2, \ldots, \alpha_n\}$ is $x = (x_1, x_2, \ldots, x_n)$, what is the coordinate of v under $\{\lambda_1 \alpha_1, \lambda_2 \alpha_2, \ldots, \lambda_n \alpha_n\}$? What are the coordinates of $w = \alpha_1 + \alpha_2 + \cdots + \alpha_n$ under $\{\alpha_1, \alpha_2, \ldots, \alpha_n\}$ and $\{\lambda_1 \alpha_1, \lambda_2 \alpha_2, \ldots, \lambda_n \alpha_n\}$?

1.11 Let $v_1, v_2, \ldots, v_n$ be vectors in a vector space V. State what is meant for $\{v_1, v_2, \ldots, v_n\}$ to be a basis of V using (i) the words "span" and "independent"; (ii) the phrase "linear combination".

1.12 Consider k vectors in $\mathbb{R}^n$ and answer the three questions in the cases of $k < n$, $k = n$, and $k > n$: (i) Are the vectors linearly independent? (ii) Do they span $\mathbb{R}^n$? (iii) Do they form a basis for $\mathbb{R}^n$?

1.13 Let $\alpha_1, \alpha_2, \ldots, \alpha_m$ be column vectors in $\mathbb{F}^n$. Let $\beta_1, \beta_2, \ldots, \beta_m$ be vectors obtained from $\alpha_1, \alpha_2, \ldots, \alpha_m$ by adding k components at bottom for each of $\alpha_1, \alpha_2, \ldots, \alpha_m$, that is, $\beta_i = \begin{pmatrix} \alpha_i \\ \gamma_i \end{pmatrix}$, $\gamma_i \in \mathbb{F}^k$, $i = 1, 2, \ldots, m$. If $\alpha_1, \alpha_2, \ldots, \alpha_m$ are linearly independent, show that $\beta_1, \beta_2, \ldots, \beta_m$ are linearly independent. What about the converse?

1.14 Let $x, y, z \in \mathbb{C}^n$ (column vectors) and let $a, b \in \mathbb{C}$. Show that

$$|a|^2 x^* x + |b|^2 y^* y \geq \pm 2 \operatorname{Re}(ab\, x^* y)$$

and

$$(x^* y)z = (zx^*)y.$$

Does this mean that z and y are proportional?

1.15 Let $p_1(x) = 1 + x$, $p_2(x) = 1 - x^2$, and let $W = \operatorname{Span}\{p_1(x), p_2(x)\}$.

    (a) Show that $p_1(x)$ and $p_2(x)$ are linearly independent (over $\mathbb{R}$).

    (b) Are 1, $p_1(x)$, and $p_2(x)$ linearly independent?

    (c) Show that $(1 + x)x \in W$.

    (d) Show that $x^2 + 2x + 1 \in W$, $x^2 + 2x - 1 \notin W$.

    (e) Do $p_1(x)$ and $p_2(x)$ span $\mathbb{P}_3[x]$?

    (f) Find explicit expressions of the elements in W.

1.16 Show that $\{1, (x - 1), (x - 1)(x - 2)\}$ is a basis of $\mathbb{P}_3[x]$. Let $W_1 = \{p(x) \in \mathbb{P}_3[x] \mid p(1) = 0\}$ and $W_2 = \{p(x) \in \mathbb{P}_3[x] \mid p(2) = p(3) = 0\}$. Show that W_1 and W_2 are subspaces of $\mathbb{P}_3[x]$. Find their bases.

1.17 If $p(x) \in \mathbb{P}_3[x]$ has coordinate $(1, 2, 3)$ with respect to the basis $\{1, x, x^2\}$, what are the respective coordinates of $p(x)$ with respect to the bases $\{2, 3x, 4x^2\}$ and $\{1, (x - 1), (x - 1)(x - 2)\}$?

1.18 Let $t \in \mathbb{R}$. Discuss the linear independence of the vectors over $\mathbb{R}$:

$$\alpha_1 = (1,\, 1,\, 0), \quad \alpha_2 = (1,\, 3,\, -1), \quad \alpha_3 = (5,\, 3,\, t).$$

1.19 Consider the subsets of $\mathbb{R}^2$ over $\mathbb{R}$. Answer true or false:

(a) $\{(x,y) \mid x^2 + y^2 = 0,\ x,\ y \in \mathbb{R}\}$ is a subspace of $\mathbb{R}^2$.

(b) $\{(x,y) \mid x^2 + y^2 \leq 1,\ x,\ y \in \mathbb{R}\}$ is a subspace of $\mathbb{R}^2$.

(c) $\{(x,y) \mid (x+y)^2 = 0,\ x,\ y \in \mathbb{R}\}$ is a subspace of $\mathbb{R}^2$.

(d) $\{(x,y) \mid x^2 - y^2 = 0,\ x,\ y \in \mathbb{R}\}$ is a subspace of $\mathbb{R}^2$.

(e) $\{(x,y) \mid x - y = 0,\ x,\ y \in \mathbb{R}\}$ is a subspace of $\mathbb{R}^2$.

(f) $\{(x,y) \mid x + y = 0,\ x,\ y \in \mathbb{R}\}$ is a subspace of $\mathbb{R}^2$.

(g) $\{(x,y) \mid xy = 0,\ x,\ y \in \mathbb{R}\}$ is a subspace of $\mathbb{R}^2$.

(h) $\{(x,y) \mid xy \geq 0,\ x,\ y \in \mathbb{R}\}$ is a subspace of $\mathbb{R}^2$.

(i) $\{(x,y) \mid x > 0,\ y > 0\}$ is a subspace of $\mathbb{R}^2$.

(j) $\{(x,y) \mid x,\ y$ are integers $\}$ is a subspace of $\mathbb{R}^2$.

(k) $\{(x,y) \mid x/y = 1,\ x,\ y \in \mathbb{R}\}$ is a subspace of $\mathbb{R}^2$.

(l) $\{(x,y) \mid y = 3x,\ x,\ y \in \mathbb{R}\}$ is a subspace of $\mathbb{R}^2$.

(m) $\{(x,y) \mid x - y = 1,\ x,\ y \in \mathbb{R}\}$ is a subspace of $\mathbb{R}^2$.

1.20 Consider the subsets of $\mathbb{P}_n[x]$ and $\mathbb{P}[x]$ over $\mathbb{R}$. Answer true or false:

(a) $\{p(x) \mid p(x) = ax + b,\ a,\ b \in \mathbb{R}\}$ is a subspace of $\mathbb{P}_3[x]$.

(b) $\{p(x) \mid p(x) = ax^2,\ a \in \mathbb{R}\}$ is a subspace of $\mathbb{P}_3[x]$.

(c) $\{p(x) \mid p(x) = a + x^2,\ a \in \mathbb{R}\}$ is a subspace of $\mathbb{P}_3[x]$.

(d) $\{p(x) \in \mathbb{P}[x] \mid p(x)$ has degree $3\}$ is a subspace of $\mathbb{P}[x]$.

(e) $\{p(x) \in \mathbb{P}[x] \mid p(1) = 0\}$ is a subspace of $\mathbb{P}[x]$.

(f) $\{p(x) \in \mathbb{P}[x] \mid p(0) = 1\}$ is a subspace of $\mathbb{P}[x]$.

(g) $\{p(x) \in \mathbb{P}[x] \mid p(0) = p(1) = 0\}$ is a subspace of $\mathbb{P}[x]$.

(h) $\{p(x) \in \mathbb{P}[x] \mid 2p(0) = p(1)\}$ is a subspace of $\mathbb{P}[x]$.

(i) $\{p(x) \in \mathbb{P}[x] \mid p(1) + p(2) = 0\}$ is a subspace of $\mathbb{P}[x]$.

(j) $\{p(x) \in \mathbb{P}[x] \mid p(1)p(2) = 0\}$ is a subspace of $\mathbb{P}[x]$.

(k) $\{p(x) \in \mathbb{P}[x] \mid p(x) \geq 0\}$ is a subspace of $\mathbb{P}[x]$.

(l) $\{p(x) \in \mathbb{P}[x] \mid p(-x) = p(x)\}$ is a subspace of $\mathbb{P}[x]$.

(m) $\{p(x) \in \mathbb{P}[x] \mid p(-x) = -p(x)\}$ is a subspace of $\mathbb{P}[x]$.

1.21   Consider the real vector space $\mathbb{R}^4$. Let

$$\alpha_1 = (1, -3, 0, 2), \quad \alpha_2 = (-2, 1, 1, 1), \quad \alpha_3 = (-1, -2, 1, 3).$$

Determine whether α_1, α_2, and α_3 are linearly dependent. Find the dimension and a basis for the subspace $\text{Span}\{\alpha_1, \alpha_2, \alpha_3\}$.

1.22   Let V be the subspace of $\mathbb{R}^4$ spanned by the 4-tuples

$$\alpha_1 = (1, 2, 3, 4), \quad \alpha_2 = (2, 3, 4, 5), \quad \alpha_3 = (3, 4, 5, 6), \quad \alpha_4 = (4, 5, 6, 7).$$

Find the dimension and a basis of V.

1.23   Let $\alpha_1 = (1, 0, -1, 1)$, $\alpha_2 = (2, 1, -2, 0)$, $\alpha_3 = (-2, -1, 0, 1)$, and $\alpha_4 = (0, -1, 2, 1)$. Determine if α_3 is a linear combination of the other three vectors. If so, write α_3 as a linear combination of $\alpha_1, \alpha_2, \alpha_4$.

1.24   Let $\alpha_1, \alpha_2, \alpha_3$ be linearly independent vectors. For what value(s) of k are the vectors $\alpha_2 - \alpha_1$, $k\alpha_3 - \alpha_2$, $\alpha_1 - \alpha_3$ linearly independent?

1.25   Consider all real-valued functions over $\mathbb{R}$. Define the addition by $(f + g)(x) = f(x) + g(x)$ and the scalar multiplication by $(rf)(x) = rf(x)$, $r \in \mathbb{R}$. Determine if each of the following sets is a vector space:

   (a) The set of all functions f such that $f(0) > 0$.

   (b) The set of all functions f such that $f(0) = 0$.

   (c) The set of all functions f such that $f(1) = 0$.

   (d) The set of all functions f such that $f(0) = 1$.

   (e) The set of all functions f such that $f(1) = 1$.

   (f) The set of all functions f such that $f(1) = f(-1)$.

   (g) The set of all continuous functions f such that $f(0) = 0$.

   (h) The set of all discontinuous functions f such that $f(0) = 0$.

   (i) The set of all *even functions* f, i.e., $f(-x) = f(x)$, $x \in \mathbb{R}$.

   (j) The set of all *odd functions* f, i.e., $f(-x) = -f(x)$, $x \in \mathbb{R}$.

   (k) The set of all functions f that are neither even nor odd.

   (l) The set of all functions f such that $f(0) = f(1)$.

   (m) The set of all functions f such that $f(0) = -f(1)$.

   (n) The set of all functions f such that $f(0) = f(1) \neq 0$.

   (o) The set of all functions f such that $f(x + \pi) = f(x)$, $x \in \mathbb{R}$.

1.26 Determine whether each of the following sets is linearly independent in the vector space $\mathcal{C}(\mathbb{R})$ of real-valued continuous functions over $\mathbb{R}$:

    (a) $\{x, e^x\}$.

    (b) $\{x, \sin x\}$.

    (c) $\{x, \sin x, 1 + x, e^x\}$.

    (d) $\{x, \sin x, 2x, e^x\}$.

    (e) $\{1, \sin x, \sin 2x\}$.

    (f) $\{1, \sin x, \sin^2 x\}$.

    (g) $\{1, \sin x, \cos x\}$.

    (h) $\{1, \sin x, \cos^2 x\}$.

    (i) $\{1, \sin^2 x, \cos^2 x\}$.

    (j) $\{1, \sin x, \cos 2x\}$.

    (k) $\{1, \sin^2 x, \cos 2x\}$.

    (l) $\{1, \sin x \cos x, \sin 2x\}$.

    (m) $\{1, \sin^2 x, e^x \cos 2x\}$.

    (n) $\{1, e^x, e^{2x}\}$.

    (o) $\{1, e, e^x, e^{2x}\}$.

    (p) $\{1, e^x, e^{2x}, \ldots, e^{nx}\}$, where n is a positive integer.

1.27 Let $\alpha_1, \alpha_2, \alpha_3, \alpha_4$ be linearly independent vectors. Answer true or false:

    (a) $\alpha_1 + \alpha_2$, $\alpha_2 + \alpha_3$, $\alpha_3 + \alpha_4$, $\alpha_4 + \alpha_1$ are linearly independent.

    (b) $\alpha_1 - \alpha_2$, $\alpha_2 - \alpha_3$, $\alpha_3 - \alpha_4$, $\alpha_4 - \alpha_1$ are linearly independent.

    (c) $\alpha_1 + \alpha_2$, $\alpha_2 + \alpha_3$, $\alpha_3 + \alpha_4$, $\alpha_4 - \alpha_1$ are linearly independent.

    (d) $\alpha_1 + \alpha_2$, $\alpha_2 + \alpha_3$, $\alpha_3 - \alpha_4$, $\alpha_4 - \alpha_1$ are linearly independent.

1.28 Let $\{\alpha_1, \alpha_2, \alpha_3\}$ be a basis for $\mathbb{R}^3$ and let $\alpha_4 = -\alpha_1 - \alpha_2 - \alpha_3$. Show that every vector v in $\mathbb{R}^3$ can be written as $v = a_1\alpha_1 + a_2\alpha_2 + a_3\alpha_3 + a_4\alpha_4$, where a_1, a_2, a_3, a_4 are unique real numbers such that $a_1 + a_2 + a_3 + a_4 = 0$. Generalize this to $\mathbb{R}^n$ (including $\mathbb{R}^2$).

1.29 If $\alpha_1, \alpha_2, \alpha_3$ are linearly dependent and $\alpha_2, \alpha_3, \alpha_4$ are linearly independent, show that (i) α_1 is a linear combination of α_2 and α_3, and (ii) α_4 is not a linear combination of $\alpha_1, \alpha_2,$ and α_3.

1.30   Let $A, B \in M_n(\mathbb{C})$ and $u, v \in \mathbb{C}^n$ (in columns). Answer true or false:

    (a) If $AB = 0$, then $A = 0$ or $B = 0$.

    (b) If $AB = 0$, then $BA = 0$.

    (c) $AB = BA$.

    (d) $(AB)^2 = A^2 B^2$.

    (e) $(A + I)^2 = A^2 + 2A + I$.

    (f) $(A + B)^2 = A^2 + 2AB + B^2$.

    (g) $u^t v = v^t u$.

    (h) $u^* v = v^* u$.

    (i) $u^* v = \overline{v^* u}$.

    (j) $uv^* = \overline{vu^*}$.

    (k) $(u^* Av)^* = v^* A^* u$.

    (l) $Au^* vB = u^* vAB$.

    (m) $Auv^* B = uv^* AB$.

    (n) If $A^2 = 0$, then $A = 0$.

    (o) If $A^* A = 0$, then $A = 0$.

    (p) If $A^2 = A$, then $A = 0$ or I.

    (q) $(A^* A)^* = A^* A$.

    (r) $(A^* + A)^* = A^* + A$.

    (s) $u^* A^* Au \geq 0$.

    (t) $u^* (A^* + A)u \geq 0$.

1.31   Let $V = \left\{ \left(\begin{smallmatrix} a+bi & 0 \\ 0 & a-bi \end{smallmatrix} \right) \mid a, b \in \mathbb{R} \right\}$ and $W = \left\{ \left(\begin{smallmatrix} a & -b \\ b & a \end{smallmatrix} \right) \mid a, b \in \mathbb{R} \right\}$.

    (a) Show that V and W are vector spaces over $\mathbb{R}$ but not over $\mathbb{C}$.

    (b) Verify the one-to-one correspondence between V and W via T:

$$T^* \begin{pmatrix} a + bi & 0 \\ 0 & a - bi \end{pmatrix} T = \begin{pmatrix} a & -b \\ b & a \end{pmatrix}, \quad \text{where } T = \begin{pmatrix} \frac{1}{\sqrt{2}} & \frac{i}{\sqrt{2}} \\ \frac{1}{\sqrt{2}} & -\frac{i}{\sqrt{2}} \end{pmatrix}.$$

    (c) Find the dimensions and bases of V and W over $\mathbb{R}$.

    (d) Show that a basis of V corresponds to a basis of W via T.

1.32 Let $W = \left\{ \begin{pmatrix} a & b \\ b & c \end{pmatrix} \mid a,\, b,\, c \in \mathbb{R} \right\}$. Show that W is a subspace of $M_2(\mathbb{R})$ over $\mathbb{R}$ and that the following matrices comprise a basis for W:

$$\begin{pmatrix} 1 & 0 \\ 0 & 0 \end{pmatrix}, \quad \begin{pmatrix} 0 & 1 \\ 1 & 0 \end{pmatrix}, \quad \begin{pmatrix} 0 & 0 \\ 0 & 1 \end{pmatrix}.$$

Find the coordinates of the matrix $\begin{pmatrix} 1 & -2 \\ -2 & 3 \end{pmatrix}$ with respect to the basis.

1.33 Let $A, B, C \in M_n(\mathbb{C})$ and $u, v, w \in \mathbb{C}^n$ (in columns) be given. Determine whether the following sets are subspaces of $\mathbb{C}^n$ or $M_n(\mathbb{C})$:

(a) $\{x \in \mathbb{C}^n \mid Ax = 0\}$.

(b) $\{x \in \mathbb{C}^n \mid A^2 x = 0\}$.

(c) $\{x \in \mathbb{C}^n \mid Ax = 2x\}$.

(d) $\{x \in \mathbb{C}^n \mid Ax = 2e, \text{ where } e = (1, \dots, 1)^t\}$.

(e) $\{x \in \mathbb{C}^n \mid Ax = Bx\}$.

(f) $\{x \in \mathbb{C}^n \mid Ax = Bx = Cx\}$.

(g) $\{x \in \mathbb{C}^n \mid (A^*A)x = 0\}$.

(h) $\{x \in \mathbb{C}^n \mid x^*u = 0\}$.

(i) $\{x \in \mathbb{C}^n \mid ux^* = 0\}$.

(j) $\{x \in \mathbb{C}^n \mid A(x - u) = 0\}$.

(k) $\{x - u \in \mathbb{C}^n \mid Ax = Au\}$.

(l) $\{X \in M_n(\mathbb{C}) \mid Xu = 0\}$.

(m) $\{X \in M_n(\mathbb{C}) \mid u^*Xu = 0\}$.

(n) $\{X \in M_n(\mathbb{C}) \mid u^*Xu = u^*u\}$.

(o) $\{X \in M_n(\mathbb{C}) \mid (X^*X)u = 0\}$.

(p) $\{X \in M_n(\mathbb{C}) \mid Xu = Xv + w\}$.

(q) $\{X \in M_n(\mathbb{C}) \mid w^*Xu = w^*Xv\}$.

1.34 Consider the following set V of polynomials in $\mathbb{P}_n[x]$:

$$V = \{f \in \mathbb{P}_n[x] \mid f(0) = f(1) = 0\}.$$

Show that V is a subspace of $\mathbb{P}_n[x]$ and find a basis of V.

1.35 Let $V = \{p(x) \in \mathbb{P}_3[x] \mid \int_{-1}^{1} p(x)dx = 0\}$. Show that V is a subspace of $\mathbb{P}_3[x]$. Find a basis for V and the elements of $\mathbb{P}_3[x] \setminus V$.

1.36   Show that $V = \{p(x) \in \mathbb{P}_4[x] \mid p(1) = 0\}$ is a subspace of $\mathbb{P}_4[x]$ and that the polynomials $1 - x, 1 - x^2, 1 - x^3$ comprise a basis for V. Find the coordinate of $1 - 2x + 2x^2 - x^3$ with respect to the basis.

1.37   Let $p_i(x) \in \mathbb{P}_4[x]$, $i = 1, 2, 3, 4$. If all $p_i(x)$ vanish at a point, say at a, i.e., $p_i(a) = 0$, $i = 1, 2, 3, 4$, show that $p_1(x), p_2(x), p_3(x), p_4(x)$ are linearly dependent. Does this hold if $p_i(a) = b$, $i = 1, 2, 3, 4$, for some nonzero number b? What if $\mathbb{P}_4[x]$ is replaced by $\mathbb{P}_5[x]$ or $\mathcal{C}(\mathbb{R})$?

1.38   Consider the vector spaces $\mathbb{P}_n[x]$ and $\mathbb{P}[x]$.

(a)   Show that $\{1, x, x^2, \ldots, x^{n-1}\}$ is a (standard) basis of $\mathbb{P}_n[x]$.

(b)   Let $a \in \mathbb{R}$. Show that the following set is also a basis of $\mathbb{P}_n[x]$:

$$\{1, (x - a), (x - a)^2, \ldots, (x - a)^{n-1}\}.$$

(c)   Find the coordinates of $f(x) \in \mathbb{P}_n[x]$ with respect to the basis

$$\{1, (x - a), (x - a)^2, \ldots, (x - a)^{n-1}\}, \quad a \in \mathbb{R}.$$

(d)   Let $a_1, a_2, \ldots, a_n \in \mathbb{R}$ be distinct. For $i = 1, 2, \ldots, n$, let

$$f_i(x) = (x - a_1) \cdots (x - a_{i-1})(x - a_{i+1}) \cdots (x - a_n).$$

Show that $\{f_1(x), \ldots, f_n(x)\}$ is also a basis for $\mathbb{P}_n[x]$.

(e)   Show that $W = \{f(x) \in \mathbb{P}_n[x] \mid f(1) = 0\}$ is a subspace of $\mathbb{P}_n[x]$. Find its dimension and a basis.

(f)   Let V be the set of all polynomials in $\mathbb{P}_n[x]$ that have degree exactly $n - 1$. Is V a subspace of $\mathbb{P}_n[x]$?

(g)   Show that $\mathbb{P}[x]$ has infinite dimension and the infinite set $\{1, x, x^2, \ldots\}$ is a basis of $\mathbb{P}[x]$. Thus, $\mathbb{P}[x] = \text{Span}\{1, x, x^2, \ldots\}$.

(h)   Can we say that $1 + x + x^2 + \cdots$ is contained in $\mathbb{P}[x]$?

(i)   Show that $\mathbb{P}_n[x]$ is a proper subspace of $\mathbb{P}[x]$.

1.39   Let S be a subspace of a finite dimensional vector space V. Show that

(a) $\dim S \leq \dim V$.

(b) $\dim S = \dim V$ if and only if $S = V$.

(c)   Every basis for S can be extended to some basis for V.

(d)   A basis for V need not contain a basis for S.

1.40 (1) Show that the following sets of infinite real sequences are vector spaces (under the usual operations) of infinite dimension:

   (a) $V_0 = \{(v_1, v_2, \dots) \mid v_i \in \mathbb{R}, i = 1, 2, \dots\}.$
   (b) $V_1 = \{(v_1, v_2, \dots) \mid v_i \neq 0 \text{ for at most finitely many } i\}.$
   (c) $V_2 = \{(v_1, v_2, \dots) \mid \sum_{i=1}^{\infty} |v_i|^2 < \infty\}.$
   (d) $V_3 = \{(v_1, v_2, \dots) \mid \sum_{i=1}^{\infty} |v_i| < \infty\}.$
   (e) $V_4 = \{(v_1, v_2, \dots) \mid \sum_{i=1}^{\infty} v_i \text{ converges}\}.$
   (f) $V_5 = \{(v_1, v_2, \dots) \mid \lim_{i \to \infty} v_i \text{ exists}\}.$

(2) Which space(s) in (1) contains $(1, 1, \dots)$?

(3) Which space(s) in (1) contains $(1, -1, 1, -1, \dots)$?

(4) Which space(s) in (1) contains $(1, 0, \dots), (0, 1, 0, \dots), \dots$?

(5) Which space(s) in (1) is spanned by $(1, 0, \dots), (0, 1, 0, \dots), \dots$?

(6) Find inclusion relations of the above vector spaces.

1.41 Let $\mathcal{C}(\mathbb{R})$ be the vector space of real-valued continuous functions on $\mathbb{R}$ (under the usual operations). Show that $\sin x$ and $\cos x$ are linearly independent and that the space spanned by $\sin x$ and $\cos x$

$$\text{Span}\{\sin x,\ \cos x\} = \{\, a \sin x + b \cos x \mid a,\, b \in \mathbb{R} \,\}$$

is contained in the solution set to the differential equation

$$y'' + y = 0.$$

Are $\sin^2 x$ and $\cos^2 x$ linearly independent? How about 1, $\sin^2 x$, and $\cos^2 x$? Find $\mathbb{R} \cap \text{Span}\{\sin x, \cos x\}$ and $\mathbb{R} \cap \text{Span}\{\sin^2 x, \cos^2 x\}$.

1.42 Show that $\alpha_1 = (1, 1, 0)$, $\alpha_2 = (1, 0, 1)$, and $\alpha_3 = (0, 1, 1)$ form a basis for $\mathbb{R}^3$. Find the coordinates of the vectors $u = (2, 0, 0)$, $v = (1, 0, 0)$, and $w = (1, 1, 1) \in \mathbb{R}^3$ with respect to the basis $\{\alpha_1, \alpha_2, \alpha_3\}$.

1.43 Show that $\alpha = \{\alpha_1, \alpha_2, \alpha_3\}$ and $\beta = \{\beta_1, \beta_2, \beta_3\}$ are bases of $\mathbb{R}^3$, where

$$\alpha_1 = (1, 0, 1), \quad \alpha_2 = (1, 1, 0), \quad \alpha_3 = (0, 1, 1);$$

$$\beta_1 = (1, 0, 3), \quad \beta_2 = (2, 2, 2), \quad \beta_3 = (-1, 1, 4).$$

If $v \in \mathbb{R}^3$ has coordinate $(1, 2, 3)$ with respect to the basis α, find the coordinate of v with respect to the basis β.

1.44   Consider the vector space $\mathbb{R}^n$ over $\mathbb{R}$ with the usual operations.

(a) Show that $\{e_1, \ldots, e_n\}$ and $\{\epsilon_1, \ldots, \epsilon_n\}$ are two bases, where

$$e_i = (0, \ldots, \overset{i}{1}, \ldots 0)^t, \quad \epsilon_i = (\overbrace{1, \ldots 1}^{i}, 0, \ldots, 0)^t, \quad i = 1, \ldots, n.$$

Are they also bases for $\mathbb{C}^n$ over $\mathbb{C}$? over $\mathbb{R}$?

(b) Find a matrix A such that $A(\epsilon_1, \epsilon_2, \ldots, \epsilon_n) = (e_1, e_2, \ldots, e_n)$.

(c) Find a matrix B such that $(\epsilon_1, \epsilon_2, \ldots, \epsilon_n) = B(e_1, e_2, \ldots, e_n)$.

(d) If $v \in \mathbb{R}^n$ has the coordinate $(1, 2, \ldots, n)$ under the basis $\{e_1, \ldots, e_n\}$, what is the coordinate of v under $\{\epsilon_1, \ldots, \epsilon_n\}$?

(e) Why are any $n+1$ vectors in $\mathbb{R}^n$ linearly dependent over $\mathbb{R}$?

(f) Find $n+1$ vectors in $\mathbb{C}^n$ that are linearly independent over $\mathbb{R}$.

1.45   Explain what is wrong with the statement: Since $\mathbb{R}^n$ is a vector space of dimension n and $\mathbb{C}^n$ is also a vector space of dimension n, and since $\mathbb{R}^n$ is contained in $\mathbb{C}^n$, we conclude that $\mathbb{R}^n = \mathbb{C}^n$.

1.46   Show that $\alpha = \{\alpha_1, \alpha_2, \alpha_3\}$ and $\beta = \{\beta_1, \beta_2, \beta_3\}$ are bases of $\mathbb{R}^3$, where

$$\alpha_1 = \begin{pmatrix} 1 \\ 1 \\ 1 \end{pmatrix}, \quad \alpha_2 = \begin{pmatrix} 1 \\ 0 \\ -1 \end{pmatrix}, \quad \alpha_3 = \begin{pmatrix} 1 \\ 0 \\ 1 \end{pmatrix};$$

$$\beta_1 = \begin{pmatrix} 1 \\ 2 \\ 1 \end{pmatrix}, \quad \beta_2 = \begin{pmatrix} 2 \\ 3 \\ 4 \end{pmatrix}, \quad \beta_3 = \begin{pmatrix} 3 \\ 4 \\ 3 \end{pmatrix}.$$

Find the matrix A such that

$$(\beta_1, \beta_2, \beta_3) = (\alpha_1, \alpha_2, \alpha_3)A.$$

If a vector $u \in \mathbb{R}^3$ has coordinate $(2, 0, -1)$ with respect to the basis α, what is the coordinate of u with respect to β? What is the coordinate of u with respect to the standard basis $\{e_1, e_2, e_3\}$ of $\mathbb{R}^3$?

1.47   If $\{\alpha_1, \alpha_2, \ldots, \alpha_n\}$ is a basis of a vector space V, $n \geq 2$, show that $\{\alpha_1, \alpha_1 + \alpha_2, \ldots, \alpha_1 + \alpha_2 + \cdots + \alpha_n\}$ is also a basis of V. Discuss whether $\{\alpha_1 + \alpha_2, \alpha_2 + \alpha_3, \ldots, \alpha_{n-1} + \alpha_n, \alpha_n + \alpha_1\}$ is a basis for V.

1.48   If $\alpha_1, \alpha_2, \ldots, \alpha_n$ are linearly independent in a vector space V and if $\alpha_1, \alpha_2, \ldots, \alpha_n, \beta$ are linearly dependent, show that β can be uniquely expressed as a linear combination of $\alpha_1, \alpha_2, \ldots, \alpha_n$.

1.49   Show that the vectors $\alpha_1(\neq 0), \alpha_2, \ldots, \alpha_n$ in a vector space V are linearly dependent if and only if there exists an integer k, $1 < k \leq n$, such that α_k is a linear combination of $\alpha_1, \alpha_2, \ldots, \alpha_{k-1}$.

1.50   Consider the vectors in a vector space. Answer true or false:

(a) If one of the vectors $\alpha_1, \alpha_2, \ldots, \alpha_r$ is the zero vector 0, then $\alpha_1, \alpha_2, \ldots, \alpha_r$ are linearly dependent.

(b) If $\alpha_1, \alpha_2, \ldots, \alpha_r$ are linearly independent and α_{r+1} is not a linear combination of $\alpha_1, \alpha_2, \ldots, \alpha_r$, then the vectors $\alpha_1, \alpha_2, \ldots, \alpha_r, \alpha_{r+1}$ are also linearly independent.

(c) If α is a linear combination of $\beta_1, \beta_2, \ldots, \beta_m$, and each β_i, $i = 1, 2, \ldots, m$, is a linear combination of $\gamma_1, \gamma_2, \ldots, \gamma_n$, then α is a linear combination of $\gamma_1, \gamma_2, \ldots, \gamma_n$.

(d) If $\alpha_1, \alpha_2, \ldots, \alpha_r$ are linearly independent, then no α_i is a linear combination of the rest α_j's. How about the converse?

(e) If $\alpha_1, \alpha_2, \ldots, \alpha_r$ are linearly dependent, then any α_i of these vectors is a linear combination of the rest α_j's.

(f) If β is not a linear combination of $\alpha_1, \alpha_2, \ldots, \alpha_r$, then $\beta, \alpha_1, \alpha_2, \ldots, \alpha_r$ are linearly independent.

(g) If any $r - 1$ vectors of $\alpha_1, \alpha_2, \ldots, \alpha_r$ are linearly independent, then $\alpha_1, \alpha_2, \ldots, \alpha_r$ are linearly independent.

(h) If $V = \text{Span}\{\alpha_1, \alpha_2, \ldots, \alpha_n\}$ and if every α_i is a linear combination of no more than r vectors in $\{\alpha_1, \alpha_2, \ldots, \alpha_n\}$ excluding α_i, then $\dim V \leq r$.

1.51   Let $\alpha_1, \ldots, \alpha_s$ and $\beta_1, \ldots, \beta_t$ be vectors in a vector space. Show that

$$\text{Span}\{\alpha_1, \ldots, \alpha_s, \beta_1, \ldots, \beta_t\} = \text{Span}\{\alpha_1, \ldots, \alpha_s\} + \text{Span}\{\beta_1, \ldots, \beta_t\}.$$

1.52   Let U and V be subspaces of $\mathbb{R}^n$ spanned by vectors $\alpha_1, \alpha_2, \ldots, \alpha_p$ and $\beta_1, \beta_2, \ldots, \beta_q$, respectively. Let W be the space spanned by $\alpha_i + \beta_j$, $i = 1, 2, \ldots, p$, $j = 1, 2, \ldots, q$. If $\dim U = s$ and $\dim V = t$, show that

$$\dim W \leq s + t.$$

1.53   Let $v_1, \ldots, v_n \in V$, where V is a vector space over a field $\mathbb{F}$, and let

$$W = \{(x_1, \ldots, x_n) \in \mathbb{F}^n \mid x_1 v_1 + \cdots + x_n v_n = 0\}.$$

Show that W is a subspace of $\mathbb{F}^n$ and that

$$\dim W = n - \dim \text{Span}\{v_1, \ldots, v_n\}.$$

1.54 Let $\alpha = \{\alpha_1, \ldots, \alpha_n\}$ and $\beta = \{\beta_1, \ldots, \beta_n\}$ be bases of a vector space V of dimension n over a field $\mathbb{F}$. Let W be the set of all vectors w in V that have the same coordinates with respect to α and β, i.e.,

$$w = x_1\alpha_1 + \cdots + x_n\alpha_n = x_1\beta_1 + \cdots + x_n\beta_n, \quad x_1, \ldots, x_n \in \mathbb{F}.$$

Show that W is a subspace of V and that

$$\dim W = n - \dim \operatorname{Span}\{\alpha_1 - \beta_1, \ldots, \alpha_n - \beta_n\}.$$

1.55 Let $\alpha_1, \alpha_2, \ldots, \alpha_r$ be linearly independent. If vector u is a linear combination of $\alpha_1, \alpha_2, \ldots, \alpha_r$, but vector v is not, show that the vectors $tu + v, \alpha_1, \ldots, \alpha_r$ are linearly independent for any scalar t.

1.56 Let V and W be vector spaces over a field $\mathbb{F}$. Denote by $V \times W$ the collection of ordered pairs (v, w), where $v \in V$ and $w \in W$, and define

$$(v_1, w_1) + (v_2, w_2) = (v_1 + v_2, w_1 + w_2), \quad k(v, w) = (kv, kw), \quad k \in \mathbb{F}.$$

  (a) Show that $V \times W$ is a vector space over $\mathbb{F}$.

  (b) Find $\dim(V \times W)$, given that $\dim V = m$ and $\dim W = n$.

  (c) Explain why $\mathbb{R} \times \mathbb{R}^2$ can be identified with $\mathbb{R}^3$ (as vector spaces).

  (d) Let $\alpha = 1 \in \mathbb{R}$, $e_1 = (1, 0), e_2 = (0, 1) \in \mathbb{R}^2$. Find $\mathbb{R} \times \mathbb{R}^2$ and $\operatorname{Span}\{(\alpha, e_1), (\alpha, e_2)\}$. Does $\operatorname{Span}\{(\alpha, e_1), (\alpha, e_2)\} = \mathbb{R} \times \mathbb{R}^2$?

  (e) Let $f_1 = (1, 0, 0), f_2 = (0, 1, 0), f_3 = (0, 0, 1) \in \mathbb{R}^3$. Show that $(e_i, f_j) \in \mathbb{R}^2 \times \mathbb{R}^3$, $i = 1, 2, j = 1, 2, 3$, are linearly dependent.

  (f) Find a basis for $\mathbb{R}^2 \times M_2(\mathbb{R})$.

  (g) What is the dimension of $M_2(\mathbb{R}) \times M_2(\mathbb{R})$?

1.57 Find all 2×2 matrices over a field $\mathbb{F}$ that commute with $\left(\begin{smallmatrix} 1 & 2 \\ 0 & 3 \end{smallmatrix}\right)$. Do these matrices form a vector space (under the usual matrix addition and scalar multiplication)? If so, find the dimension and a basis.

1.58 Show that the 2×2 matrices over a field $\mathbb{F}$ that commute with $A = \left(\begin{smallmatrix} 1 & 2 \\ 3 & 4 \end{smallmatrix}\right)$ form a vector space over $\mathbb{F}$. Find a basis of the space. Show that if $AB = BA$, then $B = xI + yA$ for some $x, y \in \mathbb{F}$.

1.59 Find the 3×3 matrices over a field $\mathbb{F}$ that commute with all 3×3 upper-triangular matrices with ones on the main diagonal. Do these matrices form a vector space (under the usual matrix addition and scalar multiplication)? If so, find the dimension and a basis.

1.60 Construct a 3×3 real matrix A with no zero columns or rows such that the columns of A are mutually orthogonal but the rows are not. Which of the following statements about such a matrix are true?

(a) The columns are linearly independent.

(b) The rows are linearly independent.

(c) If the columns have the same length, then so do the rows.

(d) If the rows have the same length, then so do the columns.

Is it possible to construct a 4×3 real matrix that has mutually orthogonal nonzero columns? How about a 3×4 such matrix?

1.61 Let V_n be the set of all $n \times n$ strictly upper-triangular matrices (i.e., the matrices that have (i, j)-entries equal to zero whenever $i \geq j$).

(a) Show that V_n is a subspace of $M_n(\mathbb{F})$.

(b) Find a basis and the dimension for V_n.

(c) Prove that the product of any n matrices in V_n is zero.

1.62 Let $u \in \mathbb{C}^n$ be a unit column vector, $P = uu^*$, $Q = 2P - I$. Show that

$$P^2 = P^* = P \quad \text{and} \quad Q^2 = Q^*Q = I.$$

1.63 Determine whether each of the following matrices is unitary, real orthogonal, Hermitian, skew-Hermitian, symmetric, and (or) normal:

$$A = \begin{pmatrix} 0 & 1 \\ i & 0 \end{pmatrix}, \quad B = \begin{pmatrix} 1 & i \\ -i & 1 \end{pmatrix}, \quad C = \begin{pmatrix} 1 & i \\ i & 1 \end{pmatrix}, \quad D = \begin{pmatrix} 1 & i \\ i & 0 \end{pmatrix},$$

$$E = \begin{pmatrix} 0 & i \\ i & 0 \end{pmatrix}, \quad F = \begin{pmatrix} i & i \\ i & -1 \end{pmatrix}, \quad G = \begin{pmatrix} \frac{1}{\sqrt{2}} & \frac{1}{\sqrt{2}} \\ -\frac{1}{\sqrt{2}} & \frac{1}{\sqrt{2}} \end{pmatrix}, \quad H = \begin{pmatrix} \frac{1}{\sqrt{2}} & \frac{i}{\sqrt{2}} \\ \frac{i}{\sqrt{2}} & \frac{1}{\sqrt{2}} \end{pmatrix}.$$

1.64 Find the space of all matrices commuting with matrix A, where

(a) $A = I_n$.

(b) $A = \begin{pmatrix} 1 & 1 \\ 0 & 1 \end{pmatrix}$.

(c) $A = \begin{pmatrix} a & 0 \\ 0 & b \end{pmatrix}$, $a \neq b$.

(d) $A = \begin{pmatrix} 0 & 1 & 0 & 0 \\ 0 & 0 & 1 & 0 \\ 0 & 0 & 0 & 1 \\ 0 & 0 & 0 & 0 \end{pmatrix}$.

(e) A is any (arbitrary) $n \times n$ matrix.

1.65 Let $A \in M_n(\mathbb{F})$ and let $C(A) = \{X \in M_n(\mathbb{F}) \mid AX = XA\}$.

(a) Show that $C(A)$ is a vector space over $\mathbb{F}$.

(b) Show that $\mathrm{Span}\{I, A, A^2, \ldots\} \subseteq C(A)$.

(c) Let $A = \left(\begin{smallmatrix} 1 & 0 & 0 \\ 0 & 1 & 0 \\ 3 & 1 & 2 \end{smallmatrix}\right)$. Find $C(A)$ and a basis of $C(A)$.

(d) Let $B = \left(\begin{smallmatrix} \lambda & 1 & 0 \\ 0 & \lambda & 1 \\ 0 & 0 & \lambda \end{smallmatrix}\right)$, where $\lambda \in \mathbb{C}$. Show that $\dim C(B) \geq 3$.

1.66 Let $A \in M_n(\mathbb{R})$ and let $T(A) = \{X \in M_n(\mathbb{R}) \mid AX + X^t A = 0\}$.

(a) Show that $T(A)$ is a subspace of $M_n(\mathbb{R})$.

(b) Show that $T(A^t) = T(A)$.

(c) What is $T(I_n)$?

(d) Find the dimension and a basis of $T(A)$ for $A = \left(\begin{smallmatrix} 1 & -1 \\ 0 & 0 \end{smallmatrix}\right)$.

1.67 Let $A, B \in M_n(\mathbb{C})$ and let $C(A, B) = \{X \in M_n(\mathbb{C}) \mid AX = XB\}$.

(a) Show that $C(A, B)$ is a subspace of $M_n(\mathbb{C})$.

(b) If $C(A, B) = M_n(\mathbb{C})$, what are A and B?

(c) Show that $C(A, B)$ is a subspace of $C(f(A), f(B))$ for any polynomial $f(x)$, that is, $AX = XB \Rightarrow f(A)X = Xf(B)$. (Note: $f(A) = a_0 I + a_1 A + \cdots + a_m A^m$ if $f(x) = a_0 + a_1 x + \cdots + a_m x^m$.)

(d) For $A = B = \left(\begin{smallmatrix} 0 & 1 \\ 1 & 0 \end{smallmatrix}\right)$, find $C(A, B)$ and a basis for $C(A, B)$.

(e) For $A = B = \left(\begin{smallmatrix} 1 & 0 \\ 0 & 2 \end{smallmatrix}\right)$, find $C(A, B)$ and a basis for $C(A, B)$.

1.68 Let A and B be $n \times n$ matrices.

(a) If A and B are unitary, show that AB is unitary. Is $A + B$ unitary?

(b) If A and B are Hermitian, show that $A + B$ is Hermitian. Is AB Hermitian?

(c) If A and B are normal, are $A + B$ and AB normal?

(d) Show that $\left(\begin{smallmatrix} 0 & A \\ A^* & 0 \end{smallmatrix}\right)$ is Hermitian and $\left(\begin{smallmatrix} A & A^* \\ A^* & A \end{smallmatrix}\right)$ is normal.

(e) Show that $\frac{1}{\sqrt{2}} \left(\begin{smallmatrix} I & I \\ I & -I \end{smallmatrix}\right)$ is real orthogonal and symmetric.

(f) Show that $\frac{1}{\sqrt{2}} \left(\begin{smallmatrix} acI & adI \\ bcI & -bdI \end{smallmatrix}\right)$ is unitary for any $a, b, c, d \in \mathbb{C}$ with absolute values equal to 1. In particular, $\frac{1}{\sqrt{2}} \left(\begin{smallmatrix} -iI & iI \\ I & I \end{smallmatrix}\right)$ is unitary.

1.69   Find the vector space spanned by the following matrices over $\mathbb{R}$:

$$\begin{pmatrix} 1 & 0 & 0 \\ 0 & 1 & 0 \\ 0 & 0 & 1 \end{pmatrix}, \quad \begin{pmatrix} 0 & 2 & 0 \\ 2 & 0 & 2 \\ 0 & 2 & 0 \end{pmatrix}, \quad \begin{pmatrix} 0 & 0 & -3 \\ 0 & 0 & 0 \\ 0 & 0 & 0 \end{pmatrix}.$$

1.70   Find a basis and the dimension for each of the following vector spaces:

(a) $M_n(\mathbb{C})$ over $\mathbb{C}$.

(b) $M_n(\mathbb{C})$ over $\mathbb{R}$.

(c) $M_n(\mathbb{R})$ over $\mathbb{R}$.

(d) $H_n(\mathbb{C})$, the $n \times n$ Hermitian matrices, over $\mathbb{R}$.

(e) $H_n(\mathbb{R})$, the $n \times n$ real symmetric matrices, over $\mathbb{R}$.

(f) $S_n(\mathbb{C})$, the $n \times n$ skew-Hermitian matrices, over $\mathbb{R}$.

(g) $S_n(\mathbb{R})$, the $n \times n$ real skew-symmetric matrices, over $\mathbb{R}$.

(h) $U_n(\mathbb{R})$, the $n \times n$ real upper-triangular matrices, over $\mathbb{R}$.

(i) $L_n(\mathbb{R})$, the $n \times n$ real lower-triangular matrices, over $\mathbb{R}$.

(j) $D_n(\mathbb{R})$, the $n \times n$ real diagonal matrices, over $\mathbb{R}$.

1.71   Find a basis of the space of all polynomials in A over $\mathbb{R}$ (or $\mathbb{C}$), where

$$A = \begin{pmatrix} 1 & 0 & 0 \\ 0 & \omega & 0 \\ 0 & 0 & \omega^2 \end{pmatrix}, \quad \omega = \tfrac{-1+\sqrt{3}\,i}{2}.$$

For example, $A^3 - A^2 + 5A + I_3$ is one of the polynomials.

1.72   Let A be the 3×3 upper-triangular matrix

$$A = \begin{pmatrix} 2 & 1 & 1 \\ 0 & 2 & 1 \\ 0 & 0 & 2 \end{pmatrix}.$$

Find $\mathrm{Span}\{I_3, A, A^2, A^3, \dots\}$, a basis, and the dimension of the span.

1.73   Let $V = \{\begin{pmatrix} a & b \\ c & d \end{pmatrix} \mid a + 2b + 3c + 4d = 0,$ where $a, b, c, d \in \mathbb{R}\}$. Show that V is a subspace of $M_2(\mathbb{R})$. Find a basis and the dimension of V.

1.74   Find the vector space spanned by $\begin{pmatrix} 1 & 0 \\ 0 & -1 \end{pmatrix}, \begin{pmatrix} 0 & 1 \\ -1 & 0 \end{pmatrix}, \begin{pmatrix} 0 & 1 \\ 0 & 0 \end{pmatrix}$, and $\begin{pmatrix} 0 & 0 \\ 1 & 0 \end{pmatrix}$ over $\mathbb{R}$. Find a basis and the dimension for the vector space.

1.75   Let M be a 2×2 (real or complex) matrix. Denote

$$W_{\mathbb{R}}(M) = \{\, Mx \mid x \in \mathbb{R}^2 \,\} \quad \text{and} \quad W_{\mathbb{C}}(M) = \{\, Mx \mid x \in \mathbb{C}^2 \,\}.$$

Answer the following questions with $A, B,$ and C in place of M, where

$$A = \begin{pmatrix} 1 & i \\ -i & 1 \end{pmatrix}, \quad B = \begin{pmatrix} 1 & 0 \\ 0 & i \end{pmatrix}, \quad C = \begin{pmatrix} 1 & -1 \\ 1 & -1 \end{pmatrix}.$$

(a) Show that the rows of A are linearly dependent over $\mathbb{C}$.

(b) Show that the rows of A are linearly independent over $\mathbb{R}$.

(c) Show that $W_{\mathbb{R}}(A) = W_{\mathbb{C}}(A) = \text{Im } A$.

(d) Show that $W_{\mathbb{R}}(B) \subsetneq W_{\mathbb{C}}(B)$.

(e) Show that $W_{\mathbb{C}}(A)$ is a subspace of $\mathbb{C}^2$ over $\mathbb{R}$ and also over $\mathbb{C}$. Find the dimensions and bases.

(f) Show that $W_{\mathbb{R}}(B)$ is a subspace of $\mathbb{C}^2$ over $\mathbb{R}$ but not over $\mathbb{C}$. Find $\dim W_{\mathbb{R}}(B)$ over $\mathbb{R}$.

(g) Show that $W_{\mathbb{R}}(C)$ is a subspace of $\mathbb{R}^2$ over $\mathbb{R}$ and $W_{\mathbb{C}}(C)$ is a subspace of $\mathbb{C}^2$ over $\mathbb{R}$ and over $\mathbb{C}$. What are their dimensions?

1.76   Show that each of the following sets is a subspace of $M_n(\mathbb{F})$ over a field $\mathbb{F}$. Find a basis and the dimension for each subspace.

(a) $S_1 = \{A = (a_{ij}) \in M_n(\mathbb{F}) \mid a_{11} = 0\}$.

(b) $S_2 = \{A = (a_{ij}) \in M_n(\mathbb{F}) \mid a_{11} = \cdots = a_{nn} = 0\}$.

(c) $S_3 = \{A = (a_{ij}) \in M_n(\mathbb{F}) \mid a_{11} + \cdots + a_{nn} = 0\}$.

(d) $S_4 = \{A = (a_{ij}) \in M_n(\mathbb{F}) \mid a_{pq} = a_{qp}$ for a fixed pair $(p, q), p \neq q\}$.

(e) $S_5 = \{A = (a_{ij}) \in M_n(\mathbb{F}) \mid a_{pq} = a_{qp}$ for all indices p and $q\}$.

(f) $S_6 = \{A = (a_{ij}) \in M_n(\mathbb{F}) \mid$ Every row sum of A is zero$\}$.

(g) $S_7 = \{A = (a_{ij}) \in M_n(\mathbb{F}) \mid$ The sum of all entries of A is zero$\}$.

1.77   Let $V_1 = \{X = (x_{ij}) \in M_n(\mathbb{C}) \mid x_{11} + \cdots + x_{nn} = 0\}$ and let $V_2 = \text{Span}\{AB - BA \mid A, B \in M_n(\mathbb{C})\}$. Show that $V_1 = V_2$.

1.78   Let $A \in M_{m \times n}(\mathbb{C})$ and $S(A) = \{X \in M_{n \times p}(\mathbb{C}) \mid AX = 0\}$. Show that $S(A)$ is a subspace of $M_{n \times p}(\mathbb{C})$. If A is an $n \times n$ matrix, show that

$$S(A) \subseteq S(A^2) \subseteq \cdots \subseteq S(A^k) \subseteq S(A^{k+1}) \quad \text{for any positive integer } k.$$

Show further that this inclusion chain must terminate, that is,

$$S(A^r) = S(A^{r+1}) = S(A^{r+2}) = \cdots \quad \text{for some positive integer } r.$$

1.79   Let $A, B \in M_n(\mathbb{F})$. If $AB = BA$, show that $\operatorname{Im}(AB) \subseteq \operatorname{Im} A \cap \operatorname{Im} B$.

1.80   Let A be an $m \times p$ matrix and let B be an $m \times q$ matrix. Show that the following statements are equivalent:

   (a) $\operatorname{Im} A \subseteq \operatorname{Im} B$.

   (b) The columns of A are linear combinations of the columns of B.

   (c) $A = BC$ for some matrix C (of size $q \times p$).

1.81   Let $A \in M_n(\mathbb{C})$ and let $x_1, \ldots, x_m \in \mathbb{C}^n$ (in columns). Determine the implication relations of the following statements:

   (a) $x_1, \ldots, x_m \in \mathbb{C}^n$ are linearly independent.

   (b) $Ax_1, \ldots, Ax_m$ are linearly independent.

   (c) $A^* x_1, \ldots, A^* x_m$ are linearly independent.

   (d) $A^* A x_1, \ldots, A^* A x_m$ are linearly independent.

   (e) $AA^* x_1, \ldots, AA^* x_m$ are linearly independent.

1.82   Let $A \in M_n(\mathbb{C})$.

   (a) Show that $\operatorname{Im}(A^2) \subseteq \operatorname{Im} A$.

   (b) Show that $\operatorname{Im}(AA^*) = \operatorname{Im} A$.

   (c) Is it true that $\operatorname{Im}(A^* A) = \operatorname{Im} A$?

1.83   Let A be an $m \times n$ matrix. Prove each of the following statements.

   (a) $\operatorname{Ker} A = \{0\}$ if and only if the columns of A are linearly independent. If the columns of A are linearly independent, are the rows of A necessarily linearly independent?

   (b) If $m < n$, then $\operatorname{Ker} A \neq \{0\}$.

   (c) $\operatorname{Ker} A \subseteq \operatorname{Ker}(A^2)$ when $m = n$.

   (d) $\operatorname{Ker}(A^* A) = \operatorname{Ker} A$. When $m = n$, is $\operatorname{Ker}(AA^*) = \operatorname{Ker} A$?

   (e) If $A = BC$, where B is $m \times m$ and C is $m \times n$, and if B is nonsingular, then $\operatorname{Ker} A = \operatorname{Ker} C$.

1.84   Let $\{v_1, v_2, \ldots, v_n\}$ be a basis of a vector space V. Suppose W is a subspace of V with dimension k, $1 < k < n$. Show that, for any subset $\{v_{i_1}, v_{i_2}, \ldots, v_{i_m}\}$ of $\{v_1, v_2, \ldots, v_n\}$, $n - k < m \leq n$, there exists a nonzero vector $w \in W$ that is a linear combination of $v_{i_1}, v_{i_2}, \ldots, v_{i_m}$.

1.85   Let V be the plane in $\mathbb{R}^3$ defined by the equation $x + 2y + 3z = 0$.

    (a) Show that V is a subspace of $\mathbb{R}^3$.

    (b) Find the dimension and a basis of V.

    (c) Show that the line $x = y = -z$ is a subspace of V.

    (d) Find a unit vector that is orthogonal (perpendicular) to V.

1.86   Let

$$
W = \left\{ \begin{pmatrix} x_1 \\ x_2 \\ x_3 \\ x_4 \end{pmatrix} \in \mathbb{C}^4 \;\middle|\; x_3 = x_1 + x_2 \text{ and } x_4 = x_1 - x_2 \right\}.
$$

    (a) Prove that W is a subspace of $\mathbb{C}^4$.

    (b) Find a basis for W. What is the dimension of W?

    (c) Prove that $\{\, k(1,0,1,1)^t \mid k \in \mathbb{C} \,\}$ is a subspace of W.

1.87   Let $A \in M_n(\mathbb{F})$, let λ and μ be scalars such that $\lambda \neq \mu$, and let $V_\lambda = \{x \in \mathbb{F}^n \mid Ax = \lambda x\}$ and $V_\mu = \{y \in \mathbb{F}^n \mid Ay = \mu y\}$. Show that

    (a) $V_\lambda \cap V_\mu = \{0\}$.

    (b) V_λ and V_μ are subspaces of $\mathbb{F}^n$.

    (c) Any two nonzero vectors, one from V_λ and one from V_μ, are linearly independent.

1.88   Let W_1 and W_2 be subspaces of a vector space V.

    (a) Show that $W_1 \cap W_2$ and $W_1 + W_2$ are subspaces of V, and

$$
W_1 \cap W_2 \subseteq W_1 \cup W_2 \subseteq W_1 + W_2.
$$

    (b) Explain the inclusions in (a) geometrically with two lines passing through the origin in the xy-plane.

    (c) When is $W_1 \cup W_2$ a subspace of V?

    (d) Show that $W_1 + W_2$ is the smallest subspace of V containing W_1 and W_2, that is, if S is a subspace of V containing W_1 and W_2, then S contains $W_1 + W_2$.

    (e) Show that $W_1 \cap W_2$ is the largest subspace of V contained in W_1 and W_2, that is, if L is a subspace of V contained in W_1 and W_2, then $W_1 \cap W_2$ contains L.

1.89 Show that each of the following sets is a subspace of $M_2(\mathbb{R})$:

$$U = \left\{ \begin{pmatrix} x & 0 \\ a & -x \end{pmatrix} \mid a, x \in \mathbb{R} \right\},$$

$$V = \left\{ \begin{pmatrix} x & b \\ a & -x \end{pmatrix} \mid a, b, x \in \mathbb{R} \right\},$$

$$W = \left\{ \begin{pmatrix} x & -(x+y) \\ 0 & y \end{pmatrix} \mid x, y \in \mathbb{R} \right\}.$$

Find $U + W$, $V \cap W$, and bases for U, V, W, $U + W$, and $V \cap W$.

1.90 Let $\{u_1, \ldots, u_m\}$ and $\{v_m, \ldots, v_n\}$ be two sets of linearly independent vectors in an n-dimensional vector space V, where $1 \leq m < n$. Show that there exists a nonzero vector w lying in $\mathrm{Span}\{u_1, \ldots, u_m\} \cap \mathrm{Span}\{v_m, \ldots, v_n\}$. Is the assertion true if the linear independence of $\{u_1, \ldots, u_m\}$ or $\{v_m, \ldots, v_n\}$ is not assumed?

1.91 Let V be a finite dimensional vector space over a field $\mathbb{F}$. Let V_1 and V_2 be subspaces of V. Show that $V_1 \subseteq V_2$ or $V_2 \subseteq V_1$ if

$$\dim(V_1 + V_2) = \dim(V_1 \cap V_2) + 1.$$

Equivalently, if $V_1 \not\subseteq V_2$ and $V_2 \not\subseteq V_1$, show that

$$\dim(V_1 + V_2) \geq \dim(V_1 \cap V_2) + 2.$$

1.92 Let V be a vector space of dimension n over a field $\mathbb{F}$ and let W be a proper subspace of V. Show that W is the intersection of all subspaces of V that contain W and have dimension $n - 1$.

1.93 Let V be a finite dimensional vector space over a field $\mathbb{F}$ and let W be a proper subspace of V. Show that there exists a basis of V such that no vector in the basis lies in W. Explain this in $\mathbb{R}^3$.

1.94 Let V be a finite dimensional vector space over a field $\mathbb{F}$ and let $W_1, \ldots, W_k$ be proper subspaces of V. Show that there exists a vector $v \in V$ that falls outside of the subspaces, i.e., $v \notin W_1 \cup \cdots \cup W_k$.

1.95 Let V be a finite dimensional vector space over a field $\mathbb{F}$ and let $W_1, \ldots, W_k$ be proper subspaces of V. Show that there exists a basis of V such that any vector in the basis is not contained in $W_1 \cup \cdots \cup W_k$.

1.96   Give an example of three subspaces of a vector space V such that

$$W_1 \cap (W_2 + W_3) \neq (W_1 \cap W_2) + (W_1 \cap W_3).$$

Why does this not contradict the following identity for any three sets

$$A \cap (B \cup C) = (A \cap B) \cup (A \cap C)?$$

1.97   Let S_1, S_2, S_3 be subspaces of a vector space V of dimension n. Show that

$$\dim(S_1 \cap S_2 \cap S_3) \geq \dim S_1 + \dim S_2 + \dim S_3 - 2n.$$

1.98   Let S_1, S_2, and S_3 be subspaces of $\mathbb{R}^4$ and they all have dimension 3. Show that $S_1 \cap S_2 \cap S_3$ contains a subspace of dimension 1.

1.99   Let V_1, V_2, and V_3 be subspaces of a vector space V of finite dimension. If $\dim V_1 = \dim V_2 = \dim V_3 = 3$ and $V_1 \cap V_2 \cap V_3 = \{0\}$, what can the smallest dimension of V be? Explain this by an example.

1.100   Let W_1 and W_2 be subspaces of a finite dimensional vector space V. Show that the following statements are equivalent:

(a) $W_1 + W_2$ is a direct sum.

(b) If $w_1 + w_2 = 0$ and $w_1 \in W_1$, $w_2 \in W_2$, then $w_1 = w_2 = 0$.

(c) $W_1 \cap W_2 = \{0\}$.

(d) $\dim(W_1 + W_2) = \dim W_1 + \dim W_2$.

Extend the direct sum of two subspaces for multiple subspaces.

1.101   If W is a nontrivial subspace of a vector space V, show that distinct subspaces W_1 and W_2 of V exist such that $V = W \oplus W_1 = W \oplus W_2$. Explain this using $\mathbb{R}^2$ and the lines (subspaces) $x = 0$, $y = 0$, $y = x$.

1.102   Let W_1, W_2, and W_3 be subspaces of a vector space V. Show that the sum $W_1 + W_2 + W_3 = \{ w_1 + w_2 + w_3 \mid w_i \in W_i, \ i = 1, 2, 3 \}$ is also a subspace of V. Show by example that $W_1 + W_2 + W_3$ need not be a direct sum (i.e., there exist elements w_1, w_2, w_3, not all zero, such that $w_1 + w_2 + w_3 = 0$ and $w_i \in W_i$, $i = 1, 2, 3$) even though

$$W_1 \cap W_2 = W_1 \cap W_3 = W_2 \cap W_3 = \{0\}.$$

1.103 Consider the vector space $\mathcal{C}(\mathbb{R})$ of all real-valued continuous functions over $\mathbb{R}$. Let W_1 and W_2 be the sets of even and odd functions in $\mathcal{C}(\mathbb{R})$, respectively. Show that W_1 and W_2 are subspaces of $\mathcal{C}(\mathbb{R})$. Show further that $\mathcal{C}(\mathbb{R}) = W_1 \oplus W_2$. What is $W_1 \cap W_2$?

1.104 Let $P = \{ax^2 + bx + c \mid a, b, c \in \mathbb{R}\}$ (i.e., $\mathbb{P}_3[x]$) and let Q be the set of continuous even functions on $\mathbb{R}$. Show that P and Q are subspaces of $\mathcal{C}(\mathbb{R})$. Determine $P \cap Q$ and $P + Q$. Is $P + Q$ a direct sum?

1.105 Let V_1 be the subspace of real symmetric matrices in $M_2(\mathbb{R})$ and let V_2 be the subspace of real skew-symmetric matrices in $M_2(\mathbb{R})$. Let

$$W_1 = \left\{ \begin{pmatrix} a & b \\ b & -a \end{pmatrix} \middle| \; a, b \in \mathbb{R} \right\}$$

and

$$W_2 = \left\{ \begin{pmatrix} c & d \\ -d & c \end{pmatrix} \middle| \; c, d \in \mathbb{R} \right\}.$$

(a) Prove that $M_2(\mathbb{R}) = V_1 \oplus V_2 = W_1 \oplus W_2$.

(b) Find the dimensions and bases for V_1, V_2, W_1, and W_2.

(c) Verify that $W_1 \subset V_1$ and $W_2 \supset V_2$.

(d) Find a subspace W_3 such that $W_3 \not\supset V_2$ and $M_2(\mathbb{R}) = W_1 \oplus W_3$.

1.106 Let $A \in M_n(\mathbb{F})$ be an *idempotent matrix*, i.e., $A^2 = A$. Show that

$$\mathbb{F}^n = \operatorname{Ker} A \oplus \operatorname{Im} A = \operatorname{Ker} A \oplus \operatorname{Ker}(I - A)$$

Give an example showing that the converse is not true in general.

1.107 Consider $M_n(\mathbb{R})$. Let H_n, S_n, V_n, and L_n be respectively the sub-spaces of symmetric matrices, skew-symmetric matrices, strictly upper-triangular matrices, and lower-triangular matrices over $\mathbb{R}$. Show that

$$M_n(\mathbb{R}) = H_n \oplus S_n = V_n \oplus L_n.$$

Discuss if the above direct sums are generalized for complex matrices.

Chapter 2

Determinants, Inverses, Ranks, and Systems of Linear Equations

Definitions, Facts, and Examples —————————————————

We summarize some basic and elementary facts about determinants, inverses, and ranks of matrices as well as systems of linear equations. We skip the proofs of these facts that are easily obtainable online or from texts.

Determinant. A *determinant* is a number assigned to a square matrix in a certain way. This number contains much information about the matrix. A very useful piece is that it tells immediately whether the matrix is invertible. For a square matrix A, we denote its determinant by

$$|A| \quad \text{or} \quad \det A.$$

Both notations are used in this book. For example, $|A| = \begin{vmatrix} X & Y \\ U & V \end{vmatrix}$ is the determinant of block matrix $A = \begin{pmatrix} X & Y \\ U & V \end{pmatrix}$, and $\det(B + C) = |B + C|$ is the determinant of $B + C$. Note that $|\cdot|$ is also used for absolute values of complex numbers (while in other books, $|A|$ may denote the modulus matrix $(A^*A)^{\frac{1}{2}}$ or the entrywise absolute value matrix $(|a_{ij}|)$). However, one can easily tell from context which use is intended.

If A is a 1×1 matrix, that is, A has one entry, say a_{11}, its determinant is defined to be $|A| = a_{11}$. If A is a 2×2 matrix, say $A = \begin{pmatrix} a_{11} & a_{12} \\ a_{21} & a_{22} \end{pmatrix}$, then $|A| = \begin{vmatrix} a_{11} & a_{12} \\ a_{21} & a_{22} \end{vmatrix} = a_{11}a_{22} - a_{12}a_{21}$. For $n \geq 3$, the determinant of an $n \times n$ matrix $A = (a_{ij})$ can be defined inductively as follows. Assume the determinant is defined for $(n-1) \times (n-1)$ matrices. Let A_{1j} denote the submatrix of A by deleting the first row and the j-th column of A. Then

$$|A| = a_{11}|A_{11}| - a_{12}|A_{12}| + \cdots + (-1)^{1+n}a_{1n}|A_{1n}|. \tag{2.1}$$

Example 2.1 Find the determinant of the matrix

$$A = \begin{pmatrix} 2 & 3 & 0 & 0 \\ 4 & a & b & 0 \\ 0 & b & a & c \\ 0 & 0 & c & a \end{pmatrix}.$$

Solution. We calculate the determinant by using the formula (2.1):

$$
\begin{aligned}
|A| &= 2 \begin{vmatrix} a & b & 0 \\ b & a & c \\ 0 & c & a \end{vmatrix} - 3 \begin{vmatrix} 4 & b & 0 \\ 0 & a & c \\ 0 & c & a \end{vmatrix} \\
&= 2 \left(a \begin{vmatrix} a & c \\ c & a \end{vmatrix} - b \begin{vmatrix} b & c \\ 0 & a \end{vmatrix} \right) - 3 \left(4 \begin{vmatrix} a & c \\ c & a \end{vmatrix} - b \begin{vmatrix} 0 & c \\ 0 & a \end{vmatrix} \right) \\
&= 2 \left(a(a^2 - c^2) - b^2 a \right) - 12(a^2 - c^2) \\
&= 2a^3 - 12a^2 - 2ab^2 - 2ac^2 + 12c^2. \quad \square
\end{aligned}
$$

Determinants can be defined in different (but equivalent) ways. Let $A = (a_{ij})$ be an $n \times n$ matrix, $n \geq 2$, and let A_{ij} denote the $(n-1) \times (n-1)$ submatrix of A by deleting row i and column j from A, $1 \leq i, j \leq n$. Like the expansion formula (2.1), the determinant of A can be computed along any row, say, row i:

$$|A| = \sum_{j=1}^{n} (-1)^{i+j} a_{ij} |A_{ij}|. \quad (Laplace\ expansion)$$

It can be proved that the value $|A|$ is independent of choices of row i. Similarly, the Laplace expansion along a column may be defined. Alternatively, determinant is the sum of all possible signed products, where every product contains one and only one element from each row and each column.

Example 2.2 Find the determinant $|A_n|$, where $A_n = (a_{ij})$ is the $n \times n$ matrix with diagonal $a_{ii} = 2$, $1 \leq i \leq n$, superdiagonal $a_{i,i+1} = -1$, subdiagonal $a_{i+1,i} = -1$, $1 \leq i \leq n-1$, and $a_{ij} = 0$ otherwise, that is,

$$A_n = \begin{pmatrix} 2 & -1 & \cdots & & 0 \\ -1 & 2 & \ddots & & \vdots \\ & \ddots & \ddots & \ddots & -1 \\ \vdots & & \ddots & \ddots & \\ 0 & \cdots & & -1 & 2 \end{pmatrix}.$$

Solution. By the Laplace formula, we expand $|A_n|$ along the first column:

$$
|A_n| = 2|A_{n-1}| - (-1) \times
\begin{vmatrix}
-1 & 0 & 0 & \cdots & 0 \\
-1 & 2 & -1 & \cdots & 0 \\
0 & -1 & 2 & \ddots & \vdots \\
\vdots & \vdots & \ddots & \ddots & -1 \\
0 & 0 & \cdots & -1 & 2
\end{vmatrix}_{(n-1)\times(n-1)}
$$

$$
= 2|A_{n-1}| - (-1)^2
\begin{vmatrix}
2 & -1 & \cdots & 0 \\
-1 & 2 & \ddots & \vdots \\
\vdots & \ddots & \ddots & -1 \\
0 & \cdots & -1 & 2
\end{vmatrix}_{(n-2)\times(n-2)}
\begin{pmatrix}
\text{the subscripts} \\
\text{indicate sizes}
\end{pmatrix}
$$

$$
= 2|A_{n-1}| - |A_{n-2}|.
$$

This gives $|A_n| - |A_{n-1}| = |A_{n-1}| - |A_{n-2}|$ for all $n \geq 3$. It follows that

$$
|A_n| = (|A_n| - |A_{n-1}|) + (|A_{n-1}| - |A_{n-2}|) + \cdots + (|A_2| - |A_1|) + |A_1|
$$
$$
= (n-1)((|A_2| - |A_1|) + |A_1| = (n-1)(3-2) + 2 = n + 1. \quad \square
$$

In the Laplace expansion formula, $|A_{ij}|$ and $(-1)^{i+j}|A_{ij}|$ are called the *minor* and *cofactor* of the (i,j)-entry a_{ij}, respectively, and the matrix whose (j,i)-entry is the cofactor of a_{ij} is referred to as the *cofactor matrix* or *adjugate* of A and is denoted by adj(A). So, adj$(A) = \left((-1)^{i+j}|A_{ji}|\right)$.

It is a very important fact that

$$
A \operatorname{adj}(A) = \operatorname{adj}(A)A = |A|I. \tag{2.2}
$$

The determinant of a square submatrix of matrix A is said to be a *minor* of A. A *principal minor* is the determinant of a principal submatrix.

Another definition of determinant in terms of permutations is concise and convenient. A permutation p on $\{1, 2, \ldots, n\}$ is said to be *even* (or *odd*) if p can be restored to natural order by an even (resp. odd) number of interchanges. For instance, consider the permutations on $\{1, 2, 3, 4\}$. (There are $4! = 24$ such permutations.) The permutation $p = (2, 1, 4, 3)$, namely, $p(1) = 2, p(2) = 1, p(3) = 4, p(4) = 3$, is even since it will become $(1, 2, 3, 4)$ after interchanging 2 and 1 and 4 and 3 (two interchanges), while $(1, 4, 3, 2)$ is odd, for interchanging 4 and 2 (one interchange) gives $(1, 2, 3, 4)$.

Let S_n be the set of all permutations of $\{1, 2, \ldots, n\}$. For $p \in S_n$, define $\operatorname{sgn}(p) = +1$ if p is even and $\operatorname{sgn}(p) = -1$ if p is odd. It can be proved that

$$
|A| = \det A = \sum_{p \in S_n} \operatorname{sgn}(p) \prod_{t=1}^{n} a_{tp(t)}. \tag{2.3}
$$

Properties of Determinants. Let A and B be $n \times n$ matrices. The following statements hold true (for rows as well as for columns).

(1) The determinant of an identity matrix is equal to 1, i.e., $|I| = 1$.

(2) If $A = \text{diag}(a_{11}, a_{22}, \dots, a_{nn})$, then $|A| = a_{11}a_{22}\cdots a_{nn}$.

(3) If $A = (a_{ij})$ is upper- or lower-triangular, then $|A| = a_{11}a_{22}\cdots a_{nn}$.

(4) If an entire row of A consists of zeros, then $|A| = 0$.

(5) If B is obtained from A by exchanging two rows of A, then $|B| = -|A|$, that is, the determinant changes its sign if two rows are exchanged.

(6) If B is obtained from A by adding (or subtracting) one row of A to (resp. from) another row of A, then $|B| = |A|$, i.e., the determinant does not change if one row is added to (subtracted from) another row.

(7) If B is obtained from A by adding one row of A multiplied by a scalar to another row of A, then $|B| = |A|$, that is, the determinant does not change if one row multiplied by a scalar is added to another row. (This fact is repeatedly used in determinant computation.)

(8) If B is obtained from A by multiplying one row of A by a scalar k, then $|B| = k|A|$. So $|kA| = k^n|A|$.

(9) Determinant is linear with respect to each row (known as *multilinear*). Take the first row as an example, $\begin{vmatrix} \alpha+k\beta \\ C \end{vmatrix} = \begin{vmatrix} \alpha \\ C \end{vmatrix} + k \begin{vmatrix} \beta \\ C \end{vmatrix}$, where $\alpha, \beta \in \mathbb{C}^n$ (in rows), k is a scalar, and C is an $(n-1) \times n$ matrix.

(10) If two rows of A are equal (identical), then $|A| = 0$.

(11) The rows of A are linearly dependent if and only if $|A| = 0$.

(12) A is singular (or $Ax = 0$ has a solution $x \neq 0$) if and only if $|A| = 0$.

(13) $|A| = |A^t|$, $|A^*| = \overline{|A|} = |\bar{A}|$, and $|A^{-1}| = |A|^{-1}$ when A is invertible.

(14) $|AB| = |A|\,|B|$; $|A^k| = |A|^k$ for any positive integer k.

(15) $|S^{-1}AS| = |A|$ for any $n \times n$ invertible matrix S.

(16) $\begin{vmatrix} A & X \\ 0 & B \end{vmatrix} = |A|\,|B|$ (for any matrix X of appropriate size).

(17) $\left| \begin{pmatrix} I & Y \\ 0 & I \end{pmatrix} \begin{pmatrix} A & C \\ D & B \end{pmatrix} \right| = \begin{vmatrix} A+YD & C+YB \\ D & B \end{vmatrix} = \begin{vmatrix} A & C \\ D & B \end{vmatrix}$.

(18) $\left| \begin{pmatrix} A & C \\ D & B \end{pmatrix} \begin{pmatrix} I & Z \\ 0 & I \end{pmatrix} \right| = \begin{vmatrix} A & AZ+C \\ D & DZ+B \end{vmatrix} = \begin{vmatrix} A & C \\ D & B \end{vmatrix}$.

Note: (16)–(17) also hold for square matrices A and B of different orders.

Example 2.3 (Vandermonde matrix) For $x_1, \ldots, x_n \in \mathbb{C}$, let

$$
V_n = \begin{pmatrix} 1 & x_1 & x_1^2 & \cdots & x_1^{n-1} \\ 1 & x_2 & x_2^2 & \cdots & x_2^{n-1} \\ \vdots & \vdots & \vdots & \ddots & \vdots \\ 1 & x_n & x_n^2 & \cdots & x_n^{n-1} \end{pmatrix}.
$$

Show that

$$
|V_n| = \prod_{1 \le i < j \le n} (x_j - x_i).
$$

Solution. We prove by induction on n. If $n = 2$, then $|V_2| = \begin{vmatrix} 1 & x_1 \\ 1 & x_2 \end{vmatrix} = x_2 - x_1$.

Suppose that the assertion is true for a Vandermonde matrix of order $n - 1$. We show the case for n. To compute $|V_n|$, multiply column $n - 1$ by x_1, then subtract it from column n; multiply column $n - 2$ by x_1, then subtract it from column $n - 1, \ldots$, multiply column 1 by x_1, then subtract it from column 2. We arrive at

$$
\begin{aligned}
|V_n| &= \begin{vmatrix} 1 & 0 & 0 & \cdots & 0 \\ 1 & x_2 - x_1 & x_2(x_2 - x_1) & \cdots & x_2^{n-2}(x_2 - x_1) \\ \vdots & \vdots & \vdots & \ddots & \vdots \\ 1 & x_n - x_1 & x_n(x_n - x_1) & \cdots & x_n^{n-2}(x_n - x_1) \end{vmatrix} \\
&= \begin{vmatrix} x_2 - x_1 & x_2(x_2 - x_1) & \cdots & x_2^{n-2}(x_2 - x_1) \\ \vdots & \vdots & \ddots & \vdots \\ x_n - x_1 & x_n(x_n - x_1) & \cdots & x_n^{n-2}(x_n - x_1) \end{vmatrix} \\
&= (x_2 - x_1) \cdots (x_n - x_1) \begin{vmatrix} 1 & x_2 & \cdots & x_2^{n-2} \\ \vdots & \vdots & \ddots & \vdots \\ 1 & x_n & \cdots & x_n^{n-2} \end{vmatrix} \\
&= (x_2 - x_1) \cdots (x_n - x_1) \prod_{2 \le i < j \le n} (x_j - x_i) \\
&= \prod_{1 \le i < j \le n} (x_j - x_i). \quad \square
\end{aligned}
$$

Elementary Row and Column Operations on (Block) Matrices.

    I. Interchange two rows (columns).

    II. Multiply a row (column) by a nonzero scalar, say $k \ne 0$.

    III. Add k times a row (column) to another row (column).

Suppose that A is a square matrix, and B, C, D are matrices obtained from A by the elementary row operations I, II, and III, respectively. Then

$$|B| = -|A|, \quad |C| = k|A|, \quad |D| = |A|.$$

Let E_I, E_{II}, and E_{III} denote the matrices obtained from the identity matrix by an application of I, II, and III, respectively, and call them *elementary matrices*. Then $B = E_I A$, $C = E_{II} A$, $D = E_{III} A$. If an elementary column operation is applied to A, the resulting matrix is A post-multiplied by the corresponding elementary matrix, i.e., AE_I, AE_{II}, or AE_{III}.

The 2×2 cases are exemplary. For instance, (I) interchange two columns:

$$\begin{pmatrix} a & b \\ c & d \end{pmatrix} \to \begin{pmatrix} b & a \\ d & c \end{pmatrix}, \quad \begin{pmatrix} a & b \\ c & d \end{pmatrix} \begin{pmatrix} 0 & 1 \\ 1 & 0 \end{pmatrix} = \begin{pmatrix} b & a \\ d & c \end{pmatrix}$$

and (III) add row one multiplied by k to row two:

$$\begin{pmatrix} a & b \\ c & d \end{pmatrix} \to \begin{pmatrix} a & b \\ c+ka & d+kb \end{pmatrix}, \quad \begin{pmatrix} 1 & 0 \\ k & 1 \end{pmatrix} \begin{pmatrix} a & b \\ c & d \end{pmatrix} = \begin{pmatrix} a & b \\ c+ka & d+kb \end{pmatrix}.$$

Elementary row and column operations for block matrices and *elementary block matrices* can be similarly defined. For example, $\left(\begin{smallmatrix} I_m & X \\ 0 & I_n \end{smallmatrix} \right)$ is a type III elementary block matrix. When applied to a block matrix, it is equivalent to applying a sequence of type III elementary matrices to the matrix.

Consider the 2×2 partitioned matrix $\left(\begin{smallmatrix} A & B \\ C & D \end{smallmatrix} \right)$. Exchanging two (block) rows and then subtracting the first (block) column post-multiplied by X (with appropriate size) from the second (block) column result in

$$\begin{pmatrix} 0 & I \\ I & 0 \end{pmatrix} \begin{pmatrix} A & B \\ C & D \end{pmatrix} \begin{pmatrix} I & -X \\ 0 & I \end{pmatrix} = \begin{pmatrix} C & D-CX \\ A & B-AX \end{pmatrix}.$$

Operation III does not change the determinant. An interesting application of this, with the fact $\left| \begin{smallmatrix} A & X \\ 0 & B \end{smallmatrix} \right| = |A||B|$ (Property (16)), is to derive the determinant identity $|AB| = |A|\,|B|$ for any $n \times n$ matrices A and B:

$$\begin{aligned} |AB| &= \begin{vmatrix} AB & A \\ 0 & I \end{vmatrix} = \left| \begin{pmatrix} AB & A \\ 0 & I \end{pmatrix} \begin{pmatrix} I & 0 \\ -B & I \end{pmatrix} \right| \quad \text{(by Property (18))} \\ &= \begin{vmatrix} 0 & A \\ -B & I \end{vmatrix} = \left| \begin{pmatrix} 0 & I \\ I & 0 \end{pmatrix} \begin{pmatrix} -B & I \\ 0 & A \end{pmatrix} \right| \quad \text{(by Property (5))} \\ &= (-1)^{n^2} |-B|\,|A| = (-1)^{n^2+n}|A|\,|B| = |A|\,|B|. \end{aligned}$$

Suppose that A is not a zero matrix. Through elementary row and column operations, A can be brought to a matrix in the form $\left(\begin{smallmatrix} I_r & 0 \\ 0 & 0 \end{smallmatrix} \right)$, that is, there are elementary matrices $R_1, \ldots, R_p$ and $C_1, \ldots, C_q$ such that

$$A = R_p \cdots R_1 \begin{pmatrix} I_r & 0 \\ 0 & 0 \end{pmatrix} C_1 \cdots C_q.$$

Inverse. Let A be an $n \times n$ matrix. A matrix B is said to be an *inverse* of A if $AB = BA = I$. If A has an inverse, then its inverse is unique, and we denote it by A^{-1}. Moreover, since $|AA^{-1}| = |A| \, |A^{-1}| = 1$, we have

$$|A^{-1}| = \frac{1}{|A|}.$$

A square matrix A is *invertible* (or *nonsingular*) if and only if $|A| \neq 0$. In addition, if A is invertible, then so is its transpose A^t, and $(A^t)^{-1} = (A^{-1})^t$; if A and B are $n \times n$ invertible matrices, then AB is invertible and

$$(AB)^{-1} = B^{-1} A^{-1}.$$

Moreover, if $A, B \in M_n(\mathbb{C})$ satisfy $AB = I$, then $BA = I$. So $B = A^{-1}$. When matrix A is invertible, the inverse can be found via its adjugate:

$$A^{-1} = \frac{1}{|A|} \operatorname{adj}(A). \tag{2.4}$$

For a 2×2 matrix, the following formula is useful and convenient:

If $A = \begin{pmatrix} a & b \\ c & d \end{pmatrix}$ and $ad - bc \neq 0$, then $A^{-1} = \frac{1}{ad - bc} \begin{pmatrix} d & -b \\ -c & a \end{pmatrix}$.

For elementary block matrix (type III), we have $\begin{pmatrix} I_m & X \\ 0 & I_n \end{pmatrix}^{-1} = \begin{pmatrix} I_m & -X \\ 0 & I_n \end{pmatrix}.$

Below are important facts on the existence of invertible or unitary matrices with the given linearly independent or orthonormal vectors.

(i) Let $v_1, \ldots, v_k \in \mathbb{C}^n$ (in columns) be k $(1 \leq k < n)$ linearly independent vectors. Then there exist vectors $v_{k+1}, \ldots, v_n \in \mathbb{C}^n$ such that $v_1, v_2, \ldots, v_n$ are linearly independent. Let $P = (v_1, v_2, \ldots, v_n)$. Then P is an $n \times n$ invertible matrix, that is, $P^{-1}P = PP^{-1} = I$.

(ii) Let $u_1, \ldots, u_k \in \mathbb{C}^n$ (in columns) be k $(1 \leq k < n)$ orthonormal vectors, i.e., $\|u_i\| = 1$, $i = 1, \ldots, k$, $u_i^* u_j = 0$ if $i \neq j$. Then there exist vectors $u_{k+1}, \ldots, u_n \in \mathbb{C}^n$ such that $u_1, u_2, \ldots, u_n$ are orthonormal. Let $U = (u_1, u_2, \ldots, u_n)$. Then U is unitary, i.e., $U^*U = UU^* = I$.

Every invertible matrix is a product of some elementary matrices. This is seen by applying on the matrix a series of elementary row operations (row reduction) to arrive at the identity matrix I. For a nonsingular matrix A, one may find the inverse of A by converting the matrix (A, I) to a matrix in the form (I, B) through elementary row operations. Let B be the product of the elementary matrices corresponding to the elementary row operations. Then B is A^{-1}, because $B(A, I) = (BA, B) = (I, B)$, revealing $BA = I$.

Example 2.4 Find the inverse of the matrix

$$A = \begin{pmatrix} 2 & -1 & 0 \\ -1 & 2 & -1 \\ 0 & -1 & 2 \end{pmatrix}.$$

Solution. We perform row operations on the matrix (A, I) to get (I, A^{-1}):

$$(A, I) = \begin{pmatrix} 2 & -1 & 0 & 1 & 0 & 0 \\ -1 & 2 & -1 & 0 & 1 & 0 \\ 0 & -1 & 2 & 0 & 0 & 1 \end{pmatrix} \quad \text{(add row 2 to row 1)}$$

$$\rightarrow \begin{pmatrix} 1 & 1 & -1 & 1 & 1 & 0 \\ -1 & 2 & -1 & 0 & 1 & 0 \\ 0 & -1 & 2 & 0 & 0 & 1 \end{pmatrix} \quad \text{(add row 1 to row 2)}$$

$$\rightarrow \begin{pmatrix} 1 & 1 & -1 & 1 & 1 & 0 \\ 0 & 3 & -2 & 1 & 2 & 0 \\ 0 & -1 & 2 & 0 & 0 & 1 \end{pmatrix} \quad \text{(add row 3 times 3 to row 2)}$$

$$\rightarrow \begin{pmatrix} 1 & 1 & -1 & 1 & 1 & 0 \\ 0 & 0 & 4 & 1 & 2 & 3 \\ 0 & -1 & 2 & 0 & 0 & 1 \end{pmatrix} \quad \text{(exchange rows 2 and 3)}$$

$$\rightarrow \begin{pmatrix} 1 & 1 & -1 & 1 & 1 & 0 \\ 0 & -1 & 2 & 0 & 0 & 1 \\ 0 & 0 & 4 & 1 & 2 & 3 \end{pmatrix} \quad \text{(add row 2 to row 1)}$$

$$\rightarrow \begin{pmatrix} 1 & 0 & 1 & 1 & 1 & 1 \\ 0 & -1 & 2 & 0 & 0 & 1 \\ 0 & 0 & 4 & 1 & 2 & 3 \end{pmatrix} \quad \text{(change the signs of row 2)}$$

$$\rightarrow \begin{pmatrix} 1 & 0 & 1 & 1 & 1 & 1 \\ 0 & 1 & -2 & 0 & 0 & -1 \\ 0 & 0 & 4 & 1 & 2 & 3 \end{pmatrix} \quad \text{(divide row 3 by 4)}$$

$$\rightarrow \begin{pmatrix} 1 & 0 & 1 & 1 & 1 & 1 \\ 0 & 1 & -2 & 0 & 0 & -1 \\ 0 & 0 & 1 & 1/4 & 1/2 & 3/4 \end{pmatrix} \quad \text{(add row 3 times 2 to row 2)}$$

$$\to \quad \begin{pmatrix} 1 & 0 & 1 & 1 & 1 & 1 \\ 0 & 1 & 0 & 1/2 & 1 & 1/2 \\ 0 & 0 & 1 & 1/4 & 1/2 & 3/4 \end{pmatrix} \quad \text{(subtract row 3 from row 1)}$$

$$\to \quad \begin{pmatrix} 1 & 0 & 0 & 3/4 & 1/2 & 1/4 \\ 0 & 1 & 0 & 1/2 & 1 & 1/2 \\ 0 & 0 & 1 & 1/4 & 1/2 & 3/4 \end{pmatrix} = (I, A^{-1}).$$

Therefore,

$$A^{-1} = \begin{pmatrix} 3/4 & 1/2 & 1/4 \\ 1/2 & 1 & 1/2 \\ 1/4 & 1/2 & 3/4 \end{pmatrix}. \quad \square$$

The next two examples demonstrate some basic skills that are often used in studying partitioned matrices. These are good exercises themselves.

Example 2.5 Let $A \in M_{m \times n}(\mathbb{C})$, $B \in M_{n \times m}(\mathbb{C})$, and $C \in M_n(\mathbb{C})$. Let

$$M = \begin{pmatrix} I_m & A \\ B & C \end{pmatrix}.$$

(1) Show that M is invertible if and only if $C - BA$ is invertible.

(2) If M and C are invertible, show that $I_m - AC^{-1}B$ is invertible, and

$$(I_m - AC^{-1}B)^{-1} = I_m + A(C - BA)^{-1}B. \tag{2.5}$$

(3) If $|M| \neq 0$, find the inverse of M in terms of A, B, and $(C - BA)^{-1}$.

Solution. (1) M is invertible if and only if $C - BA$ is invertible because

$$\begin{pmatrix} I & A \\ B & C \end{pmatrix} = \begin{pmatrix} I & 0 \\ B & I \end{pmatrix} \begin{pmatrix} I & 0 \\ 0 & C - BA \end{pmatrix} \begin{pmatrix} I & A \\ 0 & I \end{pmatrix}.$$

(2) If C is invertible, then

$$\begin{pmatrix} I & A \\ B & C \end{pmatrix} = \begin{pmatrix} I - AC^{-1}B & A \\ 0 & C \end{pmatrix} \begin{pmatrix} I & 0 \\ C^{-1}B & I \end{pmatrix}. \tag{2.6}$$

Thus, if M and C are both invertible, then $I - AC^{-1}B$ is invertible.

The identity (2.5) involving the inverses is shown in the next part.

(3) We find the inverse of M in two ways resulting two expressions. Comparing the blocks in the expressions reveals the inverse of $I - AC^{-1}B$.

Let $S = I - AC^{-1}B$. By (2.6), we have

$$
\begin{pmatrix} I & A \\ B & C \end{pmatrix}^{-1} = \begin{pmatrix} I & 0 \\ C^{-1}B & I \end{pmatrix}^{-1} \begin{pmatrix} S & A \\ 0 & C \end{pmatrix}^{-1}
$$

$$
= \begin{pmatrix} I & 0 \\ -C^{-1}B & I \end{pmatrix} \begin{pmatrix} S^{-1} & -S^{-1}AC^{-1} \\ 0 & C^{-1} \end{pmatrix}
$$

$$
= \begin{pmatrix} S^{-1} & -S^{-1}AC^{-1} \\ -C^{-1}BS^{-1} & C^{-1} + C^{-1}BS^{-1}AC^{-1} \end{pmatrix}. \tag{2.7}
$$

On the other hand, by performing row operations on (M, I), we have

$$
\begin{pmatrix} I & A & I & 0 \\ B & C & 0 & I \end{pmatrix} \rightarrow \begin{pmatrix} I & A & I & 0 \\ 0 & C - BA & -B & I \end{pmatrix}
$$

$$
\rightarrow \begin{pmatrix} I & A & I & 0 \\ 0 & I & -(C - BA)^{-1}B & (C - BA)^{-1} \end{pmatrix}
$$

$$
\rightarrow \begin{pmatrix} I & 0 & I + A(C - BA)^{-1}B & -A(C - BA)^{-1} \\ 0 & I & -(C - BA)^{-1}B & (C - BA)^{-1} \end{pmatrix},
$$

which gives

$$
\begin{pmatrix} I & A \\ B & C \end{pmatrix}^{-1} = \begin{pmatrix} I + A(C - BA)^{-1}B & -A(C - BA)^{-1} \\ -(C - BA)^{-1}B & (C - BA)^{-1} \end{pmatrix}. \tag{2.8}
$$

Comparing the $(1,1)$-blocks in (2.7) and (2.8) yields the identity (2.5):

$$
(I - AC^{-1}B)^{-1} = I + A(C - BA)^{-1}B. \quad \square
$$

In the above solution, for simplicity, we omitted the subscripts m and n for I, as we usually do. One can tell from the context which I is I_m or I_n.

Setting $C = I_n$ in (2.5), we see that $I_m - AB$ is invertible if and only if $I_n - BA$ is invertible (this is also immediate from the next example). In the case that $I_m - AB$ is invertible (then so is $I_n - BA$), we have

$$
(I_m - AB)^{-1} = I_m + A(I_n - BA)^{-1}B,
$$

which is also verified directly by checking

$$
(I - AB)\big(I + A(I - BA)^{-1}B\big)
$$

$$
= (I - AB) + (I - AB)A(I - BA)^{-1}B
$$

$$
= I - AB + A(I - BA)^{-1}B - ABA(I - BA)^{-1}B
$$

$$
= I - AB + A(I - BA)(I - BA)^{-1}B
$$

$$
= I - AB + AB = I.
$$

Example 2.6 For any $m \times n$ matrix A and $n \times m$ matrix B, show that

$$|I_m - AB| = |I_n - BA| = \begin{vmatrix} I_m & A \\ B & I_n \end{vmatrix},$$

concluding that $I_n - BA$ is invertible if and only if $I_m - AB$ is invertible.

Solution. We show the case for $|I_n - BA|$. The case of $|I_m - AB|$ is similar.

$$\begin{vmatrix} I_m & A \\ B & I_n \end{vmatrix} = \left| \begin{pmatrix} I_m & 0 \\ -B & I_n \end{pmatrix} \begin{pmatrix} I_m & A \\ B & I_n \end{pmatrix} \right| = \begin{vmatrix} I_m & A \\ 0 & I_n - BA \end{vmatrix} = |I_n - BA|. \quad \square$$

It follows immediately that $|I_m - AB| = 0 \Leftrightarrow |I_n - BA| = 0$, that is, $I_m - AB$ is nonsingular if and only if $I_n - BA$ is nonsingular.

We will see more properties of matrices AB and BA in Chapter 3. As an application of the determinant identity in Example 2.6, we show that for $A \in M_n(\mathbb{C})$, $u, v \in \mathbb{C}^n$, and $z \in \mathbb{C}$,

$$\det(A + zuv^*) = \det A + z\, v^* \operatorname{adj}(A)u. \tag{2.9}$$

To prove this, first consider the case where A is invertible. We derive

$$\begin{aligned} \det(A + zuv^*) &= \det\left(A(I + zA^{-1}uv^*)\right) \\ &= \det A \det(I + zA^{-1}uv^*) \\ &= \det A\,(1 + zv^*A^{-1}u) \\ &= \det A + zv^*(\det A \cdot A^{-1})u \\ &= \det A + zv^* \operatorname{adj}(A)u. \end{aligned}$$

If A is not invertible, use $A + tI$ in place of A. Then let $t \to 0$.

Additionally, for $A \in M_n(\mathbb{C})$, $x, y^t \in \mathbb{C}^n$, and $a \in \mathbb{C}$, we have

$$\det \begin{pmatrix} A & x \\ y & a \end{pmatrix} = a \det A - y \operatorname{adj}(A)x. \tag{2.10}$$

Computing the cofactor matrix $\operatorname{adj}(A)$ is a great amount of work. When $a \neq 0$, the formula $\det \begin{pmatrix} A & x \\ y & a \end{pmatrix} = \frac{1}{a^{n-1}} \det(aA - xy)$ is more practical.

Rank. Let A be an $m \times n$ matrix over a field $\mathbb{F}$, where $\mathbb{F} = \mathbb{R}$ or $\mathbb{C}$. The image or range of A, $\operatorname{Im} A = \{Ax \mid x \in \mathbb{F}^n\}$, is the column space of A; it is a subspace of $\mathbb{F}^m$. The *rank* of A is defined by the dimension of its image:

$$r(A) = \dim \operatorname{Im} A. \tag{2.11}$$

Since $\dim \operatorname{Im} A + \dim \operatorname{Ker} A = n$ (Chapter 1, (1.3), p. 13), we have

$$r(A) = n - \dim \operatorname{Ker} A. \tag{2.12}$$

Example 2.7 Let A be an $m \times n$ matrix. Show that

$$r(A^*A) = r(AA^*) = r(A^*) = r(A^t) = r(\bar{A}).$$

Solution. By Example 1.7 (Chapter 1, p. 13), $r(A^t) = r(A)$. $r(\bar{A}) = r(A)$ is because the columns of A are linearly independent if and only if the corresponding columns of $\bar{A}$ are linearly independent. $r(A^*A) = r(A)$ is due to the fact that $(A^*A)x = 0$ and $Ax = 0$ have the same solution space as $Ax = 0 \Leftrightarrow x^*A^*Ax = 0$. Consequently, $r(AA^*) = r(A^*) = r(A)$. $\square$

(Note: In fact, this example is a restatement of (1.5) in dimensions.)

Let A be an $m \times n$ matrix and let Q be an $n \times n$ invertible matrix. Then A and AQ have the same column space, since, obviously, $\text{Im}(AQ) \subseteq \text{Im} A$, and if $y = Ax \in \text{Im} A$, then $y = Ax = (AQ)(Q^{-1}x) \in \text{Im}(AQ)$. It follows that applications of elementary column operations do not change the rank of a matrix. This is also true for row operations, because $\text{Im} A$ and $\text{Im}(PA)$ have the same dimension for any $m \times m$ invertible matrix P. To see this, let $r(A) = r$ and take a basis $\alpha_1, \alpha_2, \ldots, \alpha_r$ (column vectors) for $\text{Im} A$, then $P\alpha_1, P\alpha_2, \ldots, P\alpha_r$ comprise a basis for $\text{Im}(PA)$, and vice versa. Thus,

$$\dim \text{Im} A = \dim \text{Im}(PAQ) \tag{2.13}$$

for any $m \times m$ invertible matrix P and any $n \times n$ invertible matrix Q.

Let $A \neq 0$. By elementary row and column operations (over the same field as the elements of A lie in), A can be brought to a matrix in the form $\begin{pmatrix} I_r & 0 \\ 0 & 0 \end{pmatrix}$, $r = r(A)$. Thus, there are invertible matrices R and C such that

$$A = R \begin{pmatrix} I_r & 0 \\ 0 & 0 \end{pmatrix} C. \tag{2.14}$$

In the following example we show how to use elementary row and/or column operations to find the rank, image, and null space of a given matrix. For rank, one may use both operations. For image, one can use row operations to find the columns of the matrix that are basis vectors of the column space, or one can use column operations to get basis vectors (not necessarily the columns of the original matrix) that span the image; one cannot use both operations to find image. For kernel (null space), only row operations can be applied. Recall that a square matrix is invertible if and only if it is a product of some elementary matrices.

The reasonings by using matrices are: For rank, if P and Q are invertible matrices, then $r(PAQ) = r(A)$. This says that row and column operations on a matrix A do not change the rank of the matrix.

For image by row operations, let $\alpha_1, \alpha_2, \ldots, \alpha_k$ be the columns of A and let $\alpha_1', \alpha_2', \ldots, \alpha_k'$ be the corresponding columns of the resulting matrix after a sequence of row operations. Then some α_i are linearly independent if and only if the corresponding α_i' are linearly independent because $PA = (P\alpha_1, P\alpha_2, \ldots, P\alpha_k) = (\alpha_1', \alpha_2', \ldots, \alpha_k')$, where P is an invertible matrix. Note that $\operatorname{Im} A$ and $\operatorname{Im}(PA)$ need not be the same. This is an effective way of selecting the largest (in number) set of linearly independent vectors from the given column vectors $\alpha_1, \alpha_2, \ldots, \alpha_k \in \mathbb{C}^n$.

For image by column operations, if $AQ = (\alpha_1'', \alpha_2'', \ldots, \alpha_k'')$, that is, $(\alpha_1, \alpha_2, \ldots, \alpha_k)Q = (\alpha_1'', \alpha_2'', \ldots, \alpha_k'')$, where Q is an invertible matrix, then $(\alpha_1, \alpha_2, \ldots, \alpha_k)x = (\alpha_1'', \alpha_2'', \ldots, \alpha_k'')y$, where $x = Qy$. It follows that

$$\operatorname{Span}\{\alpha_1, \alpha_2, \ldots, \alpha_k\} = \operatorname{Span}\{\alpha_1'', \alpha_2'', \ldots, \alpha_k''\}.$$

For the null space of A, since $Ax = 0 \Leftrightarrow (PA)x = 0$, where P is an invertible matrix, we can only perform row operations (as P is on the left).

Example 2.8 Let $A = \begin{pmatrix} 1 & 2 & 3 \\ 2 & 3 & 4 \\ 3 & 4 & 5 \\ 4 & 5 & 6 \end{pmatrix}$.

(a) Find the rank of A by row and column operations.

(b) Find the image of A (as span of basis vectors) by row operations.

(c) Find the image of A (as span of basis vectors) by column operations.

(d) Find the kernel of A (as span of basis vectors) by row operations.

Solution. (a) We perform row and column elementary operations on A:

$$A = \begin{pmatrix} 1 & 2 & 3 \\ 2 & 3 & 4 \\ 3 & 4 & 5 \\ 4 & 5 & 6 \end{pmatrix} \quad \left(\begin{array}{l}\text{from bottom to top,} \\ \text{subtract each preceding row}\end{array}\right)$$

$$\rightarrow \begin{pmatrix} 1 & 2 & 3 \\ 1 & 1 & 1 \\ 1 & 1 & 1 \\ 1 & 1 & 1 \end{pmatrix} \quad \left(\begin{array}{l}\text{from top to bottom,} \\ \text{subtract the last row,} \\ \text{then put the last row on top}\end{array}\right)$$

$$\rightarrow \begin{pmatrix} 1 & 1 & 1 \\ 0 & 1 & 2 \\ 0 & 0 & 0 \\ 0 & 0 & 0 \end{pmatrix} \quad (\text{subtract the 2nd row from the 1st row})$$

$$\rightarrow \begin{pmatrix} 1 & 0 & -1 \\ 0 & 1 & 2 \\ 0 & 0 & 0 \\ 0 & 0 & 0 \end{pmatrix} \quad \text{(by column operations)}$$

$$\rightarrow \begin{pmatrix} 1 & 0 & 0 \\ 0 & 1 & 0 \\ 0 & 0 & 0 \\ 0 & 0 & 0 \end{pmatrix} . \quad \text{Thus, the rank of } A \text{ is 2.}$$

(b) By row operations, we can find the linearly independent columns of A that span the column space of A. From (a), we have

$$A = \begin{pmatrix} 1 & 2 & 3 \\ 2 & 3 & 4 \\ 3 & 4 & 5 \\ 4 & 5 & 6 \end{pmatrix} \xrightarrow{\text{row operations}} \begin{pmatrix} 1 & 0 & -1 \\ 0 & 1 & 2 \\ 0 & 0 & 0 \\ 0 & 0 & 0 \end{pmatrix} .$$

Denote the columns of A by α_1, α_2, and α_3. Then we see any two of α_1, α_2, and α_3 are linearly independent (basis vectors). It follows that

$$\text{Im } A = \text{Span}\{\alpha_1, \alpha_2\} = \text{Span}\{\alpha_1, \alpha_3\} = \text{Span}\{\alpha_2, \alpha_3\}.$$

(c) By column operations, we can find a basis for the column space of A from the columns of the resulting matrix. The vectors in the basis are no longer the columns of A. We perform column operations on A:

$$A = \begin{pmatrix} 1 & 2 & 3 \\ 2 & 3 & 4 \\ 3 & 4 & 5 \\ 4 & 5 & 6 \end{pmatrix} \quad \begin{pmatrix} \text{from right to left,} \\ \text{subtract each previous column} \end{pmatrix}$$

$$\rightarrow \begin{pmatrix} 1 & 1 & 1 \\ 2 & 1 & 1 \\ 3 & 1 & 1 \\ 4 & 1 & 1 \end{pmatrix} \quad \begin{pmatrix} \text{subtract the last column,} \\ \text{move the last column to the left} \end{pmatrix}$$

$$\rightarrow \begin{pmatrix} 1 & 0 & 0 \\ 1 & 1 & 0 \\ 1 & 2 & 0 \\ 1 & 3 & 0 \end{pmatrix} \quad \begin{pmatrix} \text{subtract the 2nd column from} \\ \text{the first column} \end{pmatrix}$$

$$\rightarrow \begin{pmatrix} 1 & 0 & 0 \\ 0 & 1 & 0 \\ -1 & 2 & 0 \\ -2 & 3 & 0 \end{pmatrix} .$$

Thus, we have

$$\text{Im } A = \text{Span}\{\beta_1, \beta_2\}, \text{ where } \beta_1 = (1,0,-1,-2)^t, \ \beta_2 = (0,1,2,3)^t.$$

β_1 and β_2 are not columns of A, but they span the column space of A.

(d) To find the kernel (null space) of A, we can only perform row operations. From (a) or (b), we see that $Ax = 0$ if and only if $x_1 = x_3$ and $x_2 = -2x_3$. Thus, $\gamma = (1,-2,1)^t$ serves as a basis of Ker A.   □

Note that linear dependence of vectors may depend on the scalar field $\mathbb{F}$. For instance, $(1,i)$ and $(i,-1)$ are linearly dependent over $\mathbb{C}$ but not over $\mathbb{R}$. The matrix $A = \begin{pmatrix} 1 & i \\ i & -1 \end{pmatrix}$ is a complex matrix with rank 1 (not 2). For an $m \times n$ real matrix A, $W_{\mathbb{R}} = \{Ax \mid x \in \mathbb{R}^n\}$ and $W_{\mathbb{C}} = \{Ax \mid x \in \mathbb{C}^n\}$ are not the same. However, the dimension of $W_{\mathbb{R}}$ as a subspace of $\mathbb{R}^m$ over $\mathbb{R}$ is the same as the dimension of $W_{\mathbb{C}}$ as a subspace of $\mathbb{C}^m$ over $\mathbb{C}$. This can be seen by taking $A = \begin{pmatrix} I_r & 0 \\ 0 & 0 \end{pmatrix}$ (Problem 2.80).

By Example 2.7, a matrix and its transpose have the same rank, that is, $r(A^t) = r(A)$ for any matrix A. Moreover, the dimension of the column (row) space of A is r if and only if the largest number of linearly independent columns (rows) of A is r. The following fact is useful in practice.

Example 2.9 Let A be an $m \times n$ matrix and let B be a $k \times k$ submatrix of A, say lying in rows $i_1, \ldots, i_k$ and columns $j_1, \ldots, j_k$ of A. If the determinant of B is nonzero (i.e., B is nonsingular), show that the columns $j_1, \ldots, j_k$ of A are linearly independent (and the rows $i_1, \ldots, i_k$ are linearly independent). Furthermore, $r(A) = k$ if and only if A has a $k \times k$ submatrix with nonzero determinant and all $l \times l$ submatrices have zero determinant for $l > k$.

Solution. Let S_k be the $m \times k$ submatrix of A consisting of the columns $j_1, \ldots, j_k$ of A. B lies in the rows $i_1, \ldots, i_k$ of S_k. Since B is invertible, $Bx = 0$ has no nonzero solution. Thus, if $S_k x = 0$, then $Bx = 0$ (as a subsystem of $S_k x = 0$), implying $x = 0$. So, the columns $j_1, \ldots, j_k$ of A are linearly independent. Considering the columns $i_1, \ldots, i_k$ of A^t with nonsingular B^t, we see that the rows $i_1, \ldots, i_k$ of A are linearly independent.

If the rank of A is k, we assume that the columns $p_1, \ldots, p_k$ of A are linearly independent. Let C be the $m \times k$ matrix of the columns $p_1, \ldots, p_k$ of A. The rank of C is k and there are k rows of C, say rows $q_1, \ldots, q_k$, that are linearly independent. Then the submatrix of A in the intersection of rows $q_1, \ldots, q_k$ and columns $p_1, \ldots, p_k$ is nonsingular. By the above argument, any submatrix of A with size $l \times l$, $l > k$, has zero determinant. The converse is immediate from the first part.   □

For $m \times n$ matrices A and B, $\text{Im}(A + B) \subseteq \text{Im}\, A + \text{Im}\, B$ reveals

$$r(A + B) \leq r(A) + r(B). \tag{2.15}$$

For $m \times p$ matrix A and $p \times n$ matrix B, since $\text{Im}(AB) \subseteq \text{Im}\, A$, we obtain $r(AB) \leq r(A)$. By the fact that a matrix and its transpose have the same rank, we have $r(AB) = r((AB)^t) = r(B^t A^t) \leq r(B^t) = r(B)$. Thus

$$r(AB) \leq \min\{r(A), r(B)\}. \tag{2.16}$$

Example 2.10 (Sylvester's rank inequality) Let A be an $m \times p$ matrix and let B be a $p \times n$ matrix. Show that

$$r(AB) \geq r(A) + r(B) - p.$$

Solution. Assume that $A \neq 0$ (or it is trivial). Let $A = M \begin{pmatrix} I_k & 0 \\ 0 & 0 \end{pmatrix} P$, where $k = r(A)$, M has size $m \times m$, P has size $p \times p$, both invertible, and the 0 submatrix in the low-right corner is absent if $k = m$ or $k = p$. We partition $PB = \begin{pmatrix} B_1 \\ B_2 \end{pmatrix}$ such that B_1 has k rows and B_2 has $p - k$ rows. Then

$$B = P^{-1} \begin{pmatrix} B_1 \\ B_2 \end{pmatrix}, \quad r\begin{pmatrix} 0 \\ B_2 \end{pmatrix} \leq p - k = p - r(A),$$

and

$$r(AB) = r\left(M \begin{pmatrix} I_k & 0 \\ 0 & 0 \end{pmatrix} P P^{-1} \begin{pmatrix} B_1 \\ B_2 \end{pmatrix} \right) = r\left(M \begin{pmatrix} B_1 \\ 0 \end{pmatrix} \right) = r\begin{pmatrix} B_1 \\ 0 \end{pmatrix}.$$

It follows that (by (2.15))

$$r(B) = r\begin{pmatrix} B_1 \\ B_2 \end{pmatrix} \leq r\begin{pmatrix} B_1 \\ 0 \end{pmatrix} + r\begin{pmatrix} 0 \\ B_2 \end{pmatrix} \leq r(AB) + p - r(A). \quad \square$$

We now present another important rank inequality - the *Frobenius rank inequality*, with two proofs. The first proof uses the Sylvester rank inequality (which is a special case of the Frobenius inequality with $B = I$).

Example 2.11 (Frobenius rank inequality) Let A, B, and C be matrices of appropriate sizes such that the product ABC is defined. Show that

$$r(ABC) \geq r(AB) + r(BC) - r(B).$$

Solution 1. Let $r(B) = k > 0$. Let P and Q be invertible matrices such that

$$B = P \begin{pmatrix} I_k & 0 \\ 0 & 0 \end{pmatrix} Q.$$

Let B have size $p \times q$. We partition $P = (M, S)$ and $Q = \begin{pmatrix} N \\ T \end{pmatrix}$, where M has size $p \times k$ and N has size $k \times q$. Then $B = MN$ (*full rank factorization*). By Sylvester's rank inequality, we derive

$$
\begin{aligned}
r(ABC) &= r(AMNC) \\
&\geq r(AM) + r(NC) - k \\
&\geq r(AMN) + r(MNC) - k \\
&= r(AB) + r(BC) - r(B).
\end{aligned}
$$

Solution 2. We use the fact that $r(X) + r(Y) = r\left(\begin{smallmatrix} X & 0 \\ 0 & Y \end{smallmatrix}\right) \leq r\left(\begin{smallmatrix} X & 0 \\ Z & Y \end{smallmatrix}\right)$ for matrices $X, Y,$ and Z of appropriate sizes (see Problem 2.91) to derive

$$
\begin{aligned}
r(ABC) + r(B) &= r\begin{pmatrix} ABC & 0 \\ 0 & B \end{pmatrix} = r\begin{pmatrix} ABC & 0 \\ BC & B \end{pmatrix} \\
&= r\begin{pmatrix} 0 & -AB \\ BC & B \end{pmatrix} = r\begin{pmatrix} AB & 0 \\ -B & BC \end{pmatrix} \\
&\geq r\begin{pmatrix} AB & 0 \\ 0 & BC \end{pmatrix} = r(AB) + r(BC). \quad \square
\end{aligned}
$$

Systems of Linear Equations. Let $\mathbb{F}$ be a field and let A be an $m \times n$ matrix over $\mathbb{F}$. Then $Ax = 0$ represents a *homogeneous* linear equation system of m linear equations in n variables, where x is a column vector of n unknowns. The system $Ax = 0$ has the trivial solution $x = 0$. If $r(A) = n$, then $x = 0$ is the unique solution. If $r(A) < n$, then $Ax = 0$ has infinitely many solutions. The solutions constitute the vector space $\operatorname{Ker} A$ (i.e., the *solution space, null space,* or *kernel* of A). Identities (1.3) and (2.12) state

$$
\dim \operatorname{Im} A + \dim \operatorname{Ker} A = r(A) + \dim \operatorname{Ker} A = n. \tag{2.17}
$$

Let $A \in M_{m \times n}(\mathbb{F})$ and $b \in \mathbb{F}^m$. The linear system $Ax = b$, $x \in \mathbb{F}^n$, may have one solution, infinitely many solutions, or no solution. The situations can be determined by the rank of the *augmented matrix* $B = (A, b)$:

(i) If $r(B) = r(A) = n$, then $Ax = b$ has a unique solution.

(ii) If $r(B) = r(A) < n$, then $Ax = b$ has infinitely many solutions.

(iii) If $r(B) \neq r(A)$, then $Ax = b$ has no solution.

Cramer's Rule. Consider the linear equation system $Ax = b$ of n equations in n unknowns. If the $n \times n$ coefficient matrix A is invertible, i.e., $|A| \neq 0$, then the system has a unique solution $x_i = \frac{|A_i|}{|A|}$, $i = 1, 2, \ldots, n$, where A_i is the matrix obtained from A by replacing the i-th column of A with b. Note that Cramer's rule cannot be used when A is singular.

Example 2.12 Solve the system of linear equations with parameter λ:

$$\begin{cases} x_1 + x_2 + x_3 + x_4 = 0 \\ 3x_1 + 2x_2 + x_3 + x_4 = 1 \\ \lambda x_1 + x_2 + 2x_3 + 2x_4 = \lambda \\ 5x_1 + 4x_2 + 3x_3 + 3x_4 = 1. \end{cases}$$

Solution. We work on the augmented matrix:

$$\begin{pmatrix} 1 & 1 & 1 & 1 & 0 \\ 3 & 2 & 1 & 1 & 1 \\ \lambda & 1 & 2 & 2 & \lambda \\ 5 & 4 & 3 & 3 & 1 \end{pmatrix}.$$

By performing some elementary row operations, we arrive at

$$\begin{pmatrix} 1 & 0 & -1 & -1 & 1 \\ 0 & 1 & 2 & 2 & -1 \\ 0 & 0 & \lambda & \lambda & 1 \\ 0 & 0 & 0 & 0 & 0 \end{pmatrix}.$$

If $\lambda = 0$, then the 3rd row results in $0 = 1$. The system has no solution. If $\lambda \neq 0$, dividing the 3rd row by λ, we can get the row echelon form:

$$\begin{pmatrix} 1 & 0 & 0 & 0 & 1 + \frac{1}{\lambda} \\ 0 & 1 & 0 & 0 & -1 - \frac{2}{\lambda} \\ 0 & 0 & 1 & 1 & \frac{1}{\lambda} \\ 0 & 0 & 0 & 0 & 0 \end{pmatrix}.$$

Thus,

$$x_1 = 1 + \tfrac{1}{\lambda}, \quad x_2 = -1 - \tfrac{2}{\lambda}, \quad x_3 = -x_4 + \tfrac{1}{\lambda}, \quad x_4 \text{ is a free variable.} \quad \square$$

Example 2.13 Let $a_1, a_2, \ldots, a_n$ be n distinct scalars (in a field $\mathbb{F}$) and let $b_1, b_2, \ldots, b_n$ be any n scalars. Show that there exists a unique polynomial $f(x)$ with degree at most $n - 1$ such that $f(a_i) = b_i$, $i = 1, 2, \ldots, n$.

Solution. Let $f(x) = c_0 + c_1 x + c_2 x^2 + \cdots + c_{n-1} x^{n-1}$, where c_i's are scalars in $\mathbb{F}$ to be determined. Plugging a_i in $f(x)$ and equating $f(a_i)$ to b_i, we obtain a system of n linear equations in $c_0, c_1, \ldots, c_{n-1}$. The coefficient matrix of the linear system is an $n \times n$ Vandermonde matrix. By Example 2.3, the Vandermonde matrix is nonsingular because the a_i's are distinct. So, the solution to the system is unique. Thus, $f(x)$ exists and is unique. $\quad \square$

2.1 Evaluate the determinants

$$
\begin{vmatrix} 1 & 2 & 3 \\ 8 & 9 & 4 \\ 7 & 6 & 5 \end{vmatrix}, \quad
\begin{vmatrix} 1+x & 2+x & 3+x \\ 8+x & 9+x & 4+x \\ 7+x & 6+x & 5+x \end{vmatrix}, \quad
\begin{vmatrix} x^1 & x^2 & x^3 \\ x^8 & x^9 & x^4 \\ x^7 & x^6 & x^5 \end{vmatrix}.
$$

2.2 Evaluate the following determinant with $a, b, c \in \mathbb{C}$ and prove that the determinant is positive if $a, b,$ and c all lie in the open disk $|z| < \frac{1}{2}$.

$$
\begin{vmatrix} 1 & a & \bar{b} \\ \bar{a} & 1 & c \\ b & \bar{c} & 1 \end{vmatrix}.
$$

2.3 Evaluate the determinants

$$
\begin{vmatrix} 1 & 1 & 0 \\ -1 & 1 & 1 \\ 0 & -1 & 1 \end{vmatrix}, \quad
\begin{vmatrix} 1 & 1 & 0 & 0 & 0 \\ -1 & 1 & 1 & 0 & 0 \\ 0 & -1 & 1 & 1 & 0 \\ 0 & 0 & -1 & 1 & 1 \\ 0 & 0 & 0 & -1 & 1 \end{vmatrix}.
$$

In general, let $F_n = (f_{ij})$ denote the $n \times n$ matrix with $f_{ii} = 1$, $f_{i,i+1} = 1$, $f_{i+1,i} = -1$ for all possible i, and $f_{ij} = 0$ otherwise. Show that $|F_n| = |F_{n-1}| + |F_{n-2}|$ (Fibonacci sequence). Find $|F_9|$.

2.4 Explain without computation why the determinant equals zero:

$$
\begin{vmatrix} a_1 & a_2 & a_3 & a_4 & a_5 \\ b_1 & b_2 & b_3 & b_4 & b_5 \\ c_1 & c_2 & 0 & 0 & 0 \\ d_1 & d_2 & 0 & 0 & 0 \\ e_1 & e_2 & 0 & 0 & 0 \end{vmatrix}.
$$

2.5 Evaluate the determinants

$$
\begin{vmatrix} 0 & 0 & a_1 & b_1 \\ 0 & 0 & a_2 & b_2 \\ a_3 & b_3 & 0 & 0 \\ a_4 & b_4 & 0 & 0 \end{vmatrix}, \quad
\begin{vmatrix} a_1 & 0 & 0 & b_1 \\ 0 & a_2 & b_2 & 0 \\ 0 & b_3 & a_3 & 0 \\ b_4 & 0 & 0 & a_4 \end{vmatrix}, \quad
\begin{vmatrix} 0 & a_1 & 0 & 0 \\ 0 & 0 & a_2 & 0 \\ 0 & 0 & 0 & a_3 \\ a_4 & 0 & 0 & 0 \end{vmatrix}.
$$

2.6   Evaluate the determinants

$$\begin{vmatrix} 0 & 0 & x & y & a_1 \\ 0 & 0 & z & a_2 & 0 \\ 0 & 0 & a_3 & 0 & 0 \\ 0 & a_4 & 0 & 0 & 0 \\ a_5 & w & 0 & 0 & 0 \end{vmatrix}, \qquad \begin{vmatrix} 0 & 0 & 0 & 0 & 0 & a_1 \\ 0 & 0 & 0 & 0 & a_2 & b \\ 0 & 0 & 0 & a_3 & c & d \\ 0 & 0 & a_4 & e & f & g \\ 0 & a_5 & h & i & j & k \\ a_6 & l & m & n & o & p \end{vmatrix}.$$

2.7   Let A and B be $m \times n$ matrices, with $m = n$ conventionally if needed in the cases such as $|A|$ and A^2, for example. Answer true or false:

(a) $|A + B| = |A| + |B|$.

(b) If $|AB| = 0$, then $|A| = 0$ or $|B| = 0$.

(c) $|AB^t| = |B^t A|$.

(d) $|AB| = |BA|$.

(e) $|A^*| = |A|$.

(f) $|A^t| = |A|$.

(g) $|AA^t| = |A^t A|$.

(h) $|I_m - AA^t| = |I_n - A^t A|$.

(i) If AA^t is diagonal, then $A^t A$ is diagonal.

(j) If $AB = kI_n$ for some $k \neq 0$, then $BA = kI$.

(k) $|AB|^{-1} = |A^{-1}||B^{-1}|$ when A and B are invertible.

(l) $|kA| = k|A|$ for any scalar k.

(m) If $A^* A = 0$, then $A = 0$.

(n) If $A^t A = 0$, then $A = 0$.

(o) If $(A^* A)^2 = 0$, then $A = 0$.

(p) If $(A^*)^2 (A)^2 = 0$, then $A = 0$.

(q) If $AB = BA$, then $(AB)^k = A^k B^k$ for $k = 2, 3, \ldots$.

(r) If $(AB)^k = A^k B^k$ for $k = 2, 3, \ldots$, then $AB = BA$.

2.8   Compute the determinant

$$\begin{vmatrix} 1 & 1 & 1 & \cdots & 1 \\ 1 & 2 & 3 & \cdots & n \\ 1 & 2^2 & 3^2 & \cdots & n^2 \\ \vdots & \vdots & \vdots & \ddots & \vdots \\ 1 & 2^{n-1} & 3^{n-1} & \cdots & n^{n-1} \end{vmatrix}.$$

2.9 Let $p_1, \ldots, p_n, a, b \in \mathbb{C}$, define $f(x) = (p_1 - x) \cdots (p_n - x)$, and let

$$
\Delta_n = \begin{vmatrix}
p_1 & a & a & a & \cdots & a & a \\
b & p_2 & a & a & \cdots & a & a \\
b & b & p_3 & a & \cdots & a & a \\
b & b & b & p_4 & \cdots & a & a \\
\vdots & \vdots & \vdots & \vdots & \ddots & \vdots & \vdots \\
b & b & b & b & \cdots & p_{n-1} & a \\
b & b & b & b & \cdots & b & p_n
\end{vmatrix}.
$$

(a) Show that if $a \neq b$, then

$$
\Delta_n = \frac{bf(a) - af(b)}{b - a}.
$$

(b) Show that if $a = b$, then

$$
\Delta_n = a \sum_{i=1}^{n} f_i(a) + f(a),
$$

where $f_i(a)$ means $f(a)$ with the factor $(p_i - a)$ absent.

(c) Find

$$
\begin{vmatrix}
a & b & b & \cdots & b \\
b & a & b & \cdots & b \\
b & b & a & \cdots & b \\
\vdots & \vdots & \vdots & \ddots & \vdots \\
b & b & b & \cdots & a
\end{vmatrix}_{n \times n}.
$$

2.10 Show that if $a \neq b$, then

$$
\begin{vmatrix}
a+b & ab & 0 & \cdots & 0 & 0 \\
1 & a+b & ab & \cdots & 0 & 0 \\
0 & 1 & a+b & \cdots & 0 & 0 \\
\vdots & \vdots & \vdots & \ddots & \vdots & \vdots \\
0 & 0 & 0 & \cdots & a+b & ab \\
0 & 0 & 0 & \cdots & 1 & a+b
\end{vmatrix}_{n \times n} = \frac{a^{n+1} - b^{n+1}}{a - b}.
$$

Find a formula for the determinant when $a = b$.

2.11   Let $a \in \mathbb{C}$. Find the determinant $|\lambda I - A_a|$ with variable λ, where

$$
A_a = \begin{pmatrix}
0 & 1 & 0 & \cdots & 0 \\
0 & 0 & 1 & \cdots & 0 \\
\vdots & \vdots & \vdots & \ddots & \vdots \\
0 & 0 & 0 & \cdots & 1 \\
a & 0 & 0 & \cdots & 0
\end{pmatrix}_{n \times n}.
$$

2.12   Find the determinant of the matrix (with unspecified entries for 0's):

$$
A = \begin{pmatrix}
a_1 & b_1 & & & \\
& \ddots & & \ddots & \\
& & & \ddots & b_{n-1} \\
b_n & & & & a_n
\end{pmatrix}.
$$

2.13   Let A and B be bordered matrices of the identity matrix I_n:

$$
A = \begin{pmatrix} I_n & \beta \\ \alpha & a \end{pmatrix} \quad \text{and} \quad B = \begin{pmatrix} 0 & I_n \\ b & \gamma \end{pmatrix},
$$

where α, β^t, γ are row vectors of n components and $a, b \in \mathbb{C}$. Find

$$
|\lambda I_{n+1} - A| \quad \text{and} \quad |\lambda I_{n+1} - B|, \quad \text{where } \lambda \in \mathbb{C}.
$$

2.14   Let $A = (a_{ij})$ be a 3×3 real matrix and let $p(t) = \det(tI_3 - A)$. If $p(t) = t^3 - at^2 + bt - c$ and $p(t)$ has roots λ_1, λ_2, and λ_3, show that

$$
a = a_{11} + a_{22} + a_{33} = \lambda_1 + \lambda_2 + \lambda_3, \quad c = \det A = \lambda_1 \lambda_2 \lambda_3,
$$

$$
b = \begin{vmatrix} a_{11} & a_{12} \\ a_{21} & a_{22} \end{vmatrix} + \begin{vmatrix} a_{11} & a_{13} \\ a_{31} & a_{33} \end{vmatrix} + \begin{vmatrix} a_{22} & a_{23} \\ a_{32} & a_{33} \end{vmatrix} = \lambda_1 \lambda_2 + \lambda_1 \lambda_3 + \lambda_2 \lambda_3.
$$

Show that b is equal to the derivative of $p(t) = \det(tI_3 - A)$ at $t = 0$.

2.15   Let $a_{ij}(t)$ be real differentiable functions of t, $1 \leq i, j \leq n$. Show that

$$
\frac{d}{dt} \begin{vmatrix}
a_{11}(t) & \cdots & a_{1j}(t) & \cdots & a_{1n}(t) \\
a_{21}(t) & \cdots & a_{2j}(t) & \cdots & a_{2n}(t) \\
\cdots & \cdots & \cdots & \cdots & \cdots \\
a_{n1}(t) & \cdots & a_{nj}(t) & \cdots & a_{nn}(t)
\end{vmatrix}
$$

$$
= \sum_{j=1}^{n} \begin{vmatrix}
a_{11}(t) & \cdots & \frac{d}{dt} a_{1j}(t) & \cdots & a_{1n}(t) \\
a_{21}(t) & \cdots & \frac{d}{dt} a_{2j}(t) & \cdots & a_{2n}(t) \\
\cdots & \cdots & \cdots & \cdots & \cdots \\
a_{n1}(t) & \cdots & \frac{d}{dt} a_{nj}(t) & \cdots & a_{nn}(t)
\end{vmatrix}
$$

and evaluate $\frac{d}{dt} |I_n + tX|$ when $t = 0$, where $X = (x_{ij}) \in M_n(\mathbb{R})$.

2.16 If A is an $n \times n$ matrix all of whose entries are either 1 or -1, prove that $|A|$ is divisible by 2^{n-1}. If such an A contains the same number of 1's as -1's in every row (thus n is even), show that $|A| = 0$.

2.17 Let $A = (\alpha, \gamma_2, \gamma_3, \gamma_4)$ and $B = (\beta, \gamma_2, \gamma_3, \gamma_4)$ be 4×4 matrices, where α, β, γ_2, γ_3, γ_4 are column vectors in $\mathbb{R}^4$. If $\det A = 4$ and $\det B = 1$, find $\det(A + B)$. Find $\det C$, where $C = (\gamma_4, \gamma_3, \gamma_2, \alpha + \beta)$.

2.18 Let α, β, γ, and δ be column vectors in $\mathbb{R}^3$. Let $A = (\alpha, 2\gamma, 3\delta)$ and $B = (\beta, \gamma, 2\delta)$. If $\det A = 18$ and $\det B = 2$, find $\det(A \pm B)$.

2.19 If $u_1, u_2, \ldots, u_n \in \mathbb{C}^n$ (in columns) are n linearly independent vectors, show that for any nonzero scalars $a_1, a_2, \ldots, a_n \in \mathbb{C}$,

$$(a_1 u_1, a_2 u_2, \ldots, a_n u_n)^{-1} = \operatorname{diag}(a_1^{-1}, a_2^{-1}, \ldots, a_n^{-1})(u_1, u_2, \ldots, u_n)^{-1}.$$

2.20 Let $A \in M_{m \times n}(\mathbb{R})$. If the system $Ax = 0$ has a nonzero solution in the complex $\mathbb{C}^n$, show that it has a nonzero solution in the real $\mathbb{R}^n$.

2.21 Let $A \in M_{m \times n}(\mathbb{Q})$. If the system $Ax = 0$ has a nonzero solution in the complex $\mathbb{C}^n$, show that it has a nonzero solution in the rational $\mathbb{Q}^n$.

2.22 Let $\alpha_1, \alpha_2, \ldots, \alpha_m \in \mathbb{F}^n$ ($\mathbb{F} = \mathbb{R}$ or $\mathbb{C}$). Let A be the $n \times m$ matrix with $\alpha_1, \alpha_2, \ldots, \alpha_m$ as columns, i.e., $A = (\alpha_1, \alpha_2, \ldots, \alpha_m)$. Show that

(a) $\dim \operatorname{Span}\{\alpha_1, \alpha_2, \ldots, \alpha_m\} = r(A)$.

(b) $\alpha_1, \alpha_2, \ldots, \alpha_m$ are linearly dependent if and only if $r(A) < m$.

(c) If $n < m$, then $\alpha_1, \alpha_2, \ldots, \alpha_m$ are linearly dependent.

(d) If $n = m$ (i.e., A is an $n \times n$ matrix), then $\alpha_1, \alpha_2, \ldots, \alpha_m$ are linearly independent if and only if $|A| \neq 0$.

2.23 Let $\{\beta_1, \ldots, \beta_n\}$ be a basis of a vector space V of dimension n and let $\{\alpha_1, \ldots, \alpha_m\}$ be a set of vectors in V. Write each α_i as

$$\alpha_i = \sum_{j=1}^{n} a_{ij} \beta_j, \quad i = 1, \ldots, m.$$

Let $A = (a_{ij})$ be the $m \times n$ coefficient matrix and $\{i_1, \ldots, i_k\} \subseteq \{1, \ldots, m\}$. Show that $\{\alpha_{i_1}, \ldots, \alpha_{i_k}\}$ is linearly independent if and only if the corresponding rows $i_1, \ldots, i_k$ of A are linearly independent and if and only if there is a nonsingular $k \times k$ submatrix of A lying in rows $i_1, \ldots, i_k$ of A. Conclude that $\dim \operatorname{Span}\{\alpha_1, \ldots, \alpha_m\} = r(A)$.

2.24   Let $\alpha = \{\alpha_1, \ldots, \alpha_m\}$ and $\beta = \{\beta_1, \ldots, \beta_n\}$ be two sets of vectors in a vector space V such that each α_i is a linear combination of $\beta_1, \ldots, \beta_n$:

$$\alpha_i = \sum_{j=1}^{n} a_{ij}\beta_j, \quad i = 1, \ldots, m.$$

Let $A = (a_{ij})$ be the $m \times n$ coefficient matrix. Show that

(a) $\dim(\text{Span}\,\alpha) \leq \dim(\text{Span}\,\beta)$.

(b) $\dim(\text{Span}\,\alpha) \leq r(A)$.

(c) If β is linearly independent, then $\dim(\text{Span}\,\alpha) = r(A)$.

(d) If α is linearly independent, then $r(A) = m$ and $n \geq m$.

2.25   Let A be an $n \times n$ real matrix.

(a) If $A^t = -A$ and n is odd, show that $|A| = 0$.

(b) If $A^2 + I = 0$, show that n must be even.

(c) Do (a) and (b) remain true for complex matrices?

2.26   Let $A \in M_n(\mathbb{C})$. If $AA^t = I$ and $|A| < 0$, show that $|A + I| = 0$.

2.27   If A, B, C, and D are $n \times n$ matrices such that $ABCD = I$. Show that

$$ABCD = DABC = CDAB = BCDA = I.$$

What does $D^{-1}C^{-1}B^{-1}$ equal? Is it true that $BCAD = I$?

2.28   Let A be an $n \times n$ matrix. If A is invertible, show that there exists a polynomial $f(t)$ such that $A^{-1} = f(A)$.

2.29   If A is a matrix such that $A^3 = 2I$, show that $|A^2 - 2A + 2I| \neq 0$.

2.30   Show that there exist 2×2 matrices A and B such that $A^2 = B^2 = 0$ and $AB + BA = I$, but there do not exist such matrices of size 3×3.

2.31   Find the inverses of the matrices

$$\begin{pmatrix} 1 & 2 \\ 3 & 4 \end{pmatrix}, \quad \begin{pmatrix} 1 & 2 \\ 4 & 3 \end{pmatrix}, \quad \text{and} \quad \begin{pmatrix} \cos\theta & \sin\theta \\ -\sin\theta & \cos\theta \end{pmatrix}, \quad \theta \in \mathbb{R}.$$

2.32   Let $A = \begin{pmatrix} a & x \\ y & B \end{pmatrix} \in M_n(\mathbb{C})$, where $a \in \mathbb{C}$, $x^t, y \in \mathbb{C}^{n-1}$, and $B \in M_{n-1}(\mathbb{C})$. If B is nonsingular, show that $\begin{pmatrix} a - \frac{|A|}{|B|} & x \\ y & B \end{pmatrix}$ is singular.

2.33 Find the inverse of matrix M by applying elementary row operations to the augmented matrix (M, I) to get (I, M^{-1}), where M is

$$A = \begin{pmatrix} 1 & 0 & 0 \\ 0 & 1 & 0 \\ x & y & 1 \end{pmatrix} \quad \text{or} \quad B = \begin{pmatrix} 0 & 1 & 0 & 0 \\ 0 & 0 & 1 & 0 \\ 0 & 0 & 0 & 1 \\ a & b & c & d \end{pmatrix}, \quad a \neq 0.$$

2.34 Find the inverses of the matrices

$$A = \begin{pmatrix} 1 & 1 & 1 \\ 0 & 1 & 1 \\ 0 & 0 & 1 \end{pmatrix} \quad \text{and} \quad B = \begin{pmatrix} 1 & 1 & \cdots & 1 & 1 \\ 0 & 1 & \cdots & 1 & 1 \\ \vdots & \vdots & \ddots & \vdots & \vdots \\ 0 & 0 & \cdots & 1 & 1 \\ 0 & 0 & \cdots & 0 & 1 \end{pmatrix}_{n \times n}.$$

2.35 Find the determinants and inverses of the $n \times n$ matrices

$$A = \begin{pmatrix} 0 & 1 & 1 & \cdots & 1 \\ 1 & 0 & 1 & \cdots & 1 \\ 1 & 1 & 0 & \cdots & 1 \\ \vdots & \vdots & \vdots & \ddots & \vdots \\ 1 & 1 & 1 & \cdots & 0 \end{pmatrix} \quad \text{and} \quad B = \begin{pmatrix} 1 & 1 & 1 & \cdots & 1 \\ 1 & 2 & 2 & \cdots & 2 \\ 1 & 2 & 3 & \cdots & 3 \\ \vdots & \vdots & \vdots & \ddots & \vdots \\ 1 & 2 & 3 & \cdots & n \end{pmatrix}.$$

2.36 Compute $Q^t Q$, QQ^t, $(Q^t Q)^2$, and $(QQ^t)^2$, where

$$Q = \begin{pmatrix} 1 & 1 & 1 & 1 \\ 1 & -1 & 1 & 1 \\ 1 & 0 & -2 & 1 \\ 1 & 0 & 0 & -3 \end{pmatrix}.$$

2.37 Let $a_1, a_2, \ldots, a_n$ be nonzero numbers. Find the inverse of the matrix

$$A = \begin{pmatrix} 0 & a_1 & 0 & \cdots & 0 \\ 0 & 0 & a_2 & \cdots & 0 \\ \vdots & \vdots & \ddots & \ddots & \vdots \\ 0 & 0 & 0 & \cdots & a_{n-1} \\ a_n & 0 & 0 & \cdots & 0 \end{pmatrix}.$$

2.38 Find the rank of the 4×6 matrix

$$A = \begin{pmatrix} 1 & 1 & 0 & 2 & 1 \\ 0 & 1 & 1 & 1 & 0 \\ 1 & 1 & -4 & 2 & 3 \\ 2 & 2 & 4 & 0 & -2 \end{pmatrix}.$$

2.39   Let $x \in \mathbb{C}^n$ be a unit column vector. Find $(I_n + xx^*)^{-1}$.

2.40   Let $X, Y \in M_{m \times n}(\mathbb{C})$. Show that $\left(\begin{smallmatrix} I_m & X \\ 0 & I_n \end{smallmatrix}\right)$ and $\left(\begin{smallmatrix} I_m & Y \\ 0 & I_n \end{smallmatrix}\right)$ commute.

2.41   Let $X = (x_{ij})$ be an $m \times n$ matrix. For $1 \le i \le m, 1 \le j \le n$, let X_{ij} be the $m \times n$ matrix with (i, j)-entry x_{ij} and 0 elsewhere. Show that

$$\begin{pmatrix} I_m & X \\ 0 & I_n \end{pmatrix} = \prod_{i,j} \begin{pmatrix} I_m & X_{ij} \\ 0 & I_n \end{pmatrix}.$$

2.42   Let A and B be $m \times n$ and $n \times m$ matrices, respectively. If $I_n - BA$ is nonsingular, show that $\left(\begin{smallmatrix} I_m & A \\ B & I_n \end{smallmatrix}\right)$ is nonsingular, and find $\left(\begin{smallmatrix} I_m & A \\ B & I_n \end{smallmatrix}\right)^{-1}$.

2.43   Let $A, B, C, X, Y, Z \in M_n(\mathbb{C})$, where A and C are nonsingular. Find

$$\begin{pmatrix} A & B \\ 0 & C \end{pmatrix}^{-1} \quad \text{and} \quad \begin{pmatrix} I & X & Y \\ 0 & I & Z \\ 0 & 0 & I \end{pmatrix}^{-1}.$$

2.44   Let a_1, a_2, and a_3 be distinct numbers. Find the inverse of

$$V = \begin{pmatrix} 1 & 1 & 1 \\ a_1 & a_2 & a_3 \\ a_1^2 & a_2^2 & a_3^2 \end{pmatrix}.$$

2.45   Find the rank of N and a positive integer k such that $N^k = 0$, where

$$N = \begin{pmatrix} 0 & 0 & 1 \\ 0 & 0 & i \\ 1 & i & 0 \end{pmatrix}, \quad i = \sqrt{-1}.$$

2.46   Let A be an $m \times r$ matrix such that A^*A is invertible and let $M = A(A^*A)^{-1}A^*$. Show that $M^2 = M$ and $M^* = M$.

2.47   Let A and B be $m \times p$ and $m \times q$ matrices, respectively, such that $A^*B = 0$ and $p + q = m$. If $M = (A, B)$ is invertible, show that

$$M^{-1} = \begin{pmatrix} (A^*A)^{-1}A^* \\ (B^*B)^{-1}B^* \end{pmatrix} \quad \text{and} \quad A(A^*A)^{-1}A^* + B(B^*B)^{-1}B^* = I.$$

2.48   Assuming that all matrix inverses involved below exist, show that

$$(A + iB)^{-1} = (A + BA^{-1}B)^{-1} - i(B + AB^{-1}A)^{-1}.$$

2.49   Assuming that all matrix inverses involved below exist, show that

$$(A - B)^{-1} = A^{-1} + A^{-1}(B^{-1} - A^{-1})^{-1}A^{-1}.$$

Derive

(a) $(I - BA)^{-1} = I + B(I - AB)^{-1}A.$

(b) $(I + A)^{-1} = I - (A^{-1} + I)^{-1}.$

(c) $\det\big((I - A)^{-1} + (I - A^{-1})^{-1}\big) = 1.$

(d) $(I + A^*A)^{-1} = I - A^*(I + AA^*)^{-1}A.$

2.50   Let $A, B, C,$ and D be $n \times n$ matrices. If AB and CD are Hermitian and if $AD - B^*C^* = I$, show that $DA - BC = I$ and $C^*A = A^*C.$

2.51   Let A and B be 2×2 or 3×3 real matrices such that $A^2 + AB - BA + B^2 = 0$. Show that $\det(A^2 + B^2) = 0$ and $\det(A + iB) = 0.$

2.52   Let m and n be positive integers and denote $K = \begin{pmatrix} I_m & 0 \\ 0 & -I_n \end{pmatrix}$. Let S_K be the collection of all $(m + n)$-square complex matrices X such that

$$X^*KX = K.$$

(a) If $A \in S_K$, show that A^{-1} exists and A^{-1}, A^t, $\bar{A}$, $A^* \in S_K.$

(b) If $A, B \in S_K$, show that $AB \in S_K$. How about kA or $A + B$?

(c) Discuss a similar problem with $K = \begin{pmatrix} 0 & I_m \\ -I_n & 0 \end{pmatrix}$, where $m = n.$

2.53   Let A and C be $m \times m$ and $n \times n$ matrices, respectively, and let $B, D,$ and E be matrices of appropriate sizes (so that each $|\cdot|$ makes sense).

(a) Show that

$$\begin{vmatrix} A & B \\ 0 & C \end{vmatrix} = \begin{vmatrix} A & 0 \\ D & C \end{vmatrix} = |A||C|.$$

(b) Evaluate

$$\begin{vmatrix} I_m & 0 \\ 0 & I_n \end{vmatrix}, \quad \begin{vmatrix} 0 & I_m \\ I_n & 0 \end{vmatrix}, \quad \begin{vmatrix} I_m & B \\ 0 & I_n \end{vmatrix}.$$

(c) Find a formula for

$$\begin{vmatrix} 0 & A \\ C & E \end{vmatrix}.$$

2.54   Let S be the $n \times n$ *backward identity* matrix, that is,

$$
S = \begin{pmatrix}
0 & 0 & \cdots & 0 & 1 \\
0 & 0 & \cdots & 1 & 0 \\
\vdots & \vdots & & \vdots & \vdots \\
0 & 1 & \cdots & 0 & 0 \\
1 & 0 & \cdots & 0 & 0
\end{pmatrix}_{n \times n}.
$$

Show that $S^{-1} = S^t = S$. Find $|S|$ and SAS for $A = (a_{ij}) \in M_n(\mathbb{C})$.

2.55   Let $A, B, C,$ and D be $m \times p$, $m \times q$, $n \times p$, and $n \times q$ matrices, respectively, where $m + n = p + q$. Show that

$$
\begin{vmatrix} A & B \\ C & D \end{vmatrix} = (-1)^{(mn+pq)} \begin{vmatrix} D & C \\ B & A \end{vmatrix}.
$$

In particular, when A, B, C, D are square matrices of the same size,

$$
\begin{vmatrix} A & B \\ C & D \end{vmatrix} = \begin{vmatrix} D & C \\ B & A \end{vmatrix},
$$

and for a square matrix A, a column vector x, and a row vector y,

$$
\begin{vmatrix} A & x \\ y & 1 \end{vmatrix} = \begin{vmatrix} 1 & y \\ x & A \end{vmatrix}.
$$

Is it generally true that for square matrices A, B, C, D of the same size

$$
\begin{vmatrix} A & B \\ C & D \end{vmatrix} = \begin{vmatrix} A & C \\ B & D \end{vmatrix}?
$$

2.56   Let $A, B, C, D \in M_n(\mathbb{C})$.

   (a) Show that if A^{-1} exists, then

$$
\begin{vmatrix} A & B \\ C & D \end{vmatrix} = |A| \, |D - CA^{-1}B|.
$$

   (b) Show that if $AC = CA$, then

$$
\begin{vmatrix} A & B \\ C & D \end{vmatrix} = |AD - CB|.
$$

   (c) Can CB on the right-hand side of (b) be changed to BC?

   (d) Does (b) remain true if the condition $AC = CA$ is dropped?

2.57 Let A, B, C, $D \in M_n(\mathbb{C})$. If matrix $\begin{pmatrix} A & B \\ C & D \end{pmatrix}$ has rank n, show that
$$\begin{vmatrix} |A| & |B| \\ |C| & |D| \end{vmatrix} = 0.$$ Moreover, if A is invertible, show that $D = CA^{-1}B$.

2.58 Let $A, B, C, D \in M_n(\mathbb{C})$ and let $M = \begin{pmatrix} A & B \\ C & D \end{pmatrix}$. Show that

    (a) $|M| = |AD^t - BC^t|$ if $CD^t - DC^t = 0$.

    (b) $|M|^2 = |AD^t + BC^t|^2$ if $CD^t + DC^t = 0$.

    (c) $|M| = |AD^t + BC^t|$ if $CD^t + DC^t = 0$ and $|D| \neq 0$.

    (d) (c) need not be true if D is singular.

2.59 Let A, B, B_1, B_2, B_3, and B_4 be 2×2 real matrices.

    (a) Is it true that $|A + B| = |A| + |B|$ in general?

    (b) If A is nonzero, and B_1, B_2, B_3, and B_4 satisfy

$$|A + B_i| = |A| + |B_i|, \quad i = 1, 2, 3, 4,$$

    show that B_1, B_2, B_3, and B_4 are linearly dependent.

2.60 Let $M = \begin{pmatrix} A & B \\ C & D \end{pmatrix}$ be an invertible matrix with $M^{-1} = \begin{pmatrix} X & Y \\ U & V \end{pmatrix}$, where A and D are square matrices (possibly of different sizes), B and C are matrices of appropriate sizes, and X has the same size as A.

    (a) Show that $\det A = \det V \det M$.

    (b) If A is invertible, show that $X = (D - CA^{-1}B)^{-1}$.

    (c) Consider a unitary matrix W partitioned as $W = \begin{pmatrix} a & x \\ y & Z \end{pmatrix}$, where a is a number and Z is a square matrix. Show that a and $\det Z$ have the same modulus, that is, $|a| = |\det Z|$.

    (d) What conclusion can be drawn for real orthogonal matrices?

2.61 Let $A = (a_{ij}) \in M_n(\mathbb{C})$ and $X = \text{diag}(x_1, x_2, \ldots, x_n)$. Show that

$$|X + A| = x_1 x_2 \cdots x_n + a_{11} x_2 \cdots x_n + a_{22} x_1 x_3 \cdots x_n + \cdots$$
$$+ \begin{vmatrix} a_{11} & a_{12} \\ a_{21} & a_{22} \end{vmatrix} x_3 \cdots x_n + \begin{vmatrix} a_{11} & a_{13} \\ a_{31} & a_{33} \end{vmatrix} x_2 x_4 \cdots x_n + \cdots + |A|,$$

in which the coefficient of $x_i x_j \cdots x_k$ is the determinant of the submatrix of A by deleting the rows and columns $i < j < \cdots < k$.

Derive the formula for characteristic polynomial:

$$|\lambda I - A| = \lambda^n - (\text{tr } A)\lambda^{n-1} + \delta_2 \lambda^{n-2} + \cdots + (-1)^k \delta_k \lambda^{n-k} + \cdots + (-1)^n |A|,$$

where δ_k is the sum of all $k \times k$ principal minors of A, $k = 1, 2, \ldots, n$.

2.62 Let x and y be column vectors of n complex components. Show that

(a) $|I - xy^*| = 1 - y^*x$.

(b) $\begin{vmatrix} I & x \\ y^* & 1 \end{vmatrix} = \begin{vmatrix} 1 & y^* \\ x & I \end{vmatrix} = 1 - y^*x$.

(c) If $\delta = 1 - y^*x \neq 0$, then $(I - xy^*)^{-1} = I + \delta^{-1}xy^*$ and

$$\begin{pmatrix} I & x \\ y^* & 1 \end{pmatrix}^{-1} = \begin{pmatrix} I + \delta^{-1}xy^* & -\delta^{-1}x \\ -\delta^{-1}y^* & \delta^{-1} \end{pmatrix}.$$

2.63 Define the maps f from the complex numbers to real matrices and φ from the ordered pairs of complex numbers to complex matrices by

$$z = x + iy \mapsto f(z) = \begin{pmatrix} x & y \\ -y & x \end{pmatrix} \in M_2(\mathbb{R}), \text{ where } x, y \in \mathbb{R},$$

$$q = (u, v) \mapsto \varphi(q) = \begin{pmatrix} u & v \\ -\bar{v} & \bar{u} \end{pmatrix} \in M_2(\mathbb{C}), \text{ where } u, v \in \mathbb{C}.$$

(a) Find $f(0)$, $f(1)$, $f(i)$, $f(1+i)$, $\varphi(1,i)$, and $\varphi(i,1)$.

(b) Show that $f(\bar{z}) = \big(f(z)\big)^t$, where $z \in \mathbb{C}$.

(c) Show that $f(z_1 z_2) = f(z_1)f(z_2)$, where $z_1, z_2 \in \mathbb{C}$.

(d) Show that $f(kz) = kf(z)$, where $k \in \mathbb{R}, z \in \mathbb{C}$.

(e) Show that $f(z_1 + z_2) = f(z_1) + f(z_2)$, where $z_1, z_2 \in \mathbb{C}$.

(f) Show that $f(z_1)f(z_2) = f(z_2)f(z_1)$, where $z_1, z_2 \in \mathbb{C}$.

(g) Show that $f(z^{-1}) = \big(f(z)\big)^{-1}$, where $0 \neq z \in \mathbb{C}$.

(h) Find $\big(f(z)\big)^n$ for $z = r(\cos\theta + i\sin\theta)$, where $r \geq 0, \theta \in \mathbb{R}$.

(i) Show that $\big(f(z)\big)^{-1} = \frac{1}{x^2+y^2}\begin{pmatrix} x & -y \\ y & x \end{pmatrix}$, where $z = x + iy \neq 0$.

(j) Show that $f(z) = P\begin{pmatrix} x+iy & 0 \\ 0 & x-iy \end{pmatrix}P^*$, where $P = \frac{1}{\sqrt{2}}\begin{pmatrix} 1 & 1 \\ i & -i \end{pmatrix}$.

(k) Show that $f(z) = \varphi(x,y)$ if $z = x + iy$, where $x, y \in \mathbb{R}$.

(l) Show that $\det\big(\varphi(q)\big) \geq 0$. Find $\big(\varphi(q)\big)^{-1}$ when $|u|^2 + |v|^2 = 1$.

(m) Apply f to every entry of $\varphi(q)$ to obtain a 4×4 real matrix $\mathcal{R}(q) = \begin{pmatrix} f(u) & f(v) \\ f(-\bar{v}) & f(\bar{u}) \end{pmatrix}$. Show that $\det\big(\mathcal{R}(q)\big) = \big(\det(\varphi(q))\big)^2$.

(n) Show that $\mathcal{R}(q)$ is singular if and only if $\varphi(q)$ is singular, and if and only if $u = v = 0$.

(o) Show that $\mathcal{R}(q)$ is similar via permutations to a matrix of the form $\begin{pmatrix} Y & X \\ -X & Y \end{pmatrix}$, where X and Y are some 2×2 real matrices.

2.64 Let $A = A_1 + iA_2$, where A_1 and A_2 are $n \times n$ real matrices. Define

$$\chi(A) = \begin{pmatrix} A_1 & A_2 \\ -A_2 & A_1 \end{pmatrix},$$

where χ is a mapping from $M_n(\mathbb{C})$ to $M_{2n \times 2n}(\mathbb{R})$. Show that

(a) $\chi(I_n) = I_{2n}$.

(b) $\chi(A + B) = \chi(A) + \chi(B)$.

(c) $\chi(AB) = \chi(A)\chi(B)$.

(d) $\chi(A^*) = \left(\chi(A) \right)^*$.

(e) $\chi(A^{-1}) = \left(\chi(A) \right)^{-1}$ if A is invertible.

(f) If A is unitary, Hermitian, or normal, then so is $\chi(A)$.

2.65 Let $A = H + iK \in M_n(\mathbb{C})$, where H and K are Hermitian. Show that

$$U^* \begin{pmatrix} A & A^* \\ A^* & A \end{pmatrix} U = 2 \begin{pmatrix} H & 0 \\ 0 & iK \end{pmatrix}, \text{ where } U = \tfrac{1}{\sqrt{2}} \begin{pmatrix} I & -I \\ I & I \end{pmatrix}.$$

2.66 Let A and B be $n \times n$ (real or complex) matrices. Show that

(a) $P^* \begin{pmatrix} A & B \\ B & A \end{pmatrix} P = \begin{pmatrix} A+B & 0 \\ 0 & A-B \end{pmatrix}$, where $P = \tfrac{1}{\sqrt{2}} \begin{pmatrix} I & I \\ I & -I \end{pmatrix}$.

(b) $Q^* \begin{pmatrix} A & B \\ -B & A \end{pmatrix} Q = \begin{pmatrix} A+iB & 0 \\ 0 & A-iB \end{pmatrix}$, where $Q = \tfrac{1}{\sqrt{2}} \begin{pmatrix} -iI & -iI \\ I & -I \end{pmatrix}$.

(c) $Q^* \begin{pmatrix} A & B \\ B & -A \end{pmatrix} Q = \begin{pmatrix} 0 & A-iB \\ A+iB & 0 \end{pmatrix}$, where $Q = \tfrac{1}{\sqrt{2}} \begin{pmatrix} -iI & -iI \\ I & -I \end{pmatrix}$.

2.67 Let A and B be $n \times n$ (real or complex) matrices. Compute

$$U^* \begin{pmatrix} A & B \\ B & -A \end{pmatrix} U, \ U^* \begin{pmatrix} A & B \\ -B & A \end{pmatrix} U, \ V^* \begin{pmatrix} A & B \\ -B & A \end{pmatrix} V, \ V^{-1} \begin{pmatrix} A & B \\ -B & A \end{pmatrix} V,$$

where

$$U = \tfrac{1}{\sqrt{2}} \begin{pmatrix} I & iI \\ iI & I \end{pmatrix}, \quad V = \begin{pmatrix} I & iI \\ 0 & I \end{pmatrix}.$$

2.68 Let A and B be $n \times n$ real matrices. Show that

$$\begin{vmatrix} A & B \\ -B & A \end{vmatrix} \geq 0.$$

2.69 Let A and B be $n \times n$ (real or complex) matrices. Show that

$$\begin{vmatrix} A & B \\ B & A \end{vmatrix} = |A + B||A - B|.$$

2.70 Let A and B be $n \times n$ (real or complex) matrices, $C = A + B$, and $D = A - B$. If C and D are invertible, show that $\left(\begin{smallmatrix} A & B \\ B & A \end{smallmatrix} \right)^{-1}$ exists and

$$\begin{pmatrix} A & B \\ B & A \end{pmatrix}^{-1} = \frac{1}{2} \begin{pmatrix} C^{-1} + D^{-1} & C^{-1} - D^{-1} \\ C^{-1} - D^{-1} & C^{-1} + D^{-1} \end{pmatrix}.$$

2.71 Let $M = \left(\begin{smallmatrix} A & B \\ C & D \end{smallmatrix} \right)$ be a block matrix. Which of the following is M^*?

$$M_1 = \begin{pmatrix} A^* & B^* \\ C^* & D^* \end{pmatrix}, \ M_2 = \begin{pmatrix} A^* & C^* \\ B^* & D^* \end{pmatrix}, \ M_3 = \begin{pmatrix} A & B^* \\ C^* & D \end{pmatrix}, \ M_4 = \begin{pmatrix} \overline{A} & C^* \\ B^* & \overline{D} \end{pmatrix}.$$

2.72 Show that a matrix A has rank one if and only if A can be written as $A = xy^t$ for some nonzero column vectors x and y. If $A = xy^t \in M_n(\mathbb{C})$, show that $A^k = \rho^{k-1} A$ for $k = 2, 3, \ldots$, where $\rho = y^t x \in \mathbb{C}$.

2.73 If A is a Hermitian matrix of rank one, show that A can be written as $A = xx^*$ or $A = -xx^*$ for some nonzero column vector x.

2.74 Let E_n be the $n \times n$ all-ones matrix and let $k \in \mathbb{R}$. Find

    (a) an invertible matrix P such that $P^{-1} E_n P$ is diagonal,

    (b) the rank of the matrix $kI_n + E_n$, and

    (c) $\det(kI_n + E_n)$.

2.75 Let $A \in M_n(\mathbb{C})$ and let $u_1, u_2, \ldots, u_n \in \mathbb{C}^n$ be linearly independent. Show that $r(A) = n \Leftrightarrow Au_1, Au_2, \ldots, Au_n$ are linearly independent.

2.76 Let A be an $m \times n$ matrix with rank $r > 0$, show that there exist an $m \times r$ matrix M and an $r \times n$ matrix N such that $A = MN$.

2.77 For matrices A, B, and C of appropriate sizes, answer true or false:

    (a) If $A^2 = B^2$, then $A = B$ or $A = -B$.

    (b) If $r(A) = r(B)$, then $r(A^2) = r(B^2)$.

    (c) $r(A + kB) \leq r(A) + kr(B)$, where k is a positive real number.

    (d) $r(A - B) \leq r(A) - r(B)$.

    (e) $r(AB) = r(BA)$.

    (f) If $r(AB) = 0$, then $r(A) = 0$ or $r(B) = 0$.

    (g) $r(AB, AC) \leq r(A)$.

    (h) $r(BA, CA) \leq r(A)$.

2.78 Show that the inverse of a rational (real) matrix is rational (real).

2.79 Consider the 2×2 Hermitian matrix $A = \begin{pmatrix} 1 & i \\ -i & 1 \end{pmatrix}$ and let

$$W_{\mathbb{R}}(A) = \{ Ax \mid x \in \mathbb{R}^2 \} \quad \text{and} \quad W_{\mathbb{C}}(A) = \{ Ax \mid x \in \mathbb{C}^2 \}.$$

(a) Show that the rows of A are linearly dependent over $\mathbb{C}$.

(b) Show that the rows of A are linearly independent over $\mathbb{R}$.

(c) Since the rows of A are linearly independent over $\mathbb{R}$, does it follow that the matrix A is invertible?

(d) Show that U^*AU is real diagonal, where $U = \frac{1}{\sqrt{2}} \begin{pmatrix} i & -i \\ 1 & 1 \end{pmatrix}$.

(e) What is the rank of A (over $\mathbb{C}$)?

2.80 Let R be an $m \times n$ real matrix. Let $W_{\mathbb{R}}(R) = \{ Rx \mid x \in \mathbb{R}^n \}$ and $W_{\mathbb{C}}(R) = \{ Rx \mid x \in \mathbb{C}^n \}$. Show that the dimension of $W_{\mathbb{R}}(R)$ as a subspace of $\mathbb{R}^m$ over $\mathbb{R}$ is equal to the dimension of $W_{\mathbb{C}}(R)$ as a subspace of $\mathbb{C}^m$ over $\mathbb{C}$. What is the dimension of $W_{\mathbb{C}}(R)$ as a subspace of $\mathbb{C}^m$ over $\mathbb{R}$? Discuss the analogs for a complex C in place of R.

2.81 Let A be an n-square Hermitian matrix. Write $A = B + iC$, where B and C are n-square real matrices.

(a) Show that $B^t = B$ and $C^t = -C$.

(b) Show that $x^t A x = x^t B x$ and $x^t C x = 0$ for all $x \in \mathbb{R}^n$.

(c) Show that if $Ax = 0$, $x \in \mathbb{R}^n$, then $Bx = Cx = 0$.

(d) Let $A = \begin{pmatrix} 1 & i \\ -i & 1 \end{pmatrix}$. What are B and C? Find an $x \in \mathbb{C}^2$ such that $Ax = 0$ (so $x^*Ax = 0$), but $x^*Bx \neq 0$ (so $Bx \neq 0$).

(e) Let $A = \begin{pmatrix} 1 & 1+i \\ 1-i & 1 \end{pmatrix}$. Find an $x \in \mathbb{R}^2$ such that $Bx = 0$ but $Ax \neq 0$. What are the ranks of A and B?

2.82 Find the values of s and t such that the matrices have the given ranks.

(a) $A = \begin{pmatrix} 1 & 2 & -1 & -1 \\ 2 & 0 & s & 0 \\ 0 & -4 & 5 & 2 \end{pmatrix}$, $r(A) = 2$.

(b) $B = \begin{pmatrix} t & 1 & 1 & 1 \\ 1 & t & 1 & 1 \\ 1 & 1 & t & 1 \\ 1 & 1 & 1 & t \end{pmatrix}$, $r(B) = 3$.

2.83 Let A and B be n-square matrices. Show that

$$r(AB - I_n) \leq r(A - I_n) + r(B - I_n).$$

2.84   Let A and B be n-square matrices such that $AB = 0$. Show that

$$r(A) + r(B) \leq n.$$

2.85   Let A be a matrix and let B be a submatrix of A obtained by deleting s rows and t columns from A. Show that

$$r(B) \geq r(A) - (s + t).$$

2.86   Let A be a square matrix. Show that

$$r(A^2) - r(A) \leq r(A^3) - r(A^2).$$

2.87   Let A and B be n-square matrices such that $AB = BA$. Show that

$$r(A + B) \leq r(A) + r(B) - r(AB).$$

2.88   Let $A, B \in M_n(\mathbb{C})$. Show that $\dim \mathrm{Ker}(AB) \leq \dim \mathrm{Ker}\, A + \dim \mathrm{Ker}\, B$.

2.89   Let $A_1, A_2, \ldots, A_k$ be $n \times n$ matrices. If $A_1 A_2 \cdots A_k = 0$, show that

$$r(A_1) + r(A_2) + \cdots + r(A_k) \leq (k - 1)n.$$

2.90   Let matrices X, Y, and Z have the same number of rows. Show that

$$r(X, Y) \leq r(X, Z) + r(Z, Y) - r(Z).$$

2.91   Let $X \in M_{p \times q}(\mathbb{C}), Y \in M_{s \times t}(\mathbb{C})$, and $Z \in M_{s \times q}(\mathbb{C})$. Show that

$$r(X) + r(Y) = r \begin{pmatrix} X & 0 \\ 0 & Y \end{pmatrix} \leq r \begin{pmatrix} X & 0 \\ Z & Y \end{pmatrix}.$$

2.92   Let A and B be $n \times n$ matrices. Is it true that

(a) $\begin{pmatrix} A \\ B \end{pmatrix}$ and (A, B) have the same rank?

(b) $A^*A + B^*B$ and $AA^* + BB^*$ have the same rank? or

(c) $A^*A + I_n$ and $AA^* + I_n$ have the same rank?

2.93   Let $A \in M_n(\mathbb{C})$. Is $r(A^t A)$ or $r(\bar{A}A)$ equal to $r(A)$? Show that

$$r\big((A\bar{A})^k\big) = r\big((\bar{A}A)^k\big), \quad k = 1, 2, \ldots.$$

2.94 Let A and B be n-square matrices such that $A^2 = A$, $B^2 = B$, and $I - (A + B)$ is invertible. Show that $r(A) = r(B)$.

2.95 Let A and B be $n \times n$ matrices satisfying $A^2 = A$, $B^2 = B$. Prove

$$r(A - B) = r(A - AB) + r(B - AB).$$

2.96 Let $A \in M_n(\mathbb{C})$. Show that $A^2 = A$ if and only if $r(A) + r(A - I) = n$.

2.97 Let A be an $m \times n$ matrix and let B be an $n \times p$ matrix. Show that

$$r \begin{pmatrix} 0 & A \\ B & I_n \end{pmatrix} = n + r(AB).$$

2.98 Let $A \in M_n(\mathbb{C})$ and $B \in M_m(\mathbb{C})$. If A is invertible, show that

$$r \begin{pmatrix} A & C \\ 0 & B \end{pmatrix}^k = n + r(B^k), \quad k = 1, 2, \ldots.$$

2.99 Let A be an $m \times n$ matrix, $m \geq n$. Show that

$$r(I_m - AA^*) - r(I_n - A^*A) = m - n.$$

2.100 Let $\mathrm{adj}(A)$ denote the adjugate matrix (or cofactor matrix) of $A = (a_{ij}) \in M_n(\mathbb{C})$, that is, $\mathrm{adj}(A)$ is the $n \times n$ matrix whose (i, j)-entry is the cofactor $(-1)^{i+j}|A_{ji}|$ of a_{ji}, where A_{ji} is the submatrix obtained from A by deleting the j-th row and the i-th column. Show that

(a) $r(A) = n$ if and only if $r(\mathrm{adj}(A)) = n$.
(b) $r(A) = n - 1$ if and only if $r(\mathrm{adj}(A)) = 1$.
(c) $r(A) < n - 1$ if and only if $r(\mathrm{adj}(A)) = 0$.
(d) $|\mathrm{adj}(A)| = |A|^{n-1}$.
(e) $\mathrm{adj}(\mathrm{adj}(A)) = |A|^{n-2}A$.
(f) $\mathrm{adj}(AB) = \mathrm{adj}(B)\,\mathrm{adj}(A)$.
(g) $\mathrm{adj}(XAX^{-1}) = X(\mathrm{adj}(A))X^{-1}$ for any invertible $X \in M_n(\mathbb{C})$.
(h) $|\overbrace{\mathrm{adj}\cdots\mathrm{adj}}^{k}(A)| = |A|$ when A is 2×2.
(i) If A is Hermitian, so is $\mathrm{adj}(A)$.
(j) Find a formula for $\overbrace{\mathrm{adj}\cdots\mathrm{adj}}^{k}(A)$ when $|A| = 1$.

2.101   The notation A^* is used for the adjugate of matrix A in some other texts. Under what conditions is A^* in the sense of this book, i.e., $A^* = (\bar{A})^t$, the conjugate transpose, the same as $\mathrm{adj}(A)$, the adjugate matrix of A? Simply put: When does $A^* = \mathrm{adj}(A)$?

2.102   Let $A \in M_n(\mathbb{C})$. Show that

(a) $\mathrm{Im}(AA^*) = \mathrm{Im}\, A$. Does $\mathrm{Im}(A^*A) = \mathrm{Im}\, A$?

(b) $A = AA^*B$ for some matrix $B \in M_n(\mathbb{C})$.

(c) If $A^*A = AA^*$ and $A^2 x = 0$, then $Ax = 0$.

2.103   Show that A is invertible if $A = (a_{ij}) \in M_n(\mathbb{C})$ satisfies, for each i,

$$|a_{ii}| > \sum_{j=1,\, j \neq i}^{n} |a_{ij}|. \quad \text{(Row diagonal dominance)}$$

Show that $I - B$ is invertible if $B = (b_{ij}) \in M_n(\mathbb{C})$ satisfies

$$1 > \sum_{j=1}^{n} |b_{ij}|, \quad i = 1, 2, \ldots, n.$$

2.104   Let

$$A = \begin{pmatrix} 1+\lambda & 1 & 1 \\ 1 & 1+\lambda & 1 \\ 1 & 1 & 1+\lambda \end{pmatrix}, \quad \beta = \begin{pmatrix} 0 \\ \lambda \\ \lambda^2 \end{pmatrix}.$$

Find the value(s) of λ such that β belongs to the column space of A.

2.105   Find a polynomial $p(x)$ of degree 3 such that the graph of $p(x)$ contains the points $(-1, 8), (0, 4), (1, 4)$, and $(2, 14)$ in the xy-plane.

2.106   Determine the dimension of the solution space of the linear equation system of three unknowns x_1, x_2, x_3 in terms of the value(s) of λ:

$$\begin{aligned} \lambda x_1 + x_2 + x_3 &= 0 \\ x_1 + \lambda x_2 + x_3 &= 0 \\ x_1 + x_2 + \lambda x_3 &= 0. \end{aligned}$$

2.107   Determine the value(s) of λ so that the following linear equation system of the unknowns x_1, x_2, x_3 has nonzero solutions:

$$\begin{aligned} x_1 + 2x_2 - 2x_3 &= 0 \\ 2x_1 - x_2 + \lambda x_3 &= 0 \\ 3x_1 + x_2 - x_3 &= 0. \end{aligned}$$

2.108 Find the general solutions of the linear system of five unknowns:

$$x_1 + x_2 + x_5 = 0$$
$$x_1 + x_2 - x_3 = 0$$
$$x_3 + x_4 + x_5 = 0.$$

2.109 Find the dimension and a basis for the solution space of the system

$$x_1 - x_2 + 5x_3 - x_4 = 0$$
$$x_1 + x_2 - 2x_3 + 3x_4 = 0$$
$$3x_1 - x_2 + 8x_3 + x_4 = 0$$
$$x_1 + 3x_2 - 9x_3 + 7x_4 = 0.$$

2.110 Given $y \in \mathbb{C}$, find the solutions of the linear system in x_1, x_2, x_3, x_4:

$$x_4 + x_2 = yx_1$$
$$x_1 + x_3 = yx_2$$
$$x_2 + x_4 = yx_3$$
$$x_3 + x_1 = yx_4.$$

2.111 Find the solutions of the equation system in x_1, x_2, x_3 in terms of a:

$$x_1 + ax_2 + 2x_3 = 1$$
$$x_1 + (2a - 1)x_2 + 3x_3 = 1$$
$$x_1 + ax_2 + (a + 1)x_3 = 2a - 1.$$

2.112 Find a basis for the solution space of the system of $n + 1$ linear equations of $2n$ unknowns $x_1, \ldots, x_{2n}$:

$$x_1 + x_2 + \cdots + x_n \quad = 0$$
$$x_2 + x_3 + \cdots + x_{n+1} \quad = 0$$
$$\vdots$$
$$x_{n+1} + x_{n+2} + \cdots + x_{2n} = 0.$$

2.113 Let $\alpha = \{\alpha_1, \ldots, \alpha_s\}$ and $\beta = \{\beta_1, \ldots, \beta_t\}$ be two sets of linearly independent vectors in $\mathbb{R}^n$. Show that $\dim(\text{Span}\,\alpha \cap \text{Span}\,\beta) = s + t - r(A)$, where A is the $n \times (s+t)$ matrix $A = (\alpha_1, \ldots, \alpha_s, \beta_1, \ldots, \beta_t)$.

2.114 Let A_1 and A_2 be respectively $n \times s$ and $n \times t$ real matrices, where $s + t = n$. If $\text{Im}\,A_1 + \text{Im}\,A_2 = \mathbb{R}^n$, show that $\text{Im}\,A_1 + \text{Im}\,A_2$ is a direct sum and $r(A_1) + r(A_2) = r(A_1, A_2) = n$. Discuss the converse(s).

2.115 Let $A \in M_{m \times p}(\mathbb{R})$, $B \in M_{p \times n}(\mathbb{R})$, and $V = \{Bx \mid ABx = 0, x \in \mathbb{R}^n\}$. Show that V is a subspace of $\mathbb{R}^p$ and $\dim V = r(B) - r(AB)$.

2.116 Let A be an $n \times n$ matrix (over a field $\mathbb{F}$) with columns $\alpha_1, \alpha_2, \ldots, \alpha_n$.

    (a) Show that $\dim(\text{Span}\{\alpha_1, \alpha_2, \ldots, \alpha_n\}) = r(A)$.

    (b) Let P be an $n \times n$ invertible matrix and let $\beta_1, \beta_2, \ldots, \beta_n$ be the columns of PA. Show that some α_i's are linearly independent if and only if the corresponding β_i's are linearly independent.

    (c) Find the dimension and a basis of the space spanned by

$$\gamma_1 = \begin{pmatrix} 2 \\ 1 \\ 3 \\ 1 \end{pmatrix}, \quad \gamma_2 = \begin{pmatrix} 1 \\ 2 \\ 0 \\ 1 \end{pmatrix}, \quad \gamma_3 = \begin{pmatrix} 0 \\ 2 \\ -2 \\ 1 \end{pmatrix}, \quad \gamma_4 = \begin{pmatrix} 1 \\ 1 \\ 1 \\ 1 \end{pmatrix}.$$

2.117 Let W_1 and W_2 be the vector spaces over $\mathbb{R}$ spanned respectively by

$$\alpha_1 = \begin{pmatrix} 1 \\ 2 \\ -1 \\ -2 \end{pmatrix}, \quad \alpha_2 = \begin{pmatrix} 3 \\ 1 \\ 1 \\ 1 \end{pmatrix}, \quad \alpha_3 = \begin{pmatrix} -1 \\ 0 \\ 1 \\ -1 \end{pmatrix}$$

and

$$\beta_1 = \begin{pmatrix} 2 \\ 5 \\ -6 \\ -5 \end{pmatrix}, \quad \beta_2 = \begin{pmatrix} -1 \\ 2 \\ -7 \\ 3 \end{pmatrix}.$$

Find the dimensions and bases for $W_1 \cap W_2$ and $W_1 + W_2$.

2.118 Let a_{ij} be integers, $1 \le i, j \le n$. If the system of linear equations

$$\sum_{j=1}^{n} a_{ij} x_j = b_i, \quad i = 1, 2, \ldots, n,$$

has integer solutions $x_1, x_2, \ldots, x_n$ for any integers $b_1, b_2, \ldots, b_n$, show that the determinant of the coefficient matrix (a_{ij}) is either 1 or -1.

2.119 Consider the two systems of linear equations in unknowns $x_1, x_2, \ldots, x_n$

$$x_1 + x_2 + \cdots + x_n = 0 \quad \text{and} \quad x_1 = x_2 = \cdots = x_n$$

over a field $\mathbb{F}$. Let W_1 and W_2 be their solution spaces. Show that

$$\mathbb{F}^n = W_1 \oplus W_2.$$

2.120   Let $A \in M_n(\mathbb{C})$. Show that there exists an $n \times n$ nonzero matrix B such that $AB = 0$ if and only if A is singular, that is, $\det A = 0$.

2.121   Let A be a $p \times n$ matrix and let B be a $q \times n$ matrix. If $r(A) + r(B) < n$, show that there exists a nonzero column vector x of n components such that both $Ax = 0$ and $Bx = 0$.

2.122   Let A and B be $n \times n$ matrices and let $\operatorname{Ker} A$ and $\operatorname{Ker} B$ be the null spaces of A and B with dimensions l and m, respectively. Show that the null space of AB has dimension at least $\max\{l, m\}$.

2.123   Let A be a square matrix. If $r(A) = r(A^2)$, show that the equation systems $Ax = 0$ and $A^2 x = 0$ have the same solution space.

2.124   Let A and B be $m \times n$ matrices. Show that $Ax = 0$ and $Bx = 0$ have the same solution space if and only if there exists an invertible matrix C such that $A = CB$. Use this fact to show that if $r(A^2) = r(A)$, then there exists an invertible matrix D such that $A^2 = DA$.

2.125   Consider the linear equation system $Ax = b$, where $b \neq 0$. Let $\eta_1, \eta_2, \ldots, \eta_n$ be solutions to $Ax = b$. Show that a linear combination $\lambda_1 \eta_1 + \lambda_2 \eta_2 + \cdots + \lambda_n \eta_n$, where $\lambda_1, \lambda_2, \ldots, \lambda_n$ are scalars, is a solution to $Ax = b$ if and only if $\lambda_1 + \lambda_2 + \cdots + \lambda_n = 1$. Show also that if $l_1 \eta_1 + l_2 \eta_2 + \cdots + l_n \eta_n = 0$, then $l_1 + l_2 + \cdots + l_n = 0$.

2.126   Let $A \in M_{m \times n}(\mathbb{C})$. Show that for any $b \in \mathbb{C}^m$, the linear equation system $A^* Ax = A^* b$ is consistent (meaning that it has solutions).

2.127   Let $A \in M_n(\mathbb{C})$ and let b be a column vector of n components. Denote $\tilde{A} = \begin{pmatrix} A & b \\ b^* & 0 \end{pmatrix}$. If $r(\tilde{A}) = r(A)$, which of the following must be true?

(a) $Ax = b$ has infinitely many solutions.

(b) $Ax = b$ has a unique solution.

(c) $\tilde{A}x = 0$ has only solution $x = 0$.

(d) $\tilde{A}x = 0$ has nonzero solutions.

2.128   Let A be a square matrix such that the linear equation system $Ax = 0$ has a nonzero solution. For any given b, show that the linear equation system $A^t x = b$ either has no solution or has infinitely many solutions.

2.129   Let $A \in M_{m \times n}(\mathbb{R})$ and $b \in \mathbb{R}^n$. Show that every solution to $Ax = 0$ is a solution to $bx = 0$ if and only if b is in the row space of A.

2.130   Let a_{ij}, b_i, c_j, and d be scalars in a number field $\mathbb{F}$, $1 \leq i, j \leq n$. Let

$$\begin{vmatrix} a_{11} & a_{12} & \cdots & a_{1n} \\ a_{21} & a_{22} & \cdots & a_{2n} \\ \cdots & \cdots & \cdots & \cdots \\ a_{n1} & a_{n2} & \cdots & a_{nn} \end{vmatrix} \neq 0.$$

Show that

$$\begin{cases} a_{11}x_1 + a_{12}x_2 + \cdots + a_{1n}x_n = b_1 \\ a_{21}x_1 + a_{22}x_2 + \cdots + a_{2n}x_n = b_2 \\ \cdots\cdots\cdots\cdots\cdots\cdots\cdots\cdots\cdots\cdots\cdots \\ a_{n1}x_1 + a_{n2}x_2 + \cdots + a_{nn}x_n = b_n \\ c_1x_1 + c_2x_2 + \cdots + c_nx_n = d \end{cases}$$

and

$$\begin{cases} a_{11}x_1 + a_{21}x_2 + \cdots + a_{n1}x_n = c_1 \\ a_{12}x_1 + a_{22}x_2 + \cdots + a_{n2}x_n = c_2 \\ \cdots\cdots\cdots\cdots\cdots\cdots\cdots\cdots\cdots\cdots\cdots \\ a_{1n}x_1 + a_{2n}x_2 + \cdots + a_{nn}x_n = c_n \\ b_1x_1 + b_2x_2 + \cdots + b_nx_n = d \end{cases}$$

either both have a unique solution or both have no solution.

2.131   Let A and B be matrices such that $r(AB) = r(A)$. Show that

$$X_1 AB = X_2 AB \quad \Rightarrow \quad X_1 A = X_2 A.$$

2.132   Find two planes in $\mathbb{R}^3$ whose intersection is the line of the points $(1 - t, 1 + t, t), t \in \mathbb{R}$, and one of the planes contains the origin.

2.133   Let $a, b, c \in \mathbb{R}$. Show that three distinct lines in the xy-plane

$$\begin{aligned} l_1 : & \quad ax + by + c = 0 \\ l_2 : & \quad bx + cy + a = 0 \\ l_3 : & \quad cx + ay + b = 0 \end{aligned}$$

intersect at exactly one point (x, y) if and only if $a + b + c = 0$.

Chapter 3

Similarity, Eigenvalues, Matrix Decompositions, and Linear Transformations

Definitions, Facts, and Examples ⎯⎯⎯⎯⎯⎯⎯⎯⎯⎯⎯⎯⎯⎯

Similarity. Let A and B be $n \times n$ matrices over a field $\mathbb{F}$. If there exists an $n \times n$ invertible matrix P over $\mathbb{F}$ such that $P^{-1}AP = B$, we say that A and B are *similar* over $\mathbb{F}$. If P is unitary, i.e., $P^*P = PP^* = I$, and $P^{-1}AP = P^*AP = B$, we say that A and B are *unitarily similar*.

Similarity among square matrices is an *equivalence relation*, that is, i) Every square matrix is similar to itself, ii) If A is similar to B, then B is similar to A, and iii) If A is similar to B and B is similar to C, then A is similar to C. Likewise, unitary similarity is also an equivalence relation.

Note that if A and B are (unitarily) similar, then A^k and B^k are (unitarily) similar for any positive integer k because $P^{-1}A^kP = (P^{-1}AP)^k = B^k$. Similar matrices have the same determinant, for if $B = P^{-1}AP$, then

$$|B| = |P^{-1}AP| = |P^{-1}|\,|A|\,|P| = |P|^{-1}\,|A|\,|P| = |A|.$$

In addition to determinant, similar matrices share many other properties, such as the same eigenvalues which we will see soon as we proceed.

We say a matrix is *diagonalizable* if it is similar to a diagonal matrix and *unitarily diagonalizable* if it is unitarily similar to a diagonal matrix.

Example 3.1 Consider the following 2×2 matrices A, $B(t)$, and C.

(a) Let $A = \begin{pmatrix} a & a-b \\ 0 & b \end{pmatrix}$, where a and $b \in \mathbb{F}$. Show that A is always diagonalizable, but A is never unitarily diagonalizable unless $a = b$.

(b) Let $B(t) = \begin{pmatrix} 1+t & 1 \\ -t^2 & 1-t \end{pmatrix}$, where $t \in \mathbb{R}$. Show that $B(t)$ is similar to $B(0)$ for all t and that $B(t)$ is not diagonalizable for any t.

(c) If $C^2 \neq 0$, show that $C^k \neq 0$ for all positive integers k.

Solution. (a) We show that A is similar to $\begin{pmatrix} a & 0 \\ 0 & b \end{pmatrix}$ through an invertible matrix P. For this, we find an invertible matrix P such that $AP = P \begin{pmatrix} a & 0 \\ 0 & b \end{pmatrix}$. Let $P = \begin{pmatrix} x & y \\ p & q \end{pmatrix}$. We compute both sides of $AP = P \begin{pmatrix} a & 0 \\ 0 & b \end{pmatrix}$ and get

$$\begin{pmatrix} ax + (a-b)p & ay + (a-b)q \\ bp & bq \end{pmatrix} = \begin{pmatrix} ax & by \\ ap & bq \end{pmatrix}.$$

Comparing the corresponding entries on both sides, we obtain

$$(a-b)p = 0, \quad (a-b)q = (b-a)y.$$

Set $p = 0$, $q = -y$, and take $P = \begin{pmatrix} x & y \\ 0 & -y \end{pmatrix}$, $xy \neq 0$. In particular, for $x = y = 1$, $P = \begin{pmatrix} 1 & 1 \\ 0 & -1 \end{pmatrix}$. Then $P^{-1}AP = \begin{pmatrix} 1 & 1 \\ 0 & -1 \end{pmatrix}^{-1} \begin{pmatrix} a & a-b \\ 0 & b \end{pmatrix} \begin{pmatrix} 1 & 1 \\ 0 & -1 \end{pmatrix} = \begin{pmatrix} a & 0 \\ 0 & b \end{pmatrix}$.

Now we show that if $a \neq b$, then there does not exist a unitary P so that $P^{-1}AP$ is diagonal. From the above argument, we have $p = 0$. Thus, P takes the form $\begin{pmatrix} x & y \\ 0 & q \end{pmatrix}$. If P is unitary, then $|q| = 1$ which yields $y = 0$. Because $ay + (a-b)q = by$, we obtain $q = 0$, a contradiction.

(b) We need an invertible matrix $Q(t)$ such that $B(t)Q(t) = Q(t)B(0)$. As we did in (a), we can choose $Q(t)$ to be any invertible matrix in the form $\begin{pmatrix} x & y \\ -tx & x-ty \end{pmatrix}$, $x \neq 0$. In particular, we can pick $Q(t) = \begin{pmatrix} 1 & 0 \\ -t & 1 \end{pmatrix}$.

Since $B(t)$ is similar to $B(0)$, to show that $B(t)$ is not diagonalizable, it is sufficient to show that $B(0)$ is not diagonalizable. Suppose otherwise that P is such an invertible matrix that $P^{-1}B(0)P$ is diagonal. Let $P = \begin{pmatrix} x & y \\ p & q \end{pmatrix}$. Then $\begin{pmatrix} 1 & 1 \\ 0 & 1 \end{pmatrix} \begin{pmatrix} x & y \\ p & q \end{pmatrix} = \begin{pmatrix} x & y \\ p & q \end{pmatrix} \begin{pmatrix} a & 0 \\ 0 & b \end{pmatrix}$ for some a and b. A simple computation results in $a = b = 1$ and $p = q = 0$, contradicting P being invertible.

(c) If C is invertible, i.e., $|C| \neq 0$, then $C^k \neq 0$ as $|C^k| = |C|^k \neq 0$. Suppose that C is singular and let $C = \begin{pmatrix} a & b \\ c & d \end{pmatrix}$. Then $ad - bc = 0$.

If $a = 0$, then $bc = 0$. So $b = 0$ or $c = 0$. Since $C^2 \neq 0$, we obtain $d \neq 0$. A direct computation shows that the $(2,2)$-entry of C^k is $d^k \neq 0$.

If $a \neq 0$, let $P = \begin{pmatrix} 1 & 0 \\ c/a & 1 \end{pmatrix}$. Then $P^{-1} = \begin{pmatrix} 1 & 0 \\ -c/a & 1 \end{pmatrix}$. With $d - bc/a = 0$, we obtain $P^{-1}CP = \begin{pmatrix} \alpha & b \\ 0 & 0 \end{pmatrix}$, where $\alpha = a + bc/a$. It follows that

$$P^{-1}C^kP = \begin{pmatrix} \alpha & b \\ 0 & 0 \end{pmatrix}^k = \begin{pmatrix} \alpha^k & \alpha^{k-1}b \\ 0 & 0 \end{pmatrix},$$

where k is a positive integer. As $C^2 \neq 0$, $\alpha \neq 0$. Hence, $C^k \neq 0$. $\square$

Notes: In view of eigenvalues and eigenvectors (see below), if $a = b$ in Example 3.1(a), then A is already a diagonal matrix; if $a \neq b$, then A has distinct eigenvalues, consequently A is diagonalizable (see the next example). The matrix $B(t)$ in (b) always has the repeated eigenvalue 1; the eigenvectors are dependent on t. For (c), more generally, if C is an $n \times n$ matrix such that $C^n \neq 0$, then $C^k \neq 0$ for any positive integer k, equivalently, if $C^m = 0$ for some m, then $C^n = 0$.

Trace. Let $A = (a_{ij})$ be an $n \times n$ matrix. The *trace* of A, denoted by $\operatorname{tr} A$, is defined to be the sum of the entries on the main diagonal of A, that is,

$$\operatorname{tr} A = a_{11} + a_{22} + \cdots + a_{nn}.$$

Eigenvalues and Eigenvectors of a Matrix. Let A be an $n \times n$ matrix over a field $\mathbb{F}$. A scalar $\lambda \in \mathbb{F}$ is said to be an *eigenvalue* of A if

$$Ax = \lambda x \tag{3.1}$$

for some nonzero column vector $x \in \mathbb{F}^n$. Such a vector x is referred to as an *eigenvector* corresponding (belonging or assiciated) to the eigenvalue λ.

Let A be an $n \times n$ complex matrix. The fundamental theorem of algebra ensures that A has n complex eigenvalues, including the repeated ones. To see this, observe that $Ax = \lambda x$ is equivalent to $(\lambda I - A)x = 0$, which has a nonzero solution x if and only if $\lambda I - A$ is singular. This is equivalent to λ being a scalar such that $|\lambda I - A| = 0$. Thus, to find the eigenvalues of A, one needs to find the roots of the *characteristic polynomial* of A

$$p_A(\lambda) = |\lambda I - A|. \tag{3.2}$$

Let $\lambda_1(A), \lambda_2(A), \ldots, \lambda_n(A)$, or simply $\lambda_1, \lambda_2, \ldots, \lambda_n$ if A is the only matrix in discussion, be the complex roots of the characteristic polynomial

$$p_A(\lambda) = |\lambda I - A| = (\lambda - \lambda_1)(\lambda - \lambda_2) \cdots (\lambda - \lambda_n).$$

Then $\lambda_1, \lambda_2, \ldots, \lambda_n$ (not necessarily distinct) are the eigenvalues of A. Expanding the determinant, we see that the constant term of $p_A(\lambda)$ is $(-1)^n |A|$ (this is also seen by putting $\lambda = 0$) and the coefficient of λ^{n-1} is $-\operatorname{tr} A$. Multiplying out the right-hand side and comparing the coefficients give us

$$\operatorname{tr} A = a_{11} + a_{22} + \cdots + a_{nn} = \lambda_1 + \lambda_2 + \cdots + \lambda_n \tag{3.3}$$

and

$$|A| = \lambda_1 \lambda_2 \cdots \lambda_n. \tag{3.4}$$

Let $A = \left(\begin{smallmatrix} 1 & 3 \\ 4 & 2 \end{smallmatrix}\right)$. Then $p_A(\lambda) = \lambda^2 - 3\lambda - 10 = (\lambda - 5)(\lambda + 2)$. The eigenvalues of A are $5, -2$, the determinant $|A| = -10$, and trace $\operatorname{tr} A = 3$.

To find an eigenvector belonging to the eigenvalue 5, we solve the linear equation system $(5I - A)x = 0$ to get $x = \left(\begin{smallmatrix} 3 \\ 4 \end{smallmatrix}\right)$ (or any nonzero multiple of x). Similarly, $v = \left(\begin{smallmatrix} 1 \\ -1 \end{smallmatrix}\right)$ is an eigenvector of A corresponding to -2.

The eigenvectors x corresponding to the eigenvalue λ are the nontrivial solutions to the linear equation system $(\lambda I - A)x = 0$; with zero, they form the null space of $\lambda I - A$, called the *eigenspace* of A associated to λ.

Similar matrices have the same characteristic polynomial, thus the same eigenvalues and trace, but not necessarily the same corresponding eigenvectors. The eigenvalues of a diagonal matrix or an upper- (or lower-) triangular matrix are the elements on the main diagonal.

Example 3.2 Let A be an $n \times n$ matrix. Show that

(a) The eigenvectors of distinct eigenvalues are linearly independent.

(b) If the eigenvalues of A are all distinct, then A is diagonalizable.

(c) A is diagonalizable if and only if A has n linearly independent eigenvectors (which may or may not belong to different eigenvalues).

(d) If A is Hermitian (or real symmetric), then the eigenvectors of distinct eigenvalues of A are orthogonal, that is, if $Au = \lambda u$, $Av = \mu v$, where $\lambda \neq \mu$ (and u and v are nonzero column vectors), then $u^*v = 0$.

Solution. (a) Let $v_1, \ldots, v_k$ be eigenvectors of A corresponding to the eigenvalues $\lambda_1, \ldots, \lambda_k$ of A, where $\lambda_i \neq \lambda_j$ when $i \neq j$. We show that $v_1, \ldots, v_k$ are linearly independent by induction on k.

For $k = 2$, we prove that v_1 and v_2 are linearly independent. If $l_1 v_1 + l_2 v_2 = 0$, then $A(l_1 v_1 + l_2 v_2) = 0$, i.e., $l_1 \lambda_1 v_1 + l_2 \lambda_2 v_2 = 0$. Since $l_1 \lambda_1 v_1 + l_2 \lambda_1 v_2 = 0$, by subtracting, $l_2(\lambda_1 - \lambda_2)v_2 = 0$. As $\lambda_1 \neq \lambda_2$ and $v_2 \neq 0$, we have $l_2 = 0$, which yields $l_1 = 0$. So, v_1 and v_2 are linearly independent.

Suppose the statement holds for $k-1$ eigenvectors. For the case of k, let

$$l_1 v_1 + \cdots + l_k v_k = 0. \tag{3.5}$$

Applying A to the above equation, we get $l_1 A v_1 + \cdots + l_k A v_k = 0$. Thus, $l_1 \lambda_1 v_1 + \cdots + l_k \lambda_k v_k = 0$. Subtracting equation (3.5) times λ_k, we obtain

$$l_1(\lambda_1 - \lambda_k)v_1 + \cdots + l_{k-1}(\lambda_{k-1} - \lambda_k)v_{k-1} = 0.$$

By induction hypothesis, $v_1, \ldots, v_{k-1}$ are linearly independent. Thus, for $i = 1, \ldots, k-1$, $l_i(\lambda_i - \lambda_k) = 0$. As $\lambda_i \neq \lambda_k$, $i = 1, \ldots, k-1$, $l_1, \ldots, l_{k-1}$ are all zero, resulting in $l_k = 0$. So, $v_1, \ldots, v_k$ are linearly independent.

(b) Let $\lambda_1, \ldots, \lambda_n$ be the eigenvalues of A and they are all different from each other. Let $u_1, \ldots, u_n$ be eigenvectors of A corresponding to $\lambda_1, \ldots, \lambda_n$, respectively. Construct a matrix P by the column vectors $u_1, \ldots, u_n$, i.e., $P = (u_1, \ldots, u_n)$. Then P is invertible because $u_1, \ldots, u_n$ are linearly independent. It is easy to check that

$$AP = A(u_1, \ldots, u_n) = (\lambda_1 u_1, \ldots, \lambda_n u_n) = P \operatorname{diag}(\lambda_1, \ldots, \lambda_n). \qquad (3.6)$$

It follows that $P^{-1}AP = \operatorname{diag}(\lambda_1, \ldots, \lambda_n)$ is a diagonal matrix.

(c) If A is diagonalizable, then $P^{-1}AP = \operatorname{diag}(\lambda_1, \ldots, \lambda_n)$, where $\lambda_1, \ldots, \lambda_n$ are the eigenvalues of A and P is some invertible matrix. Thus, $AP = P \operatorname{diag}(\lambda_1, \ldots, \lambda_n)$. The column vectors of P are linearly independent and they are eigenvectors of A corresponding to eigenvalues $\lambda_1, \ldots, \lambda_n$. Conversely, if $u_1, \ldots, u_n$ are eigenvectors of A belonging to eigenvalues $\lambda_1, \ldots, \lambda_n$, and if $u_1, \ldots, u_n$ are linearly independent, construct a matrix P as in (3.6). Then $P^{-1}AP = \operatorname{diag}(\lambda_1, \ldots, \lambda_n)$ is diagonal.

(d) If $Au = \lambda u$, $u \neq 0$, then $u^* A u = \lambda u^* u$. By taking the conjugate transpose of both sides and using the fact that $A^* = A$, we obtain $\bar{\lambda} = \lambda$. This says that the eigenvalues of a Hermitian matrix are necessarily real.

Now suppose $Au = \lambda u$ and $Av = \mu v$, where u and v are nonzero, λ and μ are real, and $\lambda \neq \mu$. On the one hand, $u^* A v = \mu u^* v$. On the other hand,

$$u^* A v = u^* A^* v = (v^* A u)^* = (\lambda v^* u)^* = \lambda(u^* v).$$

Because $\lambda \neq \mu$, we have $u^* v = 0$, i.e., u and v are orthogonal. □

Let $A \in M_n(\mathbb{F})$. If λ is an eigenvalue of A and $u_1, \ldots, u_k$ are eigenvectors of λ that comprise a basis of the eigenspace $V_\lambda = \{x \in \mathbb{F}^n \mid Ax = \lambda x\}$, then

$$P^{-1}AP = \begin{pmatrix} \lambda I_k & \star \\ 0 & \star \end{pmatrix}, \quad (\star \text{ represents unspecified entries})$$

where P is an invertible matrix with $u_1, \ldots, u_k$ as its first k columns.

Example 3.3 Let A be an $m \times n$ matrix and let B be an $n \times m$ matrix. Show that AB and BA have the same nonzero eigenvalues, including multiplicities. Consequently, the trace cyclic property follows:

$$\operatorname{tr}(AB) = \operatorname{tr}(BA). \qquad (3.7)$$

If $m = n$, then AB and BA have the same eigenvalues. We simply put:

$$\operatorname{eig}(AB) = \operatorname{eig}(BA). \qquad (3.8)$$

Solution 1. Use determinants. Since

$$\begin{pmatrix} I_m & -A \\ 0 & \lambda I_n \end{pmatrix} \begin{pmatrix} \lambda I_m & A \\ B & I_n \end{pmatrix} = \begin{pmatrix} \lambda I_m - AB & 0 \\ \lambda B & \lambda I_n \end{pmatrix}$$

and

$$\begin{pmatrix} I_m & 0 \\ -B & \lambda I_n \end{pmatrix} \begin{pmatrix} \lambda I_m & A \\ B & I_n \end{pmatrix} = \begin{pmatrix} \lambda I_m & A \\ 0 & \lambda I_n - BA \end{pmatrix},$$

by taking determinants and equating the right-hand sides, we obtain

$$\lambda^n \det(\lambda I_m - AB) = \lambda^m \det(\lambda I_n - BA). \tag{3.9}$$

Considering the factorized forms of (3.9) over $\mathbb{C}$, we see that AB and BA have the same eigenvalues (including multiplicities), except $|m - n|$ zeros.

Solution 2. Use matrix similarity. Consider the block matrix $\begin{pmatrix} 0 & 0 \\ B & 0 \end{pmatrix}$. Adding the second row multiplied by A from the left to the first row gives $\begin{pmatrix} AB & 0 \\ B & 0 \end{pmatrix}$. Perform the similar operations for columns to get $\begin{pmatrix} 0 & 0 \\ B & BA \end{pmatrix}$. Put in equations:

$$\begin{pmatrix} I_m & A \\ 0 & I_n \end{pmatrix} \begin{pmatrix} 0 & 0 \\ B & 0 \end{pmatrix} = \begin{pmatrix} AB & 0 \\ B & 0 \end{pmatrix}$$

and

$$\begin{pmatrix} 0 & 0 \\ B & 0 \end{pmatrix} \begin{pmatrix} I_m & A \\ 0 & I_n \end{pmatrix} = \begin{pmatrix} 0 & 0 \\ B & BA \end{pmatrix}.$$

It follows that

$$\begin{pmatrix} I_m & A \\ 0 & I_n \end{pmatrix}^{-1} \begin{pmatrix} AB & 0 \\ B & 0 \end{pmatrix} \begin{pmatrix} I_m & A \\ 0 & I_n \end{pmatrix} = \begin{pmatrix} 0 & 0 \\ B & BA \end{pmatrix}.$$

Thus

$$\begin{pmatrix} AB & 0 \\ B & 0 \end{pmatrix} \quad \text{and} \quad \begin{pmatrix} 0 & 0 \\ B & BA \end{pmatrix}$$

are similar. Hence, AB and BA have the same nonzero eigenvalues, counting multiplicities. (Question: Are AB and BA similar?)

Solution 3. Use continuity argument. We first deal with the case where $m = n$. If A is nonsingular, then $BA = A^{-1}(AB)A$. Thus, AB and BA are similar and have the same eigenvalues.

If A is singular, let δ be such a positive number that $\epsilon I + A$ is nonsingular for every ϵ, $0 < \epsilon < \delta$. Then $(\epsilon I + A)B$ and $B(\epsilon I + A)$ are similar and have the same characteristic polynomials. Therefore,

$$\det \big(\lambda I_n - (\epsilon I_n + A)B \big) = \det \big(\lambda I_n - B(\epsilon I_n + A) \big), \quad 0 < \epsilon < \delta.$$

Since both sides are continuous functions of ϵ, letting $\epsilon \to 0^+$ gives

$$\det(\lambda I_n - AB) = \det(\lambda I_n - BA).$$

Thus, AB and BA have the same eigenvalues.

If $m \neq n$, say $m < n$. We add zeros to A and B so that $A_1 = \begin{pmatrix} A \\ 0 \end{pmatrix}$ and $B_1 = (B, 0)$ are $n \times n$. It follows that $A_1 B_1$ and $B_1 A_1$ have the same nonzero eigenvalues, counting multiplicities. Note that $B_1 A_1 = BA$ and $A_1 B_1 = \begin{pmatrix} AB & 0 \\ 0 & 0 \end{pmatrix}$. The desired conclusion follows at once.

The trace (as the sum of eigenvalues) identity is immediate. $\square$

The trace identity (*cyclic invariance*) can also be shown directly as follows: Write $A = (a_{ij})$ and $B = (b_{ij})$. Then the (i, i)-entry of AB and the (j, j)-entry of BA are $\sum_{j=1}^n a_{ij} b_{ji}$ and $\sum_{i=1}^m b_{ji} a_{ij}$, respectively. Thus,

$$\operatorname{tr}(AB) = \sum_{i=1}^m \sum_{j=1}^n a_{ij} b_{ji} = \sum_{j=1}^n \sum_{i=1}^m b_{ji} a_{ij} = \operatorname{tr}(BA).$$

Cayley–Hamilton Theorem and Minimal Polynomial. The vector space $M_n(\mathbb{F})$ of $n \times n$ matrices over $\mathbb{F}$ has dimension n^2. If $A \in M_n(\mathbb{F})$, then $I, A, \ldots, A^{n^2}$ are linearly dependent as vectors in $M_n(\mathbb{F})$. So, there exist $a_0, a_1, \ldots, a_{n^2} \in \mathbb{F}$, not all zero, such that $a_0 I + a_1 A + \cdots + a_{n^2} A^{n^2} = 0$. Let $p(x) = a_0 + a_1 x + \cdots + a_{n^2} x^{n^2}$. Then $p(A)$ as a polynomial in A is 0.

There are many polynomials that annihilate A. The characteristic polynomial of A is one of those. The classical *Cayley–Hamilton theorem* asserts $p_A(A) = 0$, where $p_A(\lambda) = |\lambda I - A|$ is the characteristic polynomial of A.

Of all annihilating polynomials of A, there exists a unique one that is monic (with leading coefficient 1) and has the smallest degree. Such a polynomial is referred to as the *minimal polynomial* of A, denoted by $m_A(\lambda)$. The minimal polynomial of A divides all annihilating polynomials of A, especially, it divides the characteristic polynomial. Moreover, every root of the characteristic polynomial is a root of the minimal polynomial (with possibly different multiplicities) and vice versa. For example, let

$$A = \begin{pmatrix} 1 & 1 & 0 \\ 0 & 1 & 0 \\ 0 & 0 & 1 \end{pmatrix}, \quad B = \begin{pmatrix} 1 & 1 & 0 \\ 0 & 1 & 0 \\ 0 & 0 & 2 \end{pmatrix}.$$

The characteristic polynomial and minimal polynomial of A are respectively $p_A(\lambda) = (\lambda-1)^3$ and $m_A(\lambda) = (\lambda-1)^2$, while the characteristic and minimal polynomials of B are the same: $p_B(\lambda) = m_B(\lambda) = (\lambda - 1)^2(\lambda - 2)$.

Example 3.4 Consider the following 3×3 real matrices M, N, and P.

(a) If $M^3 = I$, $M \neq I$, show that M is not diagonalizable over $\mathbb{R}$, but is diagonalizable over $\mathbb{C}$. Give an example of such a matrix.

(b) If $N = \begin{pmatrix} 1 & x & 1 \\ x & 1 & x \\ 1 & x & 1 \end{pmatrix}$, find the real value(s) of x so that N has one negative eigenvalue, one positive eigenvalue, and a zero eigenvalue.

(c) If $r(P) = 1$ and $\operatorname{tr} P = 1$, find the characteristic polynomial of P.

Solution. (a) Since $M^3 - I = 0$, $M \neq I$, the minimal polynomial of M divides $\lambda^3 - 1$, factored as $\lambda^3 - 1 = (\lambda - 1)(\lambda^2 + \lambda + 1)$ over $\mathbb{R}$. Thus, $m_M(\lambda) = p_M(\lambda) = \lambda^3 - 1$. The complex eigenvalues of M are $1, \frac{-1 \pm \sqrt{3}\,i}{2}$.

Because the eigenvalues of M are distinct, M is diagonalizable over $\mathbb{C}$ and it is similar to $\operatorname{diag}\left(1, \frac{-1+\sqrt{3}\,i}{2}, \frac{-1-\sqrt{3}\,i}{2}\right)$. As M has non-real eigenvalues, it cannot be similar to a diagonal real matrix. Below is such a matrix

$$\begin{pmatrix} 1 & 0 & 0 \\ 0 & -\frac{1}{2} & -\frac{\sqrt{3}}{2} \\ 0 & \frac{\sqrt{3}}{2} & -\frac{1}{2} \end{pmatrix}.$$

(b) The characteristic polynomial of N is $\lambda(\lambda^2 - 3\lambda + 2 - 2x^2)$. So the eigenvalues of N are $0, \frac{3 \pm \sqrt{1+8x^2}}{2}$. If $|x| > 1$, then N has one positive eigenvalue, one negative eigenvalue, and a zero eigenvalue. If $x = \pm 1$, then N has eigenvalue 0 with multiplicity 2 and one positive eigenvalue. If $|x| < 1$, then N has two positive eigenvalues and a zero eigenvalue.

(c) Since P has rank 1, the rows of P are proportional. By row and column (similarity) permutations, we may assume that the first row of P is nonzero and row 2 and row 3 are multiples of row 1. One may verify that

$$Q^{-1}PQ = \begin{pmatrix} \operatorname{tr} P & b & c \\ 0 & 0 & 0 \\ 0 & 0 & 0 \end{pmatrix} \text{ if } P = \begin{pmatrix} a & b & c \\ pa & pb & pc \\ qa & qb & qc \end{pmatrix}, \ Q = \begin{pmatrix} 1 & 0 & 0 \\ p & 1 & 0 \\ q & 0 & 1 \end{pmatrix}.$$

Since $\operatorname{tr} P = 1$, the characteristic polynomial of P is $\lambda^2(\lambda - 1)$. □

Schur Triangularization, Singular Value Decomposition (SVD), and Jordan Canonical Form. In this section, we present three fundamentally important theorems about matrix decompositions. As we proceed, more matrix decompositions, such as spectral decomposition and polar decomposition, etc., will be discussed (in Chapters 4 and 5).

- **Schur triangularization.** Let $A \in M_n(\mathbb{C})$. Then there exists an $n \times n$ unitary matrix U such that U^*AU is an upper- (or lower-) triangular matrix having the eigenvalues of A on the main diagonal. That is, every square matrix is unitarily similar to an upper- (or lower-) triangular matrix over $\mathbb{C}$. Let $\lambda_1, \ldots, \lambda_n$ be the eigenvalues of A. We have the upper-triangular form (with $\star$ for unspecified entries)

$$U^*AU = \begin{pmatrix} \lambda_1 & & \star \\ & \ddots & \\ 0 & & \lambda_n \end{pmatrix}. \quad (Schur\ decomposition)$$

Proof by induction on n: If $n = 1$, there is nothing to prove. Let $n > 1$. We reduce the problem to a matrix of order $n - 1$. Let u_1 be a unit eigenvector associated to eigenvalue λ_1. Let $P = (u_1, \ldots, u_n)$ be a unitary matrix (see (ii), p. 43, Chapter 2). Then $Au_1 = \lambda_1 u_1$, $u_i^*Au_1 = 0$, $i > 1$, and $P^*AP = \begin{pmatrix} \lambda_1 & \star \\ 0 & A_1 \end{pmatrix}$, where A_1 has order $n - 1$.

- **Singular value decomposition (SVD).** Let $A \in M_{m \times n}(\mathbb{C})$. Let λ be an eigenvalue of A^*A and let x be a corresponding eigenvector. Then $(A^*A)x = \lambda x$ implies $x^*(A^*A)x = (Ax)^*(Ax) = \lambda x^*x \geq 0$. Hence, $\lambda \geq 0$. The nonnegative square roots of the eigenvalues of A^*A are the *singular values* of A. For example, if $A = \begin{pmatrix} 0 & 10 \\ 0 & 1 \end{pmatrix}$, then the singular values of A are $0, \sqrt{101}$ (whereas the eigenvalues of A are $0, 1$).

Let A be an $m \times n$ nonzero matrix and let $\sigma_1(A) \geq \cdots \geq \sigma_r(A)$ (or leaving A's out) be the positive singular values of A. Then there exist an $m \times m$ unitary matrix U and an $n \times n$ unitary matrix V such that

$$A = UDV, \quad (Singular\ value\ decomposition)$$

where D is the $m \times n$ matrix with (i, i)-entry σ_i, $i = 1, \ldots, r$, and 0 elsewhere. This is the *singular value decomposition (SVD) theorem*. If A is a real matrix, then U and V can be chosen to be real orthogonal.

Proof by induction on the size of A. Let $A \neq 0$. If $m = 1$ or $n = 1$, say $m = 1$ and $A = (a_1, \ldots, a_n)$. Then $\sigma_1 = (|a_1|^2 + \cdots + |a_n|)^{\frac{1}{2}}$, and $\frac{1}{\sigma_1}(a_1, \ldots, a_n)$ is a unit vector. Let V be a unitary matrix with $\frac{1}{\sigma_1}(a_1, \ldots, a_n)$ as its first row. Then $A = (\sigma_1, 0, \ldots, 0)V$ is the SVD of A. Thus, the statement holds true for $m = 1$ or $n = 1$.

Let $m > 1$ and $n > 1$. We reduce the problem to a matrix of size $(m - 1) \times (n - 1)$. Let u_1 be a unit eigenvector of A^*A associated to σ_1^2, i.e., $(A^*A)u_1 = \sigma_1^2 u_1$. Let $v_1 = \frac{1}{\sigma_1}Au_1$. Then v_1 is a unit vector and $v_1^* A u_1 = \sigma_1$. Let P be an $m \times m$ unitary matrix with v_1^* as its first row and let Q be an $n \times n$ unitary matrix with u_1 as its first column. Then $PAQ = \left(\begin{smallmatrix} \sigma_1 & 0 \\ 0 & A_1 \end{smallmatrix} \right)$, where A_1 has size $(m-1) \times (n-1)$. The number of positive singular values of A equals the rank of A. If A is a square matrix, then A^*, A^t, and $\bar{A}$ have the same singular values as A. The singular values of A^{-1} (if the inverse exists) are the reciprocals of those of A. If A is an $n \times n$ matrix having eigenvalues $\lambda_1(A), \ldots, \lambda_n(A)$ and singular values $\sigma_1(A), \ldots, \sigma_n(A)$, then

$$|\det A| = \left| \prod_{i=1}^{n} \lambda_i(A) \right| = \prod_{i=1}^{n} \sigma_i(A). \qquad (3.10)$$

Singular value decomposition is one of the most important results in matrix theory; it has many applications in various scientific areas.

- **Jordan (canonical) form.** Let $A \in M_n(\mathbb{C})$. Then there exists an $n \times n$ invertible complex matrix P such that $P^{-1}AP$ is block-diagonal:

$$P^{-1}AP = \begin{pmatrix} J_1 & & & 0 \\ & J_2 & & \\ & & \ddots & \\ 0 & & & J_s \end{pmatrix}, \quad (Jordan\ decomposition)$$

where each J_t, $t = 1, 2, \ldots, s$, called a *Jordan block*, takes the form

$$J_t = \begin{pmatrix} \lambda_t & 1 & & 0 \\ & \lambda_t & \ddots & \\ & & \ddots & 1 \\ 0 & & & \lambda_t \end{pmatrix}, \quad (Jordan\ block)$$

where $\lambda_1, \ldots, \lambda_s$ are the eigenvalues of A (not necessarily different). Below are a few exemplary Jordan blocks of orders 1, 2, 3, and 4:

$$(2), \quad \begin{pmatrix} 2 & 1 \\ 0 & 2 \end{pmatrix}, \quad \begin{pmatrix} 2 & 1 & 0 \\ 0 & 2 & 1 \\ 0 & 0 & 2 \end{pmatrix}, \quad \begin{pmatrix} 2 & 1 & 0 & 0 \\ 0 & 2 & 1 & 0 \\ 0 & 0 & 2 & 1 \\ 0 & 0 & 0 & 2 \end{pmatrix}.$$

Every square matrix is similar to a Jordan canonical form (over $\mathbb{C}$). Two $n \times n$ matrices are similar if and only if they have the same Jordan canonical form (up to permutations of the Jordan blocks).

The Jordan canonical form of a matrix carries a great deal of algebraic information about the matrix; it is a useful and powerful tool for many matrix problems. For instance, if $\left(\begin{smallmatrix} 2 & 1 \\ 0 & 2 \end{smallmatrix}\right)$ is a Jordan block of a matrix, then the matrix is not diagonalizable, that is, it cannot be similar to a diagonal matrix; if $\left(\begin{smallmatrix} A & 0 \\ 0 & B \end{smallmatrix}\right)$ is similar to $\left(\begin{smallmatrix} A & 0 \\ 0 & C \end{smallmatrix}\right)$, where $A, B, C \in M_n(\mathbb{C})$, then B is similar to C (why?); and for any complex matrix F, the square block matrix $\left(\begin{smallmatrix} 0 & F \\ 0 & 0 \end{smallmatrix}\right)$ is similar to a real matrix (why?).

The determination of the Jordan blocks of a matrix needs the theory of generalized eigenvectors or elementary divisors and invariant factors (Smith normal form). The Jordan theorem itself is easily understandable. Its proof, however, requires much time and energy, and can be found in most linear algebra texts, thus is omitted here.

Example 3.5 Let $x, y, z \in \mathbb{C}$ and let

$$
A = \begin{pmatrix} 1 & 0 & x \\ 3 & 2 & 0 \\ 4 & 0 & 2 \end{pmatrix}, \quad B = \begin{pmatrix} 2 & y & 0 \\ 0 & 2 & 0 \\ 3 & 0 & 1 \end{pmatrix}, \quad C = \begin{pmatrix} 0 & 1 & 0 \\ 0 & 0 & z \\ 0 & 0 & 0 \end{pmatrix}.
$$

Show that

(a) A and B are similar if and only if $x = y = 0$.

(b) C is similar to C^t, but not unitarily similar to C^t if $|z| \neq 0, 1$.

Solution. (a) We compute the determinants to get $\det A = 4 - 8x$ and $\det B = 4$. So, if $x \neq 0$, then A and B cannot be similar.

If $x = 0$, the eigenvalues of A and B are 1, 2, 2. Their possible Jordan forms are

$$
\Lambda = \begin{pmatrix} 1 & 0 & 0 \\ 0 & 2 & 1 \\ 0 & 0 & 2 \end{pmatrix} \quad \text{or} \quad \Gamma = \begin{pmatrix} 1 & 0 & 0 \\ 0 & 2 & 0 \\ 0 & 0 & 2 \end{pmatrix}.
$$

Since the rank of $A - 2I$ is 1 for $x = 0$, Γ is the Jordan form of A. The rank of $B - 2I$ is 2 for $y \neq 0$, so Λ is the Jordan form of B when $y \neq 0$. In this case, A and B have different Jordan forms, they cannot be similar.

When $x = y = 0$, Γ is the Jordan form of A and B, so they are similar.

(b) If $z = 0$, then $P^{-1}CP = C^t$, where $P = \left(\begin{smallmatrix} 0 & 1 & 0 \\ 1 & 0 & 0 \\ 0 & 0 & 1 \end{smallmatrix}\right)$. If $z \neq 0$, the rank of C is 2. Since the eigenvalues of C are all equal to zero, the Jordan canonical form of C is $J = \left(\begin{smallmatrix} 0 & 1 & 0 \\ 0 & 0 & 1 \\ 0 & 0 & 0 \end{smallmatrix}\right)$. C is similar to J. Thus, C^t is similar to J^t. On the other hand, one checks that $Q^{-1}JQ = J^t$, where $Q = \left(\begin{smallmatrix} 0 & 0 & 1 \\ 0 & 1 & 0 \\ 1 & 0 & 0 \end{smallmatrix}\right)$.

If U is a unitary matrix such that $U^*CU = C^t$, then $CU = UC^t$. A computation shows that such a U does not exist unless $|z| = 0$ or 1. □

Example 3.6 Let A and B be $n \times n$ matrices. Show that $AA^* = BB^*$ if and only if $A = BU$ for some $n \times n$ unitary matrix U.

Solution. If $A = BU$, where U is unitary, then $AA^* = BUU^*B^* = BB^*$.

Conversely, since $AA^* = BB^*$, A and B have the same singular values. By singular value decomposition, let $A = SDT$ and $B = PDQ$, where S, T, P, Q are unitary, and $D = \text{diag}(d_1, \ldots, d_n)$ is nonnegative diagonal.

$AA^* = BB^*$ reveals $SD^2S^* = PD^2P^*$. Setting $W = (w_{ij}) = P^*S$, we have $WD^2 = D^2W$, that is, $w_{ij}d_j^2 = d_i^2w_{ij}$ for all i, j, which imply $w_{ij}d_j = d_iw_{ij}$ for all i, j, i.e., $WD = DW$. Thus, $SDS^* = PDP^*$. So

$$A = SDT = SDS^*ST = PDP^*ST = PDQ\,Q^*P^*ST = BU,$$

where $U = Q^*P^*ST$ is unitary.   □

Example 3.7 Answer the following questions about singular values.

(a) Find the singular values of matrices $\begin{pmatrix} 0 & 1 \\ -1 & 0 \end{pmatrix}$, $\begin{pmatrix} 1 & 2 & 0 \\ 0 & 1 & 2 \end{pmatrix}$, and $\begin{pmatrix} 1 & 0 \\ 2 & 1 \\ 0 & 2 \end{pmatrix}$.

(b) Give an example of a matrix that has all eigenvalues equal to zero and has a singular value s, where s is any given positive number.

(c) Show that a matrix has a zero eigenvalue if and only if it has a zero singular value; show further that if a matrix has k nonzero eigenvalues, then it has at least k nonzero singular values. Is the converse true?

(d) For a Hermitian matrix $H = (h_{ij})$, show that the singular values are the absolute values of its eigenvalues, and $\max_i |h_{ii}| \leq \max_j |\lambda_j(H)|$.

(e) For a square matrix M, show that $\max_i |\lambda_i(M)| \leq \max_j \sigma_j(M)$.

Solution. (a) The singular values of $\begin{pmatrix} 0 & 1 \\ -1 & 0 \end{pmatrix}$ are 1 and 1. For the second matrix, let $A = \begin{pmatrix} 1 & 2 & 0 \\ 0 & 1 & 2 \end{pmatrix}$. Then $A^*A = \begin{pmatrix} 1 & 2 & 0 \\ 2 & 5 & 2 \\ 0 & 2 & 4 \end{pmatrix}$. The eigenvalues of A^*A are 0, 3, and 7. Thus, the singular values of A are 0, $\sqrt{3}$, and $\sqrt{7}$. The third matrix is the transpose of A. So, its singular values are $\sqrt{3}$ and $\sqrt{7}$.

(b) Take $B = \begin{pmatrix} 0 & s \\ 0 & 0 \end{pmatrix}$, $s > 0$. Then the eigenvalues of B are 0 and 0, and the singular values of B are 0 and s. This problem says that the singular value(s) can get arbitrary large while eigenvalues remain unchanged.

(c) Let C be an $n \times n$ matrix with eigenvalues $\lambda_1, \ldots, \lambda_n$ and singular values $\sigma_1, \ldots, \sigma_n$. Note that $\det(C^*C) = \det C^* \det C$. This is the same as $(\prod_{i=1}^n \sigma_i)^2 = |\prod_{i=1}^n \lambda_i|^2$. So, a λ is zero if and only if some σ is zero.

If C has k nonzero eigenvalues, by Schur decomposition, the rank of C is at least k (as it has a nonsingular $k \times k$ submatrix). On the other hand, since $r(C^*C) = r(C)$, the rank of C^*C is at least k. Thus, C has at least k nonzero singular values. The converse is not true. For example, $C = \begin{pmatrix} 0 & 1 \\ 0 & 0 \end{pmatrix}$ has one nonzero singular, but no nonzero eigenvalues.

(d) Let $H \in M_n(\mathbb{C})$, $H^* = H$. By Schur decomposition, we assume $H = U^*DU$, where U is unitary and D is upper-triangular. As $H^* = H$, $D^* = D$. Thus, D is diagonal. Let $D = \text{diag}(\lambda_1(H), \ldots, \lambda_n(H))$. It is immediate that the singular values of H are $|\lambda_1(H)|, \ldots, |\lambda_n(H)|$, and that

$$
\begin{aligned}
|h_{ii}| &= \left| \sum_{j=1}^{n} \bar{u}_{ji} \lambda_j(H) u_{ji} \right| = \left| \sum_{j=1}^{n} \lambda_j(H) |u_{ji}|^2 \right| \\
&\leq \sum_{j=1}^{n} |\lambda_j(H)| \, |u_{ji}|^2 \leq \max_j |\lambda_j(H)| \left(\sum_{j=1}^{n} |u_{ji}|^2 \right) \\
&= \max_j |\lambda_j(H)| \quad \text{(which is equal to } \sigma_{\max}(H)\text{)}.
\end{aligned}
$$

Note that this need not be true if H is not Hermitian.

(e) Let $M \in M_n(\mathbb{C})$. By Schur decomposition, let $M = V^*TV$, where V is unitary and $T = (t_{ij})$ is upper-triangular, and the diagonal entries of T are $\lambda_1(M), \ldots, \lambda_n(M)$. Since $M^*M = V^*T^*TV$, M^*M and T^*T have the same eigenvalues. A computation shows the diagonal entries of T^*T:

$$
|\lambda_1(M)|^2, \; |\lambda_2(M)|^2 + |t_{12}|^2, \; \ldots, \; |\lambda_n(M)|^2 + \sum_{j<n} |t_{jn}|^2.
$$

Because T^*T is Hermitian, by (d), we obtain that $\max_i |\lambda_i(M)|^2 \leq \max_j \lambda_j(T^*T)$. So, $\max_i |\lambda_i(M)| \leq \max_j \sigma_j(M) \; (= \sigma_{\max}(M))$. □

Hadamard Product and Kronecker Product. Matrices can be "multiplied" in different ways. Let $A = (a_{ij}) \in M_{m \times n}(\mathbb{F})$ and $B = (b_{ij}) \in M_{p \times q}(\mathbb{F})$. If $n = p$, AB is the usual product given in Chapter 1; if $m = p$ and $n = q$, we can multiply A and B entrywise to get $A \circ B = (a_{ij}b_{ij})$, referred to as the *Hadamard product* or *Schur product*. Another basic and important product of matrices is the *Kronecker product*, also known as *tensor product* or *direct product*, which is defined by the block matrix

$$
A \otimes B = \begin{pmatrix} a_{11}B & a_{12}B & \cdots & a_{1n}B \\ a_{21}B & a_{22}B & \cdots & a_{2n}B \\ \vdots & \vdots & \ddots & \vdots \\ a_{m1}B & a_{m2}B & \cdots & a_{mn}B \end{pmatrix}.
$$

For example, if $A = \begin{pmatrix} a & b \\ c & d \end{pmatrix}$ and $B = \begin{pmatrix} 7 & i \\ 0 & x \end{pmatrix}$, then $A \circ B = \begin{pmatrix} 7a & bi \\ 0 & dx \end{pmatrix}$,

$$
A \otimes B = \begin{pmatrix} 7a & ai & 7b & bi \\ 0 & ax & 0 & bx \\ 7c & ci & 7d & di \\ 0 & cx & 0 & dx \end{pmatrix} \text{ and } B \otimes A = \begin{pmatrix} 7a & 7b & ai & bi \\ 7c & 7d & ci & di \\ 0 & 0 & ax & bx \\ 0 & 0 & cx & dx \end{pmatrix}.
$$

A very important relation between the two products is that $A \circ B$ is a principal submatrix of $A \otimes B$ (Problem 3.112).

Among many properties of the Kronecker product, we single out the following fundamental one with a proof, leaving others as Problem 3.113.

$$
(A \otimes B)(C \otimes D) = (AC) \otimes (BD). \tag{3.11}
$$

Proof: Write $A \otimes B = (a_{ij}B)$, $C \otimes D = (c_{ij}D)$. Then the (i, j)-block of $(A \otimes B)(C \otimes D)$ is $\sum_{t=1}^{n} a_{it}B c_{tj}D = \left(\sum_{t=1}^{n} a_{it}c_{tj} \right)(BD)$. This is also the (i, j)-entry of AC times BD, namely, the (i, j)-block of $(AC) \otimes (BD)$.

Linear Transformation. Let V and W be vector spaces over a field $\mathbb{F}$. A transformation (also referred to as map) $\mathcal{A}$ from V to W is said to be *linear* if

$$
\mathcal{A}(u + v) = \mathcal{A}(u) + \mathcal{A}(v), \quad u, v \in V
$$

and

$$
\mathcal{A}(kv) = k\mathcal{A}(v), \quad k \in \mathbb{F}, \ v \in V.
$$

It follows at once that $\mathcal{A}(0) = 0$. We could have written $\mathcal{A}(0_v) = 0_w$, where 0_v and 0_w stand for the zero vectors of V and W, respectively. However, from the context one can easily tell which is which. For simplicity, we use 0 for both. Sometimes we write $\mathcal{A}v$ for $\mathcal{A}(v)$ for convenience.

The zero map from V to W is defined by $0(v) = 0$, $v \in V$; the *identity map* $\mathcal{I}_V$ is the map from V to V for which $\mathcal{I}_V(v) = v$ for all $v \in V$.

The linear transformations from V to V are also called *linear operators*.

Let A be an $m \times n$ matrix over field $\mathbb{F}$ and define $\mathcal{A} : \mathbb{F}^n \to \mathbb{F}^m$ by

$$
\mathcal{A}(x) = Ax.
$$

Then $\mathcal{A}$ is a linear transformation because, for all $x, y \in \mathbb{F}^n$ and $k \in \mathbb{F}$,

$$
\mathcal{A}(x + ky) = A(x + ky) = Ax + A(ky) = \mathcal{A}(x) + k\mathcal{A}(y).
$$

If $\{e_1, \ldots, e_n\}$ is the standard basis of $\mathbb{F}^n$ and $\alpha_1, \ldots, \alpha_n$ are the columns of A, then $\mathcal{A}(e_i) = \alpha_i$, $i = 1, \ldots, n$. We may write $(\mathcal{A}(e_1), \ldots, \mathcal{A}(e_n)) = A$.

The Vector Space of Linear Transformations. Let $L(V, W)$ denote the set of all linear transformations from a vector space V to a vector space W. We define addition and scalar multiplication for $L(V, W)$ as follows:

$$(\mathcal{A} + \mathcal{B})(v) = \mathcal{A}(v) + \mathcal{B}(v), \quad (k\mathcal{A})(v) = k(\mathcal{A}(v)).$$

Then $L(V, W)$ is a vector space with respect to the addition and scalar multiplication. The zero vector in $L(V, W)$ is the zero transformation, and for every $\mathcal{A} \in L(V, W)$, $-\mathcal{A}$ is the linear transformation

$$(-\mathcal{A})(v) = -(\mathcal{A}(v)).$$

Let U, V, W be vector spaces over the same field. Let $\mathcal{A} \in L(V, W)$ and $\mathcal{B} \in L(U, V)$. The *product* of $\mathcal{A}$ and $\mathcal{B}$ is defined by the composite mapping

$$(\mathcal{A}\mathcal{B})(u) = \mathcal{A}(\mathcal{B}(u)), \quad u \in U.$$

The product $\mathcal{A}\mathcal{B}$ is a linear transformation from U to W. For three (or more) linear transformations $\mathcal{A}$, $\mathcal{B}$, and $\mathcal{C}$, one may verify that

$$\mathcal{A}(\mathcal{B}\mathcal{C}) = (\mathcal{A}\mathcal{B})\mathcal{C}.$$

The Inverses of a Linear Transformation. Let V and W be vector spaces over the same field $\mathbb{F}$. Let $\mathcal{A} : V \to W$ be a linear transformation. If $\mathcal{B} : W \to V$ is a linear transformation such that $\mathcal{A}\mathcal{B} = \mathcal{I}_W$, then $\mathcal{B}$ is called a *right-inverse* of $\mathcal{A}$. If $\mathcal{B}\mathcal{A} = \mathcal{I}_V$, then $\mathcal{B}$ is called a *left-inverse* of $\mathcal{A}$. If both $\mathcal{A}\mathcal{B} = \mathcal{I}_W$ and $\mathcal{B}\mathcal{A} = \mathcal{I}_V$, then $\mathcal{B}$ is called the *inverse* of $\mathcal{A}$, denoted by $\mathcal{A}^{-1}$. The inverse (if it exists) is unique. Suppose $\mathcal{X} : W \mapsto V$ is also a map such that $\mathcal{A}\mathcal{X} = \mathcal{I}_W$ and $\mathcal{X}\mathcal{A} = \mathcal{I}_V$. Then

$$\mathcal{X} = \mathcal{X}\mathcal{I}_W = \mathcal{X}(\mathcal{A}\mathcal{B}) = (\mathcal{X}\mathcal{A})\mathcal{B} = \mathcal{I}_V\mathcal{B} = \mathcal{B}.$$

Note: Inverses (including right- and left-inverses) are in fact defined for functions or maps between sets in pre-calculus. Linearity is not required.

A linear *isomorphism* is a bijective linear map between vector spaces that preserves algebraic structures (addition and scalar multiplication). Two vector spaces are *isomorphic* if there is an isomorphism between them.

Let $\mathbb{F}$ be a field. Then $\mathbb{F}^m$ and $\mathbb{F}^n$ are isomorphic if and only if $m = n$. Let V and W be finite dimensional vector spaces. Then V and W are isomorphic if and only if $\dim V = \dim W$.

Example 3.8 Let V be a finite dimensional vector space and let $\mathcal{A}$ be a linear map from V to V. Show that the following statements are equivalent:

(1) $\mathcal{A}$ has an inverse, i.e., $\mathcal{A}\mathcal{B} = \mathcal{B}\mathcal{A} = \mathcal{I}_V$ for some linear map $\mathcal{B}$ on V.

(2) $\mathcal{A}$ maps a basis to a basis.

(3) $\mathcal{A}$ is one-to-one.

(4) $\mathcal{A}$ is onto.

Solution. Let $\dim V = n$ and let $\{\alpha_1, \ldots, \alpha_n\}$ be a basis of V.

$(1) \Rightarrow (2)$: We need to show that $\{\mathcal{A}(\alpha_1), \ldots, \mathcal{A}(\alpha_n)\}$ is a basis. It is sufficient to show that $\mathcal{A}(\alpha_1), \ldots, \mathcal{A}(\alpha_n)$ are linearly independent. Suppose $x_1\mathcal{A}(\alpha_1) + \cdots + x_n\mathcal{A}(\alpha_n) = 0$. Then $\mathcal{A}(x_1\alpha_1 + \cdots + x_n\alpha_n) = 0$ as $\mathcal{A}$ is linear. Applying $\mathcal{B}$ to both sides from the left, we obtain $(\mathcal{B}\mathcal{A})(x_1\alpha_1 + \cdots + x_n\alpha_n) = \mathcal{I}_V(x_1\alpha_1 + \cdots + x_n\alpha_n) = x_1\alpha_1 + \cdots + x_n\alpha_n = 0$. Because $\alpha_1, \ldots, \alpha_n$ are linearly independent, we see that $x_1 = \cdots = x_n = 0$. It follows that $\mathcal{A}(\alpha_1), \ldots, \mathcal{A}(\alpha_n)$ are linearly independent.

$(2) \Rightarrow (3)$: If $\mathcal{A}$ is not one-to-one, let $v_1 \neq v_2$ and $\mathcal{A}(v_1) = \mathcal{A}(v_2)$. Let $v_1 = x_1\alpha_1 + \cdots + x_n\alpha_n$ and $v_2 = y_1\alpha_1 + \cdots + y_n\alpha_n$. $v_1 \neq v_2$ means $x_i \neq y_i$ for some i. $\mathcal{A}(v_1) = \mathcal{A}(v_2)$ gives $x_1\mathcal{A}(\alpha_1) + \cdots + x_n\mathcal{A}(\alpha_n) = y_1\mathcal{A}(\alpha_1) + \cdots + y_n\mathcal{A}(\alpha_n)$. Thus, $z_1\mathcal{A}(\alpha_1) + \cdots + z_n\mathcal{A}(\alpha_n) = 0$, where $z_i = x_i - y_i$, $i = 1, \ldots, n$. As $z_i = x_i - y_i \neq 0$ for some i, $\mathcal{A}(\alpha_1), \ldots, \mathcal{A}(\alpha_n)$ are linearly dependent, contradicting $\mathcal{A}(\alpha_1), \ldots, \mathcal{A}(\alpha_n)$ being a basis.

$(3) \Rightarrow (2)$: Suppose $\mathcal{A}$ is one-to-one. We show that $\{\mathcal{A}(\alpha_1), \ldots, \mathcal{A}(\alpha_n)\}$ is a basis. It suffices to show that $\mathcal{A}(\alpha_1), \ldots, \mathcal{A}(\alpha_n)$ are linearly independent. Let $s_1\mathcal{A}(\alpha_1) + \cdots + s_n\mathcal{A}(\alpha_n) = 0$. Set $w_0 = s_1\alpha_1 + \cdots + s_n\alpha_n$. Then $\mathcal{A}(w_0) = 0$. Because $\mathcal{A}$ is one-to-one, $w_0 = 0$ as $\mathcal{A}(0) = 0$. Now $s_1\alpha_1 + \cdots + s_n\alpha_n = 0$ and $\{\alpha_1, \ldots, \alpha_n\}$ is a basis, all s_i's are equal to zero.

$(2) \Rightarrow (4)$: Suppose $\{\mathcal{A}(\alpha_1), \ldots, \mathcal{A}(\alpha_n)\}$ is a basis. Let $v \in V$ and $v = t_1\mathcal{A}(\alpha_1) + \cdots + t_n\mathcal{A}(\alpha_n)$. Then $v = \mathcal{A}(t_1\alpha_1 + \cdots + t_n\alpha_n)$. So $\mathcal{A}$ is onto.

$(4) \Rightarrow (2)$: If $\{\mathcal{A}(\alpha_1), \ldots, \mathcal{A}(\alpha_n)\}$ is not a basis, then the span by $\mathcal{A}(\alpha_1), \ldots, \mathcal{A}(\alpha_n)$ has dimension at most $n - 1$. There exists a vector that does not lie in the span. So $\mathcal{A}$ cannot be onto.

$(3) \Rightarrow (1)$: By the previous parts, we know that $\mathcal{A}$ is one-to-one if and only if $\mathcal{A}$ is onto. If $\mathcal{A}$ is one-to-one, we show that $\mathcal{A}$ is invertible.

For $v \in V$, denote $v' = \mathcal{A}(v)$. Note that $\mathcal{A}$ is one-to-one and onto. Define $\mathcal{B}$ by $\mathcal{B}(v') = v$. Then $\mathcal{B}$ is a well-defined transformation on V. It is immediate that $\mathcal{A}\mathcal{B} = \mathcal{B}\mathcal{A} = \mathcal{I}_V$. We are left to show that $\mathcal{B}$ is linear.

On the one hand, $\mathcal{B}(u') + k\mathcal{B}(v') = u + kv$. On the other hand, $\mathcal{A}(u + kv) = \mathcal{A}(u) + k\mathcal{A}(v) = u' + kv'$. Thus, $\mathcal{B}(u' + kv') = u + kv$. It follows that $\mathcal{B}(u' + kv') = \mathcal{B}(u') + k\mathcal{B}(v')$. (See also Problem 3.119.) $\quad\square$

Note: The statements in Example 3.8 need not be true for an infinite dimensional vector space V or for vector spaces V and W of different finite dimensions. For instance, consider the infinite dimensional space $\mathbb{P}[x]$ and define a linear map $\mathcal{L}$ by $\mathcal{L}(f(x)) = xf(x)$. Then $\mathcal{L}$ is one-to-one, but not invertible. Let $\mathcal{S} : x \in \mathbb{R} \mapsto \left(\begin{smallmatrix} x \\ x \end{smallmatrix}\right) \in \mathbb{R}^2$. Then $\mathcal{S}$ is one-to-one, but not onto; while $\mathcal{T}(X) = \operatorname{tr} X$ from $M_n(\mathbb{R})$ to $\mathbb{R}$ is onto, but not one-to-one. Moreover, in the example, $\mathcal{A}\mathcal{B} = \mathcal{I}_V$ if and only if $\mathcal{B}\mathcal{A} = \mathcal{I}_V$, because if $\mathcal{B}\mathcal{A} = \mathcal{I}_V$, say, then $\mathcal{A}$ is one-to-one, thus $\mathcal{A}$ is invertible as $(3) \Rightarrow (1)$.

Example 3.9 Let $A, B, C, D \in M_n(\mathbb{C})$ and define a map $\mathcal{T}$ on $M_n(\mathbb{C})$ by

$$\mathcal{T}(X) = AXB + CX + XD, \quad X \in M_n(\mathbb{C}).$$

Show that

(1) $\mathcal{T}$ is a linear transformation on $M_n(\mathbb{C})$.
(2) If $C = D = 0$, then $\mathcal{T}$ is invertible if and only if A and B are invertible.
(3) If $B = I, C = 0$, and A and $-D$ have no common eigenvalues, then $\mathcal{T}$ is one-to-one and onto. Thus $AX + XD = E$ has a unique solution for any given $E \in M_n(\mathbb{C})$ (*Sylvester's matrix equation theorem*).

Solution. (1) It is routine to show that $\mathcal{T}(Y + kZ) = \mathcal{T}(Y) + k\mathcal{T}(Z)$.

(2) $\mathcal{T}(X) = AXB$. If A and B are invertible, then $\mathcal{T}\mathcal{S} = \mathcal{S}\mathcal{T} = \mathcal{I}$, where $\mathcal{S}$ is defined by $\mathcal{S}(X) = A^{-1}XB^{-1}$, which is also a linear map on $M_n(\mathbb{C})$. Conversely, suppose $\mathcal{T}$ is invertible and let $\mathcal{T}\mathcal{L} = \mathcal{L}\mathcal{T} = \mathcal{I}$ for a linear map $\mathcal{L}$. Then $I_n = \mathcal{T}\mathcal{L}(I_n) = A(\mathcal{L}(I_n))B$. So A and B are nonsingular.

(3) As $\mathcal{T}$ is a linear map on $M_n(\mathbb{C})$, $\mathcal{T}$ is onto if and only if it is one-to-one (by the previous example), and if and only if $\mathcal{T}$ maps only 0 to 0. Let $\mathcal{T}(X) = 0$. Then $AX + XD = 0$, or $AX = -XD$. For each positive integer k, $A^k X = X(-D)^k$. It follows that $p(A)X = Xp(-D)$ for every polynomial $p(t)$. In particular, for the characteristic polynomial p of $-D$, $p(-D) = 0$, we get $p(A)X = 0$. Because A and $-D$ have no common eigenvalues, $p(A)$ is nonsingular. Thus $X = 0$. Since $\mathcal{T}$ is one-to-one and onto, for every given $E \in M_n(\mathbb{C})$, there is a unique solution X to $AX + XD = E$. □

Kernel and Image. Let $\mathcal{A}$ be a linear transformation from a vector space V to a vector space W. The *kernel* or *null space* of $\mathcal{A}$ is defined to be

$$\operatorname{Ker} \mathcal{A} = \{\, u \in V \mid \mathcal{A}(u) = 0 \,\}$$

and *image* or *range* of $\mathcal{A}$ is the set

$$\operatorname{Im} \mathcal{A} = \{\, \mathcal{A}(u) \mid u \in V \,\}.$$

The kernel is a subspace of V and the image is a subspace of W. If V is finite dimensional, say n, then both $\operatorname{Ker}\mathcal{A}$ and $\operatorname{Im}\mathcal{A}$ have finite dimensions. If $\{u_1, u_2, \ldots, u_s\}$ is a basis for $\operatorname{Ker}\mathcal{A}$ and is extended to a basis $\{u_1, u_2, \ldots, u_s, u_{s+1}, \ldots, u_n\}$ for V, then $\{\mathcal{A}(u_{s+1}), \ldots, \mathcal{A}(u_n)\}$ is a basis for $\operatorname{Im}\mathcal{A}$. We arrive at the theorem about dimension:

$$\dim V = \dim \operatorname{Ker}\mathcal{A} + \dim \operatorname{Im}\mathcal{A}. \tag{3.12}$$

For the linear map from $\mathbb{F}^n$ to $\mathbb{F}^m$ defined by $\mathcal{A}(x) = Ax$, where $A \in M_{m \times n}(\mathbb{F})$, the kernel and image of $\mathcal{A}$ are the null space and column space of A, respectively. By (2.12) (Chapter 2), $\dim \operatorname{Im}\mathcal{A} = r(A)$, the rank of A.

Example 3.10 Let $\mathcal{A}$ be a linear transformation from $\mathbb{R}^3$ to $\mathbb{R}^4$ defined by

$$\mathcal{A}\begin{pmatrix} x \\ y \\ z \end{pmatrix} = \begin{pmatrix} x + 2y + 3z \\ 2x + 3y + 4z \\ 3x + 4y + 5z \\ 4x + 5y + 6z \end{pmatrix}.$$

Find the image of $\mathcal{A}$ (using span of basis vectors) and the kernel of $\mathcal{A}$. What are the dimensions of the image and the kernel of $\mathcal{A}$?

Solution. Let $\alpha_1 = (1, 2, 3, 4)^t, \alpha_2 = (2, 3, 4, 5)^t$, and $\alpha_3 = (3, 4, 5, 6)^t$. Then

$$\mathcal{A}\begin{pmatrix} x \\ y \\ z \end{pmatrix} = \begin{pmatrix} 1 & 2 & 3 \\ 2 & 3 & 4 \\ 3 & 4 & 5 \\ 4 & 5 & 6 \end{pmatrix}\begin{pmatrix} x \\ y \\ z \end{pmatrix} = (\alpha_1, \alpha_2, \alpha_3)\begin{pmatrix} x \\ y \\ z \end{pmatrix} = x\alpha_1 + y\alpha_2 + z\alpha_3.$$

We have

$$\operatorname{Im}\mathcal{A} = \{x\alpha_1 + y\alpha_2 + z\alpha_3 \mid x, y, z \in \mathbb{R}\}$$

and

$$\operatorname{Ker}\mathcal{A} = \{(x, y, z)^t \mid x\alpha_1 + y\alpha_2 + z\alpha_3 = 0, \ x, y, z \in \mathbb{R}\}.$$

By row operations, we obtain

$$\begin{pmatrix} 1 & 2 & 3 \\ 2 & 3 & 4 \\ 3 & 4 & 5 \\ 4 & 5 & 6 \end{pmatrix} \to \begin{pmatrix} 1 & 2 & 3 \\ 1 & 1 & 1 \\ 1 & 1 & 1 \\ 1 & 1 & 1 \end{pmatrix} \to \begin{pmatrix} 1 & 2 & 3 \\ 0 & 1 & 2 \\ 0 & 0 & 0 \\ 0 & 0 & 0 \end{pmatrix} \to \begin{pmatrix} 1 & 0 & -1 \\ 0 & 1 & 2 \\ 0 & 0 & 0 \\ 0 & 0 & 0 \end{pmatrix}.$$

Thus, the kernel of $\mathcal{A}$ consists of all $(x, y, z)^t$ satisfying $x - z = 0$ and $y + 2z = 0$. Putting $z = 1$ reveals $\operatorname{Ker}\mathcal{A} = \{k(1, -2, 1)^t \mid k \in \mathbb{R}\}$, and $\dim \operatorname{Ker}\mathcal{A} = 1$. For the image of $\mathcal{A}$, noticing that any two of α_i's are linearly independent, we arrive at $\operatorname{Im}\mathcal{A} = \operatorname{Span}\{\alpha_1, \alpha_2\} = \operatorname{Span}\{\alpha_2, \alpha_3\} = \operatorname{Span}\{\alpha_1, \alpha_3\}$, and $\dim \operatorname{Im}\mathcal{A} = 2$. (See Example 2.8.) □

The matrix on which we performed row operations plays an important role in the problem; it is called the *matrix representation* of the linear transformation. Because of the matrix representations of linear transformations, (finite dimensional) linear algebra is largely basic matrix theory.

Matrix Representation of a Linear Transformation. Let V be a vector space of dimension m over a field $\mathbb{F}$ with a (ordered) basis $\alpha = \{\alpha_1, \alpha_2, \ldots, \alpha_m\}$. It is a basic fact that every vector $v \in V$ can be uniquely expressed as $v = x_1\alpha_1 + \cdots + x_m\alpha_m$, where $x_1, x_2, \ldots, x_m \in \mathbb{F}$. For the sake of convenience, we write $v = \alpha x$, where $x = (x_1, \ldots, x_m)^t$ is the coordinate of v relative to the basis α. It follows that $\alpha x = \alpha y$ if and only if $x = y$.

Let $\mathcal{A}$ be a linear transformation from V to W, where W is a vector space of dimension n over $\mathbb{F}$ with a basis $\beta = \{\beta_1, \beta_2, \ldots, \beta_n\}$. Then $\mathcal{A}$ is determined by its action on α relative to β. To be precise, let

$$
\begin{aligned}
\mathcal{A}(\alpha_i) &= a_{1i}\beta_1 + a_{2i}\beta_2 + \cdots + a_{ni}\beta_n \\
&= (\beta_1, \beta_2, \ldots, \beta_n)(a_{1i}, a_{2i}, \ldots, a_{ni})^t \\
&= (\beta_1, \beta_2, \ldots, \beta_n)a_i = \beta a_i,
\end{aligned}
$$

where $a_i = (a_{1i}, a_{2i}, \ldots, a_{ni})^t$, $i = 1, 2, \ldots, m$. Then we write

$$\mathcal{A}(\alpha) = \big(\mathcal{A}(\alpha_1), \mathcal{A}(\alpha_2), \ldots, \mathcal{A}(\alpha_m)\big) = (\beta_1, \beta_2, \ldots, \beta_n)A = \beta A, \quad (3.13)$$

where A is the unique $n \times m$ matrix given by $A = (a_1, a_2, \ldots, a_m) = (a_{ij})$.

In (3.13), the "product" βA of the basis β and matrix A is understood as the row "vector" $\beta = (\beta_1, \beta_2, \ldots, \beta_n)$ "times" each column of A, that is,

$$\beta A = (\beta a_1, \beta a_2, \ldots, \beta a_m) = \left(\sum_{j=1}^{n} a_{j1}\beta_j, \ \sum_{j=1}^{n} a_{j2}\beta_j, \ \ldots, \ \sum_{j=1}^{n} a_{jm}\beta_j \right).$$

Note that $\beta A = \beta B$ if and only if $A = B$. Moreover, for $v = \alpha x$,

$$\mathcal{A}(v) = \mathcal{A}(\alpha x) = (\mathcal{A}(\alpha))x = (\beta A)x = \beta(Ax). \quad (3.14)$$

This says Ax is the coordinate of $\mathcal{A}(v) \in W$ relative to the basis β. Thus, the linear transformation $\mathcal{A}$ is determined by the matrix A. Such a matrix A associated with $\mathcal{A}$ is called the *matrix representation* of the linear transformation $\mathcal{A}$ relative to the (ordered) bases α of V and β of W.

Given an $n \times m$ matrix A over a field $\mathbb{F}$, define $\mathcal{A} : \mathbb{F}^m \to \mathbb{F}^n$ by

$$\mathcal{A}(x) = Ax. \quad (3.15)$$

Then the matrix representation of $\mathcal{A}$ relative to the standard basis $\alpha = \{e_1, \ldots, e_m\}$ of $\mathbb{F}^m$ and the standard basis $\beta = \{\epsilon_1, \ldots, \epsilon_n\}$ of $\mathbb{F}^n$ is A, because

$$(\mathcal{A}(e_1), \ldots, \mathcal{A}(e_m)) = (Ae_1, \ldots, Ae_m) = AI_m = A = I_n A = (\epsilon_1, \ldots, \epsilon_n)A.$$

If V has dimension m and W has dimension n, then the vector space $L(V, W)$ of all linear maps from V to W is a vector space of dimension mn. $L(V, W)$ has the same algebraic structures as $M_{m \times n}(\mathbb{F})$ (or $M_{n \times m}(\mathbb{F})$).

If $V = W$ and $\alpha = \beta$, then $\mathcal{A}(\alpha) = \alpha A$ by (3.13). In this case, we can simply call A the matrix of $\mathcal{A}$ under, or with respect to, the basis α.

It is easy to see that the matrix of the zero transformation is 0 (the zero matrix), the matrix of the identity transformation is I. Let A, B be respectively the matrices of $\mathcal{A}, \mathcal{B}$ with respect to the basis α. Then

$$(\mathcal{A} + \mathcal{B})(\alpha) = \mathcal{A}(\alpha) + \mathcal{B}(\alpha) = \alpha A + \alpha B = \alpha(A + B),$$

$$(\mathcal{A}\mathcal{B})(\alpha) = \mathcal{A}(\mathcal{B}(\alpha)) = \mathcal{A}(\alpha B) = (\mathcal{A}\alpha)B = (\alpha A)B = \alpha(AB).$$

Example 3.8 states that a linear map $\mathcal{A}$ on a finite dimensional vector space V is invertible if and only if $\mathcal{A}$ is one-to-one, i.e., $\operatorname{Ker} \mathcal{A} = \{0\}$, if and only if $\mathcal{A}$ is onto, i.e., $\operatorname{Im} \mathcal{A} = V$. Now we add another equivalent statement: $\mathcal{A}$ is invertible if and only if the matrix of $\mathcal{A}$ under a basis is invertible.

Proof: If $\mathcal{A}$ is invertible, then $\mathcal{A}\mathcal{B} = \mathcal{I}_V$, where $\mathcal{B} = \mathcal{A}^{-1}$. So, $\alpha = \alpha(AB)$. Since α is a basis, $AB = I$ and A is invertible. Conversely, suppose that A is invertible and B is the inverse of A. Define $\mathcal{B}$ by $\mathcal{B}(\alpha) = \alpha B$. Then $\mathcal{B}$ is a linear map on V and $(\mathcal{A}\mathcal{B})(\alpha) = \alpha(AB) = \alpha$. Thus, $\mathcal{A}\mathcal{B} = \mathcal{I}_V$. Similarly, one proves that $\mathcal{B}\mathcal{A} = \mathcal{I}_V$. Note that the matrix of $\mathcal{A}^{-1}$ is A^{-1}.

Example 3.11 Let $A = \begin{pmatrix} a & b \\ c & d \end{pmatrix} \in M_2(\mathbb{R})$. Define a map $\mathcal{A}$ on $M_2(\mathbb{R})$ by

$$\mathcal{A}(X) = AX, \quad X \in M_2(\mathbb{R}).$$

Show that $\mathcal{A}$ is a linear map. Find the matrix representation of $\mathcal{A}$ with respect to the (ordered) basis $\{E_{11}, E_{21}, E_{12}, E_{22}\}$, where E_{ij} is the 2×2 matrix with (i, j)-entry 1 and 0 elsewhere for $i = 1, 2$, and $j = 1, 2$.

Solution. We compute

$$\mathcal{A}(E_{11}) = \begin{pmatrix} a & b \\ c & d \end{pmatrix} \begin{pmatrix} 1 & 0 \\ 0 & 0 \end{pmatrix} = \begin{pmatrix} a & 0 \\ c & 0 \end{pmatrix} = aE_{11} + cE_{21},$$

$$\mathcal{A}(E_{21}) = \begin{pmatrix} a & b \\ c & d \end{pmatrix} \begin{pmatrix} 0 & 0 \\ 1 & 0 \end{pmatrix} = \begin{pmatrix} b & 0 \\ d & 0 \end{pmatrix} = bE_{11} + dE_{21},$$

$$A(E_{12}) = \begin{pmatrix} a & b \\ c & d \end{pmatrix} \begin{pmatrix} 0 & 1 \\ 0 & 0 \end{pmatrix} = \begin{pmatrix} 0 & a \\ 0 & c \end{pmatrix} = aE_{12} + cE_{22},$$

$$A(E_{22}) = \begin{pmatrix} a & b \\ c & d \end{pmatrix} \begin{pmatrix} 0 & 0 \\ 0 & 1 \end{pmatrix} = \begin{pmatrix} 0 & b \\ 0 & d \end{pmatrix} = bE_{12} + dE_{22}.$$

Thus,

$$(A(E_{11}), A(E_{21}), A(E_{12}), A(E_{22})) = (E_{11}, E_{21}, E_{12}, E_{22}) \begin{pmatrix} a & b & 0 & 0 \\ c & d & 0 & 0 \\ 0 & 0 & a & b \\ 0 & 0 & c & d \end{pmatrix}.$$

The matrix representation of A under $\{E_{11}, E_{21}, E_{12}, E_{22}\}$ is $\begin{pmatrix} A & 0 \\ 0 & A \end{pmatrix}$. ☐

Change-of-Basis Matrix. Let V be a finite dimensional vector space. Vectors of V generally have different coordinates with respect to different bases of V. If u is a vector in V that has coordinate x with respect to basis α and coordinate y with respect to basis β, that is, $u = \alpha x = \beta y$. What can we say about x and y? In other words, how x and y are related?

Suppose $\beta = \alpha T$, where T is an invertible matrix (which is uniquely determined by α and β). Then $u = \beta y = (\alpha T)y = \alpha(Ty)$. So $x = Ty$ and $y = T^{-1}x$. Such a matrix T is called the *change-of-basis matrix* from β to α. (Note: It is also referred to as the change-of-basis matrix from α to β in some other texts.) If the change-of-basis matrix is T, then the change of the corresponding coordinates can be computed through T.

Example 3.12 Find the change-of-basis matrix from basis $\{\alpha_1, \alpha_2\}$ to basis $\{\beta_1, \beta_2\}$ for $\mathbb{R}^2$, where $\alpha_1 = (1, 0)^t$, $\alpha_2 = (1, 1)^t$ and $\beta_1 = (2, 3)^t$, $\beta_2 = (3, 4)^t$. If v is a vector of coordinate $(5, -6)$ with respect to $\{\alpha_1, \alpha_2\}$, find the coordinate of v with respect to $\{\beta_1, \beta_2\}$.

Solution. Let S be the change-of-basis matrix from $\{\alpha_1, \alpha_2\}$ to $\{\beta_1, \beta_2\}$, i.e., $\alpha = \beta S$, where $\alpha = (\alpha_1, \alpha_2)$, $\beta = (\beta_1, \beta_2)$. Then $S = \beta^{-1}\alpha$. Compute

$$S = \begin{pmatrix} 2 & 3 \\ 3 & 4 \end{pmatrix}^{-1} \begin{pmatrix} 1 & 1 \\ 0 & 1 \end{pmatrix} = \begin{pmatrix} -4 & 3 \\ 3 & -2 \end{pmatrix} \begin{pmatrix} 1 & 1 \\ 0 & 1 \end{pmatrix} = \begin{pmatrix} -4 & -1 \\ 3 & 1 \end{pmatrix}.$$

If $v = 5\alpha_1 - 6\alpha_2$, then we have $v = -14\beta_1 + 9\beta_2$ because

$$v = (\alpha_1, \alpha_2) \begin{pmatrix} 5 \\ -6 \end{pmatrix} = (\beta_1, \beta_2) \begin{pmatrix} -4 & -1 \\ 3 & 1 \end{pmatrix} \begin{pmatrix} 5 \\ -6 \end{pmatrix} = (\beta_1, \beta_2) \begin{pmatrix} -14 \\ 9 \end{pmatrix}. \quad ☐$$

Note: The change-of-basis matrix from β to α is $S^{-1} = \begin{pmatrix} -1 & -1 \\ 3 & 4 \end{pmatrix}$.

Matrices of a Linear Operator Are Similar. Consider a vector space V of dimension n. Let α and β be two bases for V and let T be the change-of-basis matrix such that $\beta = \alpha T$. Let A_1 be the matrix of $\mathcal{A}$ under the basis α, that is, $\mathcal{A}(\alpha) = \alpha A_1$. Let A_2 be the matrix of $\mathcal{A}$ under β, i.e., $\mathcal{A}(\beta) = \beta A_2$. Then A_1 and A_2 are similar, because

$$\mathcal{A}(\beta) = \mathcal{A}(\alpha T) = (\mathcal{A}(\alpha))T = (\alpha A_1)T = (\beta T^{-1})A_1 T = \beta(T^{-1}A_1 T).$$

Thus, $A_2 = T^{-1}A_1 T$. This says that the matrices of the same linear operator under different bases are similar (via the change-of-basis matrix).

Eigenvalues of a Linear Operator. Let $\mathcal{A}$ be a linear transformation on a vector space V over $\mathbb{F}$. A scalar $\lambda \in \mathbb{F}$ is an *eigenvalue* of $\mathcal{A}$ if

$$\mathcal{A}(u) = \lambda u \tag{3.16}$$

for some nonzero vector u. Such a vector u is called an *eigenvector* of $\mathcal{A}$ corresponding (belonging or associated) to the eigenvalue λ.

Let A be the matrix representation of $\mathcal{A}$ under a basis α of V and let x be the coordinate of the vector u under α, that is, $u = \alpha x$. Then

$$\alpha(\lambda x) = \lambda(\alpha x) = \lambda u = \mathcal{A}(u) = \mathcal{A}(\alpha x) = (\mathcal{A}(\alpha))x = (\alpha A)x = \alpha(Ax).$$

Thus, $\mathcal{A}(u) = \lambda u$ if and only if $Ax = \lambda x$. So, the eigenvalues of the linear transformation $\mathcal{A}$ are exactly the eigenvalues of its matrix A under α. Note that similar matrices have the same eigenvalues. The eigenvalues of $\mathcal{A}$ through its matrices are independent of the choices of the bases.

Example 3.13 Let $\mathcal{D} = \frac{d}{dx}$ be the differential operator on the space of differentiable functions in x on $\mathbb{R}$. Let

$$U = \mathrm{Span}\{1, x, x^2, x^3\},$$
$$V = \mathrm{Span}\{1, \sin x, \cos x, e^x\},$$
$$W = \mathrm{Span}\{1, e^x, e^{2x}, e^{3x}\}.$$

Find (i) the matrix representation relative to the basis, (ii) the real eigenvalues, and (iii) the kernel of $\mathcal{D}$ as a linear map on U, V, and W.

Solution. For each spanned space, the given vectors are linearly independent, thus they comprise a basis for the span. For example, $\{1, x, x^2, x^3\}$ is a basis of U. For $\mathcal{D}$ as a linear map on U, V, and W, we compute

$$
\begin{aligned}
\mathcal{D}(1, x, x^2, x^3) &= \left(\mathcal{D}(1), \mathcal{D}(x), \mathcal{D}(x^2), \mathcal{D}(x^3)\right) \\
&= (0, 1, 2x, 3x^2) \\
&= (1, x, x^2, x^3) \begin{pmatrix} 0 & 1 & 0 & 0 \\ 0 & 0 & 2 & 0 \\ 0 & 0 & 0 & 3 \\ 0 & 0 & 0 & 0 \end{pmatrix},
\end{aligned}
$$

$$
\begin{aligned}
\mathcal{D}(1, \sin x, \cos x, e^x) &= \left(\mathcal{D}(1), \mathcal{D}(\sin x), \mathcal{D}(\cos x), \mathcal{D}(e^x)\right) \\
&= (0, \cos x, -\sin x, e^x) \\
&= \left(1, \sin x, \cos x, e^x\right) \begin{pmatrix} 0 & 0 & 0 & 0 \\ 0 & 0 & -1 & 0 \\ 0 & 1 & 0 & 0 \\ 0 & 0 & 0 & 1 \end{pmatrix},
\end{aligned}
$$

$$
\begin{aligned}
\mathcal{D}(1, e^x, e^{2x}, e^{3x}) &= \left(\mathcal{D}(1), \mathcal{D}(e^x), \mathcal{D}(e^{2x}), \mathcal{D}(e^{3x})\right) \\
&= (0, e^x, 2e^{2x}, 3e^{3x}) \\
&= \left(1, e^x, e^{2x}, e^{3x}\right) \begin{pmatrix} 0 & 0 & 0 & 0 \\ 0 & 1 & 0 & 0 \\ 0 & 0 & 2 & 0 \\ 0 & 0 & 0 & 3 \end{pmatrix}.
\end{aligned}
$$

(i) The matrices of $\mathcal{D}$ under the bases of U, V, and W are respectively

$$
\begin{pmatrix} 0 & 1 & 0 & 0 \\ 0 & 0 & 2 & 0 \\ 0 & 0 & 0 & 3 \\ 0 & 0 & 0 & 0 \end{pmatrix}, \quad \begin{pmatrix} 0 & 0 & 0 & 0 \\ 0 & 0 & -1 & 0 \\ 0 & 1 & 0 & 0 \\ 0 & 0 & 0 & 1 \end{pmatrix}, \quad \begin{pmatrix} 0 & 0 & 0 & 0 \\ 0 & 1 & 0 & 0 \\ 0 & 0 & 2 & 0 \\ 0 & 0 & 0 & 3 \end{pmatrix}.
$$

(ii) The real eigenvalues of $\mathcal{D}$ on U are $0, 0, 0, 0$; on V are 0 and 1 (plus two complex eigenvalues i and $-i$); and on W are $0, 1, 2, 3$.

(iii) $\operatorname{Ker}\mathcal{D} = \mathbb{R}$ for U, V, and W. $\operatorname{Im}\mathcal{D} = \operatorname{Span}\{1, x, x^2\}$ for U, $\operatorname{Im}\mathcal{D} = \operatorname{Span}\{\sin x, \cos x, e^x\}$ for V, and $\operatorname{Im}\mathcal{D} = \operatorname{Span}\{e^x, e^{2x}, e^{3x}\}$ for W. □

Invariant Subspaces. Let $\mathcal{A}$ be a linear operator on a vector space V. If W is a subspace of V such that $\mathcal{A}(w) \in W$ for all $w \in W$, that is, $\mathcal{A}(W) \subseteq W$, then we say that W is *invariant* under $\mathcal{A}$ (or $\mathcal{A}$-invariant). Both $\operatorname{Ker}\mathcal{A}$ and $\operatorname{Im}\mathcal{A}$ are invariant subspaces under any linear operator $\mathcal{A}$.

Example 3.14 Let A and B be $n \times n$ complex matrices such that $AB = BA$. Show that (i) A and B have a common eigenvector. (ii) There exists a unitary matrix U such that U^*AU and U^*BU are both upper-triangular.

Solution. (i) We view the matrices as linear maps on $\mathbb{C}^n$ (or on subspaces of $\mathbb{C}^n$). Note that if A is a linear map on a finite dimensional vector space V over $\mathbb{C}$, then A has at least one eigenvector in V because

$$Ax = \lambda x, \text{ for some } x \neq 0, \text{ if and only if } \det(\lambda I - A) = 0.$$

As a polynomial in λ, $\det(\lambda I - A) = 0$ always has a solution in $\mathbb{C}$. For each eigenvalue μ of B, consider the eigenspace of B for μ

$$V_\mu = \{v \in \mathbb{C}^n \mid Bv = \mu v\}.$$

If A and B commute, then for every $v \in V_\mu$,

$$B(Av) = (BA)v = (AB)v = A(Bv) = A(\mu v) = \mu(Av).$$

Thus, $Av \in V_\mu$, namely, V_μ is an invariant subspace of A. Let $\tilde{A} = A|_{V_\mu}$ be the restriction of A on V_μ. As a linear transformation on V_μ, $\tilde{A}$ has an eigenvalue λ and a corresponding eigenvector, say, v_1 in V_μ. It is a common eigenvector of $\tilde{A}$ (thus of A) and B:

$$\tilde{A}v_1 = Av_1 = \lambda v_1, \quad Bv_1 = \mu v_1, \quad v_1 \in V_\mu.$$

(ii) Use induction on n. If $n = 1$, we have nothing to show. Suppose that the assertion is true for $n - 1$. For the case of n, we use part (i). We may assume that v_1 is a unit vector. Extend v_1 to a unitary matrix U_1, that is, U_1 is a unitary matrix whose first column is v_1. Then we arrive at

$$U_1^*AU_1 = \begin{pmatrix} \lambda & \alpha \\ 0 & C \end{pmatrix} \quad \text{and} \quad U_1^*BU_1 = \begin{pmatrix} \mu & \beta \\ 0 & D \end{pmatrix},$$

where $C, D \in M_{n-1}(\mathbb{C})$, and α and β are some row vectors.

Since $AB = BA$, we have $CD = DC$. The induction hypothesis guarantees the existence of a unitary matrix $U_2 \in M_{n-1}(\mathbb{C})$ such that $U_2^*CU_2$ and $U_2^*DU_2$ are both upper-triangular. Let $U = U_1 \begin{pmatrix} 1 & 0 \\ 0 & U_2 \end{pmatrix}$. As a product of two unitary matrices, U is a unitary matrix. It is immediate that U^*AU and U^*BU are both upper-triangular. $\square$

Note that the special case of Example 3.14 with $B = I$ gives the Schur triangularization theorem. Furthermore, if A is a real matrix and if all the eigenvalues of A are real, then there exists a real orthogonal matrix R such that R^tAR is a real upper-triangular matrix (Problem 3.70).

3.1 Let A be an $n \times n$ (real or complex) matrix. Answer true or false:

    (a) If $A^2 = 0$ and λ is an eigenvalue of A, then $\lambda = 0$.

    (b) If $A^2 = 0$ and $n = 3$, then the rank of A is 0 or 1.

    (c) If $A^2 = 0$ and $n = 5$, then the rank of A is at most 2.

    (d) If $A^2 = 0$, then the rank of A is at most 2.

    (e) The determinant of A^2 is nonnegative, i.e., $\det(A^2) \geq 0$.

    (f) If A is real and λ is an eigenvalue of A, then $\lambda^2 \geq 0$.

    (g) If λ is an eigenvalue of A^2, then $\lambda \geq 0$.

    (h) If λ is an eigenvalue of AA^*, then $\lambda \geq 0$.

    (i) If $\operatorname{tr}(A^2) = 0$, then A is singular.

    (j) If $\operatorname{tr}(A^*A) = 0$, then $A = 0$.

    (k) If $A^2 \neq 0$, then $A^k \neq 0$ for all positive integers k.

    (l) If $A^2 = 0$ and A is Hermitian, then $A = 0$.

    (m) If $(A^* - A)^2 = 0$, then A has only real eigenvalues.

    (n) If $(A^* - A)^2 = 0$ and $A^3 = 0$, then $A = 0$.

3.2 Let A and B be $n \times n$ matrices. Show that

$$(A + B)^2 = A^2 + 2AB + B^2 \quad \Leftrightarrow \quad AB = BA.$$

3.3 Let A and B be $n \times n$ matrices. Show that

$$AB = A \pm B \quad \Rightarrow \quad AB = BA.$$

3.4 Show that matrices A and B are similar for any scalars x, y, z, a, b, c:

$$A = \begin{pmatrix} 1 & 0 & 0 \\ x & 2 & z \\ y & 0 & 3 \end{pmatrix}, \quad B = \begin{pmatrix} 3 & a & b \\ 0 & 2 & 0 \\ 0 & c & 1 \end{pmatrix}.$$

3.5 Show that A and B are similar for any values x, y, p, q, z and r, $zr \neq 0$:

$$A = \begin{pmatrix} 1 & 0 & 0 \\ x & 0 & 0 \\ y & z & 0 \end{pmatrix}, \quad B = \begin{pmatrix} 1 & p & q \\ 0 & 0 & r \\ 0 & 0 & 0 \end{pmatrix}.$$

3.6 Find the values of a and b such that the following matrices are similar:

$$A = \begin{pmatrix} -2 & 0 & 0 \\ 2 & a & 2 \\ 3 & 1 & 1 \end{pmatrix}, \quad B = \begin{pmatrix} -1 & 0 & 0 \\ 0 & 2 & 0 \\ 0 & 0 & b \end{pmatrix}.$$

3.7 Let A be the matrix given below. Find a matrix B that contains as many zeros as possible such that B is similar to A and compute A^{10}.

$$A = \begin{pmatrix} 1 & \sqrt{2} & \sqrt{3} & 2 \\ \sqrt{2} & 2 & \sqrt{6} & 2\sqrt{2} \\ \sqrt{3} & \sqrt{6} & 3 & 2\sqrt{3} \\ 2 & 2\sqrt{2} & 2\sqrt{3} & 4 \end{pmatrix}.$$

3.8 What are the matrices that are similar to themselves only?

3.9 If A and B commute, i.e., $AB = BA$, what can be said about the eigenvalues of $A + B$ and AB in terms of the eigenvalues of A and B?

3.10 Let A be a real matrix. A real matrix X is said to be *equivalent* to A if $PXQ = A$ for some real invertible matrices P and Q.

   (a) Show that this is an equivalence relation of matrices.

   (b) Let $A = \operatorname{diag}(1, 2, -1)$. Determine whether the matrices

$$B = \begin{pmatrix} -1 & -1 & 0 \\ 0 & 1 & 0 \\ 0 & 3 & 2 \end{pmatrix}, \; C = \begin{pmatrix} 1 & 1 & 0 \\ 1 & 2 & 0 \\ 0 & 0 & -1 \end{pmatrix}, \; D = \begin{pmatrix} 0 & 1 & 0 \\ 1 & 0 & 0 \\ 0 & 0 & 2 \end{pmatrix}, \; E = \begin{pmatrix} 0 & 1 & 1 \\ 1 & 0 & 1 \\ 1 & 1 & 2 \end{pmatrix}$$

   are equivalent or similar to A. Explain why.

3.11 Which of the following matrices are similar to $A = \operatorname{diag}(1, 4, 6)$?

$$B = \begin{pmatrix} 1 & 2 & 3 \\ 0 & 4 & 5 \\ 0 & 0 & 6 \end{pmatrix}, \; C = \begin{pmatrix} 4 & 0 & 0 \\ 7 & 1 & 0 \\ 8 & 9 & 6 \end{pmatrix}, \; D = \begin{pmatrix} 1 & 2 & 0 \\ 3 & 4 & 5 \\ 0 & 7 & 6 \end{pmatrix},$$

$$E = \begin{pmatrix} 4 & 7 & 0 \\ 0 & 1 & 0 \\ 8 & 9 & 6 \end{pmatrix}, \; F = \begin{pmatrix} 1 & 2 & 0 \\ 3 & 4 & 0 \\ 0 & 5 & 6 \end{pmatrix}, \; G = \begin{pmatrix} 1 & 0 & 0 \\ 2 & 5 & 1 \\ 3 & 1 & 5 \end{pmatrix}.$$

3.12 Find all values $a, b, c,$ and $x \in \mathbb{R}$ so that A and B are diagonalizable:

$$A = \begin{pmatrix} 2 & 0 & 0 \\ a & 2 & 0 \\ b & c & -1 \end{pmatrix}, \quad B = \begin{pmatrix} 1 & 0 & 0 & x \\ 0 & 1 & 0 & 0 \\ 0 & 0 & 0 & 1 \\ 0 & 0 & 1 & 0 \end{pmatrix}.$$

3.13   Find the eigenvalues and corresponding eigenvectors of the matrix

$$A = \begin{pmatrix} 1 & 2 & 2 \\ 2 & 1 & 2 \\ 2 & 2 & 1 \end{pmatrix},$$

then find an invertible matrix P such that $P^{-1}AP$ is diagonal.

3.14   Show that the following matrix A is not similar to a diagonal matrix:

$$A = \begin{pmatrix} 2 & 3 & 2 \\ 1 & 4 & 2 \\ 1 & -3 & 1 \end{pmatrix}.$$

3.15   Find the eigenvalues, corresponding eigenvectors, and eigenspaces of

$$A = \begin{pmatrix} 0 & 1 & 1 \\ 1 & 0 & -1 \\ 0 & 0 & 1 \end{pmatrix}.$$

3.16   Find the eigenvalues and eigenvectors of $\left(\begin{smallmatrix} 1 & a \\ a & 0 \end{smallmatrix}\right)$ in terms of $a \in \mathbb{R}$.

3.17   Let E_{ij} be the n-square matrix with the (i,j)-entry 1 and 0 elsewhere, $i,j = 1,2,\ldots,n$. For $A \in M_n(\mathbb{C})$, find AE_{ij}, $E_{ij}A$, and $E_{ij}AE_{st}$.

3.18   Compute A^2 and A^6, where

$$A = \begin{pmatrix} -1 & 1 & 1 & -1 \\ 1 & -1 & -1 & 1 \\ 1 & -1 & -1 & 1 \\ -1 & 1 & 1 & -1 \end{pmatrix}.$$

3.19   Find A^{100} and B^{100}, where

$$A = \begin{pmatrix} 1 & 2 \\ 4 & 3 \end{pmatrix}, \quad B = \begin{pmatrix} 1 & -2 & 2 \\ -2 & 1 & -2 \\ -2 & 2 & -3 \end{pmatrix}.$$

3.20   For positive integer $k \geq 2$, compute

$$\begin{pmatrix} 2 & 1 \\ 2 & 3 \end{pmatrix}^k, \quad \begin{pmatrix} 0 & 1 & 0 \\ 0 & 0 & 1 \\ 0 & 0 & 0 \end{pmatrix}^k, \quad \begin{pmatrix} 0 & 1 & 0 \\ 0 & 0 & 1 \\ 1 & 0 & 0 \end{pmatrix}^k, \quad \begin{pmatrix} \lambda & 1 & 0 \\ 0 & \lambda & 1 \\ 0 & 0 & \lambda \end{pmatrix}^k.$$

3.21   Let $A = \left(\begin{smallmatrix} 1 & 1 \\ 0 & 1 \end{smallmatrix}\right)$. Show that A^k is similar to A for all positive integers k. Show that this holds for any matrix with all eigenvalues equal to 1.

3.22   Let $u = (1, 2, 3)$, $v = (1, \frac{1}{2}, \frac{1}{3})$, and $A = u^t v$. Find A^k, $k = 1, 2, \ldots$.

3.23   Let A be a square matrix such that $\det A = 0$. Show that there exists a positive number δ such that $\det(A + \epsilon I) \neq 0$ for any $\epsilon \in (0, \delta)$.

3.24   Let $A, B \in M_n(\mathbb{C})$. If $A + zB$ is nonsingular for some $z \in \mathbb{C}$. Show that $A + rB$ is nonsingular for some $r \in \mathbb{R}$. In particular, if A and B are real matrices such that $A + iB$ is nonsingular, then there exists a real number t such that the real matrix $A + tB$ is nonsingular.

3.25   Let $A \in M_n(\mathbb{C})$. If $\lambda_1, \ldots, \lambda_n$ are the eigenvalues of A, show that $(A - \lambda_1 I) \cdots (A - \lambda_n I) = 0$.

3.26   Let $A, B \in M_n(\mathbb{C})$. Prove or disprove (by a counterexample) that

$$\text{if}\quad f(\lambda) = \det(A + \lambda I), \quad \text{then}\quad \det\big(f(B)\big) = \det(A + B).$$

3.27   Let $A, B \in M_n(\mathbb{C})$ and let $f(x)$ be a polynomial over $\mathbb{C}$. Show that

(a) If λ is an eigenvalue of A, then λ^k is an eigenvalue of A^k, where k is a positive integer. Is this true for singular values?

(b) If λ is an eigenvalue of A, then $f(\lambda)$ is an eigenvalue of $f(A)$.

(c) If A and B are similar, then $f(A)$ and $f(B)$ are similar.

(d) If $AP = QA$ for diagonal P and Q, then $Af(P) = f(Q)A$.

3.28   Show that for any 2×2 matrix A and 3×3 matrix B,

$$A^2 - (\operatorname{tr} A)A + |A|I = 0,$$

$$|\lambda I - B| = \lambda^3 - \lambda^2 \operatorname{tr} B + \lambda \operatorname{tr}\big(\operatorname{adj}(B)\big) - |B|.$$

3.29   Let $A, B \in M_n(\mathbb{C})$. Let $f(\lambda)$ be the characteristic polynomial of B:

$$f(\lambda) = |\lambda I - B|.$$

(a) Provide an example showing that $f(|A|)$, $|f(A)|$, and $|A - B|$ are all different from each other.

(b) What is wrong with $f(A) = |A - B|$?

(c) Show that the matrix $f(A)$ is invertible if and only if A and B have no common eigenvalues.

3.30   Let $A = \begin{pmatrix} a+b & a-b \\ a-b & a+b \end{pmatrix}$, $a, b \in \mathbb{C}$. Find the eigenvalues and corresponding eigenvectors of A, then find a matrix P such that $P^{-1}AP$ is diagonal.

3.31 Let $A = \left(\begin{smallmatrix} a & -b \\ b & a \end{smallmatrix}\right)$, $a, b \in \mathbb{C}$. Find the eigenvalues and corresponding eigenvectors of A, then find a matrix P such that P^*AP is diagonal.

3.32 Let $B \in M_n(\mathbb{C})$, u and v be row and column vectors, respectively. Let

$$A = \begin{pmatrix} B & -Bv \\ -uB & uBv \end{pmatrix}.$$

(a) Show that $|A| = 0$.

(b) If $|B| = 0$, show that λ^2 divides $|\lambda I - A|$.

(c) Discuss the converse of (b).

3.33 Let $A = uv^*$, where $u, v \in \mathbb{C}^n$ are nonzero column vectors. Show that

(a) $\operatorname{tr} A = v^*u$.

(b) $A^2 = (\operatorname{tr} A)A$.

(c) The eigenvalues of A are $v^*u, 0, \ldots, 0$.

(d) The singular values of A are $\|u\|\|v\|, 0, \ldots, 0$.

(e) If $w^*v = 0$, where $0 \neq w \in \mathbb{C}^n$, then w is an eigenvector of A.

(f) A is similar to $\operatorname{diag}(\operatorname{tr} A, 0, \ldots, 0)$.

(g) $\det(I + A) = 1 + \operatorname{tr} A$.

(h) If $v^*u \neq -1$, then $(I + A)^{-1} = I - \frac{1}{1+v^*u}A$.

3.34 Let $A \in M_n(\mathbb{C})$ and let $u \in \mathbb{C}^n$ be an eigenvector associated with the eigenvalue λ of A. ((λ, u) is called an *eigenpair* of A.) Show that

(a) $(\lambda + v^*u, u)$ is an eigenpair of $A + uv^*$ for any $v \in \mathbb{C}^n$.

(b) If $\lambda, \lambda_2, \ldots, \lambda_n$ are the eigenvalues of A, then $\lambda+v^*u, \lambda_2, \ldots, \lambda_n$ are the eigenvalues of $A + uv^*$ for any $v \in \mathbb{C}^n$.

3.35 Consider $\mathcal{C}(\mathbb{R})$, the space of real-valued continuous functions over $\mathbb{R}$. Let $\alpha = \{1, \cos x, \cos 2x\}$ and $\beta = \{1, \cos x, \cos^2 x\}$. Show that both α and β are linearly independent sets, and they span the same subspace of $\mathcal{C}(\mathbb{R})$. Find the change-of-basis matrix from α to β.

3.36 Let A and B be n-square matrices. Show that the characteristic polynomial of the block matrix $M = \left(\begin{smallmatrix} 0 & A \\ B & 0 \end{smallmatrix}\right)$ is an even function. So,

$$\text{if} \quad \det(\lambda I - M) = 0, \quad \text{then} \quad \det(\lambda I + M) = 0.$$

3.37 Let A and B be n-square matrices, $n \geq 2$. Answer true or false:

(a) If $A^k = 0$ for all positive integers $k \geq 2$, then $A = 0$.

(b) If $A^k = 0$ for some positive integer k, then $\operatorname{tr} A = 0$.

(c) If $A^k = 0$ for some positive integer k, then $\det A = 0$.

(d) If $A^k = 0$ for some positive integer k, then $r(A) = 0$.

(e) If $A^k = 0$ for some positive integer k, then $A^n = 0$.

(f) If $\operatorname{tr} A = 0$, then $\det A = 0$.

(g) If the rank of A is r, then A has r nonzero eigenvalues.

(h) If A has s nonzero eigenvalues, then $r(A) \geq s$.

(i) If A and B are similar, then they have the same eigenvalues.

(j) If A and B are similar, then they have the same singular values.

(k) If A and B have the same eigenvalues, then they are similar.

(l) If A and B have the same characteristic polynomial, then they have the same eigenvalues; and vice versa.

(m) If A and B have the same characteristic polynomial and also the same minimal polynomial, then they are similar.

(n) If all eigenvalues of A are zero, then $A = 0$.

(o) If all eigenvalues of A are zero, then $A^n = 0$.

(p) If all singular values of A are zero, then $A = 0$.

(q) If $\operatorname{tr} A^k = \operatorname{tr} B^k$ for all positive integers k, then $A = B$.

(r) If the eigenvalues of A are $\lambda_1, \lambda_2, \ldots, \lambda_n$, then A is similar to the diagonal matrix $\operatorname{diag}(\lambda_1, \lambda_2, \ldots, \lambda_n)$.

(s) $\operatorname{diag}(1, 2, \ldots, n)$ is similar to $\operatorname{diag}(n, \ldots, 2, 1)$.

(t) If A has a repeated eigenvalue, then A is not diagonalizable.

(u) If $A^m = I$ for some positive integer m, then A is diagonalizable.

(v) If A^m is diagonalizable for some positive integer m, so is A.

(w) If $a + bi$ is an eigenvalue of a real square matrix A, then $a - bi$ is also an eigenvalue of the matrix A, where $a, b \in \mathbb{R}$.

(x) If A is a real square matrix, then all eigenvalues of A are real.

(y) If $A = U^*TU$ is a Schur decomposition of A, where U is unitary and T is upper-triangular, then $A^2 = U^*T^2U$.

(z) If $A = UDV$ is a singular value decomposition (SVD) of A, where U, V are unitary, and D is diagonal, then $A^2 = UD^2V$.

3.38 Let $A = \left(\begin{smallmatrix} 0 & 1 \\ -1 & 0 \end{smallmatrix} \right)$. Show that A is diagonalizable over $\mathbb{C}$, not over $\mathbb{R}$. Find an invertible matrix $P \in M_2(\mathbb{C})$ such that $P^{-1}AP$ is diagonal.

3.39 Let $A = \operatorname{diag}(a_1, \ldots, a_n)$ be a diagonal matrix with $a_i \neq a_j$ for $i \neq j$. If B is a matrix commuting with A, show that B is a diagonal matrix.

3.40 Let A be an $n \times n$ nonsingular matrix of distinct eigenvalues. If B is a matrix satisfying $ABA = B$, show that B^2 is diagonalizable.

3.41 If A is a nonzero matrix such that $A^2 = A$, show that $\operatorname{tr} A > 0$.

3.42 Let $A \in M_n(\mathbb{C})$. If A is symmetric, i.e., $A = A^t$, show that

$$\operatorname{tr}\left(A^p \bar{A}^p (A\bar{A})^q \right) = \operatorname{tr}\left((\bar{A}A)^q \bar{A}^p A^p \right)$$

for any positive integers p and q.

3.43 Let A be a square complex matrix. If A is unitarily similar to its transpose A^t, i.e., $A = U^* A^t U$ for some unitary matrix U, show that

$$\operatorname{tr}\left(A^a A^{*b} A^c A^{*d} \right) = \operatorname{tr}\left(A^{*d} A^c A^{*b} A^a \right)$$

for any nonnegative integers a, b, c, and d, where A^{*k} denotes $(A^*)^k$. Derive that for such an A and for any nonnegative integers p and q,

$$\operatorname{tr}\left(A^p A^{*p} A^q A^{*q} \right) = \operatorname{tr}\left(A^{*p} A^p A^{*q} A^q \right).$$

3.44 Show that if all the eigenvalues of $A \in M_n(\mathbb{C})$ are real and if

$$\operatorname{tr} A^2 = \operatorname{tr} A^3 = \operatorname{tr} A^4 = c$$

for some constant c, then for every positive integer k,

$$\operatorname{tr} A^k = c,$$

and moreover, c must be an integer. Prove that the same conclusion can be drawn if $A^m = A^{m+1}$ for some positive integer m.

3.45 Let $A \in M_n(\mathbb{C})$. If $\operatorname{tr} A^k = 0$, $k = 1, 2, \ldots, n$, show that all eigenvalues of A are equal to zero and $A^n = 0$, that is, A is nilpotent.

3.46 Let $A, B \in M_n(\mathbb{C})$. If $AB = 0$, show that for any positive integer k,

$$\operatorname{tr}(A + B)^k = \operatorname{tr} A^k + \operatorname{tr} B^k.$$

3.47 Let $A = \left(\begin{smallmatrix} 0 & 1 \\ 1 & 1 \end{smallmatrix} \right)$ and $A^0 = I_2$. Show that for any positive integer k,

$$\operatorname{tr} A^k = \operatorname{tr} A^{k-1} + \operatorname{tr} A^{k-2}.$$

3.48   Let a, b, and c be scalars of a field $\mathbb{F}$ (where $\mathbb{F} = \mathbb{R}$ or $\mathbb{C}$), and let

$$A = \begin{pmatrix} a & b & c \\ b & c & a \\ c & a & b \end{pmatrix}, \quad B = \begin{pmatrix} b & c & a \\ c & a & b \\ a & b & c \end{pmatrix}, \quad C = \begin{pmatrix} c & a & b \\ a & b & c \\ b & c & a \end{pmatrix}.$$

(a) Show that A, B, and C are similar over $\mathbb{F}$.

(b) Over $\mathbb{R}$, show that the following statements are equivalent:

  (i) two of A, B, and C commute (say $AC = CA$).

  (ii) A, B, and C mutually commute.

  (iii) $a = b = c$.

  (iv) $A = B = C$.

  (v) A, B, and C each have at least two zero eigenvalues.
    In this case, what is the possible nonzero eigenvalue of A?

(c) Over $\mathbb{C}$, (i) give an example that $AC = CA$, but a, b, c are distinct, (ii) if two of A, B, C commute, show that they all commute, (iii) if $AC = CA$, show that they each have at least two zero eigenvalues. Then, what are possible ranks of A, B, C?

3.49   Find the eigenvalues of $A = \begin{pmatrix} 0 & 1 & 0 \\ 0 & 0 & 1 \\ 1 & 0 & 0 \end{pmatrix}$ and $B = \begin{pmatrix} a & b & c \\ c & a & b \\ b & c & a \end{pmatrix}$, $a, b, c \in \mathbb{C}$.

3.50   Find the values of $a, b, c \in \mathbb{C}$ such that matrices

$$A = \begin{pmatrix} 1 & a & 1 \\ a & 1 & b \\ 1 & b & 1 \end{pmatrix} \quad \text{and} \quad B = \begin{pmatrix} 0 & 0 & 0 \\ 0 & 1 & 0 \\ 0 & 0 & c \end{pmatrix}$$

are similar. Find a real orthogonal matrix T such that $T^{-1}AT = B$.

3.51   Let $A = \begin{pmatrix} 1 & 1 & 1 \\ 1 & 1 & 1 \\ 1 & 1 & 1 \end{pmatrix}$ and $B = \begin{pmatrix} 1 & 1 & 1 \\ 1 & 0 & 0 \\ 1 & 0 & 0 \end{pmatrix}$. Explain why A and B are diagonalizable and what the diagonal matrices are. Find real orthogonal matrices P and Q such that $P^t A P$ and $Q^t B Q$ are diagonal.

3.52   Let $a, b \in \mathbb{C}$. If matrix

$$A = \begin{pmatrix} 0 & 0 & 1 \\ a & 1 & b \\ 1 & 0 & 0 \end{pmatrix}$$

has three linearly independent eigenvectors, show that $a + b = 0$.

3.53 Find the values of $a, b \in \mathbb{C}$ so that matrix A is diagonalizable, where

$$A = \begin{pmatrix} 0 & a & 0 \\ b & 0 & a \\ 0 & 0 & b \end{pmatrix}.$$

3.54 Find the minimal and characteristic polynomials of the matrices:

$$A = \begin{pmatrix} 1 & 1 & 0 & 0 \\ 0 & 1 & 0 & 0 \\ 0 & 0 & 1 & 0 \\ 0 & 0 & 0 & 1 \end{pmatrix}, B = \begin{pmatrix} 1 & 1 & 0 & 0 \\ 0 & 1 & 0 & 0 \\ 0 & 0 & 1 & 1 \\ 0 & 0 & 0 & 1 \end{pmatrix}, C = \begin{pmatrix} 1 & 1 & 0 & 0 \\ 0 & 1 & 1 & 0 \\ 0 & 0 & 1 & 0 \\ 0 & 0 & 0 & 1 \end{pmatrix}.$$

3.55 Find the minimal and characteristic polynomials of the matrices:

$$P = \begin{pmatrix} 1 & 1 & 0 & 0 \\ 1 & 1 & 0 & 0 \\ 0 & 0 & 1 & 1 \\ 0 & 0 & 0 & 1 \end{pmatrix}, Q = \begin{pmatrix} 0 & 0 & 0 & 1 \\ 1 & 0 & 0 & 2 \\ 0 & 1 & 0 & 3 \\ 0 & 0 & 1 & 4 \end{pmatrix}, R = \begin{pmatrix} 1 & 0 & 0 & 1 \\ 0 & 1 & 0 & 2 \\ 0 & 0 & 1 & 3 \\ 0 & 0 & 0 & 4 \end{pmatrix}.$$

3.56 Show that a square matrix is diagonalizable over the complex number field $\mathbb{C}$ if and only if its minimal polynomial has no repeated roots.

3.57 Let A be an $n \times n$ real matrix with eigenvalues $\lambda_1, \lambda_2, \ldots, \lambda_n$ (which need not be real). For each k, let $\lambda_k = x_k + iy_k$, $x_k, y_k \in \mathbb{R}$. Show that

(a) $y_1 + y_2 + \cdots + y_n = 0.$

(b) $x_1 y_1 + x_2 y_2 + \cdots + x_n y_n = 0.$

(c) $\operatorname{tr} A^2 = (x_1^2 + x_2^2 + \cdots + x_n^2) - (y_1^2 + y_2^2 + \cdots + y_n^2).$

3.58 Let λ be an eigenvalue of an n-square matrix $A = (a_{ij})$. Show that there exists a positive integer k such that

$$|\lambda - a_{kk}| \leq \sum_{j=1, j \neq k}^{n} |a_{kj}|. \quad (Gershgorin\ circle\ theorem)$$

3.59 If the eigenvalues of $A = (a_{ij}) \in M_n(\mathbb{C})$ are $\lambda_1, \lambda_2, \ldots, \lambda_n$, show that

$$\sum_{i=1}^{n} |\lambda_i|^2 \leq \sum_{i, j=1}^{n} |a_{ij}|^2 \quad (Schur\ inequality)$$

and equality holds if and only if A is unitarily diagonalizable.

3.60 Let A be an $n \times n$ matrix and $P = \begin{pmatrix} A & I_n - A \\ A & I_n - A \end{pmatrix}$. Show that $P^2 = P$ and P is similar to $\begin{pmatrix} I_n & I_n - A \\ 0 & 0 \end{pmatrix}$. Thus, the eigenvalues of P are 0's and 1's.

3.61   Let A and B be n-square matrices. Show that the eigenvalues of the $2n \times 2n$ matrix $\begin{pmatrix} A & A \\ B & B \end{pmatrix}$ are the eigenvalues of $A + B$ plus n zeros.

3.62   Let λ_1 and λ_2 be two different eigenvalues of a matrix A and let u_1 and u_2 be eigenvectors of A corresponding to λ_1 and λ_2, respectively. Show that $u_1 + u_2$ is not an eigenvector of A (for any eigenvalue).

3.63   Find a 3×3 real matrix A such that

$$Au_1 = u_1, \quad Au_2 = 2u_2, \quad Au_3 = 3u_3,$$

where

$$u_1 = \begin{pmatrix} 1 \\ 2 \\ 2 \end{pmatrix}, \quad u_2 = \begin{pmatrix} 2 \\ -2 \\ 1 \end{pmatrix}, \quad u_3 = \begin{pmatrix} -2 \\ -1 \\ 2 \end{pmatrix}.$$

3.64   Let $A = \begin{pmatrix} a & b \\ c & d \end{pmatrix}$ be a 2×2 real matrix. If $\begin{pmatrix} x \\ 1 \end{pmatrix}$ is an eigenvector of A for some eigenvalue, find the value of x in terms of a, b, c, and d.

3.65   Let $A = \begin{pmatrix} a & b \\ c & d \end{pmatrix}$ be a 2×2 complex matrix with determinant 1.

    (a)   Find A^{-1}.

    (b)   Write A as a product of matrices of the forms $\begin{pmatrix} 1 & x \\ 0 & 1 \end{pmatrix}$ and $\begin{pmatrix} 1 & 0 \\ x & 1 \end{pmatrix}$.

    (c)   If $|a + d| > 2$, show that A is similar to $\begin{pmatrix} \lambda & 0 \\ 0 & \lambda^{-1} \end{pmatrix}$, $\lambda \neq 0, 1, -1$.

    (d)   If $|a + d| < 2$, show that A is similar to $\begin{pmatrix} \lambda & 0 \\ 0 & \lambda^{-1} \end{pmatrix}$, $\lambda \notin \mathbb{R} \cup i\mathbb{R}$.

    (e)   If $|a + d| = 2$ and A has real eigenvalues, what are the possible real matrices to which A is similar?

    (f)   If $|a + d| \neq 2$, show that A is similar to $\begin{pmatrix} \frac{a+d}{2} & x \\ x & \frac{a+d}{2} \end{pmatrix}$, $x \in \mathbb{C}$.

    (g)   Does (f) remain true if $|a + d| = 2$?

3.66   Find the eigenvalues of the following matrices (with $i = \sqrt{-1}$):

$$\begin{pmatrix} 1 & i \\ i & 1 \end{pmatrix}, \quad \begin{pmatrix} 1 & i \\ -i & 1 \end{pmatrix}, \quad \begin{pmatrix} 1 & i \\ i & 0 \end{pmatrix}, \quad \begin{pmatrix} 1+i & i \\ i & 1+i \end{pmatrix}.$$

3.67   Show that matrices A and B are similar but not unitarily similar:

$$A = \begin{pmatrix} 1 & 1 \\ 0 & 0 \end{pmatrix}, \quad B = \begin{pmatrix} 1 & 0 \\ 0 & 0 \end{pmatrix}.$$

3.68 Let $A \in M_n(\mathbb{C})$. Show that A^t is similar to A (but A^t and A need not be unitarily similar). Give an example showing that A^* and A are not similar even though A^* and A have the same eigenvalues.

3.69 Show that two real matrices similar over $\mathbb{C}$ are similar over $\mathbb{R}$.

3.70 If all eigenvalues of real matrix A are real, show that there exists a real invertible matrix R such that $R^{-1}AR$ is real and upper-triangular.

3.71 (a) Let $A = \left(\begin{smallmatrix} 0 & -1 \\ 1 & 0 \end{smallmatrix}\right)$. Show that $A^2 + I = 0$.

(b) If $B^2 + I = 0$, where $B \in M_2(\mathbb{C})$, show that B is similar to A.

(c) For any given $\epsilon > 0$, show that there does not exist a 2×2 real matrix C such that $C^2 + I + \epsilon \left(\begin{smallmatrix} 0 & 0 \\ 0 & 1 \end{smallmatrix}\right) = 0$.

3.72 Find the eigenvalues and corresponding eigenvectors of the matrix

$$\begin{pmatrix} 2 & 1 & 0 \\ 1 & 3 & 1 \\ 0 & 1 & 2 \end{pmatrix}.$$

Verify that the eigenvectors of distinct eigenvalues are orthogonal.

3.73 Find a real symmetric orthogonal matrix T, i.e., $T^t = T^{-1} = T$, such that $T^t AT$ is diagonal for any (real or complex) values x and y, where

$$A = \begin{pmatrix} x & y & 0 \\ y & x & y \\ 0 & y & x \end{pmatrix}.$$

3.74 Find singular value decompositions (SVD) for A and B (not unique):

$$A = \begin{pmatrix} 1 & 0 & 1 \\ 0 & 1 & 0 \end{pmatrix} \quad \text{and} \quad B = \begin{pmatrix} 1 & 1 \\ 1 & 1 \\ 1 & 1 \end{pmatrix}.$$

3.75 Let J_k be the $k \times k$ Jordan block of eigenvalue 0. What is the minimal polynomial of J_k? Find $\dim(\text{Ker } J_k)$ and $\dim(\text{Im } J_k)$.

3.76 Let $A \in M_n(\mathbb{C})$. What are the eigenvalues of $\text{adj}(A)$ and $\text{adj}(\text{adj}(A))$?

3.77 Let $A = (a_{ij}) \in M_n(\mathbb{C})$. Prove the min-max inequalities for diagonal entries, eigenvalues, and singular values: $\max_{i,j} |a_{ij}| \leq \sigma_{\max}(A)$, and

$$\sigma_{\min}(A) \leq \min_i |\lambda_i(A)| \leq \max_i |\lambda_i(A)| \leq \sigma_{\max}(A).$$

Is it true that $\max_i |a_{ii}| \leq \max_i |\lambda_i(A)|$? or $\sigma_{\min}(A) \leq \min_i |a_{ii}|$?

3.78 Let $A \in M_n(\mathbb{C})$ and $\lambda \in \mathbb{R}$. If $\lambda \geq 0$, show that λ is an eigenvalue of $A\bar{A}$ if and only if $A\bar{x} = \sqrt{\lambda}\,x$ for some nonzero $x \in \mathbb{C}^n$.

3.79 Let $p(x)$ be a polynomial in variable x and let A be a square matrix. If $p(A) = 0$, show, with or without using minimal polynomial, that $p(\lambda) = 0$, where λ is any eigenvalue of A. Is the converse true?

3.80 Let $a_1, a_2, \ldots, a_n \in \mathbb{R}$ satisfy $a_1 + a_2 + \cdots + a_n = 0$ and denote

$$A = \begin{pmatrix} a_1^2 + 1 & a_1 a_2 + 1 & \cdots & a_1 a_n + 1 \\ a_2 a_1 + 1 & a_2^2 + 1 & \cdots & a_2 a_n + 1 \\ \vdots & \vdots & \vdots & \vdots \\ a_n a_1 + 1 & a_n a_2 + 1 & \cdots & a_n^2 + 1 \end{pmatrix}.$$

Show that $A = BB^t$ for some matrix B. Find the eigenvalues of A.

3.81 Let $A, B \in M_n(\mathbb{C})$.

  (a) Show that $\mathrm{tr}(AB) = \mathrm{tr}(BA)$.

  (b) Is it true that $r(AB) = r(BA)$?

  (c) Show that $\mathrm{tr}(AB)^k = \mathrm{tr}(BA)^k$.

  (d) Is it true that $\mathrm{tr}(AB)^k = \mathrm{tr}(A^k B^k)$?

  (e) Why is A singular if $AB - BA = A$?

  (f) Show that $\mathrm{tr}(ABC) = \mathrm{tr}(BCA)$ for every $C \in M_n(\mathbb{C})$.

  (g) Is it true that $\mathrm{tr}(ABC) = \mathrm{tr}(ACB)$ in general?

  (h) Show that $\mathrm{tr}\big((AB - BA)(AB + BA)\big) = 0$.

  (i) Show that AB and BA are similar if A or B is nonsingular.

  (j) Are AB and BA similar in general?

  (k) Show that AA^* and A^*A are (unitarily) similar.

  (l) Are AA^t and A^tA similar in general?

3.82 Let m and j be positive integers with $m \geq j$. Let $S_{m,j}(X, Y)$ denote the sum of all matrix products of the form $A_1 \cdots A_m$, where each A_i is either X or Y, with Y in exactly j cases. For instance, $S_{5,2}(X, Y) = X^3 Y^2 + Y^2 X^3 + XYXYX + \cdots$. Show that for $m = 5$ and $j = 3$

$$\mathrm{tr}\,\big(S_{5,3}(X, Y)\big) = \tfrac{5}{2}\,\mathrm{tr}\,\big(XS_{4,3}(X, Y)\big).$$

3.83   Let $u, v \in \mathbb{R}^n$ be nonzero orthogonal column vectors, namely, $v^t u = 0$. Find the eigenvalues of $A = uv^t$ and corresponding eigenvectors. Is A similar to a diagonal matrix? What are the singular values of A?

3.84   (a) Let J_n denote the n-square matrix all of whose entries are 1. Find the eigenvalues and the corresponding eigenvectors of J_n.

      (b) Let $K = \left(\begin{smallmatrix} 0 & J_n \\ J_n & 0 \end{smallmatrix} \right)$. Find the eigenvalues and eigenvectors of K.

3.85   Let

$$A = \begin{pmatrix} 0 & 1 \\ 0 & 0 \end{pmatrix}, \ B = \begin{pmatrix} 0 & 0 & 1 \\ 0 & 0 & 0 \\ 0 & 0 & 0 \end{pmatrix}, \ C = \begin{pmatrix} 0 & 0 & 0 & 1 \\ 0 & 0 & 0 & 0 \\ 0 & 0 & 0 & 0 \\ 0 & 0 & 0 & 0 \end{pmatrix}.$$

      (a) Show that there does not exist a matrix X such that $X^2 = A$.

      (b) Show that there exists a matrix Y such that $Y^2 = B$.

      (c) If $Z^k = C$ for some matrix Z and integer $k \geq 2$, find Z and k.

3.86   Let $A \in M_n(\mathbb{C})$. If all eigenvalues of A are real, and s of the eigenvalues are nonzero, show that

      (a) $(\operatorname{tr} A)^2 \leq s \operatorname{tr} A^2$. When does equality hold?

      (b) $(\operatorname{tr} A)^2 \leq r(A) \operatorname{tr} A^2$ when A is Hermitian. Moreover, equality holds if and only if $A^2 = cA$ for some (real) scalar c, equivalently, $A = 0$ or all nonzero eigenvalues of A are the same.

      (c) If $(\operatorname{tr} A)^2 > (n - 1) \operatorname{tr} A^2$, then A is nonsingular.

3.87   Let A be a square matrix. If $A^3 = A$, show that $r(A) = \operatorname{tr} A^2$.

3.88   Let A and B be 3×2 and 2×3 matrices, respectively, such that

$$AB = \begin{pmatrix} 8 & 2 & -2 \\ 2 & 5 & 4 \\ -2 & 4 & 5 \end{pmatrix}.$$

Show that

$$BA = \begin{pmatrix} 9 & 0 \\ 0 & 9 \end{pmatrix}.$$

3.89 Find the Jordan blocks and Jordan canonical form of the matrix

$$A = \begin{pmatrix} 1 & 2 & 0 & 1 \\ 1 & 4 & 3 & 2 \\ 3 & 0 & 7 & 2 \\ -4 & -1 & -7 & -1 \end{pmatrix}.$$

3.90 It is given that 1, 2, and 3 are the eigenvalues of a 3×3 real symmetric matrix A and that $\alpha_1 = (-1, -1, 1)^t$ and $\alpha_2 = (1, -2, -1)^t$ are eigenvectors of A belonging to the eigenvalues 1 and 2, respectively. Find an eigenvector of A corresponding to the eigenvalue 3 and then find A. Show that such a real symmetric matrix A is unique up to row and column permutations (permutation similarity). If A is not required to be symmetric, is A unique (up to permutation similarity)?

3.91 (a) Explain why there does not exist a real symmetric matrix that has eigenvalues 0, 1, and -1, and $\alpha = (1, 1, 1)^t$ and $\beta = (2, 2, 1)^t$ as eigenvectors corresponding to the eigenvalues 0 and 1, respectively.

(b) Construct a 3×3 real symmetric matrix A such that the eigenvalues of A are 1, 1, and -1, and $\alpha = (1, 1, 1)^t$ and $\beta = (2, 2, 1)^t$ are eigenvectors corresponding to the eigenvalue 1. Is such an A unique?

3.92 Construct a 3×3 complex symmetric matrix A such that A has two zero eigenvalues and one nonzero eigenvalue, and the rank of A is 2. Explain why such a matrix of real entries does not exist.

3.93 Let T be the $n \times n$ real symmetric and tridiagonal matrix:

$$T = \begin{pmatrix} a_1 & b_1 & 0 & 0 & \dots & 0 & 0 \\ b_1 & a_2 & b_2 & 0 & \dots & 0 & 0 \\ 0 & b_2 & a_3 & b_3 & \dots & 0 & 0 \\ 0 & 0 & b_3 & a_4 & \dots & 0 & 0 \\ \vdots & \vdots & \vdots & \vdots & \ddots & \vdots & \vdots \\ 0 & 0 & 0 & 0 & \cdots & a_{n-1} & b_{n-1} \\ 0 & 0 & 0 & 0 & \dots & b_{n-1} & a_n \end{pmatrix}.$$

If all b_i's are nonzero, show that

  (a) the rank of T is at least $n - 1$,
  (b) T is similar to a diagonal matrix, and
  (c) T has n distinct eigenvalues.

3.94 Let $A \in M_n(\mathbb{C})$. Of the matrices $\bar{A}$, A^t, A^*, $\text{adj}(A)$, $\frac{A+A^*}{2}$, $(A^*A)^{\frac{1}{2}}$, which one(s) has the same eigenvalues or singular values as A?

3.95   Let $A, B \in M_n(\mathbb{C})$. The *commutator* or *Lie bracket* of A and B is $AB - BA$, denoted by $[A, B]$, i.e. $[A, B] = AB - BA$. Show that

(a) $[A, B] = [-A, -B] = -[B, A]$.

(b) $[A, B + C] = [A, B] + [A, C]$ and $[aA, bB] = ab[A, B]$, $a, b \in \mathbb{C}$.

(c) $[A, B]^* = [B^*, A^*]$.

(d) $[PXP^{-1}, Y] = 0$ if and only if $[X, P^{-1}YP] = 0$.

(e) $\operatorname{tr}[A, B] = 0$.

(f) $I - [A, B]$ is not nilpotent, i.e., $\left(I - [A, B]\right)^k \neq 0$ for any k.

(g) $[A, B]$ is never similar to the identity matrix.

(h) If the diagonal entries of A are all equal to zero, then there exist matrices X and Y such that $A = [X, Y]$.

(i) If $[A, B] = 0$, then $[A^p, B^q] = 0$ for all positive integers p, q.

(j) If $[A, B] = A$, then A is singular.

(k) If A and B are both Hermitian or both skew-Hermitian, then $[A, B]$ is skew-Hermitian.

(l) If one of A and B is Hermitian and the other one is skew-Hermitian, then $[A, B]$ is Hermitian.

(m) If A is skew-Hermitian, then $A = [B, C]$ for some Hermitian matrices B and C.

(n) If A and B are Hermitian, then the real part of every eigenvalue of $[A, B]$ is zero.

(o) If $[A, [A, A^*]] = 0$, then A is normal, i.e., $A^*A = AA^*$.

(p) $[A, [B, C]] + [B, [C, A]] + [C, [A, B]] = 0$.

(q) If $[A, B]$ commutes with A or B, then $[A, B]$ has no eigenvalues other than 0. Equivalently, $[A, B]$ is nilpotent.

3.96   Let $A \in M_n(\mathbb{C})$. Show that the following statements are equivalent:

(a) $A^2 = BA$ for some nonsingular matrix $B \in M_n(\mathbb{C})$.

(b) $r(A^2) = r(A)$.

(c) $\operatorname{Im} A \cap \operatorname{Ker} A = \{0\}$.

(d) There exist nonsingular matrices P and D of sizes $n \times n$ and $r(A) \times r(A)$, respectively, such that $A = P \begin{pmatrix} D & 0 \\ 0 & 0 \end{pmatrix} P^{-1}$.

3.97   Show that a square matrix A is diagonalizable over $\mathbb{C}$ if and only if for every eigenvalue λ of A, the following rank identity holds:

$$r(\lambda I - A) = r\big((\lambda I - A)^2\big).$$

Equivalently, A is diagonalizable if and only if $(\lambda I - A)x = 0$ whenever $(\lambda I - A)^2 x = 0$, where $x \in \mathbb{C}^n$ (that is, they have the same null space).

3.98   Let $A \in M_n(\mathbb{C})$. Denote by $\sigma_1(A)$ the largest singular value of A and by $\sigma_n(A)$ the smallest singular value of A. Show that

(a) $\sigma_1(A) = \max_{x^*x=1}(x^*A^*Ax)^{\frac{1}{2}}$.

(b) $\sigma_n(A) = \min_{x^*x=1}(x^*A^*Ax)^{\frac{1}{2}}$.

(c) $\sigma_1(A) = \max_{x^*x=1}|x^*Ax|$ if A is Hermitian.

(d) $\sigma_n(A) \neq \min_{x^*x=1}|x^*Ax|$ in general though A is Hermitian.

3.99   Let A, B, and C be $n \times n$ complex matrices. Show that

$$AB = AC \quad \text{if and only if} \quad A^*AB = A^*AC.$$

3.100   Let A and B be $n \times n$ complex matrices of the same rank. Show that

$$A^2B = A \quad \text{if and only if} \quad B^2A = B.$$

3.101   Let $A = I_n - (x^*x)^{-1}(xx^*)$, where $x \in \mathbb{C}^n$ is a nonzero vector. Find

(a) $r(A)$.   (b) $\operatorname{Im} A$.   (c) $\operatorname{Ker} A$.

3.102   Let A be a square matrix with rational entries. Show that there exists a polynomial $f(x)$ of integer coefficients such that $f(A) = 0$. Find such a polynomial of the lowest degree for $A = \operatorname{diag}(\frac{1}{2}, \frac{2}{3}, \frac{3}{4})$.

3.103   Let $A, B \in M_n(\mathbb{C})$. If the range of B is contained in that of A, i.e., $\operatorname{Im} B \subseteq \operatorname{Im} A$, show that $\left(\begin{smallmatrix} A & B \\ 0 & 0 \end{smallmatrix}\right)$ is similar to $\left(\begin{smallmatrix} A & 0 \\ 0 & 0 \end{smallmatrix}\right)$.

3.104   Let R be an $n \times n$ real matrix. Show that

(a) R is similar to an $n \times n$ real block matrix $\left(\begin{smallmatrix} D & 0 \\ 0 & E \end{smallmatrix}\right)$, where D is invertible and E is nilpotent. (D or E is allowed to be absent.)

(b)   If C is an $n \times m$ complex matrix, then the $(m + n) \times (m + n)$ block matrix $\left(\begin{smallmatrix} R & C \\ 0 & 0 \end{smallmatrix}\right)$ is similar to a real matrix.

3.105   Let A and B be $m \times m$ and $n \times n$ (real or complex) matrices, respectively. If A and B have no common eigenvalues over $\mathbb{C}$, show that the matrix equation $AX = XB$ has only the zero solution $X = 0$.

3.106   Let $A, B \in M_n(\mathbb{C})$. Show that for any scalar λ and positive integer k

$$r \begin{pmatrix} \lambda I - AB & -A \\ 0 & \lambda I \end{pmatrix}^k = r \begin{pmatrix} \lambda I & -A \\ 0 & \lambda I - BA \end{pmatrix}^k.$$

Derive that if $\lambda \neq 0$, then $r(\lambda I - AB)^k = r(\lambda I - BA)^k$. What if $\lambda = 0$?

3.107   Let $A, B, C \in M_n(\mathbb{C})$. If the matrix equation $AX - XB = C$ has a solution X, show that $\begin{pmatrix} A & C \\ 0 & B \end{pmatrix}$ and $\begin{pmatrix} A & 0 \\ 0 & B \end{pmatrix}$ are similar.

3.108   Let $A \in M_n(\mathbb{C})$. Prove that

(a)   If $\lambda \neq 0$ is an eigenvalue of A, $\frac{1}{\lambda}|A|$ is an eigenvalue of $\mathrm{adj}(A)$.

(b)   If v is an eigenvector of A, v is an eigenvector of $\mathrm{adj}(A)$.

3.109   Let A and B be $n \times n$ matrices such that $AB = BA$. Show that

(a)   If A has n distinct eigenvalues, then A, B, and AB are all diagonalizable.

(b)   If A and B are diagonalizable, then there exists an invertible matrix T such that $T^{-1}AT$ and $T^{-1}BT$ are both diagonal.

3.110   Let A be an $n \times n$ *idempotent* matrix, i.e., $A^2 = A$. Show that

(a)   If $x^*y = 0$ for all $x \in \mathrm{Ker}\, A$ and $y \in \mathrm{Im}\, A$, then $A^* = A$.

(b)   If $z^*(A^*A)z \leq z^*z$ for all $z \in \mathbb{C}^n$, then $A^* = A$.

3.111   Let $x, y, u, v \in \mathbb{C}^n$ (in columns), $\langle x, y \rangle = y^*x$, $\|x\|^2 = x^*x$. Show that

(a) $(xx^*) \circ (yy^*) = (x \circ y)(x \circ y)^*$.

(b) $(xx^*) \otimes (yy^*) = (x \otimes y)(x \otimes y)^*$.

(c) $\langle x \otimes u, y \otimes v \rangle = \langle x, y \rangle \langle u, v \rangle$.

(d) $\|x \otimes y\| = \|x\| \, \|y\|$.

3.112   Let $A, B \in M_n(\mathbb{C})$. Show that the Hadamard product $A \circ B$ is the principal submatrix of the Kronecker product $A \otimes B$ lying in the intersections of rows and columns $1, n+2, 2n+3, \ldots, n^2$. Simply put:

$$A \circ B = (A \otimes B)[\alpha], \quad \alpha = \{1, n+2, 2n+3, \ldots, n^2\}.$$

3.113   Let A, B, C, and D be matrices of appropriate sizes. Besides the mixed-product property $(A \otimes B)(C \otimes D) = (AC) \otimes (BD)$, prove that

(a) $c(A \otimes B) = (cA) \otimes B = A \otimes (cB)$, where $c \in \mathbb{C}$.

(b) $(A + B) \otimes C = A \otimes C + B \otimes C$.

(c) $A \otimes (B + C) = A \otimes B + A \otimes C$.

(d) $(A \otimes B) \otimes C = A \otimes (B \otimes C)$.

(e) $I_m \otimes I_n = I_{mn}$.

(f) $(A \otimes B)^* = A^* \otimes B^*$.

(g) $(A \otimes B)^t = A^t \otimes B^t$.

(h) $\overline{A \otimes B} = \bar{A} \otimes \bar{B}$.

(i) $(A \otimes B)^{-1} = A^{-1} \otimes B^{-1}$ if A and B are invertible.

(j) $A \otimes B$ is upper-triangular if A and B are upper-triangular.

(k) $A \otimes B$ is Hermitian if A and B are Hermitian.

(l) $A \otimes B$ is unitary if A and B are unitary.

(m) $A \otimes B$ is normal if A and B are normal.

(n) $(A \otimes B)^k = A^k \otimes B^k$ for any positive integer k.

(o) $A \otimes B = 0$ if and only if $A = 0$ or $B = 0$.

(p) If $A \otimes B = C \otimes D \neq 0$, where A and C have the same size (so do B and D), then $A = aC$ and $B = bD$ with $ab = 1$.

(q) $(A \otimes B) \circ (C \otimes D) = (A \circ C) \otimes (B \circ D)$.

(r) $(A \otimes I)(I \otimes B) = A \otimes B = (I \otimes B)(A \otimes I)$.

(s) If $A \sim B$ and $C \sim D$, then $A \otimes C \sim B \otimes D$ ($\sim$ for similarity).

(t) $(A \otimes B)(x \otimes y) = (Ax) \otimes (By)$, where x, y are column vectors.

(u) $r(A \otimes B) = r(A) r(B)$.

(v) $\operatorname{tr}(A \otimes B) = (\operatorname{tr} A)(\operatorname{tr} B)$.

(w) $\det(A \otimes B) = (\det A)^n (\det B)^m$, where A is $m \times m$, B is $n \times n$.

(x) $B \otimes A$ can be obtained from $A \otimes B$ through row and column permutations. In particular, if A and B are $n \times n$ matrices, then $P^{-1}(A \otimes B)P = B \otimes A$ for some permutation matrix P.

3.114   Let $A \in M_m(\mathbb{C})$ and $B \in M_n(\mathbb{C})$ have eigenvalues λ_i and μ_j, respectively, where $i = 1, \ldots, m$, $j = 1, \ldots, n$. Show that the eigenvalues of $A \otimes B$ are $\lambda_i \mu_j$, $i = 1, \ldots, m$, $j = 1, \ldots, n$, and the eigenvalues of $A \otimes I_n + I_m \otimes B$ are $\lambda_i + \mu_j$, $i = 1, \ldots, m$, $j = 1, \ldots, n$.

3.115 If $A = UCV$ and $B = RDS$ are singular value decompositions of A and B, respectively, where C and D are nonnegative diagonal, U, V, R, S are unitary, what is a singular value decomposition of $A \otimes B$?

3.116 Which of the following $\mathcal{A}$ are linear transformations on $\mathbb{C}^n$ over $\mathbb{C}$?

(a) $\mathcal{A}(u) = v$, where v is a fixed nonzero vector in $\mathbb{C}^n$.

(b) $\mathcal{A}(u) = 0$.

(c) $\mathcal{A}(u) = \bar{u}$.

(d) $\mathcal{A}(u) = ku$, where k is a fixed complex number.

(e) $\mathcal{A}(u) = \|u\|u$, where $\|u\|$ is the length of vector u.

(f) $\mathcal{A}(u) = \frac{1}{\|u\|}u$ if $u \neq 0$ and $\mathcal{A}(0) = 0$.

(g) $\mathcal{A}(u) = (u_1, 2u_2, \ldots, nu_n)$, where $u = (u_1, u_2, \ldots, u_n) \in \mathbb{C}^n$.

Show that $\mathcal{T}(u) = \frac{u+\bar{u}}{2}$ is a linear transformation on $\mathbb{C}^n$ as a vector space over $\mathbb{R}$, however, $\mathcal{T}$ is not a linear transformation on $\mathbb{C}^n$ over $\mathbb{C}$.

3.117 Determine whether each of the following maps $\mathcal{T}$ is (i) linear, (ii) one-to-one, (iii) onto, and (iv) invertible (i.e., one-to-one and onto).

(a) $\mathcal{T} : \mathbb{R} \to \mathbb{R}$ defined by $\mathcal{T}(r) = |r|$.

(b) $\mathcal{T} : \mathbb{R} \to \mathbb{R}$ defined by $\mathcal{T}(r) = r^2$.

(c) $\mathcal{T} : \mathbb{R} \to \mathbb{R}$ defined by $\mathcal{T}(r) = -r$.

(d) $\mathcal{T} : \mathbb{R} \to \mathbb{R}$ defined by $\mathcal{T}(r) = r$ if r is not an integer, and $\mathcal{T}(r) = -r$ if r is an integer.

(e) $\mathcal{T} : \mathbb{R} \to \mathbb{R}^2$ defined by $\mathcal{T}(r) = (r, -r)$.

(f) $\mathcal{T} : \mathbb{R}^2 \to \mathbb{R}$ defined by $\mathcal{T}(x, y) = x$.

(g) $\mathcal{T} : \mathbb{R}^2 \to \mathbb{R}$ defined by $\mathcal{T}(x, y) = x + y$.

(h) $\mathcal{T} : \mathbb{R}^2 \to \mathbb{R}$ defined by $\mathcal{T}(v) = \|v\|$.

(i) $\mathcal{T} : \mathbb{C} \to \mathbb{C}$ defined by $\mathcal{T}(z) = e^{i\theta}z$, where $\theta \in \mathbb{R}$ is fixed.

(j) $\mathcal{T} : \mathbb{C} \to \mathbb{C}$ defined by $\mathcal{T}(z) = z + z_0$, where $0 \neq z_0 \in \mathbb{C}$ is fixed.

(k) $\mathcal{T} : \mathbb{R}^2 \to \mathbb{R}^2$ defined by $\mathcal{T}(x, y) = (y, x)$.

(l) $\mathcal{T} : \mathbb{R}^2 \to \mathbb{R}^2$ defined by $\mathcal{T}(x, y) = (x + 2y, 3x + 4y)$.

(m) $\mathcal{T} : \mathbb{R}^2 \to \mathbb{R}^2$ defined by $\mathcal{T}(x, y) = (x + 2y, 2x + 4y)$.

(n) $\mathcal{T} : \mathbb{P}[x] \to \mathbb{P}[x]$ defined by $\mathcal{T}(p(x)) = \frac{d}{dx}(p(x))$.

(o) $\mathcal{T} : \mathbb{P}[x] \to \mathbb{P}[x]$ defined by $\mathcal{T}(p(x)) = xp(x)$.

3.118 Let $\mathcal{A}$ be a (possibly nonlinear) map from (vector) space V to space W.

(a) Show that there exists a map $\mathcal{L}$ from W to V such that $\mathcal{L}\mathcal{A} = \mathcal{I}_V$ (i.e., $\mathcal{A}$ is left-invertible) if and only if $\mathcal{A}$ is one-to-one.

(b) If $\mathcal{A}$ is linear and one-to-one, define $\mathcal{T}(w) = v$ if $\mathcal{A}(v) = w$ and $\mathcal{T}(w) = 0$ if $w \in W \setminus \operatorname{Im}\mathcal{A}$. Is $\mathcal{T}$ necessarily linear?

(c) Let $\mathcal{A}$ be left-invertible. Give an example showing that $\mathcal{A}$ may have different left-inverses $\mathcal{L}_1 \neq \mathcal{L}_2$: $\mathcal{L}_1\mathcal{A} = \mathcal{L}_2\mathcal{A} = \mathcal{I}_V$.

(d) Show that there exists a map $\mathcal{R}$ from W to V such that $\mathcal{A}\mathcal{R} = \mathcal{I}_W$ (i.e., $\mathcal{A}$ is right-invertible) if and only if $\mathcal{A}$ is onto.

(e) Show that $\mathcal{A}$ is invertible (i.e., $\mathcal{A}$ is left- and right-invertible) if and only if $\mathcal{A}$ is bijective (i.e., one-to-one and onto).

(f) If $\mathcal{A}$ has a left-inverse and it also has a right-inverse, then the left- and right-inverses are unique, and they are equal (to $\mathcal{A}^{-1}$).

3.119 Let V and W be vector spaces over a field $\mathbb{F}$ and let $\mathcal{L} : V \to W$ be a bijective (one-to-one and onto) map. Define the inverse $\mathcal{L}^{-1}$ of $\mathcal{L}$ by $\mathcal{L}^{-1}(w) = v$ if $\mathcal{L}(v) = w$. If $\mathcal{L}$ is linear, show that $\mathcal{L}^{-1}$ is also linear.

3.120 Let A be an $m \times n$ matrix over $\mathbb{F}$ ($= \mathbb{R}$ or $\mathbb{C}$) and $\mathcal{T}(x) = Ax$, $x \in \mathbb{F}^n$.

(a) (One-to-one case) Show that the following are equivalent:

(i) $\mathcal{T}$ is one-to-one.

(ii) $Ax = 0$ has only the trivial solution $x = 0$.

(iii) The columns of A are linearly independent.

(iv) The rank of A is n.

(v) A has a left inverse, i.e., $LA = I_n$ for some matrix L.

(vi) $\dim \operatorname{Im}\mathcal{T} = n$.

(b) (Onto case) Show that the following are equivalent:

(i) $\mathcal{T}$ is onto.

(ii) $Ax = b$ has a solution(s) for every $b \in \mathbb{F}^m$.

(iii) The rows of A are linearly independent.

(iv) The rank of A is m.

(v) A has a right inverse, i.e., $AR = I_m$ for some matrix R.

(vi) $\dim \operatorname{Im}\mathcal{T} = m$.

3.121 Let $\mathcal{A}$ be a linear transformation on a vector space. Show that

$$\operatorname{Ker}\mathcal{A} \subseteq \operatorname{Im}(\mathcal{I} - \mathcal{A}) \quad \text{and} \quad \operatorname{Im}\mathcal{A} \supseteq \operatorname{Ker}(\mathcal{I} - \mathcal{A}).$$

3.122   For the vector space V of real-valued functions on $\mathbb{R}$, define

$$\mathcal{A}(f) = \tfrac{1}{2}\big(f(x) + f(-x)\big), \quad f \in V.$$

(a) Show that $\mathcal{A}$ is a linear map.
(b) Show that $\mathcal{A}^2 = \mathcal{A}$.
(c) Find $\operatorname{Im}\mathcal{A}$ and $\operatorname{Ker}\mathcal{A}$.
(d) Find a matrix representation of $\mathcal{A}$ when $V = \mathbb{P}_5[x]$.

3.123   Let $\mathcal{A}$ be a linear transformation on a finite dimensional vector space V and let V_1 and V_2 be subspaces of V. Answer true or false:

(a) $\mathcal{A}(V_1 \cap V_2) = \mathcal{A}(V_1) \cap \mathcal{A}(V_2)$.
(b) $\mathcal{A}(V_1 \cup V_2) = \mathcal{A}(V_1) \cup \mathcal{A}(V_2)$.
(c) $\mathcal{A}(V_1 + V_2) = \mathcal{A}(V_1) + \mathcal{A}(V_2)$.
(d) $\mathcal{A}(V_1 \oplus V_2) = \mathcal{A}(V_1) \oplus \mathcal{A}(V_2)$.

3.124   Let V and W be finite dimensional vector spaces and let $\mathcal{A}$ be an arbitrary linear transformation from V to W. Answer true or false:

(a) $\operatorname{Ker}\mathcal{A} = \{0\}$.
(b) If $V = W$ and $\operatorname{Im}\mathcal{A} \subseteq \operatorname{Ker}\mathcal{A}$, then $\mathcal{A} = 0$.
(c) If $V = W$ and $\operatorname{Im}\mathcal{A} \subseteq \operatorname{Ker}\mathcal{A}$, then $\mathcal{A}^2 = 0$.
(d) $\operatorname{Ker}\mathcal{A}^2 \supseteq \operatorname{Ker}\mathcal{A}$.
(e) $\operatorname{Ker}\mathcal{A}^2 \subseteq \operatorname{Ker}\mathcal{A}$.
(f) $\operatorname{Im}\mathcal{A}^2 \supseteq \operatorname{Im}\mathcal{A}$.
(g) $\operatorname{Im}\mathcal{A}^2 \subseteq \operatorname{Im}\mathcal{A}$.
(h) $\dim\operatorname{Ker}\mathcal{A} \le \dim\operatorname{Im}\mathcal{A}$.
(i) $\dim\operatorname{Ker}\mathcal{A} \le \dim V$.
(j) $\mathcal{A}$ is one-to-one $\Leftrightarrow \operatorname{Ker}\mathcal{A} = \{0\} \Leftrightarrow \dim\operatorname{Im} A = \dim V$.
(k) $\mathcal{A}$ is onto $\Leftrightarrow \operatorname{Im}\mathcal{A} = W \Leftrightarrow \dim\operatorname{Im} A = \dim W$.
(l) If $\dim V \le \dim W$, then one-to-one maps from V to W exist.
(m) If $\dim V \le \dim W$, then every map from V to W is one-to-one.
(n) If $\mathcal{B}$ is a linear map from W to V and $\mathcal{AB} = I_W$, then $\mathcal{BA} = \mathcal{I}_V$.
(o) If $\dim V = \dim W$ and if $\mathcal{B}$ is a map from W to V such that $\mathcal{AB} = I_W$, then $\mathcal{BA} = \mathcal{I}_V$ and B is linear (as $\mathcal{A}$ is).
(p) If $\dim V = \dim W$ and if $\mathcal{A}$ and $\mathcal{B}$ are maps, but not assumed to be linear, such that $\mathcal{AB} = I_W$, then $\mathcal{BA} = \mathcal{I}_V$.

3.125   Let $A, B \in M_n(\mathbb{C})$ be given and let $n > 1$. Which of the following maps on $M_n(\mathbb{C})$ are linear? Which ones are invertible? (That is, which ones are one-to-one and onto?)

    (a) $T_1(X) = X^t$.

    (b) $T_2(X) = X \pm X^t$.

    (c) $T_3(X) = \operatorname{tr}(X)I_n \pm X$.

    (d) $T_4(X) = \operatorname{tr}(X)X$.

    (e) $T_5(X) = \operatorname{tr}(X)I_n$.

    (f) $T_6(X) = \operatorname{diag}(X)$.

    (g) $\mathcal{L}_1(X) = AX$.

    (h) $\mathcal{L}_2(X) = AXA^*$.

    (i) $\mathcal{L}_3(X) = AX - X^tB$.

    (j) $\mathcal{L}_4(X) = AX + B$.

    (k) $\mathcal{L}_5(X) = AXB$.

    (l) $\mathcal{L}_6(X) = XAX$.

3.126   Let $A = \left(\begin{smallmatrix} 0 & 1 \\ 1 & 0 \end{smallmatrix}\right)$ and define a transformation T on $M_2(\mathbb{C})$ by

$$T(X) = AX^t - X, \quad X \in M_2(\mathbb{C}).$$

    (a)   If $X = \left(\begin{smallmatrix} a & b \\ c & d \end{smallmatrix}\right)$, find $T(X)$ explicitly.

    (b)   Show that T is a linear transformation.

    (c)   Find a matrix representation of T.

    (d)   Find $\operatorname{Im} T$ and $\operatorname{Ker} T$.

    (e)   Find all matrices X such that $T(X)$ is singular.

3.127   Let $A = \left(\begin{smallmatrix} 0 & 1 \\ 1 & 0 \end{smallmatrix}\right)$ and define a transformation T on $M_2(\mathbb{C})$ by

$$T(X) = AX - XA, \quad X \in M_2(\mathbb{C}).$$

    (a)   Show that T is a linear transformation.

    (b)   If $X = \left(\begin{smallmatrix} a & b \\ c & d \end{smallmatrix}\right)$, find $T(X)$ and $T^2(X)$ explicitly.

    (c)   Show that $AT(X) + T(X)A = 0$.

    (d)   Find $\operatorname{Im} T$ and $\operatorname{Im} T^2$.

    (e)   Find a matrix representation of T.

3.128   Let $A, B \in M_n(\mathbb{C})$. Define a transformation on $M_n(\mathbb{C})$ by

$$T(X) = AX - XA, \quad X \in M_n(\mathbb{C}).$$

Show that

(a) T is a linear transformation.

(b)   Zero is an eigenvalue of T.

(c)   If $A^k = 0$, then $T^{2k} = 0$.

(d) $T(XY) = XT(Y) + T(X)Y$.

(e) $n \leq \dim \operatorname{Ker} T$ and $\dim \operatorname{Im} T \leq n^2 - n$.

(f)   If A is diagonalizable, so is a matrix representation of T.

(g)   If A and B commute, so do T and $\mathcal{L}$, where $\mathcal{L}$ is defined as

$$\mathcal{L}(X) = BX - XB, \quad X \in M_n(\mathbb{C}).$$

Find all matrices A such that $T = 0$, and discuss the converse of (g).

3.129   Let W be a subspace of a finite dimensional vector space V. If $\mathcal{A}$ is a linear transformation from W to V, show that $\mathcal{A}$ can be extended to a linear transformation $\mathcal{L}$ on V, i.e., the restriction of $\mathcal{L}$ on W is $\mathcal{A}$.

3.130   Let $\mathcal{A}$ be a linear transformation from a vector space V to a vector space W and let $\dim V = n$. If $\{\alpha_1, \ldots, \alpha_s, \alpha_{s+1}, \ldots, \alpha_n\}$ is a basis for V such that $\{\alpha_1, \ldots, \alpha_s\}$ is a basis for $\operatorname{Ker} \mathcal{A}$, show that

(a) $\{\mathcal{A}(\alpha_{s+1}), \ldots, \mathcal{A}(\alpha_n)\}$ is a basis for $\operatorname{Im} \mathcal{A}$.

(b) $\dim \operatorname{Ker} \mathcal{A} + \dim \operatorname{Im} \mathcal{A} = n$.

(c) $V = \operatorname{Ker} \mathcal{A} \oplus \operatorname{Span}\{\alpha_{s+1}, \ldots, \alpha_n\}$.

Is $\operatorname{Ker} \mathcal{A} + \operatorname{Im} \mathcal{A}$ necessarily a direct sum when $V = W$? If $\{\beta_1, \ldots, \beta_n\}$ is a basis for V, does it follow that some β_i's fall in $\operatorname{Ker} \mathcal{A}$?

3.131   Let $\mathcal{A}$ be a linear transformation on a vector space V of dimension n.

(a)   If for some vector v, the vectors $v, \mathcal{A}(v), \mathcal{A}^2(v), \ldots, \mathcal{A}^{n-1}(v)$ are linearly independent, show that every eigenvalue of $\mathcal{A}$ has only one corresponding eigenvector up to a scalar multiplication.

(b)   If $\mathcal{A}$ has n distinct eigenvalues, show that there is a vector u such that $u, \mathcal{A}(u), \mathcal{A}^2(u), \ldots, \mathcal{A}^{n-1}(u)$ are linearly independent.

3.132   Let $\mathcal{A}$ be a linear map on a vector space V of dimension n. If

$$\mathcal{A}^{n-1}(x) \neq 0, \quad \text{but} \quad \mathcal{A}^n(x) = 0, \quad \text{for some} \quad x \in V,$$

show that

$$x, \mathcal{A}(x), \ldots, \mathcal{A}^{n-1}(x)$$

are linearly independent, and thus form a basis of V. What are the eigenvalues of $\mathcal{A}$? Find the matrix representation of $\mathcal{A}$ under the basis.

3.133   Let V be a vector space of dimension n with basis $\{\alpha_1, \alpha_2, \ldots, \alpha_n\}$. Let $\mathcal{A}$ be the linear map on V defined by

$$\mathcal{A}(\alpha_1) = \alpha_2, \ \mathcal{A}(\alpha_2) = \alpha_3, \ \ldots, \ \mathcal{A}(\alpha_{n-1}) = \alpha_n, \ \mathcal{A}(\alpha_n) = 0.$$

(a)   Show that $\mathcal{A}^{n-1} \neq 0$ and $\mathcal{A}^n = 0$.

(b)   If $\mathcal{B}$ is a linear map on V such that $\mathcal{B}^{n-1} \neq 0$ and $\mathcal{B}^n = 0$, show that $\mathcal{A}$ and $\mathcal{B}$ are similar, that is, their matrices with respect to some (possibly different) bases are similar.

(c)   Let W be an $\mathcal{A}$-invariant subspace of V, i.e., $\mathcal{A}(w) \in W$ for all $w \in W$. If $u \in W$ and $\mathcal{A}^{n-1}(u) \neq 0$, show that $W = V$.

(d)   Find the matrix of $\mathcal{A}$ with respect to the basis $\{\alpha_1, \alpha_2, \ldots, \alpha_n\}$.

3.134   Let V and W be finite dimensional vector spaces over a field $\mathbb{F}$ and let $\mathcal{A}$ be a linear transformation from V to W. Prove or disprove:

(a)   If the vectors $\alpha_1, \alpha_2, \ldots, \alpha_n$ in V are linearly independent, then $\mathcal{A}(\alpha_1), \mathcal{A}(\alpha_2), \ldots, \mathcal{A}(\alpha_n)$ in W are linearly independent.

(b)   If the vectors $\mathcal{A}(\alpha_1), \mathcal{A}(\alpha_2), \ldots, \mathcal{A}(\alpha_n)$ in W are linearly independent, then $\alpha_1, \alpha_2, \ldots, \alpha_n$ in V are linearly independent.

3.135   Let $\{\alpha_1, \alpha_2, \alpha_3\}$ be a basis for a vector space V of dimension three. Let $\mathcal{A}$ be the linear transformation on V such that

$$\mathcal{A}(\alpha_1) = \alpha_1, \quad \mathcal{A}(\alpha_2) = \alpha_1 + \alpha_2, \quad \mathcal{A}(\alpha_3) = \alpha_1 + \alpha_2 + \alpha_3.$$

(a)   Show that $\mathcal{A}$ is invertible.

(b)   Find $\mathcal{A}^{-1}$.

(c)   Find $2\mathcal{A} - \mathcal{A}^{-1}$.

3.136   Let $\mathcal{A}$ be the linear transformation defined on $\mathbb{R}^3$ (in rows) by

$$\mathcal{A}(x,\, y,\, z) = (0,\, x,\, y).$$

Find the characteristic and minimal polynomials of $\mathcal{A}$, $\mathcal{A}^2$, and $\mathcal{A}^3$, and find the images and kernels of $\mathcal{A}$, $\mathcal{A}^2$, and $\mathcal{A}^3$.

3.137   If $\mathcal{A}$ is a linear transformation on $\mathbb{R}^3$ (in columns) such that

$$\mathcal{A}\begin{pmatrix} 1 \\ 0 \\ 1 \end{pmatrix} = \begin{pmatrix} 2 \\ 3 \\ -1 \end{pmatrix}, \quad \mathcal{A}\begin{pmatrix} 1 \\ -1 \\ 1 \end{pmatrix} = \begin{pmatrix} 3 \\ 0 \\ -2 \end{pmatrix}, \quad \mathcal{A}\begin{pmatrix} -2 \\ 7 \\ -1 \end{pmatrix} = \begin{pmatrix} 2 \\ 3 \\ -1 \end{pmatrix},$$

find Im$\mathcal{A}$ (as span of basis vectors) and a matrix A so that $\mathcal{A}(x) = Ax$.

3.138   Let $\mathcal{A}$ and $\mathcal{B}$ be linear transformations on $\mathbb{R}^2$. It is given that the matrix representation of $\mathcal{A}$ under the basis $\{\alpha_1 = (1,2),\ \alpha_2 = (2,1)\}$ is $\begin{pmatrix} 1 & 2 \\ 2 & 3 \end{pmatrix}$, and that the matrix representation of $\mathcal{B}$ under the basis $\{\beta_1 = (1,1),\ \beta_2 = (1,2)\}$ is $\begin{pmatrix} 3 & 3 \\ 2 & 4 \end{pmatrix}$. Let $u = (3,3) \in \mathbb{R}^2$. Find

   (a)   The matrix representation of $\mathcal{A} + \mathcal{B}$ under $\{\beta_1, \beta_2\}$.

   (b)   The matrix representation of $\mathcal{A}\mathcal{B}$ under $\{\alpha_1, \alpha_2\}$.

   (c)   The coordinate of $\mathcal{A}(u)$ with respect to $\{\alpha_1, \alpha_2\}$.

   (d)   The coordinate of $\mathcal{B}(u)$ with respect to $\{\beta_1, \beta_2\}$.

3.139   Let V be a vector space of dimension 4 and let $\mathcal{A}$ be a linear map on V having matrix representation A under a basis $\{\epsilon_1, \epsilon_2, \epsilon_3, \epsilon_4\}$, where

$$A = \begin{pmatrix} 1 & 0 & 2 & 1 \\ -1 & 2 & 1 & 3 \\ 1 & 2 & 5 & 5 \\ 2 & -2 & 1 & -2 \end{pmatrix}.$$

   (a)   Find Ker $\mathcal{A}$.

   (b)   Find Im $\mathcal{A}$.

   (c)   Take a basis for Ker $\mathcal{A}$, extend it to a basis of V, and then find the matrix representation of $\mathcal{A}$ under this basis.

3.140   Let $\{\epsilon_1, \epsilon_2, \epsilon_3, \epsilon_4\}$ be a basis for a vector space V of dimension four. Define a linear transformation on V such that

$$\mathcal{A}(\epsilon_1) = \mathcal{A}(\epsilon_2) = \mathcal{A}(\epsilon_3) = \epsilon_1, \quad \mathcal{A}(\epsilon_4) = \epsilon_2.$$

Find Im $\mathcal{A}$, Ker $\mathcal{A}$, Ker $\mathcal{A}$ + Im $\mathcal{A}$, and Ker $\mathcal{A} \cap$ Im $\mathcal{A}$.

3.141 Let W be an invariant subspace of a linear transformation $\mathcal{A}$ on a finite dimensional vector space V, that is, $\mathcal{A}(w) \in W$ for all $w \in W$.

(a) If $\mathcal{A}$ is invertible, show that W is also invariant under $\mathcal{A}^{-1}$.

(b) If $V = W \oplus W'$, is W' necessarily invariant under $\mathcal{A}$?

3.142 Let $\mathcal{A}$ be a linear map on $\mathbb{R}^2$ with the matrix $A = \left(\begin{smallmatrix} 2 & 1 \\ 0 & 2 \end{smallmatrix}\right)$ under the basis $\{\alpha_1 = (1,0), \alpha_2 = (0,1)\}$. Let W_1 be the subspace of $\mathbb{R}^2$ spanned by α_1. Show that W_1 is invariant under $\mathcal{A}$ and that there does not exist a subspace W_2 invariant under $\mathcal{A}$ such that $\mathbb{R}^2 = W_1 \oplus W_2$.

3.143 Consider the vector space of all 2×2 real matrices. Let E_{ij} be the 2×2 matrix with (i,j)-entry 1 and other entries 0, i, $j = 1$, 2. Let

$$A = \begin{pmatrix} 1 & -1 \\ -1 & 1 \end{pmatrix}$$

and define

$$\mathcal{A}(X) = AX, \quad X \in M_2(\mathbb{R}).$$

(a) Show that $\mathcal{A}$ is a linear transformation on $M_2(\mathbb{R})$.

(b) Find the matrix of $\mathcal{A}$ under the basis $E_{11}, E_{12}, E_{21}, E_{22}$.

(c) Find $\operatorname{Im}\mathcal{A}$, its dimension, and a basis.

(d) Find $\operatorname{Ker}A$, its dimension, and a basis.

3.144 Let $A \in M_{m \times n}(\mathbb{C})$. Define a transformation $\mathcal{A} : M_{n \times p}(\mathbb{C}) \to M_{m \times p}(\mathbb{C})$ by

$$\mathcal{A}(X) = AX, \quad X \in M_{n \times p}(\mathbb{C}).$$

(a) Show that $\mathcal{A}$ is a linear transformation.

(b) When $p = 1$, find a matrix representation of $\mathcal{A}$.

(c) When $p = 2$, find a matrix representation of $\mathcal{A}$.

(d) When $m = n = p$, find a matrix representation of $\mathcal{A}$.

(e) For $m = n$, show that a scalar λ is an eigenvalue of the linear transformation $\mathcal{A}$ if and only if λ is an eigenvalue of the matrix A (possibly of different multiplicities).

(f) If $m = n$, how are the characteristic polynomials of $\mathcal{A}$ and A related? Find the characteristic polynomial of $\mathcal{A}$ in terms of A.

3.145   Let $\mathcal{A}$ and $\mathcal{B}$ be transformations on $\mathbb{R}^2$ (in rows) defined by

$$\mathcal{A}(x,\,y) = (y,\,x) \quad \text{and} \quad \mathcal{B}(x,\,y) = (x - y,\,x - y).$$

(a) Show that $\mathcal{A}$ and $\mathcal{B}$ are linear transformations.
(b) Find the nontrivial invariant subspace(s) of $\mathcal{A}$.
(c) Find $\operatorname{Ker} \mathcal{B}$ and $\operatorname{Im} \mathcal{B}$.
(d) Show that $\dim \operatorname{Ker} \mathcal{B} + \dim \operatorname{Im} \mathcal{B} = 2$, but $\operatorname{Ker} \mathcal{B} + \operatorname{Im} \mathcal{B}$ is not a direct sum. Is the sum $\operatorname{Ker} \mathcal{B} + \operatorname{Im} \mathcal{B}^*$ a direct sum, where $\mathcal{B}^*$ is defined by $\mathcal{B}^*(x, y) = (x + y, -x - y)$?

3.146   Define transformations $\mathcal{A}$ and $\mathcal{B}$ on $\mathbb{R}^n$ (in rows) by

$$\mathcal{A}(x_1, x_2, \ldots, x_n) = (0, x_1, x_2, \ldots, x_{n-1})$$

and

$$\mathcal{B}(x_1, x_2, \ldots, x_n) = (x_n, x_1, x_2, \ldots, x_{n-1}).$$

(a) Show that $\mathcal{A}$ and $\mathcal{B}$ are linear transformations.
(b) Find $\mathcal{AB}$, $\mathcal{BA}$, $\mathcal{A}^n$, and $\mathcal{B}^n$.
(c) Find matrix representations of $\mathcal{A}$ and $\mathcal{B}$.
(d) Find $\operatorname{Ker} \mathcal{A}$, $\operatorname{Ker} \mathcal{B}$, and their dimensions.

3.147   Let $\mathcal{A}$ be a linear transformation on a finite dimensional vector space V. If $\{\alpha_1, \ldots, \alpha_m\}$ is a basis for $\operatorname{Im} \mathcal{A}$ and if $\{\beta_1, \ldots, \beta_m\}$ is a set of vectors of V such that $\mathcal{A}(\beta_i) = \alpha_i$, $i = 1, \ldots, m$, show that

$$V = \operatorname{Span}\{\beta_1, \ldots, \beta_m\} \oplus \operatorname{Ker} \mathcal{A}.$$

3.148   Let $\mathcal{A}$ be a linear transformation on a finite dimensional vector space V. Show that the following statements are equivalent:

(a) $\dim(\operatorname{Im} \mathcal{A}^2) = \dim \operatorname{Im} \mathcal{A}$.
(b) $\operatorname{Im} \mathcal{A}^2 = \operatorname{Im} \mathcal{A}$.
(c) $V = \operatorname{Im} \mathcal{A} \oplus \operatorname{Ker} \mathcal{A}$.

Specifically, if $\mathcal{A}^2 = \mathcal{A}$, then $V = \operatorname{Im} \mathcal{A} \oplus \operatorname{Ker} \mathcal{A}$. Is the converse true?

3.149   A linear transformation $\mathcal{L}$ on a vector space V is said to be *idempotent* or a *projection* if $\mathcal{L}^2 = \mathcal{L}$. Let $\mathcal{A}$ and $\mathcal{B}$ be projections on a vector space V. Show that $\mathcal{A}$ and $\mathcal{B}$ commute with $(\mathcal{A} - \mathcal{B})^2$ and show that

$$(\mathcal{A} - \mathcal{B})^2 + (\mathcal{I} - \mathcal{A} - \mathcal{B})^2 = \mathcal{I}.$$

3.150   Let $\mathcal{A}$ be an idempotent linear map (i.e., $\mathcal{A}^2 = \mathcal{A}$) on a vector space V of dimension n. Let A be the matrix of $\mathcal{A}$ under some basis. Prove:

    (a) $\mathcal{I} - \mathcal{A}$ is idempotent.

    (b) $(\mathcal{I} - \mathcal{A})(\mathcal{I} - t\mathcal{A}) = \mathcal{I} - \mathcal{A}$ for any scalar t.

    (c) $(2\mathcal{A} - \mathcal{I})^2 = \mathcal{I}$.

    (d) $\mathcal{A} + \mathcal{I}$ is invertible. Find $(\mathcal{A} + \mathcal{I})^{-1}$.

    (e) $\operatorname{Ker}\mathcal{A} = \{\, x - \mathcal{A}(x) \mid x \in V \,\} = \operatorname{Im}(\mathcal{I} - \mathcal{A})$.

    (f) $V = \operatorname{Im}\mathcal{A} \oplus \operatorname{Ker}\mathcal{A}$, and $\operatorname{Im}\mathcal{A}$ and $\operatorname{Ker}\mathcal{A}$ are $\mathcal{A}$-invariant.

    (g) $\mathcal{A}(x) = x$ for every $x \in \operatorname{Im}\mathcal{A}$.

    (h) $\mathcal{A}$ has only eigenvalues 1's and 0's. Moreover, $\operatorname{Im}\mathcal{A}$ equals the eigenspace V_1 of 1 and $\operatorname{Ker}\mathcal{A}$ equals the eigenspace V_0 of 0.

    (i) If $V = M \oplus L$, where M and L are subspaces of V, then there exists a unique linear transformation $\mathcal{B}$ on V such that $\mathcal{B}^2 = \mathcal{B}$, $\operatorname{Im}\mathcal{B} = M$, and $\operatorname{Ker}\mathcal{B} = L$.

    (j) The matrix of $\mathcal{A}$ under some basis is in the form $\left(\begin{smallmatrix} I_r & 0 \\ 0 & 0 \end{smallmatrix}\right)$.

    (k) $r(A) + r(A - I) = n$.

    (l) $r(A) = \operatorname{tr} A = \dim \operatorname{Im}\mathcal{A}$.

    (m) $\det(A + I) = 2^{r(A)}$.

3.151   Let $A \in M_n(\mathbb{C})$ and let λ be an eigenvalue of A. The eigenspace of A associated to λ consists of all solutions to the equation system $Ax = \lambda x$; it is the null space (or kernel) of $\lambda I - A$; that is, it is the set of the zero vector and all eigenvectors of A associated to λ. Let

$$V_\lambda = \{x \in \mathbb{C}^n \mid Ax = \lambda x\}.$$

The dimension of V_λ is called the *geometric multiplicity* of λ, while the *algebraic multiplicity* of λ as an eigenvalue of A is the power of $t - \lambda$ in the factored characteristic polynomial $p(t) = \det(tI - A)$.

    (a) Let $A = \left(\begin{smallmatrix} 1 & 1 \\ 0 & 0 \end{smallmatrix}\right)$. Find the geometric multiplicity and algebraic multiplicity of 1 as an eigenvalue of A.

    (b) Let $A = \left(\begin{smallmatrix} 1 & 1 \\ 0 & 1 \end{smallmatrix}\right)$. Find the geometric multiplicity and algebraic multiplicity of 1 as an eigenvalue of A.

    (c) Show that the geometric multiplicity of λ is $n - r(\lambda I - A)$.

    (d) Show that an eigenvalue's algebraic multiplicity is at least as large as its geometric multiplicity.

    (e) Show that a matrix is diagonalizable if and only if algebraic and geometric multiplicities are equal for every eigenvalue.

3.152   Let $\mathcal{A}$ and $\mathcal{B}$ be linear transformations on a finite dimensional vector space V over the complex field $\mathbb{C}$ satisfying $\mathcal{AB} = \mathcal{BA}$. Show that

(a) If λ is an eigenvalue of $\mathcal{A}$, then the eigenspace

$$V_\lambda(\mathcal{A}) = \{\, x \in V \mid \mathcal{A}x = \lambda x \,\}$$

is invariant under $\mathcal{B}$.

(b) $\operatorname{Im}\mathcal{A}$ and $\operatorname{Ker}\mathcal{A}$ are invariant under $\mathcal{B}$.

(c) $\mathcal{A}$ and $\mathcal{B}$ have at least one common eigenvector (not necessarily belonging to the same eigenvalue).

(d) The matrix representations of $\mathcal{A}$ and $\mathcal{B}$ are both upper-triangular with respect to some basis.

If $\mathbb{C}$ is replaced with $\mathbb{R}$, which of the above remain(s) true?

3.153   Let $t \in \mathbb{R}$. Define a map $\mathcal{S}_t$ on $\mathcal{C}(\mathbb{R})$ by $\big(\mathcal{S}_t(f)\big)(x) = f(x+t)$ (known as the *shift operator*). Show that $\mathcal{S}_t$ is linear. Find $\operatorname{Ker}\mathcal{S}_t$ and $\operatorname{Im}\mathcal{S}_t$.

3.154   Let $\mathcal{D}$ be the *differential operator* on $\mathbb{P}_n[x]$ defined by $\mathcal{D}(p) = p'$, i.e.,

$$\mathcal{D}\big(p(x)\big) = a_1 + 2a_2x + \cdots + ia_ix^{i-1} + \cdots + (n-1)a_{n-1}x^{n-2},$$

where

$$p(x) = a_0 + a_1x + a_2x^2 + \cdots + a_{n-1}x^{n-1} \in \mathbb{P}_n[x].$$

(a) Show that $\mathcal{D}$ is a linear transformation on $\mathbb{P}_n[x]$.

(b) Find the eigenvalues and eigenvectors of $\mathcal{D}$ and $\mathcal{I} + \mathcal{D}$.

(c) Find the matrix representations of $\mathcal{D}$ under the bases $\{\, 1,\, x,\, x^2,\, \ldots,\, x^{n-1} \,\}$ and $\{\, 1, x, \frac{x^2}{2!}, \frac{x^3}{3!}, \ldots, \frac{x^{n-1}}{(n-1)!} \,\}$.

(d) Is any matrix representation of $\mathcal{D}$ diagonalizable?

3.155   For $\mathbb{P}_4[x]$, the real polynomials of degree less than 4, define a map

$$\mathcal{L}(p) = p' - p'', \quad p \in \mathbb{P}_4[x].$$

(a) Show that $\mathcal{L}$ is a linear transformation.

(b) Find a matrix representation of $\mathcal{L}$.

(c) Find the image $\operatorname{Im}\mathcal{L}$ and the kernel $\operatorname{Ker}\mathcal{L}$.

3.156   Let $\mathcal{C}_\infty(\mathbb{R})$ be the vector space of real-valued functions on $\mathbb{R}$ having derivatives of all orders (known as infinitely differentiable functions).

(a)   Consider the differential operator

$$\mathcal{D}_1(y) = y'' + ay' + by, \quad y \in \mathcal{C}_\infty(\mathbb{R}),$$

where a and b are real constants. Show that $y = e^{\lambda x}$ lies in $\operatorname{Ker}\mathcal{D}_1$ if and only if λ is a root of the quadratic equation

$$t^2 + at + b = 0.$$

(b)   Consider the second differential operator

$$\mathcal{D}_2(y) = y'', \quad y \in \mathcal{C}_\infty(\mathbb{R}).$$

Show that $y = ce^{\lambda x}$ is an eigenvector of $\mathcal{D}_2$ for any nonzero $c \in \mathbb{R}$ and that every positive number is an eigenvalue of $\mathcal{D}_2$.

3.157   For $\mathbb{P}[x]$, the space of all real polynomials (of finite degrees), define

$$\mathcal{A}\big(f(x)\big) = f'(x) \quad \text{and} \quad \mathcal{B}\big(f(x)\big) = xf(x), \quad f(x) \in \mathbb{P}[x].$$

Show that

(a) $\mathcal{A}$ and $\mathcal{B}$ are linear transformations.
(b) $\operatorname{Im}\mathcal{A} = \mathbb{P}[x]$ and $\operatorname{Ker}\mathcal{A} \neq \{0\}$.
(c) $\operatorname{Ker}\mathcal{B} = \{0\}$ and $\mathcal{B}$ does not have an inverse.
(d) $\mathcal{AB} - \mathcal{BA} = \mathcal{I}$.
(e) $\mathcal{A}^k\mathcal{B} - \mathcal{B}\mathcal{A}^k = k\mathcal{A}^{k-1}$ for every positive integer k.

3.158   Consider $\mathbb{P}_n[x]$, the real polynomials of degrees less than n. Define

$$\mathcal{A}\big(p(x)\big) = xp'(x) - p(x), \quad p(x) \in \mathbb{P}_n[x].$$

(a)   Show that $\mathcal{A}$ is a linear transformation.
(b)   Find $\operatorname{Ker}\mathcal{A}$ and $\operatorname{Im}\mathcal{A}$.
(c)   Show that $\mathbb{P}_n[x] = \operatorname{Ker}\mathcal{A} \oplus \operatorname{Im}\mathcal{A}$.

3.159   Consider $\mathbb{P}_n[x]$, the real polynomials of degrees less than n. Define

$$\mathcal{A}\big(p(x)\big) = x\big(p'(x) - p'(0)\big), \quad p(x) \in \mathbb{P}_n[x].$$

(a)   Show that $\mathcal{A}$ is a linear transformation.
(b)   Find the eigenvalues and eigenvectors of $\mathcal{A}$.
(c)   Show that a matrix of $\mathcal{A}$ is diagonalizable.

3.160 Let V be an n-dimensional vector space over $\mathbb{C}$ and let $\mathcal{A}$ be a linear map with the matrix representation A under a basis $\{u_1, u_2, \ldots, u_n\}$:

$$A = \begin{pmatrix} \lambda & 0 & 0 & \ldots & 0 & 0 \\ 1 & \lambda & 0 & \ldots & 0 & 0 \\ 0 & 1 & \lambda & \ldots & 0 & 0 \\ \vdots & \vdots & \vdots & \vdots & \vdots & \vdots \\ 0 & 0 & 0 & \ldots & \lambda & 0 \\ 0 & 0 & 0 & \ldots & 1 & \lambda \end{pmatrix},$$

that is,

$$\mathcal{A}(u_1, \ldots, u_n) = \big(\mathcal{A}(u_1), \ldots, \mathcal{A}(u_n)\big) = (u_1, \ldots, u_n)A.$$

Show that

(a) V is the only invariant subspace of $\mathcal{A}$ containing u_1.

(b) Every nonzero invariant subspace of $\mathcal{A}$ contains u_n.

(c) Every subspace

$$V_i = \text{Span}\{u_{n-i+1}, \ldots, u_n\}, \quad i = 1, 2, \ldots, n,$$

is invariant under $\mathcal{A}$, and $x \in V$ lies in V_i if and only if

$$(\mathcal{A} - \lambda \mathcal{I})^i(x) = 0.$$

(d) $V_1, V_2, \ldots, V_n$ are the only nonzero invariant subspaces of $\mathcal{A}$.

(e) $\text{Span}\{u_n\}$ is the only eigenspace of $\mathcal{A}$.

(f) V cannot be equal to a direct sum of two nontrivial invariant subspaces of $\mathcal{A}$.

3.161 Let V be a finite dimensional vector space and let $\mathcal{A}$ be a linear map on V. Show that there exists a positive integer k such that

$$V = \text{Im}\, \mathcal{A}^k \oplus \text{Ker}\, \mathcal{A}^k.$$

Chapter 4

Special Matrices

Definitions, Facts, and Examples _____

There are many special types of matrices. We begin with unitary matrices, focusing on Hermitian, positive semidefinite, and normal matrices. We present spectral decompositions and basic properties for these matrices.

Unitary Matrix. An $n \times n$ complex matrix $U = (u_{ij})$ is called a *unitary* matrix if $U^*U(= UU^*) = I_n$, i.e., $\sum_{k=1}^n \overline{u_{ki}}u_{kj} = 0$ if $i \neq j$ and $\sum_{k=1}^n |u_{kj}|^2 = 1$ for every j. So, U is unitary if and only if the columns (rows) of U are *orthonormal*, meaning that they are mutually orthogonal unit vectors.

If U is unitary, it is immediate that $U^* = U^{-1}$ and $|\det U| = 1$.

A real unitary matrix is referred to as a *real orthogonal* matrix. An identity matrix is obviously real orthogonal. Let $\theta, \theta_1, \ldots, \theta_n \in \mathbb{R}$. The first three matrices below are real orthogonal, the last one is unitary:

$$\begin{pmatrix} 0 & 1 \\ 1 & 0 \end{pmatrix}, \quad \begin{pmatrix} \sin\theta & \cos\theta \\ -\cos\theta & \sin\theta \end{pmatrix}, \quad \begin{pmatrix} \frac{1}{\sqrt{2}} & 0 & \frac{1}{\sqrt{2}} \\ 0 & 1 & 0 \\ -\frac{1}{\sqrt{2}} & 0 & \frac{1}{\sqrt{2}} \end{pmatrix}, \quad \begin{pmatrix} e^{i\theta_1} & & 0 \\ & \ddots & \\ 0 & & e^{i\theta_n} \end{pmatrix}.$$

Example 4.1 If $U = (u_{ij})$ is a unitary matrix partitioned as $U = \begin{pmatrix} A & B \\ C & D \end{pmatrix}$, where A and D are square matrices (possibly of different sizes), show that

$$|\det A| = |\det D|.$$

Derive that for any entry of U, say u_{ij}, $|u_{ij}| = |\det U_{ij}|$, where $\det U_{ij}$ is the minor of u_{ij}, concluding that $u_{ij} = 0$ if and only if U_{ij} is singular.

Solution. Let A have size $p \times p$ and let D have size $q \times q$, $p + q = n$. Since $UU^* = I_n$, we have $|\det U| = 1$ and $\begin{pmatrix} A & B \\ C & D \end{pmatrix} \begin{pmatrix} A^* & C^* \\ B^* & D^* \end{pmatrix} = \begin{pmatrix} I_p & 0 \\ 0 & I_q \end{pmatrix}$.

Thus, $AC^* + BD^* = 0$ and $CC^* + DD^* = I_q$. A computation reveals $\begin{pmatrix} A & B \\ C & D \end{pmatrix} \begin{pmatrix} I & C^* \\ 0 & D^* \end{pmatrix} = \begin{pmatrix} A & 0 \\ C & I_q \end{pmatrix}$. Taking determinants for both sides, we obtain

$$\det U \det(D^*) = \det A.$$

Since $|\det U| = 1$ and $\det(D^*) = \overline{\det D}$, we get $|\det A| = |\det D|$. The remaining parts are immediate through row and column permutations.   □

Example 4.2 (Cayley transform) Prove the following statements.

(1) If A is an $n \times n$ real orthogonal matrix whose eigenvalues are all different from -1, then $S = (I - A)(I + A)^{-1}$ is skew-symmetric.

(2) If S is an $n \times n$ real skew-symmetric matrix, then $A = (I - S)(I + S)^{-1}$ is a real orthogonal matrix with no eigenvalue equal to -1.

(3) The correspondence $A \leftrightarrow S$ in (1) and (2) is one-to-one.

(4) Find $S = (I - A)(I + A)^{-1}$ for $A = \begin{pmatrix} \cos\theta & -\sin\theta \\ \sin\theta & \cos\theta \end{pmatrix}$, $0 \leq \theta < 2\pi$, $\theta \neq \pi$.

Solution. (1) If A has no eigenvalue equal to -1, then $I + A$ is nonsingular. Thus, $S = (I - A)(I + A)^{-1}$ is well-defined. Moreover, it is easy to check that $S = (I - A)(I + A)^{-1} = (I + A)^{-1}(I - A)$. We show $S^t = -S$:

$$
\begin{aligned}
S^t &= \left((I - A)(I + A)^{-1}\right)^t \\
&= \left((I + A)^{-1}\right)^t (I - A)^t \\
&= (I + A^t)^{-1}(I - A^t) \\
&= (A^t A + A^t)^{-1}(A^t A - A^t) \\
&= (A + I)^{-1}(A^t)^{-1} A^t (A - I) \\
&= -(A + I)^{-1}(I - A) = -S.
\end{aligned}
$$

(2) If S is an $n \times n$ real skew-symmetric matrix, then the eigenvalues of S are pure imaginary (or zero) and $I + S$ is nonsingular. We show $A^t A = I$:

$$
\begin{aligned}
A^t A &= \left((I - S)(I + S)^{-1}\right)^t (I - S)(I + S)^{-1} \\
&= (I + S^t)^{-1}(I - S^t)(I - S)(I + S)^{-1} \\
&= (I - S)^{-1}(I + S)(I - S)(I + S)^{-1} \\
&= (I + S)(I - S)^{-1}(I - S)(I + S)^{-1} = I.
\end{aligned}
$$

To show that A has no eigenvalue equal to -1, we prove $\det(A + I) \neq 0$:

$$
\begin{aligned}
\det(A + I) &= \det\left((I - S)(I + S)^{-1} + I\right) \\
&= \det\left((I - S)(I + S)^{-1} + (I + S)(I + S)^{-1}\right) \\
&= \det(2I)\det\left((I + S)^{-1}\right) = 2^n \det\left((I + S)^{-1}\right) \neq 0.
\end{aligned}
$$

(3) It is a one-to-one correspondence due to the equivalence relations:

$$A = B \quad \Leftrightarrow \quad (I + B)(I - A) = (I - B)(I + A)$$
$$\Leftrightarrow \quad (I - A)(I + A)^{-1} = (I + B)^{-1}(I - B)$$
$$\Leftrightarrow \quad (I - A)(I + A)^{-1} = (I - B)(I + B)^{-1}.$$

(4) For $A = \begin{pmatrix} \cos\theta & -\sin\theta \\ \sin\theta & \cos\theta \end{pmatrix}$, $\theta \neq \pi$, -1 is not an eigenvalue. We compute

$$
\begin{aligned}
S &= \begin{pmatrix} 1 - \cos\theta & \sin\theta \\ -\sin\theta & 1 - \cos\theta \end{pmatrix} \begin{pmatrix} 1 + \cos\theta & -\sin\theta \\ \sin\theta & 1 + \cos\theta \end{pmatrix}^{-1} \\
&= \frac{1}{2 + 2\cos\theta} \begin{pmatrix} 1 - \cos\theta & \sin\theta \\ -\sin\theta & 1 - \cos\theta \end{pmatrix} \begin{pmatrix} 1 + \cos\theta & \sin\theta \\ -\sin\theta & 1 + \cos\theta \end{pmatrix} \\
&= \begin{pmatrix} 0 & \frac{\sin\theta}{1 + \cos\theta} \\ -\frac{\sin\theta}{1 + \cos\theta} & 0 \end{pmatrix} = \begin{pmatrix} 0 & \tan\frac{\theta}{2} \\ -\tan\frac{\theta}{2} & 0 \end{pmatrix}. \quad \square
\end{aligned}
$$

Hermitian Matrix. A square complex matrix $A = (a_{st})$ is *Hermitian* if $A^* = A$, i.e., $\bar{A}^t = A$, or $\overline{a_{ts}} = a_{st}$ for all s and t. It is immediate that the entries $a_{11}, \ldots, a_{nn}$ on the main diagonal of a Hermitian matrix are necessarily real. A matrix A is *skew-Hermitian* if and only if iA is Hermitian. A real Hermitian matrix is a real symmetric matrix.

Recall Schur's triangularization theorem: For every $n \times n$ matrix A, there exists an $n \times n$ unitary matrix U such that $A = U^*DU$, where D is upper-triangular. If A is Hermitian, then $D^* = D$. So, D is real diagonal.

The *spectral decomposition* for Hermitian matrices follows:

A Hermitian matrix A can be decomposed as $A = U^*DU$, where U is unitary, $D = \mathrm{diag}(\lambda_1, \ldots, \lambda_n)$, and $\lambda_1, \ldots, \lambda_n$ are the eigenvalues of A.

If A is real symmetric, U can be chosen to be real. So, every real symmetric matrix A can be written as $A = Q^tDQ$, where Q is real orthogonal and D is real diagonal with the eigenvalues of A on the main diagonal.

The eigenvalues of a Hermitian matrix are necessarily real as the main diagonal entries of D are the eigenvalues of A. This is also seen as follows. Let λ be an eigenvalue of Hermitian A and $Ax = \lambda x$ for some nonzero vector x. Taking conjugate transpose yields $x^*A^* = \bar{\lambda}x^*$. Thus,

$$\lambda x^*x = x^*(\lambda x) = x^*Ax = x^*A^*x = \bar{\lambda}x^*x.$$

Therefore, $\lambda = \bar{\lambda}$ and λ is real. Similarly, one proves that the eigenvalues of a skew-Hermitian matrix are pure imaginary (or zero).

Let $A \in M_n(\mathbb{C})$ be a Hermitian matrix. The eigenvalues $\lambda_i(A)$ of A (abbreviated to λ_i sometimes for simplicity) are usually decreasingly ordered:

$$\lambda_{\max}(A) = \lambda_1(A) \geq \lambda_2(A) \geq \cdots \geq \lambda_n(A) = \lambda_{\min}(A).$$

More can be said about the spectral decomposition of a Hermitian matrix. Let H be an $n \times n$ Hermitian matrix with eigenvalues $\lambda_1, \lambda_2, \ldots, \lambda_n$. Then there is a unitary matrix U such that $U^* H U = \mathrm{diag}(\lambda_1, \lambda_2, \ldots, \lambda_n)$ or $HU = U \,\mathrm{diag}(\lambda_1, \lambda_2, \ldots, \lambda_n)$. The column vectors $u_1, u_2, \ldots, u_n$ of U are orthonormal eigenvectors of H corresponding to $\lambda_1, \lambda_2, \ldots, \lambda_n$, respectively,

$$Hu_i = \lambda_i u_i, \quad u_i^* u_j = \delta_{ij}, \quad i, j = 1, 2, \ldots, n,$$

where $\delta_{ij} = 1$ if $i = j$ and 0 otherwise (the *Kronecker delta*).

Example 4.3 Let H be an $n \times n$ Hermitian matrix with eigenvalues $\lambda_1 \geq \lambda_2 \geq \cdots \geq \lambda_n$ and corresponding orthonormal eigenvectors $u_1, u_2, \ldots, u_n$. Let $W = \mathrm{Span}\{u_p, \ldots, u_q\}$, $1 \leq p \leq q \leq n$. Then for any unit vector $x \in W$,

$$\lambda_q \leq x^* H x \leq \lambda_p.$$

Solution. Let $x = x_p u_p + \cdots + x_q u_q$. Since $u_p, \ldots, u_q$ are orthonormal,

$$
\begin{aligned}
x^* H x &= x^*(x_q H u_p + \cdots + x_q H u_q) \\
&= x^*(\lambda_p x_p u_p + \cdots + \lambda_q x_q u_q) \\
&= \lambda_p x_p x^* u_p + \cdots + \lambda_q x_q x^* u_q \\
&= \lambda_p |x_p|^2 + \cdots + \lambda_q |x_q|^2.
\end{aligned}
$$

The inequality follows because $\sum_{i=p}^{q} |x_i|^2 = 1$ as x is a unit vector. □

The cases of λ_1 and λ_n in the example are well-known results. Note that $Hu_1 = \lambda_1 u_1$ and $Hu_n = \lambda_n u_n$. Therefore, $u_1^* H u_1 = \lambda_1$ and $u_n^* H u_n = \lambda_n$.

Rayleigh–Ritz Min-Max Theorem. Let H be an $n \times n$ Hermitian matrix. Let $\lambda_{\min}(H)$ and $\lambda_{\max}(H)$ denote the smallest eigenvalue and the largest eigenvalue of H, respectively. Then

$$\lambda_{\min}(H) = \min\{x^* H x \mid x \in \mathbb{C}^n, \|x\| = 1\},$$

$$\lambda_{\max}(H) = \max\{x^* H x \mid x \in \mathbb{C}^n, \|x\| = 1\}.$$

It is immediate that for Hermitian matrices H and K of the same size

$$\lambda_{\min}(H) + \lambda_{\min}(K) \leq \lambda_{\min}(H+K) \leq \lambda_{\max}(H+K) \leq \lambda_{\max}(H) + \lambda_{\max}(K).$$

If H is an $n \times n$ Hermitian matrix such that $x^* H x \geq 0$ for all $x \in \mathbb{C}^n$, then H is said to be *positive semidefinite*, denoted by $H \geq 0$. If $x^* H x > 0$ for all nonzero $x \in \mathbb{C}^n$, then H is *positive definite*, written as $H > 0$. Equivalently, $H \geq 0$ (or > 0) means $H = H^*$ and $\lambda_{\min}(H) \geq 0$ (resp. > 0). For Hermitian matrices A and B of the same size, if $B - A \geq 0$, we write $B \geq A$ or $A \leq B$, with strict inequalities for the positive definite case.

Example 4.4 Let A and B be $n \times n$ Hermitian matrices with eigenvalues $\lambda_1(A) \geq \lambda_2(A) \geq \cdots \geq \lambda_n(A)$ and $\lambda_1(B) \geq \lambda_2(B) \geq \cdots \geq \lambda_n(B)$, respectively. If $A \leq B$, i.e., $B - A$ is positive semidefinite, show that

$$\lambda_i(A) \leq \lambda_i(B), \quad i = 1, 2, \ldots, n. \tag{4.1}$$

Solution. Let $E = B - A \geq 0$. Let $x_1, \ldots, x_n$ and $y_1, \ldots, y_n$ be orthonormal eigenvectors of A and B corresponding to $\lambda_1(A), \ldots, \lambda_n(A)$ and $\lambda_1(B), \ldots, \lambda_n(B)$, respectively. For each fixed i, let $S_1 = \text{Span}\{x_1, \ldots, x_i\}$ and $S_2 = \{y_i, \ldots, y_n\}$. The dimension identity shows

$$\dim(S_1 \cap S_2) = \dim S_1 + \dim S_2 - \dim(S_1 + S_2) \geq i + (n - i + 1) - n \geq 1.$$

So, there is a nonzero vector $v \in S_1 \cap S_2$. Thus, by Example 4.3, we have

$$\lambda_i(A) \leq v^* A v = v^*(B - E)v = v^* B v - v^* E v \leq \lambda_i(B) - v^* E v \leq \lambda_i(B). \quad \square$$

Example 4.5 (Cauchy's Eigenvalue Interlacing Theorem) Let H be an $n \times n$ Hermitian matrix with eigenvalues $\lambda_1 \geq \cdots \geq \lambda_n$ and let A be an $m \times m$ principal submatrix of H with eigenvalues $\mu_1 \geq \cdots \geq \mu_m$. Then

$$\lambda_{k+n-m} \leq \mu_k \leq \lambda_k, \quad k = 1, 2, \ldots, m.$$

In particular, when $m = n - 1$, the eigenvalues of A interlace those of H:

$$\lambda_n \leq \mu_{n-1} \leq \lambda_{n-1} \leq \cdots \leq \lambda_2 \leq \mu_1 \leq \lambda_1.$$

Solution. It suffices to show the case of $m = n-1$ by considering a sequence of principal submatrices, two consecutive ones differing in size by one. Simultaneously permuting rows and columns of H, we can place A in the upper-left corner. Thus, without loss of generality, we assume that

$$H = \begin{pmatrix} A & \alpha \\ \alpha^* & a \end{pmatrix}, \quad \alpha \in \mathbb{C}^{n-1}, \ a \in \mathbb{R}.$$

Let $H = U \operatorname{diag}(\lambda_1, \ldots, \lambda_n) U^*$ and $A = V \operatorname{diag}(\mu_1, \ldots, \mu_{n-1}) V^*$, where U and V are unitary matrices of orders n and $n - 1$, respectively. Let

$u_1, \ldots, u_n$ be the (from left to right) columns of U and let $v_1, \ldots, v_{n-1}$ be the columns of V. Let $w_i \in \mathbb{C}^n$ be the vector obtained from v_i by adding a zero at the bottom, that is, $w_i = \binom{v_i}{0}$, $i = 1, 2, \ldots, n-1$. We claim that

$$\mu_k \leq \lambda_k, \quad k = 1, 2, \ldots, n-1.$$

Let

$$S_1 = \mathrm{Span}\{w_1, \ldots, w_k\}, \quad S_2 = \mathrm{Span}\{u_k, \ldots, u_n\}.$$

The dimension identity ensures $S_1 \cap S_2 \neq \{0\}$. Let $x = \binom{y}{0} \in S_1 \cap S_2$, where y is a unit vector in $\mathrm{Span}\{v_1, \ldots, v_k\}$. By Example 4.3, we obtain

$$\mu_k \leq y^* A y = x^* H x \leq \lambda_k.$$

Applying this to $-H$ reveals the inequalities $\mu_k \geq \lambda_{k+1}$ for each k. □

Inertia. The eigenvalues of a Hermitian matrix are all real. How many of them are positive, negative, and zero? And how can these numbers be determined? Let H be a Hermitian matrix. Since the multi-set of the eigenvalues of H is uniquely determined by the characteristic polynomial of H, the numbers of positive, negative, and zero eigenvalues (if any) depend solely on H. We denote them by $i_+(H), i_-(H)$, and $i_0(H)$, respectively, and call the triple $(i_+(H), i_-(H), i_0(H))$ the *inertia* of H.

By the spectral decomposition, if H is a Hermitian matrix, then there exists a unitary matrix U such that $H = U^* \mathrm{diag}(\lambda_1, \ldots, \lambda_n) U$.

Permuting simultaneously rows and columns, we can arrange the positive, nonnegative, and zero eigenvalues along the diagonal. Then we can write $H = U^* F^* \mathrm{diag}(1, \ldots, 1, -1, \ldots, -1, 0, \ldots, 0) F U$, where F is a nonsingular matrix. Thus, there exists an invertible matrix S such that

$$H = S^* \begin{pmatrix} I_p & 0 & 0 \\ 0 & -I_q & 0 \\ 0 & 0 & 0 \end{pmatrix} S, \tag{4.2}$$

where $p = i_+(H)$, $q = i_-(H)$, and $p + q = r(H)$. (The diagonal block $I_p, -I_q$, or 0 is absent if there is no positive, negative, or zero eigenvalue.)

Sylvester's law of inertia states that the inertia of a Hermitian matrix does not change under simultaneous row and column operations. To be precise, if H is an $n \times n$ Hermitian matrix, then for any $n \times n$ invertible matrix S, $S^* H S$ and H have the same *inertia*. It follows that two Hermitian matrices are *-congruent* if and only if they have the same inertia.

Example 4.6 If A and B are diagonal unitary matrices such that $XAX^* = B$ for some nonsingular matrix X, show that A and B are permutation-similar, that is, $PAP^t = B$ for some permutation matrix P.

Solution. Let $M \in M_n(\mathbb{C})$. The *Cartesian decomposition of M* is

$$M = \Re(M) + \Im(M)i, \quad \Re(M) = \tfrac{1}{2}(M + M^*), \quad \Im(M) = \tfrac{1}{2i}(M - M^*).$$

Then $\Re(M)$ and $\Im(M)$ are Hermitian, and for any invertible $S \in M_n(\mathbb{C})$,

$$\Re(SMS^*) = S\Re(M)S^*, \quad \Im(SMS^*) = S\Im(M)S^*.$$

We show that each diagonal element of A appears on the diagonal of B exactly the same number of times. Since $XAX^* = B$ if and only if $X(e^{i\theta}A)X^* = e^{i\theta}B$, $\theta \in \mathbb{R}$, it is sufficient to show that A and B have the same number of 1's on their diagonals. Note that for any $\theta \in \mathbb{R}$,

$$i_+\big(\Im(e^{i\theta}A)\big) = i_+\big(X\Im(e^{i\theta}A)X^*\big) = i_+\big(\Im(X(e^{i\theta}A)X^*)\big) = i_+\big(\Im(e^{i\theta}B)\big).$$

If A is real, then B is real (vice versa). In this case, it is obvious.

Suppose some diagonal entries of A are nonreal. Let 1 appear k times in A ($0 < k < n$), and k' times in B. Place all nonreal diagonal elements of A and B on the unit circle. Note that the inertia of the imaginary part $\Im(A)$ of A is determined accordingly by the numbers of the points above, below, and on the x-axis (i.e., the y-coordinates of the points).

Let γ be the point (if more than one, take any one) of all these points that has the smallest y-coordinate in absolute value. Let $\delta > 0$ be the angle between γ (through the origin) and the x-axis. Then for all $0 < \theta < \delta$,

$$i_+\big(\Im(e^{i\theta}A)\big) = i_+\big(\Im(A)\big) + k, \quad i_+\big(\Im(e^{i\theta}B)\big) = i_+\big(\Im(B)\big) + k'.$$

By $i_+\big(\Im(e^{i\theta}A)\big) = i_+\big(\Im(e^{i\theta}B)\big)$ and $i_+\big(\Im(A)\big) = i_+\big(\Im(B)\big)$, we get $k = k'$. $\square$

The above proof is simple intuitively. The inertias of the imaginary parts of A and B are determined respectively by the numbers of the points located in the (open) upper-half and lower-half planes, and on the x-axis. Rotate (counter clockwise) the unit circle a bit (equivalently, multiply the matrices by $e^{i\theta}$ for a small θ) so that the points not on the x-axis are left in their planes, but the 1's on the x-axis are "lifted" a bit. Such a rotation increases the positive inertia of $\Im(A)$ by the number of 1's. The same is true for B. This shows that A and B have the same number of 1's.

The statement in the above example can be rewritten as follows: If $\lambda_1, \ldots, \lambda_n, \mu_1, \ldots, \mu_n$ are $2n$ complex numbers located on the unit circle and if X is an invertible matrix such that $X \operatorname{diag}(\lambda_1, \ldots, \lambda_n)X^* = \operatorname{diag}(\mu_1, \ldots, \mu_n)$, then $\{\lambda_1, \ldots, \lambda_n\} = \{\mu_1, \ldots, \mu_n\}$ as multi-sets.

Positive Semidefinite Matrices. Positivity (positive definiteness and positive semidefiniteness) of a matrix is a central topic in linear algebra and matrix theory. We say $A \in M_n(\mathbb{C})$ is positive semidefinite and write

$$A \geq 0 \quad \text{if } x^*Ax \geq 0 \text{ for all } x \in \mathbb{C}^n.$$

We write $A > 0$ to mean that A is positive definite, i.e., $A \geq 0$ and A is nonsingular. Note that some texts use $A \geq 0$ for entrywise nonnegativity of A. Moreover, the hermicity of a square matrix is inherent in the definition of positive semidefiniteness, that is, a positive semidefinite matrix is necessarily Hermitian. This is seen as follows. Since x^*Ax is real, $x^*Ax = (x^*Ax)^* = x^*A^*x$, implying $x^*(A^* - A)x = 0$ for all $x \in \mathbb{C}^n$. Thus, all eigenvalues of $A^* - A$ are zero. On the other hand, $A^* - A$ is skew-Hermitian, so $A^* - A = 0$, and $A^* = A$ is Hermitian (Problem 4.3c).

Note that it is possible that for some real square matrix A, $x^tAx \geq 0$ for all real vectors x, but A is not positive semidefinite in our sense. Take $A = \begin{pmatrix} 0 & 1 \\ -1 & 0 \end{pmatrix}$. It is easy to verify that $x^tAx = 0$ for all $x \in \mathbb{R}^2$.

The eigenvalues of a positive semidefinite matrix are necessarily nonnegative. To see this, let λ be an eigenvalue of A and $Ax = \lambda x$ for some nonzero vector x. Then $x^*Ax = \lambda x^*x \geq 0$. Therefore, $\lambda \geq 0$.

As to *spectral decomposition*, a positive semidefinite matrix A can be decomposed as $A = U^*DU$, where U is unitary, $D = \text{diag}(\lambda_1, \lambda_2, \ldots, \lambda_n)$, and $\lambda_1, \lambda_2, \ldots, \lambda_n$ are the nonnegative eigenvalues of A.

Below are some basic properties of positive semidefinite matrices. (1)–(4) are straightforward; (5) requires some work (Example 4.7). They remain valid for positive definite matrices with "positive" for "nonnegative".

(1) All main diagonal entries are nonnegative.

(2) All eigenvalues are nonnegative.

(3) The determinant is nonnegative.

(4) The trace is nonnegative.

(5) Every positive semidefinite matrix has a unique positive semidefinite square root. In other words, if $A \geq 0$, then there exists a unique $B \geq 0$ such that $A = B^2$. Such a square root of A is denoted by $A^{\frac{1}{2}}$.

Example 4.7 Let $A \geq 0$. There exists a unique $B \geq 0$ such that $A = B^2$.

Solution. Let $\lambda_1, \ldots, \lambda_n$ be the eigenvalues of A, all $\lambda_i \geq 0$. By spectral decomposition, let $A = U^*\Lambda U$, where U is unitary and $\Lambda = \text{diag}(\lambda_1, \ldots, \lambda_n)$. Take $B = U^*\Gamma U$, where $\Gamma = \text{diag}(\lambda_1^{\frac{1}{2}}, \ldots, \lambda_n^{\frac{1}{2}})$. Then $B \geq 0$ and $B^2 = A$.

For uniqueness, suppose $C \geq 0$ and $C^2 = B^2 = A$. The eigenvalues of C are $\lambda_1^{\frac{1}{2}}, \ldots, \lambda_n^{\frac{1}{2}}$. Let $C = V\Gamma V^*$, where V is unitary. Put $T = (t_{ij}) = UV$. Then $C^2 = B^2$ yields $T\Lambda = \Lambda T$, implying $\lambda_j t_{ij} = \lambda_i t_{ij}$, so $\lambda_j^{\frac{1}{2}} t_{ij} = \lambda_i^{\frac{1}{2}} t_{ij}$ for all i, j. Thus $T\Gamma = \Gamma T$, revealing $U^*\Gamma U = V\Gamma V^*$, i.e., $B = C$.   □

Below are necessary and sufficient conditions for a Hermitian matrix A to be positive semidefinite. The positive definite case is similarly stated.

(6) All principal minors of A are nonnegative.

(7) All principal submatrices of A are positive semidefinite.

(8) $A = S^*S$ for some matrix S.

(9) T^*AT is positive semidefinite for any matrix $T \in M_n(\mathbb{C})$.

The condition (6) for positive definite matrices can be relaxed a bit. To be exact, a Hermitian matrix A is positive definite if and only if all leading principal minors (including $\det A$) are positive. We show the sufficiency by induction on the matrix size and the Cauchy eigenvalue interlacing theorem.

Denote by A_k the leading principal submatrix of A in the intersection of first k rows and columns. If $n = k = 1$, we have nothing to show. Suppose that A_k is positive definite, we show that A_{k+1} is positive definite.

Since A_k is a principal submatrix of A_{k+1}, the eigenvalues of A_k interlace those of A_{k+1}. The assumption asserts that the eigenvalues of A_k are all positive. Thus, A_{k+1} has at least k positive eigenvalues. On the other hand, $\det(A_{k+1}) > 0$ is the product of all eigenvalues of A_{k+1}. Hence, all eigenvalues of A_{k+1} are positive. By induction, $A_n = A$ is positive definite.

Example 4.8 (Hadamard determinantal inequality) Let A be an $n \times n$ positive semidefinite matrix with diagonal entries $a_{11}, \ldots, a_{nn}$. Show that

$$\det A \leq a_{11} \cdots a_{nn}. \tag{4.3}$$

Equality occurs if and only if some row (and column) is zero or A is diagonal. *Solution 1.* If $\det A = 0$, the inequality is trivial. Assume that A is nonsingular. Partition $A = \begin{pmatrix} a_{11} & \alpha^* \\ \alpha & A_{11} \end{pmatrix}$, where $\alpha \in \mathbb{C}^{n-1}$ is a column vector and A_{11} is a positive definite matrix of size $n-1$. We show that $\det A \leq a_{11} \det A_{11}$.

Note that

$$\begin{pmatrix} 1 & -\alpha^* A_{11}^{-1} \\ 0 & I_{n-1} \end{pmatrix} \begin{pmatrix} a_{11} & \alpha^* \\ \alpha & A_{11} \end{pmatrix} = \begin{pmatrix} a_{11} - \alpha^* A_{11}^{-1}\alpha & 0 \\ \alpha & A_{11} \end{pmatrix}.$$

Taking the determinants of both sides, we obtain

$$\det A = (a_{11} - \alpha^* A_{11}^{-1}\alpha) \det A_{11} \leq a_{11} \det A_{11}.$$

Repeat the discussion on $\det A_{11}$. Inductively, inequality (4.3) follows.

For the equality case, if some row (column) is zero or if A is diagonal, then equality obviously holds. Conversely, suppose equality occurs and no diagonal entry is zero. We show that A is diagonal by induction on n.

If $n = 2$, then $\det A = a_{11}a_{22} - |a_{12}|^2 = a_{11}a_{22}$ gives $a_{12} = 0$. Suppose that the claim is true for positive semidefinite matrices of size less than n. For the case of n, $\det A = a_{11} \det A_{11}$ reveals $a_{22} \cdots a_{nn} = \det A_{11}$. By induction hypothesis, A_{11} is diagonal (and nonzero). Then $a_{11} - \alpha^* A_{11}^{-1} \alpha = a_{11}$. Thus, $\alpha^* A_{11}^{-1} \alpha = 0$, which yields $\alpha = 0$. Therefore, A is diagonal.

Solution 2. Without loss of generality, we assume that no diagonal entry of A is zero. We first show that if $a_{11} = \cdots = a_{nn} = 1$ then $\det A \le 1$. Let $\lambda_1, \ldots, \lambda_n$ be the eigenvalues of A. The inequality of arithmetic and geometric means yields

$$\det A = \lambda_1 \cdots \lambda_n \le \left(\frac{\lambda_1 + \cdots + \lambda_n}{n} \right)^n = \left(\frac{\operatorname{tr} A}{n} \right)^n = 1^n = 1.$$

Equality occurs if and only if $\lambda_1 = \cdots = \lambda_n = 1$, that is, $A = I_n$.

For a general A, we write $A = DBD$, where $D = \operatorname{diag}(\sqrt{a_{11}}, \ldots, \sqrt{a_{nn}})$ and B is a positive semidefinite matrix with 1's on the main diagonal. Then

$$\det A = (\det D)^2 \det B = a_{11} \cdots a_{nn} \det B \le a_{11} \cdots a_{nn}.$$

Equality occurs if and only if all the eigenvalues of B are equal to 1, that is, B is the identity matrix. It follows that A is a diagonal matrix.

Solution 3. Assume that A is nonsingular. We use (or borrow Problem 5.64) the *Cholesky decomposition* $A = C^*C$, where $C = (c_{ij})$ is an upper-triangular matrix with all $c_{ii} > 0$. Then $a_{ii} = c_{ii}^2 + \sum_{k<i} |c_{ki}|^2$. Thus

$$\det A = |\det C|^2 = \prod_{i=1}^{n} c_{ii}^2 \le \prod_{i=1}^{n} \left(c_{ii}^2 + \sum_{k<i} |c_{ki}|^2 \right) = \prod_{i=1}^{n} a_{ii}. \tag{4.4}$$

If some $a_{ii} = 0$ or A is diagonal, then equality holds. For the converse, if $\det A = \prod_{i=1}^{n} a_{ii}$ and no a_{ii} is zero, then $\det A \ne 0$ and A is nonsingular. Equality holds in (4.4) if and only if all $c_{ki} = 0$, $k \ne i$. So A is diagonal. □

Note that the Hadamard inequality follows from the Fischer inequality concerning partitioned positive semidefinite matrices (Problem 4.63).

Polar decomposition. Let $A \in M_{m \times n}(\mathbb{C})$. Then there exists an $m \times n$ matrix U satisfying $U^*U = I_n$ if $m \ge n$ or $UU^* = I_m$ if $m < n$ such that

$$A = U(A^*A)^{\frac{1}{2}} = (AA^*)^{\frac{1}{2}} U. \tag{4.5}$$

To show this, consider the case $m = n$ first. Let $A = PDQ$ be a singular value decomposition of A, where P and Q are unitary, and $D = \operatorname{diag}(\sigma_1(A), \ldots, \sigma_n(A))$. Then $(A^*A)^{\frac{1}{2}} = Q^*DQ$, $(AA^*)^{\frac{1}{2}} = PDP^*$, and

$$A = PDQ = (PQ)(Q^*DQ) = (PDP^*)(PQ) = U(A^*A)^{\frac{1}{2}} = (AA^*)^{\frac{1}{2}}U,$$

where $U = PQ$ is $n \times n$ unitary as P and Q are $n \times n$ unitary.

If $m > n$, we add zero columns to make $(A, 0)$ an $m \times m$ matrix. By the preceding argument, $(A, 0) = V\big((A,0)^*(A,0)\big)^{\frac{1}{2}} = \big((A,0)(A,0)^*\big)^{\frac{1}{2}}V$ for some $m \times m$ unitary V. Observe that $\big((A,0)^*(A,0)\big)^{\frac{1}{2}} = \begin{pmatrix} (A^*A)^{\frac{1}{2}} & 0 \\ 0 & 0 \end{pmatrix}$ and that $\big((A,0)(A,0)^*\big)^{\frac{1}{2}} = (AA^*)^{\frac{1}{2}}$. Take U to be the first n columns of V. Then U is an $m \times n$ matrix whose columns are orthonormal, i.e., $U^*U = I_n$. A computation reveals $(A, 0) = \big(U(A^*A)^{\frac{1}{2}}, 0\big) = \big((AA^*)^{\frac{1}{2}}U, 0\big)$. Comparing the first n columns, we see that U satisfies (4.5) as desired.

If $m < n$, an application of the above argument to A^* ensures an $n \times m$ matrix W such that $W^*W = I_m$ and $A^* = W(AA^*)^{\frac{1}{2}} = (A^*A)^{\frac{1}{2}}W$. So, $A = (AA^*)^{\frac{1}{2}}W^* = W^*(A^*A)^{\frac{1}{2}}$. Let $U = W^*$. Then U satisfies (4.5).

Example 4.9 Let $A = (a_{ij}) \geq 0$, $B = (b_{ij}) \geq 0$, both $n \times n$. Show that

(1) The eigenvalues of AB are all nonnegative.

(2) $\lambda_{\min}(A)\lambda_{\min}(B) \leq \lambda_{\min}(AB) \leq \lambda_{\max}(AB) \leq \lambda_{\max}(A)\lambda_{\max}(B)$.

(3) P^*AP and P^*BP are both diagonal for some invertible matrix P.

(4) The Krokecker product $A \otimes B = (a_{ij}B)$ is positive semidefinite.

(5) The Hadamard product $A \circ B = (a_{ij}b_{ij})$ is positive semidefinite.

Solution. (1) Since A is positive semidefinite, there exists a matrix C such that $A = C^*C$. Because $AB = C^*CB$ has the same eigenvalues as CBC^* which is positive semidefinite, the eigenvalues of AB are all nonnegative.

(2) We show $\lambda_{\max}(AB) \leq \lambda_{\max}(A)\lambda_{\max}(B)$. Note that $A \leq \lambda_{\max}(A)I$.

$$
\begin{aligned}
\lambda_{\max}(AB) &= \lambda_{\max}(B^{\frac{1}{2}}AB^{\frac{1}{2}}) \\
&= \max_{\|x\|=1} x^*B^{\frac{1}{2}}AB^{\frac{1}{2}}x \\
&\leq \max_{\|x\|=1} x^*B^{\frac{1}{2}}\big(\lambda_{\max}(A)I\big)B^{\frac{1}{2}}x \\
&= \lambda_{\max}(A)\max_{\|x\|=1} x^*B^{\frac{1}{2}}B^{\frac{1}{2}}x \\
&= \lambda_{\max}(A)\max_{\|x\|=1} x^*Bx \\
&= \lambda_{\max}(A)\lambda_{\max}(B).
\end{aligned}
$$

The inequality for minimal eigenvalues is similarly proved.

(3) Since A and B are positive semidefinite matrices, so is $A + B$. Let r be the rank of $A + B$ and let S be a nonsingular matrix such that

$$S^*(A+B)S = \begin{pmatrix} I_r & 0 \\ 0 & 0 \end{pmatrix}.$$

(Note: If $r = 0$, there is nothing to show.) Conformally partition S^*BS as

$$S^*BS = \begin{pmatrix} B_{11} & B_{12} \\ B_{21} & B_{22} \end{pmatrix}.$$

(Note: If $r = n$, then B_{12}, B_{21}, and B_{22} are absent.) Since $S^*(A+B)S \geq S^*BS$, we obtain $B_{12} = 0, B_{21} = 0$, and $B_{22} = 0$. Because $B_{11} \geq 0$, there exists an $r \times r$ unitary matrix T such that $T^*B_{11}T$ is diagonal. Put

$$P = S\begin{pmatrix} T & 0 \\ 0 & I_{n-r} \end{pmatrix}.$$

Then P^*BP and $P^*AP = P^*(A+B)P - P^*BP$ are both diagonal.

(4) Let $A = U^*CU$ and $B = V^*DV$, where U, V are unitary, C, D are nonnegative diagonal. Recall the Kronecker product from Chapter 3, p. 89, $A \otimes B = (U^*CU) \otimes (V^*DV) = (U \otimes V)^*(C \otimes D)(U \otimes V) \geq 0$.

(5) Because $A \circ B$ is a principal submatrix of $A \otimes B$ (Problem 3.112).

Alternative proof. Let $A = U^* \operatorname{diag}(\lambda_1, \ldots, \lambda_n)U$ be a spectral decomposition of A, where λ_i's are the (nonnegative) eigenvalues of A and U is unitary. Let u_i be the i-th row of U. Then $A = \sum_{i=1}^n \lambda_i u_i^* u_i$. Denote by U_i the diagonal matrix with the components of u_i on the main diagonal of U_i (in the natural order). Observe that $(u_i^* u_i) \circ B = U_i^* B U_i$. We have

$$A \circ B = \left(\sum_{i=1}^n \lambda_i u_i^* u_i \right) \circ B = \sum_{i=1}^n \lambda_i (u_i^* u_i) \circ B = \sum_{i=1}^n \lambda_i U_i^* B U_i \geq 0. \quad \square$$

In fact, more can be said about (3) regarding the simultaneous *-congruent diagonalization of A and B if one (or both) of A or B is invertible. If, say, A is invertible, we write $A = C^*C$ for some nonsingular matrix C. Consider matrix $(C^{-1})^*BC^{-1}$. Since it is positive semidefinite, there exists a unitary matrix U such that $(C^{-1})^*BC^{-1} = UDU^*$, where D is nonnegative diagonal. Let $P = C^{-1}U$. Then $P^*AP = I$ and $P^*BP = D$.

From the proofs of (4) and (5), we see that if A and B are positive definite, then so are their Kronecker product and Hadamard product, i.e.,

$$A > 0, \, B > 0 \; \Rightarrow \; A \otimes B > 0, \, A \circ B > 0. \tag{4.6}$$

Example 4.10 Let $A \in M_n(\mathbb{C})$ and $A > 0$. Consider the matrix equation

$$AX + XA = Y. \quad (Lyapunov\ equation)$$

Show that

(1) For every given $Y \in M_n(\mathbb{C})$, there exists a unique solution X.

(2) If Y is Hermitian, then X is Hermitian.

(3) If Y is positive semidefinite, then X is positive semidefinite.

(4) If Y is positive definite, then X is positive definite.

Solution. (1) Existence. First consider the case where A is a diagonal matrix, say $A = \operatorname{diag}(\lambda_1, \ldots, \lambda_n)$ with all $\lambda_i > 0$. A simple computation gives $x_{ij} = \frac{1}{\lambda_i + \lambda_j} y_{ij}$, $1 \leq i, j \leq n$, and $X = (x_{ij})$ is a solution to the matrix equation. For a general $A > 0$, let $A = U^* D U$ be a spectral decomposition of A, where $D = \operatorname{diag}(\lambda_1, \ldots, \lambda_n)$ and U is unitary. Then $AX + XA = Y$ is equivalent to $D\tilde{X} + \tilde{X}D = \tilde{Y}$, where $\tilde{X} = UXU^*$ and $\tilde{Y} = UYU^*$. The above discussion ensures a solution $\tilde{X}$ to the latter equation, which in turn reveals a solution $X = U^* \tilde{X} U$ to the original equation.

Uniqueness. If Z is also a matrix such that $AZ + ZA = Y$, subtracting the equations gives $A(X - Z) + (X - Z)A = 0$, implying $DW + WD = 0$, where $W = (w_{ij}) = U(X - Z)U^*$. It follows that $(\lambda_i + \lambda_j)w_{ij} = 0$ for all i, j, resulting $w_{ij} = 0$ for all i, j. So $W = 0$ and $Z = X$.

(2) Let $Y^* = Y$. Then $(AX + XA)^* = Y^*$ gives $X^*A^* + A^*X^* = Y^*$, i.e., $AX^* + X^*A = Y$. Because of the uniqueness of the solution, $X^* = X$.

(3) Let $Y \geq 0$. We claim that the eigenvalues of X are nonnegative. Let λ be an (real) eigenvalue of X with a corresponding eigenvector v: $Xv = \lambda v$, $v \neq 0$. Then $v^*X = \lambda v^*$, and $v^*(AX)v + v^*(XA)v = v^*Yv$ gives $2\lambda(v^*Av) = v^*Yv$. Since $A > 0$ and $Y \geq 0$, we see $\lambda \geq 0$, so $X \geq 0$.

(4) If $Y > 0$, the proof of (3) immediately gives $\lambda > 0$, i.e., $X > 0$. $\square$

One may use Sylvester's matrix equation theorem (Example 3.9) to show the existence and uniqueness of the solution of the Lyapunov equation.

Example 4.11 Let $M = \begin{pmatrix} A & B \\ B^* & C \end{pmatrix} \geq 0$, where $A, B, C \in M_n(\mathbb{C})$. Show that

(1) $C - B^*A^{-1}B \geq 0$ if A is nonsingular.

(2) $B = A^{\frac{1}{2}}X$ for some matrix X.

(3) $\lambda_{\max}(M) \leq \lambda_{\max}(A) + \lambda_{\max}(C)$.

(4) $\begin{pmatrix} \operatorname{tr} A & \operatorname{tr} B \\ \operatorname{tr} B^* & \operatorname{tr} C \end{pmatrix} \geq 0.$

(5) $\begin{pmatrix} \det A & \det B \\ \det B^* & \det C \end{pmatrix} \geq 0.$

(6) $\begin{pmatrix} \|A\| & \|B\| \\ \|B^*\| & \|C\| \end{pmatrix} \geq 0$, where $\| \cdot \| = \sigma_{\max}(\cdot)$ (the largest singular value).

Solution. (1) $\begin{pmatrix} I & 0 \\ -B^*A^{-1} & I \end{pmatrix} \begin{pmatrix} A & B \\ B^* & C \end{pmatrix} \begin{pmatrix} I & -A^{-1}B \\ 0 & I \end{pmatrix} = \begin{pmatrix} A & 0 \\ 0 & C - B^*A^{-1}B \end{pmatrix} \geq 0.$

(2) Since $M \geq 0$, we can write $M = P^*P$ for some P. Let $P = (S,T)$, where S and T are $2n \times n$ matrices. Then

$$M = \begin{pmatrix} S^* \\ T^* \end{pmatrix} (S,T) = \begin{pmatrix} S^*S & S^*T \\ T^*S & T^*T \end{pmatrix}.$$

Thus, $A = S^*S$ and $B = S^*T$. By polar decomposition of S^*, there exists an $n \times 2n$ (partial unitary) matrix U such that $S^* = (S^*S)^{\frac{1}{2}}U = A^{\frac{1}{2}}U$. Therefore, $B = S^*T = A^{\frac{1}{2}}UT = A^{\frac{1}{2}}X$, where $X = UT$.

(3) Following the proof of (2) and using the fact $\lambda_{\max}(XY) = \lambda_{\max}(YX)$,

$$\begin{aligned}
\lambda_{\max}(M) &= \lambda_{\max}\left(\begin{pmatrix} S^* \\ T^* \end{pmatrix}(S,T)\right) \\
&= \lambda_{\max}\left((S,T)\begin{pmatrix} S^* \\ T^* \end{pmatrix}\right) \\
&= \lambda_{\max}(SS^* + TT^*) \\
&\leq \lambda_{\max}(SS^*) + \lambda_{\max}(TT^*) \\
&= \lambda_{\max}(S^*S) + \lambda_{\max}(T^*T) \\
&= \lambda_{\max}(A) + \lambda_{\max}(C).
\end{aligned}$$

(4) $M = \begin{pmatrix} A & B \\ B^* & C \end{pmatrix} \geq 0$ implies the 2×2 principal submatrices $\begin{pmatrix} a_{ii} & b_{ii} \\ \overline{b_{ii}} & c_{ii} \end{pmatrix} \geq 0$, $i = 1, \ldots, n$. Taking the sum of these 2×2 matrices reveals

$$\sum_{i=1}^{n} \begin{pmatrix} a_{ii} & b_{ii} \\ \overline{b_{ii}} & c_{ii} \end{pmatrix} = \begin{pmatrix} \sum_i a_{ii} & \sum_i b_{ii} \\ \sum_i \overline{b_{ii}} & \sum_i c_{ii} \end{pmatrix} = \begin{pmatrix} \operatorname{tr} A & \operatorname{tr} B \\ \operatorname{tr} B^* & \operatorname{tr} C \end{pmatrix}.$$

(5) For $\begin{pmatrix} \det A & \det B \\ \det B^* & \det C \end{pmatrix} \geq 0$, it suffices to show that $\det A \det C \geq \det(B^*B)$. This is the same as $\det C \geq \det(B^*A^{-1}B)$ when A is nonsingular, which is true because $M \geq 0$ implies $C \geq B^*A^{-1}B$. So for nonsingular A, $\det A \det C \geq \det(B^*B)$. If A is singular, then $A + \epsilon I > 0$ for $\epsilon > 0$. Use $A + \epsilon I$ in place of A in the inequality, then let $\epsilon \to 0^+$.

(6) To show $\begin{pmatrix} \|A\| & \|B\| \\ \|B^*\| & \|C\| \end{pmatrix} \geq 0$, let $B = USV$ be a singular value decomposition of B, where U and V are unitary, and S has $\|B\|$ in the $(1,1)$ position. We have $\begin{pmatrix} U^* & 0 \\ 0 & V \end{pmatrix} \begin{pmatrix} A & B \\ B^* & C \end{pmatrix} \begin{pmatrix} U & 0 \\ 0 & V^* \end{pmatrix} = \begin{pmatrix} U^*AU & S \\ S^t & VCV^* \end{pmatrix} \geq 0.$ By

extracting the (1,1)-entry in each block, we get $\begin{pmatrix} a & \|B\| \\ \|B^*\| & c \end{pmatrix} \geq 0$, where a is the (1,1)-entry of U^*AU and c is the (1,1)-entry of VCV^*. As $a \leq \|U^*AU\| = \|A\|$ and $c \leq \|VCV^*\| = \|C\|$, we are done. □

The statements (1), (2), (3), and (6) hold for square matrices A and C of different sizes, while B is required to be square in (4) and (5). The matrix $C - B^*A^{-1}B$ in (1) is known as the *Schur complement* of the block matrix M with respect to A. The proof of (1) shows that if $A > 0$, then

$$\begin{pmatrix} A & B \\ B^* & C \end{pmatrix} \geq 0 \quad \Leftrightarrow \quad C - B^*A^{-1}B \geq 0. \tag{4.7}$$

Moreover, $\| \cdot \| = \sigma_{\max}(\cdot)$ is also known as the spectral norm.

Example 4.12 Let A and B be $n \times n$ positive definite matrices and let X and Y be any $m \times n$ complex matrices. Denote by $M[\alpha]$ the principal submatrix of matrix M with rows and columns indexed by α. Show that

(1) $(A[\alpha])^{-1} \leq A^{-1}[\alpha]$.

(2) $X[\alpha](A[\alpha])^{-1}X^*[\alpha] \leq (XA^{-1}X^*)[\alpha]$.

(3) $(X \circ Y)(A \circ B)^{-1}(X \circ Y)^* \leq (XA^{-1}X^*) \circ (YB^{-1}Y^*)$.

Solution. (1) By Schur complement, $\begin{pmatrix} A & I_n \\ I_n & A^{-1} \end{pmatrix} \geq 0$. Extracting the principal submatrix indexed by α from each block, we get $\begin{pmatrix} A[\alpha] & I_n[\alpha] \\ I_n[\alpha] & A^{-1}[\alpha] \end{pmatrix} \geq 0$. Thus, $A^{-1}[\alpha] \geq (A[\alpha])^{-1}$. (2) is similarly proven with $\begin{pmatrix} A & X^* \\ X & XA^{-1}X^* \end{pmatrix} \geq 0$.
(3) is derived by applying (2) and (1) as follows: (α is an index set)

$$\begin{aligned}
(XA^{-1}X^*) \circ (YB^{-1}Y^*) &= \big((XA^{-1}X^*) \otimes (YB^{-1}Y^*)\big)[\alpha] \\
&= \big((X \otimes Y)(A \otimes B)^{-1}(X \otimes Y)^*\big)[\alpha] \\
&\geq (X \otimes Y)[\alpha]\,(A \otimes B)^{-1}[\alpha]\,(X \otimes Y)^*[\alpha] \\
&\geq (X \otimes Y)[\alpha]\,\big((A \otimes B)[\alpha]\big)^{-1}(X \otimes Y)^*[\alpha] \\
&= (X \circ Y)(A \circ B)^{-1}(X \circ Y)^*. \quad \square
\end{aligned}$$

A similar inequality (Problem 6.130) is obtained by Schur complement:

$$(X + Y)(A + B)^{-1}(X + Y)^* \leq XA^{-1}X^* + YB^{-1}Y^*.$$

In the above example, (1) is a special case of (2) with $X = I$, and (3) yields some interesting inequalities by setting $X = Y = I$ or $A = B = I$.

Let f be a linear map from $M_n(\mathbb{C})$ to $M_m(\mathbb{C})$, where $n, m \geq 1$. We say that f is *positive* (and) *unital* if $f(A) \geq 0$ whenever $A \geq 0$ and $f(I_n) = I_m$. For example, a map that sends a square matrix to its principal submatrix (in a fixed position) is positive unital; $f(A) = \frac{1}{n} \operatorname{tr} A$ is positive unital (from $M_n(\mathbb{C})$ to $\mathbb{C}$); and $f(A) = A \circ C$ is positive unital, where C is a fixed *correlation* matrix (i.e., $C \geq 0$ and all main diagonal entries of C are equal to 1). In particular, setting $C = I_n$, $f(A) = \operatorname{diag}(A)$ is positive unital.

Example 4.13 Let A be an $n \times n$ positive semidefinite matrix with the largest eigenvalue λ_1 and the smallest eigenvalue λ_n. Let f be a positive unital linear map from $M_n(\mathbb{C})$ to $M_m(\mathbb{C})$. Show that

(1) If $A > 0$ (i.e., A is positive definite), then $f(A) > 0$.

(2) f is monotone, that is, if $X \leq Y$ then $f(X) \leq f(Y)$.

(3) $f(A^2) \leq \big(f(A)\big)^2 + \frac{1}{4}(\lambda_1 - \lambda_n)^2 I_m$.

(4) $f(A^{-1}) \leq \frac{(\lambda_1 + \lambda_n)^2}{4\lambda_1 \lambda_n} \big(f(A)\big)^{-1}$ if A is nonsingular.

Solution. (1) Let $A = P + \epsilon I_n$, where $P \geq 0$ and $\epsilon > 0$. Then

$$f(A) = f(P + \epsilon I_n) = f(P) + \epsilon f(I_n) = f(P) + \epsilon I_m \geq \epsilon I_m > 0.$$

(2) $Y - X \geq 0 \Rightarrow f(Y - X) = f(Y) - f(X) \geq 0 \Rightarrow f(Y) \geq f(X)$. In fact, by (1), we see that f is strictly monotone, namely, $Y > X \Rightarrow f(Y) > f(X)$.

(3) $(\lambda_1 I_n - A)(A - \lambda_n I_n) \geq 0 \Rightarrow \lambda_1 A + \lambda_n A - \lambda_1 \lambda_n I_n - A^2 \geq 0$. Thus,

$$\lambda_1 f(A) + \lambda_n f(A) - \lambda_1 \lambda_n I_m \geq f(A^2).$$

By spectral decomposition, let $f(A) = U^* D U$, where U is unitary and D is diagonal. Since $4(\lambda_1 - x)(x - \lambda_n) \leq (\lambda_1 - \lambda_n)^2$ for any real x, we have

$$\begin{aligned}
f(A^2) - \big(f(A)\big)^2 &\leq \lambda_1 f(A) + \lambda_n f(A) - \lambda_1 \lambda_n I_m - \big(f(A)\big)^2 \\
&= \big(\lambda_1 I_m - f(A)\big)\big(f(A) - \lambda_n I_m\big) \\
&= U^*\big(\lambda_1 I_m - D\big)\big(D - \lambda_n I_m\big) U \\
&\leq \tfrac{1}{4}(\lambda_1 - \lambda_n)^2 I_m.
\end{aligned}$$

(4) $(\lambda_1 A^{-1} - I_n)(A - \lambda_n I_n) \geq 0 \Rightarrow (\lambda_1 + \lambda_n) I_n - \lambda_1 \lambda_n A^{-1} - A \geq 0$. Thus,

$$(\lambda_1 + \lambda_n) I_m - f(A) \geq \lambda_1 \lambda_n f(A^{-1}). \tag{4.8}$$

Now notice that for all real λ and $x \neq 0$, $(\lambda - x)^2 \geq 0 \Rightarrow \lambda - x \leq \frac{1}{4}\lambda^2 x^{-1}$. Replacing λ by $\lambda_1 + \lambda_n$, we get $(\lambda_1 + \lambda_n) - x \leq \frac{1}{4}(\lambda_1 + \lambda_n)^2 x^{-1}$, which yields $(\lambda_1 + \lambda_n)I_m - f(A) \leq \frac{1}{4}(\lambda_1 + \lambda_n)^2 (f(A))^{-1}$. It follows, by (4.8), that

$$f(A^{-1}) \leq \frac{1}{\lambda_1 \lambda_n}\big((\lambda_1 + \lambda_n)I_m - f(A)\big) \leq \frac{(\lambda_1 + \lambda_n)^2}{4\lambda_1 \lambda_n}(f(A))^{-1}. \quad \square$$

A variety of inequalities may be derived from the example. For instance, setting $f(X) = X[\alpha]$ in (4), we get a reverse inequality of Example 4.12(1):

$$A^{-1}[\alpha] \leq \frac{(\lambda_1 + \lambda_n)^2}{4\lambda_1 \lambda_n}(A[\alpha])^{-1}. \tag{4.9}$$

Diagonal elements, eigenvalues, and singular values of a matrix carry much information about the matrix. How are they related? Let A be a $p \times q$ submatrix of a matrix M. Say, A is located in the upper-left corner of M. Write $M = \begin{pmatrix} A & B \\ C & D \end{pmatrix}$. Then $M^*M = \begin{pmatrix} A^*A + C^*C & \star \\ \star & \star \end{pmatrix}$. Example 4.4 reveals

$$\lambda_i(A^*A) \leq \lambda_i(A^*A + C^*C) \leq \lambda_i(M^*M).$$

It is immediate that the singular values of A are dominated by the corresponding singular values of M, i.e., for each $i = 1, \ldots, r(A) \leq \min\{p, q\}$,

$$\sigma_i(A) \leq \sigma_i(M). \tag{4.10}$$

Example 4.14 Let A be an $n \times n$ complex matrix with main diagonal entries d_i's, eigenvalues λ_i's, and singular values σ_i's ordered respectively as

$$|d_1| \geq \cdots \geq |d_n|, \quad |\lambda_1| \geq \cdots \geq |\lambda_n|, \quad \sigma_1 \geq \cdots \geq \sigma_n.$$

Show that for each $k = 1, \ldots, n$,

$$|d_1| + \cdots + |d_k| \leq \sigma_1 + \cdots + \sigma_k \tag{4.11}$$

and

$$|\lambda_1| + \cdots + |\lambda_k| \leq \sigma_1 + \cdots + \sigma_k. \tag{4.12}$$

Solution. For (4.11), through permutations (which do not change singular values) we can place $d_1, \ldots, d_k, \ldots, d_n$ along the main diagonal of A.

Let A_k be the $k \times k$ leading principal submatrix of A. Let $A_k = UDV$ be a singular value decomposition of A_k, where $D = \mathrm{diag}(\sigma_1(A_k), \ldots, \sigma_k(A_k))$, and $U = (u_{ij})$, $V = (v_{ij})$ are $k \times k$ unitary matrices. We compute to get

$$
\begin{aligned}
|d_1| + \cdots + |d_k| &= \sum_{i=1}^{k} \left| \sum_{j=1}^{k} u_{ij} \sigma_j(A_k) v_{ji} \right| \\
&\leq \sum_{j=1}^{k} \sum_{i=1}^{k} \left| u_{ij} v_{ji} \right| \sigma_j(A_k) \\
&\leq \sum_{j=1}^{k} \sigma_j(A_k) \\
&\leq \sum_{j=1}^{k} \sigma_j(A). \text{ (By (4.10))}
\end{aligned}
$$

The second-to-last inequality is due to the Cauchy–Schwartz inequality:

$$
\sum_{i=1}^{k} |u_{ij} v_{ji}| \leq \sqrt{\sum_{i=1}^{k} |u_{ij}|^2 \sum_{i=1}^{k} |v_{ji}|^2} = 1.
$$

For the eigenvalue inequalities (4.12), by Schur triangularization, let $W^*AW = T$, where W is unitary and T is upper-triangular with $\lambda_1, \ldots, \lambda_n$ on the main diagonal. Note that A and T have the same singular values. An application of inequalities (4.11) to T gives inequalities (4.12). $\square$

As a side product of (4.11), $|\operatorname{tr} A| \leq \sigma_1 + \cdots + \sigma_n$. One may show by examples that $|d_1| + \cdots + |d_k|$ and $|\lambda_1| + \cdots + |\lambda_k|$ need not be comparable.

Normal Matrices. An $n \times n$ complex matrix A is called a *normal* matrix if $A^*A = AA^*$, that is, A and A^* commute. Unitary, Hermitian, skew-Hermitian, and positive semidefinite matrices are normal matrices.

Spectral Decomposition. Let A be an $n \times n$ matrix with eigenvalues $\lambda_1, \lambda_2, \ldots, \lambda_n$ over $\mathbb{C}$. Then A is normal (Hermitian, positive semidefinite, or positive definite) if and only if A is unitarily diagonalizable, that is, $U^*AU = \operatorname{diag}(\lambda_1, \lambda_2, \ldots, \lambda_n)$ for some unitary matrix U, where $\lambda_1, \lambda_2, \ldots, \lambda_n$ are complex (real, nonnegative, or positive, respectively) numbers. Moreover, the columns u_i's of U are corresponding eigenvectors of λ_i's that are orthonormal and comprise a basis for $\mathbb{C}^n$ (or $\mathbb{R}^n$ for the real case). We have

$$
A = \sum_{i=1}^{n} \lambda_i u_i u_i^*. \tag{4.13}
$$

Example 4.15 Let $A = (a_{ij}) \in M_n(\mathbb{C})$. Prove the *Schur inequality*

$$\sum_{i=1}^{n} |\lambda_i(A)|^2 \leq \sum_{i,j=1}^{n} |a_{ij}|^2 = \operatorname{tr}(A^*A)$$

with equality if and only if A is normal. Use this result to show that if $A, B \in M_n(\mathbb{C})$ are normal and if AB is normal, then BA is normal.

Solution. (The first part, i.e., Schur's inequality, is Problem 3.59. Below is a recap of its solution.) Let $A = U^*TU$ be a Schur decomposition of A, where U is unitary, $T = (t_{ij})$ is upper-triangular, and $t_{ii} = \lambda_i(A)$, $i = 1, 2, \ldots, n$. Then $A^*A = U^*T^*TU$, and $\operatorname{tr}(A^*A) = \operatorname{tr}(T^*T)$. We compute

$$\operatorname{tr}(A^*A) = \sum_{i,j=1}^{n} |a_{ij}|^2, \quad \operatorname{tr}(T^*T) = \sum_{i=1}^{n} |\lambda_i(A)|^2 + \sum_{i<j} |t_{ij}|^2.$$

The inequality is immediate.

For the equality case, notice that each $t_{ij} = 0$, $i < j$, that is, T is diagonal. Hence, A is unitarily diagonalizable, thus normal. Conversely, if A is normal, then T is diagonal, and $\operatorname{tr}(A^*A) = \operatorname{tr}(T^*T) = \sum_{i=1}^{n} |\lambda_i(A)|^2$.

To show the normality of BA, let $C = BA$. It is sufficient to prove

$$\operatorname{tr}(C^*C) = \sum_{i=1}^{n} |\lambda_i(C)|^2.$$

We compute

$$
\begin{aligned}
\operatorname{tr}(C^*C) &= \operatorname{tr}\big((BA)^*(BA)\big) = \operatorname{tr}(A^*B^*BA) \\
&= \operatorname{tr}(B^*BAA^*) \quad (\text{by } \operatorname{tr}(XY) = \operatorname{tr}(YX)) \\
&= \operatorname{tr}(BB^*A^*A) \quad (A \text{ and } B \text{ are normal}) \\
&= \operatorname{tr}(B^*A^*AB) \quad (\text{by } \operatorname{tr}(XY) = \operatorname{tr}(YX)) \\
&= \sum_{i=1}^{n} |\lambda_i(AB)|^2 \quad (AB \text{ is normal}) \\
&= \sum_{i=1}^{n} |\lambda_i(BA)|^2 \quad (\lambda_i(AB) = \lambda_i(BA)) \\
&= \sum_{i=1}^{n} |\lambda_i(C)|^2. \quad \square
\end{aligned}
$$

There are many more special types of matrices one may explore. For instance, the Toeplitz matrix, stochastic matrix, nonnegative matrix, and M-matrix, etc. All these matrices are useful in various fields.

Chapter 4 Problems

4.1 Let A and B be $n \times n$ Hermitian matrices. Answer true or false:

(a) $A + B$ is Hermitian.

(b) cA is Hermitian for every $c \in \mathbb{C}$.

(c) AB is Hermitian.

(d) ABA is Hermitian.

(e) If $AB = 0$, then $A = 0$ or $B = 0$.

(f) If $AB = 0$, then $BA = 0$.

(g) If $A^2 = 0$, then $A = 0$.

(h) If $A^2 = I$, then $A = \pm I$.

(i) If $A^3 = I$, then $A = I$.

(j) $-A$, A^t, $\bar{A}$, and A^{-1} (if A is invertible) are all Hermitian.

(k) The main diagonal entries of A are all real.

(l) The eigenvalues of A are all real.

(m) The eigenvalues of AB are all real.

(n) The eigenvalues of ABA are all real.

(o) The eigenvalues of AB^2 are all real.

(p) The eigenvalues of A^2B^2 are all nonnegative.

(q) The eigenvalues of $(AB)^2$ are all nonnegative.

(r) The determinant $\det A$ is real.

(s) The trace $\operatorname{tr}(AB)$ is real.

4.2 Let $A \in M_n(\mathbb{C})$. Show that the following statements are equivalent:

(a) A is Hermitian, that is, $A^* = A$.

(b) There is a unitary matrix U such that U^*AU is real diagonal.

(c) x^*Ax is real for every $x \in \mathbb{C}^n$.

(d) $A^2 = A^*A$.

(e) $A^2 = AA^*$.

(f) $\operatorname{tr} A^2 = \operatorname{tr}(A^*A)$.

(g) $\operatorname{tr} A^2 = \operatorname{tr}(AA^*)$.

4.3   Let $A, B, C \in M_n(\mathbb{C})$. Show that

(a) $x^t A x = 0$ for all $x \in \mathbb{R}^n$ if and only if $A^t = -A$.

(b) $x^t A y = 0$ for all $x, y \in \mathbb{R}^n$ if and only if $A = 0$.

(c) $x^* A x = 0$ for all $x \in \mathbb{C}^n$ if and only if $A = 0$.

(d) $x^* A x$ is real for all $x \in \mathbb{C}^n$ if and only if A is Hermitian.

(e) $x^* A x$ is a fixed constant for all unit vectors $x \in \mathbb{C}^n$ if and only if A is a scalar matrix, i.e., $A = cI$ for some $c \in \mathbb{C}$.

(f) If A is real symmetric, then (i) $x^t A x = 0$ for all $x \in \mathbb{R}^n$ if and only if $A = 0$, (ii) $x^t A x > 0$ for all nonzero $x \in \mathbb{R}^n$ implies $z^* A z > 0$ for all nonzero $z \in \mathbb{C}^n$ (i.e., $A > 0$), (iii) $x^t A x \geq 0$ for all $x \in \mathbb{R}^n$ implies $z^* A z \geq 0$ for all $z \in \mathbb{C}^n$ (i.e., $A \geq 0$).

(g) A need not be equal to B when A and B are real and satisfying

$$x^t A x = x^t B x \quad \text{for all } x \in \mathbb{R}^n.$$

(h) $y^* A y = 0$ for some $y \in \mathbb{C}^n$ does not necessarily imply $A y = 0$.

(i) A real $B \neq 0$ exists such that $x^t B x = 0$ for all $x \in \mathbb{R}^n$.

(j) A Hermitian $C \neq 0$ exists such that $x^t C x = 0$ for all $x \in \mathbb{R}^n$.

4.4   Show that the matrix U is unitary, then find its eigenvalues, where

$$U = \frac{1}{\sqrt{2}} \begin{pmatrix} 1 & 0 & 0 & i \\ 0 & i & 1 & 0 \\ 0 & i & -1 & 0 \\ 1 & 0 & 0 & -i \end{pmatrix}. \quad (i = \sqrt{-1})$$

4.5   Let $\lambda_1, \ldots, \lambda_n$ be the eigenvalues of $A = (a_{ij}) \in M_n(\mathbb{C})$. Show that

$$\sum_{i=1}^{n} \lambda_i^2 = \sum_{i,j=1}^{n} a_{ij} a_{ji} \quad \text{and} \quad \sum_{i=1}^{n} \lambda_i^2 = \sum_{i,j=1}^{n} |a_{ij}|^2 \text{ if } A \text{ is Hermitian.}$$

4.6   (a) Let $u \in \mathbb{C}^n$ be a unit vector and let $H = I_n - 2uu^*$. Show that

$$H = H^* = H^{-1}.$$

(b) Let $x \neq y \in \mathbb{C}^n$, $\|x\| = \|y\|$, $y^* x \in \mathbb{R}$, and $w = \frac{x-y}{\|x-y\|}$. Show that

$$Sx = y, \quad \text{where } S = I_n - 2ww^*.$$

(c) For $e = (1, 1, 1)^t$, find a 3×3 real orthogonal matrix Q such that

$$Qe = \sqrt{3}\,(1, 0, 0)^t.$$

4.7 Let A be an $n \times n$ Hermitian matrix. Show that if $\det A < 0$, then $x^* A x < 0$ for some $x \in \mathbb{C}^n$, equivalently, if A is positive semidefinite, then $\det A \geq 0$. If $\det A \geq 0$, is it true that $y^* A y \geq 0$ for some $y \in \mathbb{C}^n$?

4.8 Let A and B be $n \times n$ Hermitian matrices. Show that $A + B$ is always Hermitian and that AB is Hermitian if and only if $AB = BA$.

4.9 Let A and B be Hermitian matrices of the same size. If AB is Hermitian, show that every eigenvalue λ of AB can be written as $\lambda = ab$, where a is an eigenvalue of A and b is an eigenvalue of B.

4.10 Let Y be a square matrix. A matrix X is said to be a k-th *root* of Y if $X^k = Y$. Let A, B, C, and D be respectively the following matrices

$$
\begin{pmatrix} 1 & 2 & 1 \\ 2 & 4 & 2 \\ 1 & 2 & 1 \end{pmatrix}, \quad
\begin{pmatrix} 0 & 1 & 0 \\ 0 & 0 & 0 \\ 0 & 0 & 0 \end{pmatrix}, \quad
\begin{pmatrix} 0 & 1 \\ 0 & 0 \end{pmatrix}, \quad
\begin{pmatrix} C & 0 \\ 0 & C \end{pmatrix}.
$$

Show that

(a) A has a real symmetric cubic root.

(b) B does not have a complex cubic root.

(c) B has a square root.

(d) C does not have a square root.

(e) D has a square root.

(f) Every normal matrix has k-th roots that are also normal.

(g) If $X^2 = Y$, then $\det(\lambda I - X)$ is a divisor of $\det(\lambda^2 I - Y)$.

Find a 2×2 matrix X such that $X^3 = I_2$ and $X \neq I_2$.

4.11 Let $a, b, c \in \mathbb{R}$, $d, e \in \mathbb{C}$, and let $A = \begin{pmatrix} a & 0 & d \\ 0 & b & e \\ \bar{d} & \bar{e} & c \end{pmatrix}$. Let $p(t)$ be the characteristic polynomial of A with roots r_1, r_2, and r_3. Show that

(a) $\det A$ is real.

(b) The roots r_1, r_2, r_3 of $p(t)$ are all real.

(c) If $a = b$, then $a = b$ is a root of $p(t)$.

(d) If $a \neq b$, then a and b lie in different intervals of the roots, i.e., if $a < b$ and $r_1 \leq r_2 \leq r_3$, then $r_1 \leq a \leq r_2 \leq b \leq r_3$.

4.12 Show that for any nonzero real number x, the 2×2 real matrix

$$X = \begin{pmatrix} 2 & 3x \\ \frac{3}{x} & 2 \end{pmatrix}$$

satisfies the matrix equation $X^2 - 4X - 5I_2 = 0$. (Thus, the equation has an infinite number of distinct 2×2 matrices as its solutions.)

4.13 Explain (without computation) why matrix A has one positive eigenvalue and one negative eigenvalue, and matrix B has two positive eigenvalues and two negative eigenvalues (counting multiplicity), where

$$A = \begin{pmatrix} 1 & 1.01 & 1 \\ 1.01 & 1 & 1.01 \\ 1 & 1.01 & 1 \end{pmatrix}, \quad B = \begin{pmatrix} 0 & 1.01 & 1 & 0 \\ 1.01 & 0 & 1.01 & 0 \\ 1 & 1.01 & 0 & 1.01 \\ 0 & 0 & 1.01 & 0 \end{pmatrix}.$$

4.14 If A is an invertible Hermitian matrix, show that A and A^{-1} are *-congruent, that is, $P^*AP = A^{-1}$ for some invertible matrix P.

4.15 Let $A, B \in M_n(\mathbb{F})$. We say that A and B are ^{t}-congruent (versus the *-congruence) over $\mathbb{F}$ if $T^tBT = A$ for some invertible $T \in M_n(\mathbb{F})$.

(a) Show that ^{t}-congruence is an equivalence relation.

(b) Show that $A = \begin{pmatrix} 1 & 0 & 0 \\ 0 & 2 & 0 \\ 0 & 0 & -1 \end{pmatrix}$ and $B = \begin{pmatrix} 1 & -1 & 0 \\ -1 & 2 & 0 \\ 0 & 0 & 3 \end{pmatrix}$ are ^{t}-congruent over $\mathbb{C}$, i.e., there exists a complex invertible matrix C such that $C^tBC = A$, but A and B are not ^{t}-congruent over $\mathbb{R}$, i.e., there does not exist a real invertible matrix R such that $R^tBR = A$. Explain why a complex invertible matrix S does not exist for which $S^*BS = A$.

4.16 Let $A = \operatorname{diag}(1, 2, -1)$. Determine whether the matrices

$$B = \begin{pmatrix} 1 & -1 & 0 \\ -1 & 2 & 0 \\ 0 & 0 & 3 \end{pmatrix}, \quad C = \begin{pmatrix} -2 & 0 & 0 \\ 0 & 1 & 0 \\ 0 & 0 & 1 \end{pmatrix}, \quad D = \begin{pmatrix} 0 & 1 & 0 \\ 1 & 0 & 0 \\ 0 & 0 & 2 \end{pmatrix}$$

are equivalent (i.e., $PBQ = A$), ^{t}-congruent (i.e., $T^tBT = A$), *-congruent (i.e., $S^*BS = A$), or similar (i.e., $R^{-1}BR = A$) to A over $\mathbb{R}$ and $\mathbb{C}$ (where P, Q, T, S, R are nonsingular matrices). Explain why.

4.17 Show that HK and KH are similar for any Hermitian $H, K \in M_n(\mathbb{C})$.

4.18 Let A be an $n \times n$ nonzero Hermitian matrix with rank r. Show that all nonzero $r \times r$ principal minors of A have the same sign.

4.19 Let $A = (a_{ij})$ be a 3×3 Hermitian matrix. Let $M = (|a_{ij}|)$, i.e., M is the matrix whose entries are absolute values of the entries of A. Show that if A is positive semidefinite then M is positive semidefinite.

4.20 Let $A = (a_{ij})$ be an $m \times n$ matrix and $1 \le p \ne q \le m$. Show that

$$\max_{i,j} |a_{ij}| \le \sigma_{\max}(A),$$

$$\max_{j} |a_{pj} + a_{qj}| \le \sqrt{2}\, \sigma_{\max}(A),$$

$$\max_{j} \left(|a_{pj}|^2 + |a_{qj}|^2 \right)^{\frac{1}{2}} \le \sigma_{\max}(A),$$

$$\max_{j} |\omega a_{1j} + \omega^2 a_{2j} + \cdots + \omega^m a_{mj}| \le \sqrt{m}\, \sigma_{\max}(A),$$

where ω is a primitive m-th root of unity.

4.21 Let A, B, and C be $n \times n$ Hermitian matrices. Show that

(a) Neither $\operatorname{tr}(A^2) \le (\operatorname{tr} A)^2$ nor $\operatorname{tr}(A^2) \ge (\operatorname{tr} A)^2$ holds in general.

(b) $\operatorname{tr}(AB)^k$ is real for every positive integer k.

(c) $\operatorname{tr}(AB)^2 \le \operatorname{tr}(A^2 B^2)$. Equality holds if and only if $AB = BA$.

(d) It is not true that $\operatorname{tr}(ABC)^2 \le \operatorname{tr}(A^2 B^2 C^2)$ in general.

(e) $\left(\operatorname{tr}(AB) \right)^2 \le (\operatorname{tr} A^2)(\operatorname{tr} B^2)$. Equality holds if and only if $A = kB$ or $B = kA$ for a scalar k.

4.22 Let A be an $n \times n$ Hermitian matrix. (Note: This precondition is important.) Show that the following statements are equivalent:

(a) A is positive semidefinite, that is, $x^* A x \ge 0$ for all $x \in \mathbb{C}^n$.

(b) All eigenvalues of A are nonnegative.

(c) $A = U^* \operatorname{diag}(\lambda_1, \lambda_2, \ldots, \lambda_n) U$ for some unitary matrix U, where $\lambda_1, \lambda_2, \ldots, \lambda_n$ are all nonnegative (eigenvalues of A).

(d) $A = B^* B$ for some matrix B.

(e) $A = T^* T$ for some $r \times n$ matrix T with rank $r = r(T) = r(A)$.

(f) All principal minors of A are nonnegative.

(g) $\operatorname{tr}(AX) \ge 0$ for all positive semidefinite matrices X.

(h) $X^* A X \ge 0$ for all $n \times m$ matrices X.

4.23 Let A and B be $n \times n$ positive semidefinite matrices.

(a) Show that $A^2 + AB + BA + B^2$ is positive semidefinite.

(b) Construct an example showing that neither $A^2 + AB + BA$ nor $AB + BA$ is positive semidefinite.

(c) If A and $AB + BA$ are both positive definite, show that B is positive definite.

4.24 Show that for any $m \times n$ matrices A and B,

$$A^*A + B^*B \geq \pm(A^*B + B^*A).$$

4.25 Let A and B be positive semidefinite matrices of the same size. If the largest eigenvalues of A and B are less than or equal to 1, show that

$$AB + BA \geq -\tfrac{1}{4}I.$$

4.26 If the following matrices are positive definite, what are $x, y, p, q \in \mathbb{C}$?

$$\begin{pmatrix} 1 & x & \frac{1}{2} \\ \bar{x} & 1 & 0 \\ \frac{1}{2} & 0 & 1 \end{pmatrix}, \quad \begin{pmatrix} 1 & y & -1 \\ y & 4 & 2 \\ -1 & 2 & 4 \end{pmatrix}, \quad \begin{pmatrix} 1 & 0 & 0 & 1 \\ 0 & p & 0 & 0 \\ 0 & 0 & 1 & q \\ 1 & 0 & \bar{q} & p \end{pmatrix}.$$

4.27 Find the conditions on the real numbers x and y such that

$$\begin{pmatrix} x & 1 & -1 \\ 1 & 2 & y \\ -1 & y & 3 \end{pmatrix} \geq 0.$$

4.28 Let $A = (a_{ij})$ be an $n \times n$ Hermitian matrix such that

$$a_{ii} = 1 \quad \text{and} \quad \sum_{j=1}^{n} |a_{ij}| \leq 2, \quad i = 1, 2, \ldots, n.$$

Show that $0 \leq \lambda \leq 2$ for any eigenvalue λ of A and $0 \leq \det A \leq 1$.

4.29 (i) Give an example of a Hermitian matrix whose leading principal minors are all nonnegative, but the matrix is not positive semidefinite. (ii) Find a non-Hermitian matrix whose principal minors and eigenvalues are all nonnegative, but the matrix is not positive semidefinite. (iii) Is it possible for some non-Hermitian matrix $A \in M_n(\mathbb{C})$ to satisfy $x^t A x \geq 0$ for all $x \in \mathbb{R}^n$? or $x^* A x \geq 0$ for all $x \in \mathbb{C}^n$?

4.30 Let $A = (a_{ij})$ be an $n \times n$ positive semidefinite matrix.

(a) Show that a solution to $X^2 = A$ is not unique in general.

(b) Show that $\operatorname{Im} A = \operatorname{Im} A^{\frac{1}{2}} = \operatorname{Im} A^{\frac{1}{4}} = \cdots$.

(c) Find $A^{\frac{1}{2}}$ when A is $\begin{pmatrix} 2 & 0 \\ 0 & 0 \end{pmatrix}$, $\begin{pmatrix} 1 & 1 \\ 1 & 1 \end{pmatrix}$, or $\frac{1}{2}\begin{pmatrix} 5 & -3 \\ -3 & 5 \end{pmatrix}$.

4.31 Let $A = (a_{ij})$ be an $n \times n$ positive semidefinite matrix. Show that

(a) $a_{ii} \geq 0$, $i = 1, 2, \ldots, n$, and if $a_{ii} = 0$, then the i-th row and i-th column of A consist entirely of 0.

(b) $a_{ii}a_{jj} \geq |a_{ij}|^2$ for each pair of i and j. Thus, of all entries of A in absolute values, i.e., $|a_{ij}|$, the largest one is nonnegative and is on the main diagonal, say a_{kk}, $a_{kk} \geq |a_{ij}|$ for all i, j.

(c) $\det A = 0$ if a principal submatrix of A is singular.

(d) There exists an $n \times n$ invertible matrix P such that

$$A = P^* \begin{pmatrix} I_{r(A)} & 0 \\ 0 & 0 \end{pmatrix} P.$$

Is it always possible to choose a unitary P in the above?

(e) The transpose A^t and the conjugate $\bar{A}$ are positive semidefinite.

(f) A has a unique positive semidefinite k-th root, $k = 1, 2, 3, \ldots$.

4.32 Let A be an $m \times n$ matrix and let $x \in \mathbb{C}^n$ (in columns). Show that

$$(A^*A)x = 0 \quad \Leftrightarrow \quad Ax = 0;$$

$$\operatorname{tr}(A^*A) = 0 \iff A^*A = 0 \iff A = 0.$$

4.33 Let A be an $n \times n$ positive definite matrix. Show that for every $x \in \mathbb{C}^n$

$$x^*A^{-1}x = \max_{y \in \mathbb{C}^n}(x^*y + y^*x - y^*Ay).$$

4.34 Let A and B be $n \times n$ positive semidefinite matrices. Show that

$$\operatorname{Im}(AB) \cap \operatorname{Ker}(AB) = \{0\}.$$

In particular, setting $B = I$ gives that for any positive semidefinite A,

$$\operatorname{Im} A \cap \operatorname{Ker} A = \{0\}.$$

4.35 Let $A, B \in M_n(\mathbb{C})$ and let A be positive semidefinite. Show that

$$AB + BA = 0 \quad \Leftrightarrow \quad AB = BA = 0.$$

4.36   Let H and G be square matrices. Prove the following statements:

(a)   If H is Hermitian and $H^5 + H^3 + 2H = 4I$, then $H = I$.

(b)   If G is positive semidefinite and $G^3 + 3G^2 + G = 0$, then $G = 0$.

4.37   Let A be an $n \times n$ Hermitian matrix. Show that

(a) $A^2 + I_n \geq kA$, where $-2 \leq k \leq 2$.

(b) $\left(A - \lambda_{\min}(A)I_n\right)\left(\lambda_{\max}(A)I_n - A\right) \geq 0$.

4.38   Let $A \in M_n(\mathbb{C})$.

(a)   If A is Hermitian, show that $A^2 \geq 0$. Give an example of a non-Hermitian matrix B such that B^2 is positive definite.

(b)   If A is skew-Hermitian, show that $A^2 \leq 0$.

(c)   If A is upper- (or lower-) triangular, show that the eigenvalues of A are exactly the main diagonal entries of A.

(d)   Show that a nonzero normal matrix has nonzero eigenvalue(s).

(e)   If A is a Hermitian matrix whose eigenvalues are exactly the entries on the main diagonal of A, show that A is diagonal.

(f)   If the eigenvalues of a matrix A are exactly the singular values of A, show that A is positive semidefinite.

4.39   For $X \in M_n(\mathbb{C})$, define $f(X) = X^*X$. Show that f is a *convex function* on $M_n(\mathbb{C})$, that is, for any $t \in [0,1]$, with $\tilde{t} = 1 - t$,

$$f(tA + \tilde{t}B) \leq tf(A) + \tilde{t}f(B), \quad A, B \in M_n(\mathbb{C}).$$

4.40   Let A and B be $n \times n$ positive semidefinite matrices. If the eigenvalues of A and B are all contained in the interval $[a, b]$, where $0 \leq a < b$, show that for any $t \in [0,1]$, with $\tilde{t} = 1 - t$,

$$0 \leq tA^2 + \tilde{t}B^2 - (tA + \tilde{t}B)^2 \leq \tfrac{1}{4}(a - b)^2 I.$$

4.41   Assume that the eigenvalues, if all real, of a square matrix are ordered decreasingly. Let A be an $n \times n$ positive definite matrix. Show that

$$\lambda_i(A^{\frac{1}{2}}) = \sqrt{\lambda_i(A)}, \quad \lambda_i(I - A) = 1 - \lambda_{n-i+1}(A), \quad \text{and}$$

$$\lambda_i(A^{-1}) = \frac{1}{\lambda_{n-i+1}(A)}, \quad i = 1, 2, \dots, n.$$

If $\sigma_i(X)$ denotes the i-th singular value of $X \in M_n(\mathbb{C})$, is it true that

$$\sigma_i(I - X) = 1 - \sigma_{n-i+1}(X)?$$

4.42   Let A and B be $n \times n$ positive semidefinite matrices. Show that

(a) $|A + B| \geq |A| + |B|$. Equality holds if and only if $n = 1$ or $|A + B| = 0$ or $A = 0$ or $B = 0$. As a result, if $A > 0$ and $B \neq 0$, then $|A + B| > |A| + |B|$.

(b) $|A + B|^{\frac{1}{n}} \geq |A|^{\frac{1}{n}} + |B|^{\frac{1}{n}}$.

(c) $|A|^t |B|^{\tilde{t}} \leq |tA + \tilde{t}B|$, where $t \in [0, 1]$, $\tilde{t} = 1 - t$. In particular, for every positive integer $k \leq n$, $\sqrt{|A||B|} \leq \frac{|A+B|}{2^k} \leq \frac{|A+B|}{2}$.

(d) $\sqrt{|A||B|} \leq \frac{|A|+|B|}{2} \leq \frac{|A+B|}{2}$.

4.43   Let $A, B \in M_n(\mathbb{C})$ and $A \geq B \geq 0$. Show that

(a) $C^* A C \geq C^* B C$ for every $C \in M_{n \times m}(\mathbb{C})$.

(b) $A + C \geq B + D$, where $C, D \in M_n(\mathbb{C})$, and $C \geq D$.

(c) $\operatorname{tr} A \geq \operatorname{tr} B$.

(d) $\lambda_{\max}(A) \geq \lambda_{\max}(B)$.

(e) $\det A \geq \det B$.

(f) $r(A) \geq r(B)$.

(g) $B^{-1} \geq A^{-1}$ (when the inverses exist).

(h) $A^{\frac{1}{2}} \geq B^{\frac{1}{2}}$. Does $A \geq B$ imply $A^2 \geq B^2$?

4.44   Let A be positive definite and let B be Hermitian, both $n \times n$. Prove

(a) The eigenvalues of AB and $A^{-1}B$ are necessarily real.

(b) $A + B \geq 0$ if and only if $\lambda(A^{-1}B) \geq -1$, where $\lambda(A^{-1}B)$ denotes any eigenvalue of $A^{-1}B$.

(c) The rank of AB is equal to the number of nonzero eigenvalues of AB. Is this true if $A \geq 0$ and B is Hermitian?

4.45   Let A and B be $n \times n$ Hermitian matrices.

(a) Give an example showing that the eigenvalues of AB are not real.

(b) If A or B is positive semidefinite, show that all the eigenvalues of AB are necessarily real.

(c) If A or B is positive definite, show that AB is diagonalizable.

(d) Give an example showing that the positive definiteness in (c) is necessary, that is, if one of A and B is positive semidefinite and the other is Hermitian, then AB need not be diagonalizable.

4.46   Let $\sigma_{\max}(A)$ be the largest singular value of a square matrix A and let $\lambda(\frac{A+A^*}{2})$ denote any eigenvalue of $\frac{A+A^*}{2}$. Show that

$$\left|\lambda\left(\frac{A+A^*}{2}\right)\right| \leq \sigma_{\max}(A) \quad \text{and} \quad \operatorname{tr}\left(\frac{A+A^*}{2}\right)^2 \leq \operatorname{tr}(A^*A).$$

4.47   For $n \times n$ positive semidefinite matrices A and B, show that

(a) $A^{\frac{1}{2}}BA^{\frac{1}{2}} \geq 0$.

(b) The eigenvalues of AB and BA are all nonnegative.

(c) AB need not be positive semidefinite.

(d) AB is positive semidefinite if and only if $AB = BA$.

(e) $\operatorname{tr}(AB^2A) = \operatorname{tr}(BA^2B)$.

(f) $\operatorname{tr}(AB^2A)^{\frac{1}{2}} = \operatorname{tr}(BA^2B)^{\frac{1}{2}}$.

(g) $\operatorname{tr}(AB) \leq \operatorname{tr}A\operatorname{tr}B \leq \frac{1}{2}\big((\operatorname{tr}A)^2 + (\operatorname{tr}B)^2\big)$.

(h) $\operatorname{tr}(AB) \leq \lambda_{\max}(A)\operatorname{tr}B$.

(i) $\operatorname{tr}(AB) \leq \frac{1}{4}(\operatorname{tr}A + \operatorname{tr}B)^2$.

(j) $\operatorname{tr}(AB) \leq \frac{1}{2}(\operatorname{tr}A^2 + \operatorname{tr}B^2)$.

(k) $\big((\operatorname{tr}A^2)(\operatorname{tr}B^2)\big)^{\frac{1}{2}} \leq \frac{1}{2}(\operatorname{tr}A^2 + \operatorname{tr}B^2) \leq \frac{1}{2}\big((\operatorname{tr}A)^2 + (\operatorname{tr}B)^2\big)$.

4.48   Let $A, B, C,$ and D be $n \times n$ positive semidefinite matrices.

(a) Show that $AB + BA$ is Hermitian.

(b) Is it true that $AB + BA \geq 0$?

(c) Is it true that $A^2 + B^2 \geq 2AB$?

(d) Is it true that $A^2 + B^2 \geq AB + BA$?

(e) Is it true that $\operatorname{tr}A\operatorname{tr}B \leq \frac{1}{2}(\operatorname{tr}A^2 + \operatorname{tr}B^2)$?

(f) Does $\operatorname{tr}A = \operatorname{tr}B$ imply $\operatorname{tr}A^{\frac{1}{2}} = \operatorname{tr}B^{\frac{1}{2}}$ or $\operatorname{tr}A^2 = \operatorname{tr}B^2$?

(g) Show that $\operatorname{tr}(AB) \leq \operatorname{tr}(CD)$ if $A \leq C$ and $B \leq D$.

(h) Show that $\lambda_{\max}(AB) \leq \lambda_{\max}(A)\lambda_{\max}(B)$.

(i) Is it true that $\lambda_{\max}(ABC) \leq \lambda_{\max}(A)\lambda_{\max}(B)\lambda_{\max}(C)$?

(j) Show that for any nonnegative real numbers p and q,

$$\lambda_{\max}(pA + qB) \leq p\lambda_{\max}(A) + q\lambda_{\max}(B).$$

In particular,

$$\lambda_{\max}(A + B) \leq \lambda_{\max}(A) + \lambda_{\max}(B).$$

4.49   Construct examples or answer questions.

(a) A 3×3 Hermitian matrix H contains a 2×2 principal submatrix K such that $\lambda = 1$ is an eigenvalue of H with multiplicity 1 and $\lambda = 1$ is an eigenvalue of K with multiplicity 2.

(b) Non-Hermitian matrices A and B have only positive eigenvalues, while AB has only negative eigenvalues.

(c) Is it possible that $A + B$, where $A, B \in M_n(\mathbb{R})$, has only negative eigenvalues while A and B have only positive eigenvalues?

(d) $A, B, C \in M_2(\mathbb{R})$ are positive definite (thus their eigenvalues are all positive), while ABC has only negative eigenvalues.

(e) Can the matrices in (d) be 3×3 or of any odd numbered size?

(f) $\left(\begin{smallmatrix} A & B \\ B^* & C \end{smallmatrix} \right)$ is positive semidefinite, but $\left(\begin{smallmatrix} A & B^* \\ B & C \end{smallmatrix} \right)$ is not.

4.50   Prove the following statements:

(a) If $A > 0$ and B is Hermitian, then there exists an invertible matrix P such that $P^*AP = I$ and P^*BP is diagonal.

(b) If $A \geq 0$ and $B \geq 0$, then there exists an invertible matrix P such that both P^*AP and P^*BP are diagonal matrices. Can the condition $B \geq 0$ be weakened so that B is Hermitian?

(c) If $A > 0$, $B > 0$, and $\det(\lambda A - B) = 0$, then $\lambda > 0$.

(d) If $A > 0$, $B > 0$, and $\det(\lambda A - B) = 0$ has no other real solutions than $\lambda = 1$, then $A = B$.

(e) If $A \geq B > 0$, and $\det(\lambda B - A) = 0$, then $\lambda \geq 1$.

4.51   Let $A, B \in M_n(\mathbb{C})$ and let A be positive semidefinite. Show that
$$AB = BA \;\Leftrightarrow\; A^2B = BA^2 \;\Leftrightarrow\; A^{\frac{1}{2}}B = BA^{\frac{1}{2}}.$$
If both A and B are positive semidefinite, show that
$$AB = BA \;\Leftrightarrow\; A^2B^2 = B^2A^2 \;\Leftrightarrow\; A^{\frac{1}{2}}B^{\frac{1}{2}} = B^{\frac{1}{2}}A^{\frac{1}{2}}.$$

4.52   Let A, B, and C be $n \times n$ positive semidefinite matrices. If C commutes with AB and $A - B$, show that C commutes with A and B.

4.53   Let A be an $n \times n$ positive definite matrix. If B is an $n \times n$ matrix such that $A - B^*AB$ is positive definite, show that $|\lambda| < 1$ for all eigenvalues λ of B. Is it true that $\sigma < 1$ for all singular values σ of B?

4.54 Let $M = \left(\begin{smallmatrix} A & B \\ B^* & C \end{smallmatrix}\right) \geq 0$, where $A, B, C \in M_n(\mathbb{C})$. Show that

(a) $r(B) \leq r(A)$.

(b) $\operatorname{Im} B \subseteq \operatorname{Im} A$.

(c) $|\operatorname{tr} B|^2 \leq \operatorname{tr} A \operatorname{tr} C$.

(d) $|\det B|^2 \leq \det A \det C$.

(e) $A \pm (B + B^*) + C \geq 0$.

(f) The sum of all entries of M is nonnegative.

(g) $\left(\begin{smallmatrix} \sum(A) & \sum(B) \\ \sum(B^*) & \sum(C) \end{smallmatrix}\right) \geq 0$, where $\sum(X) = \sum_{i,j} x_{ij}$.

(h) $\left(\begin{smallmatrix} \sum_2(A) & \sum_2(B) \\ \sum_2(B^*) & \sum_2(C) \end{smallmatrix}\right) \geq 0$, where $\sum_2(X) = \sum_{i,j} |x_{ij}|^2$.

4.55 Let A and B be $n \times n$ Hermitian matrices. Show that

$$A \pm B \geq 0 \quad \Leftrightarrow \quad \begin{pmatrix} A & B \\ B & A \end{pmatrix} \geq 0.$$

4.56 Let A and B be $n \times n$ real matrices. Show that

$$A + iB \geq 0 \quad \Leftrightarrow \quad A - iB \geq 0 \quad \Leftrightarrow \quad \begin{pmatrix} A & -B \\ B & A \end{pmatrix} \geq 0.$$

4.57 Let A, B, and C be $n \times n$ matrices such that $\left(\begin{smallmatrix} A & B \\ B^* & C \end{smallmatrix}\right) \geq 0$. Show that

$$\begin{pmatrix} A + C & B + B^* \\ B + B^* & A + C \end{pmatrix} \geq 0 \quad \text{and} \quad \begin{pmatrix} A + C & 0 \\ 0 & A + C \end{pmatrix} \geq \begin{pmatrix} A & B^* \\ B & C \end{pmatrix}.$$

4.58 Let A, B, and C be $n \times n$ matrices such that $\left(\begin{smallmatrix} A & B \\ B^* & C \end{smallmatrix}\right) \geq 0$. Show that

$$|\det(B + B^*)| \leq \det(A + C) \quad \text{and} \quad |\det(B \circ B^*)| \leq \det(A \circ C).$$

4.59 Let $A \in M_n(\mathbb{C})$ and let $\sigma_1, \ldots, \sigma_n$ be the singular values of A. Let $M = \left(\begin{smallmatrix} 0 & A \\ A^* & 0 \end{smallmatrix}\right)$, $N = \left(\begin{smallmatrix} I & A \\ A^* & I \end{smallmatrix}\right)$, and $K = \left(\begin{smallmatrix} kI & A \\ A^* & kI \end{smallmatrix}\right)$, where $k \in \mathbb{R}$.

(a) Show that $\det M = (-1)^n |\det A|^2$.

(b) If $A \neq 0$, why is M never positive semidefinite?

(c) Show that the eigenvalues of M are $\pm \sigma_1, \ldots, \pm \sigma_n$.

(d) Find the eigenvalues and singular values of N.

(e) Find the eigenvalues and singular values of K.

4.60 Let A be an $n \times n$ positive definite matrix with eigenvalues $\lambda_1, \ldots, \lambda_n$. Find the eigenvalues of the partitioned matrix $M = \begin{pmatrix} A & I \\ I & A^{-1} \end{pmatrix}$.

4.61 Find the singular values of the $n \times n$ real symmetric matrix

$$A = \begin{pmatrix} 1 & 1 & \cdots & 1 \\ 1 & -1 & \cdots & 0 \\ \vdots & \vdots & \ddots & \vdots \\ 1 & 0 & \cdots & -1 \end{pmatrix}, \quad a_{ij} = 0 \text{ if } i, j \geq 2, \ i \neq j.$$

What are the eigenvalues of A? and what is the rank of A?

4.62 Let $\sigma_{\max}(X)$ denote the largest singular value of the matrix X (which is also called the *spectral norm* of X). For $A, B \in M_n(\mathbb{C})$, show that

$$\sigma_{\max}(AB) \leq \sigma_{\max}(A)\sigma_{\max}(B),$$

$$\sigma_{\max}(A + B) \leq \sigma_{\max}(A) + \sigma_{\max}(B),$$

$$\sigma_{\max}(A^2 - B^2) \leq \sigma_{\max}(A + B)\sigma_{\max}(A - B).$$

4.63 Let $A = \begin{pmatrix} B & C \\ C^* & D \end{pmatrix} \geq 0$, where $B \in M_m(\mathbb{C})$ and $D \in M_n(\mathbb{C})$. Show that

$$\det A \leq \det B \det D. \quad (\textit{Fischer's inequality})$$

Equality holds if and only if $C = 0$ or $\det B = 0$ or $\det D = 0$.

4.64 (a) Show that for any $m \times n$ complex matrix $E = (e_{ij})$

$$\det(E^*E) \leq \prod_{j=1}^{n} \sum_{i=1}^{m} |e_{ij}|^2.$$

(b) Let F be an $m \times n$ matrix. If G consists of some columns of F and H contains the remaining columns of F, show that

$$\det(F^*F) \leq \det(G^*G) \det(H^*H).$$

4.65 Let X and Y be square matrices of the same size. Show that

$$|\det(I - YX^*)|^2 \leq \det(I + XX^*)\det(I + Y^*Y)$$

and

$$|\det(X + Y)|^2 \leq \det(I + XX^*)\det(I + Y^*Y).$$

4.66   Let H be an $n \times n$ positive semidefinite matrix and write $H = A + iB$, where $A = (a_{st})$ and $B = (b_{st})$ are $n \times n$ real matrices. Show that

(a) A is positive semidefinite and $B^t = -B$.

(b) $a_{ss} a_{tt} \geq a_{st}^2 + b_{st}^2$ for each pair of s, t.

(c) $\det H \leq \det A$. When does equality hold?

(d)   If A is singular, then H is singular. Is the converse true?

4.67   Let $A \in M_n(\mathbb{C})$. Show that

(a) $A = AA^* S$ for some matrix S.

(b)   If A is nonsingular, then matrix S in (a) is unique.

(c) $A = (AA^*)^{\frac{1}{2}} T$ for some matrix T.

(d)   If A is nonsingular, then T in (c) is unique and it is unitary.

4.68   Let $X \in M_{m \times n}(\mathbb{C})$. The matrix $(X^* X)^{\frac{1}{2}}$ is called the *modulus* of X, denoted by $\ell(X)$. (Note: In some texts, $|X|$ is used for modulus.)

(I) Find $\ell(X)$ and $\ell(X^*)$, where X is $\left(\begin{smallmatrix} 0 & 1 \\ 0 & 0 \end{smallmatrix}\right)$, $\left(\begin{smallmatrix} 1 & 1 \\ 1 & 1 \end{smallmatrix}\right)$, $\left(\begin{smallmatrix} 0 & 1 \\ 1 & 0 \end{smallmatrix}\right)$, or $\left(\begin{smallmatrix} 1 & 1 \\ 0 & 0 \end{smallmatrix}\right)$.

(II) Let A be an $n \times n$ matrix. Prove each of the following statements:

(a) $\det(\ell(A)) = |\det A|$.

(b) $A = \ell(A)$ if $A \geq 0$.

(c) $\ell(\ell(A)) = \ell(A)$.

(d)   If $A = UDV$ is a singular value decomposition of A, then

$$\ell(A) = V^* D V \quad \text{and} \quad \ell(A^*) = U D U^*.$$

(e) $\ell(A)$ and $\ell(A^*)$ are unitarily similar.

(f) $\ell(A) = \ell(A^*)$ if and only if A is normal.

(g) $\left(\begin{smallmatrix} \ell(A) & A^* \\ A & \ell(A^*) \end{smallmatrix}\right)$ is positive semidefinite.

(h) $A\ell(A) = \ell(A^*)A$; A need not commute with $\ell(A)$.

(i) $\ell(A)H = H\ell(A)$ if $AH = HA$ and H is Hermitian.

(j) $A = \ell(A^*)W = W\ell(A)$ for some unitary matrix W.

(k) $\ell(A)A^{-1}$ is unitary and $A(\ell(A))^{-1}A^* = \ell(A^*)$ if A^{-1} exists.

(l) $\ell(AB)$ and $\ell\big(\ell(A)\ell(B^*)\big)$ have the same eigenvalues.

4.69   Let $A \in M_{m \times n}(\mathbb{C})$ and $B \in M_{p \times n}(\mathbb{C})$. If $r(B) = p$, show that

$$AA^* \geq AB^*(BB^*)^{-1}BA^*.$$

4.70   Let A and B be both $m \times n$ complex matrices. Show that

$$\begin{pmatrix} A^*A & A^*B \\ B^*A & B^*B \end{pmatrix} \geq 0 \quad \text{and} \quad \begin{pmatrix} |A^*A| & |A^*B| \\ |B^*A| & |B^*B| \end{pmatrix} \geq 0.$$

Determine whether the following are true:

$$\begin{pmatrix} A^*A & B^*A \\ A^*B & B^*B \end{pmatrix} \geq 0 \quad \text{and} \quad \begin{pmatrix} |A^*A| & |B^*A| \\ |A^*B| & |B^*B| \end{pmatrix} \geq 0.$$

4.71   Show that a matrix A satisfies $\frac{A+A^*}{2} = (AA^*)^{\frac{1}{2}}$ if and only if $A \geq 0$.

4.72   Show that a matrix A satisfies $\frac{A+A^*}{2} = AA^*$ if and only if A is normal and its eigenvalues all lie on the circle $|z - \frac{1}{2}| = \frac{1}{2}$ in the complex plane.

4.73   Let A and B be $n \times n$ positive semidefinite matrices.

(a)   Find $A \circ I$, where I is the $n \times n$ identity matrix.

(b)   Find $A \circ J$, where J is the $n \times n$ all-ones matrix.

(c)   Show that $A^p \circ B^q \geq 0$ for all positive integers p and q.

(d)   Show that $(a_{ij}^k) \geq 0$ and $(|a_{ij}|^{2k}) \geq 0$ for all positive integers k.

(e)   Show that $\lambda_{\max}(A \circ B) \leq \lambda_{\max}(A)\lambda_{\max}(B)$.

(f)   Is it true that $A \circ B$ is singular when A or B is singular?

(g)   Show that $\operatorname{tr}(A \circ B) \leq \frac{1}{2}\operatorname{tr}(A \circ A + B \circ B)$.

4.74   Let $A, B \in M_{m \times n}(\mathbb{C})$. Let A_i and B_j denote the i-th and the j-th columns of A and B, respectively, $i, j = 1, 2, \ldots, n$. Show that

$$(AA^*) \circ (BB^*) = (A \circ B)(A^* \circ B^*) + \sum_{i \neq j}(A_i \circ B_j)(A_i^* \circ B_j^*)$$

and

$$(A \circ B)(A^* \circ B^*) \leq (AA^*) \circ (BB^*).$$

In particular, for Hermitian matrices A and B of the same size,

$$(A \circ B)^2 \leq A^2 \circ B^2.$$

4.75 Let A and B be $n \times n$ correlation matrices, i.e., A and B are positive semidefinite and all entries on the main diagonals are 1. Show that

$$A^{\frac{1}{2}} \circ B^{\frac{1}{2}} \leq I.$$

4.76 Let A be a positive definite matrix. Partition A and A^{-1} conformably as

$$A = \begin{pmatrix} B & C \\ C^* & D \end{pmatrix}, \quad A^{-1} = \begin{pmatrix} U & V \\ V^* & W \end{pmatrix}.$$

(a) Show that U and W can be expressed, respectively, as

$$U = (B - CD^{-1}C^*)^{-1} = B^{-1} + B^{-1}CWC^*B^{-1},$$
$$W = (D - C^*B^{-1}C)^{-1} = D^{-1} + D^{-1}C^*UCD^{-1}.$$

(b) Show that U is nonsingular and that

$$A - \begin{pmatrix} U^{-1} & 0 \\ 0 & 0 \end{pmatrix} \geq 0.$$

4.77 (a) Find the eigenvalues of the block matrices

$$\begin{pmatrix} I_2 & I_2 \\ I_2 & I_2 \end{pmatrix} \quad \text{and} \quad \begin{pmatrix} I_n & I_n \\ I_n & I_n \end{pmatrix}.$$

(b) If A is an $n \times n$ positive semidefinite matrix, show that

$$\begin{pmatrix} \operatorname{diag}(A) & \operatorname{diag}(A) \\ \operatorname{diag}(A) & \operatorname{diag}(A) \end{pmatrix} \geq 0.$$

4.78 Let $A > 0$. Show that

(a) $A + A^{-1} \geq 2I$.

(b) $A \circ A^{-1} \geq I$.

4.79 Let $A = (a_{ij})$ be an $n \times n$ matrix of nonnegative entries. If each row sum of A is equal to 1, namely, $\sum_{j=1}^{n} a_{ij} = 1$ for each i, show that

(a) $|\lambda| \leq 1$ for every eigenvalue λ of A.

(b) 1 is an eigenvalue of A. Find a corresponding eigenvector.

(c) If A^{-1} exists, then each row sum of A^{-1} is also equal to 1.

(d) If $A = U \circ \overline{U}$, where U is a unitary matrix, then (i) each row (column) sum of A is 1, (ii) 1 is an eigenvalue of A, and (iii) A^t and A have a common eigenvalue and a common eigenvector.

4.80 Let $A \in M_n(\mathbb{C})$.

(a) If $A \geq 0$, show that $\left| \begin{smallmatrix} A & x \\ x^* & 0 \end{smallmatrix} \right| \leq 0$ for every column vector $x \in \mathbb{C}^n$. The inequality is strict if A is nonsingular and $x \neq 0$.

(b) If $A > 0$ and $x \neq 0$, find the inverse of $\left(\begin{smallmatrix} A & x \\ x^* & 0 \end{smallmatrix} \right)$.

4.81 Let A be an $n \times n$ real orthogonal matrix, i.e., A is real and $A^t A = AA^t = I_n$. Let $\lambda = a + ib$ be an eigenvalue of A with a corresponding eigenvector $u = x + iy$, where $a, b \in \mathbb{R}$, $x, y \in \mathbb{R}^n$. If $b \neq 0$, show that

$$x^t y = 0 \quad \text{and} \quad x^t x = y^t y.$$

4.82 Let U be an $n \times n$ unitary matrix, i.e., $U^* U = UU^* = I$. Show that

(a) $U^* = U^{-1}$.

(b) U^t and $\overline{U}$ are unitary.

(c) UV is unitary for every $n \times n$ unitary matrix V.

(d) The eigenvalues of U are all equal to 1 in absolute value.

(e) If λ is an eigenvalue of U, then $\frac{1}{\lambda}$ is an eigenvalue of U^*.

(f) $|x^* U x| \leq 1$ for every unit vector $x \in \mathbb{C}^n$.

(g) $\|Ux\| = 1$ for every unit vector $x \in \mathbb{C}^n$.

(h) Each column (row) sum of $U \circ \overline{U} = (|u_{ij}|^2)$ is equal to 1.

(i) If x and y are eigenvectors of U belonging to distinct eigenvalues, then $x^* y = 0$.

(j) The columns (rows) of U form an orthonormal basis for $\mathbb{C}^n$.

(k) For any k rows of U, $1 \leq k \leq n$, there exist k columns such that the submatrix on these rows and columns is nonsingular.

(l) $|\operatorname{tr}(UA)| \leq \operatorname{tr} A$ for every $n \times n$ matrix $A \geq 0$.

Which of the above statements imply that U is unitary?

4.83 Let $A \in M_{m \times n}(\mathbb{C})$. If the spectral norm $\sigma_{\max}(A) \leq 1$, show that

$$U = \begin{pmatrix} A & (I - AA^*)^{\frac{1}{2}} \\ (I - A^*A)^{\frac{1}{2}} & -A^* \end{pmatrix} \quad \text{is unitary.}$$

4.84 Show that a square complex matrix U is unitary if and only if the column (row) vectors of U all have length 1 and $|\det U| = 1$.

4.85 If the eigenvalues of $A \in M_n(\mathbb{C})$ are all equal to 1 in absolute value and if $\|Ax\| \leq 1$ for all unit vectors $x \in \mathbb{C}^n$, show that A is unitary.

4.86 Show that the $n \times n$ matrix U below with the (i, j)-entry $\frac{1}{\sqrt{n}}\omega^{(i-1)(j-1)}$, where ω is a primitive n-th root of unity, i.e., $\omega^n = 1$, $\omega^k \neq 1$, $1 \leq k < n$, is Vandermonde, symmetric, and unitary:

$$U = \frac{1}{\sqrt{n}} \begin{pmatrix} 1 & 1 & 1 & \cdots & 1 \\ 1 & \omega & \omega^2 & \cdots & \omega^{n-1} \\ 1 & \omega^2 & \omega^4 & \cdots & \omega^{2n-2} \\ \vdots & \vdots & \vdots & \vdots & \vdots \\ 1 & \omega^{n-1} & \omega^{2n-2} & \cdots & \omega^{(n-1)^2} \end{pmatrix}.$$

4.87 Let $A \in M_n(\mathbb{C})$ and let $U \in M_n(\mathbb{C})$ be a unitary matrix. Show that

$$\min_U \sigma_{\max}(U \circ A) \leq \frac{1}{\sqrt{n}} \Big(\sum_{i,j=1}^n |a_{ij}|^2 \Big)^{\frac{1}{2}}.$$

4.88 Let A be an $n \times n$ real symmetric matrix. Show that $I - iA$ and $I + iA$ are nonsingular and that $(I - iA)(I + iA)^{-1}$ is unitary.

4.89 Let A be an $n \times n$ invertible matrix. Show that A is normal, i.e., $A^*A = AA^*$, if and only if $A^{-1}A^*$ is unitary.

4.90 Find all 2×2 real orthogonal matrices.

4.91 Let $A = (a_{ij}) \in M_3(\mathbb{R})$ and $A \neq 0$. If the transpose A^t of A is equal to the adjugate $\mathrm{adj}(A)$ of A, show that A is an orthogonal matrix.

4.92 Let A be a Hermitian matrix. Show that the eigenvectors of A belonging to distinct eigenvalues are orthogonal, i.e., if $Au = \lambda u$ and $Av = \mu v$, $\lambda \neq \mu$, then $u^*v = 0$. What if A is non-Hermitian?

4.93 Show that there are no real orthogonal matrices A and B satisfying $A^2 - B^2 = AB$. Do unitary A and B exist such that $A^2 - B^2 = AB$?

4.94 If A and B are $n \times n$ real orthogonal matrices satisfying

$$\det A + \det B = 0,$$

show that

$$\det(A + B) = 0.$$

Can this be generalized to (complex) unitary matrices?

4.95 Let A and B be $n \times n$ real matrices. If $A > 0$ and $B = -B^t$, that is, A is real positive definite and B is real skew-symmetric, show that

$$\det(A + B) > 0.$$

Does this hold true for complex $A > 0$ and skew-Hermitian B?

4.96 Let $A \in M_n(\mathbb{C})$. Show that if A is unitary, then so is the matrix

$$\frac{1}{\sqrt{2}} \begin{pmatrix} A & -A \\ A & A \end{pmatrix}.$$

4.97 Let $A, B \in M_n(\mathbb{R})$. Show the rank identity:

$$r\begin{pmatrix} A & B \\ -B & A \end{pmatrix} = 2r(A + iB).$$

4.98 Let A be an $n \times n$ complex matrix. Show that A can be written as

$$A = R_1 + iR_2 = H_1 + iH_2,$$

where R_1 and R_2 are real; H_1 and H_2 are Hermitian. Find R_1, R_2, H_1, and H_2 in terms of A, and show that if $A \geq 0$, then $H_1 \geq 0$, $R_1 \geq 0$.

4.99 Let $A \in M_n(\mathbb{C})$ and $A \neq I$.

  (a) Can A be positive definite and unitary?

  (b) Can A be Hermitian and unitary?

  (c) Can A be upper-triangular, non-diagonal, and unitary?

4.100 Let V be an $n \times k$ (partial unitary) matrix such that $V^*V = I_k, k \leq n$. Define $f : M_n(\mathbb{C}) \to M_k(\mathbb{C})$ by $f(X) = V^*XV, X \in M_n(\mathbb{C})$. Show that

  (a) f is a positive unital linear map.

  (b) For any $n \times n$ positive semidefinite A,

$$V^*A^2V - (V^*AV)^2 \leq \tfrac{1}{4}\big(\lambda_{\max}(A) - \lambda_{\min}(A)\big)^2 I_k.$$

  (c) For any unit $x \in \mathbb{C}^n$ and $n \times n$ positive definite matrix A,

$$(x^*Ax)(x^*A^{-1}x) \leq \frac{\big(\lambda_{\max}(A) + \lambda_{\min}(A)\big)^2}{4\lambda_{\max}(A)\lambda_{\min}(A)}.$$

4.101   Let A and B be $n \times n$ positive definite matrices. Show that

$$(\lambda A + \tilde{\lambda} B)^{-1} \leq \lambda A^{-1} + \tilde{\lambda} B^{-1}, \quad \text{where } \lambda \in [0,1], \ \tilde{\lambda} = 1 - \lambda.$$

4.102   Let $A = (a_{ij}) \in M_n(\mathbb{C})$ and let $\lambda_1, \ldots, \lambda_n$ be the eigenvalues of A. Show that the following statements are equivalent:

  (a) A is normal, that is, $A^*A = AA^*$.

  (b) $I - A$ is normal.

  (c) A is unitarily diagonalizable, that is, there exists a unitary matrix U such that $U^*AU = \text{diag}(\lambda_1, \ldots, \lambda_n)$.

  (d) $\mathbb{C}^n$ has an orthonormal basis consisting of eigenvectors of A.

  (e) $Ax = \lambda x \Leftrightarrow A^*x = \bar{\lambda}x$, where $\lambda \in \mathbb{C}$, i.e., an eigenvector of A associated with λ is an eigenvector of A^* associated with $\bar{\lambda}$.

  (f) $A^* = AU$ for some unitary U.

  (g) $A^* = VA$ for some unitary V.

  (h) $\text{tr}(A^*A) = \sum_{i,j=1}^n |a_{ij}|^2 = \sum_{i=1}^n |\lambda_i|^2$.

  (i) The singular values of A are $|\lambda_1|, \ldots, |\lambda_n|$.

  (j) $\text{tr}(A^*A)^2 = \text{tr}\left((A^*)^2 A^2\right)$.

  (k) $\|Ax\| = \|A^*x\|$ for every $x \in \mathbb{C}^n$, where $\|y\| = \sqrt{y^*y}$, $y \in \mathbb{C}^n$.

  (l) $A + A^*$ and $A - A^*$ commute.

  (m) $A^*A - AA^*$ is positive semidefinite.

  (n) A commutes with A^*A.

  (o) A commutes with $AA^* - A^*A$.

4.103   Show that A is a normal matrix if and only if the following hold:

  (a) $A = B + iC$, where B and C are Hermitian and commute.

  (b) $A = HP = PH$, where $H \geq 0$ and P is unitary.

4.104   Let $A \in M_n(\mathbb{C})$ be a normal matrix. Show that

  (a) $\text{Ker}\, A^* = \text{Ker}\, A$.

  (b) $\text{Im}\, A^* = \text{Im}\, A$.

  (c) $\mathbb{C}^n = \text{Im}\, A \oplus \text{Ker}\, A$.

4.105   If $A = \begin{pmatrix} B & C \\ 0 & D \end{pmatrix}$ is a normal matrix, where B and D are square matrices (possibly of different sizes), what can be said about B, C, and D?

4.106   If A is a normal matrix and it commutes with matrix B, show that the transpose conjugate A^* of A also commutes with B.

4.107   Let A be a normal matrix. Show that

$$A\bar{A} = 0 \Leftrightarrow \bar{A}A = 0 \Leftrightarrow AA^t = 0 \Leftrightarrow A^tA = 0.$$

4.108   Let A and B be $n \times n$ normal matrices. Prove each of the statements:

(a) If $AB = BA$, then there exists a unitary matrix U such that U^*AU and U^*BU are both diagonal.

(b) If $AB = BA$, then AB and $A + B$ are normal. Give examples showing that the condition that $AB = BA$ is necessary.

(c) If $AB^* = B^*A$, then both AB and BA are normal.

(d) $A + iB$ is normal if and only if $AB^* + A^*B$ is Hermitian.

4.109   If A is a 3×3 matrix such that $A^2 = I$ and $A \neq \pm I$, show that the rank of one of $A + I$ and $A - I$ is 1 and the rank of the other is 2.

4.110   If A is a real matrix such that $A^3 + A = 0$, show that $\operatorname{tr} A = 0$.

4.111   Let A be an $n \times n$ matrix. If $A^k = I$ for some positive integer k, show that $\operatorname{tr}(A^{-1}) = \overline{\operatorname{tr}(A)}$. If such a k is less than n, show that A has repeated eigenvalues, that is, A cannot have n distinct eigenvalues.

4.112   (a) Let A be an $n \times n$ matrix. If $A^k = I$ for some positive integer k, show that $T^{-1}AT$ is diagonal for some complex matrix T.

(b) For $B = \begin{pmatrix} 0 & -1 \\ 1 & 0 \end{pmatrix}$, show that $B^4 = I_2$ and there does not exist a real invertible matrix P such that $P^{-1}BP$ is diagonal.

4.113   A matrix $A \in M_n(\mathbb{C})$ is *nilpotent* if $A^k = 0$ for some positive integer k; *idempotent* or a *projection* if $A^2 = A$; *involutory* if $A^2 = I$. Show that

(a) A is nilpotent if and only if all eigenvalues of A are zero.

(b) A is idempotent if and only if A is similar to a diagonal matrix of the form $\operatorname{diag}(1, \ldots, 1, 0, \ldots, 0)$.

(c) A is involutory if and only if A is similar to a diagonal matrix of the form $\operatorname{diag}(1, \ldots, 1, -1, \ldots, -1)$.

(d) A is idempotent if and only if $r(A) + r(A - I) = n$.

(e) A is involutory if and only if $r(A + I) + r(A - I) = n$.

4.114 Let A and B be nilpotent matrices of the same size. If $AB = BA$, show that $A + B$ and AB are also nilpotent. What if $AB \neq BA$?

4.115 Let $A, B \in M_n(\mathbb{C})$. If $AB + BA = 0$ and if B is not nilpotent, show that the matrix equation $AX + XA = B$ has no solution.

4.116 If A is a nilpotent matrix, show that

    (a) $I - A$ is invertible. Find $(I - A)^{-1}$.

    (b) $I + A$ is also invertible.

    (c) $\operatorname{tr} A = 0$.

    (d) A is not diagonalizable unless $A = 0$.

4.117 Let A and B be idempotent matrices of the same size. Show that $A + B$ is idempotent if and only if $AB = BA = 0$.

4.118 Let A and B be Hermitian. Show that if AB is idempotent, so is BA.

4.119 Let A be an $n \times n$ matrix of rank $r > 0$. Show that $A^2 = A$ if and only if there exist an $r \times n$ matrix B and an $n \times r$ matrix C, both of rank r, such that $A = BC$ and $CB = I_r$. Show also that if $A^2 = A$ then

$$\det(2I_n - A) = 2^{n-r} \quad \text{and} \quad \det(A + I_n) = 2^r.$$

4.120 Let A be an $n \times n$ matrix of rank $r > 0$. If A satisfies $A^2 = A$ and $A \neq I$, show that for every positive integer k, $1 < k \leq n - r$, there exists an $n \times n$ matrix B such that $AB = BA = 0$, and

$$(A + B)^{k+1} = (A + B)^k \neq (A + B)^{k-1}.$$

4.121 Let $A \in M_n(\mathbb{C})$ be an idempotent matrix.

    (a) Find $\det(A + I)$ and $\det(A - I)$.

    (b) Show that $r(A) = \operatorname{tr} A$.

    (c) Show that $\dim \operatorname{Im} A = \operatorname{tr} A$.

4.122 (a) Let A be an $n \times n$ idempotent matrix. Show that

$$x \in \operatorname{Im} A \quad \text{if and only if} \quad x = Ax.$$

    (b) Let A and B be $n \times n$ idempotent Hermitian matrices. Show that

$$\operatorname{Im} A = \operatorname{Im} B \quad \text{if and only if} \quad A = B.$$

4.123 A square complex matrix X is said to be an *orthogonal projection* if it is idempotent and Hermitian (i.e., $X^2 = X$ and $X^* = X$). Let A and B be orthogonal projections of the same size. Show that

$$B \leq A \qquad \text{if and only if} \qquad AB = B.$$

4.124 Let P and Q be orthogonal projections of the same size. Show that the following statements are equivalent:

(a) $P + Q$ is an orthogonal projection.

(b) $PQ + QP = 0$.

(c) $PQ = 0$.

(d) $QP = 0$.

(e) $\operatorname{Im} P \subseteq \operatorname{Ker} Q$.

(f) $\operatorname{Im} Q \subseteq \operatorname{Ker} P$.

4.125 Let $A \in M_n(\mathbb{C})$ be an involutory matrix, i.e., $A^2 = I$. Show that the following statements are equivalent:

(a) A is Hermitian.

(b) A is normal.

(c) A is unitary.

(d) All singular values of A are equal to 1.

4.126 Let $A \in M_n(\mathbb{C})$ be an involutory matrix, i.e., $A^2 = I$. Show that

(a) $\frac{1}{2}(I + A)$ and $\frac{1}{2}(I - A)$ are idempotent, and their product is 0.

(b) $r(I + A) + r(I - A) = n$.

(c) A has only eigenvalues ± 1.

(d) $\mathbb{C}^n = V_1 \oplus V_{-1}$, where V_1 and V_{-1} are the eigenspaces of the eigenvalues 1 and -1, respectively.

(e) $\operatorname{Im}(I - A) \subseteq \operatorname{Ker}(I + A)$.

Conversely, which of the above statements imply $A^2 = I$?

4.127 Let A and B be $n \times n$ nonsingular matrices such that $ABA = B$ and $BAB = A$. Show that $A^2 = B^2$ and $A^4 = B^4 = I$.

4.128 Let A and B be $n \times n$ involutory matrices, i.e., $A^2 = B^2 = I$. Show that

$$\text{Im}(AB - BA) = \text{Im}(A - B) \cap \text{Im}(A + B).$$

4.129 Let $A, B \in M_n(\mathbb{C})$ be such that $A = \frac{1}{2}(B + I)$. Show that A is idempotent, i.e., $A^2 = A$, if and only if B is involutory, i.e. $B^2 = I$.

4.130 Let A be a square matrix and let λ be a nonzero scalar. Show that

(a) $\begin{pmatrix} I & A \\ A^* & I \end{pmatrix}$ is Hermitian.

(b) $\begin{pmatrix} A & A^* \\ A^* & A \end{pmatrix}$ is normal.

(c) $\begin{pmatrix} A & \frac{1}{\lambda}A \\ \lambda(I - A) & I - A \end{pmatrix}$ is idempotent.

(d) $\begin{pmatrix} A & \frac{1}{\lambda}A \\ -\lambda A & -A \end{pmatrix}$ is nilpotent.

(e) $\begin{pmatrix} (A^*A)^{\frac{1}{2}} & A^* \\ A & (AA^*)^{\frac{1}{2}} \end{pmatrix}$ is positive semidefinite.

4.131 For $A \in M_n(\mathbb{C})$, denote the *field of values* or *numerical range* of A by

$$F(A) = \{x^*Ax \mid x \in \mathbb{C}^n, \, x^*x = 1\} = \left\{ \frac{x^*Ax}{x^*x} \mid x \in \mathbb{C}^n, \, x \neq 0 \right\}.$$

(a) Show that $F(A + cI) = F(A) + c$ for every $c \in \mathbb{C}$.

(b) Show that $F(cA) = cF(A)$ for every $c \in \mathbb{C}$.

(c) Show that the diagonal entries of A contained in $F(A)$.

(d) Show that the eigenvalues of A are contained in $F(A)$.

(e) Are the singular values of A contained in $F(A)$?

(f) Show that $F(U^*AU) = F(A)$ for any unitary $U \in M_n(\mathbb{C})$.

(g) If A is singular, show that $0 \in F(A)$. Is the converse true?

(h) Show that $0 \notin F(A)$ if and only if $0 \notin F(A^{-1})$.

(i) Describe $F(A)$ when A is Hermitian.

(j) Describe $F(A)$ when A is positive semidefinite.

(k) Determine $F(A)$, where A is a matrix in the following:

$$\begin{pmatrix} 0 & 1 \\ 1 & 0 \end{pmatrix}, \begin{pmatrix} 1 & 0 \\ 0 & 0 \end{pmatrix}, \begin{pmatrix} 0 & 1 \\ 0 & 0 \end{pmatrix}, \begin{pmatrix} 0 & 0 \\ 1 & 1 \end{pmatrix}, \begin{pmatrix} 1 & 0 \\ 0 & 1+i \end{pmatrix}, \begin{pmatrix} 0 & 0 & 0 \\ 0 & 1 & 0 \\ 0 & 0 & 1+i \end{pmatrix}.$$

4.132   Let $A \in M_n(\mathbb{C})$ be a normal matrix with eigenvalues $\lambda_1, \ldots, \lambda_n$ (not necessarily distinct). Show that the numerical range $F(A)$ is the convex hull of $\lambda_1, \ldots, \lambda_n$, i.e., $F(A)$ is the smallest polygon in the complex plane containing all $\lambda_1, \ldots, \lambda_n$. Is the converse true?

4.133   Let $A \in M_n(\mathbb{C})$. If $0 \in F(A)$ and $1 \in F(A)$, show that $[0, 1] \subseteq F(A)$.

4.134   (a) Let $A \in M_n(\mathbb{C})$ be positive definite. Show that it is impossible that both $x^* A x < 1$ and $x^* A^{-1} x < 1$ hold for a unit $x \in \mathbb{C}^n$.

   (b) Show by example that it is possible that both $x^* A x > 1$ and $x^* A^{-1} x > 1$ hold for some positive definite A and unit $x \in \mathbb{C}^n$.

4.135   A *permutation matrix* is a square matrix that has exactly one 1 in each row and each column, and all other entries are equal to 0.

   (a) How many $n \times n$ permutation matrices are there?

   (b) Show that the product of two $n \times n$ permutation matrices is again a permutation matrix. How about the sum?

   (c) Show that a permutation matrix is an orthogonal matrix.

   (d) Show that two permutation matrices are similar if and only if they have the same eigenvalues.

   (e) Show that every permutation matrix is invertible and its inverse equals its transpose (which is also a permutation matrix).

   (f) Let R, P, and Q be permutation matrices. If $R = t_1 P + t_2 Q$, where $0 < t_1, t_2 < 1$, $t_1 + t_2 = 1$, show that $R = P = Q$.

   (g) Find the permutation matrices P such that $P^2 = I$.

   (h) Show that 1 is an eigenvalue of any permutation matrix.

   (i) Find all 4×4 symmetric permutation matrices that have just two positive eigenvalues. What are these positive eigenvalues?

   (j) Does there exist a 5×5 symmetric permutation matrix that has just two positive eigenvalues?

   (k) Show that two symmetric permutation matrices are similar if and only if they have the same number of positive eigenvalues.

   (l) Let $P = \begin{pmatrix} A & B \\ C & D \end{pmatrix}$ be a partitioned permutation matrix, where A, B, C, and D have size $n \times n$. Show that $B = 0$ if and only if $C = 0$ and that $A = 0$ if and only if $D = 0$.

4.136   Let P be the following $n \times n$ permutation matrix, $n \geq 2$,

$$P = \begin{pmatrix} 0 & 1 & 0 & \cdots & 0 & 0 \\ 0 & 0 & 1 & \cdots & 0 & 0 \\ 0 & 0 & 0 & \cdots & 0 & 0 \\ \vdots & \vdots & \vdots & \ddots & \vdots & \vdots \\ 0 & 0 & 0 & \cdots & 0 & 1 \\ 1 & 0 & 0 & \cdots & 0 & 0 \end{pmatrix} = \begin{pmatrix} 0 & I_{n-1} \\ 1 & 0 \end{pmatrix}.$$

(a)   Show that for any positive integer k, $1 \leq k < n$,

$$P^k = \begin{pmatrix} 0 & I_{n-k} \\ I_k & 0 \end{pmatrix}$$

and

$$P^{n-1} = P^t, \quad P^n = I_n, \quad P^t = P^{-1}.$$

(b)   For what k, $1 \leq k < n$, is P^k a symmetric matrix?

(c)   What are the eigenvalues of P and P^k?

(d)   Show that $P, P^2, \ldots, P^n$ are linearly independent.

(e)   Show that $P^i + P^j$ is a normal matrix for $1 \leq i, j < n$.

(f)   When is $P^i + P^j$ a symmetric matrix for $1 \leq i, j < n$?

(g)   If $n \geq 3$, show that P is diagonalizable over $\mathbb{C}$ but not over $\mathbb{R}$.

(h)   If $1 < i < n$ and $(i, n) = 1$, show that P^i is similar to P.

(i)   Give an example showing that if $(i, n) > 1$, then (h) is false, that is, P^i and P need not be similar.

   (Note: (p, q) denotes the greatest common divisor of p and q.)

4.137   Let A be an $m \times m$ invertible matrix with nonnegative integer entries. Let $\Sigma(A^n)$ be the sum of all entries of A^n. If there exists a constant N such that $\Sigma(A^n) \leq N$ for all n, show that A is a permutation matrix. Equivalently, if the union over all n of the sets of entries of A^n is a finite set, show that A is a permutation matrix.

4.138   Let A be an $n \times n$ matrix such that every row and every column have one and only one nonzero entry that is either 1 or -1 and all other entries are 0. Show that $A^k = I$ for some positive integer k.

4.139   Let A be an $(n-1) \times n$ matrix of integers such that the row sums are all equal to zero. Show that $\det(AA^t) = nk^2$ for some integer k.

4.140   Let A be an $n \times n$ matrix whose entries are equal to either 1 or -1 and whose rows are mutually orthogonal. Suppose that A has an $s \times t$ submatrix having entries all equal to 1. Show that $st \leq n$.

4.141   Let A be an $n \times n$ nonzero real symmetric matrix. Denote the sum of all entries of A by $S(A)$, i.e., $S(A) = \sum_{i,j=1}^{n} a_{ij}$. Show that

$$\left(S(A)\right)^2 \leq nS(A^2).$$

4.142   Let

$$A = \begin{pmatrix} 0 & 0 & 1 \\ 1 & 0 & 0 \\ 0 & -2 & 0 \end{pmatrix} \quad \text{and} \quad B = \begin{pmatrix} 0 & 1 & 0 \\ 0 & 0 & 1 \\ 1 & 0 & 0 \end{pmatrix}.$$

Show that

$$(A + tB)^3 = A^3 + t^3 B^3, \quad t \in \mathbb{R}.$$

4.143   Show that the matrix equation $X^4 + Y^4 = Z^4$ has nontrivial solutions:

$$\begin{pmatrix} 0 & x \\ 1 & 0 \end{pmatrix}^4 + \begin{pmatrix} 0 & y \\ 1 & 0 \end{pmatrix}^4 = \begin{pmatrix} 0 & z \\ 1 & 0 \end{pmatrix}^4,$$

where $x = 2pq$, $y = p^2 - q^2$, and $z = p^2 + q^2$, p and q are integers.

4.144   (a) If A is a normal matrix whose eigenvalues are exactly the main diagonal entries of A, show that A is a diagonal matrix.

        (b) Construct a 3×3 real non-symmetric matrix A that satisfies (i) the diagonal entries are $1, 2, 3$, (ii) the eigenvalues are $1, 2, 3$, and (iii) no entry of A is zero.

        (c) If $A = \begin{pmatrix} a & b \\ c & d \end{pmatrix} \in M_2(\mathbb{C})$ is a normal matrix and a is an eigenvalue of A, show that $b = c = 0$, thus $A = \text{diag}(a, d)$.

        (d) Construct a 3×3 real positive definite matrix $A = \begin{pmatrix} a & b & c \\ b & d & e \\ c & e & f \end{pmatrix}$ such that a is an eigenvalue of A, $b^2 + c^2 \neq 0$ (A is not diagonal).

        (e) What are the matrices whose singular values are eigenvalues?

4.145   Let $A \in M_n(\mathbb{C})$ and $H = \frac{1}{2}(A + A^*)$. Let $\lambda_1, \ldots, \lambda_n$ and $\mu_1, \ldots, \mu_n$ be respectively the eigenvalues of A and H arranged in such an order that $|\operatorname{Re}\lambda_1| \geq \cdots \geq |\operatorname{Re}\lambda_n|$ and $|\mu_1| \geq \cdots \geq |\mu_n|$. Show that

$$\sum_{t=1}^{k} |\operatorname{Re}\lambda_t| \leq \sum_{t=1}^{k} |\mu_t|, \quad k = 1, \ldots, n.$$

Chapter 5

Inner Product Spaces

Definitions, Facts, and Examples _____

Inner Product Space. Let V be a vector space over a scalar field $\mathbb{F}$ ($= \mathbb{C}$ or $\mathbb{R}$). An *inner product* on V is a mapping that assigns every ordered pair of vectors u and v in V a unique scalar in $\mathbb{F}$, denoted by $\langle u, v \rangle$, and it possesses the properties (i)–(iv): for all vectors $u, v, w \in V$ and scalars $\lambda \in \mathbb{F}$,

(i) Positivity: $\langle u, u \rangle \geq 0$. Equality holds if and only if $u = 0$.

(ii) Homogeneity: $\langle \lambda u, v \rangle = \lambda \langle u, v \rangle$.

(iii) Additivity: $\langle u, v + w \rangle = \langle u, v \rangle + \langle u, w \rangle$.

(iv) Hermitian Symmetry: $\langle u, v \rangle = \overline{\langle v, u \rangle}$.

A vector space equipped with an inner product is called an *inner product space*. An inner product space over $\mathbb{R}$ is also referred to as a *Euclidean space*.

Let W be a subspace of an inner product space V. With the restriction of the inner product of V on W, W is also an inner product space.

One may derive additional properties for an inner product space:

$$\langle u, \lambda v \rangle = \bar{\lambda} \langle u, v \rangle,$$
$$\langle \lambda u + \mu v, w \rangle = \lambda \langle u, w \rangle + \mu \langle v, w \rangle,$$
$$\langle u + v, u + v \rangle = \langle u, u \rangle + 2 \operatorname{Re}\langle u, v \rangle + \langle v, v \rangle.$$

Note that in some texts, the homogeneity is defined with respect to the second variable v, i.e., $\langle u, \lambda v \rangle = \lambda \langle u, v \rangle$. In general, this is different from (but equivalent to) our definition. In the case of Euclidean space, there is no difference, because $\langle u, v \rangle = \langle v, u \rangle$ for Euclidean spaces.

A few examples of inner product spaces we often see:

- $\mathbb{R}^n$ is an inner product space with the standard inner product

$$\langle u, v \rangle = v^t u = u_1 v_1 + u_2 v_2 + \cdots + u_n v_n.$$

- $\mathbb{C}^n$ is an inner product space with the standard inner product

$$\langle u, v \rangle = v^* u = u_1 \bar{v}_1 + u_2 \bar{v}_2 + \cdots + u_n \bar{v}_n.$$

- $M_{m \times n}(\mathbb{C})$ is an inner product space with the *Frobenius inner product*

$$\langle A, B \rangle = \text{tr}(B^* A).$$

- Let l^2 be the vector space of all infinite complex sequences $u = (u_1, u_2, \dots)$ with the convergence property $\sum_{i=1}^{\infty} |u_i|^2 < \infty$. Then l^2 is an inner product space with respect to the inner product

$$\langle u, v \rangle = \sum_{i=1}^{\infty} u_i \bar{v}_i.$$

- The vector space $C[a, b]$ of all real-valued continuous functions on the closed interval $[a, b]$ is an inner product space with the inner product

$$\langle f, g \rangle = \int_a^b f(t) g(t) dt.$$

Cauchy–Schwarz Inequality. The Cauchy–Schwartz inequality is one of the most fundamental results in mathematics with a great amount of applications. In previous chapters we used its version for $\mathbb{R}^n$ (and $\mathbb{C}^n$). It states: for any two vectors $u = (u_1, u_2, \dots, u_n)^t, v = (v_1, v_2, \dots, v_n)^t \in \mathbb{R}^n$,

$$\left(\sum_{i=1}^{n} u_i v_i \right)^2 \le \sum_{i=1}^{n} u_i^2 \sum_{i=1}^{n} v_i^2. \tag{5.1}$$

Equality occurs if and only if $u = kv$ or $v = ku$ for some scalar $k \in \mathbb{R}$. The inequality (5.1) follows immediately from the identity

$$\sum_{i=1}^{n} u_i^2 \sum_{i=1}^{n} v_i^2 - \left(\sum_{i=1}^{n} u_i v_i \right)^2 = \frac{1}{2} \sum_{i, j} \left(u_i v_j - u_j v_i \right)^2. \tag{5.2}$$

(5.1) can also be shown by considering the quadratic function in t:

$$\sum_{i=1}^{n} (tu_i + v_i)^2 = t^2 \sum_{i=1}^{n} u_i^2 + 2t \sum_{i=1}^{n} u_i v_i + \sum_{i=1}^{n} v_i^2 \tag{5.3}$$

whose discriminant is nonpositive. (Note: in (5.3) we assumed $u \ne 0$.)

Slight modifications of (5.2) reveal the inequality for vectors in $\mathbb{C}^n$:

$$\left| \sum_{i=1}^{n} u_i v_i \right|^2 \leq \sum_{i=1}^{n} |u_i|^2 \sum_{i=1}^{n} |v_i|^2. \tag{5.4}$$

The inequality (5.4) can be proved in various ways. Using matrices, if we set $A = (u, v)$, then $A^*A = \begin{pmatrix} u^* \\ v^* \end{pmatrix} (u, v) = \begin{pmatrix} u^*u & u^*v \\ v^*u & v^*v \end{pmatrix}$ is a positive semidefinite matrix. Taking the determinant of A^*A, we obtain

$$\begin{vmatrix} u^*u & u^*v \\ v^*u & v^*v \end{vmatrix} = (u^*u)(v^*v) - |u^*v|^2 \geq 0. \tag{5.5}$$

"=" holds if and only if A^*A is singular, i.e., u and v are linearly dependent. Exercise: Let $A = uv^* - vu^*$. Use $\operatorname{tr}(A^*A) \geq 0$ to show (5.4).

The Cauchy–Schwartz inequality holds in more general settings of inner product spaces. Let V be an inner product space over $\mathbb{C}$ or $\mathbb{R}$ with inner product $\langle \cdot, \cdot \rangle$. The *Cauchy–Schwartz inequality* states that for all $u, v \in V$,

$$|\langle u, v \rangle|^2 \leq \langle u, u \rangle \langle v, v \rangle. \tag{5.6}$$

Equality holds if and only if $u = kv$ or $v = ku$ for some scalar k.

Proof. We may assume that u and v are nonzero and that $\langle u, v \rangle$ is real (otherwise, replace u by $e^{i\theta} u$ and choose $\theta \in \mathbb{R}$ so that $\langle e^{i\theta} u, v \rangle$ is real). Let $f(t) = \langle u + tv, u + tv \rangle$, $t \in \mathbb{R}$. Then $f(t) \geq 0$ for all real t. On the other hand, $f(t) = \langle u, u \rangle + 2t\langle u, v \rangle + t^2 \langle v, v \rangle$. The discriminant of the quadratic function in t is $4|\langle u, v \rangle|^2 - 4\langle u, u \rangle \langle v, v \rangle$, and it is nonpositive, revealing the inequality (5.6). Equality occurs if and only if $f(t) = 0$ for some $t \in \mathbb{R}$, i.e., $u + tv = 0$, in other words, u and v are linearly dependent.

Another proof. Suppose $v \neq 0$ and let $r = \frac{\langle u, v \rangle}{\langle v, v \rangle}$. We compute

$$
\begin{aligned}
\langle u - rv, u - rv \rangle &= \langle u, u \rangle - \langle rv, u \rangle - \langle u, rv \rangle + \langle rv, rv \rangle \\
&= \langle u, u \rangle - r\langle v, u \rangle - \bar{r}\langle u, v \rangle + |r|^2 \langle v, v \rangle \\
&= \langle u, u \rangle - r\overline{\langle u, v \rangle} - \bar{r}\langle u, v \rangle + |r|^2 \langle v, v \rangle \\
&= \langle u, u \rangle - \frac{|\langle u, v \rangle|^2}{\langle v, v \rangle} - \frac{|\langle u, v \rangle|^2}{\langle v, v \rangle} + \frac{|\langle u, v \rangle|^2}{\langle v, v \rangle} \\
&= \langle u, u \rangle - \frac{|\langle u, v \rangle|^2}{\langle v, v \rangle}.
\end{aligned}
$$

Since $0 \leq \langle u - rv, u - rv \rangle$, it follows that $|\langle u, v \rangle|^2 \leq \langle u, u \rangle \langle v, v \rangle$. Equality occurs if and only if $u - rv = 0$, that is, u and v are linearly dependent.

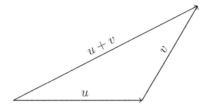

Figure 5.1: The triangle inequality

Length or Norm of a Vector. Let V be a vector space over $\mathbb{F}$. A *norm* $\|\cdot\|$ on V is a function from V to $[0, \infty)$ that has the following properties:

(1) $\|u\| \geq 0$ for all $u \in V$. Equality holds if and only if $u = 0$.

(2) $\|\lambda u\| = |\lambda| \, \|u\|$ for all $u \in V$ and $\lambda \in \mathbb{F}$.

(3) Triangle inequality: $\|u + v\| \leq \|u\| + \|v\|$ for all $u, v \in V$.

A *normed (vector) space* is a vector space on which a norm is defined. For a vector u in a normed space, $\|u\|$ is called the *length* or *norm* of u. The zero vector has length 0. A *unit vector* is a vector of norm 1.

If V is an inner product space, then the inner product naturally induces a norm: $\|u\| = \sqrt{\langle u, u \rangle}$. (The triangle inequality is derived by the Cauchy–Schwartz inequality below.) However, a norm need not be induced by an inner product (Problem 5.51). We focus mainly on inner product spaces. By saying that V is an inner product space, we mean that V is already equipped with $\langle \cdot, \cdot \rangle$ as well as the *induced norm* $\|\cdot\| = \sqrt{\langle \cdot, \cdot \rangle}$. In terms of the induced norm, the Cauchy–Schwartz inequality is written as

$$|\langle u, v \rangle| \leq \|u\| \, \|v\|. \tag{5.7}$$

The triangle inequality (3) is a consequence of the Cauchy–Schwarz inequality for any inner product space with the induced norm, because

$$
\begin{aligned}
\|u + v\|^2 &= \langle u + v, u + v \rangle \\
&= \langle u, u \rangle + \langle u, v \rangle + \langle v, u \rangle + \langle v, v \rangle \\
&= \|u\|^2 + 2 \operatorname{Re}\langle u, v \rangle + \|v\|^2 \\
&\leq \|u\|^2 + 2|\langle u, v \rangle| + \|v\|^2 \\
&\leq \|u\|^2 + 2\|u\| \, \|v\| + \|v\|^2 \\
&= (\|u\| + \|v\|)^2.
\end{aligned}
$$

The triangle inequality for an inner product space becomes an equality, i.e., $\|u+v\| = \|u\| + \|v\|$, if and only if $v = 0$ or $u = 0$ or $u = kv$ for some $k > 0$.

A few common norms on $\mathbb{C}^n$ (and $\mathbb{R}^n$):

(i) ℓ_1-norm: $\|u\|_1 = |u_1| + |u_2| + \cdots + |u_n|$.

(ii) ℓ_2-norm: $\|u\|_2 = (|u_1|^2 + |u_2|^2 + \cdots + |u_n|^2)^{\frac{1}{2}}$ (*Euclidean norm*).

(iii) ℓ_p-norm: $\|u\|_p = (|u_1|^p + |u_2|^p + \cdots + |u_n|^p)^{\frac{1}{p}}$, $p \geq 1$.

(iv) ℓ_∞-norm: $\|u\|_\infty = \max\{|u_1|, |u_2|, \ldots, |u_n|\}$.

Angles between vectors and between one dimensional subspaces may be introduced through the inner product. For nonzero vectors u and v in a Euclidean space, say $\mathbb{R}^n$, we may define the angle α between u and v, and the angle β between the spaces Span$\{u\}$ and Span$\{v\}$ respectively by

$$\alpha = \cos^{-1} \frac{\langle u, v \rangle}{\|u\| \, \|v\|}, \quad \alpha \in [0, \pi], \quad \beta = \cos^{-1} \frac{|\langle u, v \rangle|}{\|u\| \, \|v\|}, \quad \beta \in [0, \tfrac{\pi}{2}].$$

Let $A \in M_n(\mathbb{C})$. Applying the Cauchy–Schwartz inequality gives

$$|\langle Ax, y \rangle| \leq \|Ax\| \, \|y\|, \quad x, y \in \mathbb{C}^n.$$

In particular, for unit $x \in \mathbb{C}^n$, as $x^* A x = \langle Ax, x \rangle$ and $\|Ax\|^2 = x^* A^* A x$,

$$|x^* A x| \leq \|Ax\| = \left(x^* (A^* A) x\right)^{\frac{1}{2}}.$$

So the eigenvalues $\lambda(A)$ of A and the largest singular value of A obey

$$\max |\lambda(A)| \leq \sigma_{\max}(A). \tag{5.8}$$

Example 5.1 Let A and B be $m \times n$ complex matrices. Show that

$$|\operatorname{tr}(AB^*)|^2 \leq \operatorname{tr}(AA^*) \operatorname{tr}(BB^*).$$

Solution. Note that $M_{m \times n}(\mathbb{C})$ is an inner product space with respect to

$$\langle A, B \rangle = \operatorname{tr}(B^* A) = \operatorname{tr}(AB^*).$$

The desired inequality follows from the Cauchy–Schwartz inequality. □

Two common norms of matrices:

- The *Frobenius norm*, also known as the *Schur norm* or the *Hilbert–Schmidt norm*, is derived from the usual inner product:

$$\|A\|_F = \sqrt{\langle A, A \rangle} = \left(\operatorname{tr}(A^* A)\right)^{\frac{1}{2}} = \left(\sum_{i,\, j} |a_{ij}|^2\right)^{\frac{1}{2}} = \left(\sum_i \sigma_i^2(A)\right)^{\frac{1}{2}}.$$

- The *spectral norm*, also called the *operator norm*, is induced by the ℓ_2-norm on $\mathbb{C}^n$, and it is equal to the largest singular value:

$$\begin{aligned} \|A\|_{\mathrm{sp}} &= \|A\|_{\mathrm{op}} = \|A\|_2 = \sigma_{\max}(A) \\ &= \max_{\|x\|_2=1} \|Ax\|_2 = \max_{x \neq 0} \frac{\|Ax\|_2}{\|x\|_2}. \end{aligned} \quad (5.9)$$

It is immediate that for any $m \times n$ matrix A and any vector $x \in \mathbb{C}^n$,

$$\|A\|_{\mathrm{sp}} \leq \|A\|_F \quad \text{and} \quad \|Ax\|_2 \leq \|A\|_{\mathrm{sp}} \|x\|_2.$$

There are other norms on matrices such as the Schatten p-norms, Fan norms, etc. This is a huge area of matrix analysis that one may explore.

Unless otherwise stated, for $x \in \mathbb{C}^n$ (or $\mathbb{R}^n$), $\|x\| = \sqrt{x^*x} = \sqrt{\langle x, x \rangle}$, which is the ℓ_2-norm of x with respect to the standard inner product, and for $A \in M_{m \times n}(\mathbb{C})$, $\|A\| = \|A\|_F = \sqrt{\mathrm{tr}(A^*A)} = \sqrt{\langle A, A \rangle}$, which is the Frobenius norm of A with respect to the usual (Frobenius) inner product.

We have seen that the Cauchy–Schwartz inequality easily follows from the nonnegativity of the determinant of the 2×2 matrix $\begin{pmatrix} \langle u, u \rangle & \langle v, u \rangle \\ \langle u, v \rangle & \langle v, v \rangle \end{pmatrix}$. More inequalities can be derived similarly via matrices of higher dimensions.

Example 5.2 Let u, v, and w be any three vectors in $\mathbb{C}^n$. Show that

$$\|u\|^2\|v\|^2\|w\|^2 + 2|\langle u, v \rangle \langle v, w \rangle \langle w, u \rangle|$$

$$\geq \|u\|^2\|v\|^2\|w\|^2 + 2\,\mathrm{Re}\left(\langle u, v \rangle \langle v, w \rangle \langle w, u \rangle\right) \quad (5.10)$$

$$\geq \|u\|^2|\langle v, w \rangle|^2 + \|v\|^2|\langle w, u \rangle|^2 + \|w\|^2|\langle u, v \rangle|^2 \quad (5.11)$$

and then derive the following inequalities:

(1) For any vectors $u, v \in \mathbb{C}^n$ (the Cauchy–Schwartz inequality),

$$\|u\|\|v\| \geq |\langle u, v \rangle|.$$

(2) For all unit vectors $u, v, w \in \mathbb{C}^n$,

$$\begin{aligned} &|\langle u, v \rangle|^2 + |\langle v, w \rangle|^2 + |\langle w, u \rangle|^2 \\ \leq\ & 1 + 2\,\mathrm{Re}\left(\langle u, v \rangle \langle v, w \rangle \langle w, u \rangle\right) \\ \leq\ & 1 + 2|\langle u, v \rangle \langle v, w \rangle \langle w, u \rangle|. \end{aligned}$$

(3) For any $u = (u_1, \ldots, u_n)^t, v = (v_1, \ldots, v_n)^t \in \mathbb{C}^n$ and for $i = 1, \ldots, n$,

$$\|u\|^2\|v\|^2 - |\langle u, v \rangle|^2 \geq 2|u_i v_i|\left(\|u\|\|v\| - |\langle u, v \rangle|\right)$$

and

$$\|u\|^2\|v\|^2 - |\langle u, v\rangle|^2 \geq \tfrac{2}{n}\langle |u|, |v|\rangle(\|u\|\|v\| - |\langle u, v\rangle|),$$

where $|x| = (|x_1|, \ldots, |x_n|)^t$ for $x = (x_1, \ldots, x_n)^t \in \mathbb{C}^n$,

(4) For any vectors $u, v \in \mathbb{C}^n$, if u and v are not linearly dependent, then

$$\|u\|\|v\| - |\langle u, v\rangle| \geq 2\big(\tfrac{1}{n}\langle |u|, |v|\rangle - |\langle u, v\rangle|\big).$$

(5) For any vectors $u, v \in \mathbb{C}^n$

$$\|u\|^2\|v\|^2 - |\langle u, v\rangle|^2 \geq \tfrac{1}{n}\big(s_v\|u\| - s_u\|v\|\big)^2,$$

where $s_x = |x_1 + \cdots + x_n|$ for $x = (x_1, \ldots, x_n)^t \in \mathbb{C}^n$.

Solution. Since $\operatorname{Re} c \leq |c|$ for $c \in \mathbb{C}$, inequality (5.10) is immediate. (5.11) is obtained by taking the determinant of the positive semidefinite matrix

$$\begin{pmatrix} u^* \\ v^* \\ w^* \end{pmatrix} (u, v, w) = \begin{pmatrix} \langle u, u\rangle & \langle v, u\rangle & \langle w, u\rangle \\ \langle u, v\rangle & \langle v, v\rangle & \langle w, v\rangle \\ \langle u, w\rangle & \langle v, w\rangle & \langle w, w\rangle \end{pmatrix}.$$

(1) In (5.11), let w be a unit vector that is orthogonal to u, i.e., $w^*u = 0$.

(2) This is immediate from (5.10) and (5.11) because $\|u\| = \|v\| = \|w\| = 1$.

(3) Let $w = (0, \ldots, 1, \ldots 0)^t \in \mathbb{C}^n$ with 1 in the i-th position and 0 elsewhere. Then $\langle v, w\rangle = v_i$ and $\langle w, u\rangle = \overline{u_i}$. Inequality (5.11) reveals

$$\|u\|^2\|v\|^2 + 2|u_i v_i||\langle u, v\rangle| \geq |v_i|^2\|u\|^2 + |u_i|^2\|v\|^2 + |\langle u, v\rangle|^2.$$

It follows that

$$\begin{aligned} \|u\|^2\|v\|^2 - |\langle u, v\rangle|^2 &\geq |v_i|^2\|u\|^2 + |u_i|^2\|v\|^2 - 2|u_i v_i||\langle u, v\rangle| \\ &\geq 2|u_i v_i|\big(\|u\|\|v\| - |\langle u, v\rangle|\big). \end{aligned}$$

By taking sum over i and dividing by n, we arrive at

$$\|u\|^2\|v\|^2 - |\langle u, v\rangle|^2 \geq \tfrac{2}{n}\langle |u|, |v|\rangle(\|u\|\|v\| - |\langle u, v\rangle|).$$

(4) If u and v are linearly independent, then $\|u\|\|v\| - |\langle u, v\rangle| \neq 0$. So

$$\|u\|\|v\| + |\langle u, v\rangle| \geq \tfrac{2}{n}\langle |u|, |v|\rangle.$$

Subtracting $2|\langle u, v\rangle|$ from both sides implies the desired inequality.

(5) Let $w = (\frac{1}{\sqrt{n}}, \ldots, \frac{1}{\sqrt{n}})^t$. Compute the expression in (5.11) to get

$$|\langle u, v \rangle|^2 + \tfrac{1}{n} s_u^2 \|v\|^2 + \tfrac{1}{n} s_v^2 \|u\|^2.$$

Observe that

$$|\langle u, v \rangle \langle v, w \rangle \langle w, u \rangle| = \tfrac{1}{n} s_u s_v |\langle u, v \rangle| \leq \tfrac{1}{n} s_u s_v \|u\| \|v\|.$$

By inequality (5.11), we arrive at

$$\begin{aligned}
\|u\|^2 \|v\|^2 &\geq |\langle u, v \rangle|^2 + \tfrac{1}{n} s_u^2 \|v\|^2 + \tfrac{1}{n} s_v^2 \|u\|^2 - \tfrac{2}{n} s_u s_v \|u\| \|v\| \\
&= |\langle u, v \rangle|^2 + \tfrac{1}{n}(s_v \|u\| - s_u \|v\|)^2. \quad \square
\end{aligned}$$

Orthogonal Vectors. Let u and v be vectors in an inner product space V. If the inner product $\langle u, v \rangle = 0$, then we say that u and v are *orthogonal* and write $u \perp v$. Nonzero orthogonal vectors are necessarily linearly independent. Suppose, say, $\lambda u + \mu v = 0$. Then $\lambda = 0$, thus $\mu = 0$, due to

$$0 = \langle \lambda u + \mu v, u \rangle = \lambda \langle u, u \rangle + \mu \langle v, u \rangle = \lambda \langle u, u \rangle.$$

A subset S of V is called an *orthogonal set* if $u \perp v$ for all $u, v \in S$, $u \neq v$. S is further said to be an *orthonormal set* if S is an orthogonal set and all vectors in S are unit vectors. Two subsets S and T of V are said to be *orthogonal*, denoted by $S \perp T$, if $u \perp v$ for all $u \in S$ and all $v \in T$.

We denote by $u^{\perp}$ and $S^{\perp}$ the collections of all vectors in V that are orthogonal to the vector u and the subset S, respectively, that is,

$$u^{\perp} = \{ v \in V \mid \langle u, v \rangle = 0 \}$$

and

$$S^{\perp} = \{ v \in V \mid \langle u, v \rangle = 0 \text{ for all } u \in S \}.$$

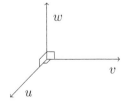
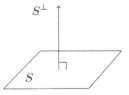

Figure 5.2: Orthogonality

These sets are called *orthogonal complements* of u and S, respectively. One may check that an orthogonal complement is always a subspace of V. Obviously, $S \cap S^{\perp} = \{0\}$, $\{0\}^{\perp} = V$, and $V^{\perp} = \{0\}$. Moreover, $S \subseteq (S^{\perp})^{\perp}$. If S is a subspace of a finite dimensional space, then $(S^{\perp})^{\perp} = S$, and

$$V = S \oplus S^{\perp}. \tag{5.12}$$

Given vectors u and v (which are assumed to be linearly independent), how do we use u and v to find or construct vectors that are orthogonal? On the xy-plane, draw vectors u and v with the same initial point, draw a line containing v, project u onto the line to get a vector, say w, and make a right triangle with u as the hypotenuse and w as the adjacent side, then the third side is the vector $u - w$, which is orthogonal to v. Let $w = kv$, where k is a scalar. Then $0 = \langle u - kv, v \rangle = \langle u, v \rangle - k\langle v, v \rangle$. So $k = \frac{\langle u,v \rangle}{\langle v,v \rangle}$. w is called the *projection* of u onto v, denoted by $\text{Proj}_v(u)$. Thus,

$$\text{Proj}_v(u) = \frac{\langle u, v \rangle}{\langle v, v \rangle} v.$$

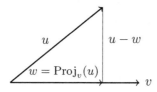

Figure 5.3: Projection of u onto v

Example 5.3 Let V be an inner product space. Let u and v be two given vectors in V and $v \neq 0$. Show that

(1) $u - \text{Proj}_v(u)$ is orthogonal to v.

(2) $\|\text{Proj}_v(u)\| \leq \|u\|$.

(3) $\mathcal{V} : x \mapsto \text{Proj}_v(x)$ is a linear map on V.

(4) $\mathcal{V}^2 = \mathcal{V}$, where $\mathcal{V}$ is defined as in (3).

Solution. (1) By a direct computation, we get

$$\langle u - \text{Proj}_v(u), v \rangle = \langle u, v \rangle - \left\langle \frac{\langle u, v \rangle}{\langle v, v \rangle} v, v \right\rangle = \langle u, v \rangle - \langle u, v \rangle = 0.$$

(2) By the Cauchy–Schwartz inequality, we have

$$\|\text{Proj}_v(u)\|^2 = \left\langle \frac{\langle u, v\rangle}{\langle v, v\rangle}v, \frac{\langle u, v\rangle}{\langle v, v\rangle}v \right\rangle = \frac{|\langle u, v\rangle|^2}{\langle v, v\rangle} \leq \langle u, u\rangle = \|u\|^2.$$

(3) Let x and y be arbitrary vectors in V and let k be any scalar. Then

$$
\begin{aligned}
\mathcal{V}(x + ky) &= \text{Proj}_v(x + ky) = \frac{\langle x + ky, v\rangle}{\langle v, v\rangle}v \\
&= \frac{\langle x, v\rangle + k\langle y, v\rangle}{\langle v, v\rangle}v = \frac{\langle x, v\rangle}{\langle v, v\rangle}v + k\frac{\langle y, v\rangle}{\langle v, v\rangle}v \\
&= \mathcal{V}(x) + k\mathcal{V}(y).
\end{aligned}
$$

(4) Denote $\tilde{x} = \text{Proj}_v(x)$. It is easy to verify that $\langle \tilde{x}, v\rangle = \langle x, v\rangle$. Then

$$\mathcal{V}^2(x) = \mathcal{V}\big(\text{Proj}_v(x)\big) = \mathcal{V}(\tilde{x}) = \frac{\langle \tilde{x}, v\rangle}{\langle v, v\rangle}v = \frac{\langle x, v\rangle}{\langle v, v\rangle}v = \mathcal{V}(x). \quad \square$$

Orthogonal Basis; Orthonormal Basis. Let $\{\alpha_1, \ldots, \alpha_n\}$ be a basis for an inner product space V. If $\alpha_1, \ldots, \alpha_n$ are mutually orthogonal, that is, $\langle \alpha_i, \alpha_j\rangle = 0$ whenever $i \neq j$, then we say that the basis is an *orthogonal basis*. If, in addition, every vector in the basis has length 1, we call such a basis an *orthonormal basis*. Thus, $\alpha_1, \ldots, \alpha_n$ comprise an orthonormal basis for an n-dimensional inner product space if and only if

$$\langle \alpha_i, \alpha_j\rangle = \begin{cases} 0 & \text{if } i \neq j, \\ 1 & \text{if } i = j. \end{cases}$$

The standard basis $\{e_1, \ldots, e_n\}$ is an orthonormal basis for $\mathbb{R}^n$ as well as for $\mathbb{C}^n$ under the usual inner product. The column (row) vectors of any $n \times n$ unitary matrix is also an orthonormal basis for $\mathbb{C}^n$.

If $\{u_1, \ldots, u_k\}$ is an orthogonal set, then $\langle u_i, u_j\rangle = 0$ for all $i \neq j$. Let $u = u_1 + \cdots + u_k$. Computing $\|u\|$, we see that

$$\|u\|^2 = \left\|\sum_{i=1}^{k} u_i\right\|^2 = \sum_{i=1}^{k} \|u_i\|^2. \tag{5.13}$$

In particular, for orthogonal vectors u and v (*Pythagorean theorem*)

$$\|u + v\|^2 = \|u\|^2 + \|v\|^2.$$

If $\{\alpha_1, \ldots, \alpha_n\}$ is an orthonormal basis for an inner product space V over $\mathbb{C}$ (or $\mathbb{R}$), then every vector u in V can be uniquely expressed as

$$u = \sum_{i=1}^{n} \langle u, \alpha_i \rangle \alpha_i. \tag{5.14}$$

It follows that

$$\|u\|^2 = \langle u, u \rangle = \sum_{i=1}^{n} |\langle u, \alpha_i \rangle|^2.$$

Let u and v be vectors with respective coordinates $x = (x_1, \ldots, x_n)^t$ and $y = (y_1, \ldots, y_n)^t$ with respect to the orthonormal basis $\{\alpha_1, \ldots, \alpha_n\}$, i.e.,

$$u = x_1 \alpha + \cdots + x_n \alpha_n, \quad v = y_1 \alpha_1 + \cdots + y_n \alpha_n.$$

Then the inner products for V and $\mathbb{C}^n$ (both denoted by $\langle \cdot, \cdot \rangle$) are equal:

$$\langle u, v \rangle = \Big\langle \sum_{i=1}^{n} x_i \alpha_i, \sum_{j=1}^{n} y_j \alpha_j \Big\rangle = \sum_{i=1}^{n} \overline{y_i} x_i = y^* x = \langle x, y \rangle. \tag{5.15}$$

Let $\mathcal{A}$ be a linear transformation on V and let A be the matrix representation of $\mathcal{A}$ with respect to the orthonormal basis $\{\alpha_1, \ldots, \alpha_n\}$. Then the coordinate of $\mathcal{A}(u)$ under the basis is Ax, because

$$\mathcal{A}(u) = \sum_{i=1}^{n} x_i \mathcal{A}(\alpha_i) = (\mathcal{A}(\alpha_1), \ldots, \mathcal{A}(\alpha_n))x = (\alpha_1, \ldots, \alpha_n)Ax.$$

We have

$$\langle \mathcal{A}(u), v \rangle = y^* A x = \langle Ax, y \rangle. \tag{5.16}$$

Does every finite dimensional inner product space possess an orthonormal basis? The answer is affirmative. If V is an inner product space with a basis $\{\alpha_1, \alpha_2, \ldots, \alpha_n\}$, the *Gram–Schmidt process* is a method for obtaining an orthonormal basis from the basis. It goes as follows:

With the given linearly independent vectors $\alpha_1, \alpha_2, \ldots, \alpha_n$, we first derive mutually orthogonal vectors $\beta_1, \beta_2, \ldots, \beta_n$ by the Gram–Schmidt process, then obtain orthonormal vectors $\gamma_1, \gamma_2, \ldots, \gamma_n$ by normalization.

$$\beta_1 = \alpha_1 \qquad\qquad\qquad\qquad\qquad \gamma_1 = \tfrac{1}{\|\beta_1\|}\beta_1$$
$$\beta_2 = \alpha_2 - \operatorname{Proj}_{\beta_1}(\alpha_2) \qquad\qquad \gamma_2 = \tfrac{1}{\|\beta_2\|}\beta_2$$
$$\beta_3 = \alpha_3 - \operatorname{Proj}_{\beta_1}(\alpha_3) - \operatorname{Proj}_{\beta_2}(\alpha_3) \qquad \gamma_3 = \tfrac{1}{\|\beta_3\|}\beta_3$$

$$\cdots\cdots\cdots\cdots\cdots\cdots\cdots\cdots\cdots\cdots\cdots\cdots \qquad\qquad \cdots\cdots\cdots\cdots$$

$$\beta_n = \alpha_n - \sum_{i=1}^{n-1} \operatorname{Proj}_{\beta_i}(\alpha_n) \qquad\qquad \gamma_n = \tfrac{1}{\|\beta_n\|}\beta_n.$$

Note that $\{\alpha_1, \alpha_2, \ldots, \alpha_n\}$ is a linearly independent set, $\{\beta_1, \beta_2, \ldots, \beta_n\}$ is an orthogonal set, $\{\gamma_1, \gamma_2, \ldots, \gamma_n\}$ is an orthonormal set. For each $k \leq n$,

$$\text{Span}\{\alpha_1, \alpha_2, \ldots, \alpha_k\} = \text{Span}\{\beta_1, \beta_2, \ldots, \beta_k\} = \text{Span}\{\gamma_1, \gamma_2, \ldots, \gamma_k\}.$$

Example 5.4 Let $\alpha_1 = (1,1,0)^t$, $\alpha_2 = (1,0,1)^t$, $\alpha_3 = (0,1,1)^t$ be a basis of $\mathbb{R}^3$. Use the Gram–Schmidt process to obtain an orthonormal basis for $\mathbb{R}^3$. If A is the 3×3 matrix with $\alpha_1, \alpha_2, \alpha_3$ as its columns, write $A = QR$ as a product of an orthogonal matrix Q and an upper-triangular matrix R.

Solution. Use the Gram–Schmidt process to get orthogonal vectors:

$$\beta_1 = \alpha_1 = (1,1,0)^t$$
$$\beta_2 = \alpha_2 - \frac{\langle \alpha_2, \beta_1 \rangle}{\langle \beta_1, \beta_1 \rangle} \beta_1 = \tfrac{1}{2}(1,-1,2)^t$$
$$\beta_3 = \alpha_3 - \frac{\langle \alpha_3, \beta_2 \rangle}{\langle \beta_2, \beta_2 \rangle} \beta_2 - \frac{\langle \alpha_3, \beta_1 \rangle}{\langle \beta_1, \beta_1 \rangle} \beta_1 = \tfrac{2}{3}(-1,1,1)^t.$$

Normalize β_1, β_2, and β_3 to get orthonormal vectors γ_1, γ_2, and γ_3:

$$\|\beta_1\| = \sqrt{2}, \qquad \gamma_1 = \tfrac{1}{\|\beta_1\|}\beta_1 = \tfrac{1}{\sqrt{2}}(1,1,0)^t$$
$$\|\beta_2\| = \sqrt{\tfrac{3}{2}}, \qquad \gamma_2 = \tfrac{1}{\|\beta_2\|}\beta_2 = \tfrac{1}{\sqrt{6}}(1,-1,2)^t$$
$$\|\beta_3\| = \tfrac{2}{\sqrt{3}}, \qquad \gamma_3 = \tfrac{1}{\|\beta_3\|}\beta_3 = \tfrac{1}{\sqrt{3}}(-1,1,1)^t.$$

Thus

$$(\alpha_1, \alpha_2, \alpha_3) = (\beta_1, \beta_2, \beta_3) \begin{pmatrix} 1 & \frac{\langle \alpha_2, \beta_1 \rangle}{\langle \beta_1, \beta_1 \rangle} & \frac{\langle \alpha_3, \beta_1 \rangle}{\langle \beta_1, \beta_1 \rangle} \\ 0 & 1 & \frac{\langle \alpha_3, \beta_2 \rangle}{\langle \beta_2, \beta_2 \rangle} \\ 0 & 0 & 1 \end{pmatrix}$$

$$= (\gamma_1, \gamma_2, \gamma_3) \begin{pmatrix} \|\beta_1\| & \frac{\langle \alpha_2, \beta_1 \rangle}{\|\beta_1\|} & \frac{\langle \alpha_3, \beta_1 \rangle}{\|\beta_1\|} \\ 0 & \|\beta_2\| & \frac{\langle \alpha_3, \beta_2 \rangle}{\|\beta_2\|} \\ 0 & 0 & \|\beta_3\| \end{pmatrix}.$$

It follows that $A = QR$, where Q is orthogonal and R is upper-triangular:

$$Q = \begin{pmatrix} \frac{1}{\sqrt{2}} & \frac{1}{\sqrt{6}} & -\frac{1}{\sqrt{3}} \\ \frac{1}{\sqrt{2}} & -\frac{1}{\sqrt{6}} & \frac{1}{\sqrt{3}} \\ 0 & \frac{2}{\sqrt{6}} & \frac{1}{\sqrt{3}} \end{pmatrix}, \quad R = \begin{pmatrix} \sqrt{2} & \frac{1}{\sqrt{2}} & \frac{1}{\sqrt{2}} \\ 0 & \frac{3}{\sqrt{6}} & \frac{1}{\sqrt{6}} \\ 0 & 0 & \frac{2}{\sqrt{3}} \end{pmatrix}. \quad \square$$

Example 5.5 Let $u_1, u_2, \ldots, u_n$ be vectors in an inner product space V. Let $G = (g_{ij})$ be the $n \times n$ matrix with $g_{ij} = \langle u_j, u_i \rangle$ (called *a Gram matrix*)

$$G = \begin{pmatrix} \langle u_1, u_1 \rangle & \langle u_2, u_1 \rangle & \cdots & \langle u_n, u_1 \rangle \\ \langle u_1, u_2 \rangle & \langle u_2, u_2 \rangle & \cdots & \langle u_n, u_2 \rangle \\ \vdots & \vdots & \ddots & \vdots \\ \langle u_1, u_n \rangle & \langle u_2, u_n \rangle & \cdots & \langle u_n, u_n \rangle \end{pmatrix}.$$

Show that

(1) G is positive semidefinite.

(2) G is singular if and only if $u_1, u_2, \ldots, u_n$ are linearly dependent.

(3) The rank of G is equal to the dimension of $\text{Span}\{u_1, u_2, \ldots, u_n\}$.

(4) $\det G \le \prod_{i=1}^{n} \|u_i\|^2$. When does equality occur?

(5) $G = B^*B$ for some $n \times n$ matrix B.

(6) If $\{u_1, u_2, \ldots, u_n\}$ is a basis for the inner product space V, x and y are respectively the coordinates of u and $v \in V$ under the basis, then

$$\langle u, v \rangle = y^* G x.$$

(7) If $\mathcal{L}$ is a linear map with matrix L under the basis $\{u_1, u_2, \ldots, u_n\}$, then

$$\langle \mathcal{L}(u), v \rangle = y^* (GL) x.$$

Solution.

(1) We show that $x^* G x \ge 0$ for all $x = (x_1, x_2, \ldots, x_n)^t \in \mathbb{C}^n$.

$$\begin{aligned} x^* G x &= \sum_{i,j} \overline{x}_i g_{ij} x_j = \sum_{i,j} \overline{x}_i \langle u_j, u_i \rangle x_j \\ &= \sum_{i,j} \langle x_j u_j, x_i u_i \rangle = \sum_j \left(\sum_i \langle x_j u_j, x_i u_i \rangle \right) \\ &= \sum_j \left\langle x_j u_j, \sum_i x_i u_i \right\rangle = \left\langle \sum_j x_j u_j, \sum_i x_i u_i \right\rangle \\ &= \left\| \sum_i x_i u_i \right\|^2 \ge 0. \end{aligned}$$

(2) This is immediate from (1): if G is singular, then $Gx = 0$ for some nonzero x which gives $\sum_i x_i u_i = 0$, i.e., $u_1, \ldots, u_n$ are linearly dependent. For the converse, note that when G is positive semidefinite, $Gx = 0$ if and only if $x^* G x = 0$, where x is a nonzero column vector.

(3) Since G is positive semidefinite, the rank of G is r if and only if G has a principal nonzero minor of order r and all minors of order $k > r$ are equal to zero. Based on this, (2) implies (3).

(4) This is immediate from the Hadamard inequality (Example 4.8, Chapter 4, p. 140). Equality occurs if and only if some $u_i = 0$, or G is diagonal, equivalently, $u_1, u_2, \ldots, u_n$ are mutually orthogonal.

(5) Choose an orthonormal basis (of n vectors) and let $\widetilde{u}_i$ be the coordinate of u_i under the basis. Then $\langle u_j, u_i \rangle = \langle \widetilde{u}_j, \widetilde{u}_i \rangle = \widetilde{u}_i^{\,*} \widetilde{u}_j$ for all i, j. Thus, $G = (\langle u_j, u_i \rangle) = (\widetilde{u}_i^{\,*} \widetilde{u}_j) = B^* B$, where $B = (\widetilde{u}_1, \widetilde{u}_2, \ldots, \widetilde{u}_n)$.

(6) For $u = x_1 u_1 + x_2 u_2 + \cdots + x_n u_n$ and $v = y_1 u_1 + y_2 u_2 + \cdots + y_n u_n$,

$$\langle u, v \rangle = \left\langle \sum_{j=1}^{n} x_j u_j, \sum_{i=1}^{n} y_i u_i \right\rangle = \sum_{j=1}^{n} \sum_{i=1}^{n} x_j \langle u_j, u_i \rangle \overline{y_i} = y^* G x.$$

(7) Note that the coordinate of $\mathcal{L}(u)$ under the basis is Lx. Thus, by (6),

$$\langle \mathcal{L}(u), v \rangle = y^* G(Lx) = y^* (GL)x. \qquad \square$$

It is immediate that the Gram matrix G is a nonnegative diagonal matrix if and only if $u_1, u_2, \ldots, u_n$ are orthogonal, and $G = I_n$ if and only if $u_1, u_2, \ldots, u_n$ are orthonormal. The above (7) shows that $\langle \mathcal{L}(u), v \rangle = y^* Lx$ if the basis is orthonormal. The case of $n = 2$ in Example 5.5(1) results in the Cauchy–Schwartz inequality; the case of $n = 3$ has been used to generate variants of the Cauchy–Schwartz inequality in Example 5.2. Moreover, since a matrix is positive semidefinite if and only if its transpose is positive semidefinite, it is immaterial to arrange the inner products in rows or in columns in a Gram (or Gramian) matrix.

As an application, consider the matrix space $M_n(\mathbb{C})$ with the usual inner product $\langle X, Y \rangle = \operatorname{tr}(Y^* X)$. The following matrix of traces involving $A_1, \ldots, A_m \in M_n(\mathbb{C})$ is Gramian, thus is positive semidefinite:

$$\begin{pmatrix} \operatorname{tr}(A_1^* A_1) & \operatorname{tr}(A_1^* A_2) & \cdots & \operatorname{tr}(A_1^* A_m) \\ \operatorname{tr}(A_2^* A_1) & \operatorname{tr}(A_2^* A_2) & \cdots & \operatorname{tr}(A_2^* A_m) \\ \vdots & \vdots & \ddots & \vdots \\ \operatorname{tr}(A_m^* A_1) & \operatorname{tr}(A_m^* A_2) & \cdots & \operatorname{tr}(A_m^* A_m) \end{pmatrix}.$$

Linear transformations (maps or operators) are a central topic of linear algebra. Orthogonal transformations are of particular importance; they preserve the inner product (i.e., the inner product of two vectors before the transformation is equal to their inner product after the transformation) as well as the lengths of vectors (and angles between them).

Example 5.6 Let $\mathcal{A}$ be a linear transformation on a Euclidean space V of dimension n. Show that the following statements are equivalent:

(1) $\langle \mathcal{A}(u), \mathcal{A}(v) \rangle = \langle u, v \rangle$ for all u and v in V, that is, $\mathcal{A}$ preserves the inner product. (Such an $\mathcal{A}$ is called an *orthogonal transformation*.)

(2) $\|\mathcal{A}(u)\| = \|u\|$ for all $u \in V$, that is, $\mathcal{A}$ preserves a vector's length. (Such an $\mathcal{A}$ is called an *isometry*.)

(3) $\mathcal{A}$ maps an orthonormal basis to an orthonormal basis, that is, if $\{\alpha_1, \alpha_2, \ldots, \alpha_n\}$ is an orthonormal basis for V, then so is $\{\mathcal{A}(\alpha_1), \mathcal{A}(\alpha_2), \ldots, \mathcal{A}(\alpha_n)\}$, i.e., $\langle \mathcal{A}(\alpha_i), \mathcal{A}(\alpha_j) \rangle = \langle \alpha_i, \alpha_j \rangle$ for all i, j.

(4) The matrix representation A of $\mathcal{A}$ with respect to some orthonormal basis is a real orthogonal matrix, that is, $A^t A = A A^t = I$.

Solution. We show that $(1) \Rightarrow (2) \Rightarrow (3) \Rightarrow (4) \Rightarrow (1)$.

$(1) \Rightarrow (2)$: This is obvious as (2) is equivalent to $\langle \mathcal{A}(u), \mathcal{A}(u) \rangle = \langle u, u \rangle$.

$(2) \Rightarrow (3)$: Over the reals $\mathbb{R}$, on the one hand,

$$\|\alpha_i + \alpha_j\|^2 = \langle \alpha_i + \alpha_j, \alpha_i + \alpha_j \rangle = \langle \alpha_i, \alpha_i \rangle + 2\langle \alpha_i, \alpha_j \rangle + \langle \alpha_j, \alpha_j \rangle.$$

On the other hand,

$$
\begin{aligned}
\|\mathcal{A}(\alpha_i + \alpha_j)\|^2 &= \langle \mathcal{A}(\alpha_i + \alpha_j), \mathcal{A}(\alpha_i + \alpha_j) \rangle \\
&= \langle \mathcal{A}(\alpha_i), \mathcal{A}(\alpha_i) \rangle + \langle \mathcal{A}(\alpha_i), \mathcal{A}(\alpha_j) \rangle \\
&\quad + \langle \mathcal{A}(\alpha_j), \mathcal{A}(\alpha_i) \rangle + \langle \mathcal{A}(\alpha_j), \mathcal{A}(\alpha_j) \rangle \\
&= \langle \alpha_i, \alpha_i \rangle + 2\langle \mathcal{A}(\alpha_i), \mathcal{A}(\alpha_j) \rangle + \langle \alpha_j, \alpha_j \rangle.
\end{aligned}
$$

It follows that for all i and j, $\langle \mathcal{A}(\alpha_i), \mathcal{A}(\alpha_j) \rangle = \langle \alpha_i, \alpha_j \rangle$.

$(3) \Rightarrow (4)$: Let $A = (a_{ij})$ be the matrix of $\mathcal{A}$ under the basis $\{\alpha_1, \ldots, \alpha_n\}$. Let a_k be the k-th column of A. Then $\mathcal{A}(\alpha_k) = \sum_{t=1}^n a_{tk} \alpha_t$. We compute

$$\langle \alpha_i, \alpha_j \rangle = \langle \mathcal{A}(\alpha_i), \mathcal{A}(\alpha_j) \rangle = \Big\langle \sum_{t=1}^n a_{ti} \alpha_t, \sum_{t=1}^n a_{tj} \alpha_t \Big\rangle = \sum_{t=1}^n a_{ti} a_{tj} = \langle a_i, a_j \rangle.$$

This says that the columns of A are orthonormal, that is, $A^t A = A A^t = I_n$.

$(4) \Rightarrow (1)$: Let x and y be respectively the coordinates of u and v under the basis $\{\alpha_1, \alpha_2, \ldots, \alpha_n\}$. Since the basis is orthonormal, we have $\langle u, v \rangle = y^t x$ and $\langle \mathcal{A}(u), \mathcal{A}(v) \rangle = (Ay)^t (Ax) = y^t A^t A x = y^t x = \langle u, v \rangle$. $\square$

A few remarks about the example are in order. Linearity of $\mathcal{A}$ is a precondition. Otherwise (2) does not imply (1). In other words, (1) and (2) are not equivalent for general (nonlinear) maps. (See Problems 5.83 and 5.85.) In (3), the word "orthonormal" cannot be replaced by "orthogonal".

Exercise: Work out the example over $\mathbb{C}$ (for unitary transformations).

Example 5.7 (Riesz Representation Theorem) Let V be an inner product space of dimension n over $\mathbb{F}$. A transformation φ from V to $\mathbb{F}$ is linear (called *linear functional*) if and only if there exists a vector $y \in V$ such that

$$\varphi(x) = \langle x, y \rangle \text{ for all } x \in V.$$

Solution 1. For any fixed $y \in V$, the function φ defined by $\varphi(x) = \langle x, y \rangle$ is a linear transformation from V to $\mathbb{F}$ because for all $u, v \in V$ and $k \in \mathbb{F}$,

$$\varphi(u + kv) = \langle u + kv, y \rangle = \langle u, y \rangle + k\langle v, y \rangle = \varphi(u) + k\varphi(v).$$

Conversely, let φ be a linear functional. We show that $\varphi(x) = \langle x, y \rangle$ for some y. To this end, let $\alpha_1, \alpha_2, \ldots, \alpha_n$ be an orthonormal basis of V. Let

$$y = \overline{\varphi(\alpha_1)}\,\alpha_1 + \overline{\varphi(\alpha_2)}\,\alpha_2 + \cdots + \overline{\varphi(\alpha_n)}\,\alpha_n.$$

For $x \in V$, we write $x = \langle x, \alpha_1 \rangle \alpha_1 + \langle x, \alpha_2 \rangle \alpha_2 + \cdots + \langle x, \alpha_n \rangle \alpha_n$. Thus

$$
\begin{aligned}
\varphi(x) &= \varphi\big(\langle x, \alpha_1 \rangle \alpha_1 + \langle x, \alpha_2 \rangle \alpha_2 + \cdots + \langle x, \alpha_n \rangle \alpha_n\big) \\
&= \langle x, \alpha_1 \rangle \varphi(\alpha_1) + \langle x, \alpha_2 \rangle \varphi(\alpha_2) + \cdots + \langle x, \alpha_n \rangle \varphi(\alpha_n) \\
&= \langle x, \overline{\varphi(\alpha_1)}\,\alpha_1 \rangle + \langle x, \overline{\varphi(\alpha_2)}\,\alpha_2 \rangle + \cdots + \langle x, \overline{\varphi(\alpha_n)}\,\alpha_n \rangle \\
&= \langle x, \overline{\varphi(\alpha_1)}\,\alpha_1 + \overline{\varphi(\alpha_2)}\,\alpha_2 + \cdots + \overline{\varphi(\alpha_n)}\,\alpha_n \rangle = \langle x, y \rangle.
\end{aligned}
$$

Solution 2. We only show that if φ is a linear functional, then it can be represented by the inner product $\varphi(x) = \langle x, y \rangle$ for some fixed $y \in V$.

Let $W = \{w \in V \mid \varphi(w) = 0\}$. Then W is a subspace of V. If $W = V$, then $\varphi = 0$ and we take $y = 0$. Let W be a proper subspace of V. Then the orthogonal complement $W^{\perp}$ of W has dimension at least one.

We show it is in fact one. For this, we show that any two vectors in $W^{\perp}$ are linearly dependent. Let $u_1, u_2 \in W^{\perp}$. Since $\varphi(u_1), \varphi(u_2) \in \mathbb{F}$, there is a scalar $k \in \mathbb{F}$ such that $\varphi(u_1) = k\varphi(u_2)$ (or similarly, $\varphi(u_2) = k\varphi(u_1)$). It follows that $\varphi(u_1 - ku_2) = 0$ and $u_1 - ku_2 \in W$. On the other hand, $W^{\perp}$ is a subspace, so $u_1 - ku_2 \in W^{\perp}$. Therefore, $u_1 - ku_2 = 0$ and $u_1 = ku_2$.

Take a unit vector $z \in W^{\perp}$. Since V is the direct sum of $W^{\perp}$ and W, if $x \in V$, we write $x = \lambda z + w$, where $\lambda \in \mathbb{F}$, $w \in W$. Then $\langle x, z \rangle = \lambda$ and

$$\varphi(x) = \varphi(\lambda z + w) = \varphi(\lambda z) + \varphi(w) = \lambda\varphi(z) = \langle x, z \rangle \varphi(z) = \langle x, \overline{\varphi(z)}\,z \rangle.$$

Now we set $y = \overline{\varphi(z)}z$. Then $\varphi(x) = \langle x, y \rangle$ for all $x \in V$. □

Note that the vector $y \in V$ for the representation of the linear functional φ is unique because $\langle x, y_1 \rangle = \langle x, y_2 \rangle$ for all $x \in V$ implies $y_1 = y_2$.

Chapter 5 Problems

5.1 $\mathbb{C}^2$ is a vector space over $\mathbb{R}$ (of dimension 4). Is $\mathbb{C}^2$ an inner product space over $\mathbb{R}$ with respect to the usual inner product $\langle u, v \rangle = v^* u$?

5.2   Let V be an inner product space over $\mathbb{C}$. Show that for any nonzero vector $u \in V$, $\frac{1}{\|u\|} u$ is a unit vector, and for any vectors $v, w \in V$,

$$\langle v, \langle v, w \rangle w \rangle = |\langle v, w \rangle|^2 = \langle v, w \rangle \langle w, v \rangle.$$

Is it true that

$$\langle \langle v, w \rangle v, w \rangle = |\langle v, w \rangle|^2?$$

5.3   Let V be an inner product space over $\mathbb{C}$. Show that for all vectors $u, v, w \in V$ and scalars $\lambda, \mu \in \mathbb{C}$, the following identities hold:

$$\langle \lambda u + \mu v, w \rangle = \lambda \langle u, w \rangle + \mu \langle v, w \rangle$$

and

$$\langle w, \lambda u + \mu v \rangle = \overline{\lambda} \langle w, u \rangle + \overline{\mu} \langle w, v \rangle.$$

5.4   Let V be an inner product space and $u, v, w \in V$. Answer true or false:

(a) $|\langle u, v \rangle| \leq \|u\| + \|v\|$.

(b) $|\langle u, v \rangle| \leq \frac{1}{2}(\|u\|^2 + \|v\|^2)$.

(c) $|\langle u, v \rangle| \leq |\langle u, w \rangle| + |\langle w, v \rangle|$.

(d) $\|u + v\| \leq \|u + w\| + \|w + v\|$.

(e) $\|u + v\| \leq \|u + w\| + \|w - v\|$.

5.5   Let $x = (x_1, x_2, \dots, x_n) \in \mathbb{C}^n$ and $\|x\|_\infty = \max\{ |x_1|, |x_2|, \dots, |x_n| \}$. For $x, y \in \mathbb{C}^n$, define $\langle x, y \rangle_\infty = \|x\|_\infty \|y\|_\infty$. Determine whether or not each of the following statements is true:

(a) $\langle x, x \rangle_\infty \geq 0$. Equality holds if and only if $x = 0$,

(b) $\langle \lambda x, y \rangle_\infty = \lambda \langle x, y \rangle_\infty$, $\lambda \in \mathbb{C}$,

(c) $\langle x, y + z \rangle_\infty = \langle x, y \rangle_\infty + \langle x, z \rangle_\infty$,

(d) $\langle x, y \rangle_\infty = \overline{\langle y, x \rangle_\infty}$, and

(e) x and y are linearly dependent if $\|x + y\|_\infty = \|x\|_\infty + \|y\|_\infty$.

5.6   Let V be a vector space with inner products $\langle \cdot, \cdot \rangle_1$ and $\langle \cdot, \cdot \rangle_2$. Show that $\langle \cdot, \cdot \rangle_3 = \langle \cdot, \cdot \rangle_1 + \langle \cdot, \cdot \rangle_2$ is also an inner product of V.

5.7   Let V be a vector space with norms $\|\cdot\|^{(1)}$ and $\|\cdot\|^{(2)}$. Show that $\|\cdot\|^{(3)} = 4\|\cdot\|^{(1)} + 5\|\cdot\|^{(2)}$ is also a norm on V. What if -5 replaces 5?

5.8   Consider the following three norms on $\mathbb{R}^2$: for $x = (x_1, x_2) \in \mathbb{R}^2$,

$$\|x\|_1 = |x_1| + |x_2|, \ \|x\|_2 = (x_1^2 + x_2^2)^{\frac{1}{2}}, \ \|x\|_\infty = \max\{|x_1|, |x_2|\}.$$

Describe and sketch the sets in the xy-plane:

(a) $S_1 = \{x \in \mathbb{R}^2 \mid \|x\|_1 \leq 1\}$.

(b) $S_2 = \{x \in \mathbb{R}^2 \mid \|x\|_2 \leq 1\}$.

(c) $S_3 = \{x \in \mathbb{R}^2 \mid \|x\|_\infty \leq 1\}$.

5.9   Show that the function η defined below is a norm on $\mathbb{R}^2$:

$$\eta(x, y) = \sqrt{x^2 + 4y^2}, \ (x, y) \in \mathbb{R}^2.$$

Describe and sketch the set $\{(x, y) \in \mathbb{R}^2 \mid \eta(x, y) \leq 1\}$ in the xy-plane.

5.10 Show that η is a norm on $\mathbb{R}^2$, where $\eta : \mathbb{R}^2 \mapsto [0, \infty)$ is defined by

$$\eta(x, y) = \sqrt{(x - y)^2 + ky^2}, \ (x, y) \in \mathbb{R}^2, \text{ where } k > 0 \text{ is fixed.}$$

5.11 Let $p_1, \ldots, p_n > 0$ and $t_1, \ldots, t_n > 0$ with $t_1 + \cdots + t_n = 1$. Show that

(a) $\sum_{i=1}^n p_i^{\frac{1}{2}} \leq (n \sum_{i=1}^n p_i)^{\frac{1}{2}}$. When does equality occur?

(b) $\left(\sum_{i=1}^n t_i p_i\right)^2 \leq \sum_{i=1}^n t_i p_i^2$. When does equality occur?

(c) $n^2 \leq \left(\sum_{i=1}^n \frac{1}{p_i}\right)\left(\sum_{i=1}^n p_i\right)$. When does equality occur?

5.12 Let A be an $n \times n$ positive semidefinite matrix. Show that

$$(x^* A x)^k \leq x^* A^k x, \text{ for any unit } x \in \mathbb{C}^n \text{ and } k = 1, 2, \ldots.$$

5.13 For each pair of vectors x and y in $\mathbb{C}^3$, assign a scalar $\lfloor x, y \rfloor$ as follows:

$$\lfloor x, y \rfloor = y^* A x, \quad \text{where } A = \begin{pmatrix} 1 & 0 & 1 \\ 0 & 2 & 0 \\ 1 & 0 & 2 \end{pmatrix}.$$

Show that $\mathbb{C}^3$ is an inner product space with respect to $\lfloor \cdot, \cdot \rfloor$. If the entry 2 in the $(2, 2)$ position is replaced by -2, is it still an inner product space? If x and y on the right-hand side are switched, i.e., $\lfloor x, y \rfloor = x^* A y$, is $\mathbb{C}^3$ still an inner product space over $\mathbb{C}$? or $\mathbb{R}$?

5.14 Let $x = (1, 0)$ and $y = (0, 1)$. Compute the eigenvalues of

$$A = \begin{pmatrix} xx^t & xy^t \\ yx^t & yy^t \end{pmatrix}, \quad B = \begin{pmatrix} x^t x & x^t y \\ y^t x & y^t y \end{pmatrix}, \quad C = \begin{pmatrix} x^t x & y^t x \\ x^t y & y^t y \end{pmatrix}.$$

5.15 Let $u, v, x, y \in \mathbb{C}^n$ (in columns). Prove the matrix inequality

$$(x^* y) uv^* + (y^* x) vu^* \le (x^* x) uu^* + (y^* y) vv^*.$$

5.16 Let $p, q \in \mathbb{R}$. Define a function $f : \mathbb{R}^2 \times \mathbb{R}^2 \mapsto \mathbb{R}$ as follows:

$$f(x, y) = x_1 y_1 + 2 x_1 y_2 + p x_2 y_1 + q x_2 y_2,$$

where $x = (x_1, x_2)^t$, $y = (y_1, y_2)^t \in \mathbb{R}^2$. Find the values of p and q such that each of the following holds for all $x, y, z \in \mathbb{R}^2$ and $\lambda \in \mathbb{R}$:

  (a) $f(x, x) \ge 0$.
  (b) $f(x, y) = f(y, x)$.
  (c) $f(\lambda x, y) = \lambda f(x, y)$.
  (d) $f(x + y, z) = f(x, z) + f(y, z)$.
  (e) f defines an inner product on $\mathbb{R}^2$.

5.17 Let

$$V = \left\{ x = \begin{pmatrix} x_1 \\ x_2 \\ x_3 \\ x_4 \end{pmatrix} \in \mathbb{R}^4 \ \middle| \ x_1 = x_3 + x_4, \ x_2 = x_3 - x_4 \right\}.$$

Show that V is a subspace of $\mathbb{R}^4$. Find a basis for V and for $V^\perp$.

5.18 Let $\alpha_1 = (1, 0, 0, 0)$ and $\alpha_2 = \left(0, \frac{1}{2}, \frac{1}{2}, \frac{1}{\sqrt{2}} \right)$. Find vectors α_3 and α_4 in $\mathbb{R}^4$ so that $\alpha_1, \alpha_2, \alpha_3, \alpha_4$ comprise an orthonormal basis for $\mathbb{R}^4$.

5.19 Let $S = \{ (1, 1, 1, 1), (0, 1, -1, 3) \}$. Find $S^\perp$ in $\mathbb{R}^4$.

5.20 Let $u = (1, 1, 0), v = (1, -2, 1) \in \mathbb{R}^3$. Find the angle between the vectors u and v and the angle between the spaces $\mathrm{Span}\{u\}$ and $\mathrm{Span}\{v\}$.

5.21 Find $\mathbb{R}^\perp$, $1^\perp$, and $x^\perp$ with respect to the inner product on $\mathbb{P}_4[x]$

$$\langle f, g \rangle = \int_0^1 f(x) g(x) dx.$$

5.22   For the real polynomial space $\mathbb{P}_4[x]$, define an inner product as

$$\langle f, g \rangle = \int_{-1}^{1} f(x)g(x)dx.$$

(a) Find $\|u(x)\|$, where $u(x) = 1$. Is $u(x) = 1$ a unit vector? Find all numbers (i.e., constant polynomials) that are unit vectors.

(b) Find $1^{\perp}$, i.e., all polynomials that are orthogonal to $u(x) = 1$.

(c) Find an orthonormal basis for the subspace spanned by x and x^2. Show that there is no nonzero number (i.e., constant polynomial) as a vector that is orthogonal to $\mathrm{Span}\{x, x^2\}$.

(d) Complete the basis in (c) to an orthonormal basis for $\mathbb{P}_4[x]$ with respect to the inner product.

(e) Show that $\langle f, g \rangle = 0$ for $f \in V_1$, $g \in V_2$, i.e., $V_1 \perp V_2$, where

$$V_1 = \mathrm{Span}\{1, x\} \quad \text{and} \quad V_2 = \mathrm{Span}\left\{x^2 - \tfrac{1}{3}, x^3 - \tfrac{3}{5}x\right\}.$$

(f) Show that $\mathbb{P}_4[x] = V_1 \oplus V_2$.

5.23   For the real polynomial space $\mathbb{P}_4[x]$, define two inner products

$$\langle f, g \rangle_1 = \int_{-1}^{1} f(x)g(x)\, dx \quad \text{and} \quad \langle f, g \rangle_2 = \int_{0}^{1} f(x)g(x)\, dx.$$

(a) Find the norms of $f(x) = 1$ with respect to $\langle \cdot, \cdot \rangle_1$ and $\langle \cdot, \cdot \rangle_2$.

(b) Find a nonzero $p(x) \in \mathbb{P}_4[x]$ such that $\langle 1, p \rangle_1 = \langle 1, p \rangle_2 = 0$.

(c) Find a pair $v(x), w(x) \in \mathbb{P}_4[x]$ such that $\langle v, w \rangle_1 = 0$, $\langle v, w \rangle_2 \neq 0$.

(d) Find all $p(x) \in \mathbb{P}_4[x]$ such that $\langle p, p \rangle_1 = \langle p, p \rangle_2$.

(e) Can an orthonormal basis of $\mathbb{P}_4[x]$ with respect to $\langle \cdot, \cdot \rangle_1$ be an orthonormal basis with respect to $\langle \cdot, \cdot \rangle_2$?

(f) Extend $\{1, x\}$ to an orthogonal basis with respect to $\langle \cdot, \cdot \rangle_1$.

(g) Find linearly independent $r(x), s(x), t(x) \in \mathbb{P}_4[x]$ such that they are all orthogonal to $f(x) = 1$ with respect to $\langle \cdot, \cdot \rangle_2$.

(h) Find a subspace W of $\mathbb{P}_4[x]$ such that every element in W is orthogonal to x with respect to $\langle \cdot, \cdot \rangle_1$ but no element except zero in W is orthogonal to x with respect to $\langle \cdot, \cdot \rangle_2$.

5.24   Let $f_1, \ldots, f_n$ be real-valued continuous functions on $[a, b]$. Show that $f_1, \ldots, f_n$ are linearly dependent on $[a, b]$ if and only if the determinant of matrix $A = (a_{ij})$ is equal to zero, where $a_{ij} = \int_{a}^{b} f_i(x) f_j(x)\, dx$.

5.25 Define an inner product on the vector space $\mathcal{C}[-\pi, \pi]$ of real-valued continuous functions on $[-\pi, \pi]$ by

$$\langle f, g \rangle = \int_{-\pi}^{\pi} f(x)g(x)dx.$$

(a) Find the length of $p(x) = 1$.

(b) Find the length of $q(x) = \frac{1}{\sqrt{2\pi}}$.

(c) Find the length of $r(x) = \sin x$.

(d) Show that every odd function is orthogonal to 1.

(e) Show that $\langle \sin mx, \cos nx \rangle = 0$ for any $m \neq 0$ and any n.

5.26 Let $u_1, u_2, \ldots, u_n \in \mathbb{C}^m$ (i.e., n column vectors of m components). Determine if each of the following matrices is positive semidefinite:

$$\begin{pmatrix} u_1^*u_1 & u_1^*u_2 & \cdots & u_1^*u_n \\ u_2^*u_1 & u_2^*u_2 & \cdots & u_2^*u_n \\ \cdots & \cdots & \cdots & \cdots \\ u_n^*u_1 & u_n^*u_2 & \cdots & u_n^*u_n \end{pmatrix}, \quad \begin{pmatrix} u_1^*u_1 & u_2^*u_1 & \cdots & u_n^*u_1 \\ u_1^*u_2 & u_2^*u_2 & \cdots & u_n^*u_2 \\ \cdots & \cdots & \cdots & \cdots \\ u_1^*u_n & u_2^*u_n & \cdots & u_n^*u_n \end{pmatrix},$$

$$\begin{pmatrix} u_1u_1^* & u_1u_2^* & \cdots & u_1u_n^* \\ u_2u_1^* & u_2u_2^* & \cdots & u_2u_n^* \\ \cdots & \cdots & \cdots & \cdots \\ u_nu_1^* & u_nu_2^* & \cdots & u_nu_n^* \end{pmatrix}, \quad \begin{pmatrix} u_1u_1^* & u_2u_1^* & \cdots & u_nu_1^* \\ u_1u_2^* & u_2u_2^* & \cdots & u_nu_2^* \\ \cdots & \cdots & \cdots & \cdots \\ u_1u_n^* & u_2u_n^* & \cdots & u_nu_n^* \end{pmatrix}.$$

5.27 Let V be an inner product space over $\mathbb{C}$ and let $u_1, u_2, \ldots, u_n$ be n vectors in V. Show that matrices G and H are positive semidefinite:

$$G = \begin{pmatrix} \langle u_1, u_1 \rangle^2 & \langle u_2, u_1 \rangle^2 & \cdots & \langle u_n, u_1 \rangle^2 \\ \langle u_1, u_2 \rangle^2 & \langle u_2, u_2 \rangle^2 & \cdots & \langle u_n, u_2 \rangle^2 \\ \vdots & \vdots & \ddots & \vdots \\ \langle u_1, u_n \rangle^2 & \langle u_2, u_n \rangle^2 & \cdots & \langle u_n, u_n \rangle^2 \end{pmatrix}$$

and

$$H = \begin{pmatrix} \langle u_1, u_1 \rangle^2 & |\langle u_2, u_1 \rangle|^2 & \cdots & |\langle u_n, u_1 \rangle|^2 \\ |\langle u_1, u_2 \rangle|^2 & \langle u_2, u_2 \rangle^2 & \cdots & |\langle u_n, u_2 \rangle|^2 \\ \vdots & \vdots & \ddots & \vdots \\ |\langle u_1, u_n \rangle|^2 & |\langle u_2, u_n \rangle|^2 & \cdots & \langle u_n, u_n \rangle^2 \end{pmatrix}.$$

5.28 Let $u, v \in \mathbb{C}^n$ have norms less than 1, i.e., $\|u\| < 1, \|v\| < 1$. Show that

$$\begin{pmatrix} \frac{1}{1-\langle u,u \rangle} & \frac{1}{1-\langle u,v \rangle} \\ \frac{1}{1-\langle v,u \rangle} & \frac{1}{1-\langle v,v \rangle} \end{pmatrix} \geq 0.$$

5.29 Show that a square matrix is positive semidefinite if and only if it is a Gram matrix. To be precise, $A \in M_n(\mathbb{C})$ is positive semidefinite if and only if there exist n vectors $u_1, u_2, \ldots, u_n \in \mathbb{C}^n$ such that

$$A = (a_{ij}), \quad \text{where } a_{ij} = \langle u_j, u_i \rangle = u_i^* u_j.$$

5.30 Show that $\langle \cdot, \cdot \rangle$ is an inner product for $\mathbb{C}^n$ if and only if there exists an n-square positive definite matrix A such that for all $x, y \in \mathbb{C}^n$

$$\langle x, y \rangle = y^* A x.$$

5.31 Let V be a vector space over $\mathbb{F}$ ($= \mathbb{C}$ or $\mathbb{R}$). A mapping $\langle\!\langle \cdot, \cdot \rangle\!\rangle$: $V \times V \mapsto \mathbb{F}$ is called an *indefinite inner product* if for all vectors $u, v, w \in V$ and scalars $\lambda, \mu \in \mathbb{F}$, (i) $\langle\!\langle \lambda u, v \rangle\!\rangle = \lambda \langle\!\langle u, v \rangle\!\rangle$, (ii) $\langle\!\langle u, v + w \rangle\!\rangle = \langle\!\langle u, v \rangle\!\rangle + \langle\!\langle u, w \rangle\!\rangle$, and (iii) $\langle\!\langle u, v \rangle\!\rangle = \overline{\langle\!\langle v, u \rangle\!\rangle}$. (I.e., the positivity condition in the definition of the usual inner product is dropped.)

(a) Give an example of 2×2 matrix A such that $\mathbb{C}^2$ is an indefinite inner product space with $\langle\!\langle u, v \rangle\!\rangle = v^* A u$, $u, v \in \mathbb{C}^2$.

(b) If V is an indefinite inner product space over $\mathbb{C}$, show that

$$\mathrm{Re}\langle\!\langle u, v \rangle\!\rangle = \frac{1}{4}(\langle\!\langle u + v, u + v \rangle\!\rangle - \langle\!\langle u - v, u - v \rangle\!\rangle), \quad u, v \in V.$$

(c) If $A = \left(\begin{smallmatrix} 1 & 0 \\ 0 & -1 \end{smallmatrix} \right)$ and $V = \mathbb{R}^2$ (over $\mathbb{F} = \mathbb{R}$) is equipped with $\langle\!\langle u, v \rangle\!\rangle = u^t A v$, $u, v \in \mathbb{R}^2$, find a nonzero vector u such that $\langle\!\langle u, u \rangle\!\rangle = 0$, find the sets $K_0 = \{v \in V \mid \langle\!\langle v, v \rangle\!\rangle = 0\}$, $K_+ = \{v \in V \mid \langle\!\langle v, v \rangle\!\rangle > 0\}$, and $K_- = \{v \in V \mid \langle\!\langle v, v \rangle\!\rangle < 0\}$.

(d) Given a vector $w \in V$, show that $W = \{v \in V \mid \langle\!\langle v, w \rangle\!\rangle = 0\}$ is a subspace of V, and that $w \in K_0$ if and only if $w \in W$.

5.32 Show that pairwise orthogonal nonzero vectors are linearly independent, i.e., if $u_1, \ldots, u_n$ are nonzero vectors and $\langle u_i, u_j \rangle = 0$ whenever $i \neq j$, then $u_1, \ldots, u_n$ are linearly independent. Is the converse true?

5.33 Let $\mathcal{A}$ be a linear map on an inner product space V. Show that

$$\langle \mathcal{A}(x), x \rangle \langle x, \mathcal{A}(x) \rangle \le \langle \mathcal{A}(x), \mathcal{A}(x) \rangle \text{ for any unit } x \in V.$$

5.34 Let $\mathcal{A}$ be a linear map on an inner product space V of dimension n over $\mathbb{C}$ such that $\langle \mathcal{A}(v), v \rangle \ge 0$ for all $v \in V$ (called a *positive map*). Show that a matrix of $\mathcal{A}$ under an orthonormal basis is positive semidefinite and $\langle \mathcal{A}(x), x \rangle^k \le \langle \mathcal{A}^k(x), x \rangle$ for any unit $x \in V$ and $k = 1, 2, \ldots$.

5.35   Let A be an $n \times n$ positive semidefinite matrix and $x \in \mathbb{C}^n$. Show that

    (a) $(I - A)(I + A)^{-1}x = x$ if and only if $x \in \operatorname{Ker} A$.

    (b) $\|(I - A)(I + A)^{-1}x\| \le \|x\|$. When does equality occur?

5.36   Let V be an inner product space over $\mathbb{R}$ with the inner product $\langle \cdot, \cdot \rangle$.

    (a) If $v_1, v_2, v_3, v_4 \in V$ are pairwise product negative, that is,

$$\langle v_i, v_j \rangle < 0, \quad i, j = 1, 2, 3, 4, \ i \ne j,$$

        show that any three of v_1, v_2, v_3, v_4 are linearly independent.

    (b) Is it possible for four vectors in the xy-plane to have pairwise negative products? How about three vectors?

    (c) Are v_1, v_2, v_3, v_4 in (a) necessarily linearly independent?

    (d) Suppose that u, v, and w are three unit vectors in the xy-plane. What are the possible maximum and minimum values of

$$\langle u, v \rangle + \langle v, w \rangle + \langle w, u \rangle?$$

5.37   Let $\{v_1, v_2, \ldots, v_n\}$ be an orthonormal basis for an inner product space V with the inner product $\langle \cdot, \cdot \rangle$. Show that for any $x \in V$,

$$x = \sum_{i=1}^{n} \langle x, v_i \rangle v_i \quad \text{and} \quad \langle x, x \rangle \ge \sum_{i=1}^{k} |\langle x, v_i \rangle|^2, \ 1 \le k \le n.$$

When does equality occur in the last inequality?

5.38   Let $\{v_1, v_2, \ldots, v_n\}$ be an orthonormal basis for an inner product space V with the inner product $\langle \cdot, \cdot \rangle$. Show that for any $x, y \in V$,

$$\langle x, y \rangle = \sum_{i=1}^{n} \langle x, v_i \rangle \langle v_i, y \rangle.$$

5.39   Show that the Frobenius inner product $\langle A, B \rangle = \operatorname{tr}(B^t A)$ on $M_n(\mathbb{R})$ can be expressed in terms of the entrywise product $A \circ B = (a_{ij}b_{ij})$ (i.e., the Hadamard or Schur product) of $A = (a_{ij})$ and $B = (b_{ij})$ as

$$\langle A, B \rangle = e^t(A \circ B)e, \quad \text{where } e = (1, \ldots, 1)^t \in \mathbb{R}^n.$$

5.40   Consider the matrix space $M_n(\mathbb{C})$ with the Frobenius inner product

$$\langle A, B \rangle = \mathrm{tr}(B^*A), \quad A, B \in M_n(\mathbb{C}).$$

Show that

(a) $M_n(\mathbb{C})$ is an inner product space.

(b) $\mathrm{tr}(A^*A) = 0$ if and only if $A = 0$.

(c) $|\mathrm{tr}(AB)|^2 \leq \mathrm{tr}(A^*A)\,\mathrm{tr}(B^*B)$.

(d) $\mathrm{tr}(ABB^*A^*) \leq \mathrm{tr}(A^*A)\,\mathrm{tr}(B^*B)$, i.e., $\|AB\| \leq \|A\|\,\|B\|$.

(e) $\|A^*A - AA^*\| \leq \sqrt{2}\,\|A\|^2$.

(f) If $\mathrm{tr}(AX) = 0$ for every $X \in M_n(\mathbb{C})$, then $A = 0$.

(g) $W = \{\, X \in M_n(\mathbb{C}) \mid \mathrm{tr}\,X = 0 \,\}$ is a subspace of $M_n(\mathbb{C})$ and its orthogonal complement $W^\perp$ consists of all scalar matrices, that is, if $\mathrm{tr}(AX) = 0$ for all $X \in M_n(\mathbb{C})$ with $\mathrm{tr}\,X = 0$, then $A = \lambda I$ for some scalar λ. Find the dimensions of W and $W^\perp$.

(h) If $\langle A, X \rangle \geq 0$ for all $X \geq 0$ in $M_n(\mathbb{C})$, then $A \geq 0$.

5.41   For $M_n(\mathbb{C})$, $\langle A, B \rangle = \mathrm{tr}(B^*A)$ is the standard inner product. Determine if each of the following also defines an inner product for $M_n(\mathbb{C})$:

(a) $\lfloor A, B \rfloor = \mathrm{tr}(A^*B)$.

(b) $\lfloor A, B \rfloor = \mathrm{tr}(AB^*)$.

(c) $\lfloor A, B \rfloor = \mathrm{tr}(BA)$.

(d) $\lfloor A, B \rfloor = \mathrm{tr}(B^*AA^*B)$.

(e) $\lfloor A, B \rfloor = \sigma_{\max}(AB)$ (the largest singular value of AB).

(f) $\lfloor A, B \rfloor = \mathrm{tr}(B^*A) + \mathrm{tr}(A^*B)$.

(g) $\lfloor A, B \rfloor = 2\,\mathrm{tr}(B^*A) + 3\,\mathrm{tr}(AB^*)$.

(h) $\lfloor A, B \rfloor = \mathrm{tr}(B^*DA)$, where D is a diagonal matrix.

(i) $\lfloor A, B \rfloor = \mathrm{tr}(B^*DA)$, where D is a positive diagonal matrix.

5.42   If the 2×2 matrix $\left(\begin{smallmatrix} a & b \\ c & d \end{smallmatrix}\right)$ is unitarily similar to $\left(\begin{smallmatrix} \lambda_1 & x \\ 0 & \lambda_2 \end{smallmatrix}\right)$, show that

$$|x| = \left(|a|^2 + |b|^2 + |c|^2 + |d|^2 - |\lambda_1|^2 - |\lambda_2|^2\right)^{\frac{1}{2}}.$$

5.43 Let $A = (a_{ij})$ and $B = (b_{ij})$ be $n \times n$ matrices, and let $A \circ B = (a_{ij}b_{ij})$ be the Hadamard (entrywise) product of A and B. Show that

$$\|A \circ B\|_F \leq \|A\|_F \|B\|_F$$

and

$$\|A \circ B\|_{\mathrm{sp}} \leq \|(A^*A) \circ (B^*B)\|_{\mathrm{sp}}^{\frac{1}{2}} \leq \|A\|_{\mathrm{sp}} \|B\|_{\mathrm{sp}}.$$

5.44 A norm $\| \cdot \|$ on the vector space $M_n(\mathbb{C})$ is called a *submultiplicative norm* if $\|AB\| \leq \|A\|\|B\|$ for all $n \times n$ matrices A and B; a norm $\| \cdot \|$ on $M_n(\mathbb{C})$ is *unitarily invariant* if $\|UAV\| = \|A\|$ for all $n \times n$ unitary matrices U and V, and $n \times n$ matrices $A = (a_{ij})$. Show that

  (a) $\|A\|_F$, the Frobenius norm, is submultiplicative and unitarily invariant, i.e., $\|AB\|_F \leq \|A\|_F \|B\|_F$ and $\|UAV\|_F = \|A\|_F$.

  (b) $\|A\|_{\mathrm{sp}}$, the spectral norm, is submultiplicative and unitarily invariant, i.e., $\|AB\|_{\mathrm{sp}} \leq \|A\|_{\mathrm{sp}} \|B\|_{\mathrm{sp}}$ and $\|UAV\|_{\mathrm{sp}} = \|A\|_{\mathrm{sp}}$.

  (c) $\|A\|_1 = \sum_{i,j=1}^n |a_{ij}|$, the ℓ_1-norm, is submultiplicative, but not unitarily invariant.

  (d) $\|A\|_\infty = \max_{1 \leq i, j \leq n} |a_{ij}|$, the ℓ_∞-norm (or the max norm), is not submultiplicative, nor unitarily invariant.

Note: The definitions of *norms* (such as vector norm, matrix norm, etc) and their notations may vary from text to text.

5.45 Let A be an n-square complex matrix and denote

$$\rho(A) = \max\{ |\lambda| \mid \lambda \text{ is an eigenvalue of } A \} \quad (\textit{spectral radius})$$
$$\omega(A) = \max\{ |x^*Ax| \mid x^*x = 1, \, x \in \mathbb{C}^n \} \quad (\textit{numerical radius})$$
$$\sigma(A) = \max\{ \|Ax\| \mid x^*x = 1, \, x \in \mathbb{C}^n \} \quad (\textit{spectral norm}).$$

  (a) Show that $\rho(A) \leq \omega(A) \leq \sigma(A)$.

  (b) Which of ρ, ω, and σ is a norm on the vector space $M_n(\mathbb{C})$?

  (c) Which of ρ, ω, and σ has the submultiplicative property? i.e., $\xi(AB) \leq \xi(A)\xi(B)$ for all $A, B \in M_n(\mathbb{C})$, where ξ is ρ, ω, or σ.

  (d) Which of ρ, ω, and σ has the unitarily invariant property? i.e., $\xi(UAV) = \xi(A)$ for all $A \in M_n(\mathbb{C})$ and unitary $U, V \in M_n(\mathbb{C})$.

5.46 Find all 2×2 complex matrices that are orthogonal to both $\left(\begin{smallmatrix} 1 & 0 \\ 0 & -1 \end{smallmatrix}\right)$ and $\left(\begin{smallmatrix} 0 & 1 \\ 1 & 0 \end{smallmatrix}\right)$ with respect to the Frobenius inner product $\langle A, B \rangle = \mathrm{tr}(B^*A)$.

5.47   Let $u_1, \ldots, u_n \in \mathbb{R}^n$ be linearly independent and $\lambda_1, \ldots, \lambda_n \in \mathbb{R}$. Show that there exists a unique real matrix A having the eigenpairs $(\lambda_1, u_1), \ldots, (\lambda_n, u_n)$. If $u_1, \ldots, u_n$ are orthogonal, A is symmetric.

5.48   For $u = (1,0), v = (1,-1) \in \mathbb{R}^2$ with the standard inner product, find

    (a) $u^\perp, v^\perp$.

    (b) $u^\perp \cap v^\perp$.

    (c) $\{u, v\}^\perp$.

    (d) $(\mathrm{Span}\{u, v\})^\perp$.

    (e) $\mathrm{Span}\{u^\perp, v^\perp\}$.

5.49   Let V be an inner product space over field $\mathbb{F}$ (where $\mathbb{F} = \mathbb{R}$ or $\mathbb{C}$) with the inner product $\langle \cdot, \cdot \rangle$ and the induced norm $\| \cdot \|$. Show that

    (a) $\|u + v\| \leq \|u\| + \|v\|$.

    (b) $|\, \|u\| - \|v\| \,| \leq \|u - v\|$.

    (c) $\|u + v\| \, \|u - v\| \leq \|u\|^2 + \|v\|^2$.

    (d) $\langle u, v \rangle = \frac{1}{4}\left(\|u + v\|^2 - \|u - v\|^2\right)$ if $\mathbb{F} = \mathbb{R}$.

    (e) $\langle u, v \rangle = \frac{1}{4}\left(\|u+v\|^2 - \|u-v\|^2 + i\|u+iv\|^2 - i\|u-iv\|^2\right)$ if $\mathbb{F} = \mathbb{C}$.

5.50   Let V be a vector space equipped with an inner product $\langle \cdot, \cdot \rangle$ and let $\|\cdot\|$ be the induced norm by the inner product, that is, $\|v\| = \sqrt{\langle v, v \rangle}$.

    (a) Show that (*Parallelogram identity*) for all $x, y \in V$,

$$\|x + y\|^2 + \|x - y\|^2 = 2\|x\|^2 + 2\|y\|^2.$$

    (b) Show that $\|x + y\| = \|x\| + \|y\|$ if and only if

$$\|sx + ty\| = s\|x\| + t\|y\|, \quad \text{for all } s, t \geq 0.$$

    (c) If $\|x\| = \|y\|$, show that $x+y$ and $x-y$ are orthogonal. Explain this with a geometric graph in the xy-plane.

    (d) If x and y are orthogonal, i.e., $\langle x, y \rangle = 0$, show that

$$\|x + y\|^2 = \|x\|^2 + \|y\|^2.$$

    (e) Is the converse of (d) true over $\mathbb{C}$? over $\mathbb{R}$?

    (f) Show by examples in $\mathbb{R}^2$ that neither the ℓ_1-norm nor the ℓ_∞-norm on $\mathbb{R}^2$ satisfies the parallelogram identity in (a).

5.51   An inner product $\langle \cdot, \cdot \rangle$ on a vector space produces a norm $\| \cdot \| = \sqrt{\langle \cdot, \cdot \rangle}$. However, a norm need not be derived by an inner product. Consider $\mathbb{R}^n$ with 1-norm and 2-norm defined for $x = (x_1, \ldots, x_n)$ by

$$\|x\|_1 = \sum_{i=1}^{n} |x_i|, \quad \|x\|_2 = \left(\sum_{i=1}^{n} |x_i|^2 \right)^{\frac{1}{2}}.$$

(a) Show that $\| \cdot \|_2$ is a norm on $\mathbb{R}^n$ induced by an inner product.

(b) Show that $\| \cdot \|_1$ is a norm on $\mathbb{R}^n$ which is not induced by any inner product, that is, there does not exist an inner product $\langle \cdot, \cdot \rangle$ on $\mathbb{R}^n$ such that $\|x\|_1 = \sqrt{\langle x, x \rangle}$ for all $x \in \mathbb{R}^n$.

(c) Show that there exist positive constants a and b such that

$$a\|x\|_1 \leq \|x\|_2 \leq b\|x\|_1 \quad \text{for all } x \in \mathbb{R}^n.$$

(d) For what x does $\|x\|_1 = \|x\|_2$?

5.52   For $x, y \in \mathbb{R}$, define

$$d(x, y) = \frac{|x - y|}{1 + |x - y|}.$$

(a) Show that $d(x, y) \leq d(x, z) + d(z, y)$ for all $x, y, z \in \mathbb{R}$.

(b) Let $g(x) = d(x, 0)$. Show that $g(x)$ is not a norm on $\mathbb{R}$.

5.53   Let V be a normed space with norm $\| \cdot \|$. Show that for $x, y \in V$

$$1 \leq \frac{\|x + y\|^2 + \|x - y\|^2}{\|x\|^2 + \|y\|^2} \leq 4, \quad \text{where } x \neq 0 \text{ or } y \neq 0.$$

5.54   Let A, B, and C be $n \times n$ matrices such that $\begin{pmatrix} A & B^* \\ B & C \end{pmatrix} \geq 0$. Show that

$$|\langle Bx, y \rangle|^2 \leq \langle Ax, x \rangle \langle Cy, y \rangle$$

for all $x, y \in \mathbb{C}^n$. In particular, for any $A \geq 0$,

$$|\langle Ax, y \rangle|^2 \leq \langle Ax, x \rangle \langle Ay, y \rangle$$

and for any $A > 0$,

$$|\langle x, y \rangle|^2 \leq \langle Ax, x \rangle \langle A^{-1}y, y \rangle.$$

Derive for any $M \in M_n(\mathbb{C})$ with $\ell(M) = (M^*M)^{\frac{1}{2}}$ and all $x, y \in \mathbb{C}^n$,

$$|\langle Mx, y \rangle|^2 \leq \langle \ell(M)x, x \rangle \langle \ell(M^*)y, y \rangle.$$

5.55 Let W be a subspace of an inner product space V of finite dimension.

  (a) Show that $V = W \oplus W^\perp$.

  Consequently, $\dim V = \dim W + \dim W^\perp$ and the decomposition of every vector is unique, that is, if $v = w_1 + w_1' = w_2 + w_2'$, where $w_1, w_2 \in W$, $w_1', w_2' \in W^\perp$, then $w_1 = w_2$, $w_1' = w_2'$.

  (b) Show that $V = \operatorname{Im} \mathcal{A} \oplus (\operatorname{Im} \mathcal{A})^\perp$, where $\mathcal{A}$ is a linear map on V.

5.56 Let W be a subspace of an inner product space V and let $W^\perp$ be the orthogonal complement of W in V. (Thus $V = W \oplus W^\perp$.) Every $v \in V$ can be uniquely written as $v = w_0 + w'$, where $w_0 \in W$, $w' \in W^\perp$. Such w_0 is called the *orthogonal projection* of v onto W.

Show that $w_0 \in W$ is the orthogonal projection of $v \in V$ onto W if and only if the distance between v and w_0 is the shortest in the sense

$$\|v - w_0\| \leq \|v - w\|, \quad \text{for all } w \in W.$$

5.57 Let W be a subspace of an inner product space V. Answer true or false:

  (a) There is a unique subspace W' such that $W' + W = V$.

  (b) There is a unique subspace W' such that $W' \oplus W = V$.

  (c) There is a unique subspace W' such that $W' \oplus W = V$ and $W' \perp W$ if V is finite dimensional.

  (d) There is a unique subspace W' such that $W' \oplus W = V$ and $W' \perp W$ if V is infinite dimensional.

  (e) $(W^\perp)^\perp = W$.

5.58 Let P be the plane defined by $\{(x, y, z) \in \mathbb{R}^3 \mid x - 2y + z = 0\}$ in $\mathbb{R}^3$. Find the (shortest) distance from the point $(1, 0, 1)$ to the plane P.

5.59 In $\mathbb{R}^3$, let V be the subspace spanned by vectors $u = (1, 2, 3)$ and $v = (1, 0, -1)$. Find the distance from the vector $w = (-\frac{1}{2}, 1, 1)$ to V.

5.60 Let W_1 and W_2 be subspaces of an inner product space V. If W_1 and W_2 are finite dimensional, show that

  (a) $(W_1 + W_2)^\perp = W_1^\perp \cap W_2^\perp$.

  (b) $(W_1 \cap W_2)^\perp = W_1^\perp + W_2^\perp$.

5.61 Let $A \in M_{m \times n}(\mathbb{R})$. Show that (for row, column, and null spaces)

$$(\operatorname{Im} A)^\perp = \operatorname{Ker} A^t \quad \text{and} \quad (\operatorname{Im} A^t)^\perp = \operatorname{Ker} A.$$

5.62 Let $S = \{u_1, \ldots, u_p\}$ be an orthogonal set of nonzero vectors in an n-dimensional inner product space V, that is, $\langle u_i, u_j \rangle = 0$ if $i \neq j$. Let $T = \{v_1, \ldots, v_q\}$ be a set of vectors in V that are all orthogonal to S, namely, $\langle v_i, u_j \rangle = 0$ for all i and j. If $p + q > n$, show that T is a linearly dependent set, i.e., $v_1, \ldots, v_q$ are linearly dependent.

5.63 Let V be an inner product space of finite dimension and let $S^\perp$ be the orthogonal complement of (nonempty) subset S of V. Show that

(a) $S = \{0\} \Leftrightarrow S^\perp = V$.

(b) $S = V \Leftrightarrow S^\perp = \{0\}$.

(c) $S \cap S^\perp = \{0\}$ or $\emptyset$ (the empty set).

(d) $S^\perp \cap \operatorname{Span} S = \{0\}$.

(e) $S^\perp$ is a subspace of V.

(f) $S^\perp = \{0\}$ if and only if $\operatorname{Span} S = V$.

(g) $S \subseteq \operatorname{Span} S \subseteq (S^\perp)^\perp$.

(h) $S \subseteq T \Rightarrow T^\perp \subseteq S^\perp$.

(i) $S^\perp = (\operatorname{Span} S)^\perp$.

(j) $(S^\perp)^\perp = \operatorname{Span} S$.

(k) $(S^\perp)^\perp = S$ if and only if S is a subspace of V.

(l) $((S^\perp)^\perp)^\perp = S^\perp$.

(m) $S^\perp \oplus \operatorname{Span} S = V$.

(n) $\dim S^\perp + \dim(S^\perp)^\perp = \dim V$.

5.64 (a) (**QR factorization**) If A is a nonsingular (square) matrix, show that there exist a unique unitary matrix Q and a unique upper-triangular matrix R with positive diagonal entries such that

$$A = QR.$$

(b) (**Cholesky factorization**) If A is a positive definite matrix, show that there is a unique upper-triangular matrix C with positive diagonal entries such that

$$A = C^*C.$$

5.65 Find a real orthogonal matrix T such that $T^t A T$ is diagonal, where

$$A = \begin{pmatrix} 4 & 2 & 2 \\ 2 & 4 & 2 \\ 2 & 2 & 4 \end{pmatrix}.$$

5.66 Find an orthonormal basis for the null space of the linear equation

$$x_1 + 2x_2 + 3x_3 = 0.$$

5.67 Find the null space S for the system of linear equations and find $S^\perp$:

$$\begin{aligned} x_1 - 2x_2 + 3x_3 - 4x_4 &= 0 \\ x_1 + 5x_2 + 3x_3 + 3x_4 &= 0. \end{aligned}$$

5.68 Find an orthonormal basis for the solution space of the equations:

$$\begin{aligned} x_1 + x_2 - x_3 + x_5 &= 0 \\ 2x_1 + x_2 - x_3 + x_4 - 3x_5 &= 0. \end{aligned}$$

5.69 Find an orthonormal basis for $\mathrm{Span}\{\alpha_1, \alpha_2, \alpha_3, \alpha_4\} \subseteq \mathbb{R}^4$, where

$$\alpha_1 = \begin{pmatrix} 1 \\ -1 \\ 0 \\ 1 \end{pmatrix}, \quad \alpha_2 = \begin{pmatrix} 4 \\ 1 \\ 1 \\ 0 \end{pmatrix}, \quad \alpha_3 = \begin{pmatrix} 3 \\ 1 \\ 2 \\ -2 \end{pmatrix}, \quad \alpha_4 = \begin{pmatrix} 0 \\ 1 \\ -1 \\ 1 \end{pmatrix}.$$

5.70 Find an orthonormal basis for the row space of the matrix

$$A = \begin{pmatrix} 0 & 1 & 0 & 0 \\ 1 & 0 & 2 & 1 \\ 1 & 0 & 0 & 0 \\ -1 & 0 & 2 & 1 \end{pmatrix}.$$

5.71 Let V_1 and V_2 be two subspaces of an inner product space V of finite dimension. If $\dim V_1 < \dim V_2$, show that there exists a nonzero vector in V_2 that is orthogonal to all vectors in V_1.

5.72 Let $\mathcal{A}$ be a linear transformation on an inner product space V equipped with the inner product $\langle \cdot, \cdot \rangle$ over a field $\mathbb{F}$. If one defines, for $x, y \in V$,

$$[x, y] = \langle \mathcal{A}(x), \mathcal{A}(y) \rangle,$$

what $\mathcal{A}$ will make $[\cdot, \cdot]$ an inner product for V?

5.73 Let V be a Euclidean vector space with a basis $\{u_1, \ldots, u_n\}$. Define

$$\mathcal{L}(v) = \sum_{i=1}^{n} \langle v, u_i \rangle u_i, \quad v \in V.$$

Show that $\mathcal{L}$ is linear, positive (i.e., $\langle \mathcal{L}(v), v \rangle \geq 0$), and invertible.

5.74   Let V be an inner product space of dimension n and let $\{u_1, \ldots, u_n\}$ be an orthonormal basis for V. If $\mathcal{A}$ is a linear transformation on V and if $A = (a_{ij})$ is the matrix representation of $\mathcal{A}$ with respect to the basis $\{u_1, \ldots, u_n\}$, show that $a_{ij} = \langle \mathcal{A}(u_j), u_i \rangle$, $1 \le i, j \le n$, that is,

$$A = \begin{pmatrix} \langle \mathcal{A}(u_1), u_1 \rangle & \cdots & \langle \mathcal{A}(u_n), u_1 \rangle \\ \vdots & \ddots & \vdots \\ \langle \mathcal{A}(u_1), u_n \rangle & \cdots & \langle \mathcal{A}(u_n), u_n \rangle \end{pmatrix}.$$

5.75   Let V and W be inner product spaces over a field $\mathbb{F}$. Let $\mathcal{A}$ and $\mathcal{B}$ be transformations from V to W, and $\mathcal{C}$ a transformation from W to V.

   (a)   Show that $\mathcal{A} = \mathcal{B}$ if and only if

$$\langle \mathcal{A}(v), w \rangle = \langle \mathcal{B}(v), w \rangle, \quad \text{for all } v \in V \text{ and } w \in W.$$

   (b)   If $\mathcal{A}$ is linear, show that $\mathcal{C}$ is linear if $\mathcal{C}$ satisfies

$$\langle \mathcal{A}(v), w \rangle = \langle v, \mathcal{C}(w) \rangle, \quad \text{for all } v \in V \text{ and } w \in W.$$

5.76   Let $\mathcal{A}$ be a linear map on an inner product space V of dimension n. Let $\mathcal{A}^*$ be the *adjoint* of $\mathcal{A}$, i.e., $\mathcal{A}^*$ is the linear map on V such that

$$\langle \mathcal{A}(x), y \rangle = \langle x, \mathcal{A}^*(y) \rangle, \quad \text{for all } x, y \in V.$$

Show that

   (a) $\mathcal{A}^*$ exists and is unique.
   (b) $(\mathcal{A}^*)^* = \mathcal{A}$.
   (c) $\operatorname{Ker} \mathcal{A}^* = (\operatorname{Im} \mathcal{A})^\perp$.
   (d) $\operatorname{Im} \mathcal{A}^* = (\operatorname{Ker} \mathcal{A})^\perp$.
   (e) $\dim \operatorname{Im} \mathcal{A}^* = \dim \operatorname{Im} \mathcal{A}$.
   (f) $V = \operatorname{Ker} \mathcal{A}^* \oplus \operatorname{Im} \mathcal{A} = \operatorname{Im} \mathcal{A}^* \oplus \operatorname{Ker} \mathcal{A}$.
   (g)   If the matrix representation of $\mathcal{A}$ under an orthonormal basis is A, then the matrix representation of $\mathcal{A}^*$ under the basis is A^*.
   (h)   If $\mathcal{B}$ is also a linear map on V, then $(\mathcal{A}\mathcal{B})^* = \mathcal{B}^*\mathcal{A}^*$ and $(\mathcal{A} + \mathcal{B})^* = \mathcal{A}^* + \mathcal{B}^*$. So, $(\mathcal{A}^*\mathcal{A})^* = \mathcal{A}^*\mathcal{A}$ and $(\mathcal{A}^* + \mathcal{A})^* = \mathcal{A}^* + \mathcal{A}$.

5.77   Let $A \in M_{m \times n}(\mathbb{C})$. Define the matrix-induced linear map $\mathcal{A} : \mathbb{C}^n \to \mathbb{C}^m$ by $\mathcal{A}(x) = Ax$. With $\langle \cdot, \cdot \rangle$ for the usual inner product, show that

$$\langle \mathcal{A}(x), y \rangle = \langle Ax, y \rangle = y^*Ax = \langle x, A^*y \rangle = \langle x, \mathcal{A}^*(y) \rangle.$$

5.78 Let A be an $m \times n$ matrix over a field $\mathbb{F}$. Define a linear operator $\mathcal{A}$ from $M_{n \times p}(\mathbb{F})$ to $M_{m \times p}(\mathbb{F})$ by $\mathcal{A}(X) = AX$. Find the adjoint of $\mathcal{A}$ (with respect to the Frobenius inner product $\langle X, Y \rangle = \operatorname{tr}(Y^*X)$).

5.79 Let $\mathcal{A}$ be a *self-adjoint* linear operator on an inner product space V of finite dimension over $\mathbb{F}$ (where $\mathbb{F} = \mathbb{C}$ or $\mathbb{R}$), i.e., $\mathcal{A} = \mathcal{A}^*$, that is,

$$\langle \mathcal{A}(u), v \rangle = \langle u, \mathcal{A}(v) \rangle, \quad \text{for all } u, v \in V.$$

Show that

(a) The eigenvalues of $\mathcal{A}$ are all real.

(b) $\mathcal{A}$ is self-adjoint if and only if the matrix representation of $\mathcal{A}$ with respect to some orthonormal basis is Hermitian.

(c) There exists an orthonormal basis for V in which every basis vector is an eigenvector of $\mathcal{A}$. In other words, there exists a set of eigenvectors of $\mathcal{A}$ comprising an orthonormal basis for V.

5.80 Let $\mathcal{A}$ be a self-adjoint linear transformation on an inner product space V of finite dimension over $\mathbb{C}$, and let W be a k-dimensional subspace of V. If $\langle \mathcal{A}(x), x \rangle > 0$ for all nonzero vectors x in W, show that $\mathcal{A}$ has at least k positive eigenvalues (counting multiplicity).

5.81 Let $\mathcal{A}$ be a map on a Euclidean space V of dimension n defined by

$$\langle \mathcal{A}(u), v \rangle = -\langle u, \mathcal{A}(v) \rangle, \quad u, v \in V.$$

Show that $\mathcal{A}$ is linear and if λ is a real eigenvalue of $\mathcal{A}$, then $\lambda = 0$. Show also that the matrix of $\mathcal{A}^2$ with respect to some basis is diagonal.

5.82 Let $\mathcal{A}$ and $\mathcal{B}$ be linear maps on an inner product space V such that

$$\langle \mathcal{A}(u), \mathcal{B}(v) \rangle = \langle u, v \rangle, \quad \text{for all } u, v \in V.$$

Show that $\mathcal{A}$ and $\mathcal{B}$ are invertible and $\mathcal{A}^{-1} = \mathcal{B}^*$, $\mathcal{B}^{-1} = \mathcal{A}^*$.

5.83 If $\mathcal{A}$ is a map on an inner product space V over a field $\mathbb{F}$ such that

$$\langle \mathcal{A}(x), \mathcal{A}(y) \rangle = \langle x, y \rangle, \quad \text{for all } x, y \in V,$$

show that $\mathcal{A}$ is linear (i.e., a map that preserves the inner product is linear) and that $\mathcal{A}^*\mathcal{A} = \mathcal{A}\mathcal{A}^* = \mathcal{I}_V$. (Such $\mathcal{A}$ is called an *orthogonal transformation* if $\mathbb{F}$ is $\mathbb{R}$ or a *unitary transformation* if $\mathbb{F}$ is $\mathbb{C}$.)

5.84 Let $\mathcal{A}$ be a linear map on an inner product space V over $\mathbb{F}$ ($= \mathbb{C}$ or $\mathbb{R}$).

(a) If $\mathbb{F} = \mathbb{C}$ and $\langle \mathcal{A}(v), v \rangle = 0$ for all $v \in V$, show that $\mathcal{A} = 0$.

(b) If $\mathbb{F} = \mathbb{R}$, give an example $\mathcal{A} \neq 0$, $\langle \mathcal{A}(v), v \rangle = 0$ for all $v \in V$.

(c) Give an example $\mathcal{A}$, for some v, $\langle \mathcal{A}(v), v \rangle = 0$, $\langle \mathcal{A}^2(v), v \rangle \neq 0$.

(d) For $\mathbb{F} = \mathbb{C}$ or $\mathbb{R}$, if $\mathcal{A}$ is self-adjoint and $\langle \mathcal{A}(v), v \rangle = 0$ for all $v \in V$, show that $\mathcal{A} = 0$. Equivalently, if $\mathcal{A}$ and $\mathcal{B}$ are self-adjoint and $\langle \mathcal{A}(v), v \rangle = \langle \mathcal{B}(v), v \rangle$ for all $v \in V$, then $\mathcal{A} = \mathcal{B}$.

5.85 Let V be an inner product space over $\mathbb{R}$. As is known, a linear map on V preserves inner product if and only if it preserves length. To be precise, a linear map $\mathcal{A}$ on V is an orthogonal transformation (i.e., $\langle \mathcal{A}(u), \mathcal{A}(v) \rangle = \langle u, v \rangle$ for all $u, v \in V$) if and only if $\mathcal{A}$ is an isometry (i.e., $\|\mathcal{A}(u)\| = \|u\|$ for all $u \in V$). The word "linear" in the statement as a precondition is necessary. Likewise, a map on V that preserves distance need not be linear. Provide the following examples:

(a) A nonlinear transformation $\mathcal{L}$ on V that preserves length, i.e., $\|\mathcal{L}(u)\| = \|u\|$ for all $u \in V$.

(b) A nonlinear transformation $\mathcal{D}$ on V that preserves distance, i.e., $\|\mathcal{D}(u) - \mathcal{D}(v)\| = \|u - v\|$ for all $u, v \in V$.

5.86 Show that a linear operator $\mathcal{L}$ on a finite dimensional inner product space V is an isometry (i.e., $\|\mathcal{L}(v)\| = \|v\|$ for every $v \in V$) if and only if $\mathcal{L}^* \mathcal{L} = \mathcal{I}$, where $\mathcal{I}$ is the identity operator on V.

5.87 Let $\mathcal{A}$ be a linear map on an inner product space V of dimension n.

(a) If $\{\alpha_1, \alpha_2, \ldots, \alpha_n\}$ is a basis of V and $\mathcal{A}$ satisfies

$$\langle \mathcal{A}(\alpha_i), \mathcal{A}(\alpha_j) \rangle = \langle \alpha_i, \alpha_j \rangle, \quad i, j = 1, 2, \ldots, n,$$

show that $\mathcal{A}$ is an orthogonal map.

(b) If $\{\alpha_1, \alpha_2, \ldots, \alpha_n\}$ is an orthogonal basis of V and $\mathcal{A}$ satisfies

$$\langle \mathcal{A}(\alpha_i), \mathcal{A}(\alpha_i) \rangle = \langle \alpha_i, \alpha_i \rangle, \quad i = 1, 2, \ldots, n,$$

is $\mathcal{A}$ an orthogonal map? i.e., is it true that $\langle \mathcal{A}(u), \mathcal{A}(v) \rangle = \langle u, v \rangle$ for all $u, v \in V$? In other words, does $\|\mathcal{A}(u)\| = \|u\|$?

5.88 Let V be an n-dimensional inner product space over $\mathbb{R}$ equipped with an inner product $\langle \cdot, \cdot \rangle$. Let $\mathcal{A}$ be a linear transformation such that

$$\langle \mathcal{A}(x), y \rangle + \langle x, \mathcal{A}(y) \rangle = 0, \quad \text{for all } x, y \in V.$$

If n is an odd number, show that $\mathcal{A}$ is singular. Show by examples that the statement is not true if n is even or if $\mathbb{R}$ is changed to $\mathbb{C}$.

5.89 Let $\{\alpha_1, \alpha_2, \ldots, \alpha_n\}$ and $\{\beta_1, \beta_2, \ldots, \beta_n\}$ be two sets of vectors of a Euclidean space V of dimension n. (a) Does there exist a linear transformation that maps α_i to β_i for $i = 1, 2, \ldots, n$? (b) Show that

$$\langle \alpha_i, \alpha_j \rangle = \langle \beta_i, \beta_j \rangle, \quad i, j = 1, 2, \ldots, n,$$

if and only if there exists an orthogonal linear map $\mathcal{A}$ such that

$$\mathcal{A}(\alpha_i) = \beta_i, \quad i = 1, 2, \ldots, n.$$

5.90 Let $\mathcal{A}$ and $\mathcal{B}$ be linear operators on a Euclidean space V. Show that

$$\langle \mathcal{A}(v), \mathcal{A}(v) \rangle = \langle \mathcal{B}(v), \mathcal{B}(v) \rangle, \quad \text{for all } v \in V$$

if and only if there exists an orthogonal operator $\mathcal{C}$ on V such that

$$\mathcal{A} = \mathcal{C}\mathcal{B}.$$

5.91 Let V be a vector space over $\mathbb{F}$ ($= \mathbb{C}$ or $\mathbb{R}$). A linear transformation from V to $\mathbb{F}$ is referred to as a *linear functional*. Show that each of the following transformations from V to $\mathbb{F}$ is a linear functional:

(a) $f \mapsto f(0)$ from $\mathcal{C}[-\pi, \pi]$ to $\mathbb{R}$.

(b) $f \mapsto \int_0^1 f(x)dx$ from $\mathcal{C}[0, 1]$ to $\mathbb{R}$.

(c) $A \mapsto \operatorname{tr} A$ from $M_n(\mathbb{F})$ to $\mathbb{F}$.

(d) $x \mapsto \sum_{i=1}^k x_i$ from $\mathbb{F}^n$ to $\mathbb{F}$, where $x = (x_1, \ldots, x_n)$, $1 \leq k \leq n$.

(e) $x \mapsto e^t A x$ from $\mathbb{C}^n$ to $\mathbb{C}$, where $A \in M_n(\mathbb{C})$, $e^t = (1, \ldots, 1)$.

(f) $x \mapsto \langle x, x_0 \rangle$ from V with inner product $\langle \cdot, \cdot \rangle$ to $\mathbb{F}$, where $x_0 \in V$.

5.92 Let $\mathcal{T}$ be an orthogonal transformation on an inner product space V. Show that $V = W_1 \oplus W_2$, where

$$W_1 = \{\, x \in V \mid \mathcal{T}(x) = x \,\} \quad \text{and} \quad W_2 = \{\, x - \mathcal{T}(x) \mid x \in V \,\}.$$

5.93   Let V be an n-dimensional inner product space over $\mathbb{R}$. A linear functional on V is a linear map from V to $\mathbb{R}$ and the *dual space* of V, denoted by V^*, is the vector space of all linear functionals on V.

   (a)   For $v \in V$, define a map $\mathcal{L}_v$ from V to $\mathbb{R}$ by

$$\mathcal{L}_v(u) = \langle u, v \rangle, \quad \text{for all } u \in V.$$

   Show that $\mathcal{L}_v$ is a linear functional for every fixed v.

   (b)   Let $\mathcal{L}$ be the map from V to V^* defined by

$$\mathcal{L}(v) = \mathcal{L}_v, \quad \text{for all } v \in V.$$

   Show that $\mathcal{L}$ is linear.

   (c)   Show that $\mathcal{L}$ is one-to-one and onto.

   (d)   Find a basis for the dual space V^*.

   (e)   Find a basis for the dual space of $\mathbb{R}^2$.

   (f)   In view of the linear functionals $\mathcal{L}_{e_i}$ for which $\mathcal{L}_{e_i}(e_i) = \langle e_i, e_i \rangle = 1$, where $\{e_1, \ldots, e_n\}$ is an orthonormal basis of V, how would the concept of dual space of an inner product space be defined for a vector space without invoking an inner product?

5.94   Let u be a unit vector in a Euclidean space V of dimension n. Define

$$\mathcal{A}(x) = x - 2\langle x, u \rangle u, \ x \in V. \ (\textit{Householder's transformation})$$

Show that

   (a) $\mathcal{A}$ is an orthogonal transformation.

   (b) $\mathcal{A}$ is self-adjoint, i.e., $\mathcal{A}^* = \mathcal{A}$.

   (c) $\mathcal{A}$ is an involution, i.e., $\mathcal{A}^2 = \mathcal{I}$.

   (d)   If A is a matrix representation of $\mathcal{A}$, then $\det A = -1$.

   (e)   The matrix representation of $\mathcal{A}$ under any orthonormal basis has the form $I - 2vv^t$, where v is some column vector.

   (f)   If $x = ku + y$ and $\langle u, y \rangle = 0$, then $\mathcal{A}(x) = -ku + y$.

   (g)   If $\mathcal{B}$ is an orthogonal transformation having 1 as an eigenvalue with eigenspace of dimension $n - 1$, then for some unit $w \in V$,

$$\mathcal{B}(x) = x - 2\langle x, w \rangle w, \quad \text{for all } x \in V.$$

5.95 Let u be a unit vector in an inner product space V. Define

$$\mathcal{L}_u(x) = x - 2\langle x, u \rangle u, \quad x \in V.$$

Show that $\mathcal{L}_v \mathcal{L}_w + \mathcal{I} = \mathcal{L}_v + \mathcal{L}_w$ for any orthonormal vectors v and w.

5.96 Let $A = \left(\begin{smallmatrix} 1 & 1 \\ 0 & 0 \end{smallmatrix}\right)$. ($A$ can be viewed as the linear map $x \mapsto Ax$ on $\mathbb{R}^2$.)

   (a) Show that $A^2 = A$ (i.e., A is idempotent).

   (b) Show that $\operatorname{Im} A$ and $\operatorname{Ker} A$ are invariant subspaces of A.

   (c) Show that $\mathbb{R}^2 = \operatorname{Im} A \oplus \operatorname{Ker} A$ is a direct sum.

   (d) Are $\operatorname{Im} A$ and $\operatorname{Ker} A$ orthogonal to each other?

   (e) Show that A is diagonalizable, i.e., there exists an nonsingular matrix P such that $P^{-1}AP$ is diagonal. Explain this in terms of matrix representation with respect to a basis.

   (f) Show that A is not unitarily diagonalizable, that is, there does not exist a unitary (real orthogonal) matrix U such that U^*AU is diagonal. What does this mean in terms of matrix representation with respect to an orthonormal basis?

5.97 Let V be an inner product space of finite dimension and let W be a proper subspace of V (so $V = W \oplus W^\perp$). A linear transformation $\mathcal{P}$ on V is called the *orthogonal projection* from V onto W if

$$\mathcal{P}(w) = w \text{ for all } w \in W \text{ and } \mathcal{P}(w') = 0 \text{ for all } w' \in W^\perp.$$

   (a) Show the existence of orthogonal projection $\mathcal{P}$ for the given W.

   (b) Show that such a $\mathcal{P}$ is uniquely determined by W (and $W^\perp$).

   (c) Show that $\mathcal{P}^2 = \mathcal{P}$, $\mathcal{P}^* = \mathcal{P}$; and vice versa with $W = \operatorname{Im} \mathcal{P}$.

   (d) Show the decomposition $V = \operatorname{Im} \mathcal{P} \oplus (\operatorname{Im} \mathcal{P})^\perp$.

   (e) Show that there exists a linear map $\mathcal{T}$ on V such that $\mathcal{P} = \mathcal{T}$ on W, i.e., $\mathcal{P}(w) = \mathcal{T}(w)$ for all $w \in W$, but $\mathcal{P} \neq \mathcal{T}$ on V.

   (f) Show that for every $v \in V$, $\langle \mathcal{P}(v), v \rangle \geq 0$.

   (g) Show that for every $v \in V$, $\|\mathcal{P}(v)\| \leq \|v\|$.

   (h) Show that $\mathcal{I} - \mathcal{P}$ is the orthogonal projection onto $W^\perp$.

   (i) Show that for every $v \in V$, $\|v\|^2 = \|\mathcal{P}(v)\|^2 + \|(\mathcal{I} - \mathcal{P})(v)\|^2$.

   (j) Describe orthogonal projection and orthogonal transformation in terms of their actions on bases. Can a linear map be both?

5.98 Let V be an inner product space with basis $\{\alpha_1, \ldots, \alpha_n\}$ and let $\{\beta_1, \ldots, \beta_n\}$ be a set of vectors in V. Define a linear map $\mathcal{A}$ on V by

$$\mathcal{A}(\alpha_i) = \beta_i, \quad i = 1, \ldots, n.$$

For what $\{\alpha_1, \ldots, \alpha_n\}$ and $\{\beta_1, \ldots, \beta_n\}$,

(a) is $\mathcal{A}$ an invertible transformation?

(b) is $\mathcal{A}$ an orthogonal transformation?

(c) is $\mathcal{A}$ an orthogonal projection?

5.99 Let $\mathcal{A}$ be a linear map on a finite dimensional inner product space V.

(a) If $\mathcal{A}^2 = \mathcal{A}$ and $\operatorname{Im}\mathcal{A} \perp \operatorname{Ker}\mathcal{A}$, show that $\mathcal{A}^* = \mathcal{A}$.

(b) If $\mathcal{A}^2 = \mathcal{A}$ and $\|\mathcal{A}(x)\| \leq \|x\|$ for all $x \in V$, show that $\mathcal{A}^* = \mathcal{A}$.

5.100 If $\mathcal{A}$ is a self-adjoint linear map on a finite dimensional inner product space V such that $V = \operatorname{Im}\mathcal{A} \oplus \operatorname{Ker}\mathcal{A}$, does it follow that $\mathcal{A}^2 = \mathcal{A}$?

5.101 Let V be an inner product space of finite dimension and let $\mathcal{L}$ be a linear map on V. Show that the following statements are equivalent:

(a) $\mathcal{L}^2 = \mathcal{L}$ (idempotent) and $\mathcal{L}^* = \mathcal{L}$ (self-adjoint).

(b) $\mathcal{L}$ is an orthogonal projection onto the image of $\mathcal{L}$.

(c) The matrix L of $\mathcal{L}$ under an orthonormal basis of V is idempotent and Hermitian, namely, $L^2 = L$ and $L^* = L$.

(d) The matrix L of $\mathcal{L}$ under an orthonormal basis of V is unitarily similar to $\left(\begin{smallmatrix} I_r & 0 \\ 0 & 0 \end{smallmatrix} \right)$, where r is the dimension of $\operatorname{Im}\mathcal{L}$.

(e) There exists an orthonormal basis of V with respect to which the matrix of $\mathcal{L}$ is $\left(\begin{smallmatrix} I_r & 0 \\ 0 & 0 \end{smallmatrix} \right)$, where r is the dimension of $\operatorname{Im}\mathcal{L}$.

5.102 Let $\mathcal{P}$ and $\mathcal{Q}$ be orthogonal projections (onto their respective ranges) on a finite dimensional inner product space. Show that

(a) $\mathcal{P} + \mathcal{Q}$ is an orthogonal projection if and only if $\mathcal{P}\mathcal{Q} = 0$.

(b) $\mathcal{P}\mathcal{Q}$ is an orthogonal projection if and only if $\mathcal{P}\mathcal{Q} = \mathcal{Q}\mathcal{P}$.

5.103 Let $\mathcal{P}$ and $\mathcal{Q}$ be orthogonal projections (onto their respective ranges) on a finite dimensional inner product space. Show that

$$(\operatorname{Im}\mathcal{P} \cap \operatorname{Im}Q)^{\perp} = \operatorname{Ker}\mathcal{P} + \operatorname{Ker}Q.$$

5.104 Construct examples of nontrivial (i.e., nonzero and nonidentity) different linear transformations $\mathcal{L}_1$, $\mathcal{L}_2$, and $\mathcal{L}_3$ on $\mathbb{R}^3$ such that they are identical (i.e., $\mathcal{L}_1 = \mathcal{L}_2 = \mathcal{L}_3$) on $W = \{(x, 0, 0) \in \mathbb{R}^3 \mid x \in \mathbb{R}\}$ and

(a) $\mathcal{L}_1^2 = \mathcal{L}_1$, $\mathcal{L}_1^* = \mathcal{L}_1$, and $\mathcal{L}_1$ is the orthogonal projection onto W.

(b) $\mathcal{L}_2^2 = \mathcal{L}_2$, $\mathcal{L}_2^* = \mathcal{L}_2$, but $\mathcal{L}_2$ is not orthogonal projection onto W.

(c) $\mathcal{L}_3^2 = \mathcal{L}_3$ and $\mathcal{L}_3^* \neq \mathcal{L}_3$. Find $\mathcal{L}_3^*$ explicitly.

5.105 Show that a linear operator $\mathcal{L}$ on an inner product space V is *normal*, i.e., $\mathcal{L}\mathcal{L}^* = \mathcal{L}^*\mathcal{L}$, if and only if $\|\mathcal{L}(v)\| = \|\mathcal{L}^*(v)\|$ for all $v \in V$.

5.106 Let $\mathcal{P}_1$, $\mathcal{P}_2$, ..., $\mathcal{P}_m$ be idempotent linear transformations on an n-dimensional vector space V, that is, $\mathcal{P}_i^2 = \mathcal{P}_i$, $i = 1, 2, \ldots, m$.

(a) If
$$\mathcal{P}_1 + \mathcal{P}_2 + \cdots + \mathcal{P}_m = \mathcal{I},$$
show that
$$V = \operatorname{Im} \mathcal{P}_1 \oplus \operatorname{Im} \mathcal{P}_2 \oplus \cdots \oplus \operatorname{Im} \mathcal{P}_m$$
and
$$\mathcal{P}_i \mathcal{P}_j = 0, \ \ i, j = 1, 2, \ldots, m, \ \ i \neq j.$$

(b) Define an inner product for V such that each $\mathcal{P}_i$ is an orthogonal projection onto its image.

(c) If
$$\mathcal{P}_i \mathcal{P}_j = 0, \ \ i, j = 1, 2, \ldots, m, \ \ i \neq j,$$
Show that
$$V = \operatorname{Im} \mathcal{P}_1 \oplus \operatorname{Im} \mathcal{P}_2 \oplus \cdots \oplus \operatorname{Im} \mathcal{P}_m \oplus \cap_{i=1}^m \operatorname{Ker} \mathcal{P}_i.$$

5.107 Let V be a vector space of dimension n over $\mathbb{R}$. Let $\mathcal{T}$ be a linear operator on V, and let a and b be real numbers such that $a \neq b$ and
$$(a\mathcal{I} - \mathcal{T})(b\mathcal{I} - \mathcal{T}) = 0.$$

(a) Find all eigenvalues of $\mathcal{T}$.

(b) Show that $V = \operatorname{Ker}(a\mathcal{I} - \mathcal{T}) \oplus \operatorname{Ker}(b\mathcal{I} - \mathcal{T})$.

(c) Show that a matrix of $\mathcal{T}$ under some basis is $\operatorname{diag}(aI_k, bI_{n-k})$.

(d) If $\mathcal{T}$ is self-adjoint, show that $\operatorname{Ker}(a\mathcal{I} - \mathcal{T}) \perp \operatorname{Ker}(b\mathcal{I} - \mathcal{T})$.

5.108   Let V be an inner product space over a field $\mathbb{F}$ with dual space V^*
(i.e., the space of all linear functionals from V to $\mathbb{F}$) and let W be a
subspace of V. Let $W^\circ = \{f \in V^* \mid f(x) = 0 \text{ if } x \in W\}$. Show that

(a) W° is a subspace of V^*, called *annihilator* of W.

(b) If V is finite dimensional, then $\dim W^\circ = \dim V - \dim W$.

(c) $W_1^\circ \cap W_2^\circ = (W_1 + W_2)^\circ$ for subspaces W_1 and W_2 of V.

5.109   Let V be an inner product space of finite dimension. Let W_1 and W_2
be subspaces of V and let $\mathcal{A}$ be a linear transformation on V. Consider

(1) $W_1 + W_2$: the sum of W_1 and W_2.

(2) $W_1 \oplus W_2$: the direct sum of W_1 and W_2.

(3) $W_1 \oplus_\perp W_2$: the orthogonal (direct) sum in which $W_1 \perp W_2$.

(4) $W_1 \oplus_\mathcal{A} W_2$: the direct sum where W_1 and W_2 are $\mathcal{A}$-invariant.

(5) $W_1 \oplus_{\mathcal{A},\perp} W_2$: the direct sum where W_1 and W_2 are $\mathcal{A}$-invariant
and W_1 and W_2 are mutually orthogonal.

Answer the questions:

(a) State an important property of each sum.

(b) Explain the differences between the sums.

(c) Draw diagrams in the xy-plane to illustrate the concepts.

(d) What can be said about a matrix representation of $\mathcal{A}$ in (4)?

(e) Is $\mathcal{A}$ in (5) an orthogonal map or an orthonormal projection?

Chapter 6

Miscellaneous Problems

6.1 Let V be the set of all ordered real pairs (x, y). Define an operation $\boxplus$ among the elements of V as follows: for (x_1, y_1) and (x_2, y_2) in V,

$$(x_1, y_1) \boxplus (x_2, y_2) = (x_1 x_2 - y_1 y_2, x_1 y_2 + y_1 x_2).$$

  (a) Compute $(1, 2) \boxplus (3, 4)$.

  (b) Compute $(0, 0) \boxplus (x, y)$.

  (c) Compute $(x, y) \boxplus (y, x)$.

  (d) Show that $(1, 2) \boxplus (x, y) \neq (0, 0)$ unless $x = y = 0$.

  (e) Find $(a, b) \in V$ so that $(a, b) \boxplus (x, y) = (x, y)$ for all $(x, y) \in V$.

  (f) Show that $(x, y) \boxplus (p, q) = (p, q) \boxplus (x, y)$ for all $(x, y), (p, q) \in V$.

  (g) Show that the ordered pair (a, b) found in (e) is unique.

  (h) Is V a vector space with respect to $\boxplus$ as addition and the usual scalar multiplication $\lambda(x, y) = (\lambda x, \lambda y)$? why or why not?

6.2 Let V be a vector space of dimension n and let $V_1, \ldots, V_m$ be subspaces of V. If the sum of the dimensions of $V_1, \ldots, V_m$ is greater than $(m - 1)n$, show that there exists a nonzero vector in $\cap_{i=1}^m V_i$.

6.3 Show that a vector space over an infinite field $\mathbb{F}$ (say $\mathbb{F} = \mathbb{C}$, $\mathbb{R}$, or $\mathbb{Q}$) cannot be the union of a finite number of proper spaces.

6.4 Let v be a vector in $\mathbb{R}^3$ that has coordinate $(1, 2, 3)$ relative to some basis $\alpha = \{\alpha_1, \alpha_2, \alpha_3\}$. Find a basis $\beta = \{\beta_1, \beta_2, \beta_3\}$ in terms of α such that the coordinate of v relative to β is $(1, 1, 1)$ and no basis vector of β lies in a plane spanned by two basis vectors of α.

6.5   Let H_n be the set of $n \times n$ Hermitian matrices and let S_n be the set of $n \times n$ symmetric (complex) matrices. With respect to the usual matrix addition and scalar multiplication, answer the following questions. In the case of a vector space, find a basis and its dimension.

(a) Is H_n a vector space over $\mathbb{C}$?

(b) Is H_n a vector space over $\mathbb{R}$?

(c) Is S_n a vector space over $\mathbb{C}$?

(d) Is S_n a vector space over $\mathbb{R}$?

(e) Is $H_n \cap S_n$ a vector space over $\mathbb{C}$?

(f) Is $H_n \cap S_n$ a vector space over $\mathbb{R}$?

6.6   Let $A \in M_n(\mathbb{C})$ and $\operatorname{Im} A = \{Ax \mid x \in \mathbb{C}^n\}$ (the image, range, or column space of A). Assume that matrices P, Q, R, S have size $n \times n$.

(a) If $\operatorname{Im} R = \operatorname{Im} S$, show that $\operatorname{Im}(PRQ) = \operatorname{Im}(PSQ)$ for all invertible matrices P and Q.

(b) If $\operatorname{Im}(PRQ) = \operatorname{Im}(PSQ)$ for some invertible matrices P and Q, show that $\operatorname{Im} R = \operatorname{Im} S$.

(c) Is it true that $\operatorname{Im} R = \operatorname{Im}(PRQ)$ for all invertible P and Q?

(d) Is it true that $\operatorname{Im} R = \operatorname{Im}(RQ)$ for all invertible Q?

(e) Is it true that $\operatorname{Im} R = \operatorname{Im}(PR)$ for all invertible P?

6.7   Let A be a square $(0, 1, -1)$-matrix (that is, the entries of A are 0's, 1's, and (or) -1's). If $\det A \neq 0$, $|\det A| \neq 1$, and $|\det X| \leq 1$ for every proper square submatrix X of A, show that $|\det A| = 2$.

6.8   Let $A = (a_{ij}) \in M_n(\mathbb{R})$ and all $|a_{ij}| \leq 1$. If every row of A contains at most one positive entry and at most one negative entry, show that

$$|\det A| \leq 1.$$

6.9   Let $a_1, \ldots, a_n, b_1, \ldots, b_n$ be real numbers. Let $M = (m_{ij})$, where

$$m_{ij} = \begin{cases} a_i b_j, & \text{if } i < j \\ 0, & \text{if } i = j \\ -a_j b_i, & \text{if } i > j. \end{cases}$$

Prove that $\det M = 0$ if n is odd and $\det M = \prod_{i=1}^{n/2} a_{2i-1}^2 b_{2i}^2$ if n is even.

6.10 Let A be an $n \times n$ $(0,1)$-matrix partitioned as $A = (P,Q)$. If each row of P and each row of Q contain at most one 1, for example,
$A = \begin{pmatrix} 0 & 0 & 1 \\ 1 & 0 & 0 \\ 0 & 1 & 1 \end{pmatrix}$, $P = \begin{pmatrix} 0 & 0 \\ 1 & 0 \\ 0 & 1 \end{pmatrix}$, $Q = \begin{pmatrix} 1 \\ 0 \\ 1 \end{pmatrix}$, show that $\det A = 0$, 1, or -1.

6.11 Let $A = (a_{ij})$ be an $n \times n$ complex matrix. If every row of A contains at most k nonzero entries, where $1 \leq k \leq n$, show that

$$|\det A| \leq k^{\frac{n}{2}} \prod_{i=1}^{n} \max_{j} |a_{ij}|.$$

6.12 Let A and B be $n \times n$ integral matrices (i.e., having integer entries). If $\det A$ and $\det B$ are relatively prime (or coprime, i.e., GCD=1), show that there exist integral matrices X and Y such that $AX + YB = I_n$.

6.13 Let M be an $m \times n$ matrix. To extract the submatrix of M lying in the rows $i_1, i_2, \ldots, i_p$ and columns $j_1, j_2, \ldots, j_q$, what matrices should be applied to M? In other words, find matrices X and Y such that XMY is the submatrix of M in the selected rows and columns.

6.14 Let X and Y be $n \times n$ matrices such that the rank of $X - Y$ is equal to one. Show that $Y^3 = YXY$ if and only if $Y^2 = XY$ or $Y^2 = YX$.

6.15 Let A and B be idempotent matrices of the same size. Show that if $\det(A-B) \neq 0$, then $\det(A+B) \neq 0$ and $r(AB - BA) = r(AB + BA)$.

6.16 Let A and B be $n \times n$ real matrices such that $A^2 + B^2 = AB - BA$. If $A^2 + B^2$ is nonsingular, show that n is a multiple of 4.

6.17 Let $M = \begin{pmatrix} A & \beta \\ \alpha & \gamma \end{pmatrix}$, where A is an $n \times n$ matrix, α is a row of n components, β is a column of n components, and γ is a number.

    (a) What is the least integer $k > 0$ such that for all A, α, β, and γ

$$r(M) \leq r(A) + k?$$

    (b) If $r(A) = 1$ and M is invertible, what are possible sizes of M?

    (c) If $r(M) = n + 1$, what is the smallest rank of A?

6.18   (a) Let A be an $n \times (n-1)$ matrix and let b be a column vector of n components, both over a field $\mathbb{F}$. Show that $Ax = b$ has a unique solution $x \in \mathbb{F}^{n-1}$ if and only if $r(A) = r(A,b) = n - 1$.

    (b) If B is an $(n-1) \times n$ matrix and c is a column vector of $n-1$ components, show that $By = c$ never has a unique solution y.

6.19 Let $A \in M_n(\mathbb{C})$ and $\mathfrak{R}(A) = \frac{A+A^*}{2}$ (Hermitian part of A). Show that

$$\mathfrak{R}(XAX^*) = X\mathfrak{R}(A)X^*, \quad \text{for any } X \in M_{m \times n}(\mathbb{C}).$$

Derive an identity for submatrices by taking $X = \left(\begin{smallmatrix} I_k & 0 \\ 0 & 0 \end{smallmatrix} \right)$, $1 \leq k \leq n$.

6.20 Let $A \in M_n(\mathbb{C})$. Let $H = \frac{1}{2}(A+A^*)$ and $K = \frac{1}{2i}(A-A^*)$. Show that

(a) H and K are Hermitian.

(b) $H^2 + K^2 = \frac{1}{2}(A^*A + AA^*)$.

(c) $A = H + iK$ and such a (Cartesian) decomposition is unique.

(d) $\sigma_{\max}(A) \leq \max_j |\lambda_j(H)| + \max_j |\lambda_j(K)|$.

6.21 Let $A \in M_n(\mathbb{C})$ and $A = H + iK$, where H and K are Hermitian. Let $B = \frac{1}{2} \left(\begin{smallmatrix} A & A^* \\ A^* & A \end{smallmatrix} \right)$. Prove each of the following statements:

(a) B is normal.

(b) B is invertible if and only if H and K are invertible.

(c) B is unitary if and only if $H^2 = K^2 = I$.

(d) B is positive semidefinite if and only if A is positive semidefinite.

6.22 Let $x = (x_1, x_2, x_3) \in \mathbb{R}^3$ with $x_1 x_2 x_3 \neq 0$, and let $\tilde{x} = (\frac{1}{x_1}, \frac{1}{x_2}, \frac{1}{x_3})$.

(a) If α, β, γ are such vectors in $\mathbb{R}^3$ that α, β, γ are mutually orthogonal, show that $\tilde{\alpha}, \tilde{\beta}, \tilde{\gamma}$ are linearly dependent.

(b) If $U = (u_{ij})$ is a 3×3 real orthogonal matrix with no zero entries, show that the matrix $\tilde{U} = \left(u_{ij}^{-1} \right)$ is singular.

(c) Show by an example that (b) need not be true for 2×2 matrices.

6.23 Let $x, y, z \in \mathbb{C}$. Find the minimal polynomial of matrix $\begin{pmatrix} 0 & 1 & 0 \\ 0 & 0 & 1 \\ x & y & z \end{pmatrix}$.

6.24 Let $A \in M_n(\mathbb{C})$. Describe the relations among the characteristic polynomial, minimal polynomial, and Jordan canonical form of A.

6.25 Let $A \in M_n(\mathbb{C})$. Show that the degree of the minimal polynomial of A equals the dimension of $\mathrm{Span}\{I, A, A^2, \dots\}$ and equals the sum of the orders of the largest Jordan blocks for distinct eigenvalues of A.

6.26 Show that every Jordan block is a product of two symmetric matrices.

6.27   Let J be an $n \times n$ Jordan block in upper-triangular form, i.e.,

$$J = \begin{pmatrix} \lambda & 1 & & 0 \\ & \lambda & \ddots & \\ & & \ddots & 1 \\ 0 & & & \lambda \end{pmatrix}.$$

Let k be a positive integer. Compute the (i, j)-entry of J^k.

6.28   Let

$$A = \begin{pmatrix} 1 & 2 & 0 & 0 \\ 4 & 3 & 0 & 0 \\ 0 & 0 & 1 & 3 \\ 0 & 0 & 4 & 2 \end{pmatrix} \quad \text{and} \quad B = \begin{pmatrix} 1 & 2 & 0 & 1 \\ 4 & 3 & 0 & 0 \\ 0 & 0 & 1 & 3 \\ 0 & 0 & 4 & 2 \end{pmatrix}.$$

(a) Show that A and B have the same eigenvalues.
(b) Show that A is diagonalizable and B is not diagonalizable.

6.29   Construct a 3×3 real symmetric matrix A such that the eigenvalues of A are 1, 1, and -1, and $\alpha = (1, 0, 1)^t$ and $\beta = (0, 1, 1)^t$ are eigenvectors corresponding to the eigenvalue 1. Show that such a real symmetric matrix A is uniquely determined by the given eigenvalues 1, 1, -1, and the eigenspace $V_1 = \text{Span}\{\alpha, \beta\}$ associated to 1.

6.30   Show that $\max_{\|x\|=1}\{(x^*A^*Ax)^{\frac{1}{2}}\} = \max_{\|x\|=1}\{x^*(A^*A)^{\frac{1}{2}}x\}$.

6.31   Let A be an n-square complex matrix with eigenvalues $\lambda_1, \ldots, \lambda_n$ and singular values $\sigma_1, \ldots, \sigma_n$, which are decreasingly ordered such that

$$\sigma_1 \geq \cdots \geq \sigma_n, \quad |\lambda_1| \geq \cdots \geq |\lambda_n|.$$

Show that (i) $\sigma_n = 0 \Leftrightarrow \lambda_n = 0$, (ii) $\sigma_1 \geq |\lambda_1|$, and (iii) $\sigma_n \leq |\lambda_n|$.

6.32   Let $a, x, y \in \mathbb{R}$ and $A = \begin{pmatrix} a & x & 1 \\ x & 0 & y \\ 1 & y & a \end{pmatrix}$. Show that for any values x and y

$$\lambda_{\max}(A) \geq \max\{a + 1, 0\} \quad \text{and} \quad \lambda_{\min}(A) \leq \min\{a - 1, 0\}.$$

If $a = 1$, for what values of x and y, does A have a zero eigenvalue?

6.33   Let $A, B, C \in M_n(\mathbb{C})$. Define "addition" $\hat{+}$ for the matrices by

$$A\hat{+}B = \tfrac{1}{2}(AB + BA).$$

Show that $\text{tr}\left(A\hat{+}(B\hat{+}C)\right) = \text{tr}\left((A\hat{+}B)\hat{+}C\right)$ for all $A, B, C \in M_n(\mathbb{C})$.
Does it hold that $A\hat{+}(B\hat{+}C) = (A\hat{+}B)\hat{+}C$ for all $A, B, C \in M_n(\mathbb{C})$?

6.34 Find the values of y in terms of x so that the eigenvalues of the matrix $\begin{pmatrix} 1 & x & 1 \\ y & 1 & x \\ 1 & x & 1 \end{pmatrix}$ do not depend on x (nor y). What are the eigenvalues?

6.35 Determine diagonalizability of the following matrices over $\mathbb{Q}$, $\mathbb{R}$, and $\mathbb{C}$:

$$A = \begin{pmatrix} 1 & 2 \\ 4 & 3 \end{pmatrix}, \ B = \begin{pmatrix} 1 & 2 \\ 3 & 4 \end{pmatrix}, \ C = \begin{pmatrix} 1 & 3 \\ 4 & 2 \end{pmatrix}, \ D = \begin{pmatrix} 1 & -2 \\ 3 & 4 \end{pmatrix}.$$

6.36 Show that matrix $A(t) = \begin{pmatrix} \cos t & -\sin t \\ \sin t & \cos t \end{pmatrix}$ is unitarily diagonalizable over $\mathbb{C}$ for any $t \in \mathbb{R}$. When is it similar to a real diagonal matrix?

6.37 Let A be a 2×2 real matrix. Show that there exists a real invertible matrix P such that $P^{-1}AP$ is in one of the following forms:

$$\begin{pmatrix} a & 0 \\ 0 & b \end{pmatrix}, \ \begin{pmatrix} a & 1 \\ 0 & a \end{pmatrix}, \ r\begin{pmatrix} \cos\theta & -\sin\theta \\ \sin\theta & \cos\theta \end{pmatrix}, \ a, b, r, \theta \in \mathbb{R}.$$

6.38 Let $A, B \in M_n(\mathbb{C})$ be such that $AB = \lambda BA$ for some $\lambda \in \mathbb{C}$.

(a) If A and B are invertible, show that $\lambda^n = 1$.

(b) If B is invertible but A is not, show that $|\lambda| = 1$ or $A^n = 0$.

(c) If B is invertible, n is odd, and $\lambda = -1$, i.e., $AB = -BA$, show that 0 is an eigenvalue of A. Is this true if n is even?

(d) Find an example of A, B, and λ such that $AB = \lambda BA$, where n is odd, B is invertible, $\lambda^n \neq 1$, and $A^n \neq 0$.

6.39 Let A be an $n \times n$ matrix. If $\operatorname{tr} A = 0$, show that A is similar to a matrix with all main diagonal entries equal to 0.

6.40 Let A be an $n \times n$ complex matrix. Show that A has repeated eigenvalues if and only if A commutes with some nonzero nilpotent matrix, equivalently, the eigenvalues of A are all distinct if and only if there does not exist a nonzero nilpotent matrix that commutes with A.

6.41 If A is an $n \times n$ matrix such that $A^2 = 0$, prove that $r(A) \leq \frac{n}{2}$.

6.42 Show that the matrix equation $X^2 + X + I = 0$ has no solution X in $M_n(\mathbb{R})$ when n is odd and that it has solutions X in $M_n(\mathbb{R})$ when n is even. Find a 2×2 real matrix X that satisfies the equation.

6.43 Let A be an $n \times n$ complex matrix such that $A^5 + A^3 + 2A^2 + 2I = 0$. Show that $r(A^2 + I) + r(A^3 + 2I) = n$.

6.44 Given $\lambda_2, \lambda_3, \ldots, \lambda_n \in \mathbb{C}$, let $\lambda_1 = \lambda_2 + \lambda_3 + \cdots + \lambda_n$ and let

$$
A = \begin{pmatrix}
0 & \lambda_2 & \lambda_3 & \cdots & \lambda_n \\
\lambda_2 & 0 & \lambda_3 & \cdots & \lambda_n \\
\lambda_2 & \lambda_3 & 0 & \cdots & \lambda_n \\
\vdots & \vdots & \vdots & \ddots & \vdots \\
\lambda_2 & \lambda_3 & \lambda_4 & \cdots & 0
\end{pmatrix}.
$$

(a) Show that λ_1 is an eigenvalue of A.

(b) If $\lambda_2, \lambda_3, \ldots, \lambda_n$ are positive numbers, find the eigenspace of λ_1.

6.45 Show that every real orthogonal matrix of odd order (like 3×3, etc.) with determinant 1 has an eigenvalue 1.

6.46 Let $x, y \in \mathbb{C}$ and let $A = \left(\begin{smallmatrix} 1 & y \\ x & 1 \end{smallmatrix} \right)$, $B = \left(\begin{smallmatrix} 1 & 1 \\ x & y \end{smallmatrix} \right)$, $C = \left(\begin{smallmatrix} 1 & x \\ 1 & y \end{smallmatrix} \right)$, $D = \left(\begin{smallmatrix} x & 1 \\ 1 & y \end{smallmatrix} \right)$. Find the values of x and y for each matrix such that the matrix is diagonalizable over $\mathbb{C}$. Show that for all real positive numbers x and y, the four matrices are all diagonalizable over $\mathbb{R}$ (the diagonalizability need not be via the same invertible matrix).

6.47 Let $A \in M_n(\mathbb{F})$. If $X \in M_n(\mathbb{F})$ satisfies $X^2 = A$, we say that X is a (not unique) square root of A over $\mathbb{F}$. For $A \in M_2(\mathbb{C})$, prove that

(a) A has a square root unless both eigenvalues of A are 0 and $A \neq 0$.

(b) If $a + d \neq 0$ or $a^2 + bc \neq 0$, then $A = \left(\begin{smallmatrix} a & b \\ c & d \end{smallmatrix} \right)$ has a square root.

(c) If $\operatorname{tr} A \neq \det A$, then A has a square root.

(d) If $(\operatorname{tr} A)^2 \neq 4 \det A$, then A has a square root.

6.48 Let $A(z) = \left(\begin{smallmatrix} 1 & z \\ z & -1 \end{smallmatrix} \right)$, $z \in \mathbb{C}$.

(a) Find the eigenvalues of $A(z)$ in terms of z.

(b) Show that $A(z)$ is diagonalizable for all $z \in \mathbb{R}$,

(c) Over $\mathbb{C}$, show that $A(z)$ is diagonalizable or similar to $\left(\begin{smallmatrix} 0 & 1 \\ 0 & 0 \end{smallmatrix} \right)$.

(d) For what values of $z \in \mathbb{C}$ is $A(z)$ not diagonalizable? Choose such a z and find a basis of $\mathbb{C}^2$ so that the matrix representation of $A(z)$ as a linear transformation under the basis is $\left(\begin{smallmatrix} 0 & 1 \\ 0 & 0 \end{smallmatrix} \right)$.

6.49   Let a and b be positive numbers, and let $A = \left(\begin{smallmatrix} 0 & a \\ b & 0 \end{smallmatrix}\right)$ and $B = \left(\begin{smallmatrix} 1 & a \\ b & 1 \end{smallmatrix}\right)$.

   (a) Find the eigenvalues of A and B in terms of a and b.

   (b) Find a nonsingular P so that $P^{-1}AP$ and $P^{-1}BP$ are diagonal.

   (c) Evaluate A^n and B^n in terms of a, b, and positive integers n.

6.50   Let A be a 2×2 matrix with integral (i.e., integer) entries. Show that

   (a) Unless A has only zero eigenvalues, it is impossible for A to have both eigenvalues in the open disk $|z| < 1$ in the xy-plane.

   (b) If A has two distinct integral eigenvalues, then there exists an invertible integral matrix Z such that ZAZ^{-1} is diagonal.

6.51   Let X be an integral matrix. We say that X is diagonalizable over the integers $\mathbb{Z}$ if there exist an integral matrix P with $\det P \neq 0$ and an integral diagonal matrix T such that $P^{-1}XP = T$. Show that

   (a) $A = \left(\begin{smallmatrix} 1 & 1 \\ 0 & 1 \end{smallmatrix}\right)$ is not diagonalizable over $\mathbb{Z}$ (or over $\mathbb{C}$).

   (b) $B = \left(\begin{smallmatrix} 1 & 1 \\ 3z^2 & 2z+1 \end{smallmatrix}\right)$ is diagonalizable over $\mathbb{Z}$ for any integer $z \neq 0$.

   (c) $C = \left(\begin{smallmatrix} 1 & 1 \\ 1 & 1 \end{smallmatrix}\right)$ is diagonalizable over $\mathbb{Z}$. Find a nonsingular integral matrix P such that $P^{-1}CP$ is integral and diagonal.

   (d) An integral matrix is diagonalizable over $\mathbb{Q}$ if and only if it is diagonalizable over $\mathbb{Z}$.

   (e) $E = \left(\begin{smallmatrix} -1 & 1 \\ 1 & 1 \end{smallmatrix}\right)$ is diagonalizable over $\mathbb{R}$, but not over $\mathbb{Q}$ or $\mathbb{Z}$.

   (f) $F = \left(\begin{smallmatrix} 1 & 1 \\ -1 & 1 \end{smallmatrix}\right)$ is diagonalizable over $\mathbb{C}$, but not over $\mathbb{R}$, $\mathbb{Q}$, or $\mathbb{Z}$.

6.52   Prove the statements about real matrices with complex eigenvalues:

   (a) Let A be a 2×2 real matrix. If i is an eigenvalue of A, then there is an invertible real matrix R such that $R^{-1}AR = \left(\begin{smallmatrix} 0 & 1 \\ -1 & 0 \end{smallmatrix}\right)$.

   (b) Let A be a 4×4 real matrix. If $\det(A^2 + I_4) = 0$ and $\operatorname{tr} A \neq 0$, then there is a nonzero real matrix B such that $AB + BA = 0$.

   (c) Let A be an $n \times n$ real matrix. If i is a simple eigenvalue of A, then there is a nonzero real matrix B such that $AB + BA = 0$.

6.53   Let A be an $n \times n$ real matrix. If all eigenvalues of A are real, show that there exists an $n \times n$ real invertible matrix P such that $P^{-1}AP = J$, where J is the (real) Jordan canonical form of A. (Recall the Jordan decomposition theorem which states that every square matrix is similar to its Jordan canonical form over the complex number field.)

6.54 Let A be a 2×2 real matrix having eigenvalue $i = \sqrt{-1}$ and associated eigenvector $u \in \mathbb{C}^2$. If $u = u_1 + iu_2$, where $u_1, u_2 \in \mathbb{R}^2$, show that $\{u_1, u_2\}$ is a basis of $\mathbb{R}^2$. If $v = u_1 + u_2$ and $w = u_1 - u_2$, show that $\{v, w\}$ is also a basis of $\mathbb{R}^2$. What are the matrix representations of A as linear transformation $x \mapsto Ax$ on $\mathbb{R}^2$ with respect to the bases?

6.55 Show that an idempotent matrix $(A^2 = A)$ is diagonalizable and that any two $n \times n$ idempotent matrices of the same rank are similar.

6.56 Let A be an $n \times n$ matrix such that $A^n = 0$ and $A^{n-1} \neq 0$, $n > 1$. Show that there does not exist a matrix X such that $X^2 = A$.

6.57 Let $x = (x_1, x_2, \ldots, x_n) \in \mathbb{C}^n$ be a row vector and let $A = \begin{pmatrix} 1 & x \\ 0 & I_n \end{pmatrix}$. If $\lambda_{\max}$ and $\lambda_{\min}$ are respectively the largest and smallest eigenvalues of A^*A, show that $\lambda_{\max} \cdot \lambda_{\min} = 1$. Find all eigenvalues of A^*A.

6.58 Let $X, Y \in M_n(\mathbb{C})$. Assume the inverses in (b) below exist. Show that

(a) If $X + Y = I$, then $XY = YX$.

(b) If $(I + X)^{-1} + (I + Y)^{-1} = I$, then $XY = YX$.

(c) If $X > 0$, $Y > 0$, and $(I+X)^{-1}+(I+Y)^{-1} = I$, then $X = Y^{-1}$.

6.59 Let A be an $m \times n$ matrix. Let B be a matrix containing A as a submatrix and the entries of B that are not in A are all zeros. For example, $B = \begin{pmatrix} 0 & 0 \\ A & 0 \end{pmatrix}$ (a matrix obtained from A by bordering 0's) and
$$B = \begin{pmatrix} 0 & a & 0 & b \\ 0 & 0 & 0 & 0 \\ 0 & c & 0 & d \end{pmatrix} \text{ for } A = \begin{pmatrix} a & b \\ c & d \end{pmatrix} \text{ (by inserting 0's in } A\text{). Show that}$$
A and B have the same singular values except some extra zeros for B. What if B is obtained from A by "randomly" inserting 0's in A?

6.60 Let $A, B, C \in M_n(\mathbb{C})$.

(a) Show that $\begin{pmatrix} 0 & A \\ 0 & 0 \end{pmatrix}$ is similar to $\begin{pmatrix} 0 & A^t \\ 0 & 0 \end{pmatrix}$.

(b) If $\begin{pmatrix} A & 0 \\ 0 & B \end{pmatrix}$ is similar to $\begin{pmatrix} A & 0 \\ 0 & C \end{pmatrix}$, show that B is similar to C.

(c) If all eigenvalues of A are real, show that A^* is similar to A.

(d) If A and A^* have the same eigenvalues, are they similar?

(e) Show that A^*A is unitarily similar to AA^*.

(f) Is A^tA necessarily similar to AA^t?

6.61 Let A and B be arbitrary $n \times n$ complex matrices. Which of the following block matrices are similar to the $2n \times 2n$ block matrix $\left(\begin{smallmatrix} A & B \\ 0 & 0 \end{smallmatrix} \right)$?

(a) $\begin{pmatrix} 0 & 0 \\ A & B \end{pmatrix}$, (b) $\begin{pmatrix} A & 0 \\ B & 0 \end{pmatrix}$, (c) $\begin{pmatrix} 0 & 0 \\ B & A \end{pmatrix}$,

(d) $\begin{pmatrix} A & 0 \\ 0 & B \end{pmatrix}$, (e) $\begin{pmatrix} A^t & 0 \\ B^t & 0 \end{pmatrix}$, (f) $\begin{pmatrix} A^* & 0 \\ B^* & 0 \end{pmatrix}$.

If A is nonsingular, which ones are always similar to $\left(\begin{smallmatrix} A & B \\ 0 & 0 \end{smallmatrix} \right)$?

6.62 Let A and B be $n \times n$ matrices over a field $\mathbb{F}$ ($= \mathbb{C}$ or $\mathbb{R}$), and let

$$M = \begin{pmatrix} A & A \\ B & B \end{pmatrix}, \quad N = \begin{pmatrix} B & B \\ A & A \end{pmatrix}, \quad R = \begin{pmatrix} A & B \\ A & B \end{pmatrix}, \quad S = \begin{pmatrix} B & A \\ B & A \end{pmatrix}.$$

Show that

(a) M and N are similar.

(b) R and S are similar.

(c) M, N, R, and S have the same eigenvalues as $A+B$ plus n zeros, concluding that they have the same characteristic polynomial.

(d) M and R need not be similar in general.

(e) M and R are similar if A and B are invertible.

(f) M and R are similar if $n = 1$, i.e., $M, R \in M_2(\mathbb{F})$.

(g) M and R are similar if A and B are symmetric.

6.63 Let $P, Q, R, S, T \in M_n(\mathbb{C})$. Let $A = \left(\begin{smallmatrix} P & Q \\ R & S \end{smallmatrix} \right)$ and $B = \left(\begin{smallmatrix} T & 0 \\ 0 & 0 \end{smallmatrix} \right)$. If P and T are nonsingular, show that AB and BA are similar.

6.64 Let $A \in M_n(\mathbb{C})$. Show that $A = AA^*B$ for some $B \in M_n(\mathbb{C})$ and that $\left(\begin{smallmatrix} A^*A & A^*C \\ 0 & 0 \end{smallmatrix} \right)$ and $\left(\begin{smallmatrix} A^*A & 0 \\ 0 & 0 \end{smallmatrix} \right)$ are similar for any $C \in M_n(\mathbb{C})$.

6.65 Let $A, B \in M_n(\mathbb{C})$. Answer the questions or prove the statements:

(a) If A and B are similar, are AB and BA necessarily similar?

(b) AB and BA are similar $\Leftrightarrow r(AB)^k = r(BA)^k$, $k = 1, 2, \dots, n$.

(c) (i) $A\bar{A}$ is similar to $\bar{A}A$. (ii) AA^* is similar to A^*A.
(iii) $\bar{A}A^t$ is similar to $A^t\bar{A}$. (iv) Is AA^t similar to A^tA?

(d) If A and B are Hermitian, then AB is similar to BA.

6.66 Let $A = (a_{ij}) \in M_n(\mathbb{C})$, $B = (b_{ij}) = (A^*A)^{\frac{1}{2}}$, and $C = (c_{ij}) = (AA^*)^{\frac{1}{2}}$. Show that $|a_{ij}|^2 \leq b_{jj}c_{ii}$ for all i and j. In particular, $|a_{ii}| \leq \sqrt{b_{ii}c_{ii}}$ for each i. Is it true that $|a_{ij}|^2 \leq b_{ii}c_{jj}$ for all i and j?

6.67 Let $A \in M_n(\mathbb{R})$ and let $\mathbb{P}_k[A] = \mathrm{Span}\{I, A, \ldots, A^{k-1}\}$ over $\mathbb{R}$.

  (a) Show that $\mathbb{P}_k[A] = \mathbb{P}_n[A]$ for all positive integers $k \geq n$.

  (b) What is the smallest number of k for which $\mathbb{P}_k[A] = \mathbb{P}_n[A]$?

  (c) Let $A = \mathrm{diag}(\lambda_1, \lambda_2, \ldots, \lambda_n)$, where $\lambda_1, \lambda_2, \ldots, \lambda_n$ are distinct real numbers. Find the dimension and a basis of $\mathbb{P}_n[A]$.

  (d) Let $A = \left(\begin{smallmatrix} 0 & 1 & 0 \\ 0 & 0 & 1 \\ 0 & 0 & 2 \end{smallmatrix}\right)$ and $B = \left(\begin{smallmatrix} 2 & 1 & 0 \\ 0 & 2 & 0 \\ 0 & 0 & 2 \end{smallmatrix}\right)$. Find $\mathbb{P}_3[A]$ and $\mathbb{P}_3[B]$.

6.68 Let $A \in M_n(\mathbb{R})$, $f(x) \in \mathbb{P}[x]$, $f(x) = f_1(x)f_2(x)$, $V = \mathrm{Ker}\, f(A)$, and $V_i = \mathrm{Ker}\, f_i(A)$, $i = 1, 2$. If $f_1(x)$ and $f_2(x)$ are relatively prime, i.e., $(f_1(x), f_2(x)) = 1$, show that $V = V_1 \oplus V_2$, and V_1, V_2 are invariant subspaces (of $\mathbb{R}^n$) under A, that is, if $w \in V_i$, then $Aw \in V_i$, $i = 1, 2$.

6.69 Let $A \in M_n(\mathbb{R})$, $f(x) \in \mathbb{P}[x]$, and $f(x) = f_1(x)f_2(x)$. If $f_1(x)$ and $f_2(x)$ are relatively prime (i.e., coprime, $(f_1(x), f_2(x)) = 1$), show that

$$r\big(f(A)\big) + n = r\big(f_1(A)\big) + r\big(f_2(A)\big).$$

6.70 Let $\mathcal{A}$ be a linear transformation from $\mathbb{R}^2$ to $\mathbb{R}^2$ that satisfies

$$\mathcal{A}\left(\begin{smallmatrix}1\\0\end{smallmatrix}\right) = \left(\begin{smallmatrix}1\\4\end{smallmatrix}\right) \quad \text{and} \quad \mathcal{A}\left(\begin{smallmatrix}1\\-1\end{smallmatrix}\right) = \left(\begin{smallmatrix}-1\\1\end{smallmatrix}\right).$$

  (a) Find $\mathcal{A}\left(\begin{smallmatrix}-2\\3\end{smallmatrix}\right)$.

  (b) Find $\mathrm{Im}\,\mathcal{A} = \{\mathcal{A}\left(\begin{smallmatrix}x\\y\end{smallmatrix}\right) \mid x, y \in \mathbb{R}\}$.

  (c) Find the matrix representation A_e of $\mathcal{A}$ with respect to the standard ordered basis $\{e_1 = \left(\begin{smallmatrix}1\\0\end{smallmatrix}\right),\ e_2 = \left(\begin{smallmatrix}0\\1\end{smallmatrix}\right)\}$ of $\mathbb{R}^2$.

  (d) Find the matrix representation A_α of $\mathcal{A}$ with respect to the ordered basis $\{\alpha_1 = \left(\begin{smallmatrix}1\\0\end{smallmatrix}\right),\ \alpha_2 = \left(\begin{smallmatrix}1\\-1\end{smallmatrix}\right)\}$ of $\mathbb{R}^2$.

  (e) Find an invertible matrix P such that $P^{-1}A_e P = A_\alpha$.

  (f) Find the eigenvalues of $\mathcal{A}$.

6.71 Let $A = \left(\begin{smallmatrix} a & b \\ c & d \end{smallmatrix}\right) \in M_2(\mathbb{R})$. Define a transformation $\mathcal{A}$ on $M_2(\mathbb{R})$ by

$$\mathcal{A}(X) = AX, \quad X \in M_2(\mathbb{R}).$$

Show that $\mathcal{A}$ is linear. Find the matrices of $\mathcal{A}$ with respect to the ordered bases $\{E_{11}, E_{12}, E_{21}, E_{22}\}$ and $\{E_{11}, E_{21}, E_{12}, E_{22}\}$, where E_{ij} is the 2×2 matrix with (i, j)-entry 1 and 0 elsewhere for $i, j = 1, 2$.

6.72   Let $A = \left(\begin{smallmatrix} 1 & 2 \\ 3 & 4 \end{smallmatrix}\right)$ and $B = \left(\begin{smallmatrix} 0 & 1 \\ 1 & 0 \end{smallmatrix}\right)$. Define a transformation $\mathcal{L}$ on $M_2(\mathbb{R})$ by $\mathcal{L}(X) = AXB$. Show that $\mathcal{L}$ is linear. Compute the eigenvalues, trace, and determinant of a matrix representation of $\mathcal{L}$.

6.73   Let k be a given real number. Define a transformation on $M_n(\mathbb{R})$ by

$$\mathcal{L}(X) = X + kX^t, \quad X \in M_n(\mathbb{R}).$$

(a)   Show that $\mathcal{L}$ is linear.

(b)   Find the images of $\mathcal{L}$ for $k = 1, -1$, and 2.

(c)   Find all values of k such that $\mathcal{L}$ is invertible.

(d)   Find matrix representations of $\mathcal{L}$ for $n = 2$, $k = 1$, and $k = 2$.

(e)   Show that $\mathcal{L}$ is self-adjoint (as to the standard inner product).

6.74   Let $A = \mathrm{diag}(1, 1, -1)$. Define a linear transformation on $M_3(\mathbb{R})$ by $\mathcal{A}(X) = AX - XA$. Find the dimensions of $\mathrm{Ker}\,\mathcal{A}$ and $\mathrm{Im}\,\mathcal{A}$.

6.75   Let L be a 3×3 real matrix having determinant 16 and minimal polynomial $\lambda^2 - 6\lambda + 8$. Define a linear transformation on $M_3(\mathbb{R})$ by $\mathcal{L}(X) = LX$. Find the trace (of a matrix representation) of $\mathcal{L}$.

6.76   Let $\mathcal{A} : \mathbb{F}^3 \to \mathbb{F}^3$ ($\mathbb{F} = \mathbb{C}$ or $\mathbb{R}$) be a linear transformation defined by

$$\mathcal{A}: \ e_1 \mapsto \left(\tfrac{1}{2}, 0, -\tfrac{\sqrt{3}}{2}\right), \ e_2 \mapsto (0, 1, 0), \ e_3 \mapsto \left(\tfrac{\sqrt{3}}{2}, 0, \tfrac{1}{2}\right),$$

where $\{e_1, e_2, e_3\}$ is the standard basis of $\mathbb{F}^3$. Show that $\mathcal{A}$ has a diagonal matrix representation over $\mathbb{C}$ but not over $\mathbb{R}$. Find a basis of $\mathbb{C}^3$ with respect to which $\mathcal{A}$ has a diagonal matrix representation.

6.77   Let $\mathcal{A}$ and $\mathcal{B}$ be linear transformations on $\mathbb{R}^2$ such that $\mathcal{A}^2 = \mathcal{B}^2 = \mathcal{I}$ and $\mathcal{A}\mathcal{B} + \mathcal{B}\mathcal{A} = 0$. Show that

(a) $\mathcal{B}$ maps an eigenspace of $\mathcal{A}$ onto another eigenspace of $\mathcal{A}$.

(b)   There exists a basis of $\mathbb{R}^2$ such that the matrices of $\mathcal{A}$ and $\mathcal{B}$ with respect to the basis are respectively $\left(\begin{smallmatrix} 1 & 0 \\ 0 & -1 \end{smallmatrix}\right)$ and $\left(\begin{smallmatrix} 0 & 1 \\ 1 & 0 \end{smallmatrix}\right)$.

(c)   There exist two bases of $\mathbb{R}^2$, say basis I and basis II, such that the matrix of $\mathcal{A}$ under basis I is $\left(\begin{smallmatrix} 1 & 0 \\ 0 & -1 \end{smallmatrix}\right)$ and the matrix of $\mathcal{B}$ under basis II is $\left(\begin{smallmatrix} 1 & 0 \\ 0 & -1 \end{smallmatrix}\right)$. Can bases I and II be the same?

6.78   Let $\mathcal{A}$ be a linear operator on a finite dimensional vector space V and let A be the matrix representation of $\mathcal{A}$ with respect to a basis $\{\alpha_1, \ldots, \alpha_n\}$ of V. What is the relation between $\mathrm{Im}\,\mathcal{A}$ and $\mathrm{Im}\,A$?

6.79   Let V be a vector space over $\mathbb{C}$ with a basis $\{u_1, u_2, \ldots, u_n\}$. Let $\{v_1, v_2, \ldots, v_n\}$ be a permutation (re-ordering) of $\{u_1, u_2, \ldots, u_n\}$. Define a linear map $\mathcal{L}$ on V by $\mathcal{L}(u_i) = v_i$, $i = 1, 2, \ldots, n$. Show that $\mathcal{L}$ is diagonalizable over $\mathbb{C}$. Is $\mathcal{L}$ diagonalizable over $\mathbb{R}$?

6.80   Let V be a vector space of dimension n over $\mathbb{R}$ and let $\mathcal{L}$ be a linear transformation on V. If $\dim V \geq 3$, show that $\mathcal{L}$ has a non-trivial invariant subspace, i.e., there is a subspace $W \neq \{0\}, V$, such that $\mathcal{L}(W) \subset W$. Find an example showing that if $n = 2$, then $\mathcal{L}$ may not have an invariant subspace of dimension one.

6.81   Let $A \in M_n(\mathbb{C})$. Suppose that $\det(I+A) \neq 0$. Let $B = 2(I+A)^{-1}-I$. Show that $\det(I + B) \neq 0$ and $A = 2(I + B)^{-1} - I$. Show also that

$$A^* + A = 0 \quad \Leftrightarrow \quad B^*B = I.$$

6.82   Let $A \in M_n(\mathbb{C})$, $\mathfrak{R}(A) = \frac{1}{2}(A + A^*)$. If $\det(I - A) \neq 0$, show that

$$
\begin{aligned}
\mathfrak{R}\big((I + A)(I - A)^{-1}\big) &= 2\mathfrak{R}\big((I - A)^{-1}\big) - I \\
&= 2\mathfrak{R}\big((I - A)^{-1} - \tfrac{1}{2}I\big) \\
&= (I - A^*)^{-1}(I - A^*A)(I - A)^{-1}.
\end{aligned}
$$

6.83   Let $A \in M_n(\mathbb{C})$. Show that

$$\sigma_{\min}(I - A) + \sigma_{\min}(A) \leq 1 \leq \sigma_{\max}(I - A) + \sigma_{\max}(A).$$

6.84   Let $A, B, C \in M_n(\mathbb{C})$ be three positive semidefinite matrices and let $\lambda(ABC)$ denote any eigenvalue of ABC. Give an example showing that $\lambda(ABC)$ can be a nonreal complex number, and show that

$$|\lambda(ABC)| \leq \lambda_{\max}(A)\lambda_{\max}(B)\lambda_{\max}(C).$$

6.85   Let $M = \left(\begin{smallmatrix} A & B \\ B^* & C \end{smallmatrix}\right)$, where $A = \left(\begin{smallmatrix} 2 & 0 \\ 0 & 1 \end{smallmatrix}\right)$, $B = \left(\begin{smallmatrix} 1 & 1 \\ 0 & 1 \end{smallmatrix}\right)$, and $C = \left(\begin{smallmatrix} 1 & 1 \\ 1 & x \end{smallmatrix}\right)$. Show that M is positive semidefinite for all $x \in [2, \infty)$. However, the block matrix $\left(\begin{smallmatrix} A & B^* \\ B & C \end{smallmatrix}\right)$ is never positive semidefinite for any $x \in \mathbb{R}$.

6.86   Let a, b, c be positive real numbers, x, y be complex numbers, and let

$$
A = \begin{pmatrix} a & x & 0 \\ \bar{x} & b & y \\ 0 & \bar{y} & c \end{pmatrix}.
$$

Show that A is positive semidefinite if and only if $b \geq |x|^2 a^{-1} + |y|^2 c^{-1}$.

6.87   Find a 3×3 real symmetric matrix $A = (a_{ij})$ such that A is singular, but its Hadamard (entrywise) power $A^{[3]} = (a_{ij}^3)$ is nonsingular.

6.88   Let $A = (a_{ij}) \in M_n(\mathbb{C})$ and let $U \in M_n(\mathbb{C})$ be a unitary matrix. For the Hadamard (i.e., entrywise) product of U and A, show that

$$\|U \circ A\|_F \leq \sqrt[4]{n} \left(\sum_{i,j=1}^n |a_{ij}|^4 \right)^{\frac{1}{4}}.$$

6.89   Which of the following A and B (or both) is positive semidefinite?

$$A = \begin{pmatrix} 1 & a & 0 & -a \\ a & 1 & a & 0 \\ 0 & a & 1 & a \\ -a & 0 & a & 1 \end{pmatrix}, \quad B = \begin{pmatrix} 1 & a & 0 & a \\ a & 1 & a & 0 \\ 0 & a & 1 & a \\ a & 0 & a & 1 \end{pmatrix}, \quad a = \frac{1}{\sqrt{2}}.$$

6.90   Let A and B be $n \times n$ positive definite matrices. If $UA = BV$, where U and V are unitary, show that $U = V$. Does $U = V$ if $UA = VB$?

6.91   Let $A \in M_n(\mathbb{C})$. Show that A is normal (i.e., $A^*A = AA^*$) if and only if $A^* = p(A)$ for some complex-coefficient polynomial p.

6.92   Let $A, B \in M_n(\mathbb{C})$.

(a) If A is positive semidefinite, show that there exists a real-coefficient polynomial p such that $A^{\frac{1}{2}} = p(A)$.

(b) If B is Hermitian, show that there exists a real-coefficient polynomial f such that $(B^*B)^{\frac{1}{2}} = f(B)$.

(c) If A and B are positive semidefinite, show that there exists a real-coefficient polynomial g such that $A^{\frac{1}{2}} = g(A)$, $B^{\frac{1}{2}} = g(B)$.

6.93   Let $D = \operatorname{diag}(\lambda_1, \ldots, \lambda_n)$ be a real diagonal matrix and let U be an $n \times n$ unitary matrix. If $\det(U_k^* D U_k) \geq 0$ for all U_k consisting of any k columns of U, $k = 1, \ldots, n$, show that $\lambda_1, \ldots, \lambda_n$ are nonnegative.

6.94   Let A and B be positive semidefinite matrices of the same size.

(a) Find an example that $AB + BA$ has a negative eigenvalue.

(b) Show that $\pm(AB + BA) \leq A^2 + B^2$.

(c) Show that $|\det(AB + BA)| \leq \det(A^2 + B^2)$.

(d) Show that $\|AB + BA\|_{\mathrm{sp}} \leq \|A^2 + B^2\|_{\mathrm{sp}}$.

6.95 Let $0 \le A, B \le I_n$, that is, A and B are $n \times n$ positive semidefinite matrices with eigenvalues contained in the interval $[0, 1]$.

    (a) Show that $-\frac{1}{4}I_n \le AB + BA \le 2I_n$.

    (b) Give an example showing that $AB + BA$ has an eigenvalue $-\frac{1}{4}$.

    (c) Is it possible to find some A and B such that $AB + BA = -\frac{1}{4}I_n$?

    (d) Show that $A(I - B) + B(I - A) \ge 0$.

    (e) Show that $-\frac{1}{2}I_n \le i(AB - BA) \le \frac{1}{2}I_n$.

6.96 Find the eigenvalues of $\left(\begin{smallmatrix} xx^* & x \\ x^* & 0 \end{smallmatrix}\right)$ and $\left(\begin{smallmatrix} x^*x & x^* \\ x & 0 \end{smallmatrix}\right)$, where $x \in \mathbb{C}^n$, $\|x\| = 1$.

6.97 Let $A \in M_{m \times n}(\mathbb{C})$, $M_t = \left(\begin{smallmatrix} A^*A & A^* \\ A & tI_m \end{smallmatrix}\right)$, and $N_t = \left(\begin{smallmatrix} AA^* & A \\ A^* & tI_n \end{smallmatrix}\right)$, $t \in \mathbb{R}$.

    (a) For $A = (2, -1, 3i) \in \mathbb{C}^3$, compute A^*A and AA^*.

    (b) For $t = 0$, prove the rank identity $r(M_0) = r(N_0) = 2r(A)$.

    (c) If $A^*A = I_n$, find the eigenvalues of M_t and N_t.

    (d) If $m = n$, show that M_t and N_t are unitarily similar.

6.98 Let P and Q be $n \times n$ positive definite matrices and let $M = QPQ^{-1}$. If $MP = PM$, show that $QP = PQ$, and consequently, $M = P$.

6.99 Let A be an $n \times n$ real matrix such that $A + A^t$ is positive definite. Show that the real eigenvalues (if any) of A are positive and $\det A > 0$.

6.100 Let $A > 0$ and $B > 0$ (i.e., A and B are positive definite). Show that

$$\left(\begin{array}{cc} (A+A)^{-1} & (A+B)^{-1} \\ (B+A)^{-1} & (B+B)^{-1} \end{array} \right) \ge 0.$$

6.101 Let $a_1, a_2, \ldots, a_n$ be positive numbers. Show that (Cauchy matrix)

$$\left(\begin{array}{cccc} \frac{1}{a_1+a_1} & \frac{1}{a_1+a_2} & \cdots & \frac{1}{a_1+a_n} \\ \frac{1}{a_2+a_1} & \frac{1}{a_2+a_2} & \cdots & \frac{1}{a_2+a_n} \\ \cdots & \cdots & \cdots & \cdots \\ \frac{1}{a_n+a_1} & \frac{1}{a_n+a_2} & \cdots & \frac{1}{a_n+a_n} \end{array} \right) \ge 0.$$

6.102 Let $H > 0$ and let K be Hermitian, both $n \times n$. Let $\ell(K) = (K^*K)^{\frac{1}{2}}$.

    (a) Show that $\left(\begin{smallmatrix} H & K \\ K & H \end{smallmatrix}\right) \ge 0$ if and only if $H \ge \pm K$.

    (b) If $H \ge \ell(K)$, show that $H \ge \pm K$.

    (c) If $H \ge \pm K$, does it follow that $H \ge \ell(K)$?

6.103 Let A be a square matrix with rank r and let $H = A + A^*$. Denote by $i_+(H)$ the number of positive eigenvalues of H. Show that $i_+(H) \leq r$.

6.104 A matrix is called a *contraction* if its spectral norm is no more than 1. Let A be an $m \times n$ matrix. Show that the following are equivalent:

(a) A is a contraction.

(b) $I_m \geq AA^*$.

(c) $I_n \geq A^*A$.

(d) $\begin{pmatrix} I_n & A^* \\ A & I_m \end{pmatrix} \geq 0$.

(e) $\begin{pmatrix} I_m & A \\ A^* & I_n \end{pmatrix} \geq 0$.

6.105 Let $A, B, C, D \in M_n(\mathbb{C})$.

(a) Show that the product of two contractions is a contraction.

(b) Show that $\operatorname{tr}(C^*AC) \leq \operatorname{tr} A$ if $A \geq 0$ and C is a contraction. Is it true that $C^*AC \leq A$ for all $A \geq 0$ and contraction C?

(c) If $B^*B \leq A$, does it follow that $BB^* \leq A$?

(d) If $0 \leq D \leq I$ and $A \geq 0$, show that $DAD \leq A$.

6.106 Let $X \in M_n(\mathbb{C})$. Show that $\begin{pmatrix} I_n & X & 0 \\ X^* & I_n & X \\ 0 & X^* & I_n \end{pmatrix} \geq 0 \Leftrightarrow I_n \geq X^*X + XX^*$.

6.107 Let $A, B, C, X, Y \in M_n(\mathbb{C})$, where A, B, C are positive definite. Let

$$M = \begin{pmatrix} A & X & 0 \\ X^* & B & Y \\ 0 & Y^* & C \end{pmatrix}.$$

Show that M is positive semidefinite if and only if $B \geq X^*A^{-1}X + YC^{-1}Y^*$. Can the inequality be changed to $B \geq X^*A^{-1}X + Y^*C^{-1}Y$?

6.108 Let $C = (c_{ij})$ be an $n \times n$ contractive matrix. Show that

$$1 - n + \sum_{j=1}^{n}\sum_{i=1}^{n} |c_{ij}|^2 \leq |\det C|^2 \leq \prod_{j=1}^{n} \left(\sum_{i=1}^{n} |c_{ij}|^2 \right) \leq 1.$$

If all three equalities occur simultaneously, what is C?

6.109 A square matrix with nonnegative entries is said to be *doubly stochastic* if the sum of the entries on each row is 1 and the sum of the entries on each column is 1. Let M be the $n \times n$ matrix in which every entry is $\frac{1}{n}$. Of all $n \times n$ matrices that are both positive semidefinite and doubly stochastic, show that M is the smallest in the sense that if A positive semidefinite and doubly stochastic, then $A \geq M$.

6.110 If A and B are $n \times n$ normal matrices, show that $r(AB) = r(BA)$. In particular, for normal matrices A and B, if $AB = 0$ then $BA = 0$. Find a normal A and a nonnormal B such that $AB = 0$ but $BA \neq 0$.

6.111 Let $A, B \in M_n(\mathbb{C})$. Show that if AB and BA are normal, then AB and BA are similar. For the matrices C and D given below, show that C and D are both normal, but CD and DC are not similar.

$$C = \begin{pmatrix} 0 & 0 & 1 & 0 \\ 1 & 0 & 0 & 0 \\ 0 & 1 & 0 & 0 \\ 0 & 0 & 0 & 0 \end{pmatrix}, \quad D = \begin{pmatrix} 0 & 0 & 0 & 0 \\ 0 & 1 & 0 & 0 \\ 0 & 0 & 0 & 1 \\ 0 & 0 & 1 & 0 \end{pmatrix}.$$

6.112 Let $\lambda_1, \lambda_2, \dots, \lambda_n$ be positive numbers with $\lambda_1 + \lambda_2 + \dots + \lambda_n < 1$ and let $D = \text{diag}(\frac{1}{\lambda_1}, \frac{1}{\lambda_2}, \dots, \frac{1}{\lambda_n})$. Show that the matrix $D - E$ is positive definite, where E is the $n \times n$ all-ones matrix.

6.113 Let $A \in M_n(\mathbb{C})$. Show that the following statements are equivalent:

(a) A is similar to $\bar{A}$.

(b) A is similar to A^*.

(c) A is similar to a real matrix.

(d) A is the product of two Hermitian matrices.

6.114 Let A, B, and C be $n \times n$ matrices. Prove the following statements:

(a) AB is diagonalizable if A is positive definite and B is Hermitian.

(b) AB is diagonalizable if A and B are both positive semidefinite.

(c) (a) is false if A is positive semidefinite and B is Hermitian.

(d) AB is similar to a real matrix if A and B are both Hermitian.

6.115 Let A, B, and C be $n \times n$ positive semidefinite matrices. Show that

(a) AB is positive semidefinite if and only if AB is Hermitian.

(b) ABC is positive semidefinite if and only if ABC is Hermitian.

6.116 Show that a matrix P is the product of two positive semidefinite matrices if and only if P is similar to a nonnegative diagonal matrix.

6.117 Let $A \in M_n(\mathbb{C})$. If $A\bar{A} = I$, show that $A = B(\bar{B})^{-1}$ for a matrix B.

6.118 Let A, B, and C be $n \times n$ matrices such that $\begin{pmatrix} A & B \\ B^* & C \end{pmatrix} \geq 0$. Show that

$$\operatorname{tr}(AC) - \operatorname{tr}(B^*B) \leq (\operatorname{tr} A)(\operatorname{tr} C) - (\operatorname{tr} B^*)(\operatorname{tr} B).$$

6.119 Define an inner product on $\mathbb{P}_4[t]$ by $\langle f, g \rangle = \int_0^1 f(t)g(t)\,dt$. Find all $f(t) \in \mathbb{P}_4[t]$ such that $\langle t, f(t) \rangle = \langle t^2, f(t) \rangle = 0$, and $f(0) = 1$.

6.120 Let $\mathbb{P}_3[x] = \{a + bx + cx^2 \mid a, b, c \in \mathbb{R}\}$ be the inner product space with $\langle f, g \rangle = \frac{1}{2} \int_{-1}^1 f(x)g(x)\,dx$. Let $\mathcal{L} : \mathbb{P}_3[x] \to \mathbb{P}_3[x]$ be defined by

$$\mathcal{L}\big(f(x)\big) = f(x) - f(-x).$$

   (a) Show that $\mathcal{L}$ is a linear map.
   (b) Find the image $\operatorname{Im}\mathcal{L}$ of $\mathcal{L}$.
   (c) Solve $\mathcal{L}(f) = -f$ for f.
   (d) Find an orthonormal basis for $\mathbb{P}_3[x]$.
   (e) Show that $\operatorname{Im}\mathcal{L}$ is orthogonal to $\mathbb{R}$.
   (f) Find a subspace W such that $\mathbb{P}_3[x] = \operatorname{Im}\mathcal{L} \oplus W$.
   (g) Show that $\frac{1}{2}\mathcal{L}$ is an orthogonal projection onto $\operatorname{Im}\mathcal{L}$.

6.121 Consider the normed vector space $\mathcal{C}[0,1]$, i.e., the space of all real-valued continuous functions on $[0,1]$ with the *supremum norm*

$$\|f\| = \max_{x \in [0,1]} |f(x)|.$$

Show that $\| \cdot \|$ is a norm that is not derived from any inner product.

6.122 Consider $\mathbb{R}^3$ with basis $\alpha_1 = (1,0,1), \alpha_2 = (0,1,1), \alpha_3 = (-1,0,1)$. Find a dual basis f_1, f_2, f_3 for the dual space (linear functionals) of $\mathbb{R}^3$.

6.123 Let A, B, C, and D be $n \times n$ matrices. Let $M = \begin{pmatrix} A & B \\ C & D \end{pmatrix}$. Show that

   (a) If M is normal, then $\|B\|_F = \|C\|_F$. Is it true that $\|A\|_F = \|D\|_F$? ($\| \cdot \|_F$ denotes the Frobenius norm.)
   (b) If M is unitary, then A and D have the same singular values; and B and C have the same singular values. Moreover,

$$|\det A| = |\det D|, \quad |\det B| = |\det C|.$$

6.124   Let $H = (h_{ij})$ be an $n \times n$ Hermitian matrix of eigenvalues $\lambda_1 \geq \cdots \geq \lambda_n$ and corresponding orthonormal eigenvectors $u_1, \ldots, u_n$. Denote by u_{ij} the i-th component of u_j, $i, j = 1, \ldots, n$. Let H_k be the principal submatrix of H by deleting the k-th row and column from H and let the eigenvalues of H_k be $\mu_{k1} \geq \cdots \geq \mu_{k(n-1)}$, $k = 1, \ldots, n$. Show that

(a) $\lambda_k^2 = \sum_{s,i,t=1}^{n} h_{si} h_{it} u_{tk} \overline{u_{sk}}$.

(b) $\det(H_k) = \det(H) \sum\limits_{j=1}^{n} \frac{1}{\lambda_j} |u_{kj}|^2$ if $\lambda_1, \ldots, \lambda_n$ are nonzero.

(c) $|u_{ij}|^2 = \dfrac{\prod\limits_{k=1}^{n-1} (\lambda_j - \mu_{ik})}{\prod\limits_{k=1,\, k \neq j}^{n} (\lambda_j - \lambda_k)}$ if $\lambda_1, \ldots, \lambda_n$ are distinct.

6.125   Let $V = \{(x, y) \mid x, y \in \mathbb{R}^2\}$. Define $f : V \to \mathbb{R}$ and $g : V \to M_2(\mathbb{R})$ by

$$f(x, y) = x^t y \in \mathbb{R}, \quad g(x, y) = xy^t \in M_2(\mathbb{R}).$$

(a) Show that V is a vector space over $\mathbb{R}$ with respect to the natural addition $(x_1, y_1) + (x_2, y_2) = (x_1 + x_2, y_1 + y_2)$ and the scalar multiplication $k(x, y) = (kx, ky)$ (via the operations of $\mathbb{R}^2$).

(b) Show that f is linear with respect to each component, but f is not linear overall, i.e., f is not a linear map from V to $\mathbb{R}$.

(c) Show that g is linear with respect to each component, but g is not linear overall, i.e., g is not a linear map from V to $M_2(\mathbb{R})$.

(d) Determine the image (i.e., range) $\operatorname{Im} g = \{g(v) \mid v \in V\}$ of g, and find two vectors u and v in V such that $g(u) + g(v)$ is not in the range $\operatorname{Im} g$. (So, $\operatorname{Im} g$ is not a subspace of $M_2(\mathbb{R})$.)

6.126   Let $A[\alpha]$ denote the principal submatrix of $A \in M_n(\mathbb{C})$ consisting of the rows and columns indexed by the sequence α. Show that

(a) $(B^*B)[\alpha] \geq (B^*[\alpha])(B[\alpha])$ for any $B \in M_n(\mathbb{C})$.

(b) $(P^2)[\alpha] \geq (P[\alpha])^2$ for any positive semidefinite P.

(c) $(P^{\frac{1}{2}})[\alpha] \leq (P[\alpha])^{\frac{1}{2}}$ for any positive semidefinite P.

(d) $(P^{-1})[\alpha] \geq (P[\alpha])^{-1}$ for any positive definite P.

6.127   Let $A, B \in M_n(\mathbb{C})$. If $A > 0$ and $B > 0$, show that $A \circ B > 0$.

6.128   Let $A, B \in M_n(\mathbb{C})$. If $A > 0$ and $B > 0$, show that

$$(A \circ B)^{-1} \leq A^{-1} \circ B^{-1}.$$

6.129   Let $X, Y \in M_{m \times n}(\mathbb{C})$. Prove the inequality of the Hadamard product

$$(X^*X) \circ (Y^*Y) \pm (X^*Y) \circ (Y^*X) \geq 0.$$

6.130   Let A and B be $n \times n$ positive definite matrices and let X and Y be any $m \times n$ complex matrices. Show that

$$(X + Y)(A + B)^{-1}(X + Y)^* \leq (XA^{-1}X^*) + (YB^{-1}Y^*).$$

6.131   Let $i_+(X), i_-(X)$, and $i_0(X)$ be respectively the numbers of positive, negative, and zero eigenvalues of a Hermitian matrix X (known as the *inertia* of X). Let H and K be $n \times n$ Hermitian matrices. Show that

    (a) If H_1 is a principal submatrix of H, then $i_\pm(H_1) \leq i_\pm(H)$.

    (b) If $K \leq H$, then $i_+(K) \leq i_+(H)$. Is $i_-(K) \leq i_-(H)$?

    (c) For any $n \times m$ matrix X, $i_\pm(X^*HX) \leq \min\{r(H), r(X)\}$.

    (d) For any $n \times m$ matrix X, $i_\pm(X^*HX) \leq i_\pm(H)$.

    (e) For any $n \times m$ matrix X with $m \geq n$, $i_0(X^*HX) \geq i_0(H)$.

6.132   Let E be the $n \times n$ all-ones matrix and let F be the $n \times n$ matrix with ones in the first row and first column, and zeros elsewhere, i.e.,

$$E = \begin{pmatrix} 1 & \cdots & 1 \\ \vdots & \ddots & \vdots \\ 1 & \cdots & 1 \end{pmatrix}, \quad F = \begin{pmatrix} 1 & 1 & \cdots & 1 \\ 1 & 0 & \cdots & 0 \\ \vdots & \vdots & \ddots & \vdots \\ 1 & 0 & \cdots & 0 \end{pmatrix}.$$

Find $n \times n$ real orthogonal matrices P and Q such that

    (a) $Pe = \sqrt{n}\, e_1$, where $e = (1, \ldots, 1)^t$, $e_1 = (1, 0, \ldots, 0)^t \in \mathbb{R}^n$.

    (b) $P^tEP = \begin{pmatrix} n & 0 \\ 0 & 0 \end{pmatrix}$.

    (c) $Q^tFQ = \begin{pmatrix} N & 0 \\ 0 & 0 \end{pmatrix}$, where $N = \begin{pmatrix} \frac{1}{\sqrt{n-1}} & \sqrt{n-1} \\ \sqrt{n-1} & 0 \end{pmatrix}$.

6.133 Let $A = (a_{ij})$ be a positive semidefinite matrix. Let $f(x)$ be a polynomial of finite degree with nonnegative coefficients. Show that $f(A)$ and $(f(a_{ij}))$ are both positive semidefinite.

6.134 Let A, B, $C \in M_n(\mathbb{C})$. Let $M = \begin{pmatrix} A & B \\ B^* & C \end{pmatrix}$ and $N = \begin{pmatrix} \det A & \det B \\ \det B^* & \det C \end{pmatrix}$. If M is positive semidefinite, show that $\det M \leq \det N$, equivalently,

$$\det \begin{pmatrix} A & B \\ B^* & C \end{pmatrix} \leq \det A \det C - |\det B|^2.$$

6.135 Let A and B be $n \times n$ positive semidefinite matrices. Show that

$$AB + BA \leq (\operatorname{tr} B)A + (\operatorname{tr} A)B.$$

Hints and Answers for Chapter 1

1.1   (a) Yes. Dimension is 1. $\{1\}$ is a basis.

     (b) Yes. Dimension is 2. $\{1, i\}$ is a basis.

     (c) No, since $i \in \mathbb{C}$ and $1 \in \mathbb{R}$, but $i \cdot 1 = i \notin \mathbb{R}$.

     (d) Yes. Dimension is infinite, since π is a transcendental number and $1, \pi, \pi^2, \ldots$ are linearly independent over $\mathbb{Q}$.

     (e) No, since $\sqrt{2} \in \mathbb{R}$ and $1 \in \mathbb{Q}$, but $\sqrt{2} \cdot 1 = \sqrt{2} \notin \mathbb{Q}$.

     (f) No, since $\mathbb{Z}$ is not a field.

     (g) Yes only over $\mathbb{Q}$, the dimension is 3, and $\{1, \sqrt{2}, \sqrt{5}\}$ is a basis.

1.2   (a) Suppose $0'$ is also a zero for addition. Then $0 = 0 + 0' = 0'$.

     (b) Let v' be also a vector such that $v + v' = 0$. By the properties of addition, $-v = -v + 0 = -v + (v + v') = (-v + v) + v' = 0 + v' = v'$.

     (c) If $\lambda = 0$, then $\lambda v = (\lambda + \lambda)v = \lambda v + \lambda v$. If $v = 0$, then $\lambda v = \lambda(v + v) = \lambda v + \lambda v$. In either case, adding $-(\lambda v)$ to both sides reveals $\lambda v = 0$. Conversely, if $\lambda v = 0$, we show $\lambda = 0$ or/and $v = 0$. If $\lambda \neq 0$, then $v = 1v = (\lambda^{-1}\lambda)v = \lambda^{-1}(\lambda v) = \lambda^{-1}0 = 0$.

     (d) $v + (-1)v = 1v + (-1)v = (1 + (-1))v = 0v = 0$. So $(-1)v = -v$.

     (e) $(-\lambda)v + (\lambda v) = (-\lambda + \lambda)v = 0v = 0$. So $(-\lambda v) = -\lambda v$. Similarly, $\lambda(-v) + (\lambda v) = \lambda(-v + v) = \lambda 0 = 0$. So $\lambda(-v) = -(\lambda v)$.

     (f) $\lambda(u - v) = \lambda(u + (-v)) = \lambda u + \lambda(-v) = \lambda u - \lambda v$.

     (g) $u + v = w \Leftrightarrow (u + v) + (-v) = w + (-v) \Leftrightarrow u = w - v$.

1.3 $2(u + v) = 2u + 2v = (1 + 1)u + (1 + 1)v = (u + u) + (v + v) = u + (u + v) + v$. And also $2(u + v) = (1 + 1)(u + v) = (u + v) + (u + v) = u + (v + u) + v$. Thus $u + (u + v) + v = u + (v + u) + v$, implying $u + v = v + u$.

The conditions (axioms) of a vector space are not all independent. However, they are all listed for convenience in use and applications.

1.4   (a) $\{(x, y) \in \mathbb{R}^2 \mid x, y > 0\}$, i.e., all vectors with the initial point $O = (0, 0)$ and terminal points in the first quadrant.

     (b) $\{(x, y) \in \mathbb{R}^2 \mid xy > 0\}$, i.e., all vectors with the initial point $O = (0, 0)$ and terminal points in the first or third quadrants.

1.5   Yes over $\mathbb{C}$, $\mathbb{R}$, and $\mathbb{Q}$. The dimensions are 2, 4, and ∞, respectively.

1.6   Suppose that the vector space V has a nonzero element z. Then $\{\, rz \mid r \in \mathbb{F} \,\}$ is an infinite set, where $\mathbb{F} = \mathbb{C}$, $\mathbb{R}$, or $\mathbb{Q}$.

If $u + v = 0$ and $w + v = 0$, then $u = u + (w + v) = (u + v) + w = w$.

1.7   For the first part, the addition and scalar multiplication for V are defined in the same way as for $\mathbb{R}^2$. It is sufficient to notice that V is a line passing through O. If the scalar multiplication for V is defined to be $\lambda \odot (x, y) = (\lambda x, 0)$, then V is no longer a vector space.

1.8   It is easy to check that $\mathbb{H}$ is closed under the usual matrix addition. As to scalar multiplication, if λ is a real number, then

$$\lambda \begin{pmatrix} a & b \\ -\bar{b} & \bar{a} \end{pmatrix} = \begin{pmatrix} \lambda a & \lambda b \\ -\lambda \bar{b} & \lambda \bar{a} \end{pmatrix} \in \mathbb{H}.$$

If λ is a nonreal complex number, then $\lambda \bar{a} \neq \overline{\lambda a}$. So $\mathbb{H}$ is not a vector space over $\mathbb{C}$ as it is not closed under the scalar multiplication.

1.9   Check all the conditions for a vector space. For instance, the condition $(\lambda \mu) v = \lambda (\mu v)$ in the definition of vector space is satisfied, because

$$(ab) \odot x = x^{ab} = (x^b)^a = a \odot (b \odot x), \quad a,\, b \in \mathbb{R},\ x \in \mathbb{R}^+.$$

The dimension of the vector space is 1, since for any $x \in \mathbb{R}^+$,

$$x = (\log x) \odot 10.$$

Thus $\{10\}$ is a basis. Any two numbers in $\mathbb{R}^+$ are linearly dependent. $\mathbb{R}^+$ is not a vector space over $\mathbb{R}$ with respect to $\boxtimes$ and $\boxplus$, since the condition $\lambda(u + v) = \lambda u + \lambda v$ in the definition is not satisfied:

$$2 = 2 \boxtimes (1 \boxplus 1) \neq (2 \boxtimes 1) \boxplus (2 \boxtimes 1) = 4.$$

1.10   It suffices to show that $\lambda_1 \alpha_1, \lambda_2 \alpha_2, \ldots, \lambda_n \alpha_n$ are linearly independent. Let $l_1, l_2, \ldots, l_n$ be scalars. If

$$0 = l_1(\lambda_1 \alpha_1) + \cdots + l_n(\lambda_n \alpha_n) = (l_1 \lambda_1)\alpha_1 + \cdots + (l_n \lambda_n)\alpha_n,$$

then each $l_i \lambda_i = 0$, thus $l_i = 0$, $i = 1, 2, \ldots, n$, since all $\lambda_i \neq 0$. For $v = x_1 \alpha_1 + \cdots + x_n \alpha_n = (x_1/\lambda_1)(\lambda_1 \alpha_1) + \cdots + (x_n/\lambda_n)(\lambda_n \alpha_n)$, it follows that the coordinate of v under the basis $\{\lambda_1 \alpha_1, \ldots, \lambda_n \alpha_n\}$ is $(x_1/\lambda_1, \ldots, x_n/\lambda_n)$. The coordinate of $w = \alpha_1 + \cdots + \alpha_n$ under $\{\alpha_1, \ldots, \alpha_n\}$ is $(1, \ldots, 1)$, under $\{\lambda_1 \alpha_1, \ldots, \lambda_n \alpha_n\}$ is $(1/\lambda_1, \ldots, 1/\lambda_n)$.

1.11   (i) The vectors $v_1, v_2, \ldots, v_n$ form a basis of V if and only if they span V and they are linearly independent.

(ii) The vectors $v_1, v_2, \ldots, v_n$ form a basis of V if and only if every vector of V is a linear combination of $v_1, v_2, \ldots, v_n$ and any vector in $\{v_1, v_2, \ldots, v_n\}$ is not a linear combination of the remaining vectors.

1.12   (i) No if $k > n$ (i.e., linearly dependent); inconclusive for $k \leq n$.

(ii) No if $k < n$; inconclusive for $k \geq n$.

(iii) No if $k < n$ or $k > n$; inconclusive for $k = n$.

1.13   If $\beta_1, \beta_2, \ldots, \beta_m$ are linearly dependent, then $x_1\beta_1 + x_2\beta_2 + \cdots + x_m\beta_m = 0$, not all x_i's are zero. Then $x_1\alpha_1 + x_2\alpha_2 + \cdots + x_m\alpha_m = 0$, and $\alpha_1, \alpha_2, \ldots, \alpha_m$ are linearly dependent, a contradiction. The converse is not true, that is, if $\beta_1, \beta_2, \ldots, \beta_m$ are linearly independent, $\alpha_1, \alpha_2, \ldots, \alpha_m$ need not be linearly independent (by taking $\alpha_1 = 0$).

1.14   Compute $(\bar{a}x \pm by)^*(\bar{a}x \pm by) \geq 0$ to get the inequality. The identity is by a direct computation. Alternatively, x^*y is a number, $(x^*y)z$ is a multiple of z. zx^* is an $n \times n$ matrix, $(zx^*)y$ is a matrix times a vector. $(x^*y)z = z(x^*y) = (zx^*)y$. z and y need not be proportional. (Note: for matrix A and scalar λ, $A\lambda$ is understood as λA or $A(\lambda I)$.)

1.15   (a) $l_1(1+x) + l_2(1-x^2) = 0 \Rightarrow (l_1+l_2) + l_1x - l_2x^2 = 0 \Rightarrow l_1 = l_2 = 0$.

(b) Similarly, one shows that $1, p_1(x), p_2(x)$ are linearly independent.

(c) $(1+x)x = (1+x) - (1-x^2) \in W = \mathrm{Span}\{p_1(x), p_2(x)\}$.

(d) $x^2 + 2x + 1 = 2(1+x) - (1-x^2) \in W$. There do not exist numbers a and b such that $x^2 + 2x - 1 = a(1+x) + b(1-x^2)$.

(e) No. $\dim\big(\mathrm{Span}\{p_1(x), p_2(x)\}\big) = 2 < 3 = \dim \mathbb{P}_3[x]$.

(f) $\mathrm{Span}\{p_1(x), p_2(x)\} = \{ax^2 + bx + (b-a) \mid a, b \in \mathbb{R}\}$.

1.16   It is sufficient to show that $1, (x-1), (x-1)(x-2)$ are linearly independent. Let $\lambda_1 1 + \lambda_2(x-1) + \lambda_3(x-1)(x-2) = 0$. Setting $x = 1$, $x = 2$, and $x = 3$, respectively, yields $\lambda_1 = \lambda_2 = \lambda_3 = 0$.

To see that W_1 is a subspace of $\mathbb{P}_3[x]$, let $p, q \in W_1$. It follows that $(p+q)(1) = p(1) + q(1) = 0$. Thus $p + q \in W_1$. For any scalar λ, $(\lambda p)(1) = \lambda p(1) = 0$. So $\lambda p \in W_1$. Thus W_1 is a subspace of $\mathbb{P}_3[x]$. In a similar manner, one can show that W_2 is also a subspace.

$\dim W_1 = 2$, $(x-1)$ and $(x-1)(x-2)$ form a basis of W_1.

$\dim W_2 = 1$, $x^2 - 5x + 6 = (x-2)(x-3)$ is a basis.

1.17 $1+2x+3x^2 = \frac{1}{2}(2) + \frac{2}{3}(3x) + \frac{3}{4}(4x^2) = 6+11(x-1)+3(x-1)(x-2)$. So the coordinates of $p(x)$ are respectively $(\frac{1}{2}, \frac{2}{3}, \frac{3}{4})$ and $(6, 11, 3)$.

1.18   The vectors are linearly independent if and only if $t \neq 1$.

1.19   (a) True. (b) False. (c) True. (d) False. (e) True. (f) True. (g) False. (h) False. (i) False. (j) False. (k) False. (l) True. (m) False.

1.20   (a) True. (b) True. (c) False. (d) False. (e) True. (f) False. (g) True. (h) True. (i) True. (j) False. (k) False. (l) True. (m) True.

1.21 $\alpha_3 = \alpha_1 + \alpha_2 \Rightarrow \alpha_1, \alpha_2, \alpha_3$ are linearly dependent. However, α_1 and α_2 are not proportional, so they are linearly independent and thus form a basis for $\mathrm{Span}\{\alpha_1, \alpha_2, \alpha_3\}$. The dimension of the span is 2.

1.22   Since $\alpha_4 - \alpha_3 = \alpha_3 - \alpha_2 = \alpha_2 - \alpha_1$, we have $\alpha_3 = 2\alpha_2 - \alpha_1$ and $\alpha_4 = 3\alpha_2 - 2\alpha_1$. Obviously, α_1 and α_2 are linearly independent, and thus they form a basis for V and $\dim V = 2$.

1.23 $\alpha_3 = 2\alpha_1 - 2\alpha_2 - \alpha_4$.

1.24 $k \neq 1$.

1.25   (a) No. (b) Yes. (c) Yes. (d) No. (e) No. (f) Yes. (g) Yes. (h) No. (i) Yes. (j) Yes. (k) No. (l) Yes. (m) Yes. (n) No. (o) Yes.

1.26   (a) Yes (i.e., linearly independent). (b) Yes. (c) Yes. (d) No (i.e., linearly dependent). (e) Yes. (f) Yes. (g) Yes. (h) Yes. (i) No. (j) Yes. (k) No. (l) No. (m) Yes. (n) Yes. (o) No. (p) Yes.

1.27   (c) is true; others are false.

1.28   Let $v = x_1\alpha_1 + x_2\alpha_2 + x_3\alpha_3$. Set $a_4 = -\frac{1}{4}(x_1 + x_2 + x_3)$ and $a_i = x_i + a_4$, $i = 1, 2, 3$. Then $v = a_1\alpha_1 + a_2\alpha_2 + a_3\alpha_3 + a_4\alpha_4$. Suppose $v = b_1\alpha_1 + b_2\alpha_2 + b_3\alpha_3 + b_4\alpha_4$ with $b_1 + b_2 + b_3 + b_4 = 0$. Since $\{\alpha_1, \alpha_2, \alpha_3\}$ is a basis, we have $b_1 - b_4 = x_1$, $b_2 - b_4 = x_2$, $b_3 - b_4 = x_3$, implying $-4b_4 = x_1 + x_2 + x_3$ and $b_4 = a_4$. Hence, $b_i = a_i$, $i = 1, 2, 3$.

For the case of n, if $\{\alpha_1, \ldots, \alpha_n\}$ is a basis for $\mathbb{R}^n$ and $\alpha_{n+1} = -(\alpha_1 + \cdots + \alpha_n)$, then every vector in $\mathbb{R}^n$ can be uniquely written as a linear combination of the vectors $\alpha_1, \ldots, \alpha_{n+1}$ with the sum of the coefficients equal to zero. (Like a_4, $a_{n+1} = -\frac{1}{n+1}(x_1 + \cdots + x_n)$.)

One can explain this geometrically with $\mathbb{R}^2$ (i.e., in the xy-plane).

1.29   (i) Since α_1, α_2, and α_3 are linearly dependent, there are scalars x_1, x_2, x_3, not all zero, such that $x_1\alpha_1 + x_2\alpha_2 + x_2\alpha_3 = 0$. x_1 cannot be zero, otherwise α_2 and α_3 would be linearly dependent, which would contradict the linear independency of α_2, α_3, and α_4. It follows that $\alpha_1 = (-\frac{x_2}{x_1})\alpha_2 + (-\frac{x_3}{x_1})\alpha_3$, so α_1 is a linear combination of α_2 and α_3.

(ii) Suppose α_4 is a linear combination of α_1, α_2, and α_3. Let $\alpha_4 = y_1\alpha_1 + y_2\alpha_2 + y_3\alpha_3$. Substitute α_1 by the linear combination of α_2 and α_3 in (i). We see that α_4 is a linear combination of α_2 and α_3. This is a contradiction to the linear independency of α_2, α_3, and α_4.

1.30   (a) False.

(b) False. $A = \left(\begin{smallmatrix} 1 & 1 \\ 0 & 0 \end{smallmatrix}\right)$, $B = \left(\begin{smallmatrix} 1 & 0 \\ -1 & 0 \end{smallmatrix}\right)$.

(c) False.

(d) False.

(e) True.

(f) False.

(g) True.

(h) False (unless u and v are real).

(i) True. For any $c \in \mathbb{C}$, $\bar{c} = c^*$.

(j) False. uv^* is an $n \times n$ matrix.

(k) True.

(l) True. u^*v is a scalar.

(m) False. uv^* is an $n \times n$ matrix.

(n) False. $A = \left(\begin{smallmatrix} 0 & 1 \\ 0 & 0 \end{smallmatrix}\right) \neq 0$, but $A^2 = 0$.

(o) True. If $a_{ij} \neq 0$, then the (j,j)-entry of A^*A is $\sum_{k=1}^{n} |a_{kj}|^2 \neq 0$.

(p) False.

(q) True.

(r) True.

(s) True. Note that $x^*x \geq 0$ for any $x \in \mathbb{C}^n$.

(t) False. It is real, not necessarily nonnegative.

1.31   (a) Let $k \in \mathbb{R}$. If $X \in V$, $Y \in W$, then $kX \in V$, $kY \in W$. V and W are closed under the operations and are vector spaces over $\mathbb{R}$. But they are not closed under scalar multiplication over $\mathbb{C}$. For example, $\left(\begin{smallmatrix} 1 & 0 \\ 0 & 1 \end{smallmatrix}\right)$ is in V and W, but $i \cdot \left(\begin{smallmatrix} 1 & 0 \\ 0 & 1 \end{smallmatrix}\right) = \left(\begin{smallmatrix} i & 0 \\ 0 & i \end{smallmatrix}\right)$ is in neither.

(b) A direct computation reveals

$$\begin{pmatrix} \frac{1}{\sqrt{2}} & \frac{1}{\sqrt{2}} \\ -\frac{i}{\sqrt{2}} & \frac{i}{\sqrt{2}} \end{pmatrix} \begin{pmatrix} a+bi & 0 \\ 0 & a-bi \end{pmatrix} \begin{pmatrix} \frac{1}{\sqrt{2}} & \frac{i}{\sqrt{2}} \\ \frac{1}{\sqrt{2}} & -\frac{i}{\sqrt{2}} \end{pmatrix} = \begin{pmatrix} a & -b \\ b & a \end{pmatrix}.$$

(c) As $\begin{pmatrix} a+bi & 0 \\ 0 & a-bi \end{pmatrix} = a \begin{pmatrix} 1 & 0 \\ 0 & 1 \end{pmatrix} + b \begin{pmatrix} i & 0 \\ 0 & -i \end{pmatrix}$, $\{ \begin{pmatrix} 1 & 0 \\ 0 & 1 \end{pmatrix}, \begin{pmatrix} i & 0 \\ 0 & -i \end{pmatrix} \}$ is a basis of V. Likewise, $\{ \begin{pmatrix} 1 & 0 \\ 0 & 1 \end{pmatrix}, \begin{pmatrix} 0 & -1 \\ 1 & 0 \end{pmatrix} \}$ is basis of W. $\dim V = \dim W = 2$.

(d) $\begin{pmatrix} 1 & 0 \\ 0 & 1 \end{pmatrix}$ of V corresponds to $\begin{pmatrix} 1 & 0 \\ 0 & 1 \end{pmatrix}$ of W, and $\begin{pmatrix} i & 0 \\ 0 & -i \end{pmatrix}$ to $\begin{pmatrix} 0 & -1 \\ 1 & 0 \end{pmatrix}$. In general, if $\{A, B\}$ is a basis for V, then $\{T^*AT, T^*BT\}$ is a basis for W because $aA + bB = 0$ if and only if $aT^*AT + bT^*BT = 0$.

1.32 It is routine to check that W is closed under addition and scalar multiplication and that the three given matrices are linearly independent.

$$\begin{pmatrix} a & b \\ b & c \end{pmatrix} = a \begin{pmatrix} 1 & 0 \\ 0 & 0 \end{pmatrix} + b \begin{pmatrix} 0 & 1 \\ 1 & 0 \end{pmatrix} + c \begin{pmatrix} 0 & 0 \\ 0 & 1 \end{pmatrix}.$$

The coordinate of $\begin{pmatrix} 1 & -2 \\ -2 & 3 \end{pmatrix}$ with respect to the basis is $(1, -2, 3)$.

1.33 (a) Yes. This is the null space of A.

(b) Yes. This is the null space of A^2.

(c) Yes. This is the null space of $A - 2I$.

(d) No. The set contains no zero vector.

(e) Yes. This is the null space of $A - B$.

(f) Yes. This is the intersection of $\mathrm{Ker}(A - B)$ and $\mathrm{Ker}(A - C)$.

(g) Yes. This is the same as the null space of A.

(h) Yes. This is the orthogonal space of u.

(i) Yes. Note that ux^* is an $n \times n$ matrix.

(j) No unless $Au = 0$.

(k) Yes. This is $\mathrm{Ker}\, A$.

(l) Yes. $(X + kY)u = Xu + kYu$.

(m) Yes. $u^*(X + kY)u = u^*Xu + ku^*Yu$.

(n) No unless $u = 0$.

(o) Yes. $(X^*X)u = 0 \Leftrightarrow u^*(X^*X)u = 0 \Leftrightarrow Xu = 0$.

(p) No unless $w = 0$.

(q) Yes. $w^*X(u - v) = 0$.

1.34 If $n = 1$ or 2, then $V = \{0\}$, we have nothing to show. Let $n > 2$. It is routine to show that V is a subspace of $\mathbb{P}_n[x]$. To find a basis of V, let $f(x) = a_0 + a_1 x + \cdots + a_{n-1} x^{n-1}$. Since $f(0) = f(1) = 0$, we have $a_0 = a_0 + a_1 + \cdots + a_{n-1} = 0$. We can rewrite $f(x)$ in the form

$$f(x) = a_2(x^2 - x) + \cdots + a_{n-1}(x^{n-1} - x).$$

Thus, $\{x^2 - x, x^3 - x, \ldots, x^{n-1} - x\}$ is a basis of V and $\dim V = n - 2$.

1.35 By properties of integration, one shows that V is a subspace. To find a basis, let $p(x) = a + bx + cx^2$. $\int_{-1}^{1} p(x)dx = 0$ yields $3a + c = 0$. Thus $p(x) = a + bx - 3ax^2 = bx + a(1 - 3x^2)$. So $\{x, 1 - 3x^2\}$ is a basis for V. All $a + bx + cx^2$ with $c \neq -3a$ are in $\mathbb{P}_3[x]$ but not in V.

1.36 If $p(x), q(x) \in V$, then $p(1) = q(1) = 0$, and $p(1) + kq(1) = 0$ for any $k \in \mathbb{R}$. So V is a subspace of $\mathbb{P}_4[x]$. V does not contain all elements of $\mathbb{P}_4[x]$. So, $\dim V \leq 3$. It is easy to show that $1 - x, 1 - x^2, 1 - x^3$ are in V and are linearly independent. To find the coordinate, let

$$
\begin{aligned}
1 - 2x + 2x^2 - x^3 &= a(1 - x) + b(1 - x^2) + c(1 - x^3) \\
&= (a + b + c) - ax - bx^2 - cx^3.
\end{aligned}
$$

Then $a + b + c = 1, a = 2, b = -2, c = 1$. So the coordinate is $(2, -2, 1)$.

1.37 Suppose that $p_1(x), p_2(x), p_3(x), p_4(x)$ are linearly independent. Since $\mathbb{P}_4[x]$ has dimension 4, $\{p_1(x), p_2(x), p_3(x), p_4(x)\}$ is a basis for $\mathbb{P}_4[x]$. Thus, $p(x) = 1$ is a linear combination of $p_1(x), p_2(x), p_3(x), p_4(x)$, implying $p(a) = 0$, contradicting $p(a) = 1$.

If $p_i(a) = b \neq 0$, $i = 1, 2, 3, 4$, then $p_i(x)$, $i = 1, 2, 3, 4$, need not be linearly dependent. Take, $1, 1 - x, 1 - x^2, 1 - x^3$, for example. They have the same value 1 at 0, but they are linearly independent. The linear dependence conclusion does not hold for four polynomials in $\mathbb{P}_5[x]$ because $x, x - x^2, x - x^3, x - x^4$, all with value 0 at $a = 0$, are linearly independent in $\mathbb{P}_5[x]$ (or in $\mathcal{C}(\mathbb{R})$).

1.38 (a) To show that $\{1, x, \ldots, x^{n-1}\}$ is a basis, let $\lambda_0, \lambda_1, \ldots, \lambda_{n-1}$ be scalars such that $\lambda_0 + \lambda_1 x + \cdots + \lambda_{n-1} x^{n-1} = 0$. Setting $x = 0$ yields $\lambda_0 = 0$. In a similar way by factoring x each time, we see that $\lambda_1 = \cdots = \lambda_{n-1} = 0$. Thus $\{1, x, \ldots, x^{n-1}\}$ is a linearly independent set, hence, it is a basis as it spans $\mathbb{P}_n[x]$.

(b) Similar to (a). Or one may make use of Taylor's formula:

$$f(x) = f(a) + \frac{f'(a)}{1!}(x - a) + \cdots + \frac{f^{(n-1)}(a)}{(n-1)!}(x - a)^{n-1}.$$

(c) $\left(f(a), \frac{f'(a)}{1!}, \ldots, \frac{f^{(n-1)}(a)}{(n-1)!}\right).$

(d) It suffices to show that $f_1(x), \ldots, f_n(x)$ are linearly independent. Let $k_1 f_1(x) + \cdots + k_n f_n(x) = 0$. Setting $x = a_i$ gives all $k_i = 0$.

(e) If $f, g \in W$, then $f(1) = 0$ and $g(1) = 0$. Thus $(f + g)(1) = f(1) + g(1) = 0$, and $f + g \in W$. For $\lambda \in \mathbb{R}$, $(\lambda f)(1) = \lambda f(1) = 0$, so $\lambda f \in W$. Thus, W is a subspace. W consists of all $p(x) = x_1 + x_2 x + \cdots + x_n x^{n-1}$ with $x_1 + x_2 + \cdots + x_n = 0$. So, $\dim W = n-1$ and $\{1 - x, \, 1 - x^2, \, \ldots, \, 1 - x^{n-1}\}$ is a basis of W.

(f) No. $((x^{n-1} + 1) - x^{n-1} = 1$ has degree 0.)

(g) $1, x, \ldots, x^n$ are linearly independent for any n. So $\mathbb{P}[x]$ is infinite dimensional. Any polynomial in $\mathbb{P}[x]$, with degree m, say, is a linear combination of $1, x, \ldots, x^m$. So $1, x, x^2, \ldots$ span $\mathbb{P}[x]$. Thus, $\mathbb{P}[x]$ is an infinite dimensional space with a basis $\{1, x, x^2, \ldots\}$.

(h) $\mathbb{P}[x]$ consists of all polynomials of finite degrees; it is finitely generated. So, $1 + x + x^2 + \cdots \notin \mathbb{P}[x]$.

(i) Obviously, $\mathbb{P}_n[x]$ is a proper subset of $\mathbb{P}[x]$ as $x^n \notin \mathbb{P}_n[x]$.

1.39   Linearly independent vectors of S are also linearly independent vectors of V. This gives (a). If $\dim S = \dim V$, then a basis of S is also a basis of V, so $S = V$ and (b) holds. To see (c), if $\alpha = \{\alpha_1, \alpha_2, \ldots, \alpha_k\}$ is a basis of S, and if every $v \in V$ is a linear combination of the vectors in α, then α is a basis for V by definition. Otherwise, there exists a vector $\beta \in V$ such that $\alpha_1, \alpha_2, \ldots, \alpha_k, \beta$ are linearly independent. Inductively, α can be extended to a basis of V. For (d), one may take V to be the xy-plane with basis $\{(1,0), (0,1)\}$ and take S to be the line $y = x$ with basis $\{(1,1)\}$.

1.40   (1)   One verifies that V_0 is a vector space with zero vector $(0, 0, \ldots)$. Obviously, V_i, $1 \le i \le 5$, are subsets of V_0. By calculus of convergent series, we see that V_2, V_3, V_4, and V_5 are subspaces.

For V_1, if $u, v \in V_1$, i.e., u and v are infinite sequences in which only finitely many (or no) components are nonzero, then $u + kv$ has finitely many nonzero components, where k is any scalar. Thus, V_1 is also a subspace of V_0.

Apparently, V_1 is contained in all other V_i's. Let $e_1 = (1, 0, \ldots)$, $e_2 = (0, 1, 0, \ldots)$, $\ldots$. Every e_i lies in V_1 and $\{e_1, e_2, \ldots\}$ is a linearly independent set (of countably many elements). Thus V_1 is (finitely) spanned by $e_1, e_2, \ldots$ and V_1 is infinite dimensional.

(2) Only V_0 and V_5 contain $(1, 1, \ldots)$.

(3) Only V_0 contains $(1, -1, 1, -1, \ldots)$.

(4) All V_i's, $0 \leq i \leq 5$, contain $e_1, e_2, \ldots$

(5) $V_1 = \mathrm{Span}\{e_1, e_2, \ldots\}$ and $\{e_1, e_2, \ldots\}$ is a basis of V_1. V_1 is a proper subspace of V_0 because $(1, 1, \ldots) \in V_0 \setminus V_1$.

One might be tempted to write $v = (v_1, v_2, \ldots) = v_1 e_1 + v_2 e_2 + \cdots$ and to conclude that $\{e_1, e_2, \ldots\}$ is a basis of V_0. However, such an expression involves an infinite sum that makes no sense in general. For example, what would be $e_1 + e_1 + \cdots$? It cannot be a vector in V_0. (This is why infinite sums are avoided.)

It is known from advanced analysis that every vector space has a basis and that no countable set of vectors can span V_0. In general, a basis for an infinite dimensional vector space is not easy to find or to describe. (This is beyond our scope.)

(6) Of the six vector spaces, V_1 is the smallest (it is contained in other V_i's), and V_0 is the largest (it contains all other V_i's).

V_5 contains V_i, $1 \leq i \leq 4$. V_2 is the so-called ℓ^2 space. V_3 is the space of absolute convergence sequences. V_4 is the space of convergence sequences. By calculus, $V_3 \subset V_2$ and $V_3 \subset V_4$.

1.41 Let $a \sin x + b \cos x = 0$. Setting $x = 0$ gives $b = 0$; putting $x = \frac{\pi}{2}$ yields $a = 0$. So $\sin x$ and $\cos x$ are linearly independent. In the same way, we see that $\sin^2 x$ and $\cos^2 x$ are linearly independent.

For $y = a \sin x + b \cos x$, it is easy to check that $y'' = -y$.

1, $\sin^2 x$, and $\cos^2 x$ are linearly dependent since $\sin^2 x + \cos^2 x = 1$.

$\mathrm{Span}\{\sin x, \cos x\} \cap \mathbb{R} = \{0\}$ and $\mathrm{Span}\{\sin^2 x, \cos^2 x\} \cap \mathbb{R} = \mathbb{R}$.

1.42 Let $x_1 \alpha_1 + x_2 \alpha_2 + x_3 \alpha_3 = 0$. Then $x_1 + x_2 = 0$, $x_1 + x_3 = 0$, and $x_2 + x_3 = 0$. Thus $x_1 = x_2 = x_3 = 0$. The coordinates of u, v, and w under the basis are $(1, 1, -1)$, $\frac{1}{2}(1, 1, -1)$, and $\frac{1}{2}(1, 1, 1)$, respectively.

1.43 By elementary operations, we see that $\alpha_1, \alpha_2, \alpha_3$ are linearly independent. Thus, they comprise a basis for $\mathbb{R}^3$. So do $\beta_1, \beta_2, \beta_3$.

$v = \alpha_1 + 2\alpha_2 + 3\alpha_3 = 3e_1 + 5e_2 + 4e_3 = -\frac{8}{9}\beta_1 + \frac{20}{9}\beta_2 + \frac{5}{9}\beta_3$.

1.44 (a) $\{e_1, \ldots, e_n\}$ and $\{\epsilon_1, \ldots, \epsilon_n\}$ are linearly independent sets on $\mathbb{C}$ (and also on $\mathbb{R}$). e_i's and ϵ_i's are bases of $\mathbb{C}^n$ over $\mathbb{C}$; they are not bases of $\mathbb{C}^n$ over $\mathbb{R}$. The dimension of $\mathbb{C}^n$ over $\mathbb{R}$ is $2n$.

(b) $A = \begin{pmatrix} 1 & -1 & 0 & \cdots & 0 & 0 \\ 0 & 1 & -1 & \cdots & 0 & 0 \\ 0 & 0 & 1 & \cdots & 0 & 0 \\ \vdots & \vdots & \vdots & & \vdots & \vdots \\ 0 & 0 & 0 & \cdots & 1 & -1 \\ 0 & 0 & 0 & \cdots & 0 & 1 \end{pmatrix}.$

(c) $A(\epsilon_1, \epsilon_2, \ldots, \epsilon_n) = I_n$. $B = (\epsilon_1, \epsilon_2, \ldots, \epsilon_n) = A^{-1}$.

(d) $(-1, \ldots, -1, n)$.

(e) The dimension of $\mathbb{R}^n$ over $\mathbb{R}$ is n.

(f) $e_1, e_2, \ldots, e_n$ and $(i, 0, \ldots, 0)^t$, where $i = \sqrt{-1}$.

1.45 $\mathbb{R}^n$ is a vector space of dimension n over $\mathbb{R}$, while $\mathbb{C}^n$ is a vector space of dimension n over $\mathbb{C}$. $\mathbb{C}^n$ as a vector space over $\mathbb{R}$ has dimension $2n$. $\mathbb{R}^n$ is a subspace of $\mathbb{C}^n$ over $\mathbb{R}$; $\mathbb{R}^n$ is not a subspace of $\mathbb{C}^n$ over $\mathbb{C}$.

1.46 Let $x_1\alpha_1 + x_2\alpha_2 + x_3\alpha_3 = 0$. By solving the system of linear equations, we get $x_1 = x_2 = x_3 = 0$. So α is a basis for $\mathbb{R}^3$. Similarly, β is also a basis. (Note: The easiest way to see that α or β is a basis is to show that the determinant $\det(\alpha_1, \alpha_2, \alpha_3) \neq 0$. See Chapter 2.)

If A is a matrix such that $\beta = \alpha A$, then $A = \alpha^{-1}\beta$. This gives

$$A = \begin{pmatrix} 1 & 1 & 1 \\ 1 & 0 & 0 \\ 1 & -1 & 1 \end{pmatrix}^{-1} \begin{pmatrix} 1 & 2 & 3 \\ 2 & 3 & 4 \\ 1 & 4 & 3 \end{pmatrix} = \begin{pmatrix} 2 & 3 & 4 \\ 0 & -1 & 0 \\ -1 & 0 & -1 \end{pmatrix}.$$

If the coordinate of u under α is $(2, 0, -1)$ (the first column of A), then

$$u = (\alpha_1, \alpha_2, \alpha_3)(2, 0, -1)^t = \beta A^{-1}(2, 0, -1)^t = \beta(1, 0, 0)^t.$$

The coordinate of u under the basis β is $(1, 0, 0)$. $u = \beta_1 = (1, 2, 1)^t$.

1.47 Let $l_1\alpha_1 + l_2(\alpha_1 + \alpha_2) + \cdots + l_n(\alpha_1 + \alpha_2 + \cdots + \alpha_n) = 0$. Then

$$(l_1 + l_2 + \cdots + l_n)\alpha_1 + (l_2 + \cdots + l_n)\alpha_2 + \cdots + l_n\alpha_n = 0.$$

Since $\alpha_1, \alpha_2, \ldots, \alpha_n$ are linearly independent and the coefficient of α_n is l_n, we have $l_n = 0$. The coefficient of α_{n-1} is $l_{n-1} + l_n$, so $l_{n-1} = 0$. Inductively, $l_1 = l_2 = \cdots = l_{n-2} = 0$. So, $\alpha_1, \alpha_1 + \alpha_2, \ldots, \alpha_1 + \alpha_2 + \cdots + \alpha_n$ are linearly independent, thus form a basis (for dim $V = n$).

Let $x_1(\alpha_1 + \alpha_2) + x_2(\alpha_2 + \alpha_3) + \cdots + x_n(\alpha_n + \alpha_1) = 0$. Then

$$(x_1 + x_n)\alpha_1 + (x_1 + x_2)\alpha_2 + \cdots + (x_{n-1} + x_n)\alpha_n = 0$$

and
$$x_1 + x_n = 0, \ x_1 + x_2 = 0, \ \ldots, \ x_{n-1} + x_n = 0.$$

The system of these equations has a nonzero solution if and only

if the rows (or columns) of $\begin{pmatrix} 1 & 0 & 0 & \cdots & 0 & 1 \\ 1 & 1 & 0 & \cdots & 0 & 0 \\ 0 & 1 & 1 & \cdots & 0 & 0 \\ \vdots & \vdots & \vdots & & \vdots & \vdots \\ 0 & 0 & 0 & \cdots & 1 & 0 \\ 0 & 0 & 0 & \cdots & 1 & 1 \end{pmatrix}$ are linearly de-

pendent (or use determinant $1 + (-1)^{n+1}$). If n is even, the vectors are linearly dependent. If n is odd, they are linearly independent and thus form a basis. The converse is also true.

1.48 Let $a_1\alpha_1 + \cdots + a_n\alpha_n - b\beta = 0$, where $a_1, a_2, \ldots, a_n, b$ are not all zero. We claim $b \neq 0$. Suppose $b = 0$. Linear independence of $\alpha_1, \alpha_2, \ldots, \alpha_n$ yields $a_1 = a_2 = \cdots = a_n = 0$. So $b \neq 0$. Thus,

$$\beta = \tfrac{a_1}{b}\alpha_1 + \cdots + \tfrac{a_n}{b}\alpha_n.$$

To see the uniqueness, let $\beta = c_1\alpha_1 + \cdots + c_n\alpha_n$. Then

$$\left(\tfrac{a_1}{b} - c_1\right)\alpha_1 + \cdots + \left(\tfrac{a_n}{b} - c_n\right)\alpha_n = 0.$$

Since $\alpha_1, \alpha_2, \ldots, \alpha_n$ are linearly independent, we have

$$\tfrac{a_i}{b} - c_i = 0 \quad \text{or} \quad c_i = \tfrac{a_i}{b}, \quad i = 1, 2, \ldots, n.$$

1.49 Let $\alpha_1, \ldots, \alpha_n$ be linearly dependent and $a_1\alpha_1 + \cdots + a_n\alpha_n = 0$, where not all a's are zero. Let k be the largest index such that $a_k \neq 0$. Then α_k is a linear combination of $\alpha_1, \ldots, \alpha_{k-1}$. The converse is obvious.

1.50 (a) True. (b) True. (c) True. (d) True. The converse is also true. (e) False. (f) False. (g) False. (h) False. For an example, take $\alpha_1 = (1,0,0), \alpha_2 = (0,1,0), \alpha_3 = (0,0,1), \alpha_4 = \alpha_1+\alpha_2, \alpha_5 = \alpha_2+\alpha_3, \alpha_6 = \alpha_3+\alpha_1$ in $\mathbb{R}^3$. Then any α_i is a linear combination of other two α_j's. $r = 2$, but $\dim \mathbb{R}^3 = 3$.

1.51 Let $\alpha = \{\alpha_1, \ldots, \alpha_s\}$, $\beta = \{\beta_1, \ldots, \beta_t\}$. Since $\text{Span}\,\alpha \subseteq \text{Span}\{\alpha, \beta\}$ and $\text{Span}\,\beta \subseteq \text{Span}\{\alpha, \beta\}$, we have $\text{Span}\,\alpha + \text{Span}\,\beta \subseteq \text{Span}\{\alpha, \beta\}$. Conversely, $\{\alpha, \beta\} \subseteq \text{Span}\,\alpha + \text{Span}\,\beta$, so $\text{Span}\{\alpha, \beta\} \subseteq \text{Span}\,\alpha + \text{Span}\,\beta$, concluding $\text{Span}\{\alpha, \beta\} = \text{Span}\,\alpha + \text{Span}\,\beta$.

1.52   If $s + t \geq n$, we have nothing to prove. Without loss of generality, let $\alpha_1, \ldots, \alpha_s$ be a basis for U and let $\beta_1, \ldots, \beta_t$ be a basis for V. Then every vector $\alpha_i + \beta_j$, thus every vector in W, is a linear combination of $\alpha_1, \ldots, \alpha_s, \beta_1, \ldots, \beta_t$ ($s + t$ vectors). It follows that $\dim W \leq s + t$.

1.53   It is straightforward to show that W is a (nonempty) subspace of $\mathbb{F}^n$. We may assume that $V = \mathrm{Span}\{v_1, \ldots, v_n\}$. Let $\dim V = k$. If $k = n$, then $W = \{0\}$. Suppose that $k < n$ and $\{v_1, \ldots, v_k\}$ is a basis of V. Then each v_{k+j}, $j \geq 1$, is a linear combination of $v_1, \ldots, v_k$. Let $v_{k+j} = \sum_{i=1}^{k} c_{ij} v_i$, $j = 1, \ldots, n - k$. Let $A = (I_k, C) \in M_{k \times n}(\mathbb{F})$, where $C = (c_{ij}) \in M_{k \times (n-k)}(\mathbb{F})$. Then $x_1 v_1 + \cdots + x_n v_n = 0 \Leftrightarrow y_1 v_1 + \cdots + y_k v_k = 0$, where $y = (y_1, \ldots, y_k)^t = Ax$. Or one may write $(v_1, \ldots, v_n)x = 0 \Leftrightarrow (v_1, \ldots, v_k)y = 0$, $y = Ax$. Since $v_1, \ldots, v_k$ are linearly independent, the solutions to $x_1 v_1 + + \cdots + x_n v_n = 0$ are the solutions to $Ax = 0$. Thus, $W = \mathrm{Ker}\, A$. The first k columns (i.e., I_k) of A are linearly independent and A has only k rows, so $\mathrm{Im}\, A = \mathbb{F}^k$. It follows that $\dim W = \dim \mathrm{Ker}\, A = n - \dim \mathrm{Im}\, A = n - k$.

1.54 W consists of all vectors $w = x_1 \alpha_1 + \cdots + x_n \alpha_n = x_1 \beta_1 + \cdots + x_n \beta_n$. $0 \in W$. We may write $W = \{ w \in V \mid w = \alpha x = \beta x, x \in \mathbb{F}^n \}$. For $u, v \in W$, let $u = \alpha x = \beta x$ and $v = \alpha y = \beta y$. Then $u + kv = \alpha(x + ky) = \beta(x + ky) \in W$ for scalars k. So W is a subspace of V.

Note that $x_1 \alpha_1 + \cdots + x_n \alpha_n = x_1 \beta_1 + \cdots + x_n \beta_n$ if and only if $x_1(\alpha_1 - \beta_1) + \cdots + x_n(\alpha_n - \beta_n) = 0$. Denote $v_i = \alpha_i - \beta_i$ for each i. Then the dimension identity follows immediately from Problem 1.53.

1.55   Suppose $a_0(tu + v) + a_1 \alpha_1 + \cdots + a_r \alpha_r = 0$ for scalars $a_0, a_1, \ldots, a_r$. We first claim $a_0 = 0$. Otherwise, dividing both sides by a_0, since u is a linear combination of $\alpha_1, \ldots, \alpha_r$, we see that v is a linear combination of $\alpha_1, \ldots, \alpha_r$, a contradiction. Now that $a_0 = 0$, the linear independence of $\alpha_1, \ldots, \alpha_r$ implies $a_1 = a_2 = \cdots = a_r = 0$.

1.56   (a) It is obvious that $V \times W$ is closed under the addition and scalar multiplication. If 0_v and 0_w are zero vectors of V and W, respectively, then $(0_v, 0_w)$ is the zero vector of $V \times W$. It is routine to check that other conditions for a vector space are also satisfied.

     (b) If $\{\alpha_1, \alpha_2, \ldots, \alpha_m\}$ is a basis for V and $\{\beta_1, \beta_2, \ldots, \beta_n\}$ is a basis for W, one may show that $(\alpha_i, 0), (0, \beta_j)$, $i = 1, 2, \ldots, m$, $j = 1, 2, \ldots, n$, form a basis for $V \times W$. So $\dim(V \times W) = m + n$.

     Note: $V \times W$ is known as the *Cartesian product* of V and W. In contrast, $V \otimes W$ is the *tensor product* of V and W which is spanned by the tensor vectors $\alpha_i \otimes \beta_j$. $\dim(V \otimes W) = mn$.

(c) Identify $(x, (y, z)) \in \mathbb{R} \times \mathbb{R}^2$ with $(x, y, z) \in \mathbb{R}^3$.

(d) $\alpha = 1$ is a basis for $\mathbb{R}$, $\{e_1, e_2\}$ is a basis for $\mathbb{R}^2$. $\mathbb{R} \times \mathbb{R}^2 = \{(x, (y, z)) \mid x, y, z \in \mathbb{R}\}$ is spanned by $(1, (0, 0)), (0, (1, 0))$, and $(0, (0, 1))$. The dimension of $\mathbb{R} \times \mathbb{R}^2$ is 3. $\mathrm{Span}\{(\alpha, e_1), (\alpha, e_2)\}$ is spanned by two vectors, its dimension is at most 2, so they cannot be the same. In fact, $\mathrm{Span}\{(\alpha, e_1), (\alpha, e_2)\}$ consists of all elements $(x, (y, z))$ satisfying $x = y + z$.

(e) The dimension of $\mathbb{R}^2 \times \mathbb{R}^3$ is $2 + 3 = 5$. There are six vectors in $\{(e_i, f_j)\}$. So, they are linearly dependent. In fact,

$$(e_1, f_1) + (e_1, f_2) - 2(e_1, f_3) - (e_2, f_1) - (e_2, f_2) + 2(e_2, f_3) = 0.$$

(f) Let E_{ij} be the 2×2 matrix with (i, j)-entry 1 and all other entries 0, $i, j = 1, 2$. $\{E_{11}, E_{12}, E_{21}, E_{22}\}$ is a basis for $M_2(\mathbb{R})$. The six vectors $(e_t, 0), (0, E_{ij}), t, i, j = 1, 2$, form a basis for $\mathbb{R}^2 \times M_2(\mathbb{R})$.

(g) $4 + 4 = 8$.

1.57   It is a vector space of the matrices $\left(\begin{smallmatrix} a & b \\ 0 & a+b \end{smallmatrix} \right) = a \left(\begin{smallmatrix} 1 & 0 \\ 0 & 1 \end{smallmatrix} \right) + b \left(\begin{smallmatrix} 0 & 1 \\ 0 & 1 \end{smallmatrix} \right)$, where $a, b \in \mathbb{F}$. Dimension is 2, and $\{\left(\begin{smallmatrix} 1 & 0 \\ 0 & 1 \end{smallmatrix} \right), \left(\begin{smallmatrix} 0 & 1 \\ 0 & 1 \end{smallmatrix} \right)\}$ serves as a basis.

1.58   If $AX = XA$ and $AY = YA$, then $A(X + kY) = AX + kAY = XA + kYA = (X + kY)A$. Thus, the matrices commuting with A form a subspace. Letting $\left(\begin{smallmatrix} a & b \\ c & d \end{smallmatrix} \right) \left(\begin{smallmatrix} 1 & 2 \\ 3 & 4 \end{smallmatrix} \right) = \left(\begin{smallmatrix} 1 & 2 \\ 3 & 4 \end{smallmatrix} \right) \left(\begin{smallmatrix} a & b \\ c & d \end{smallmatrix} \right)$, we solve $3b = 2c, 2a + 3b = 2d, d = a + c$ to get $c = \frac{3}{2}b$ and $d = a + \frac{3}{2}b$. Thus the matrices have the form $\left(\begin{smallmatrix} a & b \\ \frac{3}{2}b & a+\frac{3}{2}b \end{smallmatrix} \right)$. Replacing b with $2b$, we get $\left(\begin{smallmatrix} a & 2b \\ 3b & a+3b \end{smallmatrix} \right) = \left(\begin{smallmatrix} (a-b)+b & 2b \\ 3b & (a-b)+4b \end{smallmatrix} \right) = xI + yA$, where $x = a - b$, $y = b$. So $\{I, A\}$ is a basis for the subspace. $\{\left(\begin{smallmatrix} 1 & 0 \\ 0 & 1 \end{smallmatrix} \right), \left(\begin{smallmatrix} 0 & 2 \\ 3 & 3 \end{smallmatrix} \right)\}$ is another basis.

1.59   It is a vector space of the matrices $\left(\begin{smallmatrix} a & 0 & b \\ 0 & a & 0 \\ 0 & 0 & a \end{smallmatrix} \right) = a \left(\begin{smallmatrix} 1 & 0 & 0 \\ 0 & 1 & 0 \\ 0 & 0 & 1 \end{smallmatrix} \right) + b \left(\begin{smallmatrix} 0 & 0 & 1 \\ 0 & 0 & 0 \\ 0 & 0 & 0 \end{smallmatrix} \right)$, where $a, b \in \mathbb{F}$. Dimension is 2; $\left(\begin{smallmatrix} 1 & 0 & 0 \\ 0 & 1 & 0 \\ 0 & 0 & 1 \end{smallmatrix} \right)$ and $\left(\begin{smallmatrix} 0 & 0 & 1 \\ 0 & 0 & 0 \\ 0 & 0 & 0 \end{smallmatrix} \right)$ serve as a basis.

1.60   Take $A = \left(\begin{smallmatrix} 1 & 0 & -\sqrt{3} \\ 0 & 2 & 0 \\ 1 & 0 & \sqrt{3} \end{smallmatrix} \right)$. Note that $A^t A = \left(\begin{smallmatrix} 2 & 0 & 0 \\ 0 & 4 & 0 \\ 0 & 0 & 6 \end{smallmatrix} \right)$, $AA^t = \left(\begin{smallmatrix} 4 & 0 & -2 \\ 0 & 4 & 0 \\ -2 & 0 & 4 \end{smallmatrix} \right)$.

(a) True. Orthogonal nonzero vectors are linearly independent.

(b) True. ("Row rank = column rank".)

(c) True. If the columns of A have the same length, say k, then $A^t A = k^2 I$, implying $AA^t = k^2 I$ (one may use inverse here). So the rows of A (are orthogonal and) each have length k also.

(d) False. See the above A.

One can construct a 4×3 real matrix that has mutually orthogonal nonzero columns. However, there does not exist a 3×4 real matrix that has mutually orthogonal nonzero columns.

1.61 (a) It is routine to verify that V_n is a subspace of $M_n(\mathbb{F})$.

(b) The matrices E_{ij} with (i,j)-entry 1 and 0 elsewhere, $1 \le i < j \le n$, comprise a basis for V_n and the dimension is $\frac{n(n-1)}{2}$.

(c) Examine the product of two matrices of V_n. Let $A = (a_{ij})$, $a_{ij} = 0$ if $i \ge j$, and $B = (b_{ij})$, $b_{ij} = 0$ if $i \ge j$. Let $C = (c_{ij}) = AB$. Then $c_{ij} = \sum_{k=1}^n a_{ik}b_{kj}$. If $i \ge j - 1$, then $i \ge k$ or $k \ge j$, so $c_{ij} = 0$. This shows that C has 0's in $(i, i+1)$ positions, $i = 1, \ldots, n - 1$. In this way, multiplying n strictly upper-triangular matrices, all (possible) nonzero entries are "pushed" out.

Or one computes $F^2, F^3, \ldots, F^n$, where F is a strictly upper-triangular 0 - 1 matrix, to calculate the number of 0's each time.

1.62 By direct verifications.

1.63 A: unitary and normal. B: Hermitian and normal. C: symmetric and normal. D: symmetric, not normal. E: symmetric, skew-Hermitian, unitary and normal. F: symmetric, not normal. G: real orthogonal and normal. H: unitary and normal.

1.64 (a) All $n \times n$ matrices.

(b) All matrices of the form $\left(\begin{smallmatrix} a & b \\ 0 & a \end{smallmatrix}\right)$, $a, b \in \mathbb{F}$.

(c) All matrices of the form $\left(\begin{smallmatrix} c & 0 \\ 0 & d \end{smallmatrix}\right)$, $c, d \in \mathbb{F}$.

(d) All matrices of the form $\left(\begin{smallmatrix} a & b & c & d \\ 0 & a & b & c \\ 0 & 0 & a & b \\ 0 & 0 & 0 & a \end{smallmatrix}\right)$, $a, b, c, d \in \mathbb{F}$.

(e) All $n \times n$ scalar matrices cI_n, $c \in \mathbb{F}$.

1.65 (a) $0 \in C(A)$. If $X, Y \in C(A)$, then $AX = XA$ and $AY = YA$. Thus $A(X + Y) = AX + AY = XA + YA = (X + Y)A$, that is, $X + Y \in C(A)$. For scalars k, $A(kX) = k(AX) = k(XA) = (kX)A$, so $kX \in C(A)$. Thus, $C(A)$ is closed under the matrix addition and scalar multiplication, and $C(A)$ is a vector space.

(b) This is nearly obvious because $I, A, A^2, \ldots$ all commute with A.

(c)　The matrices that commute with A are in the form

$$\begin{pmatrix} a & b & 0 \\ c & d & 0 \\ -3a-c+3e & -3b-d+e & e \end{pmatrix} = a \begin{pmatrix} 1 & 0 & 0 \\ 0 & 0 & 0 \\ -3 & 0 & 0 \end{pmatrix} + b \begin{pmatrix} 0 & 1 & 0 \\ 0 & 0 & 0 \\ 0 & -3 & 0 \end{pmatrix} + c \begin{pmatrix} 0 & 0 & 0 \\ 1 & 0 & 0 \\ -1 & 0 & 0 \end{pmatrix}$$
$$+ d \begin{pmatrix} 0 & 0 & 0 \\ 0 & 1 & 0 \\ 0 & -1 & 0 \end{pmatrix} + e \begin{pmatrix} 0 & 0 & 0 \\ 0 & 0 & 0 \\ 3 & 1 & 1 \end{pmatrix}.$$

$\dim C(A) = 5$ (5 independent variables a, b, c, d, e). The five matrices on the right-hand side comprise a basis of $C(A)$.

(d)　One can verify that $\begin{pmatrix} a & b & c \\ 0 & a & b \\ 0 & 0 & a \end{pmatrix}$ commutes with B. There are 3 independent variables a, b, c. So $\dim C(B) \geq 3$.

1.66　(a)　$0 \in T(A)$. If $X, Y \in T(A)$, then $A(X + kY) + (X + kY)^t A = AX + X^t A + k(AY + Y^t A) = 0$. So, $T(A)$ is a subspace of $M_n(\mathbb{R})$.

(b)　Since $A^t X + X^t A^t = 0 \Leftrightarrow X^t A + AX = 0$, $T(A^t) = T(A)$.

(c)　$T(I_n)$ consists of all $n \times n$ real skew-symmetric matrices.

(d)　For $A = \begin{pmatrix} 1 & -1 \\ 0 & 0 \end{pmatrix}$, $T(A) = \{\begin{pmatrix} a & 0 \\ 2a & -a \end{pmatrix} \mid a \in \mathbb{R}\}$. $\begin{pmatrix} 1 & 0 \\ 2 & -1 \end{pmatrix}$ serves as a basis of $T(A)$; the dimension is 1.

1.67　(a)　If $X, Y \in C(A, B)$, then $A(X + kY) = AX + kAY = (X + kY)B$.

(b)　$A = B = kI$, where k is any scalar.

(c)　If $X \in C(A, B)$, then $A^i X = X B^i$ for every positive integer i. It follows that $f(A)X = X f(B)$ for any polynomial f.

(d)　$C(A, B) = \{\begin{pmatrix} a & b \\ b & a \end{pmatrix}\}$ and $\{\begin{pmatrix} 1 & 0 \\ 0 & 1 \end{pmatrix}, \begin{pmatrix} 0 & 1 \\ 1 & 0 \end{pmatrix}\}$ is a basis for $C(A, B)$.

(e)　$C(A, B) = \{\begin{pmatrix} a & 0 \\ 0 & b \end{pmatrix}\}$ and $\{\begin{pmatrix} 1 & 0 \\ 0 & 0 \end{pmatrix}, \begin{pmatrix} 0 & 0 \\ 0 & 1 \end{pmatrix}\}$ is a basis for $C(A, B)$.

1.68　(a)　$(AB)^*(AB) = B^* A^* AB = I$. $A + B$ need not be unitary.

(b)　$(A + B)^* = A^* + B^* = A + B$. AB need not be Hermitian.

(c)　$A + B$ and AB need not be normal.

(d)　By direct verifications.

(e)　By direct verifications.

(f)　$\frac{1}{2} \begin{pmatrix} \overline{a}cI & \overline{b}cI \\ \overline{a}dI & -\overline{b}dI \end{pmatrix} \begin{pmatrix} acI & adI \\ bcI & -bdI \end{pmatrix} = \begin{pmatrix} I & 0 \\ 0 & I \end{pmatrix}$. Set $a = -i, b = c = 1, d = -1$.

1.69　$\begin{pmatrix} x & y & z \\ y & x & y \\ 0 & y & x \end{pmatrix}$, where $x, y, z \in \mathbb{R}$.

1.70　Let E_{st} denote the $n \times n$ matrix with the (s, t)-entry 1 and 0 elsewhere.

(a)　E_{st}, $1 \leq s, t \leq n$, form a basis. Dimension is n^2.

(b) $E_{st}, iE_{st}, 1 \le s, t \le n$, form a basis. Dimension is $2n^2$.

(c) $E_{st}, 1 \le s, t \le n$, form a basis. Dimension is n^2.

(d) $E_{st}+E_{ts}, s \le t, i(E_{st}-E_{ts}), s < t$, form a basis. Dimension is n^2.

(e) $E_{st} + E_{ts}, s \le t$, form a basis. Dimension is $\frac{n(n+1)}{2}$.

(f) $E_{st}-E_{ts}, s < t, i(E_{st}+E_{ts}), s \le t$, form a basis. Dimension is n^2.

(g) $E_{st} - E_{ts}, s < t$, form a basis. Dimension is $\frac{n(n-1)}{2}$.

(h) $E_{st}, 1 \le s \le t \le n$, form a basis. Dimension is $\frac{n(n+1)}{2}$.

(i) $E_{st}, 1 \le t \le s \le n$, form a basis. Dimension is $\frac{n(n+1)}{2}$.

(j) $E_{st}, 1 \le s = t \le n$, form a basis. Dimension is n.

1.71 Using vandermonde matrix, one can show that $\{I, A, A^2\}$ is a basis (over $\mathbb{R}$ or $\mathbb{C}$). The dimension of the span is 3. Note that $A^3 = I$.

1.72 Write $A = 2I_3+B+C$, where $B = \left(\begin{smallmatrix} 0 & 1 & 0 \\ 0 & 0 & 1 \\ 0 & 0 & 0 \end{smallmatrix}\right)$ and $C = \left(\begin{smallmatrix} 0 & 0 & 1 \\ 0 & 0 & 0 \\ 0 & 0 & 0 \end{smallmatrix}\right)$. By computations, we see that $A^2, A^3, \ldots$ are linear combinations of I_3, B, C. Thus, $\text{Span}\{I_3, A, A^2, A^3, \ldots\} = \text{Span}\{I_3, B, C\}$, and $\{I_3, B, C\}$ is a basis. The dimension of the span is 3.

1.73 It is not difficult to show that V is a subspace of $M_2(\mathbb{R})$. Note that $a = -2b - 3c - 4d$. $\dim V = 3$. $\{ \left(\begin{smallmatrix} -2 & 1 \\ 0 & 0 \end{smallmatrix}\right), \left(\begin{smallmatrix} -3 & 0 \\ 1 & 0 \end{smallmatrix}\right), \left(\begin{smallmatrix} -4 & 0 \\ 0 & 1 \end{smallmatrix}\right) \}$ is a basis:

$$\left(\begin{smallmatrix} a & b \\ c & d \end{smallmatrix}\right) = \left(\begin{smallmatrix} -2b-3c-4d & b \\ c & d \end{smallmatrix}\right) = b \left(\begin{smallmatrix} -2 & 1 \\ 0 & 0 \end{smallmatrix}\right) + c \left(\begin{smallmatrix} -3 & 0 \\ 1 & 0 \end{smallmatrix}\right) + d \left(\begin{smallmatrix} -4 & 0 \\ 0 & 1 \end{smallmatrix}\right).$$

1.74 $\text{Span}\{ \left(\begin{smallmatrix} 1 & 0 \\ 0 & -1 \end{smallmatrix}\right), \left(\begin{smallmatrix} 0 & 1 \\ -1 & 0 \end{smallmatrix}\right), \left(\begin{smallmatrix} 0 & 1 \\ 0 & 0 \end{smallmatrix}\right), \left(\begin{smallmatrix} 0 & 0 \\ 1 & 0 \end{smallmatrix}\right) \} = \{ \left(\begin{smallmatrix} a & b \\ c & -a \end{smallmatrix}\right) \mid a, b, c \in \mathbb{R} \}$. $\{ \left(\begin{smallmatrix} 1 & 0 \\ 0 & -1 \end{smallmatrix}\right), \left(\begin{smallmatrix} 0 & 1 \\ 0 & 0 \end{smallmatrix}\right), \left(\begin{smallmatrix} 0 & 0 \\ 1 & 0 \end{smallmatrix}\right) \}$ is a basis and the dimension is 3.

1.75 (a) $(1, i) = i(-i, 1)$.

(b) If $a(1, i) + b(-i, 1) = 0$, where $a, b \in \mathbb{R}$, then $a - bi = 0$. Thus $a = b = 0$. (Linear dependence is related to the underlying field.)

(c) Obviously, $W_{\mathbb{R}}(A) \subseteq W_{\mathbb{C}}(A)$. For the reversal inclusion, let $A \left(\begin{smallmatrix} a \\ b \end{smallmatrix}\right) = \left(\begin{smallmatrix} a+ib \\ b-ia \end{smallmatrix}\right) \in W_{\mathbb{C}}(A)$, where $a, b \in \mathbb{C}$. Let $a = a_1 + ia_2$, $b = b_1 + ib_2$, where $a_1, a_2, b_1, b_2 \in \mathbb{R}$. Then $\left(\begin{smallmatrix} a+ib \\ b-ia \end{smallmatrix}\right) = \left(\begin{smallmatrix} p+iq \\ q-ip \end{smallmatrix}\right) = A \left(\begin{smallmatrix} p \\ q \end{smallmatrix}\right)$, where $p = a_1 - b_2$, $q = a_2 + b_1 \in \mathbb{R}$. So $A \left(\begin{smallmatrix} a \\ b \end{smallmatrix}\right) \in W_{\mathbb{R}}(A)$. Therefore, $W_{\mathbb{R}}(A) = W_{\mathbb{C}}(A)$. $W_{\mathbb{C}}(A) = \text{Im } A$ by definition.

(d) $W_{\mathbb{R}}(B) = \{ \left(\begin{smallmatrix} a \\ bi \end{smallmatrix}\right) \mid a, b \in \mathbb{R} \} \subsetneqq W_{\mathbb{C}}(B) = \{ \left(\begin{smallmatrix} a \\ b \end{smallmatrix}\right) \mid a, b \in \mathbb{C} \}$.

(e) $W_{\mathbb{C}}(A) = \{a \begin{pmatrix} 1 \\ -i \end{pmatrix} + b \begin{pmatrix} i \\ 1 \end{pmatrix} \mid a, b \in \mathbb{R}\} = \{c \begin{pmatrix} 1 \\ -i \end{pmatrix} \mid c \in \mathbb{C}\} = \operatorname{Im} A.$

Over $\mathbb{R}$, $W_{\mathbb{C}}(A)$ has dimension 2 with basis $\{\begin{pmatrix} 1 \\ -i \end{pmatrix}, \begin{pmatrix} i \\ 1 \end{pmatrix}\}$; over $\mathbb{C}$, $W_{\mathbb{C}}(A)$ has dimension 1 with basis $\begin{pmatrix} 1 \\ -i \end{pmatrix}$.

(f) It is routine to show that $W_{\mathbb{R}}(B) = \{\begin{pmatrix} a \\ bi \end{pmatrix} \mid a, b \in \mathbb{R}\}$ is a subspace of $\mathbb{C}^2$ over $\mathbb{R}$, but not over $\mathbb{C}$ because it is not closed under the complex scalar multiplication. $\dim W_{\mathbb{R}}(B)$ over $\mathbb{R}$ is 2 with basis $\{\begin{pmatrix} 1 \\ 0 \end{pmatrix}, \begin{pmatrix} 0 \\ i \end{pmatrix}\}$. Note that $W_{\mathbb{C}}(B) = \mathbb{C}^2$ over $\mathbb{C}$ has dimension 2.

(g) $W_{\mathbb{R}}(C) = \{\begin{pmatrix} a \\ a \end{pmatrix} \mid a \in \mathbb{R}\}$ and $W_{\mathbb{C}}(C) = \{\begin{pmatrix} a \\ a \end{pmatrix} \mid a \in \mathbb{C}\}$. $W_{\mathbb{C}}(C)$ is a subspace of $\mathbb{C}^2$ over $\mathbb{R}$ and over $\mathbb{C}$. The dimension of $W_{\mathbb{R}}(C)$ over $\mathbb{R}$ is 1. The dimension of $W_{\mathbb{C}}(C)$ over $\mathbb{R}$ is 2; over $\mathbb{C}$ is 1.

1.76 One checks that each set is closed under the matrix addition and scalar multiplication. An easy way to find dimensions is to count the number of "free" entries in the matrix. For example, S_1 has dimension $n^2 - 1$ because, of the n^2 entries, only $(1,1)$ is fixed to be 0. Let E_{ij}, $1 \le i, j \le n$, be the standard basis of $M_n(\mathbb{C})$ (see (1.2), p. 9).

(a) $\dim(S_1) = n^2 - 1$. All E_{ij}, except E_{11}, form a basis.

(b) $\dim(S_2) = n^2 - n$. All E_{ij}, except $E_{11}, \ldots, E_{nn}$, form a basis.

(c) $\dim(S_3) = n^2 - 1$. All E_{ij}, $i \ne j$, $E_{ii} - E_{nn}$, $i < n$, form a basis.

(d) $\dim(S_4) = n^2 - 1$. All E_{ij}, $(i,j) \ne (p,q), (q,p)$, and $E_{pq} + E_{qp}$, form a basis. ($n^2 - 1$ free positions.)

(e) $\dim(S_5) = \frac{n(n+1)}{2}$. All E_{ii} and $E_{ij} + E_{ji}$, $i < j$ form a basis.

(f) $\dim(S_6) = n^2 - n$. All $E_{ij} - E_{in}$, $j < n$, form a basis.

(g) $\dim(S_7) = n^2 - 1$. All $E_{ij} - E_{nn}$, $(i,j) \ne (n,n)$, form a basis.

1.77 We call $x_{11} + \cdots + x_{nn}$ the trace of $X = (x_{ij})$. V_1 is the space of trace 0 matrices, $\dim V_1 = n^2 - 1$. $V_2 = \operatorname{Span}\{AB - BA \mid A, B \in M_n(\mathbb{C})\}$. One verifies that the trace of $AB - BA$ is 0. So $V_2 \subseteq V_1$.

Let E_{ij}, $1 \le i, j \le n$, be the standard basis of $M_n(\mathbb{C})$ (see (1.2), p. 9). Then for $i \ne j$, $E_{ij} = E_{is}E_{sj} - E_{sj}E_{is}$. So $E_{ij} \in V_2$ if $i \ne j$. Observe that V_2 contains all the diagonal matrices with one 1 in the (i,i) position, $1 \le i \le n-1$ and -1 in the (n,n) position. Thus, the dimension of V_2 is $n^2 - 1$. Consequently, $V_2 = V_1$.

1.78 First, we show that $S(A)$ is closed under the addition and scalar multiplication. Let $X, Y \in S(A)$ and let c be a scalar. Then $A(X+Y) = AX + AY = 0$ and $A(cX) = c(AX) = 0$. So $S(A)$ is a subspace of $M_{n \times p}(\mathbb{C})$. When $m = n$ and if $X \in S(A^k)$, i.e., $A^k X = 0$, then $A^{k+1} X = A(A^k X) = 0$. Thus, $X \in S(A^{k+1})$. Hence, $S(A^k) \subseteq S(A^{k+1})$. Since each $S(A^k)$ is a subspace of $M_{n \times p}(\mathbb{C})$ and the dimension of $M_{n \times p}(\mathbb{C})$ is finite, there must exist a positive integer r such that $\dim S(A^r) = \dim S(A^{r+1}) = \dim S(A^{r+2}) = \cdots$. It follows that

$$S(A) \subseteq S(A^2) \subseteq \cdots \subseteq S(A^r) = S(A^{r+1}) = S(A^{r+2}) = \cdots.$$

1.79   Let $AB = BA$. For any $x \in \mathbb{C}^n$, $(AB)x = A(Bx) \in \operatorname{Im} A$ and $(AB)x = (BA)x = B(Ax) \in \operatorname{Im} B$. Thus, $\operatorname{Im}(AB) \subseteq \operatorname{Im} A \cap \operatorname{Im} B$.

1.80   (a) $\Rightarrow$ (b): Since $\operatorname{Im} A \subseteq \operatorname{Im} B$, every column of A is contained in $\operatorname{Im} B$, which is spanned by the columns of B. Thus every column of A is a linear combination of the columns of B. So (a) implies (b).

(b) $\Rightarrow$ (c): For a matrix X, denote the j-th column of X by X_j. Write

$$A = (A_1, A_2, \ldots, A_p), \quad B = (B_1, B_2, \ldots, B_q).$$

If (b) holds, then for each $j = 1, 2, \ldots, p$, $A_j = c_{1j} B_1 + \cdots + c_{qj} B_q$ for some scalars c_{ij}, $i = 1, 2, \ldots, q$. Letting $C = (c_{ij})$ gives $A = BC$.

(c) $\Rightarrow$ (a): If $x \in \operatorname{Im} A$, let $x = Ay$ for some y. Since $A = BC$, we have $x = Ay = (BC)y = B(Cy)$. Thus $x \in \operatorname{Im} B$, and $\operatorname{Im} A \subseteq \operatorname{Im} B$. Alternatively, $A = BC$ yields $A_i = BC_i = (B_1, \ldots, B_q)C_i$ for each $i = 1, 2, \ldots, p$, that is, each column A_i of A is a linear combination of the columns of B. Thus any linear combination of A_i is also a linear combination of the columns of B. Hence $\operatorname{Im} A \subseteq \operatorname{Im} B$.

1.81   Each of (b), (c), (d), and (e) implies (a). (Converse is false for $A = 0$.)

(b) $\not\Rightarrow$ (c): Take $A = \left(\begin{smallmatrix} 1 & 0 \\ 1 & 0 \end{smallmatrix}\right)$ and $x_1 = \left(\begin{smallmatrix} 1 \\ -1 \end{smallmatrix}\right)$. Then $Ax_1 \neq 0$, $A^* x_1 = 0$. (Note: $AA^* x_1 = 0$, $A^* A x_1 \neq 0$.) Similarly, (c) $\not\Rightarrow$ (b).

(b) $\Leftrightarrow$ (d): We only show that if $Ax_1, \ldots, Ax_m$ are linearly independent, then $A^* Ax_1, \ldots, A^* Ax_m$ are linearly independent. Let $l_1 A^* Ax_1 + \cdots + l_m A^* Ax_m = 0$. Then $A^* A(l_1 x_1 + \cdots + l_m x_m) = 0$, so $u^* A^* Au = 0$, where $u = l_1 x_1 + \cdots + l_m x_m$. By the fact $y^* y = 0 \Rightarrow y = 0$, we have $Au = 0$, i.e., $l_1 Ax_1 + \cdots + l_m Ax_m = 0$. Because $Ax_1, \ldots, Ax_m$ are linearly independent, $l_1 = \cdots = l_m = 0$, as desired.

(c) $\Leftrightarrow$ (e): Replace A with A^* in (b) and (d).

Note that if A is nonsingular, then all statements are equivalent.

1.82   (a) For any $x \in \mathbb{C}^n$, $A^2 x = A(Ax) \in \operatorname{Im} A$. So $\operatorname{Im}(A^2) \subseteq \operatorname{Im} A$.

(b) Since $(AA^*)x = A(A^*x) \in \text{Im} A$, we have $\text{Im}(AA^*) \subseteq \text{Im} A$. We claim that $\text{Im}(AA^*)$ and $\text{Im} A$ have the same dimension. Since $\dim \text{Im} A = \dim \text{Im} A^*$ (see (1.5), p. 13), $\dim \text{Ker} A = \dim \text{Ker} A^*$ (see (1.3), p. 13). As $\text{Ker}(AA^*) = \text{Ker} A^*$ and $\dim \text{Ker}(AA^*) = \dim \text{Ker} A$, we get $\dim \text{Im}(AA^*) = \dim \text{Im} A$. (Or use the rank identity $r(AA^*) = r(A) = r(A^*)$, Example 2.7, p. 48.)

(c) $\text{Im}(A^*A)$ need not be equal to $\text{Im} A$. Take $A = \left(\begin{smallmatrix} 1 & 1 \\ 0 & 0 \end{smallmatrix}\right)$. Then $\text{Im} A = \{\left(\begin{smallmatrix} a \\ 0 \end{smallmatrix}\right) \mid a \in \mathbb{C}\}$, but $\text{Im}(A^*A) = \{\left(\begin{smallmatrix} a \\ a \end{smallmatrix}\right) \mid a \in \mathbb{C}\}$.

1.83   (a) It is routine to check that $\text{Ker} A$ is a subspace of $\mathbb{F}^n$. Write $A = (\alpha_1, \alpha_2, \ldots, \alpha_n)$, where α_i's are columns in $\mathbb{F}^m$. Then $\text{Ker} A = \{0\}$ if and only if $Ax = 0$ has the unique solution $x = 0$, i.e., $x_1\alpha_1 + x_2\alpha_2 + \cdots + x_n\alpha_n = 0$ if and only if $x_1 = x_2 = \cdots = x_n = 0$. If the columns are linearly independent, then $m \geq n$. In the case of $m > n$, the rows of A are linearly dependent. In the case of $m = n$, the rows of A are linearly independent.

(b) If $m < n$, then $\dim \text{Im} A \leq m < n$. So the columns of A are linearly dependent. Thus, $Ax = 0$ for some nonzero x. (Alternatively, use rank $r(A) \leq m < n$ to get $\dim \text{Ker} A = n - r(A) > 0$.)

(c) If $Ax = 0$, then $A^2x = 0$.

(d) $Ax = 0 \Rightarrow A^*Ax = 0 \Rightarrow x^*A^*Ax = 0 \Rightarrow Ax = 0$. So, $Ax = 0 \Leftrightarrow A^*Ax = 0$. It follows that $\text{Ker} A = \text{Ker}(A^*A)$. $\text{Ker}(AA^*)$ need not be equal to $\text{Ker} A$. Take $A = \left(\begin{smallmatrix} 1 & 1 \\ 0 & 0 \end{smallmatrix}\right)$ and $x = \left(\begin{smallmatrix} 1 \\ -1 \end{smallmatrix}\right)$. Then $x \in \text{Ker} A$ but $x \notin \text{Ker}(AA^*)$. However, $\dim \text{Ker}(AA^*) = \dim \text{Ker} A$.

(e) Let $A = BC$. Clearly, $\text{Ker} C \subseteq \text{Ker} A$. If $Ax = (BC)x = 0$ and B is invertible, then $Cx = B^{-1}(BC)x = 0$, so $\text{Ker} A \subseteq \text{Ker} C$.

1.84   Let W_1 and W_2 be subspaces of a finite dimensional vector space. If $\dim W_1 + \dim W_2 > \dim(W_1 + W_2)$, then, by the dimension identity, $W_1 \cap W_2 \neq \{0\}$. Note that $\dim W + \dim(\text{Span}\{v_{i_1}, \ldots, v_{i_m}\}) = k + m > n$. There must be a nonzero vector in $W \cap \text{Span}\{v_{i_1}, \ldots, v_{i_m}\}$.

1.85   (a) $V = \{(x, y, z) \in \mathbb{R}^3 \mid x + 2y + 3z = 0\}$ and V is closed under the addition and scalar multiplication. So V is a subspace of $\mathbb{R}^3$.

(b) The dimension of V is 2 and $\{v_1 = (2, -1, 0), v_2 = (3, 0, -1)\}$ serves as a basis (not unique).

(c) Let W be the line; it contains all triples in the form $(t, t, -t)$, $t \in \mathbb{R}$. W is a subspace with $\{w_1 = (1, 1, -1)\}$ as a basis.

(d) Let $u = (r, s, t)$ be orthogonal to V, i.e., $(x, y, z)u^t = rx + sy + tz = 0$, where $x + 2y + 3z = 0$. Simply take $u = (1, 2, 3)$. Then $v = \frac{1}{\sqrt{14}}(1, 2, 3)$ is a unit vector perpendicular to the plane V.

1.86 (a) Let $u = (x_1, x_2, x_3, x_4)^t$, $v = (y_1, y_2, y_3, y_4)^t \in W$. Then, for any scalar λ, $\lambda u + v = (\lambda x_1 + y_1, \lambda x_2 + y_2, \lambda x_3 + y_3, \lambda x_4 + y_4)^t$, and $\lambda x_3 + y_3 = \lambda(x_1 + x_2) + (y_1 + y_2) = (\lambda x_1 + y_1) + (\lambda x_2 + y_2)$ and $\lambda x_4 + y_4 = \lambda(x_1 - x_2) + (y_1 - y_2) = (\lambda x_1 + y_1) - (\lambda x_2 + y_2)$. It follows that $\lambda u + v \in W$ and thus W is a subspace of $\mathbb{C}^4$.

(b) $(1, 0, 1, 1)^t$ and $(0, 1, 1, -1)^t$ form a basis of W. $\dim W = 2$.

(c) It is sufficient to notice that $(1, 0, 1, 1)^t \in W$.

1.87 (a) Let $u \in V_\lambda \cap V_\mu$. Then $Au = \lambda u$ and $Au = \mu u$. By subtraction, we have $0 = (\lambda - \mu)u$. Since $\lambda \neq \mu$, $u = 0$.

(b) If $u, v \in V_\lambda$, then $A(u + kv) = Au + kAv = \lambda u + k\lambda v = \lambda(u + kv)$. So $u + kv \in V_\lambda$ and V_λ is a subspace. Similarly, V_μ is a subspace.

(c) Let $x \in V_\lambda$ and $y \in V_\mu$, $x \neq 0$ and $y \neq 0$. Let $ax + by = 0$. We show $a = b = 0$. Applying A, we have $a\lambda x + b\mu y = 0$. As $\lambda \neq \mu$, we may assume that $\lambda \neq 0$. So $\lambda(ax + by) = a\lambda x + b\lambda y = 0$. Thus, $b(\mu - \lambda)y = 0$, implying $b = 0$, consequently $a = 0$.

1.88 (a) By the definitions, the inclusions are nearly trivial.

(b) Take W_1 and W_2 to be the x-axis and the line $y = x$, respectively. Then $W_1 \cap W_2 = \{0\}$, while $W_1 + W_2$ is the entire xy-plane.

(c) In general, $W_1 \cup W_2$ is not a subspace (take the x- and y-axes). $W_1 \cup W_2$ is a subspace if and only if one of W_1 and W_2 is contained in the other: $W_1 \subseteq W_2$ or $W_2 \subseteq W_1$, i.e., $W_1 \cup W_2 = W_1 + W_2$.

(d) If S is a subspace containing W_1 and W_2, then every vector in the form $w_1 + w_2$, $w_1 \in W_1$, $w_2 \in W_2$, is contained in S. Thus $W_1 + W_2$ is contained in S.

(e) If L is contained in W_1 and W_2, then $L \subseteq W_1 \cap W_2$. So $W_1 \cap W_2$ contains L and it is the largest in this sense.

1.89 It is routine to check that each of U, V, and W is closed under the addition and scalar multiplication of $M_2(\mathbb{R})$ over $\mathbb{R}$. So they are subspaces of $M_2(\mathbb{R})$. Clearly, U is a subspace of V (by setting $b = 0$).

$U + W = \left\{ \begin{pmatrix} x_1 + x_2 & -(x_2 + y_2) \\ a_1 & -x_1 + y_2 \end{pmatrix} \right\} = \left\{ \begin{pmatrix} x & -y \\ z & -(x + y) \end{pmatrix} \right\}$ has three free variables. So $\dim(U + W) = 3$. $\left\{ \begin{pmatrix} 1 & 0 \\ 0 & -1 \end{pmatrix}, \begin{pmatrix} 0 & 1 \\ 0 & -1 \end{pmatrix}, \begin{pmatrix} 0 & 0 \\ 1 & 0 \end{pmatrix} \right\}$ is a basis.

$V \cap W = \left\{ \left(\begin{smallmatrix} x & 0 \\ 0 & -x \end{smallmatrix} \right) \right\}$, and $\left\{ \left(\begin{smallmatrix} 1 & 0 \\ 0 & -1 \end{smallmatrix} \right) \right\}$ is a basis.

$\left(\begin{smallmatrix} x & 0 \\ a & -x \end{smallmatrix} \right) = a \left(\begin{smallmatrix} 0 & 0 \\ 1 & 0 \end{smallmatrix} \right) + x \left(\begin{smallmatrix} 1 & 0 \\ 0 & -1 \end{smallmatrix} \right)$. So $\left\{ \left(\begin{smallmatrix} 0 & 0 \\ 1 & 0 \end{smallmatrix} \right), \left(\begin{smallmatrix} 1 & 0 \\ 0 & -1 \end{smallmatrix} \right) \right\}$ is a basis for U.

$\left\{ \left(\begin{smallmatrix} 0 & 0 \\ 1 & 0 \end{smallmatrix} \right), \left(\begin{smallmatrix} 1 & 0 \\ 0 & -1 \end{smallmatrix} \right), \left(\begin{smallmatrix} 0 & 1 \\ 0 & 0 \end{smallmatrix} \right) \right\}$ is a basis for V.

$\left\{ \left(\begin{smallmatrix} 1 & -1 \\ 0 & 0 \end{smallmatrix} \right), \left(\begin{smallmatrix} 0 & -1 \\ 0 & 1 \end{smallmatrix} \right) \right\}$ is a basis for W.

1.90 $\dim(\mathrm{Span}\{u_1, \ldots, u_m\} \cap \mathrm{Span}\{v_m, \ldots, v_n\}) \geq m + (n - m + 1) - n \geq 1$.
If the linear independence is not assumed, then the statement is false.

1.91   Since $V_1 \cap V_2 \subseteq V_1 \subseteq V_1 + V_2$, $\dim(V_1 \cap V_2) \leq \dim V_1 \leq \dim(V_1 + V_2)$.
If $\dim(V_1 \cap V_2) + 1 = \dim(V_1 + V_2)$, then either $\dim V_1 = \dim(V_1 \cap V_2)$
or $\dim V_1 = \dim(V_1 + V_2)$. The former says $V_1 = V_1 \cap V_2$. Thus
$V_1 \subseteq V_2$. The latter ensures $V_1 = V_1 + V_2$. As a result, $V_2 \subseteq V_1$.

1.92   Let $S = \cap\{U \mid W \subseteq U, U$ is a subspace with $\dim U = n - 1\}$. Then
S is a subspace of V and $W \subseteq S$. We now show that $S \subseteq W$.
Suppose otherwise $s \in S \setminus W$. Let $\{s, w_2, \ldots, w_k, v_{k+1}, \ldots, v_n\}$ be a
basis for V, where $\{w_2, \ldots, w_k\}$ is a basis for W (the w's are absent
if $W = \{0\}$). Consider $Y = \mathrm{Span}\{w_2, \ldots, w_k, v_{k+1}, \ldots, v_n\}$. Then Y
contains W and has dimension $n - 1$, but Y contains no s. By the
definition of S, $S \subseteq Y$, s cannot be in S, a contradiction.

1.93   Fact: If W_1 and W_2 are proper subspaces of V, then there exists a
vector $v \in V$ such that $v \notin W_1 \cup W_2$. (See Example 1.5.)

   If $W = \{0\}$, there is nothing to show. Suppose $1 \leq \dim W < \dim V = n$. There exists a nonzero vector, say v_1, such that $v_1 \in V \setminus W$. Let
$V_1 = \mathrm{Span}\{v_1\}$. V_1 is a proper subspace of V. By the fact, there exists
$v_2 \notin W \cup V_1$. It is readily seen that v_1 and v_2 are linearly independent.
If $n = 2$, then we are done. If $n > 2$, let $V_2 = \mathrm{Span}\{v_1, v_2\}$. By the
fact again, there exists $v_3 \notin W \cup V_2$. v_1, v_2, and v_3 are linearly
independent. In this way, if $v_1, \ldots, v_k$ are found and $k < n$, we can
find v_{k+1} by the fact. At end, linearly independent vectors $v_1, \ldots, v_n$
outside W are obtained and $v_1, \ldots, v_n$ comprise a basis of V.

   In $\mathbb{R}^3$, take $W = W_1$ to be the xy-plane (and W_2 to be the yz-plane).
Choose a basis of $\mathbb{R}^3$ such that no basis vector lies in the plane(s).

1.94   We use induction on k. The case of $k = 1$ is clearly true. Suppose
the statement is true for k subspaces. We show the case of $k + 1$. Let
$\alpha \notin W_1 \cup \cdots \cup W_k$ by hypothesis. If $\alpha \notin W_{k+1}$, then we are done.
Otherwise, $\alpha \in W_{k+1}$. Let $\beta \notin W_{k+1}$. Then any vector in the form
$s\alpha + \beta$, where s is a positive integer, is not contained in W_{k+1}.

Now consider the vectors (that are all outside W_{k+1}):

$$v_1 = \alpha + \beta, \; v_2 = 2\alpha + \beta, \; v_3 = 3\alpha + \beta, \ldots, v_{k+1} = (k+1)\alpha + \beta.$$

The above $k + 1$ vectors cannot be all contained in $W_1 \cup \cdots \cup W_k$, because, otherwise, two would lie in one of $W_1, \ldots, W_k$, say W_t. Subtracting the two vectors would imply $\alpha \in W_t$, a contradiction.

Note: The proof works for a finite field $\mathbb{F}$ of at least $k+1$ elements.

1.95   By Problem 1.94, there exists $v_1 \notin W_1 \cup \cdots \cup W_k$. Let $W_{k+1} = \mathrm{Span}\{v_1\}$. Then W_{k+1} is a proper subspace of V. By Problem 1.94 again, there exists $v_2 \notin W_1 \cup \cdots \cup W_k \cup W_{k+1}$. v_1 and v_2 are linearly independent. Let $W_{k+2} = \mathrm{Span}\{v_1, v_2\}$. If $W_{k+2} = V$, then we are done. Otherwise, W_{k+2} is a proper subspace of V. Let $v_3 \notin W_1 \cup \cdots \cup W_k \cup W_{k+2}$. Then $v_1, v_2,$ and v_3 are linearly independent. Since V has finite dimension, repeat this process until a finite set of linearly independent vectors v_i's that span V is obtained. These vectors are not contained in $W_1 \cup \cdots \cup W_k$ and form a basis of V.

1.96   For a counterexample, consider the xy-plane and take $W_1, W_2,$ and W_3 to be the x-, y-axes, and the line $y = x$, respectively. It does not contradict the set identity as the sum is usually "bigger" than the union. The former is a subspace, while the latter is not.

1.97   Use the dimension identity twice to obtain

$$\begin{aligned}\dim(S_1 \cap S_2 \cap S_3) &= \dim S_1 + \dim(S_2 \cap S_3) - \dim(S_1 + S_2 \cap S_3) \\ &= \dim S_1 + \dim S_2 + \dim S_3 \\ &\quad - \dim(S_2 + S_3) - \dim(S_1 + S_2 \cap S_3) \\ &\geq \dim S_1 + \dim S_2 + \dim S_3 - 2n.\end{aligned}$$

1.98 $\dim(S_1 \cap S_2 \cap S_3) \geq \dim S_1 + \dim S_2 + \dim S_3 - 2n = 9 - 8 = 1.$

1.99   If $\dim V = n \leq 4$, then $\dim V_1 + \dim V_2 + \dim V_3 - 2n \geq 1$. So $n = 5$. For an example, take $\mathbb{R}^5$, $V_1 = \{(x_1, x_2, x_3, 0, 0)\}$, $V_2 = \{(0, 0, y_1, y_2, y_3)\}$, and $V_3 = \{(z_1, z_2, 0, z_3, 0)\}$.

1.100   (a) $\Leftrightarrow$ (b): If (a) holds, (b) is immediate. Conversely, let $w \in W_1 + W_2$ be written as $w = w_1 + w_2 = v_1 + v_2$, where $w_1, v_1 \in W_1$ and $w_2, v_2 \in W_2$. Then $(w_1 - v_1) + (w_2 - v_2) = 0$. By (b), $w_1 - v_1 = 0$, so $w_1 = v_1$. Likewise $w_2 = v_2$. This says the decomposition of w is unique. (b) $\Leftrightarrow$ (c): If (b) holds and $w \in W_1 \cap W_2$, then $w + (-w) = 0$. By (b),

$w = 0$. If (c) holds and $w_1 + w_2 = 0$, then $w_1 = -w_2 \in W_1 \cap W_2$. By (c), $w_1 = w_2 = 0$. (c) $\Leftrightarrow$ (d): By the dimension identity.

For multiple subspaces $W_1, W_2, \ldots, W_k$, $k \geq 3$, let $W = W_1 + W_2 + \cdots + W_k$. We say that W is a *direct sum* of W_1, W_2, $\ldots$, W_k if for each $w \in W$, w can be expressed in exactly one way as a sum of vectors in $W_1, W_2, \ldots, W_k$. The following statements are equivalent:

(i) W is a direct sum of $W_1, W_2, \ldots, W_k$.

(ii) If $0 = w_1 + w_2 + \cdots + w_k$, $w_i \in W_i$, then all $w_i = 0$.

(iii) $\dim W = \dim W_1 + \dim W_2 + \cdots + \dim W_k$.

(iv) $W_i \cap \sum_{j \neq i} W_j = \{0\}$, $i = 1, 2, \ldots, k$.

1.101   If $W \neq V$, then $\dim(W) < \dim V$. Let $\{\alpha_1, \ldots, \alpha_m, \alpha_{m+1}, \ldots, \alpha_n\}$ be a basis of V, where $\alpha_1, \ldots, \alpha_m \in W$. (This is possible since one may choose a basis for W then extend it to a basis of V.) Set $W_1 = \mathrm{Span}\{\alpha_{m+1}, \ldots, \alpha_n\}$ and $W_2 = \mathrm{Span}\{\alpha_1 + \alpha_{m+1}, \alpha_{m+2}, \ldots, \alpha_n\}$. Then $W_1 \neq W_2$. One may show that $V = W \oplus W_1$ and $V = W \oplus W_2$. For example, take $V = \mathbb{R}^2$ (i.e., the xy-plane), W to be the x-axis, W_1 and W_2 to be the lines (subspaces) $x = 0$ and $y = x$, respectively. Then $\mathbb{R}^2 = W \oplus W_1 = W \oplus W_2$, $W_1 \neq W_2$.

1.102   It is sufficient to notice that when $w_i, v_i \in W_i$, $i = 1, 2, 3$, $\lambda \in \mathbb{F}$,

$$(w_1 + w_2 + w_3) + \lambda(v_1 + v_2 + v_3) = (w_1 + \lambda v_1) + (w_2 + \lambda v_2) + (w_3 + \lambda v_3)$$

belongs to $W_1 + W_2 + W_3$. For a counterexample, consider the xy-plane and take the x-axis, y-axis, and the line $y = x$.

1.103   Let f and g be even functions. Then for any $r \in \mathbb{R}$,

$$(f + rg)(-x) = f(-x) + rg(-x) = f(x) + rg(x) = (f + rg)(x),$$

that is, $f + rg \in W_1$. So W_1 is a subspace. Similarly, W_2 is a subspace. Now for any $f \in \mathcal{C}(\mathbb{R})$, we can write $f = f_e + f_o$, where

$$f_e = \tfrac{1}{2}\big(f(x) + f(-x)\big), \qquad f_o = \tfrac{1}{2}\big(f(x) - f(-x)\big).$$

Hence, $\mathcal{C}(\mathbb{R}) = W_1 + W_2$. Obviously, $W_1 \cap W_2 = \{0\}$. Thus $\mathcal{C}(\mathbb{R}) = W_1 \oplus W_2$. (Many functions like $x^2 + x^3$ are neither even nor odd.)

1.104 $P \cap Q = \{ax^2 + c \mid a, c \in \mathbb{R}\}$ and

$$P + Q = \{bx + p(x) \mid b \in \mathbb{R} \text{ and } p \text{ is an even function}\}.$$

Note that $x^2 \in P \cap Q$ and $x^3 \notin P + Q$. Moreover, $\dim(P \cap Q) = 2$ and $\dim(P + Q) = \infty$ because $\{1, x + x^2, x + x^4, \ldots, x + x^{2k}\}$ is a linearly independent set for any positive integer k. $P + Q$ is not a direct sum of P and Q, but $P + Q = L \oplus Q \subsetneq \mathcal{C}(\mathbb{R})$, where $L = \{lx \mid l \in \mathbb{R}\}$.

1.105 (a) For any real square matrix A, $A = B + C$, where $B = \frac{1}{2}(A + A^t)$ is real symmetric and $C = \frac{1}{2}(A - A^t)$ is real skew-symmetric. A matrix is real symmetric and skew-symmetric if and only if the matrix is zero. Thus, $M_2(\mathbb{R}) = V_1 \oplus V_2$. Observing that

$$\begin{pmatrix} x & y \\ u & v \end{pmatrix} = \begin{pmatrix} \frac{x-v}{2} & \frac{y+u}{2} \\ \frac{y+u}{2} & -\frac{x-v}{2} \end{pmatrix} + \begin{pmatrix} \frac{x+v}{2} & \frac{y-u}{2} \\ -\frac{y-u}{2} & \frac{x+v}{2} \end{pmatrix},$$

we have $M_2(\mathbb{R}) = W_1 + W_2$. One may check that $W_1 \cap W_2 = \{0\}$.

(b) $\left\{ \begin{pmatrix} 1 & 0 \\ 0 & 0 \end{pmatrix}, \begin{pmatrix} 0 & 0 \\ 0 & 1 \end{pmatrix}, \begin{pmatrix} 0 & 1 \\ 1 & 0 \end{pmatrix} \right\}$ is a basis for V_1. $\dim V_1 = 3$.

$\left\{ \begin{pmatrix} 0 & 1 \\ -1 & 0 \end{pmatrix} \right\}$ is a basis for V_2. $\dim V_2 = 1$.

$\left\{ \begin{pmatrix} 1 & 0 \\ 0 & -1 \end{pmatrix}, \begin{pmatrix} 0 & 1 \\ 1 & 0 \end{pmatrix} \right\}$ is a basis for W_1. $\dim W_1 = 2$.

$\left\{ \begin{pmatrix} 1 & 0 \\ 0 & 1 \end{pmatrix}, \begin{pmatrix} 0 & 1 \\ -1 & 0 \end{pmatrix} \right\}$ is a basis for W_2. $\dim W_2 = 2$.

(c) This part is nearly trivial.

(d) There are many such W_3's. Take $W_3 = \left\{ \begin{pmatrix} c & 2d \\ -d & c \end{pmatrix} \mid c, d \in \mathbb{R} \right\}$.

1.106 If $x \in \operatorname{Im} A$, then $x = Ay$ for some y, $(I - A)x = (I - A)Ay = 0$, i.e., $x \in \operatorname{Ker}(I - A)$. So, $\operatorname{Im} A \subseteq \operatorname{Ker}(I - A)$. Conversely, if $x \in \operatorname{Ker}(I - A)$, then $(I - A)x = 0$ and $x = Ax \in \operatorname{Im} A$. Thus, $\operatorname{Im} A = \operatorname{Ker}(I - A)$.

We show $\mathbb{F}^n = \operatorname{Ker} A \oplus \operatorname{Ker}(I - A)$.

Since $A^2 = A$, $A(I - A) = (I - A)A = 0$. For any $x \in \mathbb{F}^n$, we have $A(I-A)x = (I-A)Ax = 0$. This says $Ax \in \operatorname{Ker}(I-A)$ and $(I-A)x \in \operatorname{Ker} A$. Thus $x = (I - A)x + Ax \in \operatorname{Ker} A + \operatorname{Ker}(I - A)$. This reveals $\operatorname{Ker} A + \operatorname{Ker}(I - A) = \mathbb{F}^n$. For direct sum, let $y \in \operatorname{Ker} A \cap \operatorname{Ker}(I - A)$. Then $Ay = 0$ and $(I - A)y = 0$. So $y = (I - A)y + Ay = 0$.

The converse need not be true. Take $A = \begin{pmatrix} 2 & 0 \\ 0 & 0 \end{pmatrix}$. (Note that A is Hermitian.) Then $A^2 \neq A$. Compute to get $\operatorname{Im} A = \{(a, 0)^t \in \mathbb{F}^2 \mid a \in \mathbb{F}\}$ and $\operatorname{Ker} A = \{(0, b)^t \in \mathbb{F}^2 \mid b \in \mathbb{F}\}$. Obviously, $\mathbb{F}^2 = \operatorname{Im} A \oplus \operatorname{Ker} A$.

1.107 Over the real number field $\mathbb{R}$, H_n, S_n, V_n and L_n are all subspaces of $M_n(\mathbb{R})$. For every real matrix $A = (a_{ij})$, we can write

$$A = \tfrac{1}{2}(A + A^t) + \tfrac{1}{2}(A - A^t) \in H_n + S_n.$$

Thus, $M_n(\mathbb{R}) = H_n + S_n$. Since $H_n \cap S_n = \{0\}$, the sum is direct.

To see $M_n(\mathbb{R}) = V_n \oplus L_n$, let A_v be the matrix with a_{ij} in the (i, j) position for all $i < j$ and other entries zero, and let A_l be the matrix with a_{ij} in the (i, j) position for all $i \geq j$ and other entries zero. Then $A_v \in V_n$ and $A_l \in L_n$. Obviously, $A = A_v + A_l$. Thus, $M_n(\mathbb{R}) = V_n + L_n$. Since $V_n \cap L_n = \{0\}$, we arrive at $M_n(\mathbb{R}) = V_n \oplus L_n$.

Over the complex numbers $\mathbb{C}$, V_n is the set of $n \times n$ strictly upper-triangular complex matrices, and L_n is the set of $n \times n$ lower-triangular complex matrices. They are both subspaces of $M_n(\mathbb{C})$, and $M_n(\mathbb{C}) = V_n \oplus L_n$ is a direct sum. However, H_n (Hermitian matrices) and S_n (skew-Hermitian matrices) are not subspaces of $M_n(\mathbb{C})$ because they are not closed under the scalar multiplication. So a direct sum of subspaces makes no sense for them, even though every complex square matrix A can be uniquely expressed as the sum of a Hermitian matrix and a skew-Hermitian matrix: $A = \frac{1}{2}(A + A^*) + \frac{1}{2}(A - A^*)$.

Hints and Answers for Chapter 2

2.1 -48, $-48 - 12x$, $-x^{19} + x^{17} + x^{13} - x^{11}$.

2.2 $\begin{vmatrix} 1 & a & \bar{b} \\ \bar{a} & 1 & c \\ b & \bar{c} & 1 \end{vmatrix} = 1 + 2\operatorname{Re}(abc) - (|a|^2 + |b|^2 + |c|^2) > \frac{1}{4} + 2\operatorname{Re}(abc) \geq \frac{1}{4} - 2|abc| > 0$ for a, b, c in the open disk $|z| < \frac{1}{2}$.

2.3 3, 8. By expanding the $n \times n$ determinant along the first column, we obtain the recursive formula $|F_n| = |F_{n-1}| + |F_{n-2}|$. ($\{|F_n|\}$ is a so-called Fibonacci sequence.) $|F_9| = 55$.

2.4 The zero submatrix is too "big"; every expanded term contains a zero.

2.5 $(a_1 b_2 - a_2 b_1)(a_3 b_4 - a_4 b_3)$, $(a_2 a_3 - b_2 b_3)(a_1 a_4 - b_1 b_4)$, $-a_1 a_2 a_3 a_4$.

2.6 $a_1 a_2 a_3 a_4 a_5$, $-a_1 a_2 a_3 a_4 a_5 a_6$.

2.7 (a) False.

    (b) True.

    (c) True if $m = n$. False if $m \neq n$. Take $A = B = (1, 0)$.

    (d) True if $m = n$. False if $m \neq n$. Take $A = B^t = (1, 0)$.

    (e) False (unless A is real). $|A^*| = \overline{|A|}$.

    (f) True.

    (g) True if $m = n$. False if $m \neq n$. Take $A = (1, 0)$.

    (h) True for any m and n.

    (i) False.

    (j) True.

    (k) True.

    (l) False (unless $n = 1$).

    (m) True. $r(A^* A) = r(A)$.

    (n) False (unless A is real). Take $A = \left(\begin{smallmatrix} 1 & 0 \\ i & 0 \end{smallmatrix} \right)$.

    (o) True. If $(A^* A)^2 = 0$, then $A^* A = 0$, so $A = 0$.

    (p) False. Take $A = \left(\begin{smallmatrix} 0 & 1 \\ 0 & 0 \end{smallmatrix} \right)$.

    (q) True.

(r) False. Take $A = \begin{pmatrix} 0 & 1 \\ 0 & 0 \end{pmatrix}$ and $B = \begin{pmatrix} 1 & 0 \\ 0 & 0 \end{pmatrix}$.

2.8 By Vandermonde determinant, $(n-1)!(n-2)!\dots 2!$, which is equal to $(n-1)(n-2)^2 \cdots 2^{n-2}$ for $n > 2$. If $n = 1, 2$, the determinant is 1.

2.9 (a) Use induction on n. Subtract column two from column one and expand the resulting determinant along the first column:

$$\Delta_n = (p_1 - a)\Delta_{n-1} + a(p_2 - b) \cdots (p_n - b),$$

where Δ_{n-1} is the determinant with $p_2, \dots, p_n$. By induction,

$$\Delta_{n-1} = \frac{bF(a) - aF(b)}{b - a},$$

where $F(x) = (p_2 - x) \cdots (p_n - x)$. Upon simplification,

$$\Delta_n = \frac{bf(a) - af(b)}{b - a}, \quad \text{if } a \neq b.$$

(b) For the case of $a = b$, we can manipulate the above proof to get

$$
\begin{aligned}
\Delta_n &= (p_1 - a)\Delta_{n-1} + af_1(a) \\
&= (p_1 - a)\big((p_2 - a)\Delta_{n-2} + a(p_3 - a) \cdots (p_n - a)\big) + af_1(a) \\
&= (p_1 - a)(p_2 - a)\Delta_{n-2} + af_2(a) + af_1(a) \\
&= \cdots\cdots \\
&= (p_1 - a) \cdots (p_{n-2} - a)\Delta_2 + af_{n-2}(a) + \cdots + af_1(a).
\end{aligned}
$$

Then use $\Delta_2 = p_n p_{n-1} - a^2 = p_n(p_{n-1} - a) + (p_n - a)a$.

One may also compute the determinant by replacing the last column with $(a, a, \dots, a, a)^t + (0, 0, \dots, 0, p_n - a)^t$. (The determinant remains the same.) Then compute two separate determinants.

(c) $\big(a + (n-1)b\big)(a - b)^{n-1}$.

2.10 If $b = 0$, it is trivial. Assume $b \neq 0$ and $a \neq b$. Let d_n denote the $n \times n$ determinant. We prove by induction on n. For $n = 1$, it is true. Expanding along the first row, we have $d_n = (a + b)d_{n-1} - abd_{n-2}$. By induction hypothesis, $d_{n-1} = \frac{a^n - b^n}{a - b}$ and $d_{n-2} = \frac{a^{n-1} - b^{n-1}}{a - b}$. Substituting these in $d_n = (a + b)d_{n-1} - abd_{n-2}$, we obtain the formula. If $a = b$, the determinant is equal to $(n+1)a^n$.

2.11 $\lambda^n - a$.

2.12 $\det A = a_1 \cdots a_n + (-1)^{n-1} b_1 \cdots b_n$.

2.13 $|\lambda I - A| = (\lambda-1)^{n-1}\big((\lambda-1)(\lambda-a)-\alpha\beta\big)$. In particular, if $\lambda = 1, n = 1$, then $|\lambda I - A| = -\alpha\beta$; if $\lambda = 1, n \geq 2$, then $|\lambda I - A| = 0$.

For B, let $\gamma = (b_1, b_2, \ldots, b_n)$. Expand $|\lambda I - B|$ along the first column to get $|\lambda I - B| = \lambda\Delta_n - b$, where Δ_n is the same type determinant containing $b_1, \ldots, b_n$. Thus, $|\lambda I - B| = \lambda\Delta_n - b = \lambda(\lambda\Delta_{n-1} - b_1) - b = \lambda^2\Delta_{n-1} - b_1\lambda - b = \lambda^3\Delta_{n-2} - b_2\lambda^2 - b_1\lambda - b$. Inductively, we have $|\lambda I - B| = \lambda^{n-1}\Delta_2 - b_{n-2}\lambda^{n-2} - \cdots - b_2\lambda^2 - b_1\lambda - b$. Since $\Delta_2 = \lambda(\lambda - b_n) - b_{n-1}$, we obtain $|\lambda I - B| = \lambda^{n+1} - b_n\lambda^n - \cdots - b_2\lambda^2 - b_1\lambda - b$.

2.14 By direct computations. Note that $p'(0)$ is the coefficient of t.

2.15 Expand the determinant as the sum of all products of n elements (functions of t) each of which is taken from a different row and a different column. Then use the product rule for multiple functions: Let $f_1(t), \ldots, f_n(t)$ be differentiable functions. Then $\frac{d}{dt}\big(f_1(t) \cdots f_n(t)\big)$ $= \sum_{j=1}^n f_1(t) \cdots \frac{d}{dt}f_j(t) \cdots f_n(t)$.

$\frac{d}{dt}|I_n + tX| = x_{11} + \cdots + x_{nn} = \operatorname{tr} X$ (the trace of X) when $t = 0$.

2.16 Subtract the first row from other $n - 1$ rows. The resulting matrix has entries only 0 and ± 2 except the first row, i.e., each of the last $n - 1$ rows is divisible by 2. So we can factor out $n - 1$ 2's.

If n is even, and the number of 1's and the number of -1's are the same in each row, adding all columns gives a zero column. So $|A| = 0$.

2.17 $\det(A + B) = 40$, $\det C = 5$.

2.18 $\det(A - B) = \det(\alpha - \beta, \gamma, \delta) = \det(\alpha, \gamma, \delta) - \det(\beta, \gamma, \delta) = \frac{1}{6}\det A - \frac{1}{2}\det B = 3 - 1 = 2$. $\det(A + B) = \det(\alpha + \beta, 3\gamma, 5\delta) = 45 + 15 = 60$.

2.19 $(UD)^{-1} = D^{-1}U^{-1}$, where $U = (u_1, \ldots, u_n)$, $D = \operatorname{diag}(a_1, \ldots, a_n)$.

2.20 Write the complex solution as $x = x_1 + ix_2$, where x_1 and x_2 are real, not both zero. Then $Ax_1 = 0$ and $Ax_2 = 0$.

2.21 First note that the inverse of a nonsingular rational matrix is rational. This is readily seen by using the cofactor (adjugate) matrix.

If $A = 0$, then there is nothing to prove. Let $A \neq 0$. By row and column operations on A (only invoking rational numbers), we can bring A to $D = \begin{pmatrix} I_r & 0 \\ 0 & 0 \end{pmatrix}$, $1 \leq r \leq n$, equivalently, $A = PDQ$ for some nonsingular rational matrices P and Q. So, $Ax = 0$ is the same as

$PDQx = 0$, which has the same solutions as $Dy = 0$, where $y = Qx$. Since $Ax = 0$ has a nonzero solution x in $\mathbb{C}^n$, $Dy = 0$ has a nonzero solution $y = Qx$ in $\mathbb{C}^n$. It follows that $r < n$ and $Dz = 0$ for any $z = (0, \ldots, 0, z_{r+1}, \ldots, z_n)^t$ in $\mathbb{Q}^n$. As a result, $(PDQ)(Q^{-1}z) = 0$, that is, $Aw = 0$, where $w = Q^{-1}z$ is nonzero in $\mathbb{Q}^n$ for $z \neq 0$.

(Note: For $Ax = b$, where A and b are rational, considering the ranks of A and (A, b) and modifying the above proof a bit, we see if $Ax = b$ has a solution in $\mathbb{C}^n$, then it has a solution in $\mathbb{Q}^n$.)

Another proof. Let $Ax = 0$, where $x = (x_1, \ldots, x_n)^t \in \mathbb{C}^n$. Let $V = \text{Span}\{x_1, \ldots, x_n\}$ over $\mathbb{Q}$. Then $\dim V \leq n$. Let $\dim V = k$ and let $\{y_1, \ldots, y_k\}$ be a basis of V. Then each x_i is a linear combination of $y_1, \ldots, y_k$ with rational coefficients. Thus, we can write $x = Qy$, where $y = (y_1, \ldots, y_k)^t$ and Q is a nonzero $n \times k$ rational matrix. So, $Ax = 0$ gives $AQy = 0$. Since $y_1, \ldots, y_k$ are linearly independent (over $\mathbb{Q}$), $AQ = 0$. Hence, $A\alpha = 0$ for every nonzero column α of Q.

2.22 (a) $\text{Span}\{\alpha_1, \alpha_2, \ldots, \alpha_m\} = \text{Im}\, A$ and $\dim \text{Im}\, A = r(A)$.

(b) This is immediate from (a).

(c) Because $r(A) \leq n < m$.

(d) $|A| \neq 0 \Leftrightarrow r(A) = m \Leftrightarrow A$'s columns are linearly independent.

2.23 We may write $(\alpha_1, \alpha_2, \ldots, \alpha_m) = (\beta_1, \beta_2, \ldots, \beta_n)A^t$. Let A_k consist of rows $i_1, \ldots, i_k$ of A. A_k contains a $k \times k$ nonsingular submatrix if and only if the rows of A_k are linearly independent, i.e., $r(A_k) = k$. Since $(\alpha_{i_1}, \ldots, \alpha_{i_k}) = (\beta_1, \beta_2, \ldots, \beta_n)A_k^t$, $\alpha_{i_1}, \ldots, \alpha_{i_k}$ are linearly independent $\Leftrightarrow (\alpha_{i_1}, \ldots, \alpha_{i_k})x = 0$ only if $x = 0 \Leftrightarrow (\beta_1, \beta_2, \ldots, \beta_n)A_k^t x = 0$ only if $x = 0 \Leftrightarrow A_k^t x = 0$ only if $x = 0 \Leftrightarrow r(A_k) = k$.

So the number of vectors in a basis of $\text{Span}\{a_1, a_2, \ldots, a_m\}$ is $r(A)$.

2.24 (a) Because each α_i is contained in $\text{Span}\,\beta$, we see that $\text{Span}\,\alpha$ is a subspace of $\text{Span}\,\beta$. So the dimension inequality follows.

(b) Denote $B = A^t$ for convenience and write $(\alpha_1, \alpha_2, \ldots, \alpha_m) = (\beta_1, \beta_2, \ldots, \beta_n)B$. Let $\dim(\text{Span}\,\alpha) = s$ and, without loss of generality, let $\alpha_1, \alpha_2, \ldots, \alpha_s$ be a basis of $\text{Span}\,\alpha$. Let B_s consist of the first s columns of B. So, $(\alpha_1, \alpha_2, \ldots, \alpha_s) = (\beta_1, \beta_2, \ldots, \beta_n)B_s$. As $(\alpha_1, \alpha_2, \ldots, \alpha_s)x = (\beta_1, \beta_2, \ldots, \beta_n)B_s x$, we see $B_s x = 0$ has no nonzero solutions. Thus, the columns of B_s are linearly independent and $s = r(B_s) \leq r(B) = r(A)$.

Note: Let r_i be the i-th row of A, $i = 1, 2, \ldots, m$. The above argument shows if some α_i's are linearly independent, then the corresponding r_i's are linearly independent.

(c) This is Problem 2.23.

(d) By (b), $r(A) = m$; by (a), $m \le n$.

2.25 (a) $|A| = |A^t| = |-A| = (-1)^n |A| = -|A|$ if n is odd. So $|A| = 0$.

(b) $|A|^2 = |A^2| = |-I| = (-1)^n$. If n is odd, then $|A|^2 = -1$. This is impossible when A is a real matrix. So n is even.

(c) (a) is true. (b) is not: Take $A = iI_3$. Then $A^2 + I_3 = 0$.

2.26 $|A+I| = |A + AA^t| = |A|\,|I + A^t| = |A|\,|A+I|$. $|A+I| = 0$ as $|A| < 0$.

2.27 Note that $XY = I$ implies $YX = I$ when X and Y are square. $D^{-1}C^{-1}B^{-1} = A$. $BCAD \ne I$ in general unless $AD = DA$.

2.28 Since Span$\{I, A, A^2, \ldots\}$ is a subspace of $M_n(\mathbb{C})$, its dimension is finite, say m. Then $I, A, A^2, \ldots, A^m$ are linearly dependent. Thus, there are scalars $a_0, a_1, \ldots, a_m$, not all zero, such that $a_0 I + a_1 A + a_2 A^2 + \cdots + a_m A^m = 0$. Multiplying by A^{-1} reveals $a_0 A^{-1} + a_1 I + a_2 A + \cdots + a_m A^{m-1} = 0$. If $a_0 \ne 0$, then $A^{-1} = b_1 I + b_2 A + \cdots + b_m A^{m-1}$, where $b_i = -a_i/a_0$, $i = 1, 2, \ldots, m$. If $a_0 = 0$, then $a_1 I + a_2 A + \cdots + a_m A^{m-1} = 0$, we can repeat the above process.

Note: One may also use the characteristic polynomial and the Cayley–Hamilton theorem (see Chapter 3) to prove the statement.

2.29 Let $B = A^2 - 2A + 2I$. Then $B = A^2 - 2A + A^3 = A(A^2 + A - 2I) = A(A + 2I)(A - I)$. However, $I = A^3 - I = (A - I)(A^2 + A + I)$. So $|A - I| \ne 0$. $A^3 + 8I = 10I$, also $(A + 2I)(A^2 - 4A + 4I) = 10I$. So $|A + 2I| \ne 0$. Thus, $|B| = |A|\,|A + 2I|\,|A - I| \ne 0$ and B is invertible. (Easier way: show that B has no zero eigenvalues; see Chapter 3.)

2.30 Let $A = \left(\begin{smallmatrix} 0 & 1 \\ 0 & 0 \end{smallmatrix}\right)$ and $B = \left(\begin{smallmatrix} 0 & 0 \\ 1 & 0 \end{smallmatrix}\right)$. For 3×3 matrix X, if $X^2 = 0$, then the columns of X are solutions to $Xy = 0$. Since $\dim \operatorname{Ker} X + r(X) = 3$, we see $r(X) + r(X) \le 3$. Thus, $r(X) \le 1$. If $A^2 = B^2 = 0$, then $r(A) \le 1$, $r(B) \le 1$, and $r(AB + BA) \le 2$. Thus, $AB + BA \ne I_3$.

2.31 The inverses are $\left(\begin{smallmatrix} -2 & 1 \\ \frac{3}{2} & -\frac{1}{2} \end{smallmatrix}\right)$, $\frac{1}{5}\left(\begin{smallmatrix} -3 & 2 \\ 4 & -1 \end{smallmatrix}\right)$, and $\left(\begin{smallmatrix} \cos\theta & -\sin\theta \\ \sin\theta & \cos\theta \end{smallmatrix}\right)$, respectively.

2.32 $\begin{vmatrix} a - \frac{|A|}{|B|} & x \\ y & B \end{vmatrix} = \begin{vmatrix} a & x \\ y & B \end{vmatrix} - \begin{vmatrix} \frac{|A|}{|B|} & x \\ 0 & B \end{vmatrix} = |A| - |A| = 0.$

2.33 $A^{-1} = \begin{pmatrix} 1 & 0 & 0 \\ 0 & 1 & 0 \\ -x & -y & 1 \end{pmatrix}$, $B^{-1} = \begin{pmatrix} -b/a & -c/a & -d/a & 1/a \\ 1 & 0 & 0 & 0 \\ 0 & 1 & 0 & 0 \\ 0 & 0 & 1 & 0 \end{pmatrix}$.

2.34 $A^{-1} = \begin{pmatrix} 1 & -1 & 0 \\ 0 & 1 & -1 \\ 0 & 0 & 1 \end{pmatrix}$, $B^{-1} = \begin{pmatrix} 1 & -1 & 0 & \cdots & 0 & 0 \\ 0 & 1 & -1 & \cdots & 0 & 0 \\ 0 & 0 & 1 & \ddots & 0 & 0 \\ \vdots & \vdots & \vdots & \ddots & \ddots & \vdots \\ 0 & 0 & 0 & \cdots & 1 & -1 \\ 0 & 0 & 0 & \cdots & 0 & 1 \end{pmatrix}$.

2.35 $\det A = (-1)^{n-1}(n-1)$. The inverse of A is the matrix with $\frac{2-n}{n-1}$ on the main diagonal and $\frac{1}{n-1}$ as off-diagonal entries. Alternatively, let $A = J - I$, where J is the all-ones matrix. Then $(J-I)(\frac{1}{n-1}J - I) = I$. Thus, $A^{-1} = (J-I)^{-1} = \frac{1}{n-1}J - I$. For B, $\det B = 1$ and

$$B^{-1} = \begin{pmatrix} 2 & -1 & 0 & \cdots & 0 & 0 \\ -1 & 2 & -1 & \cdots & 0 & 0 \\ 0 & -1 & 2 & \ddots & 0 & 0 \\ \vdots & \vdots & \ddots & \ddots & \ddots & \vdots \\ 0 & 0 & 0 & \ddots & 2 & -1 \\ 0 & 0 & 0 & \cdots & -1 & 1 \end{pmatrix}.$$

2.36 A computation reveals $Q^t Q = \mathrm{diag}(4, 2, 6, 12)$; it is diagonal. $(Q^t Q)^2 = \mathrm{diag}(16, 4, 36, 144)$. QQ^t is not diagonal. $(QQ^t)^2 = Q(Q^t Q)Q^t$, and

$$QQ^t = \begin{pmatrix} 4 & 2 & 0 & -2 \\ 2 & 4 & 0 & -2 \\ 0 & 0 & 6 & -2 \\ -2 & -2 & -2 & 10 \end{pmatrix}, \ (QQ^t)^2 = \begin{pmatrix} 24 & 20 & 4 & -32 \\ 20 & 24 & 4 & -32 \\ 4 & 4 & 40 & -32 \\ -32 & -32 & -32 & 112 \end{pmatrix}.$$

2.37 $A^{-1} = \begin{pmatrix} 0 & 0 & \cdots & 0 & \frac{1}{a_n} \\ \frac{1}{a_1} & 0 & \cdots & 0 & 0 \\ 0 & \frac{1}{a_2} & \cdots & 0 & 0 \\ \vdots & \vdots & \ddots & \ddots & \vdots \\ 0 & 0 & \cdots & \frac{1}{a_{n-1}} & 0 \end{pmatrix}$.

2.38 4.

2.39 $(I_n + xx^*)^{-1} = I_n - \frac{1}{2}xx^*$ because $(I_n + xx^*)(I_n - \frac{1}{2}xx^*) = I_n$. See Problem 2.62c for a more general formula.

2.40   Verify directly that $\left(\begin{smallmatrix} I_m & X \\ 0 & I_n \end{smallmatrix}\right)\left(\begin{smallmatrix} I_m & Y \\ 0 & I_n \end{smallmatrix}\right) = \left(\begin{smallmatrix} I_m & X+Y \\ 0 & I_n \end{smallmatrix}\right) = \left(\begin{smallmatrix} I_m & Y \\ 0 & I_n \end{smallmatrix}\right)\left(\begin{smallmatrix} I_m & X \\ 0 & I_n \end{smallmatrix}\right)$.

2.41   Use the fact that $\left(\begin{smallmatrix} I_m & A \\ 0 & I_n \end{smallmatrix}\right)\left(\begin{smallmatrix} I_m & B \\ 0 & I_n \end{smallmatrix}\right) = \left(\begin{smallmatrix} I_m & A+B \\ 0 & I_n \end{smallmatrix}\right)$. We have

$$\prod_{i,j}\begin{pmatrix} I_m & X_{ij} \\ 0 & I_n \end{pmatrix} = \begin{pmatrix} I_m & \sum_{i,j} X_{ij} \\ 0 & I_n \end{pmatrix} = \begin{pmatrix} I_m & X \\ 0 & I_n \end{pmatrix}.$$

Alternatively, view $\begin{pmatrix} I_m & X_{ij} \\ 0 & I_n \end{pmatrix}$ as an elementary operation (type Ⅲ).

2.42   Perform elementary operations on rows:

$$\begin{pmatrix} I & A & I & 0 \\ B & I & 0 & I \end{pmatrix}$$

$$\rightarrow \begin{pmatrix} I & A & I & 0 \\ 0 & I-BA & -B & I \end{pmatrix}$$

$$\rightarrow \begin{pmatrix} I & A & I & 0 \\ 0 & I & -(I-BA)^{-1}B & (I-BA)^{-1} \end{pmatrix}$$

$$\rightarrow \begin{pmatrix} I & 0 & I+A(I-BA)^{-1}B & -A(I-BA)^{-1} \\ 0 & I & -(I-BA)^{-1}B & (I-BA)^{-1} \end{pmatrix}.$$

The first "$\rightarrow$" shows $\left(\begin{smallmatrix} I & 0 \\ -B & I \end{smallmatrix}\right)\left(\begin{smallmatrix} I & A \\ B & I \end{smallmatrix}\right) = \left(\begin{smallmatrix} I & A \\ 0 & I-BA \end{smallmatrix}\right)$. Thus $\left(\begin{smallmatrix} I & A \\ B & I \end{smallmatrix}\right)$ is invertible if and only if $I - BA$ is invertible. The last "$\rightarrow$" reveals

$$\begin{pmatrix} I & A \\ B & I \end{pmatrix}^{-1} = \begin{pmatrix} I+A(I-BA)^{-1}B & -A(I-BA)^{-1} \\ -(I-BA)^{-1}B & (I-BA)^{-1} \end{pmatrix}.$$

2.43 $\begin{pmatrix} A^{-1} & -A^{-1}BC^{-1} \\ 0 & C^{-1} \end{pmatrix}$, $\begin{pmatrix} I & -X & XZ-Y \\ 0 & I & -Z \\ 0 & 0 & I \end{pmatrix}$.

2.44   Use $V^{-1} = |V|^{-1}\operatorname{adj}(V)$. (Note: V is a 3×3 Vandermonde matrix.)

$$V^{-1} = \frac{1}{|V|}\begin{pmatrix} a_3a_2(a_3-a_2) & -(a_3^2-a_2^2) & a_3-a_2 \\ -a_3a_1(a_3-a_1) & a_3^2-a_1^2 & -(a_3-a_1) \\ a_2a_1(a_2-a_1) & -(a_2^2-a_1^2) & a_2-a_1 \end{pmatrix}.$$

2.45 $r(N) = 2$. $N^3 = 0$. Note that N is symmetric.

2.46 $M^2 = A(A^*A)^{-1}A^*A(A^*A)^{-1}A^* = A(A^*A)^{-1}A^* = M$, and
$M^* = \left(A(A^*A)^{-1}A^*\right)^* = A\left((A^*A)^{-1}\right)^*A^* = A(A^*A)^{-1}A^* = M.$

2.47   Check directly that $M^{-1}M = I$. The identity is due to $MM^{-1} = I$.

2.48   Verify the equivalent identity:

$$(A + iB)\big((A + BA^{-1}B)^{-1} - i(B + AB^{-1}A)^{-1}\big) = I.$$

Post-multiply by $B + AB^{-1}A$ to get

$$(A + iB)\big((A + BA^{-1}B)^{-1}(B + AB^{-1}A) - iI\big) = B + AB^{-1}A.$$

Since $(A + BA^{-1}B)^{-1}(B + AB^{-1}A) = B^{-1}A$, we need to show

$$(A + iB)\big(B^{-1}A - iI\big) = B + AB^{-1}A,$$

which is readily seen by a computation.

2.49   Verify that $(A - B)\big(A^{-1} + A^{-1}(B^{-1} - A^{-1})^{-1}A^{-1}\big) = I$.

(a) Set $A = I$ and replace B by BA. (b) Set $A = I$ and replace B by $-A$. (c) By (b), change A to $-A$. (d) By (a), change B to $-A^*$.

2.50   With AB and CD being Hermitian, we verify that

$$\begin{pmatrix} A & B^* \\ C & D^* \end{pmatrix} \begin{pmatrix} D & -B \\ -C^* & A^* \end{pmatrix} = \begin{pmatrix} I & 0 \\ 0 & I \end{pmatrix}.$$

It follows that

$$\begin{pmatrix} D & -B \\ -C^* & A^* \end{pmatrix} \begin{pmatrix} A & B^* \\ C & D^* \end{pmatrix} = \begin{pmatrix} I & 0 \\ 0 & I \end{pmatrix},$$

which implies $DA - BC = I$ and $A^*C - C^*A = 0$.

2.51 $(A+iB)(A-iB) = A^2 + B^2 + i(BA - AB) = (1+i)(BA - AB)$. Thus, $\det\big((A+iB)(A-iB)\big) = \det(A+iB)\det(A-iB) = |\det(A+iB)|^2 = (1+i)^n \det(BA - AB)$. Since $n = 2$ or 3, $(1+i)^n$ is nonreal. So $\det(A + iB) = 0$ and $\det(A^2 + B^2) = \det(BA - AB) = 0$.

2.52   (a) If $A^*KA = K$, then A is nonsingular and $K = (A^{-1})^*KA^{-1}$. So $A^{-1} \in S_K$. By taking conjugate for both sides of $A^*KA = K$, we see $(\bar{A})^*K\bar{A} = K$. So $\bar{A} \in S_K$. From $(A^{-1})^*KA^{-1} = K$, taking inverses of both sides gives $AK^{-1}A^* = K^{-1}$. Note that $K^{-1} = K$. Thus, $A^* \in S_K$. Consequently, $A^t = (\bar{A})^* \in S_K$.

(b)   Since $(AB)^*K(AB) = B^*A^*KAB = B^*KB = K$, $AB \in S_K$. But kA (unless $|k| = 1$) and $A + B$ are not in S_K in general.

(c) (a) and (b) hold. In fact, they hold for any real K, $K^2 = \pm I$.

2.53   (a) By the Laplace expansion theorem.

     (b) 1, $(-1)^{mn}$, 1.

     (c) $(-1)^{mn}|A||C|$.

2.54   Direct computation yields $S^2 = I$. So $S^{-1} = S$. It is obvious that $S^t = S$. $|S| = (-1)^{\frac{n(n-1)}{2}}$. The (i, j)-entry of SAS is $a_{n-i+1,n-j+1}$.

2.55   Notice that

$$
\begin{pmatrix} 0 & I_n \\ I_m & 0 \end{pmatrix} \begin{pmatrix} A & B \\ C & D \end{pmatrix} \begin{pmatrix} 0 & I_p \\ I_q & 0 \end{pmatrix} = \begin{pmatrix} D & C \\ B & A \end{pmatrix}.
$$

Taking the determinants of both sides gives the identity. When A, B, C, D are all square, say $m \times m$, $mn + pq = 2m^2$ is an even number. So $(-1)^{(mn+pq)} = 1$. For the case of column and row vectors, $m = p$ and $n = q = 1$. Thus $mn + pq = 2m$ is also even. The identity holds. When B and C are switched, the two determinants need not be equal. Take $A = \begin{pmatrix} 1 & 0 \\ 0 & 0 \end{pmatrix}$, $B = \begin{pmatrix} 0 & 0 \\ 1 & 0 \end{pmatrix}$, $C = \begin{pmatrix} 0 & 1 \\ 0 & 0 \end{pmatrix}$, $D = \begin{pmatrix} 0 & 0 \\ 0 & 1 \end{pmatrix}$, $\begin{vmatrix} A & B \\ C & D \end{vmatrix} \neq \begin{vmatrix} A & C \\ B & D \end{vmatrix}$.

2.56   (a) If A^{-1} exists, then

$$
\begin{pmatrix} I & 0 \\ -CA^{-1} & I \end{pmatrix} \begin{pmatrix} A & B \\ C & D \end{pmatrix} = \begin{pmatrix} A & B \\ 0 & D - CA^{-1}B \end{pmatrix}.
$$

By taking determinant,

$$
\begin{vmatrix} A & B \\ C & D \end{vmatrix} = |A|\,|D - CA^{-1}B|.
$$

(b) Suppose $AC = CA$. If A^{-1} exists,

$$
|A|\,|D - CA^{-1}B| = |AD - ACA^{-1}B| = |AD - CB|.
$$

If A is not invertible, we take a positive number δ such that $|A + \epsilon I| \neq 0$ for every ϵ, $0 < \epsilon < \delta$. Since $A + \epsilon I$ and C commute,

$$
\begin{vmatrix} A + \epsilon I & B \\ C & D \end{vmatrix} = |(A + \epsilon I)D - CB|.
$$

Note that both sides are continuous functions of ϵ. Letting $\epsilon \to 0^+$ results in the desired result for the case where A is singular.

(c) No. Take $A, B, C,$ and D to be, respectively,

$$\begin{pmatrix} 1 & -1 \\ 0 & 0 \end{pmatrix}, \begin{pmatrix} 1 & 1 \\ 1 & 1 \end{pmatrix}, \begin{pmatrix} 1 & -1 \\ 0 & 0 \end{pmatrix}, \begin{pmatrix} 1 & 0 \\ 0 & 0 \end{pmatrix}.$$

Then $AC = CA,$ $|AD - CB| = 0,$ but $|AD - BC| = 1.$

(d) No. Take $A, B, C,$ and D to be, respectively,

$$\begin{pmatrix} 1 & -1 \\ 0 & 0 \end{pmatrix}, \begin{pmatrix} 1 & 0 \\ -1 & 1 \end{pmatrix}, \begin{pmatrix} 1 & 1 \\ 1 & 1 \end{pmatrix}, \begin{pmatrix} 1 & 0 \\ 0 & 1 \end{pmatrix}.$$

Note that D commutes with the other three matrices.

2.57 It suffices to show that $|A|\,|D| - |B|\,|C| = 0.$ If A is invertible, then

$$\begin{pmatrix} I & 0 \\ -CA^{-1} & I \end{pmatrix} \begin{pmatrix} A & B \\ C & D \end{pmatrix} = \begin{pmatrix} A & B \\ 0 & D - CA^{-1}B \end{pmatrix}.$$

Since A has rank n when A^{-1} exists, we obtain

$$D - CA^{-1}B = 0, \text{ or } D = CA^{-1}B.$$

Thus, $|A|\,|D| = |A|\,|CA^{-1}B| = |A|\,|C|\,|A^{-1}|\,|B| = |B|\,|C|.$

If $|A| = 0,$ it must be shown that $|B| = 0$ or $|C| = 0.$ Suppose otherwise that B (or C in a similar way) is invertible. Then we have

$$\begin{pmatrix} I & 0 \\ -DB^{-1} & I \end{pmatrix} \begin{pmatrix} A & B \\ C & D \end{pmatrix} = \begin{pmatrix} A & B \\ C - DB^{-1}A & 0 \end{pmatrix}.$$

Since B is of rank $n,$

$$C - DB^{-1}A = 0, \text{ or } C = DB^{-1}A$$

and

$$|C| = |DB^{-1}A| = |D|\,|B^{-1}|\,|A| = 0.$$

2.58 (a) Note that

$$\begin{pmatrix} A & B \\ C & D \end{pmatrix} \begin{pmatrix} D^t & 0 \\ -C^t & I \end{pmatrix} = \begin{pmatrix} AD^t - BC^t & B \\ CD^t - DC^t & D \end{pmatrix}.$$

Using $CD^t - DC^t = 0$ and taking determinants, we have

$$\begin{vmatrix} A & B \\ C & D \end{vmatrix} |D^t| = |AD^t - BC^t|\,|D|.$$

If D is nonsingular, then the conclusion follows immediately by dividing both sides by $|D|$. Now suppose $|D| = 0$. If $C^t = C$,

$$C(D + \epsilon I)^t = (D + \epsilon I)C^t,$$

where $\epsilon > 0$. Using $D + \epsilon I$ for D in the above argument, we have

$$\begin{vmatrix} A & B \\ C & D + \epsilon I \end{vmatrix} = |A(D + \epsilon I)^t - BC^t|$$

if $D + \epsilon I$ is nonsingular. Note that both sides of the above identity are continuous functions of ϵ and there is a finite number of ϵ for which $|D + \epsilon I| = 0$. Thus, for some $\delta > 0$, $D + \epsilon I$ is nonsingular for $\epsilon \in (0, \delta)$. Letting $\epsilon \to 0^+$ yields the desired identity.

For a general $C \neq 0$, let P and Q be invertible matrices such that $PCQ = \begin{pmatrix} I_r & 0 \\ 0 & 0 \end{pmatrix}$, where r is the rank of C, $r > 0$. Consider

$$\begin{pmatrix} I & 0 \\ 0 & P \end{pmatrix} \begin{pmatrix} A & B \\ C & D \end{pmatrix} \begin{pmatrix} Q & 0 \\ 0 & (Q^t)^{-1} \end{pmatrix} = \begin{pmatrix} AQ & B(Q^t)^{-1} \\ \tilde{C} & \tilde{D} \end{pmatrix},$$

where $\tilde{C} = PCQ$, $\tilde{D} = PD(Q^t)^{-1}$. Note that $\tilde{C}\tilde{D}^t = \tilde{D}\tilde{C}^t$ and $\tilde{C} = PCQ$ is symmetric. The above discussion gives

$$|P||M| = |AQ\tilde{D}^t - B(Q^t)^{-1}\tilde{C}| = |AD^t - BC^t||P^t|.$$

Dividing both sides by $|P| = |P^t|$ reveals the desired identity.

(b) Recall the fact that a square matrix and its transpose have the same determinant. Take the determinants of both sides of

$$\begin{pmatrix} A & B \\ C & D \end{pmatrix} \begin{pmatrix} D^t & B^t \\ C^t & A^t \end{pmatrix} = \begin{pmatrix} AD^t + BC^t & AB^t + BA^t \\ 0 & CB^t + DA^t \end{pmatrix}.$$

(c) Since $|D| \neq 0$, we see $|M| = |AD^t + BC^t|$ by taking determinants:

$$\begin{pmatrix} A & B \\ C & D \end{pmatrix} \begin{pmatrix} D^t & 0 \\ C^t & I \end{pmatrix} = \begin{pmatrix} AD^t + BC^t & B \\ 0 & D \end{pmatrix}.$$

(d) Take A, B, C, D to be, respectively,

$$\begin{pmatrix} 1 & 0 \\ 0 & 0 \end{pmatrix}, \begin{pmatrix} 0 & 0 \\ 0 & 1 \end{pmatrix}, \begin{pmatrix} 0 & 1 \\ 0 & 0 \end{pmatrix}, \begin{pmatrix} 0 & 0 \\ 1 & 0 \end{pmatrix}.$$

Then $CD^t + DC^t = 0$, $|D| = 0$, $|M| = 1$. $|AD^t + BC^t| = -1$.

2.59   (a) No. Take $A = \begin{pmatrix} 1 & 0 \\ 0 & 0 \end{pmatrix}$ and $B = \begin{pmatrix} 0 & 0 \\ 0 & 1 \end{pmatrix}$.

(b) It must be shown that there exist real numbers k_1, k_2, k_3, k_4, not all zero, such that $k_1 B_1 + k_2 B_2 + k_3 B_3 + k_4 B_4 = 0$. Let

$$A = \begin{pmatrix} a & b \\ c & d \end{pmatrix}, \quad B_i = \begin{pmatrix} w_i & x_i \\ y_i & z_i \end{pmatrix}, \quad i = 1, 2, 3, 4.$$

Then $|A + B_i| = |A| + |B_i|$ leads to

$$dw_i - cx_i - by_i + az_i = 0, \quad i = 1, 2, 3, 4.$$

Consider the linear equation of four unknowns w, x, y, z:

$$dw - cx - by + az = 0.$$

Since $A \neq 0$, say, $d \neq 0$, there are three free variables, and the solution space of the equation is of dimension 3. Thus, any four vectors are linearly dependent; in particular, $B_1, B_2, B_3,$ and B_4 as solutions to the equation are linearly dependent.

2.60   For (a), $MM^{-1} = I$ implies $AY + BV = 0$ and $CY + DV = I$. Thus

$$\begin{pmatrix} A & B \\ C & D \end{pmatrix} \begin{pmatrix} I & 0 \\ 0 & V \end{pmatrix} = \begin{pmatrix} A & BV \\ C & DV \end{pmatrix} = \begin{pmatrix} A & -AY \\ C & I - CY \end{pmatrix}.$$

Taking determinants ($| \cdot |$ for det to save space) of both sides, we get

$$|M|\,|V| = \begin{vmatrix} A & -AY \\ C & I - CY \end{vmatrix} = \begin{vmatrix} A & 0 \\ C & I \end{vmatrix} \begin{vmatrix} I & -Y \\ 0 & I \end{vmatrix} = \begin{vmatrix} A & 0 \\ C & I \end{vmatrix} = |A|.$$

For (b), apply row-block operations to (M, I) to get (I, M^{-1}). The upper-left corner of this M^{-1} is $(D - CA^{-1}B)^{-1}$.

For (c), note that $W^{-1} = W^*$ and the absolute value of det W is 1.

For (d), if M is real orthogonal, then $|A| = |V|$ or $|A| = -|V|$.

2.61   To obtain the coefficient of $x_i x_j \cdots x_k$, deleting the rows and columns $i < j < \cdots < k$ results in a submatrix of A. Set all x values in the submatrix to zero and get the determinant of the submatrix of A.

This formula reveals the expansion of the (characteristic) polynomial

$$|\lambda I - A| = \lambda^n - (\operatorname{tr} A)\lambda^{n-1} + \delta_2 \lambda^{n-2} + \cdots + (-1)^k \delta_k \lambda^{n-k} + \cdots + (-1)^n |A|,$$

where δ_k is the sum of all $k \times k$ principal minors of A, $k = 1, 2, \ldots, n$.

2.62 (a) follows by observing

$$\begin{vmatrix} I & x \\ y^* & 1 \end{vmatrix} = \begin{vmatrix} I & x \\ 0 & 1 - y^*x \end{vmatrix} = 1 - y^*x$$

and

$$\begin{vmatrix} I & x \\ y^* & 1 \end{vmatrix} = \begin{vmatrix} I - xy^* & 0 \\ y^* & 1 \end{vmatrix} = |I - xy^*|.$$

For (b), with the proof of (a), it is sufficient to notice that

$$\begin{pmatrix} 0 & 1 \\ I & 0 \end{pmatrix} \begin{pmatrix} I & x \\ y^* & 1 \end{pmatrix} \begin{pmatrix} 0 & I \\ 1 & 0 \end{pmatrix} = \begin{pmatrix} 1 & y^* \\ x & I \end{pmatrix}.$$

(c) can be verified directly through multiplications.

2.63 (a) $f(0) = \begin{pmatrix} 0 & 0 \\ 0 & 0 \end{pmatrix}$, $f(1) = \begin{pmatrix} 1 & 0 \\ 0 & 1 \end{pmatrix}$, $f(i) = \begin{pmatrix} 0 & 1 \\ -1 & 0 \end{pmatrix}$, $f(1+i) = \begin{pmatrix} 1 & 1 \\ -1 & 1 \end{pmatrix}$,

$\varphi(1,i) = \begin{pmatrix} 1 & i \\ i & 1 \end{pmatrix}$, and $\varphi(i,1) = \begin{pmatrix} i & 1 \\ -1 & -i \end{pmatrix}$.

(b) $f(\bar{z}) = f(x - iy) = \begin{pmatrix} x & -y \\ y & x \end{pmatrix} = (f(z))^t.$

(c) Let $z_1 = x_1 + iy_1$ and $z_2 = x_2 + iy_2$. Then $z_1 z_2 = (x_1 x_2 - y_1 y_2) + (x_1 y_2 + y_1 x_2)i$ and $f(z_1 z_2) = \begin{pmatrix} x_1 x_2 - y_1 y_2 & x_1 y_2 + y_1 x_2 \\ -(x_1 y_2 + y_1 x_2) & x_1 x_2 - y_1 y_2 \end{pmatrix} = \begin{pmatrix} x_1 & y_1 \\ -y_1 & x_1 \end{pmatrix} \begin{pmatrix} x_2 & y_2 \\ -y_2 & x_2 \end{pmatrix} = f(z_1)f(z_2).$

(d) Note that if k is real, then $k(x + yi) = kx + (ky)i$, where kx and ky are real. Thus, $f(kz) = \begin{pmatrix} kx & ky \\ -ky & kx \end{pmatrix} = k \begin{pmatrix} x & y \\ -y & x \end{pmatrix} = kf(z).$

(e) Let $z_1 = x_1 + iy_1$ and $z_2 = x_2 + iy_2$. Compute $f(z_1 + z_2)$.

(f) Use (c) and the fact $z_1 z_2 = z_2 z_1$.

(g) Note that $I_2 = f(1) = f(z \cdot z^{-1}) = f(z)f(z^{-1})$.

(h) $(f(z))^n = f(z^n) = r^n \begin{pmatrix} \cos n\theta & \sin n\theta \\ -\sin n\theta & \cos n\theta \end{pmatrix}$ as $z^n = r^n(\cos n\theta + i \sin n\theta)$.

(i) It is sufficient to note that $z^{-1} = \frac{1}{x^2+y^2}(x - iy)$ if $z = x + iy \neq 0$.

(j) Verify by a direct computation.

(k) Compute $f(z)$ and $\varphi(q)$ with $z = x + iy$ and $q = (x, y)$.

(l) $\det(\varphi(q)) = |u|^2 + |v|^2 \geq 0$. $(\varphi(q))^{-1} = \begin{pmatrix} \bar{u} & -v \\ \bar{v} & u \end{pmatrix}$ if $|u|^2 + |v|^2 = 1$.

(m) Note that $f(u)$ and $f(v)$ commute. By computation, we have

$$
\begin{aligned}
\det(\mathcal{R}(q)) &= \det\big(f(u)f(\bar{u}) - f(-\bar{v})f(v)\big) \\
&= \det\big(f(u\bar{u}) + f(\bar{v}v)\big) \\
&= \det\big(f(u\bar{u} + \bar{v}v)\big) \\
&= (|u|^2 + |v|^2)^2 \\
&= \big(\det(\varphi(q))\big)^2.
\end{aligned}
$$

(n) This is obvious from (m).

(o) Exchange the last two rows and last two columns of $\mathcal{R}(q)$.

2.64 Let $A = A_1 + iA_2$, $B = B_1 + iB_2$, where A_1, A_2, B_1, and B_2 are real.

(a) Obvious.

(b) $\chi(A + B) = \begin{pmatrix} A_1 + B_1 & A_2 + B_2 \\ -A_2 - B_2 & A_1 + B_1 \end{pmatrix} = \chi(A) + \chi(B)$.

(c) $\chi(AB) = \begin{pmatrix} A_1 B_1 - A_2 B_2 & A_1 B_2 + A_2 B_1 \\ -A_1 B_2 - A_2 B_1 & A_1 B_1 - A_2 B_2 \end{pmatrix} = \chi(A)\chi(B)$.

(d) $\chi(A^*) = \chi(A_1^t - iA_2^t) = \begin{pmatrix} A_1^t & -A_2^t \\ A_2^t & A_1^t \end{pmatrix} = (\chi(A))^t$.

(e) This is because $\chi(A^{-1})\chi(A) = \chi(A^{-1}A) = I_{2n}$.

(f) If A is unitary, then $A^*A = I$. Thus $\chi(A^*A) = \chi(A^*)\chi(A) = I_{2n}$. Since $\chi(A^*) = (\chi(A))^*$, we see that $\chi(A)$ is unitary. The Hermitian and normal cases are similarly proven.

2.65 By a direct computation.

2.66 By direct verifications. See also Problem 1.68f.

2.67 Let $M = \begin{pmatrix} A & B \\ B & -A \end{pmatrix}$ and $N = \begin{pmatrix} A & B \\ -B & A \end{pmatrix}$. Then

$$
U^*MU = \begin{pmatrix} 0 & B + iA \\ B - iA & 0 \end{pmatrix}, \quad U^*NU = \begin{pmatrix} A + iB & 0 \\ 0 & A - iB \end{pmatrix},
$$

$$
V^*NV = \begin{pmatrix} A & B + Ai \\ -B - Ai & 2(A - iB) \end{pmatrix}, \quad V^{-1}NV = \begin{pmatrix} A + iB & 0 \\ -B & A - iB \end{pmatrix}.
$$

2.68 Applying a row and a column operation of type III, for real A and B,

$$
\begin{vmatrix} A & B \\ -B & A \end{vmatrix} = \begin{vmatrix} A + iB & 0 \\ -B & A - iB \end{vmatrix} = \overline{|A + iB|}\,|A + iB| \geq 0.
$$

Alternatively, by continuity, we only show the case of nonsingular A.

$$\begin{vmatrix} A & B \\ -B & A \end{vmatrix} = \begin{vmatrix} A & B \\ 0 & A + BA^{-1}B \end{vmatrix}$$
$$= |A||A + BA^{-1}B|$$
$$= |A|^2|I + A^{-1}BA^{-1}B| \geq 0.$$

Set $R = A^{-1}B$. Then R is a real matrix. We are done by observing

$$|I + R^2| = |I + iR|\,|I - iR| = |I + iR|\overline{|I + iR|} \geq 0.$$

If A and B are complex matrices, then it is expected that

$$\begin{vmatrix} A & B \\ -\overline{B} & \overline{A} \end{vmatrix} \geq 0.$$

This is true. To prove this, a more advanced result that $\overline{X}X$, where X is a complex matrix, is similar to R^2 for some real R is needed.

2.69 For the determinant identity, use Problem 2.66a. Alternatively, adding the second (block) column to the first column, and then subtracting the first row from the second row, we obtain

$$\begin{vmatrix} A & B \\ B & A \end{vmatrix} = \begin{vmatrix} A + B & B \\ B + A & A \end{vmatrix} = \begin{vmatrix} A + B & B \\ 0 & A - B \end{vmatrix} = |A + B|\,|A - B|.$$

2.70 By Problem 2.69, the block matrix is nonsingular if $A + B$ and $A-$ are nonsingular. Let $C = A + B$, $D = A - B$, and consider

$$\begin{pmatrix} A & B \\ B & A \end{pmatrix} \begin{pmatrix} X & Y \\ U & V \end{pmatrix} = \begin{pmatrix} I & 0 \\ 0 & I \end{pmatrix}.$$

Multiply this out to get

$$AX + BU = I, \quad AY + BV = 0, \quad BX + AU = 0, \quad BY + AV = I.$$

Adding the first equation to the third reveals $(A + B)(X + U) = I$. So $X + U = C^{-1}$. Subtracting the third equation from the first gives $(A - B)(X - U) = I$. Thus, $X - U = D^{-1}$. It follows that

$$X = \tfrac{1}{2}(C^{-1} + D^{-1}), \quad U = \tfrac{1}{2}(C^{-1} - D^{-1}).$$

In a similar way, one shows that V and Y are in the above forms as X and U, respectively, that is, $V = X$ and $Y = U$. Thus,

$$\begin{pmatrix} A & B \\ B & A \end{pmatrix}^{-1} = \tfrac{1}{2} \begin{pmatrix} C^{-1} + D^{-1} & C^{-1} - D^{-1} \\ C^{-1} - D^{-1} & C^{-1} + D^{-1} \end{pmatrix}.$$

Another way: Notice that

$$\begin{pmatrix} I & -I \\ 0 & I \end{pmatrix} \begin{pmatrix} A & B \\ B & A \end{pmatrix} \begin{pmatrix} I & I \\ 0 & I \end{pmatrix} = \begin{pmatrix} A-B & 0 \\ B & A+B \end{pmatrix}.$$

Taking the inverses of both sides and by multiplications, we get

$$\begin{pmatrix} A & B \\ B & A \end{pmatrix}^{-1} = \begin{pmatrix} I & I \\ 0 & I \end{pmatrix} \begin{pmatrix} A-B & 0 \\ B & A+B \end{pmatrix}^{-1} \begin{pmatrix} I & -I \\ 0 & I \end{pmatrix}.$$

Denote $C = A + B$ and $D = A - B$. Then we obtain by computation

$$\begin{pmatrix} A & B \\ B & A \end{pmatrix}^{-1} = \begin{pmatrix} I & I \\ 0 & I \end{pmatrix} \begin{pmatrix} D^{-1} & 0 \\ -C^{-1}BD^{-1} & C^{-1} \end{pmatrix} \begin{pmatrix} I & -I \\ 0 & I \end{pmatrix}$$

$$= \begin{pmatrix} D^{-1} - C^{-1}BD^{-1} & C^{-1} - D^{-1} + C^{-1}BD^{-1} \\ -C^{-1}BD^{-1} & C^{-1} + C^{-1}BD^{-1} \end{pmatrix}.$$

Now it is left to show that each of the blocks has the desired form. As an example, we show for the (2,2)-block:

$$C^{-1} + C^{-1}BD^{-1} = \tfrac{1}{2}(C^{-1} + D^{-1}).$$

By multiplying both sides by C from the left and by D from the right, this reduces to the trivial identity $D + B = \tfrac{1}{2}(D + C)$ or $C - D = 2B$.

2.71 $M^* = M_2$. The first column of M^* is composed of the first columns of A^* and B^*, and so on. Note that A and D need not be square.

2.72   If $r(A) = 1$, then any two rows of A are linearly dependent. All rows are multiples of the first row. The conclusion follows immediately.

If $A = xy^t$, then $A^2 = (xy^t)(xy^t) = x(y^t x)y^t = (y^t x)xy^t = \rho A$. $A^3 = A^2 A = \rho A A = \rho^2 A$. Inductively, $A^k = \rho^{k-1}A$, $k = 1, 2, 3, \ldots$.

2.73   Since A has rank one, we write $A = uv^*$ for some nonzero vectors $u = (u_1, \ldots, u_n)^t$ and $v = (v_1, \ldots, v_n)^t$. As A is Hermitian, i.e., $a_{ij} = \overline{a_{ji}}$, we have $u_i \overline{v_j} = \overline{u_j \overline{v_i}} = \overline{u_j} v_i$. This means $u = kv$ for some nonzero k ($k = \frac{\overline{u_j}}{\overline{v_j}}$, $v_j \neq 0$, independent of j). k is real as $A = kvv^*$ is Hermitian. If $k > 0$, then set $x = \sqrt{k}\,v$; if $k < 0$, set $x = \sqrt{|k|}\,v$. Then $A = xx^*$ when $k > 0$ and $A = -xx^*$ when $k < 0$.

Alternatively, this is immediate from the spectral decomposition theorem for a Hermitian matrix with rank one (see Chapter 4).

2.74 (a) Let $p_1 = \frac{1}{\sqrt{n}}(1, \ldots, 1)^t$. Let $p_i \in \mathbb{R}^n$ ($i > 1$) have 1 as i-th component and -1 as the last component. Set $P = (p_1, p_2, \ldots, p_n)$. P is nonsingular and $E_n P = P \left(\begin{smallmatrix} n & 0 \\ 0 & 0_{n-1} \end{smallmatrix} \right)$. So $P^{-1} E_n P$ is diagonal.

(b) Use elementary operations directly to find the rank (depending on k). Alternatively, use (a), $kI_n + E_n = P \left(\begin{smallmatrix} k+n & 0 \\ 0 & kI_{n-1} \end{smallmatrix} \right) P^{-1}$. Thus, $r(kI_n + E_n) = 1$ if $k = 0$; $r(kI_n + E_n) = n - 1$ if $k = -n$; and $r(kI_n + E_n) = n$ otherwise.

(c) $\det(kI_n + E_n) = k^{n-1}(k + n)$.

2.75 If $r(A) = n$, then A is invertible. Let $x_1(Au_1) + x_2(Au_2) + \cdots + x_n(Au_n) = 0$. Pre-multiplying both sides by A^{-1} shows $x_1 u_1 + x_2 u_2 + \cdots + x_n u_n = 0$. Thus $x_1 = x_2 = \cdots = x_n = 0$ since $u_1, u_2, \ldots, u_n$ are linearly independent. So $Au_1, Au_2, \ldots, Au_n$ are linearly independent. For the other direction, if $Au_1, Au_2, \ldots, Au_n$ are linearly dependent, let $y_1(Au_1) + y_2(Au_2) + \cdots + y_n(Au_n) = 0$, where not all y are zero. Then $A(y_1 u_1 + y_2 u_2 + \cdots + y_n u_n) = 0$. Since $u_1, u_2, \ldots, u_n$ are linearly independent, $y_1 u_1 + y_2 u_2 + \cdots + y_n u_n \neq 0$. Thus the system $Ax = 0$ has a nonzero solution, and A is singular.

2.76 There exist invertible matrices P and Q such that $A = P \left(\begin{smallmatrix} I_r & 0 \\ 0 & 0 \end{smallmatrix} \right) Q$, where P is $m \times m$, Q is $n \times n$. Write $P = (M, S)$ and $Q = \left(\begin{smallmatrix} N \\ T \end{smallmatrix} \right)$, where M is $m \times r$, N is $r \times n$. Then $A = (M, S) \left(\begin{smallmatrix} I_r & 0 \\ 0 & 0 \end{smallmatrix} \right) \left(\begin{smallmatrix} N \\ T \end{smallmatrix} \right) = MN$.

2.77 (a) False.

(b) False.

(c) True for $k \geq 1$. Inclusive for $0 < k < 1$.

(d) False.

(e) False.

(f) False.

(g) True since $r(AB, AC) = r(A(B, C)) \leq r(A)$.

(h) False. $A = \left(\begin{smallmatrix} 1 & 0 \\ 0 & 0 \end{smallmatrix} \right), B = I_2, C = \left(\begin{smallmatrix} 0 & 1 \\ 1 & 0 \end{smallmatrix} \right)$. $r(BA, CA) = 2$, $r(A) = 1$.

2.78 This is immediate from the cofactor (adjugate) matrix. Another proof for the real case: Let B be the inverse of real matrix A. Write $B = B_1 + iB_2$, where B_1 and B_2 are real. Then $AB = BA = I$ implies $AB_1 = B_1 A = I$ and $B_2 = 0$. Thus, $B = B_1$ is a real matrix.

2.79 (a) $(1, i) = i(-i, 1)$.

(b) If $x(1, i) + y(-i, 1) = 0$, where $x, y \in \mathbb{R}$, then $x - yi = 0$. Thus $x = y = 0$. (Linear independence of vectors and rank of a matrix are relative to the underlying field. See Problem 1.75.)

(c) No. A is not invertible.

(d) $U^*AU = \text{diag}(2, 0)$ is real diagonal (via a nonreal matrix U).

(e) $r(A) = 1$ (over $\mathbb{C}$ where the entries of the matrix belong).

2.80 A basic fact is that a set of real vectors is linearly independent over $\mathbb{R}$ if and only if the set of the vectors is linearly independent over $\mathbb{C}$.

Let $\{e_1, \ldots, e_n\}$ be the standard basis of $\mathbb{R}^n$ and $\mathbb{C}^n$. Then $Re_1, \ldots, Re_n$ are the columns of R, which lie in $W_{\mathbb{R}}(R)$ and $W_{\mathbb{C}}(R)$. The columns of R that span $W_{\mathbb{R}}(R)$ also span $W_{\mathbb{C}}(R)$. Thus, $W_{\mathbb{R}}(R) \subseteq \mathbb{R}^m$ over $\mathbb{R}$ and $W_{\mathbb{C}}(R) \subseteq \mathbb{C}^m$ over $\mathbb{C}$ as vector spaces have the same dimension.

One may denote $\text{Im}_{\mathbb{R}}(R) = W_{\mathbb{R}}(R)$ and $\text{Im}_{\mathbb{C}}(R) = W_{\mathbb{C}}(R)$.

$W_{\mathbb{C}}(R) = \text{Im}_{\mathbb{C}}(R)$ is a vector space over $\mathbb{C}$ with dimension $r(R)$.
$W_{\mathbb{C}}(R)$ is a vector space over $\mathbb{R}$ with dimension $2r(R)$.
$W_{\mathbb{R}}(R)$ is not a vector space over $\mathbb{C}$ in general.
$W_{\mathbb{R}}(R) = \text{Im}_{\mathbb{R}}(R)$ is a vector space over $\mathbb{R}$ with dimension $r(R)$.

Let C be an $m \times n$ complex matrix with rank $r(C)$ (over $\mathbb{C}$) and let

$$W_{\mathbb{R}}(C) = \{Cx \mid x \in \mathbb{R}^n\}, \quad W_{\mathbb{C}}(C) = \{Cx \mid x \in \mathbb{C}^n\}.$$

$W_{\mathbb{C}}(C) = \text{Im}_{\mathbb{C}}(C)$ is a vector space over $\mathbb{C}$ with dimension $r(C)$.
$W_{\mathbb{C}}(C)$ is a vector space over $\mathbb{R}$ with dimension $2r(C)$.
$W_{\mathbb{R}}(C)$ is not a vector space over $\mathbb{C}$ in general.
$W_{\mathbb{R}}(C) = \text{Im}_{\mathbb{R}}(C)$ is a vector space over $\mathbb{R}$. Its dimension can be any integer between $r(C)$ and $2r(C)$, depending on the number of real and complex columns of C. For example, if $C = \left(\begin{smallmatrix} 1 & 0 \\ 0 & i \end{smallmatrix}\right)$, then $r(C) = 2$ and the dimension of $W_{\mathbb{R}}(C)$ is 2; if $C = \left(\begin{smallmatrix} 1 & i \\ -i & 1 \end{smallmatrix}\right)$, then $r(C) = 1$ (over $\mathbb{C}$) and the dimension of $W_{\mathbb{R}}(C)$ over $\mathbb{R}$ is 2. (How the dimension of a vector space varies with the underlying fields is beyond our scope.)

2.81 (a) $A^* = B^* - iC^* = B^t - iC^t = B + iC$. So $B^t = B$ and $C^t = -C$.

(b) $x^t Ax = x^t Bx + ix^t Cx$. Since A is Hermitian, $x^t Ax$ is always real for $x \in \mathbb{R}^n$, as is $x^t Bx$. Therefore, $x^t Ax = x^t Bx$ and $x^t Cx = 0$.

(c) $Ax = Bx + iCx$. If $Ax = 0$ and x is real, then $0 = Ax = Bx + iCx$. Thus, $Bx = Cx = 0$.

(d) $B = \begin{pmatrix} 1 & 0 \\ 0 & 1 \end{pmatrix}$, $C = \begin{pmatrix} 0 & 1 \\ -1 & 0 \end{pmatrix}$. Take $x = (1, i)^t$. Then $Ax = 0$, so $x^*Ax = 0$. However, $x^*Bx = x^*x > 0$, hence $Bx \neq 0$.

(e) $B = \begin{pmatrix} 1 & 1 \\ 1 & 1 \end{pmatrix}$. Let $x = \begin{pmatrix} 1 \\ -1 \end{pmatrix}$. $Bx = 0$, $Ax \neq 0$, $r(B) = 1$, $r(A) = 2$.

2.82   (a) $s = 3$.   (b) $t = -3$.

2.83   Write $AB - I = AB - B + B - I = (A - I)B + (B - I)$.

2.84   Since $AB = 0$, the column vectors of B are contained in $\operatorname{Ker} A = \{\, x \mid Ax = 0 \,\}$. Since $r(A) + \dim \operatorname{Ker} A = n$, where n is the number of unknowns, it follows that $r(A) + r(B) \leq r(A) + \dim \operatorname{Ker} A = n$.

2.85   Let A_s be the submatrix of A by deleting s rows from A. Then $r(A) - s \leq r(A_s)$. Similarly, $r(A_s) - t \leq r(B)$. Thus $r(A) \leq s + t + r(B)$.

2.86   Use the Frobenius rank inequality (Example 2.11) for $A^3 = AAA$.

2.87   Let W_1, W_2, W_3, and W_4 be the column spaces of A, B, $A + B$, and AB, respectively. Since $W_3 \subseteq W_1 + W_2$, we have

$$\dim W_3 \leq \dim(W_1 + W_2) = \dim W_1 + \dim W_2 - \dim(W_1 \cap W_2).$$

Thus

$$r(A + B) \leq r(A) + r(B) - \dim(W_1 \cap W_2).$$

We claim that $r(AB) = \dim W_4 \leq \dim(W_1 \cap W_2)$. For this, we show $W_4 \subseteq W_1 \cap W_2$. Note that $\operatorname{Im}(AB) \subseteq \operatorname{Im} A$ and $\operatorname{Im}(BA) \subseteq \operatorname{Im} B$. As $AB = BA$, we have $\operatorname{Im}(AB) \subseteq \operatorname{Im} A \cap \operatorname{Im} B$, i.e., $W_4 \subseteq W_1 \cap W_2$.

2.88   By the Sylvester rank inequality, $r(AB) \geq r(A) + r(B) - n$. Then

$$
\begin{aligned}
\dim \operatorname{Ker}(AB) &= n - \dim \operatorname{Im}(AB) = n - r(AB) \\
&\leq n - (r(A) + r(B) - n) \\
&= (n - r(A)) + (n - r(B)) \\
&= (n - \dim \operatorname{Im} A) + (n - \dim \operatorname{Im} B) \\
&= \dim \operatorname{Ker} A + \dim \operatorname{Ker} B.
\end{aligned}
$$

2.89   Sylvester's rank inequality gives $r(AB) \geq r(A) + r(B) - n$. We derive

$$
\begin{aligned}
0 &= r(A_1 A_2 \cdots A_k) \\
&\geq r(A_1) + r(A_2 \cdots A_k) - n \\
&\geq r(A_1) + r(A_2) + r(A_3 \cdots A_k) - 2n \\
&\geq \cdots \\
&\geq r(A_1) + r(A_2) + \cdots + r(A_k) - (k - 1)n.
\end{aligned}
$$

2.90   Denote the column spaces of (X, Z), (Z, Y), and Z by W_1, W_2, and W_3, respectively. By the dimension identity,

$$\dim(W_1 + W_2) = \dim W_1 + \dim W_2 - \dim(W_1 \cap W_2)$$
$$= r(X, Z) + r(Z, Y) - \dim(W_1 \cap W_2).$$

Apparently, $W_3 \subseteq W_1 \cap W_2$. Note that the column space of (X, Y) is contained in $W_1 + W_2$. So $r(X, Y) \leq \dim(W_1 + W_2)$ and $r(Z) \leq \dim(W_1 \cap W_2)$. The desired inequality follows.

2.91   Let $\{x_{i_1}, \ldots, x_{i_a}\}$ and $\{y_{j_1}, \ldots, y_{j_b}\}$ be bases of the column spaces of X and Y, respectively. Then $\{\tilde{x}_{i_1}, \ldots, \tilde{x}_{i_a}, \tilde{y}_{j_1}, \ldots, \tilde{y}_{j_b}\}$ is a basis of the column space of $\begin{pmatrix} X & 0 \\ 0 & Y \end{pmatrix}$, where $\tilde{x}_{i_s} = \begin{pmatrix} x_{i_s} \\ 0 \end{pmatrix}$ and $\tilde{y}_{i_t} = \begin{pmatrix} 0 \\ y_{i_t} \end{pmatrix}$ for each s and t. Let $\tilde{z}_{i_r} = \begin{pmatrix} x_{i_r} \\ z_{i_r} \end{pmatrix}$. Then $\{\tilde{z}_{i_1}, \ldots, \tilde{z}_{i_a}, \tilde{y}_{j_1}, \ldots, \tilde{y}_{j_b}\}$ is a linearly independent set. Thus, $r(X) + r(Y) = r\begin{pmatrix} X & 0 \\ 0 & Y \end{pmatrix} \leq r\begin{pmatrix} X & 0 \\ Z & Y \end{pmatrix}$.

2.92   (a) and (b) are false: Take $A = \begin{pmatrix} 1 & 0 \\ 0 & 0 \end{pmatrix}$ and $B = \begin{pmatrix} 1 & 0 \\ 1 & 0 \end{pmatrix}$. (c) is true.

2.93 $r(A^t A)$ need not be equal to $r(A)$ (unless A is real). Take $A = \begin{pmatrix} 1 & i \\ i & -1 \end{pmatrix}$. Then $r(A^t A) = 0$, but $r(A) = 1$. Similarly, $r(\bar{A} A) \neq r(A)$.

For the rank identity, note that X and $\bar{X}$ have the same rank.

2.94   Let $C = I - (A + B)$. Since $AC = -AB = CB$ and C is invertible, we see that A and B have the same rank.

2.95   Using the Frobenius rank inequality, we derive

$$r(A - AB) + r(B - AB) = r(A^2 - AB) + r(B^2 - AB)$$
$$= r(A(A - B)) + r((B - A)B) = r(A(A - B)) + r((A - B)B)$$
$$\leq r(A(A - B)B) + r(A - B) = r(A^2 B - AB^2) + r(A - B)$$
$$= r(AB - AB) + r(A - B) = r(A - B).$$

On the other hand,

$$r(A - B) \leq r(A - AB) + r(AB - B) = r(A - AB) + r(B - BA).$$

2.96   Since $I = A + (I - A)$, we have $n = r(A + (I - A)) \leq r(A) + r(I - A)$. Note that $A^2 = A \Leftrightarrow A(I - A) = 0 \Leftrightarrow \text{Im}(I - A) \subseteq \text{Ker}\, A \Rightarrow r(I - A) \leq n - r(A)$. Thus, if $A^2 = A$, then $n = r(A) + r(I - A)$.

For the converse, if $n = r(A) + r(I - A)$, then $n = \dim \text{Im}\, A + \dim(\text{Im}(I - A))$. Since $x = Ax + (I - A)x$ for every $x \in \mathbb{C}^n$, we have

$\mathbb{C}^n = \operatorname{Im} A + \operatorname{Im}(I - A)$. The dimension identity reveals $\dim(\operatorname{Im} A \cap \operatorname{Im}(I - A)) = 0$. Thus, $\operatorname{Im} A \cap \operatorname{Im}(I - A) = \{0\}$. Observing that $\operatorname{Im}(A(I - A)) \subseteq \operatorname{Im} A$ and $\operatorname{Im}(A(I - A)) = \operatorname{Im}((I - A)A) \subseteq \operatorname{Im}(I - A)$, we arrive at $\operatorname{Im}((I - A)A) = \{0\}$, that is, $(I - A)A = 0$, or $A = A^2$.

Alternatively, use Jordan canonical form of A. See Problem 4.113.

2.97 It is sufficient to notice that

$$\begin{pmatrix} I_m & -A \\ 0 & I_n \end{pmatrix} \begin{pmatrix} 0 & A \\ B & I_n \end{pmatrix} \begin{pmatrix} I_p & 0 \\ -B & I_n \end{pmatrix} = \begin{pmatrix} -AB & 0 \\ 0 & I_n \end{pmatrix}.$$

2.98 Since A is invertible, $r\left(\begin{smallmatrix} A & C \\ 0 & B \end{smallmatrix}\right) = n + r(B)$. As $\left(\begin{smallmatrix} A & C \\ 0 & B \end{smallmatrix}\right)^k = \left(\begin{smallmatrix} A^k & \star \\ 0 & B^k \end{smallmatrix}\right)$ and A^k is invertible, the desired rank identity follows immediately.

2.99 Notice that

$$\begin{pmatrix} I_m & 0 \\ -A^* & I_n \end{pmatrix} \begin{pmatrix} I_m & A \\ A^* & I_n \end{pmatrix} \begin{pmatrix} I_m & -A \\ 0 & I_n \end{pmatrix} = \begin{pmatrix} I_m & 0 \\ 0 & I_n - A^*A \end{pmatrix}.$$

So

$$r\begin{pmatrix} I_m & A \\ A^* & I_n \end{pmatrix} = m + r(I_n - A^*A).$$

Similarly,

$$r\begin{pmatrix} I_m & A \\ A^* & I_n \end{pmatrix} = n + r(I_m - AA^*).$$

Thus

$$r(I_m - AA^*) - r(I_n - A^*A) = m - n.$$

2.100 There are two important facts regarding $\operatorname{adj}(A)$ (see (2.2), p. 39):

$$A\operatorname{adj}(A) = \operatorname{adj}(A)A = |A|I$$

and when A is invertible,

$$\operatorname{adj}(A) = |A|A^{-1}.$$

(a) $\operatorname{adj}(A)$ is invertible if and only if $|A| \neq 0$.

(b) $A\operatorname{adj}(A) = 0$ implies that the columns of $\operatorname{adj}(A)$ are the solutions of $Ax = 0$. If $r(A) = n - 1$, then $\operatorname{adj}(A) \neq 0$, $\dim \operatorname{Ker} A = 1$, and the columns of $\operatorname{adj}(A)$ are mutually linearly dependent. Thus $r(\operatorname{adj}(A)) = 1$. Conversely, if $r(\operatorname{adj}(A)) = 1$, then A has an $(n-1) \times (n-1)$ nonzero minor and A is singular, i.e, $r(A) = n-1$.

(c) Consider $(n-1) \times (n-1)$ minors of A.

(d) It follows from the two facts given above.

(e) By the first fact mentioned above

$$\mathrm{adj}(A) \cdot \mathrm{adj}(\mathrm{adj}(A)) = |\mathrm{adj}(A)| I = |A|^{n-1} I.$$

Replace the left-most $\mathrm{adj}(A)$ by $|A|A^{-1}$ when A is invertible. If A is singular, then both sides vanish.

(f) First consider the case where A and B are nonsingular. For the singular case, use $A + \varepsilon I$ and $B + \varepsilon I$ to substitute A and B, respectively, then apply an argument of continuity.

(g) By (f) with $\mathrm{adj}(X^{-1}) = |X^{-1}|X$.

(h) If $A = \begin{pmatrix} a & b \\ c & d \end{pmatrix}$, then $\mathrm{adj}(A) = \begin{pmatrix} d & -b \\ -c & a \end{pmatrix}$. So $|A| = |\mathrm{adj}(A)|$.

(i) $A_{ij} = A_{ji}^{*}$ when A is Hermitian.

(j) $\overbrace{\mathrm{adj} \cdots \mathrm{adj}}^{k}(A) = A$ when k is even; it is A^{-1} when k is odd.

2.101 There are three possibilities for $A^{*} = \mathrm{adj}(A)$: (1) $A = 0$, (2) A is a unitary matrix and $|A| = 1$, and (3) $A = \begin{pmatrix} a & b \\ -\bar{b} & \bar{a} \end{pmatrix}$, where $a, b \in \mathbb{C}$.

2.102 (a) This is because the range of AA^{*} is contained in the range of A and $r(AA^{*}) = r(A)$. $\mathrm{Im}(A^{*}A) \neq \mathrm{Im}\,A$ in general. See Problem 1.82.

(b) By (a), the columns of A are linear combinations of those of AA^{*}.

(c) $A^{2}x = 0$ and $A^{*}A = AA^{*} \Rightarrow A^{*}AAx = 0 \Rightarrow AA^{*}Ax = 0 \Rightarrow (x^{*}A^{*}A)(A^{*}Ax) = 0 \Rightarrow A^{*}Ax = 0 \Rightarrow x^{*}A^{*}Ax = 0 \Rightarrow Ax = 0$.

2.103 It suffices to show that $Ax = 0$ has only the trivial solution 0. Suppose $Ax = 0$ has a nonzero solution $x = (x_1, x_2, \ldots, x_n)^{t}$. For some s,

$$|x_s| = \max_{1 \leq i \leq n} \{|x_i|\}.$$

Then $|x_s| \neq 0$. Notice that the s-th equation of $Ax = 0$ is

$$a_{s1}x_1 + a_{s2}x_2 + \cdots + a_{ss}x_s + \cdots + a_{sn}x_n = 0.$$

Thus

$$a_{ss}x_s = -\sum_{j=1,\, j \neq s}^{n} a_{sj}x_j$$

and

$$|a_{ss}| \leq \sum_{j=1,\, j\neq s}^{n} \left| a_{sj} \frac{x_j}{x_s} \right| \leq \sum_{j=1,\, j\neq s}^{n} |a_{sj}|,$$

a contradiction to the given condition (row diagonal dominance). For the second part, let $A = I - B$. Then

$$|a_{ii}| = |1 - b_{ii}| \geq 1 - |b_{ii}| > \sum_{j=1,\, j\neq i}^{n} |b_{ij}| = \sum_{j=1,\, j\neq i}^{n} |a_{ij}|.$$

So A is row diagonally dominant, thus invertible, i.e., B is invertible.

2.104 If $\lambda \neq -3$, then $\beta \in \operatorname{Im} A$.

2.105 Let $p(x) = ax^3 + bx^2 + cx + d$, where a, b, c, d are constants to be determined. Plug in the x-coordinates of the given points and equate the corresponding y-coordinates to get a linear equation system in a, b, c, d. Solve the equation system to get $a = 1, b = 2, c = -3$, and $d = 4$. So $p(x) = x^3 + 2x^2 - 3x + 4$.

2.106 Let A be the coefficient matrix. If $\lambda = 1$, then $\dim \operatorname{Ker} A = 2$. If $\lambda = -2$, then $\dim \operatorname{Ker} A = 1$. Otherwise, $\dim \operatorname{Ker} A = 0$.

2.107 If $\lambda = 1$, the homogeneous equation system has nonzero solutions.

2.108 $\alpha = (-1, 1, 0, 0, 0)$ and $\beta = (-1, 0, -1, 0, 1)$ form a basis for the solution space. The general solution is $x = \lambda\alpha + \mu\beta$, where λ, μ are scalars.

2.109 The dimension is 2. $\eta_1 = (-\frac{3}{2}, \frac{7}{2}, 1, 0)$ and $\eta_2 = (-1, -2, 0, 1)$ form a basis for the solution space. The general solution is $\eta = s\eta_1 + t\eta_2$, where s and t are scalars.

2.110 If $y = 0$, then $x_1 = -x_3$, $x_2 = -x_4$, where x_3 and x_4 are free.

If $y = 2$, then $x_1 = x_2 = x_3 = x_4$, where x_4 is free.

If $y = -2$, then $x_1 = x_3 = -x_4$, $x_2 = x_4$, where x_4 is free.

If $y \neq 0, 2, -2$, then $x_1 = x_2 = x_3 = x_4 = 0$ is the unique solution.

2.111 If $a = 1$, then $x_3 = 0$ and $x_1 = 1 - x_2$, where x_2 is free.

If $a \neq 1$, then $x_1 = \frac{3-a}{a-1}$, $x_2 = \frac{2}{1-a}$, and $x_3 = 2$.

2.112   Apply elementary row operations to the coefficient matrix.

The dimension of the solution space is $n-1$ and the following vectors in $\mathbb{R}^{2n}$ form a basis for the solution space:

$$
\begin{aligned}
\epsilon_1 &= (-1,1,0,\ldots,0,0,\ -1,1,0,\ldots,0,0) \\
\epsilon_2 &= (-1,0,1,\ldots,0,0,\ -1,0,1,\ldots,0,0) \\
\cdots\quad & \quad\cdots\quad\cdots\quad\cdots\quad\cdots\quad\cdots \\
\epsilon_{n-1} &= (-1,0,0,\ldots,0,1,\ -1,0,0,\ldots,0,1).
\end{aligned}
$$

2.113   Since $\dim \operatorname{Im} A = r(A)$ and $\operatorname{Im} A = \operatorname{Span}\{\alpha,\beta\} = \operatorname{Span}\alpha + \operatorname{Span}\beta$,

$$
\begin{aligned}
\dim(\operatorname{Span}\alpha \cap \operatorname{Span}\beta) \\
= \quad & \dim(\operatorname{Span}\alpha) + \dim(\operatorname{Span}\beta) - \dim(\operatorname{Span}\alpha + \operatorname{Span}\beta) \\
= \quad & \dim(\operatorname{Span}\alpha) + \dim(\operatorname{Span}\beta) - \dim\operatorname{Im} A \\
= \quad & \dim(\operatorname{Span}\alpha) + \dim(\operatorname{Span}\beta) - r(A) \\
= \quad & s + t - r(A).
\end{aligned}
$$

2.114   Since $\operatorname{Im} A_1 + \operatorname{Im} A_2 = \mathbb{R}^n$, $n = \dim(\operatorname{Im} A_1 + \operatorname{Im} A_2) = \dim\operatorname{Im} A_1 + \dim\operatorname{Im} A_2 - \dim(\operatorname{Im} A_1 \cap \operatorname{Im} A_2) \leq \dim\operatorname{Im} A_1 + \dim\operatorname{Im} A_2 = r(A_1) + r(A_2) \leq n$. So $r(A_1) + r(A_2) = n$ and $\operatorname{Im} A_1 \cap \operatorname{Im} A_2 = \{0\}$. Thus $\operatorname{Im} A_1 + \operatorname{Im} A_2$ is a direct sum. Moreover, the columns of A_1 and A_2 are all linearly independent, yielding $r(A_1, A_2) = n$.

For the converse, if $r(A_1, A_2) = n$, then $\operatorname{Im} A_1 \oplus \operatorname{Im} A_2 = \mathbb{R}^n$.

If $r(A_1) + r(A_2) = r(A_1, A_2)$, then $\operatorname{Im} A_1 + \operatorname{Im} A_2$ is a direct sum, but it need not be equal to $\mathbb{R}^n$. Take $A_1 = \left(\begin{smallmatrix} 1 \\ 0 \\ 0 \end{smallmatrix}\right)$ and $A_2 = \left(\begin{smallmatrix} 0 & 0 \\ 1 & 0 \\ 0 & 0 \end{smallmatrix}\right)$ for $\mathbb{R}^3$.

If $r(A_1) + r(A_2) = n$, then neither $\operatorname{Im} A_1 + \operatorname{Im} A_2$ is a direct sum nor is it equal to $\mathbb{R}^n$ in general. Take $A_1 = A_2 = \left(\begin{smallmatrix} 1 \\ 0 \end{smallmatrix}\right)$ for $\mathbb{R}^2$.

2.115   Obviously V is contained in $\operatorname{Im} B$. Let $\dim V = s$ and let $Bx_1,\ldots,Bx_s$ be a basis for V. Extend it to a basis $Bx_1,\ldots,Bx_s, By_1,\ldots,By_t$ of $\operatorname{Im} B$, $s + t = r(B)$. As $ABz = A(Bz)$ and $ABx_i = 0$ for each x_i, it is not difficult to show that $ABy_1,\ldots,ABy_t$ comprise a basis of $\operatorname{Im}(AB)$. So $\dim\operatorname{Im}(AB) = t = r(B) - s = r(B) - \dim V$.

2.116   (a) $\operatorname{Span}\{\alpha_1,\alpha_2,\ldots,\alpha_n\} = \operatorname{Im} A$. $r(A) = \dim\operatorname{Im} A$ is equal to the largest number of column vectors that are linearly independent.

(b) Let $\{\alpha_{i_1},\alpha_{i_2},\ldots,\alpha_{i_r}\} \subseteq \{\alpha_1,\alpha_2,\ldots,\alpha_n\}$. Since $r(X) = r(PX)$ for any $p \times q$ matrix X and $p \times p$ invertible matrix P, we see that $r\big((\alpha_{i_1},\ldots,\alpha_{i_r})\big) = r\big((\beta_{i_1},\beta_{i_2},\ldots,\beta_{i_r})\big)$.

(c) Apply elementary row operations to $(\gamma_1, \gamma_2, \gamma_3, \gamma_4)$ to get

$$(\gamma_1, \gamma_2, \gamma_3, \gamma_4) \longrightarrow \begin{pmatrix} 1 & 0 & 0 & 1 \\ 0 & 1 & 0 & -1 \\ 0 & 0 & 1 & 1 \\ 0 & 0 & 0 & 0 \end{pmatrix}.$$

Thus the dimension is 3. $\{\gamma_1, \gamma_2, \gamma_3\}$ is a basis. In fact, any three of $\gamma_1, \gamma_2, \gamma_3, \gamma_4$ form a basis.

See also Problem 2.22.

2.117   For $W_1 \cap W_2$, consider the system of linear equations

$$x_1\alpha_1 + x_2\alpha_2 + x_3\alpha_3 = y_1\beta_1 + y_2\beta_2.$$

The dimensions of $W_1, W_2, W_1 \cap W_2$ and $W_1 + W_2$ are 3, 2, 1, 4, respectively. $\{\beta_1\}$ is a basis for $W_1 \cap W_2$ and $\{\alpha_1, \alpha_2, \alpha_3, \beta_2\}$ is a basis for $W_1 + W_2$, which is spanned by $\alpha_1, \alpha_2, \alpha_3, \beta_1$ and β_2.

2.118   Write the equation system as $Ax = b$. Take $b = e_i$, where e_i is the column vector with i-th component 1 and elsewhere 0, $i = 1, 2, \ldots, n$. Then there are column vectors C_i with integer components such that $AC_i = e_i$ for each i. Thus $AC = I$, where $C = (C_1, C_2, \ldots, C_n)$. Taking determinants, $|AC| = |A||C| = 1$, so $|A| = \pm 1$.

2.119 $\dim W_1 = n - 1, \dim W_2 = 1$, and $W_1 \cap W_2 = \{0\}$. So $\mathbb{F}^n = W_1 \oplus W_2$.

2.120   If $\det A = 0$, then $Ax = 0$ has a nonzero solution x_0. Let $B = (x_0, 0) \in M_n(\mathbb{C})$. Then $AB = 0$. Conversely, if $AB = 0$ for some nonzero matrix B, then $Ab = 0$ for any column b of B. Thus $Ax = 0$ has a nonzero solution and A is singular, that is, $\det A = 0$.

2.121   It suffices to show $\operatorname{Ker} A \cap \operatorname{Ker} B \neq \{0\}$. By the dimension identity,

$$\dim(\operatorname{Ker} A \cap \operatorname{Ker} B)$$
$$= \quad \dim \operatorname{Ker} A + \dim \operatorname{Ker} B - \dim(\operatorname{Ker} A + \operatorname{Ker} B)$$
$$= \quad n - r(A) + n - r(B) - \dim(\operatorname{Ker} A + \operatorname{Ker} B)$$
$$= \quad (n - r(A) - r(B)) + (n - \dim(\operatorname{Ker} A + \operatorname{Ker} B))$$
$$> \quad n - \dim(\operatorname{Ker} A + \operatorname{Ker} B) \geq 0.$$

2.122   Since $r(A) = n-l, r(B) = n-m$, we have $r(AB) \leq \min\{r(A), r(B)\} = \min\{n - l, n - m\}$. It follows that $\dim \operatorname{Ker}(AB) = n - r(AB) \geq n - \min\{r(A), r(B)\} \geq \max\{n - (n - l), n - (n - m)\} = \max\{l, m\}$.

2.123 First notice that $\operatorname{Ker} A \subseteq \operatorname{Ker}(A^2)$. If $r(A) = r(A^2)$, then $\dim \operatorname{Ker} A = \dim \operatorname{Ker}(A^2)$. This implies $\operatorname{Ker} A = \operatorname{Ker}(A^2)$.

2.124 We only show that if $Ax = 0$ and $Bx = 0$ have the same solution space (i.e., $\operatorname{Ker} A = \operatorname{Ker} B$), then $A = CB$ for some nonsingular C.

Note that A and B have the same rank, say r. Let $A = P \begin{pmatrix} A_r \\ 0 \end{pmatrix}$, where P is $(m \times m)$ nonsingular, and A_r consists of the rows of A that form a basis for the row space of A. Likewise, $B = Q \begin{pmatrix} B_r \\ 0 \end{pmatrix}$. Since $Ax = 0$ and $Bx = 0$ have the same solutions, $A_r x = 0$, $B_r x = 0$, and $\begin{pmatrix} A_r \\ B_r \end{pmatrix} x = 0$ all have the same solutions. So $r \begin{pmatrix} A_r \\ B_r \end{pmatrix} = r$. Thus, $A_r = RB_r$ for some nonsingular matrix R. Set $C = P \begin{pmatrix} R & 0 \\ 0 & I \end{pmatrix} Q^{-1}$. Then $A = P \begin{pmatrix} A_r \\ 0 \end{pmatrix} = P \begin{pmatrix} RB_r \\ 0 \end{pmatrix} = P \begin{pmatrix} R & 0 \\ 0 & I \end{pmatrix} \begin{pmatrix} B_r \\ 0 \end{pmatrix} = CB$.

If $r(A^2) = r(A)$, then $A^2 x = 0$ and $Ax = 0$ have the same solution space. From the above result, $A^2 = DA$ for some invertible D.

2.125 Suppose that $\lambda_1 \eta_1 + \cdots + \lambda_n \eta_n$ is a solution to $Ax = b$. Then

$$A(\lambda_1 \eta_1) + \cdots + A(\lambda_n \eta_n) = b.$$

Since $A(\lambda_i \eta_i) = \lambda_i b$, $i = 1, \ldots, n$, we have $(\lambda_1 + \cdots + \lambda_n)b = b$. Thus, $\lambda_1 + \cdots + \lambda_n = 1$ for $b \neq 0$. Conversely, if $\lambda_1 + \cdots + \lambda_n = 1$, then

$$A(\lambda_1 \eta_1 + \cdots + \lambda_n \eta_n) = \lambda_1 A \eta_1 + \cdots + \lambda_n A \eta_n = (\lambda_1 + \cdots + \lambda_n)b = b.$$

If $l_1 \eta_1 + l_2 \eta_2 + \cdots + l_n \eta_n = 0$, applying A to both sides from the left, we have $(l_1 + l_2 + \cdots + l_n)b = 0$. Since $b \neq 0$, we get $l_1 + l_2 + \cdots + l_n = 0$.

2.126 $r(A^*A) \leq r(A^*A, A^*b) = r\big(A^*(A, b)\big) \leq r(A^*)$. On the other hand, $r(A^*A) = r(A^*)$. So $r(A^*A) = r(A^*A, A^*b)$. Thus the coefficient matrix A^*A and the augmented matrix (A^*A, A^*b) have the same rank. It follows that $A^*Ax = A^*b$ is consistent.

2.127 $\tilde{A}$ has order $n + 1$. If $r(\tilde{A}) = r(A) \leq n$, then $\det \tilde{A} = 0$. (d) is right.

2.128 Let A be an $n \times n$ matrix. Since $Ax = 0$ has nonzero solutions, $r(A) < n$. Note that $r(A) = r(A^t)$. Let $B = (A^t, b)$. If $r(B) > r(A^t) = r(A)$, then $A^t x = b$ has no solution. If $r(B) = r(A^t) = r(A) < n$, then there are infinitely many solutions. In either case, if $r(A) < n$, then $A^t x = b$ cannot have a unique solution for any b.

2.129   Necessity: Since $Ax = 0$ and $\begin{pmatrix} A \\ b \end{pmatrix} x = 0$ have the same solution space, $r(A) = r\begin{pmatrix} A \\ b \end{pmatrix}$. Thus, b is a linear combination of the rows of A.

Sufficiency: If b lies in the row space of A, then we can write b as $b = cA$ for some $c = (c_1, \ldots, c_m)$, where $c_1, \ldots, c_m \in \mathbb{R}$. It is immediate that $Ax = 0$ implies $bx = cAx = 0$.

2.130   Let $A = (a_{ij})$, $c = (c_1, c_2, \ldots, c_n)$, $b = (b_1, b_2, \ldots, b_n)^t$. Then the augmented matrices of the equation systems are, respectively,

$$M = \begin{pmatrix} A & b \\ c & d \end{pmatrix}, \quad M^t = \begin{pmatrix} A^t & c^t \\ b^t & d \end{pmatrix}.$$

Since A is nonsingular, $r(A) = n$, we see $r(M) = n$ or $n+1$. The first equation system has a (unique) solution if and only if $r(M) = n$ (i.e., $d = cA^{-1}b$). The second system has a (unique) solution if and only if $r(M^t) = n$. However, $r(M) = r(M^t)$. The two systems are either both consistent or inconsistent. In the case they have a solution, the solution to the first system is $A^{-1}b$ and to the second is $(A^t)^{-1}c^t$.

2.131   This is immediate from the Frobenius rank inequality:

$$0 = r\big((X_1 - X_2)AB\big) \geq r\big((X_1 - X_2)A\big) + r(AB) - r(A).$$

Another solution: We show that $(X_1A - X_2A)x = 0$ for all vectors x. Note that $\operatorname{Im}(AB) \subseteq \operatorname{Im}(A)$ and $r(AB) = r(A)$ imply $\operatorname{Im}(AB) = \operatorname{Im}(A)$. So for every x, there is y such that $Ax = ABy$. It follows that $(X_1A - X_2A)x = X_1Ax - X_2Ax = X_1ABy - X_2ABy = 0$.

Alternatively, if B is nonsingular, then it is obvious. Let the rank of B be r and first consider the case $B = \begin{pmatrix} I_r & 0 \\ 0 & 0 \end{pmatrix}$. Since $r(AB) = r(A)$, the first r columns of A span the column space of A. Write A as $A = (A_r, A_rC)$. $X_1AB = X_2AB$ implies $X_1A_r = X_2A_r$. Thus $X_1A = (X_1A_r, X_1A_rC) = (X_2A_r, X_2A_rC) = X_2A$. For general B, let $B = P\begin{pmatrix} I_r & 0 \\ 0 & 0 \end{pmatrix}Q$ with P and Q invertible and use the above argument.

2.132 $x + y = 2$ and $x - y + 2z = 0$. The second plane contains the origin.

2.133   Let the lines intersect at (x_0, y_0). Adding the three equations gives

$$(x_0 + y_0 + 1)(a + b + c) = 0.$$

We show that $x_0 + y_0 + 1 \neq 0$, concluding that $a + b + c = 0$.

Consider lines l_1 and l_2 and view them as equations in a and b: $x_0 a + y_0 b = -c$, $a + x_0 b = -cy_0$. If $x_0 + y_0 + 1 = 0$, then the determinant of the coefficient matrix is $x_0^2 - y_0 = x_0^2 + x_0 + 1$, which is never zero for any real x_0. Solving for a and b in terms of c, we obtain that $a = b = c$ and all three lines are the same, contradicting the assumption.

Conversely, let $a + b + c = 0$. Consider the augmented matrix

$$\begin{pmatrix} a & b & -c \\ b & c & -a \\ c & a & -b \end{pmatrix}.$$

Because the lines are distinct, we may assume $b^2 \neq ac$. By row operations (adding first two rows to the last row), we see the system has a unique solution (x, y) that represents the intersection point.

Hints and Answers for Chapter 3

3.1 (a) True. If $Ax = \lambda x$, $0 = A^2 x = A(Ax) = \lambda^2 x$. As $x \neq 0$, $\lambda = 0$.

(b) True. Consider possible Jordan blocks of A.

(c) $r(A) = 0, 1, 2$. The Jordan form can have at most two $J_2 = \begin{pmatrix} 0 & 1 \\ 0 & 0 \end{pmatrix}$.

(d) False if $n \geq 6$. Take $A = \begin{pmatrix} J_2 & 0 & 0 \\ 0 & J_2 & 0 \\ 0 & 0 & J_2 \end{pmatrix}$, $J_2 = \begin{pmatrix} 0 & 1 \\ 0 & 0 \end{pmatrix}$. True if $n \leq 5$.

(e) False in general. True if A is real: $|\det(A^2)| = |\det(A)|^2$.

(f) False. Take $A = \begin{pmatrix} 0 & 1 \\ -1 & 0 \end{pmatrix}$.

(g) False. Take $A = \begin{pmatrix} 0 & 1 \\ -1 & 0 \end{pmatrix}$.

(h) True. If $AA^* v = \lambda v$, $v \neq 0$, then $v^* AA^* v = \lambda v^* v \geq 0$.

(i) False. Take $A = \begin{pmatrix} 1 & 1 \\ -\frac{1}{2} & 0 \end{pmatrix}$.

(j) True. Because $\operatorname{tr}(A^* A) = \sum_{i,j} |a_{ij}|^2 = 0$ implies all $a_{ij} = 0$.

(k) False if $n \geq 3$. True if $n \leq 2$.

(l) True.

(m) True. Fact: If N is a normal matrix and $N^2 = 0$, then $N = 0$. Since $A^* - A$ is skew-Hermitian (normal), $(A^* - A)^2 = 0 \Rightarrow A^* - A = 0$. So A is Hermitian, and all eigenvalues of A are real.

(n) True. Since $(A^* - A)^2 = 0$, A is Hermitian. $A^3 = 0$ says that the eigenvalues of A are all equal to zero. Thus, $A = 0$.

3.2 If $AB = BA$, then equality holds. Conversely, suppose $(A + B)^2 = A^2 + 2AB + B^2$. Since $(A + B)^2 = A^2 + AB + BA + B^2$, we have $AB + BA = 2AB$. Thus, $AB = BA$.

3.3 If $AB = A - B$, one may check that $(A+I)(I-B) = I$. So $I - B$ is the inverse of $A + I$. Thus $(A+I)(I-B) = (I-B)(A+I)$. This implies $AB = BA$. The case of $AB = A + B$ is similar as $(A-I)(B-I) = I$.

3.4 A and B have distinct eigenvalues 1, 2, and 3 for any a, b, c, x, y, z. So they are both diagonalizable, and they are similar to $\operatorname{diag}(1, 2, 3)$.

3.5 1, 0, 0 are the eigenvalues of A and B. Since $zr \neq 0$, A and B both have rank 2. A and B have the same Jordan form $\begin{pmatrix} 1 & 0 & 0 \\ 0 & 0 & 1 \\ 0 & 0 & 0 \end{pmatrix}$.

291

3.6 $a = 0$, $b = -2$. A and B have the same eigenvalues $-2, -1, 2$.

3.7 A is a rank one matrix. Let $u = (a, b, c, d)^t = (1, \sqrt{2}, \sqrt{3}, 2)^t$. Then $A = uu^t$. So the nonzero eigenvalue of A is $u^t u = 10$. A is similar to the 4×4 matrix B with $(1,1)$-entry $u^t u = 10$ and all other entries 0. $A^{10} = (uu^t)(uu^t) \cdots (uu^t) = u(u^t u) \cdots (u^t u)u^t = (u^t u)^9 (uu^t) = 10^9 A$.

3.8   If A is an $n \times n$ matrix only similar to itself, then $P^{-1}AP = A$ (i.e., $AP = PA$) for all $n \times n$ invertible matrices P. Taking $P = \text{diag}(1, \ldots, n)$ reveals that $A = \text{diag}(a_1, \ldots, a_n)$ is diagonal. Setting $P = I + E_{1i}$, where E_{1i} has 1 in the $(1, i)$ position and 0 elsewhere, yields $a_1 = a_i$, $i = 2, \ldots, n$. Thus, $A = aI$ is a scalar matrix.

3.9   If $AB = BA$, then there exists a unitary matrix U such that $A = U^* D_1 U$ and $B = U^* D_2 U$, where D_1 and D_2 are upper-triangular matrices with the eigenvalues of A and B on the respective diagonals, say $\lambda_1, \lambda_2, \ldots, \lambda_n$ and $\mu_1, \mu_2, \ldots, \mu_n$ (from top-left to bottom-right). Then the eigenvalues of $A + B$ and AB are respectively $\lambda_1 + \mu_1, \lambda_2 + \mu_2, \ldots, \lambda_n + \mu_n$, and $\lambda_1 \mu_1, \lambda_2 \mu_2, \ldots, \lambda_n \mu_n$. Note: This need not be true if A and B do not commute. Take $A = \left(\begin{smallmatrix} 0 & 1 \\ 0 & 0 \end{smallmatrix}\right)$ and $B = \left(\begin{smallmatrix} 0 & 0 \\ 1 & 0 \end{smallmatrix}\right)$.

3.10 (a) By direct verifications.

(b) Two matrices of the same size are equivalent if and only if they have the same rank. A, B, C, D are all invertible, having rank 3. E has rank 2. So, B, C, and D are equivalent to A. A, B, D are similar because they have the same set of distinct eigenvalues.

3.11 B, C, E, G.

3.12 The eigenvalues of A are 2, 2, -1. A is diagonalizable for $a = 0$ and arbitrary b and c. If $a \neq 0$, then $r(2I - A) = 2 \neq 1$, and A is not diagonalizable. For B, the eigenvalues of B are 1, 1, 1, -1. If $x = 0$, then B is diagonalizable. If $x \neq 0$, then B is not diagonalizable because $r(I - B) = 2 \neq 1$.

3.13 The eigenvalues of A are $-1, -1, 5$; and corresponding eigenvectors are $u_1 = (-1, 1, 0)^t$, $u_2 = (-1, 0, 1)^t$, $u_3 = (1, 1, 1)^t$. $P = (u_1, u_2, u_3)$.

3.14 The eigenvalues of A are $3, 3, 1$, $r(A - 3I) = 2$, so A is not diagonalizable (or the rank would be 1). Alternatively, 3 is a repeated eigenvalue, but its eigenspace has dimension 1. So A does not have three linearly independent eigenvectors, and is not diagonalizable.

3.15 $\det(\lambda I - A) = (\lambda + 1)(\lambda - 1)^2$. So A has eigenvalues -1 and 1 (with algebraic multiplicity 2). Solving $(I + A)x = 0$ gives the eigenvectors corresponding to -1 in the form $k\eta_1$, $\eta_1 = (1, -1, 0)^t, k \in \mathbb{F}$. Thus, $V_{-1} = \{k(1, -1, 0)^t \mid k \in \mathbb{F}\}$. For the repeated eigenvalue 1, solving $(I - A)x = 0$, we get a basis $\{\eta_2 = (1, 0, 1)^t, \eta_3 = (1, 1, 0)^t\}$ for the null space of $I - A$. Therefore, η_2 and η_3 are linearly independent eigenvectors that span the eigenspace $V_1 = \{s\eta_2 + t\eta_3 \mid s, t \in \mathbb{F}\}$.

A is diagonalizable: $P^{-1}AP = \text{diag}(-1, 1, 1)$, where $P = (\eta_1, \eta_2, \eta_3)$.

3.16 The eigenvalues are $\frac{1 \pm \sqrt{1+4a^2}}{2}$, and the corresponding eigenvectors are $(2a, -1 \pm \sqrt{1 + 4a^2})$ (not unique). Note: They are orthogonal.

3.17 $AE_{ij} = (0, \ldots, 0, Ae_i, 0, \ldots, 0)$ is the $n \times n$ matrix whose j-th column is the i-th column of A, and 0 elsewhere. $E_{ij}A$ is the $n \times n$ matrix whose i-th row is the j-th row of A, and 0 elsewhere. For $A = (a_{pq})$, $E_{ij}AE_{st}$ is the $n \times n$ matrix with the (i, t)-entry a_{js}, and 0 elsewhere.

3.18 $A^2 = -4A$, $A^6 = -2^{10}A$. The eigenvalues of A are $-4, 0, 0, 0$.

3.19 $A = P^{-1} \begin{pmatrix} 5 & 0 \\ 0 & -1 \end{pmatrix} P$, where $P = \begin{pmatrix} 1 & 1 \\ -2 & 1 \end{pmatrix}$. Thus

$$A^{100} = P^{-1} \begin{pmatrix} 5^{100} & 0 \\ 0 & 1 \end{pmatrix} P = \frac{1}{3} \begin{pmatrix} 5^{100} + 2 & 5^{100} - 1 \\ 2 \cdot 5^{100} - 2 & 2 \cdot 5^{100} + 1 \end{pmatrix}.$$

By computation, $B^2 = I$. Thus, $B^{100} = I$.

3.20 Notice that

$$\begin{pmatrix} 2 & 1 \\ 2 & 3 \end{pmatrix} = \begin{pmatrix} -1 & 1 \\ 1 & 2 \end{pmatrix} \begin{pmatrix} 1 & 0 \\ 0 & 4 \end{pmatrix} \begin{pmatrix} -1 & 1 \\ 1 & 2 \end{pmatrix}^{-1}.$$

It follows that

$$\begin{pmatrix} 2 & 1 \\ 2 & 3 \end{pmatrix}^k = \frac{1}{3} \begin{pmatrix} 2 + 2^{2k} & 2^{2k} - 1 \\ 2^{2k+1} - 2 & 2^{2k+1} + 1 \end{pmatrix}.$$

$$\begin{pmatrix} 0 & 1 & 0 \\ 0 & 0 & 1 \\ 0 & 0 & 0 \end{pmatrix}^k = \begin{pmatrix} 0 & 0 & 1 \\ 0 & 0 & 0 \\ 0 & 0 & 0 \end{pmatrix} \text{ if } k = 2, \; = 0_3 \text{ if } k > 2.$$

$$\begin{pmatrix} 0 & 1 & 0 \\ 0 & 0 & 1 \\ 1 & 0 & 0 \end{pmatrix}^k = \begin{pmatrix} 0 & 1 & 0 \\ 0 & 0 & 1 \\ 1 & 0 & 0 \end{pmatrix}, \quad \text{when } k = 3m + 1,$$

$$\begin{pmatrix} 0 & 1 & 0 \\ 0 & 0 & 1 \\ 1 & 0 & 0 \end{pmatrix}^k = \begin{pmatrix} 0 & 0 & 1 \\ 1 & 0 & 0 \\ 0 & 1 & 0 \end{pmatrix}, \quad \text{when } k = 3m + 2,$$

and I_3 otherwise. For the matrix of λ (with $\lambda^0 = 1$ for any λ),

$$\begin{pmatrix} \lambda & 1 & 0 \\ 0 & \lambda & 1 \\ 0 & 0 & \lambda \end{pmatrix}^k = \begin{pmatrix} \lambda^k & k\lambda^{k-1} & \frac{k(k-1)}{2}\lambda^{k-2} \\ 0 & \lambda^k & k\lambda^{k-1} \\ 0 & 0 & \lambda^k \end{pmatrix}.$$

3.21 Computations yield $A^k = \begin{pmatrix} 1 & k \\ 0 & 1 \end{pmatrix}$ and $PA^kP^{-1} = A$, where $P = \begin{pmatrix} 1 & 1 \\ 0 & k \end{pmatrix}$. For the general case, if A is an $n \times n$ matrix with all eigenvalues equal to 1, then A^k is similar to A. To see this, consider a Jordan form. It suffices to show the case where A itself is an $n \times n$ Jordan block. Then $r(I - A^k) = n - 1$. Suppose that the Jordan blocks of A^k are $J_1, J_2, \ldots, J_s$, $s \geq 2$, and $Q^{-1}A^kQ = \text{diag}(J_1, J_2, \ldots, J_s)$. Then

$$r(I - A^k) = r\big(Q^{-1}(I - A^k)Q\big) = r(I - Q^{-1}A^kQ) \leq n - 2,$$

a contradiction. Thus $s = 1$ and A^k is similar to A.

3.22 $vu^t = 3$. $A^k = u^tvu^tv\cdots u^tv = u^t(vu^t)^{k-1}v = 3^{k-1}u^tv = 3^{k-1}A$.

3.23 $\det(\lambda I + A) = 0$ has a finite number of roots; $\lambda = 0$ is a root. So, there is $\delta > 0$ such that $\det(\lambda I + A) = 0$ has no solution in $(0, \delta)$.

3.24 If A is nonsingular, then take $r = t = 0$. Suppose $\det A = 0$. Let $f(\lambda) = \det(A + \lambda B)$. Since $f(0) = 0$ and $f(z) \neq 0$, $f(\lambda)$ is not a constant. As a polynomial in λ, $f(\lambda) = 0$ has a finite number of zeros. Thus, $f(r) \neq 0$ for some real r, that is, $A + rB$ is nonsingular.

3.25 This is immediate by the Cayley–Hamilton Theorem. Let $p(\lambda) = \det(\lambda I - A)$ be the characteristic polynomial of A. Then $p(\lambda) = (\lambda - \lambda_1)\cdots(\lambda - \lambda_n)$ and $p(A) = (A - \lambda_1 I)\cdots(A - \lambda_n I) = 0$. Note that it is false to plug A in $\det(\lambda I - A)$ to get $p(A) = \det(A - A) = 0$.

3.26 Take $A = \begin{pmatrix} 1 & 0 \\ 0 & 0 \end{pmatrix}$ and $B = I_2$. Then $f(\lambda) = \det(A + \lambda I_2) = \lambda^2 + \lambda$ and $f(B) = 2I_2$. A computation shows $\det f(B) = 4$, but $\det(A + B) = 2$.

3.27 (a) If $Ax = \lambda x$, $x \neq 0$, then $A^2x = A(Ax) = \lambda(Ax) = \lambda^2 x$. Inductively, $A^kx = A(A^{k-1}x) = \lambda^{k-1}(Ax) = \lambda^k x$. So, λ^k is an eigenvalue of A^k. Alternatively, let U^*AU be a Schur triangular form of A. Then $(U^*AU)^k = U^*A^kU$, so λ^k is an eigenvalue of A^k. It does not hold for singular values. Take $A = \begin{pmatrix} 0 & 1 \\ 0 & 0 \end{pmatrix}$, $k = 2$.

(b) Let U^*AU be a Schur triangularization of A. Then $f(U^*AU) = U^*f(A)U$ and $f(\lambda)$ is an eigenvalue of $f(A)$.

(c) If $A = P^{-1}BP$, then $f(A) = f(P^{-1}BP) = P^{-1}f(B)P$.

(d) Let $A = (a_{ij})$, $P = \text{diag}(p_1, p_2, \ldots, p_n)$, $Q = \text{diag}(q_1, q_2, \ldots, q_n)$. Then $AP = QA \Rightarrow a_{ij}p_j = a_{ij}q_i \Rightarrow (p_j - q_i)a_{ij} = 0 \Rightarrow a_{ij} = 0$ or $p_j = q_i \Rightarrow (f(p_j) - f(q_i))a_{ij} = 0 \Rightarrow Af(P) = f(Q)A$.

3.28 By direct computations.

3.29 (a) Take $A = \left(\begin{smallmatrix} 2 & 0 \\ 0 & 2 \end{smallmatrix}\right)$ and $B = \left(\begin{smallmatrix} 1 & 0 \\ 0 & 0 \end{smallmatrix}\right)$. Then $f(\lambda) = \lambda^2 - \lambda$. Compute to get $f(|A|) = f(4) = 12$, $|f(A)| = |A^2 - A| = 4$, $|A - B| = 2$.

(b) $f(A)$ is a matrix, $|A-B|$ is a scalar; they cannot be equal if $n > 1$.

(c) If the eigenvalues of A are $\lambda_1, \ldots, \lambda_n$, then the eigenvalues of $f(A)$ are $f(\lambda_1), \ldots, f(\lambda_n)$. So $f(A)$ is invertible if and only if $f(\lambda_i) \neq 0$ for every i, i.e., A and B have no common eigenvalues.

3.30 $A\left(\begin{smallmatrix} 1 \\ 1 \end{smallmatrix}\right) = 2a\left(\begin{smallmatrix} 1 \\ 1 \end{smallmatrix}\right)$ and $A\left(\begin{smallmatrix} 1 \\ -1 \end{smallmatrix}\right) = 2b\left(\begin{smallmatrix} 1 \\ -1 \end{smallmatrix}\right)$. Thus, $2a$ and $2b$ are the eigenvalues of A, and $P^{-1}AP = \left(\begin{smallmatrix} 2a & 0 \\ 0 & 2b \end{smallmatrix}\right)$, where $P = \left(\begin{smallmatrix} 1 & 1 \\ 1 & -1 \end{smallmatrix}\right)$.

3.31 A is a normal matrix with eigenvalues $a+bi$ and $a-bi$, corresponding unit eigenvectors are (not unique) $\left(\frac{1}{\sqrt{2}}, -\frac{i}{\sqrt{2}}\right)^t$ and $\left(\frac{1}{\sqrt{2}}, \frac{i}{\sqrt{2}}\right)^t$. Thus

$$\begin{pmatrix} \frac{1}{\sqrt{2}} & \frac{i}{\sqrt{2}} \\ \frac{1}{\sqrt{2}} & -\frac{i}{\sqrt{2}} \end{pmatrix} \begin{pmatrix} a & -b \\ b & a \end{pmatrix} \begin{pmatrix} \frac{1}{\sqrt{2}} & \frac{1}{\sqrt{2}} \\ -\frac{i}{\sqrt{2}} & \frac{i}{\sqrt{2}} \end{pmatrix} = \begin{pmatrix} a+bi & 0 \\ 0 & a-bi \end{pmatrix}.$$

3.32 (a) A is singular because $\left(\begin{smallmatrix} I & 0 \\ u & 1 \end{smallmatrix}\right)A = \left(\begin{smallmatrix} B & -Bv \\ 0 & 0 \end{smallmatrix}\right)$ has a zero row.

(b) It suffices to show that A has two linearly independent eigenvectors associated to eigenvalue 0. This is seen by verifying that

$$A\begin{pmatrix} v \\ 1 \end{pmatrix} = 0 \quad \text{and} \quad A\begin{pmatrix} x \\ 0 \end{pmatrix} = 0,$$

where x is a nonzero solution to $Bx = 0$.

(c) Take $P = \left(\begin{smallmatrix} I & 0 \\ u & 1 \end{smallmatrix}\right)$. Then $PAP^{-1} = \left(\begin{smallmatrix} B(I+vu) & -Bv \\ 0 & 0 \end{smallmatrix}\right)$ and

$$|\lambda I_{n+1} - A| = \lambda|\lambda I_n - B(I_n + vu)|.$$

If λ^2 divides $|\lambda I - A|$, then B or $I+vu$ is singular. Since $|I+vu| = 1 + uv$, λ^2 divides $|\lambda I - A|$ if and only if $|B| = 0$ or $uv = -1$.

3.33   (a) $\operatorname{tr} A = \operatorname{tr}(uv^*) = \operatorname{tr}(v^*u) = v^*u$ as v^*u is a scalar.

   (b) $A^2 = (uv^*)(uv^*) = u(v^*u)v^* = (\operatorname{tr} A)A$.

   (c) $Au = uv^*u = (v^*u)u$. So, u is an eigenvector of A associated with the eigenvalue v^*u. Since the rank of A is 1, A has at most one nonzero eigenvalue. Because $\operatorname{tr} A = v^*u$ is equal to the sum of all eigenvalues of A, the eigenvalues of A are $v^*u, 0, \ldots, 0$.

   (d) $A^*A = (uv^*)^*(uv^*) = vu^*uv^* = (u^*u)vv^*$. The eigenvalues of A^*A are $(u^*u)(v^*v), 0, \ldots, 0$. Note that $x^*x = \|x\|^2$ for any column vector x. So, the singular values of A are $\|u\|\|v\|, 0, \ldots, 0$.

   (e) $w^*v = 0 \Rightarrow v^*w = 0 \Rightarrow Aw = uv^*w = 0 = 0w$. So w is an eigenvector of A corresponding to eigenvalue 0.

   (f) A has rank one with eigenvalues $\operatorname{tr} A, 0, \ldots, 0$, by Jordan decomposition, A is similar to $\operatorname{diag}(\operatorname{tr} A, 0, \ldots, 0)$.

   (g) This is nearly obvious because A is similar to $\operatorname{diag}(\operatorname{tr} A, 0, \ldots, 0)$.

   (h) With $A^2 = (v^*u)(uv^*)$, compute $(I + A)(I - \frac{1}{1+v^*u}A)$ to get I.

3.34   (a) $(A + uv^*)u = Au + uv^*u = \lambda u + (v^*u)u = (\lambda + v^*u)u$.

   (b) Let $u_1 = u/\|u\|$ (a unit vector). Let $U = (u_1, U_1)$ be an $n \times n$ unitary matrix with u_1 as the first column of U. Then $U^*U = I$ reveals $U^*u_1 = \begin{pmatrix} u_1^* \\ U_1^* \end{pmatrix} u_1 = \begin{pmatrix} 1 \\ 0 \end{pmatrix}$ and $U_1^*u_1 = 0 = U_1^*u$. Compute

$$U^*AU = \begin{pmatrix} u_1^* \\ U_1^* \end{pmatrix} (\lambda u_1, AU_1) = \begin{pmatrix} \lambda & \star \\ 0 & U_1^*AU_1 \end{pmatrix}.$$

So, the eigenvalues of A are λ and those of $U_1^*AU_1$. Thus, λ_2, $\ldots, \lambda_n$ are the eigenvalues of $U_1^*AU_1$. Since $u_1^*uv^*u = v^*u$ and

$$U^*uv^*U = \begin{pmatrix} u_1^* \\ U_1^* \end{pmatrix} uv^*(u_1, U_1) = \begin{pmatrix} v^*u & \star \\ 0 & 0 \end{pmatrix},$$

we have

$$
\begin{aligned}
U^*(A + uv^*)U &= U^*AU + U^*uv^*U \\
&= \begin{pmatrix} \lambda & \star \\ 0 & U_1^*AU_1 \end{pmatrix} + \begin{pmatrix} v^*u & \star \\ 0 & 0 \end{pmatrix} \\
&= \begin{pmatrix} \lambda + v^*u & \star \\ 0 & U_1^*AU_1 \end{pmatrix}.
\end{aligned}
$$

So, the eigenvalues of $A + uv^*$ are $\lambda + v^*u$ and $\lambda_2, \ldots, \lambda_n$.

3.35 It is routine to show that α is a linearly independent set; so is β. Since $\cos 2x = 2\cos^2 x - 1$, $\operatorname{Span}\alpha \subseteq \operatorname{Span}\beta$. Since they both have dimension 3, $\operatorname{Span}\alpha = \operatorname{Span}\beta$. $\alpha = \beta T$, where $T = \begin{pmatrix} 1 & 0 & -1 \\ 0 & 1 & 0 \\ 0 & 0 & 2 \end{pmatrix}$.

3.36 $\det(\lambda I - M) = \det(\lambda^2 I - BA)$ is an even function in λ.

3.37 (a) False. (b) True. (c) True. (d) False. (e) True. (f) False. (g) False. (h) True. (i) True. (j) False. (k) False. (l) True. (m) False. (n) False. (o) True. (p) True. (q) False. (r) False. True if λ_i's are distinct. (s) True. (t) False. (u) True. (v) False. (w) True. (x) False. (y) True. (z) False.

3.38 A has two different complex eigenvalues i and $-i$. So A is diagonalizable over $\mathbb{C}$, but not over $\mathbb{R}$. The eigenvectors of i and $-i$ are respectively $(1, i)^t$ and $(1, -i)^t$. Let $P = \begin{pmatrix} 1 & 1 \\ i & -i \end{pmatrix}$. Then $P^{-1}AP = \begin{pmatrix} i & 0 \\ 0 & -i \end{pmatrix}$.

3.39 Let $B = (b_{ij})$. Then $AB = BA$ implies $a_i b_{ij} = b_{ij} a_j$. For $i \neq j$, $a_i \neq a_j$ yields $b_{ij} = 0$. Thus, all off-diagonal entries of B are zero.

3.40 Observe $AB = BA^{-1}$ and it implies $AB^k = B^k A$ for any even positive integer k. (Note: $AB^k = B^k A^{-1}$ if k is odd.) In particular, $AB^2 = B^2 A$. Since the eigenvalues of A are distinct, B^2 is diagonalizable.

3.41 Since $A^2 = A$, the eigenvalues of A are 1's and 0's. Consider a Schur triangularization $A = U^*TU$ of A, where U is unitary and T is upper-triangular. Because $A \neq 0$ and $A^2 = A$, $T \neq 0$ and $T^2 = T$. So T has at least one 1 (the eigenvalue of A) on the diagonal. Hence, $\operatorname{tr} A > 0$.

3.42 Fact: if $\operatorname{tr} X = \operatorname{tr} X^*$ for square matrix X, then $\operatorname{tr} X$ is real. Now show $\operatorname{tr}\left((X^*)^p(XX^*)^q X^p\right)$ is real (nonnegative actually). Since $A = A^t$, $\bar{A} = A^*$. Now use the fact $\operatorname{tr}(XY) = \operatorname{tr}(YX)$ to compute

$$
\begin{aligned}
\operatorname{tr}\left(A^p \bar{A}^p (A\bar{A})^q\right) &= \operatorname{tr}\left(\bar{A}^p (A\bar{A})^q A^p\right) = \operatorname{tr}\overline{\left(\bar{A}^p (A\bar{A})^q A^p\right)} \\
&= \operatorname{tr}\left(A^p (\bar{A}A)^q \bar{A}^p\right) = \operatorname{tr}\left((\bar{A}A)^q \bar{A}^p A^p\right).
\end{aligned}
$$

3.43 Replace A with U^*A^tU on the left and cancel the U's and U^*'s to get

$$
\operatorname{tr}\left((A^t)^a \bar{A}^b (A^t)^c \bar{A}^d\right) = \operatorname{tr}\left((A^t)^a \bar{A}^b (A^t)^c \bar{A}^d\right)^t = \operatorname{tr}(A^{*d} A^c A^{*b} A^a).
$$

For the 2nd identity, set $a = b = p$, $c = d = q$. Use $\operatorname{tr}(XY) = \operatorname{tr}(YX)$.

3.44   Let $\lambda_1, \ldots, \lambda_n$ be the eigenvalues of A. Then

$$0 = \operatorname{tr} A^2 - 2 \operatorname{tr} A^3 + \operatorname{tr} A^4 = \sum_{i=1}^{n} \lambda_i^2 (1 - \lambda_i)^2.$$

Because all λ_i's are real, we arrive at $\lambda_i = 0$ or 1, $i = 1, \ldots, n$. Since

$$\operatorname{tr} A^2 = \sum_{i=1}^{n} \lambda_i^2 = \operatorname{tr} A = c,$$

c is a nonnegative integer, and c of the λ_i's equal 1, others 0.

If $A^m = A^{m+1}$ for some m, then the eigenvalues of A are 0's and 1's. So, A^k and A have the same eigenvalues, and $\operatorname{tr} A^k = \operatorname{tr} A$ for all k.

3.45   It is sufficient to show that all the eigenvalues of A are equal to zero. Let $\lambda_1, \lambda_2, \ldots, \lambda_n$ be the eigenvalues of A. Then

$$\operatorname{tr} A^k = 0, \quad k = 1, 2, \ldots, n,$$

is equivalent to

$$\lambda_1^k + \lambda_2^k + \cdots + \lambda_n^k = 0, \quad k = 1, 2, \ldots, n.$$

If all the λ_i's are the same, they are obviously equal to zero. Otherwise, suppose that $\lambda_{i_1}, \ldots, \lambda_{i_m}$ are the distinct nonzero eigenvalues of A, $2 \leq m \leq n$. The above equations can be written as

$$l_1 \lambda_{i_1}^k + l_2 \lambda_{i_2}^k + \cdots + l_m \lambda_{i_m}^k = 0, \quad k = 1, 2, \ldots, n,$$

where l_i's are positive integers. Consider the linear equation system

$$\lambda_{i_1}^k x_1 + \lambda_{i_2}^k x_2 + \cdots + \lambda_{i_m}^k x_m = 0, \quad k = 1, 2, \ldots, m.$$

An application of the Vandermonde determinant yields that the equation system has only the trivial solution 0, contradicting l_i's being positive integers. So, all the eigenvalues of A are equal to zero.

3.46   In the expansion of $(A + B)^k$, there are four kinds of terms: A^k, B^k, $B^m A^{k-m}$, and other terms each have a factor $AB = 0$. Note that

$$\operatorname{tr}(B^m A^{k-m}) = \operatorname{tr}\left((B^{m-1} A^{k-m-1})(AB)\right) = 0.$$

3.47   Proof 1: Calculate the eigenvalues of A to obtain $\alpha = \frac{1+\sqrt{5}}{2}$ and $\beta = \frac{1-\sqrt{5}}{2}$. Then $\operatorname{tr} A^k = \alpha^k + \beta^k$. One may verify directly that

$$\alpha^k + \beta^k = \alpha^{k-1} + \beta^{k-1} + \alpha^{k-2} + \beta^{k-2}.$$

Proof 2: A^k is in the form $\begin{pmatrix} x & y \\ y & x+y \end{pmatrix}$ for all k. Let $A^{k-2} = \begin{pmatrix} a & b \\ b & a+b \end{pmatrix}$.
Then $A^{k-1} = \begin{pmatrix} b & a+b \\ a+b & a+2b \end{pmatrix}$ and $A^k = \begin{pmatrix} a+b & a+2b \\ a+2b & 2a+3b \end{pmatrix}$. It follows that

$$\operatorname{tr} A^k = 3a + 4b = (a+3b) + (2a+b) = \operatorname{tr}(A^{k-1}) + \operatorname{tr}(A^{k-2}).$$

Note: Let $a_k = \operatorname{tr} A^k$. Then $\{a_k\} = \{1, 3, 4, 7, 11, 18, 29, 47, \dots\}$ is a sequence satisfying $a_k = a_{k-1} + a_{k-2}$ with initial values $a_1 = 1, a_2 = 3$.

3.48   (a)   Use the same elementary row and column operations on A to get B and C, that is, $A, B,$ and C are permutation-similar.

(b)   The product of any two of $A, B,$ and C is a matrix with entries either $a^2 + b^2 + c^2$ or $ab + ac + bc$. If (i) holds, say $AC = CA$. By comparison of entries, $a^2 + b^2 + c^2 = ab + ac + bc$. Thus, $\frac{1}{2}(a-b)^2 + \frac{1}{2}(a-c)^2 + \frac{1}{2}(b-c)^2 = 0$. If a, b, c are real, then $a = b = c$. All other parts follow immediately. Now we show (v) implies (iv). Note that A, B, C are real symmetric. (a) says they have the same rank. If the rank is 0, then $A = B = C = 0$. If the rank is one, by considering the 2×2 minors, we see $a = b = c$, so $A = B = C$. The possible nonzero eigenvalue is $\operatorname{tr} A = a + b + c$.

(c)   (i) Let μ be a nonreal number such that $\mu^2 - \mu + 1 = 0$, i.e., $\mu^2 + 1 = \mu$. Set $a = 1, b = 0, c = \mu$. One may check that $AC = CA$. Note that the rank of A (and also B and C) is 2, while A has only one nonzero eigenvalue which is $\operatorname{tr} A = c + 1$.
(ii) If two of A, B, C commute, then $a^2 + b^2 + c^2 - ab - bc - ca = 0$, and they all commute. (iii) Multiply $a^2 + b^2 + c^2 - ab - bc - ca = 0$ by $a + b + c$ and simplify to get $a^3 + b^3 + c^3 - 3abc = 0$. A computation reveals $|\lambda I - A| = \lambda^3 - (a+b+c)\lambda^2$. It follows that A (B and C also) has at least two zero eigenvalues. In the commuting case, possible ranks of A (B and C) are 0, 1, 2.

3.49   The characteristic polynomial of A is $\lambda^3 - 1$. So the eigenvalues of A are $1, \omega_1 = \frac{-1+\sqrt{3}i}{2}$, and $\omega_2 = \omega_1^2 = \frac{-1-\sqrt{3}i}{2}$. $B = aI_3 + bA + cA^2$. So the eigenvalues of B are $a + b + c$, $a + b\omega_1 + c\omega_1^2$, and $a + b\omega_2 + c\omega_2^2$.

3.50   The eigenvalues of B are $0, 1, c$. $\det A = -(a-b)^2$. If 0 is an eigenvalue of A, then $a = b$. If 1 is an eigenvalue of A, then $ab = 0$. Thus,

$a = b = 0$. The eigenvalues of A are 0, 1, 2. So, $c = 2$.

$$T = \begin{pmatrix} \frac{1}{\sqrt{2}} & 0 & \frac{1}{\sqrt{2}} \\ 0 & 1 & 0 \\ -\frac{1}{\sqrt{2}} & 0 & \frac{1}{\sqrt{2}} \end{pmatrix}.$$

3.51 Since A and B are real symmetric, Schur triangularization theorem (or the spectral decomposition theorem in Chapter 4) ensures that A and B are unitarily diagonalizable via real orthogonal matrices. A has rank one with eigenvalues $3, 0, 0$, so A is similar to $\mathrm{diag}(3, 0, 0)$. B has eigenvalues $-1, 0, 2$, so B is similar to $\mathrm{diag}(-1, 0, 2)$. Solving $(\lambda I - A)x = 0$ and $(\lambda I - B)x = 0$ for each eigenvalue of A and B, we obtain orthonormal eigenvectors of A and B, respectively. Construct P and Q. $P^t A P = \mathrm{diag}(3, 0, 0)$ and $Q^t B Q = \mathrm{diag}(-1, 0, 2)$, where

$$P = \frac{1}{\sqrt{6}} \begin{pmatrix} \sqrt{2} & \sqrt{3} & 1 \\ \sqrt{2} & 0 & -2 \\ \sqrt{2} & -\sqrt{3} & 1 \end{pmatrix}, \quad Q = \frac{1}{\sqrt{6}} \begin{pmatrix} \sqrt{2} & 0 & 2 \\ -\sqrt{2} & \sqrt{3} & 1 \\ -\sqrt{2} & -\sqrt{3} & 1 \end{pmatrix}.$$

Note that P and Q need not be unique. The following matrix H is real symmetric and orthogonal, and satisfies $H^t A H = \mathrm{diag}(3, 0, 0)$.

$$H = \frac{1}{3+\sqrt{3}} \begin{pmatrix} -1-\sqrt{3} & -1-\sqrt{3} & -1-\sqrt{3} \\ -1-\sqrt{3} & 2+\sqrt{3} & -1 \\ -1-\sqrt{3} & -1 & 2+\sqrt{3} \end{pmatrix}.$$

See Problems 4.6, 5.94, and 6.132.

3.52 The eigenvalues of A are $1, 1, -1$. A necessary condition for A to be diagonalizable is that the rank of $I - A$ is 1, which implies $a + b = 0$.

3.53 $a = b = 0$, or $a \neq b$ and $ab \neq 0$ (A has three distinct eigenvalues).

3.54 A, B, and C have the same characteristic polynomial $p(t) = (t-1)^4$. A and B have the same minimal polynomial $m(t) = (t-1)^2$. The minimal polynomial of C is $m(t) = (t-1)^3$. Note: they are not similar.

3.55 For P, $p(t) = m(t) = t(t-2)(t-1)^2$. For Q, $p(t) = m(t) = t^4 - 4t^3 - 3t^2 - 2t - 1$. For R, $p(t) = (t-1)^3(t-4)$, $m(t) = (t-1)(t-4)$.

3.56 Let $A \in M_n(\mathbb{C})$. If A is diagonalizable over $\mathbb{C}$ and if $\mu_1, \ldots, \mu_k$ are the distinct eigenvalues of A, then $m_A(\lambda) = (\lambda - \mu_1) \cdots (\lambda - \mu_k)$ is the minimal polynomial of A; it has no repeated roots. Conversely, if the minimal polynomial of A has no repeated roots, then A has no Jordan blocks of orders more than one. So the Jordan canonical form of A is a diagonal matrix, consequently A is diagonalizable.

3.57 (a) $\operatorname{tr} A = \sum_{k=1}^{n} x_k + i \sum_{k=1}^{n} y_k$ is real. So $\sum_{k=1}^{n} y_k = 0$.

(b) Compute

$$\operatorname{tr} A^2 = \sum_{k=1}^{n} \lambda_k^2 = \sum_{k=1}^{n} x_k^2 - \sum_{k=1}^{n} y_k^2 + 2i \sum_{k=1}^{n} x_k y_k.$$

Since A is real, $\operatorname{tr} A^2$ is real, so $\sum_{k=1}^{n} x_k y_k = 0$.

(c) See (b).

3.58 Let $Ax = \lambda x$, where $x = (x_1, x_2, \ldots, x_n)^t \neq 0$. Then for each i, $\sum_{i \neq j} a_{ij} x_j = (\lambda - a_{ii}) x_i$. Let $|x_k| = \max\{|x_1|, |x_2|, \ldots, |x_n|\} > 0$, where k is some positive integer. Then $(\lambda - a_{kk}) x_k = \sum_{j \neq k} a_{kj} x_j$. So

$$|\lambda - a_{kk}| \leq \sum_{j \neq k} |a_{kj}(x_j/x_k)| \leq \sum_{j \neq k} |a_{kj}|.$$

3.59 By Schur decomposition, let U be a unitary matrix such that $T = U^* A U$ is upper-triangular with the eigenvalues λ_i's of A on the main diagonal. Compute the trace of $A^* A$ by adding the diagonal entries:

$$\operatorname{tr}(A^* A) = \sum_{i,j} |a_{ij}|^2 = \operatorname{tr}(T^* T) \geq \sum_{i} |\lambda_i|^2.$$

Equality occurs if and only if all $t_{ij} = 0$, $i < j$, i.e., T is diagonal.

3.60 By computation, $P^2 = P$, $\left(\begin{smallmatrix} I_n & 0 \\ -I_n & I_n \end{smallmatrix} \right) \left(\begin{smallmatrix} A & I_n-A \\ A & I_n-A \end{smallmatrix} \right) \left(\begin{smallmatrix} I_n & 0 \\ I_n & I_n \end{smallmatrix} \right) = \left(\begin{smallmatrix} I_n & I_n-A \\ 0 & 0 \end{smallmatrix} \right)$.

3.61 It is sufficient to notice that $\left(\begin{smallmatrix} I_n & I_n \\ 0 & I_n \end{smallmatrix} \right) \left(\begin{smallmatrix} A & A \\ B & B \end{smallmatrix} \right) \left(\begin{smallmatrix} I_n & -I_n \\ 0 & I_n \end{smallmatrix} \right) = \left(\begin{smallmatrix} A+B & 0 \\ B & 0 \end{smallmatrix} \right)$. Alternatively, compute $A + B = (I, I) \left(\begin{smallmatrix} A & 0 \\ 0 & B \end{smallmatrix} \right) \left(\begin{smallmatrix} I \\ I \end{smallmatrix} \right)$ which has the same nonzero eigenvalues as $\left(\begin{smallmatrix} A & 0 \\ 0 & B \end{smallmatrix} \right) \left(\begin{smallmatrix} I \\ I \end{smallmatrix} \right) (I, I) = \left(\begin{smallmatrix} A & 0 \\ 0 & B \end{smallmatrix} \right) \left(\begin{smallmatrix} I & I \\ I & I \end{smallmatrix} \right) = \left(\begin{smallmatrix} A & A \\ B & B \end{smallmatrix} \right)$.

3.62 If $u_1 + u_2$ is an eigenvector of A for an eigenvalue μ, we have $A(u_1 + u_2) = \mu(u_1 + u_2)$. However, $A(u_1 + u_2) = \lambda_1 u_1 + \lambda_2 u_2$. Subtracting these equations, we have $0 = (\mu - \lambda_1) u_1 + (\mu - \lambda_2) u_2$. So, u_1 and u_2 are linearly dependent. This is impossible because the eigenvectors belonging to distinct eigenvalues are linearly independent.

3.63 From $A(u_1, u_2, u_3) = (u_1, 2u_2, 3u_3)$, we have $A = (u_1, 2u_2, 3u_3)P^{-1}$, where $P = (u_1, u_2, u_3)$. We compute P^{-1} and multiply to get

$$A = \begin{pmatrix} \frac{7}{3} & 0 & -\frac{2}{3} \\ 0 & \frac{5}{3} & -\frac{2}{3} \\ -\frac{2}{3} & -\frac{2}{3} & 2 \end{pmatrix}.$$

3.64 If $c = 0$ and $a \neq d$, then $x = -b/(a - d)$. If $c = 0$ and $a = d$, then $b = 0$ and A is the scalar matrix aI_2, x is free. If $c \neq 0$, then

$$x = \left((a - d) \pm \sqrt{(a - d)^2 + 4bc} \right)/(2c).$$

3.65 (a) $A^{-1} = \begin{pmatrix} d & -b \\ -c & a \end{pmatrix}$.

(b) If $c \neq 0$, then $\begin{pmatrix} a & b \\ c & d \end{pmatrix} = \begin{pmatrix} 1 & \frac{a-1}{c} \\ 0 & 1 \end{pmatrix} \begin{pmatrix} 1 & 0 \\ c & 1 \end{pmatrix} \begin{pmatrix} 1 & \frac{d-1}{c} \\ 0 & 1 \end{pmatrix}$.

If $c = 0$, then $a \neq 0$. Consider $\begin{pmatrix} 1 & 0 \\ 1 & 1 \end{pmatrix} \begin{pmatrix} a & b \\ 0 & d \end{pmatrix} = \begin{pmatrix} a & b \\ a & b+d \end{pmatrix}$.

(c) Let λ_1 and λ_2 be the eigenvalues of A. Then

$$\lambda_1 \lambda_2 = |A| = 1 \quad \text{and} \quad \lambda_2 = \lambda_1^{-1}.$$

We claim $\lambda_1 \neq \lambda_2$. Otherwise $|\lambda_1| = |\lambda_2| = 1$, and

$$2 \geq |\lambda_1 + \lambda_2| = |a + d| > 2.$$

Thus, A is similar to the diagonal matrix $\begin{pmatrix} \lambda_1 & 0 \\ 0 & \lambda_1^{-1} \end{pmatrix}$, $\lambda_1 \neq 0, \pm 1$.

(d) If $|a + d| < 2$, then $|a + d| = |\lambda_1 + \lambda_2| = |\lambda_1 + \lambda_1^{-1}| < 2$, so λ_1 is neither real nor pure imaginary (as $|r + \frac{1}{r}| \geq 2$ for real $r \neq 0$).

(e) If $|a+d| = 2$, the possible real eigenvalues of A are $1, 1$ or $-1, -1$. The possible real matrices to which A is similar are

$$I, \ -I, \ \begin{pmatrix} 1 & 1 \\ 0 & 1 \end{pmatrix}, \ \begin{pmatrix} -1 & 1 \\ 0 & -1 \end{pmatrix}.$$

(f) If $|a + d| \neq 2$, then A has two distinct eigenvalues λ_1, λ_2. Thus A is similar to $\operatorname{diag}(\lambda_1, \lambda_2)$. On the other hand, the matrix

$$\begin{pmatrix} \frac{\lambda_1 + \lambda_2}{2} & \frac{\lambda_1 - \lambda_2}{2} \\ \frac{\lambda_1 - \lambda_2}{2} & \frac{\lambda_1 + \lambda_2}{2} \end{pmatrix}$$

has eigenvalues λ_1 and λ_2 with corresponding eigenvectors $(1, 1)^t$ and $(1, -1)^t$, respectively. So, it is also similar to $\operatorname{diag}(\lambda_1, \lambda_2)$.

(g) No. Take $A = \begin{pmatrix} 1 & 1 \\ 0 & 1 \end{pmatrix}$.

3.66 The eigenvalues are respectively $1 + i, 1 - i$; $2, 0$; $\frac{1 \pm \sqrt{3}i}{2}$; $1 + 2i, 1$.

3.67 The eigenvalues of A and B are 1 and 0. Thus both A and B are diagonalizable and they are similar. Let $U = \begin{pmatrix} a & c \\ b & d \end{pmatrix}$ be such that $U^* B U = A$. A computation shows that such U cannot be unitary. Alternatively, because A and B have different singular values.

3.68 A and A^t have the same Jordan canonical form by considering Jordan blocks in canonical forms. Every Jordan block J of size k, say, is similar to its transpose J^t via a backward identity matrix S:

$$SJ = J^t S \quad \text{or} \quad SJS^{-1} = J^t,$$

where S has $(i, k - i + 1)$-entry 1 and 0 elsewhere, $i = 1, 2, \ldots, k$. (Note: It is also proved by Smith normal forms of A and A^t, not covered in this book.) A and A^t need not be unitarily similar. See Example 3.5, or take $A = \begin{pmatrix} 0 & 1 & 0 \\ 0 & 0 & 2 \\ 0 & 0 & 0 \end{pmatrix}$.

If $A = \begin{pmatrix} i & 1 & 0 & 0 \\ 0 & i & 0 & 0 \\ 0 & 0 & -i & 0 \\ 0 & 0 & 0 & -i \end{pmatrix}$, then A and A^* have the same eigenvalues, but they are not similar because they have different Jordan blocks.

3.69 Use the fact that if $R_1 + iR_2$ is nonsingular, where $R_1, R_2 \in M_n(\mathbb{C})$, then $R_1 + tR_2$ is nonsingular for some real t (see Problem 3.24).

Reason: Let $d(\lambda) = \det(R_1 + \lambda R_2)$. Then $d(i) \neq 0$. So $d(\lambda)$ is not identical to zero, and it has at most n roots. Let t be such a real number that $d(t) \neq 0$. Then $R_1 + tR_2$ is invertible.

Let $P^{-1}AP = B$, where A and B are real, and P is complex. Let $P = R_1 + iR_2$, where R_1 and R_2 are real. Then $AR_1 = R_1B$ and $AR_2 = R_2B$. Since $P = R_1 + iR_2$ is nonsingular, $R = R_1 + tR_2$ is real invertible for some real t, and $AR = RB$. Thus, $R^{-1}AR = B$.

In general, if two matrices A and B with entries in a field $\mathbb{F}$ are similar over an extended field $\mathbb{E}$ (containing $\mathbb{F}$), then A and B are similar over $\mathbb{F}$. An explanation is that $X^{-1}AX = B$ if and only if $AX = XB$ for an invertible X. Viewed as a system of linear equations $AX = XB$ has a solution X with $\det X \neq 0$ in $\mathbb{E}$ if and only if it has such a solution in $\mathbb{F}$, because solving the linear equations $AX = XB$ only invokes the operations of the field $\mathbb{F}$. A rigorous proof requires abstract algebra. So, two rational matrices similar over $\mathbb{C}$ are similar over $\mathbb{Q}$. This is usually shown by using Smith normal form.

3.70 By Jordan decomposition theorem, A is similar to J, a Jordan form of A, which is upper-triangular and real because the eigenvalues of A are all real. So A and J are similar over $\mathbb{R}$ (by Problem 3.69).

3.71 (a) $\begin{pmatrix} 0 & -1 \\ 1 & 0 \end{pmatrix} \begin{pmatrix} 0 & -1 \\ 1 & 0 \end{pmatrix} = -I$.

(b) A and B have two distinct eigenvalues $\pm i$. So they are similar over $\mathbb{C}$. Moreover, if B is real, then B is similar to A over $\mathbb{R}$.

(c) If $C^2 = \begin{pmatrix} -1 & 0 \\ 0 & -1-\epsilon \end{pmatrix}$, then C has two distinct eigenvalues i or $-i$, and $\sqrt{1+\epsilon}\,i$ or $-\sqrt{1+\epsilon}\,i$, which results in a nonreal trace of C, contradicting C being real.

3.72 The eigenvalues of A are $1, 2, 4$, and corresponding eigenvectors (up to nonzero scalar multiplications) are $(1, -1, 1)$, $(1, 0, -1)$, and $(1, 2, 1)$, respectively. One checks that these vectors are mutually orthogonal.

3.73 The eigenvalues of A are $\lambda_1 = x + \sqrt{2}y$, $\lambda_2 = x$, and $\lambda_3 = x - \sqrt{2}y$. Find corresponding unit eigenvectors by solving $(\lambda I - A)v = 0$. Set

$$T = \begin{pmatrix} \frac{1}{2} & \frac{\sqrt{2}}{2} & \frac{1}{2} \\ \frac{\sqrt{2}}{2} & 0 & -\frac{\sqrt{2}}{2} \\ \frac{1}{2} & -\frac{\sqrt{2}}{2} & \frac{1}{2} \end{pmatrix}.$$

3.74 $A = \begin{pmatrix} \sqrt{2} & 0 & 0 \\ 0 & 1 & 0 \end{pmatrix} \begin{pmatrix} \frac{1}{\sqrt{2}} & 0 & \frac{1}{\sqrt{2}} \\ 0 & 1 & 0 \\ \frac{1}{\sqrt{2}} & 0 & -\frac{1}{\sqrt{2}} \end{pmatrix}.$

$B = \begin{pmatrix} \frac{1}{\sqrt{3}} & \frac{1}{\sqrt{2}} & \frac{1}{\sqrt{6}} \\ \frac{1}{\sqrt{3}} & 0 & -\frac{2}{\sqrt{6}} \\ \frac{1}{\sqrt{3}} & -\frac{1}{\sqrt{2}} & \frac{1}{\sqrt{6}} \end{pmatrix} \begin{pmatrix} \sqrt{6} & 0 \\ 0 & 0 \\ 0 & 0 \end{pmatrix} \begin{pmatrix} \frac{1}{\sqrt{2}} & \frac{1}{\sqrt{2}} \\ \frac{1}{\sqrt{2}} & -\frac{1}{\sqrt{2}} \end{pmatrix}.$

3.75 $m_{J_k}(t) = t^k$, $r(J_k) = k - 1$, $\dim \mathrm{Ker}\, J_k = 1$, and $\dim \mathrm{Im}\, J_k = k - 1$.

3.76 Let $n \geq 2$. Note that $A\,\mathrm{adj}(A) = |A|I$ and $|\,\mathrm{adj}(A)| = |A|^{n-1}$.

If $r(A) = n$, then A and $\mathrm{adj}(A)$ are invertible. Let A have eigenvalues $\lambda_1, \lambda_2, \ldots, \lambda_n$. The eigenvalues of $\mathrm{adj}(A)$ are $\frac{1}{\lambda_1}|A|, \frac{1}{\lambda_2}|A|, \ldots, \frac{1}{\lambda_n}|A|$; the eigenvalues of $\mathrm{adj}(\mathrm{adj}(A))$ are $\lambda_1|A|^{n-2}, \lambda_2|A|^{n-2}, \ldots, \lambda_n|A|^{n-2}$.

If $r(A) = n - 1$, then $r(\mathrm{adj}(A)) = 1$. The eigenvalues of $\mathrm{adj}(A)$ are $0, \ldots, 0$, and $\mathrm{tr}(\mathrm{adj}(A)) = \sum_{i=1}^{n} |A_{ii}|$, where $|A_{ii}|$ is the minor by deleting the i-th row and i-th column of A, $i = 1, 2, \ldots, n$.

If $r(A) < n - 1$, $\mathrm{adj}(A)) = 0$. The eigenvalues of $\mathrm{adj}(A)$ are all zero.

If $r(A) \leq n - 1$, then $\mathrm{adj}(\mathrm{adj}(A)) = 0$.

3.77 Let $A = UDV$ be a singular value decomposition of A, where D has the singular values $\sigma_1, \ldots, \sigma_n$ of A on the diagonal. Let $W = UV$. By the Cauchy–Schwartz inequality, $\sum_s |u_{is}v_{sj}| \leq 1$ for all i, j. Thus,

$$|a_{ij}| = \left| \sum_s u_{is}\sigma_s v_{sj} \right| \leq \sum_s \sigma_s |u_{is}v_{sj}| \leq \sigma_{\max} \sum_s |u_{is}v_{sj}| \leq \sigma_{\max}(A).$$

Suppose λ_1 is an eigenvalue of A for which $|\lambda_1| = \max_i |\lambda_i(A)|$. Let $Au = \lambda_1 u$, u is a unit vector. Then $|\lambda_1|^2 = |u^*\overline{\lambda_1}\lambda_1 u| = |u^* A^* A u| \leq (\sigma_{\max}(A))^2$. So, $|\lambda_1| \leq \sigma_{\max}(A)$. For $\sigma_{\min}(A) \leq \min_i |\lambda_i(A)|$, if A is singular, then $\sigma_{\min}(A) = 0$, we have nothing to show. Otherwise, A^{-1} exists. From the above argument, $\max_j |\lambda_j(A^{-1})| \leq \sigma_{\max}(A^{-1})$. On the other hand, $\max_j |\lambda_j(A^{-1})| = (\min_i |\lambda_i(A)|)^{-1}$ and $\sigma_{\max}(A^{-1}) = (\sigma_{\min}(A))^{-1}$. Thus, $\sigma_{\min}(A) \leq \min_i |\lambda_i(A)|$. (See Example 3.7.)

It is not true that $\max_i |a_{ii}| \leq \max_i |\lambda_i(A)|$. Take $A = \begin{pmatrix} 1 & -1 \\ 1 & -1 \end{pmatrix}$. The eigenvalues of A are 0, 0; singular values of A are 2, 0, while diagonal entries of A are 1, -1; nor is $\sigma_{\min}(A) \leq \min_i |a_{ii}|$ by setting $A = \begin{pmatrix} 0 & 1 \\ 1 & 0 \end{pmatrix}$.

3.78 If $A\bar{x} = \sqrt{\lambda}x$, $x \neq 0$, then $\bar{A}x = \sqrt{\lambda}\bar{x}$. Thus, $(A\bar{A})x = A(\bar{A}x) = \sqrt{\lambda}(A\bar{x}) = \lambda x$, i.e., λ is an eigenvalue of $A\bar{A}$. Conversely, suppose that $A\bar{A}u = \lambda u$, $u \neq 0$. Case (i) If $\lambda = 0$ and $\bar{A}u = 0$, then $A\bar{u} = 0 = \sqrt{\lambda}u$. Case (ii) If $\lambda = 0$ and $\bar{A}u \neq 0$, let $x = A\bar{u}$. Then $A\bar{x} = A\bar{A}u = 0 = \sqrt{\lambda}x$. Case (iii) Let $\lambda > 0$. If $A\bar{u} = -\sqrt{\lambda}u$, set $x = iu$. If $A\bar{u} \neq -\sqrt{\lambda}u$, set $x = A\bar{u}+\sqrt{\lambda}u$. In either case, $A\bar{x} = \sqrt{\lambda}x$.

3.79 Let $Av = \lambda v$, $v \neq 0$. Then $A^k v = \lambda^k v$ for any positive integer k. If $p(A) = 0$, then $p(A)v = 0$, which implies $p(\lambda)v = 0$, thus $p(\lambda) = 0$.

For the minimal polynomial $m_A(x)$ of A, since $m_A(A) = 0$, we have $m_A(\lambda) = 0$, that is, the minimal polynomial is the unique monic polynomial that annihilates A: it has the smallest degree (of all the polynomials annihilating A) and it has all distinct eigenvalues (with certain multiplicities) of A as its roots. The minimal polynomial divides all polynomials that annihilate A. If $p(A) = 0$, then $m_A(x)|p(x)$. The converse is false. Take $A = \begin{pmatrix} 0 & 1 \\ 0 & 0 \end{pmatrix}$, $p(x) = x$, $\lambda = 0$, $m_A(x) = x^2$.

3.80 Let B be the $n \times 2$ matrix with $a_1, a_2, \ldots, a_n$ in the 1st column and 1's in the 2nd column. Then $A = BB^t$. As $a_1 + \cdots + a_n = 0$, we get

$$|\lambda I_n - A| = |\lambda I_n - BB^t| = \lambda^{n-2}|\lambda I_2 - B^t B| = \lambda^{n-2}\left(\lambda - \sum_{i=1}^{n} a_i^2\right)(\lambda - n).$$

The eigenvalues of A are $0, \ldots, 0, n, \sum a_i^2$, all nonnegative.

3.81 (a) Because AB and BA have the same eigenvalues.

(b) Take $A = \begin{pmatrix} 1 & i \\ 0 & 0 \end{pmatrix}$ and $B = A^t$. Then $AB = 0$ but $BA \neq 0$.

(c) By (a), we compute

$$\begin{aligned} \operatorname{tr}(AB)^k &= \operatorname{tr}(AB)(AB)\cdots(AB) = \operatorname{tr} A(BA)\cdots(BA)B \\ &= \operatorname{tr}(BA)(BA)\cdots(BA)B = \operatorname{tr}(BA)^k. \end{aligned}$$

(d) No, in general. Take $A = \begin{pmatrix} 0 & 1 \\ 0 & 0 \end{pmatrix}$, $B = \begin{pmatrix} 1 & 0 \\ 1 & 0 \end{pmatrix}$.

(e) If A had an inverse, then $AB - BA = A$ would imply $ABA^{-1} - B = I$. Taking trace for both sides gives $0 = n$, a contradiction.

(f) Write $ABC = A(BC)$, then use (a).

(g) No, in general. Take $A = \begin{pmatrix} 0 & 1 \\ 0 & 0 \end{pmatrix}$, $B = \begin{pmatrix} 1 & 0 \\ 0 & 0 \end{pmatrix}$, $C = \begin{pmatrix} 0 & 0 \\ 1 & 0 \end{pmatrix}$.

(h) By (a) and (c).

(i) If A or B is nonsingular, say, A, then $AB = A(BA)A^{-1}$.

(j) No. Take $A = \begin{pmatrix} 0 & 1 \\ 0 & 0 \end{pmatrix}$, $B = \begin{pmatrix} 1 & 0 \\ 0 & 0 \end{pmatrix}$. Then $AB = 0$, $BA = \begin{pmatrix} 0 & 1 \\ 0 & 0 \end{pmatrix}$.

(k) By SVD, A^*A and AA^* are unitarily similar.

(l) No in general. Take $A = \begin{pmatrix} 1 & i \\ 0 & 0 \end{pmatrix}$. Then $AA^t = 0$ but $A^tA \neq 0$.

3.82 We call a product of matrices a "word". We use the fact that $\mathrm{tr}(AB) = \mathrm{tr}(BA)$ for any square matrices A and B of the same size. For simplicity, $S_{m,j} = S_{m,j}(X,Y)$. First view $S_{5,3}$ as a collection of the words of length 5 with 3 Y's and divide (10 of) them into two groups:

$$XY^2XY, \ Y^2XYX, \ YXYXY, \ XYXY^2, \ YXY^2X$$

and

$$X^2Y^3, \ XY^3X, \ Y^3X^2, \ Y^2X^2Y, \ YX^2Y^2.$$

The words in each group all have the same trace. So

$$\tfrac{1}{5}\mathrm{tr}(S_{5,3}) = \mathrm{tr}(XY^2XY + X^2Y^3) = \mathrm{tr}\left(X(Y^2XY + XY^3)\right),$$

where $Y^2XY, \ XY^3 \in S_{4,3}$. There are two more elements in $S_{4,3}$: YXY^2 and Y^3X, which have the same trace as $Y^2XY, \ XY^3$, respectively. (In fact, all the 4 words in $S_{4,3}$ have the same trace.) The conclusion follows at once. One may generalize this to the words of length m with j copies Y and $m - j$ copies of X.

3.83 $A^2 = 0$ and 0 is the only (repeated) eigenvalue of A. Thus A cannot be similar to a diagonal matrix. The eigenvectors corresponding to 0 are the solutions to $v^t x = 0$. The singular values are $\sqrt{u^t u v^t v}, 0, \ldots, 0$.

3.84 (a) J_n has eigenvalues n and $n-1$ zeros. The eigenvectors belonging to zero are the solutions to $x_1 + x_2 + \cdots + x_n = 0$, and an eigenvector (up to a nonzero scalar) belonging to n is $(1, \ldots, 1)$.

(b) K has $2n$ eigenvalues: $0, \ldots, 0$ $(2n - 2$ copies$)$, $-n$, and n. For $\lambda = 0$, $r(\lambda I_{2n} - K) = r(K) = 2$, the eigenvectors are the solutions to $x_1 + x_2 + \cdots + x_n = 0$ and $x_{n+1} + x_{n+2} + \cdots + x_{2n} = 0$. The following $2n - 2$ vectors form a basis for the eigenspace:

$$\alpha_i = (1, 0, \ldots, 0, -1, 0, \ldots, 0), \quad i = 1, 2, \ldots, n - 1,$$

where -1 is the $(i + 1)$st component, and

$$\alpha_{n+i} = (0, \ldots, 0, 1, 0, \ldots, 0, -1, 0, \ldots, 0), \quad i = 1, 2, \ldots, n - 1,$$

where 1 is in the $(n + 1)$-position; -1 in the $(n + 1 + i)$-position. For $\lambda = -n$, $r(\lambda I - K) = 2n - 1$. Solve $(\lambda I - K)x = 0$ to get

$$x_1 = x_2 = \cdots = x_n = -x_{2n}, \quad x_{n+1} = x_{n+2} = \cdots = x_{2n},$$

where x_{2n} is free. Thus, up to scalar multiples, there is a unique eigenvector $(1, \ldots, 1, -1, \ldots, -1) \in \mathbb{R}^{2n}$ by setting $x_{2n} = -1$. For $\lambda = n$, in a similar way, we obtain the unique (up to scalar multiples) eigenvector $(1, \ldots, 1) \in \mathbb{R}^{2n}$.

3.85 (a) If X is such a matrix that $X^2 = A$, then $X \neq 0$ and X has the same Jordan form as A. For such an X, $X^2 = 0$, a contradiction.

(b) Let Y be the 3×3 matrix with $(1, 2)$- and $(2, 3)$-entries equal to 1, and other entries equal to 0, then $Y^2 = B$.

(c) If $k \geq 4$, by considering the Jordan forms, $Z^k = C$ has no solution. For $k = 2$ and $k = 3$, $Z^k = C$ has solutions, respectively,

$$Z_2 = \begin{pmatrix} 0 & 1 & 0 & 0 \\ 0 & 0 & 0 & 1 \\ 0 & 0 & 0 & 0 \\ 0 & 0 & 0 & 0 \end{pmatrix}, \quad Z_3 = \begin{pmatrix} 0 & 1 & 0 & 0 \\ 0 & 0 & 1 & 0 \\ 0 & 0 & 0 & 1 \\ 0 & 0 & 0 & 0 \end{pmatrix}.$$

3.86 (a) If $s = 0$, then the eigenvalues of A are all 0; equality holds. Let $s > 0$ and let $\lambda_1, \ldots, \lambda_s$ be the nonzero eigenvalues of A. Then $\operatorname{tr} A = \sum_{i=1}^{s} \lambda_i$. Let $\hat{\lambda} = \frac{1}{s} \operatorname{tr} A$, $S = \sum_{i=1}^{s} (\lambda_i - \hat{\lambda})^2 \geq 0$. Note that all λ_i's are real. By computation, we have

$$S = \sum_{i=1}^{s} (\lambda_i - \hat{\lambda})^2 = \sum_{i=1}^{s} \lambda_i^2 - 2 \sum_{i=1}^{s} \lambda_i \hat{\lambda} + \sum_{i=1}^{s} \hat{\lambda}^2$$

$$= \sum_{i=1}^{s} \lambda_i^2 - s\hat{\lambda}^2 = \operatorname{tr} A^2 - \frac{1}{s} (\operatorname{tr} A)^2 \geq 0.$$

The desired inequality follows. Equality holds if and only if either all eigenvalues are zero or all nonzero eigenvalues are the same.

(b) A is Hermitian. The rank of A is equal to the number of nonzero eigenvalues of A. Then apply (a). If $A^2 = cA$ for some c, then $\lambda_i^2 = c\lambda_i$ for all i. It is immediate that the nonzero eigenvalues all equal c and c is real (as A is Hermitian). The converse is easy.

(c) Let $\lambda_1, \ldots, \lambda_k$ be the nonzero eigenvalues of A. $\lambda_1^2, \ldots, \lambda_k^2$ are nonzero eigenvalues of A^2. By the Cauchy–Schwarz inequality,

$$(\operatorname{tr} A)^2 = (\lambda_1 + \cdots + \lambda_k)^2 \leq k(\lambda_1^2 + \cdots + \lambda_k^2) = k \operatorname{tr} A^2.$$

If $(\operatorname{tr} A)^2 > (n-1)\operatorname{tr} A^2$, then k must equal n. Thus, $|A| \neq 0$.

3.87 The eigenvalues of A satisfy $\lambda^3 = \lambda$. Thus, the possible eigenvalues of A are 0, 1, and -1. It follows that $\operatorname{tr} A^2$ is equal to the number of nonzero eigenvalues of A (counting multiplicities). Consider the Jordan blocks J_0, J_1, J_{-1} of A corresponding to the eigenvalues 0, 1, and -1, respectively. If the order of J_0 is more than one, then $r(A^2) < r(A)$ and it is impossible since $A^3 = A$. Thus, the Jordan blocks J_0 corresponding to 0 (if any) are 1×1. Note that J_1^2 and J_{-1}^2 have 1's on their main diagonals. The number of 1's as eigenvalues of A^2 (i.e., the number of nonzero eigenvalues of A) is exactly $\operatorname{tr} A^2$.

3.88 Computation shows that the rank of AB is 2 and $(AB)^2 = 9(AB)$. So $r(BA) = r(AB)^2 = 2$ and BA (2×2) is invertible. Observe that

$$(BA)^3 = B(AB)^2A = B(9AB)A = 9(BA)^2.$$

Since BA is invertible, it follows that $BA = 9I_2$.

3.89 $\det(\lambda I - A) = \lambda^4 - 11\lambda^3 + 45\lambda^2 - 81\lambda + 54 = (\lambda - 2)(\lambda - 3)^3$. So A has simple eigenvalue 2 and multiple eigenvalue 3 of algebraic multiplicity 3. Compute the rank of $3I - A$ to get $r(3I - A) = 3$. Thus, the Jordan block associated with the eigenvalue 3 has size 3×3, and the Jordan block for 2 has size 1×1. So the Jordan canonical form of A is

$$A = \begin{pmatrix} 3 & 1 & 0 & 0 \\ 0 & 3 & 1 & 0 \\ 0 & 0 & 3 & 0 \\ 0 & 0 & 0 & 2 \end{pmatrix}.$$

3.90 If A is real symmetric, then the eigenvectors of A for different eigenvalues are orthogonal, that is, if $Au = \lambda u$ and $Av = \mu v$, where u

and v are nonzero, then $u^t v = 0$. This is because $u^t A v = \mu u^t v$ and $v^t A u = \lambda v^t u = \lambda u^t v$, implying $u^t v = 0$ when $\lambda \neq \mu$.

Let $\alpha_3 = (x, y, z)^t$ be an eigenvector corresponding to 3. By solving $\alpha_3^t \alpha_1 = 0$ and $\alpha_3^t \alpha_2 = 0$, we can choose one of such solutions, say, $\alpha_3 = (1, 0, 1)^t$. Let $P = (\alpha_1, \alpha_2, \alpha_3)$ and $D = P^t P$. Then D is a diagonal matrix and $P^{-1} = D^{-1} P^t$. It follows that $A = P \operatorname{diag}(1, 2, 3) P^{-1} = P \operatorname{diag}(1, 2, 3) D^{-1} P^t$ is symmetric. We compute this out to get

$$A = P \operatorname{diag}(1, 2, 3) P^{-1} = \frac{1}{6} \begin{pmatrix} 13 & -2 & 5 \\ -2 & 10 & 2 \\ 5 & 2 & 13 \end{pmatrix}.$$

With the given eigenpairs $(1, \alpha_1)$ and $(2, \alpha_2)$, the eigenvector α_3 (orthogonal to α_1 and α_2 because A is symmetric) corresponding to 3 is unique up to a scalar multiple. Thus, such a symmetric matrix A is unique up to permutation similarity. A detailed proof goes as follows.

Let $Q = (a\alpha_1, b\alpha_2, c\alpha_3)$, where a, b, c are nonzero scalars. Since $P = (\alpha_1, \alpha_2, \alpha_3)$ is invertible, $Q = P \operatorname{diag}(a, b, c) = (\alpha_1, \alpha_2, \alpha_3) \operatorname{diag}(a, b, c)$ is also invertible, and $Q^{-1} = (\operatorname{diag}(a, b, c))^{-1} P^{-1}$. It follows that

$$\begin{aligned} Q \operatorname{diag}(1, 2, 3) Q^{-1} &= P \operatorname{diag}(a, b, c) \operatorname{diag}(1, 2, 3)(\operatorname{diag}(a, b, c))^{-1} P^{-1} \\ &= P \operatorname{diag}(1, 2, 3) P^{-1} = A. \end{aligned}$$

If the symmetry of A is not required, then α_3 need not be orthogonal to α_1, α_2, A is not unique. Let $\alpha_4 = (1, 0, 0)^t$, which is linearly independent with α_1 and α_2, and let $R = (\alpha_1, \alpha_2, \alpha_4)$. Then

$$A = R \operatorname{diag}(1, 2, 3) R^{-1} = \frac{1}{3} \begin{pmatrix} 9 & -1 & 5 \\ 0 & 5 & 2 \\ 0 & 1 & 4 \end{pmatrix}.$$

3.91   (a) The eigenvectors of distinct eigenvalues of a real symmetric matrix are orthogonal. $\alpha = (1, 1, 1)^t$ and $\beta = (2, 2, 1)^t$ are not orthogonal.

(b) $A = \begin{pmatrix} 0 & 1 & 0 \\ 1 & 0 & 0 \\ 0 & 0 & 1 \end{pmatrix}$. A is unique up to permutation similarity (or the order of the eigenvalues). Let B be also a real symmetric matrix with α and β as eigenvectors associated to eigenvalue 1. Let γ be an eigenvector of B for -1. Since B is symmetric, γ is orthogonal to α and β, say $\gamma = (1, -1, 0)^t$. Thus, γ is unique up to a nonzero scalar multiple. Moreover, α, β, and γ are linearly independent. Let $T = (\alpha, \beta, t\gamma)$, $t \neq 0$. Then $BT = T \operatorname{diag}(1, 1, -1)$ and $B = T \operatorname{diag}(1, 1, -1) T^{-1} = A$.

Note: $\frac{1}{3} \begin{pmatrix} 1 & -2 & 2 \\ -2 & 1 & 2 \\ 2 & 2 & 1 \end{pmatrix}$ is a symmetric matrix having eigenvalues 1, 1, and -1. $(1, 0, 1)^t$ and $(0, 1, 1)^t$ are eigenvectors associated to 1, and $(1, 1, -1)^t$ is an eigenvector corresponding to -1.

See also Problem 5.47 and Problem 6.29 .

3.92 Let $c = \frac{1}{2}(1 + \sqrt{3}i)$, i.e., c is a nonreal solution to $c^3 + 1 = 0$. Let $A = \begin{pmatrix} 1 & 0 & c \\ 0 & c & 1 \\ c & 1 & 0 \end{pmatrix}$. Then $A = A^t$, $r(A) = 2$, and the eigenvalues of A are $1 + c, 0, 0$. For real symmetric (or Hermitian) matrices, the rank of the matrix equals the number of nonzero eigenvalues. So it is impossible to construct a real symmetric matrix that satisfies the conditions.

3.93 (a) Deleting the first column and the last row in T, we obtain an $(n - 1) \times (n - 1)$ matrix whose determinant is $b_1 b_2 \cdots b_{n-1} \neq 0$. So $r(T) \geq n - 1$. For the same reason, the rank of $\lambda I - T$ is at least $n - 1$ for any real λ.

(b) Real symmetric matrices are diagonalizable with real eigenvalues.

(c) Since for any eigenvalue t of T, the rank of $tI - T$ is $n - 1$, every eigenvalue of T is simple (i.e., with algebraic multiplicity 1).

3.94 A^t always has the same eigenvalues as A, while A^t, $\bar{A}$, A^*, and $(A^*A)^{\frac{1}{2}}$ all have the same singular values as A.

3.95 (a) By direct verifications.

(b) $[A, B + C] = A(B + C) - (B + C)A = AB + AC - BA$
$-CA = (AB - BA) + (AC - CA) = [A, B] + [A, C]$.
$[aA, bB] = (aA)(bB) - (bB)(aA) = ab(AB - BA) = ab[A, B]$.

(c) $[A, B]^* = (AB - BA)^* = B^*A^* - A^*B^* = [B^*, A^*]$.

(d) Note that $P^{-1}[PXP^{-1}, Y]P = [X, P^{-1}YP]$.

(e) $\operatorname{tr}(AB - BA) = \operatorname{tr} AB - \operatorname{tr} BA = 0$.

(f) $\operatorname{tr}(I - [A, B]) = n$. If X is nilpotent, then $\operatorname{tr} X = 0$.

(g) $\operatorname{tr}[A, B] = 0 \neq \operatorname{tr} I = n$.

(h) Take $X = \operatorname{diag}(1, 2, \ldots, n)$ and $Y = (y_{ij})$, where

$$y_{ij} = \begin{cases} \frac{1}{i-j} a_{ij} & \text{if } i \neq j. \\ 0 & \text{if } i = j. \end{cases}$$

Then $A = [X, Y]$. Note that X is Hermitian.

(i) $[A, B] = 0 \Rightarrow AB = BA$. So $A^2B = A(AB) = A(BA) = (AB)A = (BA)A = BA^2$. Inductively for any positive integer p, $A^pB = BA^p$. For the same reason, $A^pB^q = B^qA^p$.

(j) If A is nonsingular, then $AB - BA = A$ implies $ABA^{-1} - B = I$. Taking trace gives $0 = n$, a contradiction.

(k) If A and B are Hermitian, then $[A, B]^* = (AB - BA)^* = B^*A^* - A^*B^* = BA - AB = -(AB - BA) = -[A, B]$. So $[A, B]$ is skew-Hermitian. The other case is similarly proved.

(l) Similar to (k).

(m) See (h). If A is skew-Hermitian, X and Y are Hermitian.

(n) Let $C = AB - BA$. Then $C^* = (AB)^* - (BA)^* = BA - AB = -C$. So C is skew-Hermitian. Thus, iC is Hermitian, and all eigenvalues of C are pure imaginary.

(o) Expanding $[A, [A, A^*]] = 0$, we obtain $A^2A^* + A^*A^2 = 2AA^*A$. Multiplying both sides by A^* from left and taking trace, we have

$$\mathrm{tr}\left((A^*)^2A^2\right) = \mathrm{tr}\left((A^*A)^2\right),$$

which implies the normality of A (see Chapter 4, Problem 4.102).

(p) By a direct verification.

(q) Let $C = [A, B]$ and let C commute with A. We show $C^n = 0$. For this, we prove $\mathrm{tr}\, C^k = 0$ for all positive integers k, then use the fact that $\mathrm{tr}\, C^k =$ for all k implies $C^n = 0$ (Problem 3.45). If $k = 1$, $\mathrm{tr}\, C = \mathrm{tr}(AB - BA) = 0$. For $k > 1$, since $AC = CA$ and $AC^{k-1} = C^{k-1}A$, by the fact $\mathrm{tr}(XY) = \mathrm{tr}(YX)$, we have

$$
\begin{aligned}
\mathrm{tr}\, C^k &= \mathrm{tr}(AB - BA)C^{k-1} = \mathrm{tr}(ABC^{k-1}) - \mathrm{tr}(BAC^{k-1}) \\
&= \mathrm{tr}(BC^{k-1}A) - \mathrm{tr}(BAC^{k-1}) = 0.
\end{aligned}
$$

3.96 (a) $\Rightarrow$ (b): Obvious.

(b) $\Rightarrow$ (c): First note that the linear systems $Ax = 0$ and $A^2x = 0$ have the same solution space when $r(A) = r(A^2)$. So $A^2z = 0 \Rightarrow Az = 0$. Let $x \in \mathrm{Im}\, A \cap \mathrm{Ker}\, A$. Then $Ax = 0$, $x = Ay$ for some y, and $0 = Ax = A(Ay) = A^2y$, implying $0 = Ay$, thus $x = Ay = 0$.

(c) $\Rightarrow$ (d): Choose bases for $\mathrm{Im}\, A$ and $\mathrm{Ker}\, A$, together they form a basis for $\mathbb{C}^n$. View A as a linear transformation on $\mathbb{C}^n$, the matrix of A under this basis is of the form $\begin{pmatrix} D & 0 \\ 0 & 0 \end{pmatrix}$, where D is invertible.

(d) $\Rightarrow$ (a): Notice that

$$A^2 = P \begin{pmatrix} D & 0 \\ 0 & I_{n-r} \end{pmatrix} P^{-1} P \begin{pmatrix} D & 0 \\ 0 & 0 \end{pmatrix} P^{-1} = BA,$$

where $B = P \begin{pmatrix} D & 0 \\ 0 & I_{n-r} \end{pmatrix} P^{-1}$ is nonsingular.

3.97   Suppose that A is diagonalizable. Let $T^{-1}AT = \text{diag}(\lambda_1, \lambda_2, \ldots, \lambda_n)$, where $\lambda_1, \lambda_2, \ldots, \lambda_n$ are the eigenvalues of A and T is an invertible matrix. Then $T^{-1}(\lambda I - A)T = \text{diag}(\lambda - \lambda_1, \lambda - \lambda_2, \ldots, \lambda - \lambda_n)$ and

$$T^{-1}(\lambda I - A)^2 T = \text{diag}\left((\lambda - \lambda_1)^2, (\lambda - \lambda_2)^2, \ldots, (\lambda - \lambda_n)^2\right).$$

As $\lambda - \lambda_i = 0$ if and only if $(\lambda - \lambda_i)^2 = 0$, $r(\lambda I - A) = r\left((\lambda I - A)^2\right)$.

Conversely, suppose A is not diagonalizable. Let J be a $k \times k$ ($k > 1$) Jordan block of A corresponding to an eigenvalue, say λ_0. Since $k > 1$, $r(\lambda_0 I - J) = r\left((\lambda_0 I - J)^2\right) + 1$. Using Jordan form of A, we see that $r(\lambda_0 I - A) > r\left((\lambda_0 I - A)\right)^2$, a contradiction.

The equivalent statement in linear equation systems is nearly obvious.

3.98   (a) If $D = \text{diag}(d_1, d_2, \ldots, d_n)$, where $d_1 \geq d_2 \geq \cdots \geq d_n$, then for $x = (x_1, x_2, \ldots, x_n)^t \in \mathbb{C}^n$ with $|x_1|^2 + |x_2|^2 + \cdots + |x_n|^2 = 1$,

$$x^* D x = d_1 |x_1|^2 + d_2 |x_2|^2 + \cdots + d_n |x_n|^2.$$

It follows that $d_1 = \max_{\|x\|=1} x^* D x$ and $d_n = \min_{\|x\|=1} x^* D x$.
Let $A = UDV$ be a singular value decomposition of A, where U, V are unitary, and D is diagonal. Then $A^* A = V^* D^2 V$.
$\sigma_1(A)$ is the square root of the largest eigenvalue of $A^* A$. So

$$
\begin{aligned}
\sigma_1(A) &= \left(\lambda_{\max}(A^* A)\right)^{\frac{1}{2}} = \left(\max_{\|x\|=1}(x^* A^* A x)\right)^{\frac{1}{2}} \\
&= \max_{\|x\|=1}(x^* A^* A x)^{\frac{1}{2}}.
\end{aligned}
$$

(b) Similar to the case of $\sigma_1(A)$.

(c) By the Schur triangularization, let $A = U^* T U$, where U is unitary and T is upper-triangular. If A is Hermitian, then T is a diagonal matrix with the eigenvalues of A on the diagonal. Thus, $\lambda_{\min}(A) \leq x^* A x \leq \lambda_{\max}(A)$ for all $\|x\| = 1$, and $\max_{\|x\|=1} |x^* A x| = \max\{|\lambda_{\min}(A)|, |\lambda_{\max}(A)|\}$. On the other hand, the singular values of the Hermitian A are the absolute values of the eigenvalues of A. Thus, $\sigma_1(A) = \max_{\|x\|=1} |x^* A x|$.

(d) Take $A = \begin{pmatrix} 1 & 0 \\ 0 & -1 \end{pmatrix}$. Then $\sigma_n(A) = 1$, but $\min_{\|x\|=1} |x^*Ax| = 0$.

3.99 We only show that if $A^*AB = A^*AC$ then $AB = AC$. Notice that $A^*A(B-C) = 0$ implies $(B^* - C^*)A^*A(B-C) = 0$. It follows that $\left(A(B-C)\right)^*\left(A(B-C)\right) = 0$. Thus $A(B-C) = 0$ and $AB = AC$.

3.100 We show one direction; the other one is similar. Let $A^2B = A$. Then $r(A) = r(A^2B) \le \min\{r(A^2), r(B)\} \le r(A)$. So $r(A) = r(A^2) = r(B)$. Thus, the null spaces of A, A^2, and B all have the same dimension. If $Bx = 0$, then $Ax = (A^2B)x = 0$. Hence, the null spaces of A^2 and B are subspaces of the null space of A, and they all have to be the same. For any $u \in \mathbb{C}^n$, $(A^2B)(Au) = A(Au)$. So $A^2BAu = A^2u$, that is, $A^2(BAu - u) = 0$, or $BAu - u \in \operatorname{Ker} A^2$. Therefore, $B(BAu - u) = 0$, i.e., $B^2Au = Bu$ for all u, or $B^2A = B$.

3.101 (a) $n - 1$. (b) $\operatorname{Im} A = \{y \in \mathbb{C}^n \mid y^*x = 0\}$. (c) $\operatorname{Ker} A = \operatorname{Span}\{x\}$.

3.102 The dimension of $M_n(\mathbb{Q})$ ($n \times n$ rational matrices) over $\mathbb{Q}$ is n^2. Thus $I, A, A^2, \ldots, A^{n^2}$ are linearly dependent over $\mathbb{Q}$. Let

$$\frac{a_0}{b_0}I + \frac{a_1}{b_1}A + \frac{a_2}{b_2}A^2 + \cdots + \frac{a_{n^2}}{b_{n^2}}A^{n^2} = 0,$$

where a's and b's are integers and none of b's is equal to 0. Take

$$f(x) = b_0b_1\cdots b_{n^2}\left(\frac{a_0}{b_0} + \frac{a_1}{b_1}x + \frac{a_2}{b_2}x^2 + \cdots + \frac{a_{n^2}}{b_{n^2}}x^{n^2}\right).$$

Then $f(A) = 0$. For the diagonal matrix $A = \operatorname{diag}(\frac{1}{2}, \frac{2}{3}, \frac{3}{4})$,

$$f(x) = 24\left(x - \tfrac{1}{2}\right)\left(x - \tfrac{2}{3}\right)\left(x - \tfrac{3}{4}\right) = 24x^3 - 46x^2 + 29x - 6.$$

3.103 $\operatorname{Im} B \subseteq \operatorname{Im} A$ is equivalent to $B = AC$ for some matrix C. Then $\begin{pmatrix} A & B \\ 0 & 0 \end{pmatrix}$ and $\begin{pmatrix} A & 0 \\ 0 & 0 \end{pmatrix}$ are similar because $\begin{pmatrix} I & C \\ 0 & I \end{pmatrix}\begin{pmatrix} A & B \\ 0 & 0 \end{pmatrix}\begin{pmatrix} I & -C \\ 0 & I \end{pmatrix} = \begin{pmatrix} A & 0 \\ 0 & 0 \end{pmatrix}$.

3.104 (a) If $R = 0$ or R is invertible, we have nothing to show. Suppose R is singular. Let J be a Jordan form of R and let $J = \begin{pmatrix} J_I & 0 \\ 0 & E \end{pmatrix}$, where J_I contains all Jordan blocks of nonzero eigenvalues, and E contains the Jordan blocks of zero eigenvalues. J_I is absent if R has only zero eigenvalues. J_I is nonsingular, and E is nilpotent. E is similar to a real matrix via a Jordan form of 0's and (or) 1's. It is sufficient to show that J_I is similar to a real matrix.

We may only consider the Jordan blocks of R associated with nonreal eigenvalues $\lambda = x + iy$. Since R is real, if $J(\lambda)$ is a Jordan block of

R, then $J(\bar{\lambda})$ is also a Jordan block of $R = \bar{R}$, and they appear in the Jordan canonical form of R the same number of times. Thus, R is similar to $\begin{pmatrix} J(\lambda) & & \\ & J(\bar{\lambda}) & \\ & & \ddots \end{pmatrix}$. Now we convert each pair of $J(\lambda)$ and $J(\bar{\lambda})$ to a real matrix via similarity.

Exam the simple case $J(\lambda) = \begin{pmatrix} \lambda & 1 \\ 0 & \lambda \end{pmatrix}$. One checks that $\begin{pmatrix} J(\lambda) & 0 \\ 0 & J(\bar{\lambda}) \end{pmatrix}$ is permutation-similar to $\begin{pmatrix} \lambda & 0 & 1 & 0 \\ 0 & \bar{\lambda} & 0 & 1 \\ 0 & 0 & \lambda & 0 \\ 0 & 0 & 0 & \bar{\lambda} \end{pmatrix}$. Notice that $\begin{pmatrix} \lambda & 0 \\ 0 & \bar{\lambda} \end{pmatrix}$ is similar to the real matrix $\begin{pmatrix} x & y \\ -y & x \end{pmatrix}$ via $P = \frac{1}{\sqrt{2}} \begin{pmatrix} 1 & 1 \\ i & -i \end{pmatrix}$ (see Problem 1.31 or Problem 2.63). Thus, $\begin{pmatrix} J(\lambda) & 0 \\ 0 & J(\bar{\lambda}) \end{pmatrix}$ is similar to a nonsingular real matrix. This idea applies to the Jordan blocks of larger sizes. We conclude that R is similar to a real matrix in the form $\begin{pmatrix} D & 0 \\ 0 & E \end{pmatrix}$, where D is invertible and E is nilpotent.

Note: It is an immediate fact from the Smith rational canonical form (or λ-matrix theory) that: Let R be a square matrix over a field $\mathbb{F}$. Then R is similar to a matrix $\begin{pmatrix} D & 0 \\ 0 & E \end{pmatrix}$, where D is invertible and E is nilpotent, both have entries from $\mathbb{F}$. (D or E can be absent.)

One may also prove the statement by subspaces. See Problem 6.68.

(b) If $R = 0$, then $\begin{pmatrix} 0 & C \\ 0 & 0 \end{pmatrix}$ is similar to a real matrix via its Jordan form which is a matrix with entries 0's and 1's. If R is nonsingular, then $\begin{pmatrix} I & R^{-1}C \\ 0 & I \end{pmatrix} \begin{pmatrix} R & C \\ 0 & 0 \end{pmatrix} \begin{pmatrix} I & -R^{-1}C \\ 0 & I \end{pmatrix} = \begin{pmatrix} R & 0 \\ 0 & 0 \end{pmatrix}$. For the general case, by (a), R is similar to $\begin{pmatrix} D & 0 \\ 0 & E \end{pmatrix}$, where D is invertible and E is nilpotent. (D or E can be absent.) Then $\begin{pmatrix} R & C \\ 0 & 0 \end{pmatrix}$ is similar to $\begin{pmatrix} D & 0 & C_1 \\ 0 & E & C_2 \\ 0 & 0 & 0 \end{pmatrix}$, where C_1 and C_2 are some (complex) matrices. Since D is invertible, $\begin{pmatrix} D & 0 & C_1 \\ 0 & E & C_2 \\ 0 & 0 & 0 \end{pmatrix}$ is similar to $\begin{pmatrix} D & 0 & 0 \\ 0 & E & C_2 \\ 0 & 0 & 0 \end{pmatrix}$. We are done by observing that $\begin{pmatrix} E & C_2 \\ 0 & 0 \end{pmatrix}$ is a nilpotent matrix which is similar to a real matrix.

3.105 $AX = XB \Rightarrow A^2X = A(AX) = A(XB) = (AX)B = XB^2$. In general, $A^kX = XB^k$ for any positive integer k. Let $p(\lambda) = |\lambda I - A|$ be the characteristic polynomial of A. Then $p(A) = 0$. It follows that $p(A)X = Xp(B) = 0$. Since A and B have no common eigenvalues, $p(B)$ is invertible. Thus, $X = 0$. See also Example 3.9.

3.106   Note that $\begin{pmatrix} AB & A \\ 0 & 0 \end{pmatrix} = P^{-1} \begin{pmatrix} 0 & A \\ 0 & BA \end{pmatrix} P$, where $P = \begin{pmatrix} I & 0 \\ B & I \end{pmatrix}$. So $\begin{pmatrix} \lambda I - AB & -A \\ 0 & \lambda I \end{pmatrix}$ is similar to $\begin{pmatrix} \lambda I & -A \\ 0 & \lambda I - BA \end{pmatrix}$. One can verify the fact that $r\left(\begin{pmatrix} X & A \\ 0 & Y \end{pmatrix}^k \right) =$

$n + r(Y^k)$, where X is an $n \times n$ invertible matrix. The rank identity for the case $\lambda \neq 0$ follows. If $\lambda = 0$, $r(AB) \neq r(BA)$ in general.

3.107 $\begin{pmatrix} I & X \\ 0 & I \end{pmatrix} \begin{pmatrix} A & C \\ 0 & B \end{pmatrix} \begin{pmatrix} I & -X \\ 0 & I \end{pmatrix} = \begin{pmatrix} A & C-(AX-XB) \\ 0 & B \end{pmatrix}$ and $\begin{pmatrix} I & X \\ 0 & I \end{pmatrix}^{-1} = \begin{pmatrix} I & -X \\ 0 & I \end{pmatrix}$.

Thus, if there is a solution X to $AX - XB = C$, then $\begin{pmatrix} A & C \\ 0 & B \end{pmatrix}$ and $\begin{pmatrix} A & 0 \\ 0 & B \end{pmatrix}$ are similar via $\begin{pmatrix} I & X \\ 0 & I \end{pmatrix}$. Note: This is (easy) part of the so-called Roth theorem. The other part says the converse is also true.

3.108 (a) Let $Av = \lambda v$, $v \neq 0$. Multiplying both sides by $\mathrm{adj}(A)$ gives

$$\mathrm{adj}(A)Av = \lambda\,\mathrm{adj}(A)v \ \text{ or } \ |A|v = \lambda\,\mathrm{adj}(A)v.$$

It is immediate that $\mathrm{adj}(A)v = \frac{1}{\lambda}|A|v$.

(b) Let $Av = \lambda v$, where $v \neq 0$. If $\lambda \neq 0$, then from the solution of (a), v is an eigenvector of $\mathrm{adj}(A)$ for the eigenvalue $\frac{1}{\lambda}|A|$.

Suppose $\lambda = 0$. If $r(A) \leq n - 2$, then $\mathrm{adj}(A) = 0$ and $\mathrm{adj}(A)v = 0 = \lambda v$. If $r(A) = n - 1$, then the solution space to $Ax = 0$ has dimension 1 and $\{v\}$ is a basis of the solution space. However, $A(\mathrm{adj}(A)v) = 0$, that is, $\mathrm{adj}(A)v$ is a solution to $Ax = 0$. Thus, $\mathrm{adj}(A)v = \mu v$ for some μ, i.e., v is an eigenvector of $\mathrm{adj}(A)$.

3.109 (a) Since the eigenvalues of A are all distinct, A is diagonalizable (Example 3.2). Let $T^{-1}AT = \mathrm{diag}(\lambda_1, \lambda_2, \ldots, \lambda_n)$, where $\lambda_1, \lambda_2, \ldots, \lambda_n$ are the eigenvalues of A. Let $C = T^{-1}BT$. Since $AB = BA$, $\mathrm{diag}(\lambda_1, \lambda_2, \ldots, \lambda_n)C = C\,\mathrm{diag}(\lambda_1, \lambda_2, \ldots, \lambda_n)$. It follows that $\lambda_i c_{ij} = c_{ij}\lambda_j$ for all i, j. As $\lambda_i \neq \lambda_j$ when $i \neq j$, we have $c_{ij} = 0$ when $i \neq j$. Thus C is diagonal, that is, $T^{-1}BT$ is diagonal. Now $T^{-1}(AB)T = T^{-1}ATT^{-1}BT$ is also diagonal.

(b) Suppose A and B are diagonalizable. Let T be an invertible matrix such that $T^{-1}AT = \mathrm{diag}(\mu_1 I, \mu_2 I, \ldots, \mu_k I)$, where μ_i are distinct eigenvalues of A, $k \leq n$, and the I's are identity matrices of appropriate sizes. Since μ's are different, $AB = BA$ implies that $T^{-1}BT = \mathrm{diag}(B_1, B_2, \ldots, B_k)$, where each B_i is a matrix of the same size as $\mu_i I$. Since B is diagonalizable, all B_i are necessarily diagonalizable. Let $R_i^{-1}B_i R_i$ be diagonal. Set $R = \mathrm{diag}(R_1, R_2, \ldots, R_k)$. Then R is invertible and both $R^{-1}T^{-1}ATR$ and $R^{-1}T^{-1}BTR$ are diagonal matrices.

3.110 If A is nonsingular, then $A^2 = A$ gives $A = I$. If $A = 0$, then it is trivial. So we assume that A is a nonzero singular matrix.

(a) Let U be a unitary matrix. Since $x^*(U^*AU)y = 0$ for all $x \in$ $\mathrm{Ker}(U^*AU)$ and $y \in \mathbb{C}^n$ if and only if $u^*Av = 0$ for all $u \in \mathrm{Ker}\, A$ and $v \in \mathbb{C}^n$, by the Schur triangularization, we assume that A is an upper-triangular matrix with diagonal entries $\lambda_1, \lambda_2, \ldots, \lambda_n$. Since $A^2 = A$, each λ_i is either 0 or 1. We assume that $\lambda_i = 0$ for $i \leq n - r$ and $\lambda_i = 1$ for $i > n - r$. We claim that $A = \begin{pmatrix} 0 & 0 \\ 0 & I_r \end{pmatrix}$. To this end, look at the first row of A. Since $\lambda_1 = 0$, take $e_1 = (1, 0, \ldots, 0)^t$. Then $Ae_1 = 0$, so $e_1^t Ay = 0$ for all $y \in \mathbb{C}^n$. This means $e_1^t A = 0$, that is, the first row of A is zero. Repeating this for λ_2 and so on, we see that A has the form $A = \begin{pmatrix} 0 & 0 \\ 0 & \Lambda \end{pmatrix}$, where Λ is an upper-triangular matrix with diagonal entries 1. Since $A^2 = A$, we have $\Lambda^2 = \Lambda$, thus $\Lambda = I_r$.

(b) Since $A^2 = A$, if all the eigenvalues of A are equal to zero, then $A^2 = A$ reveals $A = 0$ by considering the Jordan decomposition of A.

In general, without loss of generality, by the Schur triangularization, we assume that A is an upper-triangular matrix with (first) zeros and (then) ones on the main diagonal. We show that A is diagonal.

Let 1 be in the (j, j) position of $A = (a_{st})$. We take $z = e_j$ with the j-th component 1 and 0 elsewhere to get $e_j^t A^* A e_j = 1 + \sum_{k<j} |a_{kj}|^2$. If $e_j^t A^* A e_j \leq e_j^t e_j = 1$, then all $a_{kj} = 0$, $k < j$. So A is in the form $\begin{pmatrix} B & 0 \\ 0 & I_r \end{pmatrix}$. As $A^2 = A$, we have $B^2 = B$ and B has only zero eigenvalues. Thus $B = 0$. It follows that $A = \mathrm{diag}(0, \ldots, 0, 1, \ldots, 1)$. Consequently, A is Hermitian. See Problem 3.150 and Problem 5.99.

3.111   (a)   The (i, j)-entry of $(xx^*) \circ (yy^*)$ is $(x_i \overline{x_j})(y_i \overline{y_j}) = (x_i y_i)(\overline{x_j y_j})$, which is the (i, j)-entry of $(x \circ y)(x \circ y)^*$.

(b)   It is a special case of $(AB) \otimes (CD) = (A \otimes C)(B \otimes D)$.

(c) $\langle x \otimes u, \, y \otimes v \rangle = (y \otimes v)^*(x \otimes u) = (y^* \otimes v^*)(x \otimes u) = (y^* x)(v^* u) = \langle x, y \rangle \cdot \langle u, v \rangle$.

(d)   Set $y = x$, $v = u$ in (c).

3.112   By a direct verification. Or show in details as follows. Since $a_{ij} b_{ij} = (e_i^t A e_j) \otimes (e_i^t B e_j) = (e_i \otimes e_i)^t (A \otimes B)(e_j \otimes e_j)$, we have $E^t (A \otimes B) E = A \circ B$, where $E = (e_1 \otimes e_1, \ldots, e_n \otimes e_n)$ and $\{e_1, \ldots, e_n\}$ is the standard basis of $\mathbb{R}^n$. Note that $e_p^t M$ is the p-th row of matrix M and $M e_q$ is the q-th column of M. $E^t (A \otimes B) E$ is the extraction of the rows and columns $1, n + 2, 2n + 3, \ldots, n^2$ of $A \otimes B$.

3.113   We prove some statements. The remaining ones are verified directly.

(d) Let E_{ij} be a matrix of appropriate size with (i,j)-entry 1 and 0 elsewhere. Show first that $(E_{ij} \otimes E_{pq}) \otimes E_{st} = E_{ij} \otimes (E_{pq} \otimes E_{st})$. Then write a matrix X as $X = \sum_{lk} x_{lk} E_{lk}$ to get the associative property.

(i) $(A \otimes B)(A^{-1} \otimes B^{-1}) = (AA^{-1}) \otimes (BB^{-1}) = I_{mn}$.

(p) Let $a_{ij} \neq 0$. Then $a_{ij}B = c_{ij}D$, $B \neq 0$, implying $B = bD$ for some nonzero b. Likewise, $A = aC$ for some nonzero a. Then $A \otimes B = (ab)C \otimes D = C \otimes D \neq 0$. So $ab = 1$.

(q) Compare the (i,j) blocks: $(a_{ij}B) \circ (c_{ij}D) = (a_{ij}c_{ij})(B \circ D)$.

(u) Use singular value decomposition.

(v) Use Schur decomposition.

(w) Use Schur decomposition.

(x) To start, work out a special case, say, A is 2×2 and B is 3×3. In general, $B \otimes A = P(A \otimes B)Q$ for some partial permutation matrices (also known as shuffling matrices) P and Q. In the case where A and B are $n \times n$, P and Q are full permutation matrices and $P = Q^t = Q^{-1}$; $A \otimes B$ and $B \otimes A$ are permutation-similar.

3.114 By the Schur decomposition, let U and V be unitary matrices of sizes m and n, respectively, such that $U^*AU = T_1$ and $V^*BV = T_2$, where T_1 and T_2 are upper-triangular matrices with diagonal entries λ_i and μ_j, $i = 1, \ldots, m$, $j = 1, \ldots, n$, respectively. Then

$$T_1 \otimes T_2 = (U^*AU) \otimes (V^*BV) = (U^* \otimes V^*)(A \otimes B)(U \otimes V).$$

Note that $U \otimes V$ is unitary. Thus $A \otimes B$ is unitarily similar to $T_1 \otimes T_2$, whose eigenvalues are $\lambda_i \mu_j$, $i = 1, \ldots, m$, $j = 1, \ldots, n$.

For the second part, let $W = U \otimes V$. Then

$$W^*(A \otimes I_n + I_m \otimes B)W = T_1 \otimes I_n + I_m \otimes T_2$$

is an upper-triangular matrix with eigenvalues $\lambda_i + \mu_j$ for all i,j.

3.115 $A \otimes B = (UCV) \otimes (RDS) = (U \otimes R)(C \otimes D)(V \otimes S)$. The right-hand side is an SVD of $A \otimes B$. Note that $U \otimes R$ and $V \otimes S$ are unitary, and $C \otimes D$ is nonnegative diagonal.

3.116 The maps in (b), (d), (g) are linear. Let k be a scalar, $u, v \in \mathbb{C}^n$. Then

$$\mathcal{T}(u + kv) = \tfrac{1}{2}(u + kv + \overline{u + kv}) = \tfrac{1}{2}(u + \bar{u}) + \tfrac{1}{2}(kv + \bar{k}\bar{v}).$$

If the underlying field is $\mathbb{R}$, then k is real, $\mathcal{T}$ is a linear transformation on $\mathbb{C}^n$. Over $\mathbb{C}$, $\bar{k} \neq k$ in general, so $\mathcal{T}$ is not linear on $\mathbb{C}^n$ over $\mathbb{C}$.

3.117

Map	(i) linear	(ii) one-to-one	(iii) onto	(iv) invertible
(a)	no	no	no	no
(b)	no	no	no	no
(c)	yes	yes	yes	yes
(d)	no	yes	yes	yes
(e)	yes	yes	no	left-invertible
(f)	yes	no	yes	right-invertible
(g)	yes	no	yes	right-invertible
(h)	no	no	no	no
(i)	yes	yes	yes	yes
(j)	no	yes	yes	yes
(k)	yes	yes	yes	yes
(l)	yes	yes	yes	yes
(m)	yes	no	no	no
(n)	yes	no	yes	right-invertible
(o)	yes	yes	no	left-invertible

3.118 This is in essence a problem about the maps between sets.

(a) $\Rightarrow$: Let $\mathcal{L}\mathcal{A} = \mathcal{I}_V$, $v_1, v_2 \in V$ and $v_1 \neq v_2$. The $v_1 = \mathcal{I}_V(v_1) = \mathcal{L}(\mathcal{A}(v_1))$ and $v_2 = \mathcal{I}_V(v_2) = \mathcal{L}(\mathcal{A}(v_2))$. It follows that $\mathcal{A}(v_1)$ cannot be equal to $\mathcal{A}(v_2)$, i.e., $\mathcal{A}$ is one-to-one.

$\Leftarrow$: If $\mathcal{A}$ is one-to-one, define $\mathcal{L}$ by $\mathcal{L}(w) = v$ if $\mathcal{A}(v) = w \in \operatorname{Im}\mathcal{A}$, and $\mathcal{L}(w) = 0$ if $w \in W \setminus \operatorname{Im}\mathcal{A}$. Then $\mathcal{L}\mathcal{A} = \mathcal{I}_V$. This $\mathcal{T}$ may not be linear, but easily meets the requirement $\mathcal{L}\mathcal{A} = \mathcal{I}_V$.

(b) $\mathcal{T}$ need not be linear. Let $\mathcal{A} : \mathbb{R} \to \mathbb{R}^2$ be defined by $\mathcal{A}(x) = (x, 0)$. Then $\mathcal{A}$ is linear and one-to-one. $\mathcal{T} : \mathbb{R}^2 \to \mathbb{R}$ defined by $\mathcal{T}(x, 0) = x$ and $\mathcal{T}(x, y) = 0$ if $y \neq 0$ is nonlinear by checking $0 = \mathcal{T}(1,1) = \mathcal{T}\big((1,0) + (0,1)\big) \neq \mathcal{T}(1,0) + \mathcal{T}(0,1) = 1 + 0 = 1$.

(c) Let $\mathcal{A} : \mathbb{R} \to \mathbb{R}^2$ be defined by $\mathcal{A}(x) = (x, 0)$. Then $\mathcal{A}$ is one-to-one. Define $\mathcal{L}_1 : \mathbb{R}^2 \to \mathbb{R}$ by $\mathcal{L}_1(x, y) = x + y$ and $\mathcal{L}_2 : \mathbb{R}^2 \to \mathbb{R}$ by $\mathcal{L}_2(x, y) = x - y$. Then $\mathcal{L}_1\mathcal{A} = \mathcal{L}_2\mathcal{A} = \mathcal{I}_V$ but $\mathcal{L}_1 \neq \mathcal{L}_2$. Note that $\mathcal{A}$, $\mathcal{L}_1$, and $\mathcal{L}_2$ are all linear.

(d) $\Rightarrow$: Let $\mathcal{A}\mathcal{R} = \mathcal{I}_W$. For any $w \in W$, $w = \mathcal{I}_W(w) = (\mathcal{A}\mathcal{R})(w) = \mathcal{A}(\mathcal{R}(w))$. This means $\mathcal{A}$ is onto.

$\Leftarrow$: If $\mathcal{A}$ is onto, for $w \in W$, denote $[w] = \{v \in V \mid \mathcal{A}(v) = w\}$. Define $\mathcal{R}$ by $\mathcal{R}(w) = v_1$, where v_1 is a (arbitrary) fixed element in $[w]$. Then $\mathcal{A}\mathcal{R} = \mathcal{I}_W$. (Note: This $\mathcal{R}$ need not be linear.)

(e) Let $\mathcal{A}\mathcal{R} = \mathcal{I}_W$ and $\mathcal{L}\mathcal{A} = \mathcal{I}_V$. For any $w \in W$, $w = \mathcal{I}_W(w) = (\mathcal{A}\mathcal{R})(w) = \mathcal{A}(\mathcal{R}(w))$, where $\mathcal{R}(w) \in V$. Therefore, $\mathcal{A}$ is onto.

Now we prove that $\mathcal{A}$ is one-to-one. If $\mathcal{A}(v_1) = \mathcal{A}(v_2)$, then $\mathcal{L}(\mathcal{A}(v_1)) = \mathcal{L}(\mathcal{A}(v_2))$, that is, $(\mathcal{L}\mathcal{A})(v_1) = (\mathcal{L}\mathcal{A})(v_2)$. Since $\mathcal{L}\mathcal{A} = \mathcal{I}_V$, we have $v_1 = v_2$. Conversely, if $\mathcal{A}$ is bijective, define $\mathcal{B}$ from W to V such that $\mathcal{B}(w) = v$ if $\mathcal{A}(v) = w$. Then $\mathcal{A}\mathcal{B} = \mathcal{I}_W$ and $\mathcal{B}\mathcal{A} = \mathcal{I}_V$.

(f) Let $\mathcal{A}\mathcal{R} = \mathcal{I}_W$ and $\mathcal{L}\mathcal{A} = \mathcal{I}_V$. Then for any $w \in W$, $\mathcal{R}(w) = (\mathcal{L}\mathcal{A})(\mathcal{R}(w)) = \big(\mathcal{L}(\mathcal{A}\mathcal{R})\big)(w) = \mathcal{L}\big(\mathcal{I}_W(w)\big) = \mathcal{L}(w)$. So, $\mathcal{R} = \mathcal{L}$.

3.119 $\mathcal{L}^{-1}$ is obviously bijective. For $w_1, w_2 \in W$, let $v_1, v_2 \in V$ be such that $\mathcal{L}^{-1}(w_1) = v_1$ and $\mathcal{L}^{-1}(w_2) = v_2$. Since $\mathcal{L}(v_1 + kv_2) = \mathcal{L}(v_1) + k\mathcal{L}(v_2) = w_1 + kw_2$, we have $\mathcal{L}^{-1}(w_1 + kw_2) = v_1 + kv_2 = \mathcal{L}^{-1}(w_1) + k\mathcal{L}^{-1}(w_2)$, that is, $\mathcal{L}^{-1}$ is a linear map from W to V.

3.120   (a) We show (i) $\Leftrightarrow$ (v); other parts are easy. (v) $\Rightarrow$ (i) is obvious since $Ax_1 = Ax_2 \Rightarrow BAx_1 = BAx_2 \Rightarrow x_1 = x_2$. Conversely, if T is one-to-one, then $Ax = 0 \Rightarrow x = 0$. Because $(A^*A)x = 0 \Leftrightarrow Ax = 0 \Leftrightarrow x = 0$, A^*A is invertible. Take $L = (A^*A)^{-1}A^*$. Then $LA = I_n$.

(b) We show (iii) $\Leftrightarrow$ (v) as an example. If the rows of A are linearly independent, then AA^* is invertible. Setting $R = A^*(AA^*)^{-1}$ gives $AR = I_m$. Conversely, if $AR = I_m$, then $x^*(AR) = x^*$ for any $x \in \mathbb{F}^m$. If $x^*A = 0$ then $x = 0$. So the rows of A are linearly independent.

3.121   It is sufficient to notice that $x = (\mathcal{I} - \mathcal{A})x + \mathcal{A}x$.

3.122   (a) It is routine to show that $\mathcal{A}(f + kg) = \mathcal{A}(f) + k\mathcal{A}(g)$.

(b) We show that $\mathcal{A}^2(f) = \mathcal{A}(f)$ for every f as follows:

$$\begin{aligned}
\mathcal{A}^2(f) &= \mathcal{A}\big(\tfrac{1}{2}(f(x) + f(-x))\big) \\
&= \tfrac{1}{2}\big(\tfrac{1}{2}(f(x) + f(-x)) + \tfrac{1}{2}(f(-x) + f(x))\big) \\
&= \tfrac{1}{2}\big(f(x) + f(-x)\big) = \mathcal{A}(f).
\end{aligned}$$

(c) Im $\mathcal{A}$ is the set of even functions; Ker $\mathcal{A}$ is the set of odd functions.

(d) To find a matrix for $\mathcal{A}$ on $\mathbb{P}_5[x]$, take the basis $\{1, x, x^2, x^3, x^4\}$. As $\mathcal{A}(1) = 1, \mathcal{A}(x) = 0, \mathcal{A}(x^2) = x^2, \mathcal{A}(x^3) = 0, \mathcal{A}(x^4) = x^4$, we get

$$\begin{pmatrix}
1 & 0 & 0 & 0 & 0 \\
0 & 0 & 0 & 0 & 0 \\
0 & 0 & 1 & 0 & 0 \\
0 & 0 & 0 & 0 & 0 \\
0 & 0 & 0 & 0 & 1
\end{pmatrix}.$$

3.123   (a) False. It is true that $\mathcal{A}(V_1 \cap V_2) \subseteq \mathcal{A}(V_1) \cap \mathcal{A}(V_2)$. But equality does not hold in general. In $\mathbb{R}^2$, take V_1 to be the line $y = x$, V_2 to be the x-axis, and $\mathcal{A}$ to be the projection onto x-axis.

    (b) True.

    (c) True. For every $w \in V_1 + V_2$, let $w = v_1 + v_2$. Then $\mathcal{A}(w) = \mathcal{A}(v_1) + \mathcal{A}(v_2) \in \mathcal{A}(V_1) + \mathcal{A}(V_2)$. So $\mathcal{A}(V_1 + V_2) \subseteq \mathcal{A}(V_1) + \mathcal{A}(V_2)$. On the other hand, if $z \in \mathcal{A}(V_1) + \mathcal{A}(V_2)$, then $z = \mathcal{A}(z_1) + \mathcal{A}(z_2) = \mathcal{A}(z_1 + z_2) \in \mathcal{A}(V_1 + V_2)$. So equality holds.

    (d) False. In $\mathbb{R}^3$, take V_1 to be the xy-plane, V_2 to be the line $y = z$, and $\mathcal{A}$ to be the projection onto the xy-plane (not a direct sum).

3.124   (a) False.

    (b) False.

    (c) True.

    (d) False.

    (e) True.

    (f) False.

    (g) True.

    (h) False.

    (i) True.

    (j) True.

    (k) True.

    (l) True.

    (m) False.

    (n) False (unless $\dim V = \dim W$).

    (o) True. Let $\{w_1, \ldots, w_n\}$ be a basis of W. Then $\{v_1 = \mathcal{B}(w_1), \ldots, v_n = \mathcal{B}(w_n)\}$ is a basis of V. $\mathcal{A}$ linearly maps a basis to a basis. So A is invertible. One can show that $B = A^{-1}$, and it is linear.

    (p) False. This is a pre-calculus exercise. Take $V = W = \mathbb{R}$ and let
$$\mathcal{A}(x) = \begin{cases} x, & \text{if } x \geq 0 \\ x + 1, & \text{if } x < 0 \end{cases} \quad \text{and} \quad \mathcal{B}(x) = \begin{cases} x, & \text{if } x \geq 0 \\ x - 1, & \text{if } x < 0. \end{cases}$$
Then $\mathcal{A}\mathcal{B} = \mathcal{I}$ but $\mathcal{B}\mathcal{A} \neq \mathcal{I}$ as $\mathcal{B}\mathcal{A}(-0.5) = 0.5 \neq -0.5$.

3.125   A linear map on $M_n(\mathbb{C})$ is invertible if and only if it is one-to-one and onto; if and only if it maps only 0 to 0, i.e., its kernel is $\{0\}$.

(a) $T_1(X) = X^t$ is linear and invertible.

(b) $T_2(X) = X \pm X^t$ is linear but not invertible.

(c) $T_3(X) = \text{tr}(X)I_n \pm X$ is linear and invertible.

(d) $T_4(X) = \text{tr}(X)X$ is not linear (nor onto or one-to-one).

(e) $T_5(X) = \text{tr}(X)I_n$ is linear but not invertible.

(f) $T_6(X) = \text{diag}(X)$ is linear but not invertible.

(g) $\mathcal{L}_1(X) = AX$ is linear and ivertible if A is invertible.

(h) $\mathcal{L}_2(X) = AXA^*$ is linear and invertible if A is invertible.

(i) $\mathcal{L}_3(X) = AX - X^tB$ is linear but not invertible in general.

(j) $\mathcal{L}_4(X) = AX + B$ is not linear unless $B = 0$.

(k) $\mathcal{L}_5(X) = AXB$ is linear and invertible if A and B are invertible.

(l) $\mathcal{L}_6(X) = XAX$ is not linear unless $A = 0$.

3.126   (a) If $X = \left(\begin{smallmatrix} a & b \\ c & d \end{smallmatrix}\right)$, then $T(X) = \left(\begin{smallmatrix} b-a & d-b \\ a-c & c-d \end{smallmatrix}\right)$.

      (b) $T(X+kY) = A(X+kY)^t - (X+kY) = (AX^t - X) + k(AY^t - Y)$.

      (c) Use the standard (ordered) basis $\{(\begin{smallmatrix} 1 & 0 \\ 0 & 0 \end{smallmatrix}), (\begin{smallmatrix} 0 & 1 \\ 0 & 0 \end{smallmatrix}), (\begin{smallmatrix} 0 & 0 \\ 1 & 0 \end{smallmatrix}), (\begin{smallmatrix} 0 & 0 \\ 0 & 1 \end{smallmatrix})\}$. The matrix representation of T with respect to the basis is

$$T = \begin{pmatrix} -1 & 1 & 0 & 0 \\ 0 & -1 & 0 & 1 \\ 1 & 0 & -1 & 0 \\ 0 & 0 & 1 & -1 \end{pmatrix}.$$

      (d) $\text{Im}\,T = \{(\begin{smallmatrix} p & q \\ r & s \end{smallmatrix}) \mid p+q+r+s = 0\}$, $\text{Ker}\,T = \{(\begin{smallmatrix} t & t \\ t & t \end{smallmatrix})\}$, $p, q, r, s, t \in \mathbb{C}$.

      (e) $T(X)$ is singular if and only if $X = (\begin{smallmatrix} a & b \\ c & a \end{smallmatrix})$ or $X = (\begin{smallmatrix} a & b \\ b & d \end{smallmatrix})$.

3.127   (a) T is linear because for $X, Y \in M_2(\mathbb{C})$ and $k \in \mathbb{C}$,

$$\begin{aligned} T(X + kY) &= A(X + kY) - (X + kY)A \\ &= (AX - XA) + k(AY - YA) \\ &= T(X) + kT(Y). \end{aligned}$$

      (b) $T(X) = \left(\begin{smallmatrix} c-b & d-a \\ a-d & b-c \end{smallmatrix}\right)$, $T^2(X) = 2\left(\begin{smallmatrix} a-d & b-c \\ c-b & d-a \end{smallmatrix}\right)$.

      (c) Note that $A^2 = I_2$. We compute

$$AT(X) + T(X)A = A(AX - XA) + (AX - XA)A = 0.$$

(d) $\operatorname{Im} \mathcal{T} = \operatorname{Im} \mathcal{T}^2 = \{ \left(\begin{smallmatrix} x & y \\ -y & -x \end{smallmatrix} \right) \mid x, y \in \mathbb{C} \}$.

(e) Use the basis $E_{11} = \left(\begin{smallmatrix} 1 & 0 \\ 0 & 0 \end{smallmatrix} \right), E_{12} = \left(\begin{smallmatrix} 0 & 1 \\ 0 & 0 \end{smallmatrix} \right), E_{21} = \left(\begin{smallmatrix} 0 & 0 \\ 1 & 0 \end{smallmatrix} \right), E_{22} = \left(\begin{smallmatrix} 0 & 0 \\ 0 & 1 \end{smallmatrix} \right)$.

$$
\begin{aligned}
\mathcal{T}(E_{11}) &= \left(\begin{smallmatrix} 0 & -1 \\ 1 & 0 \end{smallmatrix} \right) &=& \ 0E_{11} - E_{12} + E_{21} + 0E_{22} \\
\mathcal{T}(E_{12}) &= \left(\begin{smallmatrix} -1 & 0 \\ 0 & 1 \end{smallmatrix} \right) &=& \ -E_{11} + 0E_{12} + 0E_{21} + E_{22} \\
\mathcal{T}(E_{21}) &= \left(\begin{smallmatrix} 1 & 0 \\ 0 & -1 \end{smallmatrix} \right) &=& \ E_{11} + 0E_{12} + 0E_{21} - E_{22} \\
\mathcal{T}(E_{22}) &= \left(\begin{smallmatrix} 0 & 1 \\ -1 & 0 \end{smallmatrix} \right) &=& \ 0E_{11} + E_{12} - E_{21} + 0E_{22}.
\end{aligned}
$$

The matrix representation of $\mathcal{T}$ with respect to the basis is

$$
T = \begin{pmatrix}
0 & -1 & 1 & 0 \\
-1 & 0 & 0 & 1 \\
1 & 0 & 0 & -1 \\
0 & 1 & -1 & 0
\end{pmatrix}.
$$

3.128 (a) By a direct verification.

(b) $\mathcal{T}(I) = A - A = 0 = 0I$.

(c) Expanding $\mathcal{T}^{2k}(X)$, every term takes the form $A^p X A^q$, where $p, q \geq 0$, $p + q = 2k$. Thus, every term of $\mathcal{T}^{2k}(X)$ contains a factor A^m, $m \geq k$. If $A^k = 0$, then $\mathcal{T}^{2k}(X) = 0$ for all X.

(d) By a direct verification.

(e) $\operatorname{Ker} \mathcal{T}$ consists of all matrices X such that $AX = XA$. Let $A = Q^{-1}JQ$ be a Jordan decomposition of A, where Q is a nonsingular matrix, J is a Jordan canonical form of Jordan blocks $J_1, \ldots, J_p$ with respective sizes $k_1 \times k_1, \ldots, k_p \times k_p$, $k_1 + \cdots + k_p = n$.

Consider a generic $k \times k$ Jordan block $K = \begin{pmatrix} \lambda & 1 & 0 & \cdots & 0 \\ 0 & \lambda & 1 & \cdots & 0 \\ \cdots & \cdots & \cdots & \cdots & \cdots \\ 0 & 0 & 0 & \cdots & \lambda \end{pmatrix}$. One

can verify that any $k \times k$ matrix in the form $\begin{pmatrix} a_1 & a_2 & a_3 & \cdots & a_k \\ 0 & a_1 & a_2 & \cdots & a_{k-1} \\ \cdots & \cdots & \cdots & \cdots & \cdots \\ 0 & 0 & 0 & \cdots & a_1 \end{pmatrix}$

commutes with K. Such a matrix has k independent variables. For all Jordan blocks $J_1, \ldots, J_p$, there are $k_1 + \cdots + k_p = n$ independent variables. Thus at least n linearly independent matrices commute with J. These matrices pre-multiplied by Q and post-multiplied by Q^{-1} commute with A, and they are linearly independent. Thus, $\dim \operatorname{Ker} \mathcal{T} \geq n$, and as $\dim \left(M_n(\mathbb{C}) \right) = n^2$,

$$
\dim \operatorname{Im} \mathcal{T} = n^2 - \dim \operatorname{Ker} \mathcal{T} \leq n^2 - n.
$$

(f) Let $P^{-1}AP = \text{diag}(\lambda_1, \ldots, \lambda_n)$, where P is an invertible matrix. Let P_i be the i-th column of P. Then $AP_i = \lambda_i P_i$, $i = 1, \ldots, n$. Let B_{ij} be the $n \times n$ matrix having P_i as its j-th column and 0 as other columns. Then $B_{ij}, i, j = 1, \ldots, n$, form a basis for $M_n(\mathbb{C})$. One may compute, for example, $\mathcal{T}(B_{12}) = AB_{12} - B_{12}A = \lambda_1 B_{12} - B_{12}A = B_{12}(\lambda_1 I - A)$, and so on. It follows that $\mathcal{T}$ has the following matrix representation on the basis

$$
T = \begin{pmatrix}
\lambda_1 I - A & 0 & \cdots & 0 \\
0 & \lambda_2 I - A & \cdots & 0 \\
\vdots & \vdots & \ddots & \vdots \\
0 & 0 & \cdots & \lambda_n I - A
\end{pmatrix}.
$$

It is immediate that if A is diagonalizable, then so is T.

(g) Notice that $\mathcal{T}\mathcal{L}(X) = \mathcal{L}\mathcal{T}(X)$ is equivalent to

$$
ABX + XBA = BAX + XAB
$$

or

$$
(AB - BA)X = X(AB - BA).
$$

When A and B commute, $AB - BA = 0$. So $\mathcal{T}$ and $\mathcal{L}$ commute. $\mathcal{T} = 0 \Leftrightarrow AX = XA$ for all $X \Leftrightarrow A$ is a scalar matrix.

For the converse of (g), if $\mathcal{T}$ commutes with $\mathcal{L}$, from the above display, $AB - BA$ commutes with any matrix X in $M_n(\mathbb{C})$. Thus, $AB - BA$ is a scalar matrix. Since $\text{tr}(AB - BA) = 0$, we have $AB = BA$.

3.129   Extend a basis of W to a basis of V, then use the projection onto W.

3.130   (a) Let

$$
a_{s+1}\mathcal{A}(\alpha_{s+1}) + \cdots + a_n\mathcal{A}(\alpha_n) = 0.
$$

Then

$$
\mathcal{A}(a_{s+1}\alpha_{s+1} + \cdots + a_n\alpha_n) = 0,
$$

or

$$
a_{s+1}\alpha_{s+1} + \cdots + a_n\alpha_n \in \text{Ker}\,\mathcal{A}.
$$

Let

$$
a_{s+1}\alpha_{s+1} + \cdots + a_n\alpha_n = a_1\alpha_1 + \cdots + a_s\alpha_s.
$$

Then $a_1 = \cdots = a_n = 0$ since $\{\alpha_1, \ldots, \alpha_s, \alpha_{s+1}, \ldots, \alpha_n\}$ is a basis. It follows that $A(\alpha_{s+1}), \ldots, A(\alpha_n)$ are linearly independent.

(b) By (a). See also identity (3.12), p. 94.

(c) $\dim \operatorname{Ker} \mathcal{A} + \dim \operatorname{Im} \mathcal{A} = n$ because $\{\alpha_1, \ldots, \alpha_s, \alpha_{s+1}, \ldots, \alpha_n\}$ is a basis, $\dim \operatorname{Ker} \mathcal{A} = s$, $\dim \operatorname{Im} \mathcal{A} = n - s$. See also (3.12), p. 94. The sum is not a direct sum in general. Consider $\mathcal{A}$ on $\mathbb{R}^2$ defined by $\mathcal{A}(x, y) = (x - y, x - y)$. It is possible that no β_i falls in $\operatorname{Ker} \mathcal{A}$.

3.131   (a) The vectors v, $\mathcal{A}(v)$, $\mathcal{A}^2(v)$, $\ldots$, $\mathcal{A}^{n-1}(v)$ form a basis for V. The matrix presentation A of the linear transformation $\mathcal{A}$ under this basis has a submatrix I_{n-1} on the upper-right corner. Thus, for any eigenvalue λ, $r(\lambda I - A) = n - 1$. So $\dim \operatorname{Ker}(\lambda I - A) = 1$, and the eigenvectors belonging to λ are multiples of each other.

        (b) Let u_1, u_2, $\ldots$, u_n be eigenvectors corresponding respectively to the distinct eigenvalues λ_1, λ_2, $\ldots$, λ_n of $\mathcal{A}$. Let $u = u_1 + u_2 + \cdots + u_n$. Then $\mathcal{A}(u) = \lambda_1 u_1 + \lambda_2 u_2 + \cdots + \lambda_n u_n$, $\mathcal{A}^2(u) = \lambda_1^2 u_1 + \lambda_2^2 u_2 + \cdots + \lambda_n^2 u_n$, $\ldots$, $\mathcal{A}^{n-1}(u) = \lambda_1^{n-1} u_1 + \lambda_2^{n-1} u_2 + \cdots + \lambda_n^{n-1} u_n$. The coefficient matrix of u, $\mathcal{A}(u)$, $\ldots$, $\mathcal{A}^{n-1}(u)$ under the basis $u_1, u_2, \ldots, u_n$ is a Vandermonde matrix. This matrix is nonsingular for distinct $\lambda_1, \lambda_2, \ldots, \lambda_n$. It follows that $u, \mathcal{A}(u), \mathcal{A}^2(u), \ldots, \mathcal{A}^{n-1}(u)$ are linearly independent.

3.132   Let

$$a_1 x + a_2 \mathcal{A}(x) + \cdots + a_n \mathcal{A}^{n-1}(x) = 0.$$

Applying $\mathcal{A}^k$, $k = n - 1, \ldots, 1$, to both sides of the equation gives

$$a_1 = a_2 = \cdots = a_n = 0.$$

The eigenvalues of $\mathcal{A}$ are all equal to zero. The matrix of $\mathcal{A}$ under the basis is the matrix with $(i + 1, i)$-entries equal to 1, $i = 1, \ldots, n - 1$, and 0 elsewhere. See also Example 1.2.

3.133   For (a) and (d), it is sufficient to notice that the matrix of $\mathcal{A}$ under the basis $\{\alpha_1, \alpha_2, \ldots, \alpha_n\}$ is $J_n(0)$, the Jordan block of size $n \times n$ with all entries on the main diagonal being 0. For (b), let v be such a vector that $\mathcal{B}^{n-1}(v) \neq 0$, then $\{v, \mathcal{B}(v), \ldots, \mathcal{B}^{n-1}(v)\}$ is a basis of V and the matrix of $\mathcal{B}$ under this basis is $J_n(0)$. (c) is obvious because $u, \mathcal{A}(u), \ldots, \mathcal{A}^{n-1}(u)$ are in W and linearly independent.

3.134   (a) False. Consider $\mathcal{A} = 0$.

        (b) True. If the vectors $\alpha_1, \alpha_2, \ldots, \alpha_n$ are linearly dependent, then there exist $k_1, k_2, \ldots, k_n$, not all equal to zero, such that

$$k_1 \alpha_1 + k_2 \alpha_2 + \cdots + k_n \alpha_n = 0,$$

which leads to

$$k_1 \mathcal{A}(\alpha_1) + k_2 \mathcal{A}(\alpha_2) + \cdots + k_n \mathcal{A}(\alpha_n) = 0.$$

So $\mathcal{A}(\alpha_i)$'s are linearly dependent, contradicting the assumption.

3.135 (a) $\mathcal{A}(\alpha_1, \alpha_2, \alpha_3) = (\alpha_1, \alpha_2, \alpha_3)A$, where $A = \left(\begin{smallmatrix} 1 & 1 & 1 \\ 0 & 1 & 1 \\ 0 & 0 & 1 \end{smallmatrix}\right)$ is the matrix of $\mathcal{A}$ under $\alpha = \{\alpha_1, \alpha_2, \alpha_3\}$. Since A is invertible, $\mathcal{A}$ is invertible.

(b) $(\alpha_1, \alpha_2, \alpha_3)A^{-1} = \mathcal{A}^{-1}(\alpha_1, \alpha_2, \alpha_3)$, where $A^{-1} = \left(\begin{smallmatrix} 1 & -1 & 0 \\ 0 & 1 & -1 \\ 0 & 0 & 1 \end{smallmatrix}\right)$. So $\mathcal{A}^{-1}(\alpha_1) = \alpha_1$, $\mathcal{A}^{-1}(\alpha_2) = \alpha_2 - \alpha_1$, and $\mathcal{A}^{-1}(\alpha_3) = \alpha_3 - \alpha_2$.

(c) The matrix of $2\mathcal{A} - \mathcal{A}^{-1}$ with respect to α is $2A - A^{-1} = \left(\begin{smallmatrix} 1 & 3 & 2 \\ 0 & 1 & 3 \\ 0 & 0 & 1 \end{smallmatrix}\right)$.

3.136 The characteristic polynomials are all $p(\lambda) = \lambda^3$. The minimal polynomials for $\mathcal{A}$, $\mathcal{A}^2$, and $\mathcal{A}^3$ are λ^3, λ^2, and λ, respectively. The images of $\mathcal{A}$, $\mathcal{A}^2$, and $\mathcal{A}^3$ are the yz-plane, the z-axis, and $\{0\}$, and the kernels of $\mathcal{A}$, $\mathcal{A}^2$, and $\mathcal{A}^3$ are the z-axis, the yz-plane, and $\mathbb{R}^3$, respectively.

3.137 For convenience, denote

$$u_1 = \begin{pmatrix} 1 \\ 0 \\ 1 \end{pmatrix}, \ u_2 = \begin{pmatrix} 1 \\ -1 \\ 1 \end{pmatrix}, \ u_3 = \begin{pmatrix} -2 \\ 7 \\ -1 \end{pmatrix}.$$

To find a basis for $\operatorname{Im} \mathcal{A}$, apply row operations on $(\mathcal{A}(u_1), \mathcal{A}(u_2), \mathcal{A}(u_3))$:

$$\begin{pmatrix} 2 & 3 & 2 \\ 3 & 0 & 3 \\ -1 & -2 & -1 \end{pmatrix} \longrightarrow \begin{pmatrix} 1 & 0 & 1 \\ 0 & 1 & 0 \\ 0 & 0 & 0 \end{pmatrix}.$$

Thus, we can choose $\{\mathcal{A}(u_1), \mathcal{A}(u_2)\}$ as a basis for $\operatorname{Im} \mathcal{A}$ and

$$\operatorname{Im} \mathcal{A} = \operatorname{Span}\{\mathcal{A}(u_1), \mathcal{A}(u_2)\}.$$

To find the matrix representation for $\mathcal{A}$ such that $\mathcal{A}(x) = Ax$, let

$$x = (x_1, x_2, x_3)^t = y_1 u_1 + y_2 u_2 + y_3 u_3.$$

Then

$$\begin{pmatrix} 1 & 1 & -2 \\ 0 & -1 & 7 \\ 1 & 1 & -1 \end{pmatrix} \begin{pmatrix} y_1 \\ y_2 \\ y_3 \end{pmatrix} = \begin{pmatrix} x_1 \\ x_2 \\ x_3 \end{pmatrix}.$$

Denote by B the 3×3 matrix on the left-hand side. Then $By = x$, where $y = (y_1, y_2, y_3)^t$, and $y = B^{-1}x$, where

$$B^{-1} = \begin{pmatrix} 6 & 1 & -5 \\ -7 & -1 & 7 \\ -1 & 0 & 1 \end{pmatrix}.$$

Thus

$$
\begin{aligned}
\mathcal{A}(x) &= y_1 \mathcal{A}(u_1) + y_2 \mathcal{A}(u_2) + y_3 \mathcal{A}(u_3) \\
&= \big(\mathcal{A}(u_1), \mathcal{A}(u_2), \mathcal{A}(u_3)\big) y \\
&= \big(\mathcal{A}(u_1), \mathcal{A}(u_2), \mathcal{A}(u_3)\big) B^{-1} x \\
&= \begin{pmatrix} 2 & 3 & 2 \\ 3 & 0 & 3 \\ -1 & -2 & -1 \end{pmatrix} \begin{pmatrix} 6 & 1 & -5 \\ -7 & -1 & 7 \\ -1 & 0 & 1 \end{pmatrix} \begin{pmatrix} x_1 \\ x_2 \\ x_3 \end{pmatrix} \\
&= \begin{pmatrix} -11 & -1 & 13 \\ 15 & 3 & -12 \\ 9 & 1 & -10 \end{pmatrix} \begin{pmatrix} x_1 \\ x_2 \\ x_3 \end{pmatrix}.
\end{aligned}
$$

3.138   (a)   The matrix of $\mathcal{A} + \mathcal{B}$ under $\{\beta_1, \beta_2\}$ is $\begin{pmatrix} 8 & 9 \\ \frac{4}{3} & 3 \end{pmatrix}$.

(b)   The matrix of $\mathcal{A}\mathcal{B}$ under $\{\alpha_1, \alpha_2\}$ is $\begin{pmatrix} 7 & 8 \\ 13 & 14 \end{pmatrix}$.

(c)   The coordinate of $\mathcal{A}(u)$ under $\{\alpha_1, \alpha_2\}$ is $(3, 5)$.

(d)   The coordinate of $\mathcal{B}(u)$ under $\{\beta_1, \beta_2\}$ is $(9, 6)$.

3.139   (a)   Solve the linear equation system $Ax = 0$ to get a basis for Ker A:

$$\alpha_1 = (-2, -1.5, 1, 0)^t, \quad \alpha_2 = (-1, -2, 0, 1)^t.$$

Let $\beta_1 = -2\epsilon_1 - 1.5\epsilon_2 + \epsilon_3$ and $\beta_2 = -\epsilon_1 - 2\epsilon_2 + \epsilon_4$. Then

$$\text{Ker}\,\mathcal{A} = \text{Span}\{\beta_1, \beta_2\}.$$

(b)   The rank of A is 2, and the first two columns of A are linearly independent. This reveals $\text{Im}\,\mathcal{A} = \text{Span}\{\mathcal{A}(\epsilon_1), \mathcal{A}(\epsilon_2)\}$.

(c)   From (a), we see that $\{\beta_1, \beta_2, \epsilon_1, \epsilon_2\}$ serves as a basis for V. The matrix representation of $\mathcal{A}$ with respect to this basis is

$$\begin{pmatrix} 0 & 0 & 1 & 2 \\ 0 & 0 & 2 & -2 \\ 0 & 0 & 5 & 2 \\ 0 & 0 & 4.5 & 1 \end{pmatrix}.$$

3.140 $\dim \operatorname{Im} \mathcal{A} = 2$ and $\{\epsilon_1, \epsilon_2\}$ is a basis. $\dim \operatorname{Ker} \mathcal{A} = 2$, and $\zeta_1 = \epsilon_1 - \epsilon_2$ and $\zeta_2 = \epsilon_1 - \epsilon_3$ form a basis. $\dim(\operatorname{Im} \mathcal{A} + \operatorname{Ker} \mathcal{A}) = 3$ and $\{\epsilon_1, \epsilon_2, \epsilon_3\}$ is a basis. $\dim(\operatorname{Im} \mathcal{A} \cap \operatorname{Ker} \mathcal{A}) = 1$ and $\zeta_3 = \epsilon_1 - \epsilon_2$ is a basis.

3.141   (a)   Apply $\mathcal{A}^{-1}$ to both sides of $\mathcal{A}(W) \subseteq W$ to get $W \subseteq \mathcal{A}^{-1}(W)$. However, $\dim \mathcal{A}^{-1}(W) \leq \dim W$. Therefore, $W = \mathcal{A}^{-1}(W)$.

      (b)   No. In $\mathbb{R}^2$, take W to be the x-axis, and W' to be the line $y = x$. Let $\mathcal{A}$ be the projection onto the x-axis.

3.142   With matrix A, we see that $\mathcal{A}(\alpha_1) = 2\alpha_1$ and $\mathcal{A}(\alpha_2) = \alpha_1 + 2\alpha_2$. Let $k\alpha_1 \in W_1$. Then $\mathcal{A}(k\alpha_1) = k\mathcal{A}(\alpha_1) = 2k\alpha_1 \in W_1$. So W_1 is invariant under $\mathcal{A}$. If W_2 is an invariant subspace such that $\mathbb{R}^2 = W_1 \oplus W_2$, then the dimension of W_2 is 1. Let $\alpha_2 = p\alpha_1 + w_2$, where $w_2 \in W_2$, and let $\mathcal{A}(w_2) = qw_2$. From $\mathcal{A}(\alpha_2) = \alpha_1 + 2\alpha_2$, we have $\alpha_1 + 2\alpha_2 = 2p\alpha_1 + qw_2$. Subtracting $2\alpha_2 = 2p\alpha_1 + 2w_2$, we have $\alpha_1 = (q - 2)w_2$, which is in both W_1 and W_2. But $W_1 \cap W_2 = \{0\}$, a contradiction.

3.143   (a)   It is routine to show that $\mathcal{A}$ is a linear transformation on $M_2(\mathbb{R})$.

      (b)   With respect to the basis $E_{11}, E_{12}, E_{21}, E_{22}$, the matrix of $\mathcal{A}$ is

$$\begin{pmatrix} 1 & 0 & -1 & 0 \\ 0 & 1 & 0 & -1 \\ -1 & 0 & 1 & 0 \\ 0 & -1 & 0 & 1 \end{pmatrix}.$$

      (c) $\operatorname{Im} \mathcal{A} = \{\left(\begin{smallmatrix} x & y \\ -x & -y \end{smallmatrix}\right) \mid x, y \in \mathbb{R}\}$; $\dim \operatorname{Im} \mathcal{A} = 2$; $\{\left(\begin{smallmatrix} 1 & 0 \\ -1 & 0 \end{smallmatrix}\right), \left(\begin{smallmatrix} 0 & 1 \\ 0 & -1 \end{smallmatrix}\right)\}$.

      (d) $\operatorname{Ker} \mathcal{A} = \{\left(\begin{smallmatrix} x & y \\ x & y \end{smallmatrix}\right) \mid x, y \in \mathbb{R}\}$; $\dim \operatorname{Ker} \mathcal{A} = 2$; $\{\left(\begin{smallmatrix} 1 & 0 \\ 1 & 0 \end{smallmatrix}\right), \left(\begin{smallmatrix} 0 & 1 \\ 0 & 1 \end{smallmatrix}\right)\}$.

3.144   (a) $\mathcal{A}(X + kY) = A(X + kY) = AX + kAY = \mathcal{A}(X) + k\mathcal{A}(Y)$.

      (b)   When $p = 1$, $\mathcal{A}$ maps $\mathbb{C}^n$ to $\mathbb{C}^m$. The matrix representation of $\mathcal{A}$ with respect to the standard bases of $\mathbb{C}^n$ and $\mathbb{C}^m$ is A.

      (c)   Let E_{ij}, $i = 1, \ldots, n$, $j = 1, \ldots, p$, be the standard basis of $M_{n \times p}(\mathbb{C})$ and let F_{ij}, $i = 1, \ldots, m$, $j = 1, \ldots, p$, be the standard basis of $M_{m \times p}(\mathbb{C})$. Note that AE_{ij} is an $m \times p$ matrix whose j-th column is the i-th column of A, zeros elsewhere. For $A = (a_{rs}) \in M_{m \times n}(\mathbb{C})$ and $X = (x_{rs}) \in M_{n \times p}(\mathbb{C})$, we compute

$$\mathcal{A}(X) = AX = A \sum_{i,j} x_{ij} E_{ij} = \sum_{i,j} x_{ij} AE_{ij} = \sum_{t=1}^{m} x_{ij} a_{ti} F_{tj}.$$

Let $p = 2$, $X = E_{ij}$, i.e., $x_{ij} = 1$, $i = 1, \ldots, n$, $j = 1, 2$. We get

$$\big(\mathcal{A}(E_{11}), \ldots, \mathcal{A}(E_{n1})\big) = (F_{11}, \ldots, F_{m1})A,$$

$$\big(\mathcal{A}(E_{12}), \dots, \mathcal{A}(E_{n2})\big) = (F_{12}, \dots, F_{m2})A.$$

Note that each F_{ij} in the "product" on the right-hand sides above is understood as a basis vector. Thus, the matrix of $\mathcal{A}$ with respect to the standard bases is $\big(\begin{smallmatrix} A & 0 \\ 0 & A \end{smallmatrix}\big)$.

(d) Let $m = n = p$. Take the ordered standard basis of $M_n(\mathbb{C})$

$$\{E_{11}, E_{21}, \dots, E_{n1}, E_{12}, E_{22}, \dots, E_{n2}, \dots, E_{1n}, E_{2n}, \dots, E_{nn}\},$$

Denote the i-th column of A by a_i. A computation reveals

$$\mathcal{A}(E_{ij}) = AE_{ij} = a_{1i}E_{1j} + \cdots + a_{ni}E_{nj} = (E_{1j}, \dots, E_{nj})a_i.$$

Thus, the matrix representation of $\mathcal{A}$ under the basis (which has n^2 matrices E_{ij} as basis vectors) is the $n^2 \times n^2$ matrix

$$\mathrm{diag}(A, \dots, A) = \begin{pmatrix} A & & 0 \\ & \ddots & \\ 0 & & A \end{pmatrix} \quad (\text{ n copies of } A).$$

(e) When $m = n$, A is a square matrix. λ is an eigenvalue of $\mathcal{A}$ if and only if $\mathcal{A}(X_0) = \lambda X_0$, i.e, $AX_0 = \lambda X_0$, for some $X_0 \neq 0$. If x_0 is a nonzero column vector of X_0, then $Ax_0 = \lambda x_0$ and λ is an eigenvalue of A. Conversely, if x_0 is an eigenvector of A belonging to λ, let X_0 be the matrix with all column vectors x_0. Then λ is an eigenvalue of $\mathcal{A}$ because $\mathcal{A}(X_0) = AX_0 = \lambda X_0$.

(f) The characteristic polynomial of $\mathcal{A}$ is $(\det(\lambda I - A))^n$; it is the n-th power of the characteristic polynomial of A.

3.145   (a) It is routine to show that $\mathcal{A}$ and $\mathcal{B}$ are linear.

(b) The lines $y = \pm x$, i.e., $\{(x, x) \mid x \in \mathbb{R}\}$ and $\{(x, -x) \mid x \in \mathbb{R}\}$.

(c) $\mathrm{Ker}\,\mathcal{B}$ and $\mathrm{Im}\,\mathcal{B}$ are both the line $y = x$ of dimension 1.

(d) The dimension identity is obvious. $\mathrm{Ker}\,\mathcal{B} + \mathrm{Im}\,\mathcal{B}$ is not a direct sum as $\mathrm{Ker}\,\mathcal{B} = \mathrm{Im}\,\mathcal{B}$. $\mathrm{Ker}\,\mathcal{B} + \mathrm{Im}\,\mathcal{B}^*$ is a direct sum because $\mathrm{Ker}\,\mathcal{B} \cap \mathrm{Im}\,\mathcal{B}^* = \{0\}$. Note: $\mathcal{B}^*$ is the adjoint of $\mathcal{B}$.

3.146   (a) By direct verifications.

(b) $\mathcal{AB}(x_1, x_2, \dots, x_n) = (0, x_n, x_1, x_2, \dots, x_{n-2})$.
$\mathcal{BA}(x_1, x_2, \dots, x_n) = (x_{n-1}, 0, x_1, x_2, \dots, x_{n-2})$.
$\mathcal{A}^n = 0$ and $\mathcal{B}^n = \mathcal{I}$.

(c) $A = (e_2, e_3, \ldots, e_{n-1}, 0)$ and $B = (e_2, e_3, \ldots, e_{n-1}, e_1)$, where $e_1, e_2, \ldots, e_n$ are the standard basis vectors of $\mathbb{R}^n$.

(d) $\operatorname{Ker} \mathcal{A} = \{(0, \ldots, 0, x) \mid x \in \mathbb{R}\}$ and $\operatorname{Ker} \mathcal{B} = \{0\}$. The dimensions of $\operatorname{Ker} \mathcal{A}$ and $\operatorname{Ker} \mathcal{B}$ are 1 and 0, respectively.

3.147   Let $\{\gamma_1, \ldots, \gamma_r\}$ be a basis for $\operatorname{Ker} \mathcal{A}$. To show that V is the direct sum of the subspace spanned by $\beta_1, \ldots, \beta_m$ and $\operatorname{Ker} \mathcal{A}$, we show that $\{\beta_1, \ldots, \beta_m, \gamma_1, \ldots, \gamma_r\}$ is a basis for V.

Let $v \in V$. Then $\mathcal{A}(v) \in \operatorname{Im} \mathcal{A}$. Writing

$$\mathcal{A}(v) = a_1 \alpha_1 + \cdots + a_m \alpha_m$$

and replacing α_i by $\mathcal{A}(\beta_i)$, we have

$$\mathcal{A}(v) = a_1 \mathcal{A}(\beta_1) + \cdots + a_m \mathcal{A}(\beta_m)$$

and

$$\mathcal{A}(v - a_1 \beta_1 - \cdots - a_m \beta_m) = 0.$$

Thus

$$v - a_1 \beta_1 - \cdots - a_m \beta_m \in \operatorname{Ker} \mathcal{A}.$$

Let

$$v - a_1 \beta_1 - \cdots - a_m \beta_m = b_1 \gamma_1 + \cdots + b_r \gamma_r.$$

Then

$$v = a_1 \beta_1 + \cdots + a_m \beta_m + b_1 \gamma_1 + \cdots + b_r \gamma_r.$$

Therefore

$$V = \operatorname{Span}\{\beta_1, \ldots, \beta_m\} + \operatorname{Ker} \mathcal{A}.$$

Now we show that $\beta_1, \ldots, \beta_m, \gamma_1, \ldots, \gamma_r$ are linearly independent. Let

$$c_1 \beta_1 + \cdots + c_m \beta_m + d_1 \gamma_1 + \cdots + d_r \gamma_r = 0.$$

Applying $\mathcal{A}$ to both sides of the above identity gives

$$c_1 \mathcal{A}(\beta_1) + \cdots + c_m \mathcal{A}(\beta_m) + d_1 \mathcal{A}(\gamma_1) + \cdots + d_r \mathcal{A}(\gamma_r) = 0.$$

Hence

$$c_1 \alpha_1 + \cdots + c_m \alpha_m = 0.$$

Thus $c_1 = \cdots = c_m = 0$ as $\alpha_1, \ldots, \alpha_m$ are linearly independent. So

$$d_1 \gamma_1 + \cdots + d_r \gamma_r = 0,$$

and $d_1 = \cdots = d_r = 0$ for the similar reason. The conclusion follows.

3.148   (c) ⇒ (b) ⇒ (a): Let $V = \operatorname{Im}\mathcal{A} \oplus \operatorname{Ker}\mathcal{A}$. Obviously, $\operatorname{Im}\mathcal{A}^2 \subseteq \operatorname{Im}\mathcal{A}$. Let $u \in \operatorname{Im}\mathcal{A}$. Then $u = \mathcal{A}(v)$ for some $v \in V$. Write $v = w_1 + w_2$, where $w_1 \in \operatorname{Im}\mathcal{A}$ and $w_2 \in \operatorname{Ker}\mathcal{A}$. Let $w_1 = \mathcal{A}(z_1)$. Then $u = \mathcal{A}(v) = \mathcal{A}(w_1) + \mathcal{A}(w_2) = \mathcal{A}(w_1) = \mathcal{A}^2(z_1) \in \operatorname{Im}\mathcal{A}^2$, implying $\operatorname{Im}\mathcal{A} \subseteq \operatorname{Im}\mathcal{A}^2$. Thus, $\operatorname{Im}\mathcal{A}^2 = \operatorname{Im}\mathcal{A}$ and $\dim(\operatorname{Im}\mathcal{A}^2) = \dim\operatorname{Im}(\mathcal{A})$.

(a) ⇒ (c): Let $\dim(\operatorname{Im}\mathcal{A}^2) = \dim(\operatorname{Im}\mathcal{A})$. It is sufficient to show that $\operatorname{Im}\mathcal{A} \cap \operatorname{Ker}\mathcal{A} = \{0\}$ because $\dim V = \dim(\operatorname{Im}\mathcal{A}) + \dim(\operatorname{Ker}\mathcal{A})$. Since $\dim(\operatorname{Im}\mathcal{A}^2) = \dim(\operatorname{Im}\mathcal{A})$, we have $\dim(\operatorname{Ker}\mathcal{A}^2) = \dim(\operatorname{Ker}\mathcal{A})$, thus $\operatorname{Ker}\mathcal{A}^2 = \operatorname{Ker}\mathcal{A}$ as $\operatorname{Ker}\mathcal{A} \subseteq \operatorname{Ker}\mathcal{A}^2$. Let $v \in \operatorname{Im}\mathcal{A} \cap \operatorname{Ker}\mathcal{A}$. Then $\mathcal{A}(v) = 0$. Write $v = \mathcal{A}(w)$. Then $\mathcal{A}(v) = \mathcal{A}^2(w) = 0$. So, $w \in \operatorname{Ker}\mathcal{A}^2 = \operatorname{Ker}\mathcal{A}$. It follows that $\mathcal{A}(w) = 0$, i.e., $v = 0$.

If $\mathcal{A}^2 = \mathcal{A}$, then $V = \operatorname{Im}\mathcal{A} \oplus \operatorname{Ker}\mathcal{A}$. But the converse need not be true. Take, for example, $\mathcal{A}(x, y) = (-x, 0)$ for $(x, y) \in \mathbb{R}^2$.

3.149   With $\mathcal{A}^2 = \mathcal{A}$ and $\mathcal{B}^2 = \mathcal{B}$, one may show that $\mathcal{A}(\mathcal{A}-\mathcal{B})^2 = \mathcal{A}-\mathcal{A}\mathcal{B}\mathcal{A}$. Similarly, $(\mathcal{A}-\mathcal{B})^2\mathcal{A} = \mathcal{A}-\mathcal{A}\mathcal{B}\mathcal{A}$. So $\mathcal{A}$ (similarly $\mathcal{B}$) commutes with $(\mathcal{A}-\mathcal{B})^2$. For the second part, a computation shows

$$(\mathcal{I}-\mathcal{A}-\mathcal{B})^2 = \big(\mathcal{I}-(\mathcal{A}+\mathcal{B})\big)^2 = \mathcal{I}-2\mathcal{A}-2\mathcal{B}+(\mathcal{A}+\mathcal{B})^2 = \mathcal{I}-(\mathcal{A}-\mathcal{B})^2.$$

3.150   (a), (b), (c), and (g) are easy to check. (k), (l), and (m) are by (j).

(d) $(\mathcal{A}+\mathcal{I})^{-1} = -\frac{1}{2}\mathcal{A}+\mathcal{I}$.

(e)   Since $\mathcal{A}^2 = \mathcal{A}$, $x \in \operatorname{Ker}\mathcal{A}$ if and only if $x = x - \mathcal{A}(x)$.

(f)   Since $v = \mathcal{A}(v) + \big(v - \mathcal{A}(v)\big)$, we see $V = \operatorname{Im}\mathcal{A} + \operatorname{Ker}\mathcal{A}$. Now let

$$z \in \operatorname{Im}\mathcal{A} \cap \operatorname{Ker}\mathcal{A}, \text{ i.e., } z = \mathcal{A}(y), \ \mathcal{A}(z) = 0.$$

Then $z = \mathcal{A}(y) = \mathcal{A}(\mathcal{A}(y)) = \mathcal{A}(z) = 0$ and $\operatorname{Im}\mathcal{A} \cap \operatorname{Ker}\mathcal{A} = \{0\}$. Note that $\operatorname{Im}\mathcal{A}$ and $\operatorname{Ker}\mathcal{A}$ are $\mathcal{A}$-invariant for any $\mathcal{A}$.

(h)   If $\mathcal{A}(x) = \lambda x$, $x \neq 0$, then $\mathcal{A}^2(x) = \lambda\mathcal{A}(x) = \lambda^2 x$. Since $\mathcal{A}^2(x) = \mathcal{A}(x) = \lambda x$, we have $(\lambda^2 - \lambda)x = 0$. Thus $\lambda = 0$ or $\lambda = 1$. By (g), $\operatorname{Im}\mathcal{A} = V_1$. $\operatorname{Ker}\mathcal{A} = V_0$ is always true.

(i)   Let $\mathcal{B}$ be the transformation on V (projection on M) such that

$$\mathcal{B}(x) = x, \ x \in M \quad \text{and} \quad \mathcal{B}(y) = 0, \ y \in L.$$

Such a $\mathcal{B}$ is linear and uniquely determined by M and L.

(j)   By (f), take a basis for $\operatorname{Im}\mathcal{A}$ and a basis for $\operatorname{Ker}\mathcal{A}$ to form a basis for V. The matrix of $\mathcal{A}$ under the basis has the desired form.

3.151   (a)   The geometric multiplicity and algebraic multiplicity are both 1.

           (b)   The geometric multiplicity is 1 and algebraic multiplicity is 2.

           (c)   The geometric multiplicity of λ is the dimension of the eigenspace:

$$\dim V_\lambda = \dim \mathrm{Ker}(\lambda I - A) = n - r(\lambda I - A).$$

           (d)   Let $\dim V_\lambda = k$ and let $\{v_1, \ldots, v_k\}$ be a basis (known as *eigenbasis*) of V_λ. Extend $\{v_1, \ldots, v_k\}$ to a basis for $\mathbb{C}^n$. Construct a matrix P with the basis vectors as columns. So P is invertible. Since $Av_i = \lambda v_i$ for $i = 1, \ldots, k$, we have $AP = P\Lambda$, where Λ has the form $\left(\begin{smallmatrix} \lambda I_k & B \\ 0 & C \end{smallmatrix}\right)$. Thus $P^{-1}AP = \Lambda$, and the characteristic polynomial of A is $p(t) = (t - \lambda)^k q(t)$ for some $q(t)$. Thus, the algebraic multiplicity of λ as an eigenvalue of A is at least k.

           (e) A is diagonalizable if and only if $P^{-1}AP = D$ for an invertible matrix P, where D is a diagonal matrix of the eigenvalues of A, if and only if $AP = PD$, where the columns of P (as eigenvectors) are linearly independent, if and only if the columns of P corresponding to each of the eigenvalues form a basis of the eigenspace of the eigenvalue, if and only if the algebraic and geometric multiplicities are equal for each of the eigenvalues of A.

3.152   (a)   If $\mathcal{A}x = \lambda x$, then $\mathcal{A}(\mathcal{B}x) = \mathcal{B}(\mathcal{A}x) = \lambda(\mathcal{B}x)$, thus $\mathcal{B}x \in V_\lambda(\mathcal{A})$.

           (b)   If $x \in \mathrm{Ker}\,\mathcal{A}$, then $\mathcal{A}x = 0$. Note that $\mathcal{A}(\mathcal{B}x) = \mathcal{B}(\mathcal{A}x) = 0$. Thus $\mathcal{B}x \in \mathrm{Ker}\,\mathcal{A}$, and $\mathrm{Ker}\,\mathcal{A}$ is invariant under $\mathcal{B}$. Similarly, $\mathrm{Im}\,\mathcal{A}$ is also invariant under $\mathcal{B}$.

           (c)   Let $\mathcal{B}_\lambda$ be the restriction of $\mathcal{B}$ on $V_\lambda(\mathcal{A})$, that is, $\mathcal{B}_\lambda v = \mathcal{B}v$, $v \in V_\lambda(\mathcal{A})$. $\mathcal{B}_\lambda$ has an eigenvalue in $\mathbb{C}$ and an eigenvector in $V_\lambda(\mathcal{A})$.

           (d)   By induction on dimension. Take v to be a common eigenvector of $\mathcal{A}$ and $\mathcal{B}$. Let W be a subspace such that $V = \mathrm{Span}\{v\} \oplus W$. Let $\mathcal{A}_1$ and $\mathcal{B}_1$ be the restrictions of $\mathcal{A}$ and $\mathcal{B}$ on W, respectively. Then $\mathcal{A}_1$ and $\mathcal{B}_1$ commute. Now apply induction hypothesis.

          If $\mathbb{C}$ is replaced by $\mathbb{R}$, (a) and (b) remain true. (c) and (d) are invalid over $\mathbb{R}$ because a polynomial may not be completely factorable over the reals (not algebraically closed). For example, $A = \left(\begin{smallmatrix} 0 & 1 \\ -1 & 0 \end{smallmatrix}\right)$ is not similar to any upper-triangular matrix over $\mathbb{R}$ (see Example 3.14.)

3.153 $\big(\mathcal{S}_t(f + kg)\big)(x) = f(x + t) + kg(x + t) = \big(\mathcal{S}_t(f)\big)(x) + k\big(\mathcal{S}_t(g)\big)(x)$. So, $\mathcal{S}_t(f + kg) = \mathcal{S}_t(f) + k\mathcal{S}_t(g)$. $\mathrm{Ker}\,\mathcal{S}_t = \{0\}$ and $\mathrm{Im}\,\mathcal{S}_t = \mathcal{C}(\mathbb{R})$.

         Note that $\mathcal{T} : t \mapsto \mathcal{S}_t$ is not a linear map from $\mathbb{R}$ to the space of linear maps $L(\mathcal{C}(\mathbb{R}))$ of $\mathcal{C}(\mathbb{R})$. $\mathcal{S}_0$ is the identity map on $\mathcal{C}(\mathbb{R})$.

3.154 (a) $\mathcal{D}(p + kq) = p' + kq' = \mathcal{D}(p) + k\mathcal{D}(q)$, where $k \in \mathbb{R}$, $p, q \in \mathbb{P}_n[x]$.

(b) The eigenvalues of $\mathcal{D}$ and $\mathcal{I} + \mathcal{D}$ are 0's and 1's, respectively; the eigenvectors are nonzero real numbers for both cases.

(c) The matrices of $\mathcal{D}$ under the bases are, respectively,

$$D_1 = (0, e_1, 2e_2, \ldots, (n-1)e_{n-1}), \quad D_2 = (0, e_1, e_2, \ldots, e_{n-1}),$$

where the e_i's are the standard basis (column) vectors of $\mathbb{R}^n$.

(d) No, since all eigenvalues of $\mathcal{D}$ are 0 (or $\mathcal{D}$ would be 0).

3.155 (a) It is easy to check that $\mathcal{L}$ is linear by the rules of derivatives.

(b) With respect to the basis $\{1, x, x^2, x^3\}$, the matrix of $\mathcal{L}$ is

$$\begin{pmatrix} 0 & 1 & -2 & 0 \\ 0 & 0 & 2 & -6 \\ 0 & 0 & 0 & 3 \\ 0 & 0 & 0 & 0 \end{pmatrix}.$$

(c) $\operatorname{Im}\mathcal{L} = \mathbb{P}_3[x]$ and $\operatorname{Ker}\mathcal{L} = \mathbb{R}$.

3.156 (a) Since $(e^{\lambda x})' = \lambda e^{\lambda x}$ and $(e^{\lambda x})'' = \lambda^2 e^{\lambda x}$, it suffices to notice that

$$y'' + ay' + by = (\lambda^2 + a\lambda + b)e^{\lambda x}.$$

(b) For $c \neq 0$, $ce^{\lambda x}$ is an eigenvector of $\mathcal{D}_2$ belonging to the eigenvalue λ^2. For any positive number μ, it is easy to see that

$$\mathcal{D}_2(e^{\sqrt{\mu}x}) = (\sqrt{\mu})^2(e^{\sqrt{\mu}x}) = \mu e^{\sqrt{\mu}x}.$$

Hence μ is an eigenvalue of $\mathcal{D}_2$ with $e^{\sqrt{\mu}x}$ as an eigenvector.

3.157 (a) is directly verified. (b) is easy, $\operatorname{Ker}\mathcal{A} = \mathbb{R}$.

(c) If $p(x) \in \operatorname{Ker}\mathcal{B}$, then $xp(x) = 0$, and $p(x) = 0$. $\mathcal{B}$ is one-to-one, but not onto. $\mathcal{B}$ has no inverse. Suppose otherwise $\mathcal{B}\mathcal{C} = \mathcal{I}$. Then $1 = x(\mathcal{C}(1))$, where $\mathcal{C}(1) \in \mathbb{P}[x]$. This is impossible.

(d) For every $f(x)$, $(\mathcal{A}\mathcal{B} - \mathcal{B}\mathcal{A})(f) = (xf)' - xf' = f + xf' - xf' = f$.

(e) By induction on k.

3.158 (a) Let $p, q \in \mathbb{P}_n[x]$, $k \in \mathbb{R}$. $\mathcal{A}$ is a linear transformation because

$$\begin{aligned} \mathcal{A}\big((p + kq)(x)\big) &= \mathcal{A}\big((p(x) + kq(x)\big) \\ &= x\big(p(x) + kq(x)\big)' - \big(p(x) + kq(x)\big) \\ &= xp'(x) + xkq'(x) - p(x) - kq(x) \\ &= \big(xp'(x) - p(x)\big) + k\big(xq'(x) - q(x)\big) \\ &= \mathcal{A}\big(p(x)\big) + k\mathcal{A}\big(q(x)\big). \end{aligned}$$

(b) $\operatorname{Ker}\mathcal{A} = \{\, kx \mid k \in \mathbb{R} \,\}$.

$\operatorname{Im}\mathcal{A} = \{\, a_0 + a_2 x^2 + \cdots + a_{n-1}x^{n-1} \mid a_0, a_2, \ldots, a_{n-1} \in \mathbb{R} \,\}$.

(c) By (b).

3.159 (a) It is routine to verify that $\mathcal{A}$ is a linear map.

(b) Consider the action of $\mathcal{A}$ on the basis $\{1, x, \ldots, x^{n-1}\}$. $\mathcal{A}(1) = \mathcal{A}(x) = 0$ and $\mathcal{A}(x^i) = ix^i$ for $i > 1$. The eigenvalues of $\mathcal{A}$ are $0, 0, 2, \ldots, n-1$ and associated eigenvectors are the basis vectors.

(c) The matrix of $\mathcal{A}$ with respect to the basis is $\operatorname{diag}(0, 0, 2, \ldots, n-1)$.

3.160 (a) Let W be an invariant subspace of V under $\mathcal{A}$. Observe that

$$
\begin{aligned}
u_2 &= (\mathcal{A} - \lambda \mathcal{I})(u_1), \\
u_3 &= (\mathcal{A} - \lambda \mathcal{I})(u_2), \\
&\;\;\vdots \qquad\qquad \vdots \\
u_n &= (\mathcal{A} - \lambda \mathcal{I})(u_{n-1}).
\end{aligned}
$$

Thus, $u_i = (\mathcal{A} - \lambda \mathcal{I})^{i-1}(u_1)$, $i = 1, 2, \ldots, n$. Since W is invariant under $\mathcal{A}$ and $\mathcal{A} - \lambda \mathcal{I}$, if $u_1 \in W$, then $u_2, \ldots, u_n \in W$, so $W = V$.

(b) Let W be a nonzero invariant subspace of V under $\mathcal{A}$. Let $x \in W$, $x \neq 0$, and let $x = a_k u_k + \cdots + a_n u_n$, where $k \geq 1$ and $a_k \neq 0$. If $k = n$, it is trivial. Suppose $k < n$. Since W is invariant under $\mathcal{A}$, it is invariant under $(\mathcal{A} - \lambda \mathcal{I})^{n-k}$, consequently,

$$
(\mathcal{A} - \lambda \mathcal{I})^{n-k}(x) \in W.
$$

Note that $(\mathcal{A} - \lambda \mathcal{I})(u_n) = 0$, $(\mathcal{A} - \lambda \mathcal{I})^n = 0$, and for each $i = 1, 2, \ldots, n$, $u_i = (\mathcal{A} - \lambda \mathcal{I})^{i-1}(u_1)$. We have $u_n \in W$ due to

$$
\begin{aligned}
(\mathcal{A} - \lambda \mathcal{I})^{n-k}(x) &= (\mathcal{A} - \lambda \mathcal{I})^{n-k}(a_k u_k + \cdots + a_n u_n) \\
&= a_k(\mathcal{A} - \lambda \mathcal{I})^{n-1}(u_1) + 0 + \cdots + 0 \\
&= a_k u_n \in W.
\end{aligned}
$$

(c) If $i = 1$, then $V_1 = \operatorname{Span}\{u_n\}$ and it is invariant under $\mathcal{A}$ as $\mathcal{A}(u_n) = \lambda u_n$. Let $i > 1$. Observe that for $k = n-i+1, \ldots, n-1$,

$$
(\mathcal{A} - \lambda \mathcal{I})(u_k) = u_{k+1} \in V_i \quad \text{and} \quad (\mathcal{A} - \lambda \mathcal{I})(u_n) = 0.
$$

Thus, each V_i is invariant under $\mathcal{A} - \lambda \mathcal{I}$. Write $\mathcal{A} = (\mathcal{A} - \lambda \mathcal{I}) + \lambda \mathcal{I}$. It follows that V_i is invariant under $\mathcal{A}$.

Note that $(\mathcal{A} - \lambda \mathcal{I})^i (u_j) = u_{i+j}$ with $u_{i+j} = 0$ when $i + j > n$. To show that $(\mathcal{A} - \lambda \mathcal{I})^i (x) = 0$ if and only if $x \in V_i$, let x be a linear combination of $u_1, u_2, \ldots, u_n$

$$x = x_1 u_1 + x_2 u_2 + \cdots + x_n u_n.$$

If $(\mathcal{A} - \lambda \mathcal{I})^i (x) = 0$, applying $(\mathcal{A} - \lambda \mathcal{I})^i$ to both sides results in

$$0 = x_1 u_{i+1} + \cdots + x_{n-i} u_n.$$

Thus $x_1 = x_2 = \cdots = x_{n-i} = 0$, and $x \in V_i$.
The other direction is immediate by observing that

$$(\mathcal{A} - \lambda \mathcal{I})^i (u_k) = 0, \quad k \geq n - i + 1.$$

(d) Let W be an invariant subspace of $\mathcal{A}$ with dimension m, $m \geq 1$. Then there must exist a nonzero $x \in W$ and u_i such that

$$x = a_i u_i + \cdots + a_n u_n, \quad a_i \neq 0, \; i \leq n - m + 1.$$

(Or W would lie in a space of dimension $m - 1$.) Applying

$$(\mathcal{A} - \lambda \mathcal{I})^{n-k}, \quad k = i, i+1, \ldots, i + m - 1,$$

to both sides consecutively, we obtain

$$u_n, u_{n-1}, \ldots, u_{n-m+1} \in V_m.$$

Thus $W = V_m$ because they have the same dimension.

(e) Note that an eigenspace is invariant under $\mathcal{A}$.

(f) If $V = W \oplus U$ and W and U are nontrivial invariant subspaces, then $u_n \in W$ and $u_n \in U$. This is impossible as $u_n \neq 0$.

3.161   Obviously, $\operatorname{Ker} \mathcal{A} \subseteq \operatorname{Ker} \mathcal{A}^2 \subseteq \cdots \subseteq \operatorname{Ker} \mathcal{A}^k \subseteq \operatorname{Ker} \mathcal{A}^{k+1} \subseteq \cdots$. This inclusion chain terminates because all the kernels are subspaces of V and V is finite dimensional. Thus, $\operatorname{Ker} \mathcal{A}^m = \operatorname{Ker} \mathcal{A}^{m+k}$ for some m and for all positive integers k. We claim that $V = \operatorname{Ker} \mathcal{A}^m \oplus \operatorname{Im} \mathcal{A}^m$. Since $\dim \operatorname{Ker} \mathcal{A} + \dim \operatorname{Im} \mathcal{A} = \dim V$ and $\operatorname{Ker} \mathcal{A} + \operatorname{Im} \mathcal{A} \subseteq V$, by the dimension identity, we only show $\operatorname{Ker} \mathcal{A}^m \cap \operatorname{Im} \mathcal{A}^m = \{0\}$. Let $u \in \operatorname{Ker} \mathcal{A}^m \cap \operatorname{Im} \mathcal{A}^m$. Then $\mathcal{A}^m (u) = 0$ and $u = \mathcal{A}^m (v)$ for some v. So $\mathcal{A}^m (u) = \mathcal{A}^{2m} (v) = 0$, which implies $v \in \operatorname{Ker} \mathcal{A}^{2m} = \operatorname{Ker} \mathcal{A}^m$. It follows that $\mathcal{A}^m (v) = 0$, that is, $u = \mathcal{A}^m (v) = 0$.

Hints and Answers for Chapter 4

4.1 (a) True. $(A + B)^* = A^* + B^* = A + B$.

    (b) False in general for complex c. True if c is real.

    (c) False in general. Take $A = \left(\begin{smallmatrix} 0 & 1 \\ 1 & 0 \end{smallmatrix}\right)$ and $B = \left(\begin{smallmatrix} 0 & i \\ -i & 0 \end{smallmatrix}\right)$.

    (d) True. $(ABA)^* = A^* B^* A^* = ABA$.

    (e) False. Take $A = \left(\begin{smallmatrix} 1 & 1 \\ 1 & 1 \end{smallmatrix}\right)$, $B = \left(\begin{smallmatrix} 1 & -1 \\ -1 & 1 \end{smallmatrix}\right)$.

    (f) True. $(BA)^* = A^* B^* = AB = 0$.

    (g) True. Consider the diagonal case.

    (h) False. Take $A = \left(\begin{smallmatrix} 0 & 1 \\ 1 & 0 \end{smallmatrix}\right)$.

    (i) True. The eigenvalues of A are 1, 1, 1.

    (j) True.

    (k) True.

    (l) True.

    (m) False. Take $A = \left(\begin{smallmatrix} 0 & 1 \\ 1 & 0 \end{smallmatrix}\right)$, $B = \left(\begin{smallmatrix} 1 & 0 \\ 0 & -1 \end{smallmatrix}\right)$.

    (n) True. ABA is Hermitian.

    (o) True. AB^2 has the same eigenvalues as BAB.

    (p) True. A^2 and B^2 are both positive semidefinite.

    (q) False. Take $A = \left(\begin{smallmatrix} 0 & 1 \\ 1 & 0 \end{smallmatrix}\right)$, $B = \left(\begin{smallmatrix} 1 & 0 \\ 0 & -1 \end{smallmatrix}\right)$.

    (r) True. $\det A = \det A^* = \overline{\det A^t} = \overline{\det A}$.

    (s) True. $\overline{\mathrm{tr}(AB)} = \mathrm{tr}(AB)^* = \mathrm{tr}(B^* A^*) = \mathrm{tr}(BA) = \mathrm{tr}(AB)$.

4.2 We show that (c) $\Rightarrow$ (a) and (g) $\Rightarrow$ (a). Other implications are easy.

(c) $\Rightarrow$ (a): $x^* A x \in \mathbb{R} \Rightarrow x^* A x - (x^* A x)^* = x^*(A - A^*)x = 0$. Since $A - A^*$ is skew-Hermitian, it is diagonalizable and all its eigenvalues are pure imaginary or zero. Because $x^*(A - A^*)x = 0$ for all x, all the eigenvalues of $A - A^*$ are zero. Thus, $A - A^* = 0$, i.e., $A = A^*$.

Alternatively, it is easy to see that $a_{ss} \in \mathbb{R}$, $s = 1, 2, \ldots, n$, by taking x to be the column vector with the s-th component 1, and 0 elsewhere.

Now take x to be the column vector with the s-th component 1, the t-th component c, and 0 elsewhere, where $s \neq t$ and c is an arbitrary complex number. Then $x^*Ax = a_{ss} + a_{tt}|c|^2 + a_{ts}\bar{c} + a_{st}c \in \mathbb{R}$. Thus, $\overline{a_{ts}c} + a_{st}c \in \mathbb{R}$ for all c, implying $\overline{a_{ts}} = a_{st}$, that is, $A^* = A$.

(g) $\Rightarrow$ (a): By the Schur triangularization, let $A = U^*TU$, where U is a unitary matrix and $T = (t_{st})$ is an upper-triangular matrix. Then

$$\operatorname{tr} A^2 = \operatorname{tr}(AA^*) \ \Rightarrow \ \operatorname{tr} T^2 = \operatorname{tr}(TT^*),$$

which is

$$\sum_{s=1}^{n} t_{ss}^2 = \sum_{s=1}^{n} |t_{ss}|^2 + \sum_{s<t} |t_{st}|^2.$$

It is immediate that $t_{st} = 0$ for every pair of s and t, $s < t$, and t_{ss} is real for each s. Therefore, T is real diagonal and A is Hermitian.

4.3   For $x = (x_1, \ldots, x_n)^t$, $y = (y_1, \ldots, y_n)^t \in \mathbb{C}^n$, we have

$$x^*Ay = \sum_{i,\,j=1}^{n} \overline{x_i} a_{ij} y_j.$$

(a) If $x^tAx = 0$ for every $x \in \mathbb{R}^n$, then the diagonal entries of A are all equal to zero by taking x to be the column vector with the p-th component 1 and 0 elsewhere, $p = 1, 2, \ldots, n$.
    Now take x to be the column vector with the p-th and the q-th components 1 and 0 elsewhere. Then $x^tAx = 0 \Rightarrow a_{pq} = -a_{qp}$, $p \neq q$, so, $A^t = -A$. Conversely, if $A^t = -A$, then $x^tAx = (x^tAx)^t = x^tA^tx = -(x^tAx)$ and $x^tAx = 0$ for every $x \in \mathbb{R}^n$.

(b) To show all $a_{pq} = 0$, take x and y to be the vectors whose p-th and q-th components are 1, respectively, and 0 elsewhere.

(c) It is easy to see that the diagonal entries of A are all equal to zero. Take x to be the column vector with the p-th component 1 and the q-th component c, then $x^*Ax = 0 \Rightarrow a_{qp}\bar{c} + a_{pq}c = 0$, for every $c \in \mathbb{C}$, thus $a_{qp} = a_{pq} = 0$ for all p, q, and $A = 0$.

(d) x^*Ax is real if and only if $x^*(A^* - A)x = 0$, i.e., $A^* = A$ by (c).

(e) Let $x^*Ax = c$ be a constant for all unit $x \in \mathbb{C}^n$. Then $x^*(A - cI)x = 0$. It follows from (c) that $A - cI = 0$, i.e., $A = cI$.

(f) For real symmetric A, $(z^*Az)^* = z^*A^*z = z^*Az$, z^*Az is a real number. (i) Use (a). $A^t = -A$ and $A^t = A$ if and only if $A = 0$. (ii) Let $z = x + yi$, where $x, y \in \mathbb{R}^n$. Then $z^*Az = (x^t - y^ti)A(x + yi) = x^tAx + y^tAy + i(x^tAy - y^tAx)$. Because z^*Az is real, $x^tAy - y^tAx = 0$. It follows that $z^*Az = x^tAx + y^tAy > 0$. (iii) is similar.

(g) Let $A = \begin{pmatrix} 0 & 1 \\ -1 & 0 \end{pmatrix}$, $B = 0$. Then $x^t A x = x^t B x$ for all $x \in \mathbb{R}^2$.

(h) Let $A = \begin{pmatrix} 0 & 1 \\ 1 & 0 \end{pmatrix}$, $y = \begin{pmatrix} 1 \\ 0 \end{pmatrix}$. Then $y^* A y = 0$, but $A y \neq 0$.

(i) Let $B = \begin{pmatrix} 0 & 1 \\ -1 & 0 \end{pmatrix}$, $x^t B x = 0$ for all $x \in \mathbb{R}^2$.

(j) Let $C = iB = \begin{pmatrix} 0 & i \\ -i & 0 \end{pmatrix}$ is Hermitian, $x^t C x = 0$ for all $x \in \mathbb{R}^2$.

4.4 $U^* U = I_4$. The eigenvalues are $\pm \frac{(1+\sqrt{3})+(-1+\sqrt{3})i}{2\sqrt{2}}$, $\pm \frac{(1-\sqrt{3})+(-1-\sqrt{3})i}{2\sqrt{2}}$.

4.5   If λ is an eigenvalue of A, then λ^2 is an eigenvalue of A^2. Thus the eigenvalues of A^2 are $\lambda_1^2, \ldots, \lambda_n^2$. So $\operatorname{tr} A^2 = \lambda_1^2 + \cdots + \lambda_n^2$. However, $\operatorname{tr} A^2$ is the sum of the (i,i)-entries $\sum_j a_{ij} a_{ji}$ on the main diagonal of A^2. The desired identity follows. When A is Hermitian, $a_{ij} = \overline{a_{ji}}$.

4.6   (a) Clearly $H^* = H$. $H^2 = I - 4uu^* + 4(uu^*)^2 = I$. So $H^{-1} = H$.
(b) Note that $(x - y)^*(x + y) = x^* x + x^* y - y^* x - y^* y = 0$. With this, we derive $S(x+y) = x + y$ and $S(x - y) = -(x - y)$. So $Sx = y$. For (c), taking $x = e$ and $y = \sqrt{3}(1,0,0)^t$ in (b), one can obtain

$$
\begin{aligned}
Q &= I_3 - \frac{1}{3 - \sqrt{3}} \begin{pmatrix} 1 - \sqrt{3} \\ 1 \\ 1 \end{pmatrix} (1 - \sqrt{3}, 1, 1) \\
&= \frac{1}{3 - \sqrt{3}} \begin{pmatrix} \sqrt{3} - 1 & \sqrt{3} - 1 & \sqrt{3} - 1 \\ \sqrt{3} - 1 & 2 - \sqrt{3} & -1 \\ \sqrt{3} - 1 & -1 & 2 - \sqrt{3} \end{pmatrix}.
\end{aligned}
$$

It follows that $Q e e^t Q^t = Q^t E Q = \operatorname{diag}(3, 0, 0)$, where $E = e e^t$ is the 3×3 all-ones matrix. Note that such a Q need not be unique.

See Problems 3.51, 5.94, and 6.132.

4.7   Since A is Hermitian, all eigenvalues of A are real. If $\det A < 0$, then at least one eigenvalue is negative. Denote it by λ. Then $Ax = \lambda x$ for some nonzero x, where $\lambda < 0$. Thus $x^* A x = \lambda x^* x < 0$. Take $A = \begin{pmatrix} -1 & 0 \\ 0 & -1 \end{pmatrix}$. Then $\det A = 1 > 0$, but $y^* A y < 0$ for all $y \neq 0$.

4.8 $(A + B)^* = A^* + B^* = A + B$. So $A + B$ is Hermitian. If $AB = BA$, then $(AB)^* = B^* A^* = BA = AB$; this says AB is Hermitian. Conversely, if AB is Hermitian, then $AB = (AB)^* = B^* A^* = BA$.

4.9   Since A and B are Hermitian matrices, AB is Hermitian if and only if $AB = BA$. Thus, A and B are diagonalizable through the same unitary matrix, that is, $U^* A U$ and $U^* B U$ are diagonal for some unitary matrix U. Therefore, the eigenvalues of AB are those in the form $\lambda = ab$, where a is an eigenvalue of A and b is an eigenvalue of B.

4.10 (a) $A = uu^t$, where $u = (1, 2, 1)^t$. $A^3 = 36A$. So $\left(\frac{1}{\sqrt[3]{36}}A\right)^3 = A$.

(b) Note that the eigenvalues of B are zero. Suppose $X^3 = B$. Then the eigenvalues of X are all zero, implying $X^3 = 0 \neq B$.

(c) $B = \begin{pmatrix} 0 & 0 & 1 \\ 0 & 0 & 0 \\ 0 & 1 & 0 \end{pmatrix}^2$.

(d) Similar to (b).

(e) $D = \begin{pmatrix} 0 & 0 & 1 & 0 \\ 0 & 0 & 0 & 1 \\ 0 & 1 & 0 & 0 \\ 0 & 0 & 0 & 0 \end{pmatrix}^2$.

(f) Every normal matrix is unitarily diagonalizable. Let $N \in M_n(\mathbb{C})$ be a normal matrix and let $N = U^* \operatorname{diag}(\lambda_1, \ldots, \lambda_n)U$, where U is unitary. Then a k-th root of N is $X = U^* \operatorname{diag}(\mu_1, \ldots, \mu_n)U$, where $\mu_i^k = \lambda_i$, $i = 1, \ldots, n$. X is normal, but not unique.

(g) If $X^2 = Y$, then $\lambda^2 I - X^2 = \lambda^2 I - Y$, which yields

$$(\lambda I - X)(\lambda I + X) = \lambda^2 I - Y.$$

Taking determinants of both sides reveals the desired conclusion. One may easily verify that $\left(\begin{smallmatrix} -1 & -1 \\ 1 & 0 \end{smallmatrix}\right)^3 = I_2$.

4.11 This is an easy problem if one applies the Cauchy eigenvalue interlacing theorem to A with the principal submatrix $\left(\begin{smallmatrix} a & 0 \\ 0 & b \end{smallmatrix}\right)$. We solve the problem without using the Cauchy eigenvalue interlacing theorem.

(a) A is Hermitian, so $\det A$ is real.

(b) Since A is Hermitian, the eigenvalues of A are all real.

(c) A computation gives the characteristic polynomial of A:

$$p(t) = |tI - A| = (t - a)(t - b)(t - c) - (t - b)|d|^2 - (t - a)|e|^2.$$

If $a = b$, then $(t - a)$ is a factor of $p(t)$. So a is an eigenvalue.

(d) Let $a < b$. Then $p(a) = (b-a)|d|^2 \geq 0$ and $p(b) = (a-b)|e|^2 \leq 0$. On the other hand, $p(t) = (t - r_1)(t - r_2)(t - r_3)$, $r_1 \leq r_2 \leq r_3$. Note that a and b are contained in the interval $[r_1, r_3]$, i.e., $[a, b] \subseteq [r_1, r_3]$, because of the eigenvalue min-max expressions and

$$a = e_1^t A e_1, \ e_1 = (1, 0, 0)^t, \quad b = e_2^t A e_2, \ e_2 = (0, 1, 0)^t.$$

If $a = r_2$ or $b = r_2$, then we have nothing to show. Suppose that neither a nor b is equal to r_2. We must show $r_2 \in (a, b)$. To this end, we prove that a and b cannot be on the same side of r_2.

Assume $[a, b] \subseteq [r_1, r_2)$. If $a \neq r_1$, then $p(t)$ would have a root in (a, b), a contradiction. If $a = r_1$ and $d \neq 0$, then $p(a) = (b - a)|d|^2 = 0$ implies $a = b = r_1 = r_2$, a contradiction. If $a = r_1$ and $d = 0$, then r_2 and r_3 are the eigenvalues of $\left(\begin{smallmatrix} b & e \\ \bar{e} & c \end{smallmatrix} \right)$, yielding $r_2 \leq b$, a contradiction. Similarly, one proves $[a, b] \not\subseteq (r_2, r_3]$.

4.12 By direct computations.

4.13 Both A and B are real symmetric. So their eigenvalues are all real. The first row and the third row of A are the same, so A is singular and A has a zero eigenvalue. The 2×2 minor of A in the upper-left corner is negative. So A has a negative eigenvalue. Since the trace of A is positive (i.e., 3), A has a positive eigenvalue.

For B, dividing rows 2, 4 and columns 2, 4 by 1.01 and through row and column operations, we easily get $\det B > 0$. Let a, b, c, d be the eigenvalues of B. Then $abcd > 0$. Since $\operatorname{tr} B = a + b + c + d = 0$, B has two positive eigenvalues and two negative eigenvalues.

4.14 Let $A = U^* \operatorname{diag}(\lambda_1, \ldots, \lambda_n)U$, where U is unitary and λ_i's are nonzero real. Set $P = U^* \operatorname{diag}(1/\lambda_1, \ldots, 1/\lambda_n)U$. Then $P^*AP = P = A^{-1}$.

4.15 (a) By direct verifications. (b) Over $\mathbb{C}$, $U^t AU = I_3$, where $U = \operatorname{diag}(1, \frac{1}{\sqrt{2}}, i)$. B is real symmetric, so $V^t BV$ is diagonal (with eigenvalues on the diagonal), where V is a real orthogonal matrix. There is a diagonal complex matrix W such that $W^t(V^t BV)W = I_3$. Thus, A and B are both ^{t}-congruent to I_3. Note: two symmetric matrices are ^{t}-congruent over $\mathbb{C}$ if and only if they have the same rank. A and B are both real symmetric; but they have different inertias. Thus, there does not exist a real or complex invertible S such that $S^* BS = A$.

4.16 B, C, D are equivalent to A (over $\mathbb{R}$) as they have the same rank. All the matrices are real symmetric and have the same rank, thus they are ^{t}-congruent over $\mathbb{C}$. B has three positive eigenvalues, while A, C, D each have two positive eigenvalues and one negative eigenvalues. By inertias, only C, D are *-congruent (and ^{t}-congruent) to A over $\mathbb{R}$. A and D have the same eigenvalues, so D is similar to A over $\mathbb{R}$.

4.17 If H (or K) is nonsingular, then HK is similar to $H^{-1}(HK)H = KH$.

Let H and K be singular and set $A = HK$. Then A is singular. Note that $AH = HKH = HA^*$, $A^2H = H(A^*)^2$, and inductively, we have

$$A^k H = H(A^*)^k, \quad k = 1, 2, \ldots.$$

Without loss of generality, we can assume that $A = \begin{pmatrix} D & 0 \\ 0 & E \end{pmatrix}$ is a Jordan canonical form, where D, E contain the Jordan blocks of nonzero and zero eigenvalues, respectively. So D is invertible and E is nilpotent. Partition $H = \begin{pmatrix} H_1 & H_2 \\ H_2^* & H_3 \end{pmatrix}$ conformally as $A = \begin{pmatrix} D & 0 \\ 0 & E \end{pmatrix}$. Then $A^n H = H(A^*)^n$ yields $\begin{pmatrix} D^n & 0 \\ 0 & 0 \end{pmatrix} \begin{pmatrix} H_1 & H_2 \\ H_2^* & H_3 \end{pmatrix} = \begin{pmatrix} H_1 & H_2 \\ H_2^* & H_3 \end{pmatrix} \begin{pmatrix} (D^*)^n & 0 \\ 0 & 0 \end{pmatrix}$. Comparing the blocks on both sides gives $D^n H_1 = H_1(D^*)^n$ and $D^n H_2 = 0$. Thus, $H_2 = 0$, and $H = \begin{pmatrix} H_1 & 0 \\ 0 & H_3 \end{pmatrix}$. Partition $K = \begin{pmatrix} K_1 & K_2 \\ K_2^* & K_3 \end{pmatrix}$. Then $A = HK$ reveals $D = H_1 K_1$. So H_1 is nonsingular. $AH = HA^*$ shows that $DH_1 = H_1 D^*$. Thus, D and D^* are similar. On the other hand, $E^* = E^t$ is similar to E (as every square matrix is similar to its transpose; Problem 3.68). Therefore, $A = \begin{pmatrix} D & 0 \\ 0 & E \end{pmatrix}$ is similar to $\begin{pmatrix} D^* & 0 \\ 0 & E^* \end{pmatrix} = A^*$, i.e., HK is similar to KH. (See also Problem 6.65.)

4.18   If $r = n$, we have nothing to show. Let $0 < r < n$. By spectral decomposition, we write $A = UDU^*$, where U is unitary and $D = \operatorname{diag}(\lambda_1, \ldots, \lambda_r, 0, \ldots, 0)$, $\lambda_i \neq 0, i = 1, \ldots, r$. Let $\Lambda = \operatorname{diag}(\lambda_1, \ldots, \lambda_r)$ and let U_r be the $n \times r$ matrix of the first r columns of U. Then

$$A = UDU^* = U_r \Lambda U_r^*.$$

Let B_α be an $r \times r$ principal submatrix of A, say, lying in the intersections of the rows and columns $\alpha_1, \ldots, \alpha_r$. Denote by V_r the $r \times r$ submatrix of U_r that consists of the rows of $\alpha_1, \ldots, \alpha_r$ of U_r. Then $B_\alpha = V_r \Lambda V_r^*$ and $\det B_\alpha = (\det \Lambda)|\det V_r|^2 = \lambda_1 \cdots \lambda_r |\det V_r|^2$. If B_α is nonsingular, then $\det B_\alpha$ has the same sign as $\lambda_1 \cdots \lambda_r$. Thus, all nonzero $r \times r$ principal minors of A have the same sign.

4.19   We show that all principal minors of M are nonnegative. This is obvious for 1×1 and 2×2 principal minors. We show that $\det M \geq 0$.

$\det M =$
$$a_{11}a_{22}a_{33} + 2|a_{21}||a_{32}||a_{13}| - a_{11}|a_{23}|^2 - a_{22}|a_{13}|^2 - a_{33}|a_{12}|^2,$$
$\det A =$
$$a_{11}a_{22}a_{33} + 2\operatorname{Re}\left(a_{21}a_{32}\overline{a_{13}}\right) - a_{11}|a_{23}|^2 - a_{22}|a_{13}|^2 - a_{33}|a_{12}|^2.$$

Since $\mathrm{Re}(c) \leq |c|$ for $c \in \mathbb{C}$, we have $\mathrm{Re}\left(a_{21}a_{32}\overline{a_{13}}\right) \leq |a_{21}||a_{32}||a_{13}|$. Thus, $\det M \geq \det A \geq 0$. It follows that M is positive semidefinite.

If $n \geq 4$, the matrix of entrywise absolute values of a positive semidefinite matrix is not positive semidefinite in general. Let

$$A = \begin{pmatrix} 3 & 2 & 0 & -2 \\ 2 & 3 & 2 & 0 \\ 0 & 2 & 3 & 2 \\ -2 & 0 & 2 & 3 \end{pmatrix}, \quad M = \begin{pmatrix} 3 & 2 & 0 & 2 \\ 2 & 3 & 2 & 0 \\ 0 & 2 & 3 & 2 \\ 2 & 0 & 2 & 3 \end{pmatrix}.$$

One may verify that all four leading principal minors of A are positive. Thus, A is positive definite, but M is not because $\det M = -63 < 0$.

4.20 Recall that if B is a submatrix of A, then $\sigma_{\max}(B) \leq \sigma_{\max}(A)$ (see inequality (4.10), p. 148). Take B to be a_{ij}, we get the first inequality.

For the 2nd inequality, take U to be a unitary matrix whose first row has $\frac{1}{\sqrt{2}}$ in the p-th and q-th positions (0's in other positions). UA has the same singular values as A, and $\frac{1}{\sqrt{2}}(a_{pj} + a_{qj})$ is the $(1,j)$-entry of UA. It follows that $\frac{1}{\sqrt{2}}|a_{pj} + a_{qj}| \leq \sigma_{\max}(UA) = \sigma_{\max}(A)$.

For the 3rd inequality, take B to be the submatrix $(a_{pj}, a_{qj})^t$.

For the last inequality, take V to be a unitary matrix whose first row is the unit vector $\frac{1}{\sqrt{m}}(\omega, \omega^2, \ldots, \omega^m)$, then apply the first inequality to the $(1,j)$-entry of VA. Note that $\sigma_{\max}(VA) = \sigma_{\max}(A)$.

Note: Replacing A with A^t, one gets the similar inequalities for rows.

4.21 (a) Take $A = \mathrm{diag}(1, -1)$ and I_2, respectively.

(b) Assume that $A = U^*DU$, where D is real diagonal and U is unitary. Then $AB = U^*DUB$ is similar to $DUBU^*$ whose trace is real (as the entries on the main diagonal are real). In general,

$$\mathrm{tr}(AB)^k = \mathrm{tr}(AB \cdots AB) = \mathrm{tr}\left((AB \cdots ABA)B\right)$$

is real since both $AB \cdots ABA$ and B are Hermitian.

(c) Note that the inequality is unitarily invariant, that is, we may replace A and B by U^*AU and U^*BU for any $n \times n$ unitary matrix U. Without loss of generality, assume that A is a real diagonal matrix with diagonal entries $a_1, a_2, \ldots, a_n$. Then

$$\begin{aligned} \mathrm{tr}(A^2B^2) - \mathrm{tr}(AB)^2 &= \sum_{i,j} a_i^2|b_{ij}|^2 - \sum_{i,j} a_i a_j |b_{ij}|^2 \\ &= \sum_{i<j}(a_i - a_j)^2|b_{ij}|^2 \geq 0. \end{aligned}$$

Equality holds if and only if $(a_i - a_j)|b_{ij}| = 0$ for all $i < j$, which is true if and only if $a_i b_{ij} = a_j b_{ij}$ for all i, j, that is, $AB = BA$. Alternatively, use the fact that $AB - BA$ is skew-Hermitian.

(d) For example, take $A = \begin{pmatrix} 1 & 0 & 0 \\ 0 & 1 & 0 \\ 0 & 0 & 2 \end{pmatrix}$, $B = \begin{pmatrix} 1 & 0 & -2 \\ 0 & 1 & 0 \\ -2 & 0 & 2 \end{pmatrix}$, $C = \begin{pmatrix} 2 & 0 & 1 \\ 0 & 1 & 0 \\ 1 & 0 & 1 \end{pmatrix}$.
Computations show that $\operatorname{tr}(A^2 B^2 C^2) = 0$ and $\operatorname{tr}(ABC)^2 = 9$.

(e) Since the inequality is unitarily invariant, we may assume that A is a real diagonal matrix with diagonal entries $a_1, a_2, \ldots, a_n$. By the Cauchy–Schwartz inequality, we have

$$\left(\sum_{i=1}^n a_i b_{ii} \right)^2 \le \left(\sum_{i=1}^n a_i^2 \right) \left(\sum_{i=1}^n b_{ii}^2 \right) \le \left(\sum_{i=1}^n a_i^2 \right) \left(\sum_{i, j=1}^n |b_{ij}|^2 \right).$$

Both equalities hold if and only if the vectors $(a_1, \ldots, a_n)$ and $(b_{11}, \ldots, b_{nn})$ are linearly dependent and all $b_{ij} = 0$ when $i \neq j$. It follows that B is a diagonal matrix and $B = kA$ or $A = kB$. Conversely, if $B = kA$ or $A = kB$, then equality obviously holds.

4.22 We only show that (c) $\Rightarrow$ (e), (f) $\Rightarrow$ (b), (g) $\Rightarrow$ (a), and (h) $\Leftrightarrow$ (a).

(c) $\Rightarrow$ (e): Let $A = U^* \operatorname{diag}(D, 0)U$ be a spectral decomposition of A, where D is $r \times r$, with positive diagonal. Let U_1 be the first r rows of U. Then $A = U_1^* D^{\frac{1}{2}} \cdot D^{\frac{1}{2}} U_1 = T^*T$, where $T = D^{\frac{1}{2}} U_1$, $r(T) = r$.

(f) $\Rightarrow$ (b): Let $A = U^* \operatorname{diag}(\lambda_1, \ldots, \lambda_n)U$, where λ_i's are (necessarily real) eigenvalues of A and U is unitary, and let

$$f(\lambda) = |\lambda I - A| = \lambda^n - \delta_1 \lambda^{n-1} + \delta_2 \lambda^{n-2} - \cdots + (-1)^n \delta_n.$$

It can be shown by expanding the determinant $|\lambda I - A|$ that δ_i is the sum of all $i \times i$ principal minors of A, $i = 1, 2, \ldots, n$. Thus, if all the minors of A are nonnegative, then $f(\lambda)$ has no negative zeros.

(g) $\Rightarrow$ (a): Take $X = xx^*$, where $x \in \mathbb{C}^n$ is any column vector.

(h) $\Leftrightarrow$ (a): $x^*Ax \ge 0$ for all $x \in \mathbb{C}^n \Rightarrow y^*(X^*AX)y = (Xy)^*A(Xy) \ge 0$ for all $y \in \mathbb{C}^m$. This says $X^*AX \ge 0$. Conversely, if for some (fixed) positive integer m, $X^*AX \ge 0$ for all $n \times m$ matrices X, then for any column vector $x \in \mathbb{C}^n$, we take X to be the $n \times m$ matrix with first column x and elsewhere 0. It follows that $x^*Ax \ge 0$, i.e., $A \ge 0$.

4.23 (a) $(A + B)^*(A + B) = (A + B)^2 = A^2 + AB + BA + B^2 \ge 0$.
(b) $A^2 + AB + BA$ and $AB + BA$ are not positive semidefinite for

$$A = \begin{pmatrix} 1 & 1 \\ 1 & 1 \end{pmatrix} \ge 0, \quad B = \begin{pmatrix} 2 & 1 \\ 1 & 1 \end{pmatrix} \ge 0,$$

as one checks

$$A^2 + AB + BA = \begin{pmatrix} 8 & 7 \\ 7 & 6 \end{pmatrix} \not\geq 0, \quad AB + BA = \begin{pmatrix} 6 & 5 \\ 5 & 4 \end{pmatrix} \not\geq 0.$$

(c) If A and $AB+BA$ are positive definite, we show that B is positive definite. Let $C = AB + BA$. Multiply C by $A^{-\frac{1}{2}}$ from both sides:

$$0 < A^{-\frac{1}{2}}CA^{-\frac{1}{2}} = A^{\frac{1}{2}}BA^{-\frac{1}{2}} + A^{-\frac{1}{2}}BA^{\frac{1}{2}} = D + D^*,$$

where $D = A^{\frac{1}{2}}BA^{-\frac{1}{2}}$. It is sufficient to show that D is nonsingular. Suppose $Dx = 0$ for some nonzero column vector x. Then $x^*D^* = 0$ and $x^*(D+D^*)x = x^*Dx+x^*D^*x = 0$. This contradicts $D+D^* > 0$.

4.24　Expanding $(A \pm B)^*(A \pm B) \geq 0$ yields the desired inequalities.

4.25　Note that $0 \leq A \leq I \Rightarrow 0 \leq A^2 \leq A$. Likewise, $B^2 \leq B$. One derives

$$\begin{aligned} 0 \leq (A + B - \tfrac{1}{2}I)^2 &= (A^2 - A) + (B^2 - B) + AB + BA + \tfrac{1}{4}I \\ &\leq AB + BA + \tfrac{1}{4}I \Rightarrow AB + BA \geq -\tfrac{1}{4}I. \end{aligned}$$

It is possible that $AB + BA$ has an eigenvalue $-\frac{1}{4}$. See Problem 6.95.

4.26　$x, q \in \mathbb{C}$, $y \in \mathbb{R}$, $p > 0$, and $|x| < \frac{\sqrt{3}}{2}$, $-2 < y < 1$, $p > 1 + |q|^2$.

4.27　For the matrix to be positive semidefinite, all principal minors need be nonnegative. So we have (i) $2x \geq 1$, (ii) $3x \geq 1$, (iii) $y^2 \leq 6$, and (iv) $x \geq \frac{5+2y}{6-y^2}$. (i) yields (ii). One may show that if $y^2 < 6$, then the least value of $\frac{5+2y}{6-y^2}$ is 0.5. So (iii) and (iv) imply (i) and (ii). It follows that $y^2 < 6$ and $x \geq \frac{5+2y}{6-y^2}$ are necessary and sufficient conditions.

4.28　Let $Ax = \lambda x$, where $x = (x_1, x_2, \ldots, x_n)^t \neq 0$. Since A is Hermitian, λ is real. Let $|x_i| = \max_j |x_j|$. Then $x_i \neq 0$. From $\sum_{j=1}^{n} a_{ij}x_j = \lambda x_i$, we have $(\lambda - 1)x_i = \sum_{j=1, j\neq i}^{n} a_{ij}x_j$. Since $\sum_{j=1, j\neq i}^{n} |a_{ij}| \leq 1$, taking absolute values, we get $|\lambda - 1||x_i| \leq |x_i|$. Thus, $|\lambda - 1| \leq 1$, i.e., $0 \leq \lambda \leq 2$. So A is positive semidefinite. The determinant inequality follows from an application of the Hadamard inequality to A.

4.29　(i) $\begin{pmatrix} 0 & 0 \\ 0 & -1 \end{pmatrix}$. (ii) $\begin{pmatrix} 1 & 1 \\ 0 & 1 \end{pmatrix}$. (iii) The answer to the first part is affirmative. For instance, $A = \begin{pmatrix} 0 & 1 \\ -1 & 0 \end{pmatrix}$, $x^t Ax = 0$ for all $x \in \mathbb{R}^2$. If for all $x \in \mathbb{C}^n$, $x^* Ax \geq 0$, then A has to be positive semidefinite (Hermitian).

4.30 (a) If $A \neq 0$, then $B = A^{\frac{1}{2}} \neq 0$, and $(\pm B)^2 = A$. In fact, there are many Hermitian matrices X that satisfy $X^2 = A$. However, only one of them is positive semidefinite, which is $A^{\frac{1}{2}}$.

(b) We only show $\operatorname{Im} A = \operatorname{Im} A^{\frac{1}{2}}$. Obviously, $\operatorname{Im} A \subseteq \operatorname{Im} A^{\frac{1}{2}}$ as $A = (A^{\frac{1}{2}})^2$. On the other hand, A and $A^{\frac{1}{2}}$ have the same rank.

(c) The square roots are, respectively,

$$\begin{pmatrix} \sqrt{2} & 0 \\ 0 & 0 \end{pmatrix}, \quad \frac{1}{\sqrt{2}} \begin{pmatrix} 1 & 1 \\ 1 & 1 \end{pmatrix}, \quad \frac{1}{2} \begin{pmatrix} 3 & -1 \\ -1 & 3 \end{pmatrix}.$$

4.31 (a) Consider each diagonal entry as a minor, or take x to be the column vector whose i-th component is 1, and 0 elsewhere. Then $a_{ii} = x^* A x$. The second part follows from (b).

(b) All 2×2 principal submatrices $\left(\begin{smallmatrix} a_{ii} & a_{ij} \\ \overline{a_{ij}} & a_{jj} \end{smallmatrix} \right) \geq 0$. So, $\left| \begin{smallmatrix} a_{ii} & a_{ij} \\ \overline{a_{ij}} & a_{jj} \end{smallmatrix} \right| \geq 0$. Thus, $|a_{ij}|^2 \leq a_{ii} a_{jj}$, $|a_{ij}| \leq \max\{a_{ii}, a_{jj}\} \leq \max\{a_{11}, \ldots, a_{nn}\}$.

(c) Assume that B is a principal submatrix of A in the upper-left corner. If $\det B = 0$, then $Bv = 0$ for some $v \neq 0$. Set $x = (v^t, 0, \ldots, 0)^t \in \mathbb{C}^n$. Then $x \neq 0$, $Ax = 0$. So A is singular.

(d) A is unitarily diagonalizable. Let $A = U^* D U$, where $D = \operatorname{diag}(\lambda_1, \lambda_2, \ldots, \lambda_n)$. Split each positive λ_i as $\frac{1}{\sqrt{\lambda_i}} 1 \frac{1}{\sqrt{\lambda_i}}$. P need not be unitary in general.

(e) $x^* A x \geq 0 \Rightarrow (x^* A x)^t \geq 0$. So $x^t A^t \bar{x} \geq 0$ or $y^* A^t y \geq 0$ for all $y = \bar{x}$. Thus, $A^t \geq 0$. Likewise, $\bar{A} \geq 0$.

(f) Let $A = U^* \operatorname{diag}(\lambda_1, \ldots, \lambda_n) U$, where U is unitary. Then a k-th root of A is $X = U^* \operatorname{diag}(\mu_1, \ldots, \mu_n) U$, where $\mu_i = \sqrt[k]{\lambda_i}$, $i = 1, \ldots, n$. Since λ_i's are nonnegative, μ_i's are nonnegative. It is obvious that $X^k = A$. If Y is a positive semidefinite matrix such that $Y^k = A$, then the eigenvalues of Y are μ_i's. If $Y = V^* \operatorname{diag}(\mu_1, \ldots, \mu_n) V$, where V is unitary, $X^k = Y^k$ implies $U^* \operatorname{diag}(\lambda_1, \ldots, \lambda_n) U = V^* \operatorname{diag}(\lambda_1, \ldots, \lambda_n) V$. Let $W = (w_{ij}) == VU^*$. Then $W \operatorname{diag}(\lambda_1, \ldots, \lambda_n) = \operatorname{diag}(\lambda_1, \ldots, \lambda_n) W$, implying $w_{ij} \lambda_j = w_{ij} \lambda_i$ for all i, j. It follows that $w_{ij} \sqrt[k]{\lambda_j} = w_{ij} \sqrt[k]{\lambda_i}$ for all i, j. From this, we get $X = Y$.

4.32 $(Ax)^*(Ax) = (x^* A^*)(Ax) = 0$ if and only if $Ax = 0$. For $P \geq 0$, $\operatorname{tr} P = 0$ (sum of the entries on the main diagonal) if and only if $P = 0$. By this, $\operatorname{tr}(A^* A) = 0 \Leftrightarrow A^* A = 0 \Leftrightarrow A = 0$ as $r(A^* A) = r(A)$. Or directly, $\operatorname{tr}(A^* A) = \sum_{i,j} |a_{ij}|^2 = 0$ if and only if all $a_{ij} = 0$.

4.33   It is sufficient to show the diagonal case $A = \text{diag}(\lambda_1, \lambda_2, \ldots, \lambda_n)$. For a general case, let $A = U^*DU$ and replace x with Ux and y with Uy in the diagonal case. For $x = (x_1, \ldots, x_n)^t$ and $y = (y_1, \ldots, y_n)^t$,

$$x^*A^{-1}x = \sum_{k=1}^{n} \lambda_k^{-1}|x_k|^2, \quad y^*Ay = \sum_{k=1}^{n} \lambda_k |y_k|^2.$$

Notice that

$$x_k\overline{y_k} + y_k\overline{x_k} - \lambda_k|y_k|^2 = -\lambda_k(y_k - \lambda_k^{-1}x_k)\overline{(y_k - \lambda_k^{-1}x_k)} + \lambda_k^{-1}|x_k|^2.$$

Taking sum and maximizing both sides, we obtain

$$\max_y(x^*y + y^*x - y^*Ay) = \sum_{k=1}^{n} \lambda_k^{-1}|x_k|^2 = x^*A^{-1}x,$$

and the maximum is attained when $y_k = \lambda_k^{-1}x_k$, $k = 1, 2, \ldots, n$.

4.34   To show $\text{Im}(AB) \cap \text{Ker}(AB) = \{0\}$, let y be in the intersection and let $y = (AB)x$ for some x. Since $y \in \text{Ker}(AB)$, $(AB)y = (AB)^2x = 0$. We prove $y = (AB)x = 0$ as follows:

$$
\begin{aligned}
(AB)^2x = 0 \quad &\Rightarrow \quad (ABAB)x = 0 \\
&\Rightarrow \quad (x^*B)(ABAB)x = 0 \\
&\Rightarrow \quad (x^*BAB^{\frac{1}{2}})(B^{\frac{1}{2}}ABx) = 0 \\
&\Rightarrow \quad (B^{\frac{1}{2}}AB)x = 0 \\
&\Rightarrow \quad B^{\frac{1}{2}}(B^{\frac{1}{2}}AB)x = (BAB)x = 0 \\
&\Rightarrow \quad (x^*BA^{\frac{1}{2}})(A^{\frac{1}{2}}Bx) = 0 \\
&\Rightarrow \quad (A^{\frac{1}{2}}B)x = 0 \\
&\Rightarrow \quad (AB)x = 0.
\end{aligned}
$$

4.35   " $\Leftarrow$ " is obvious. For " $\Rightarrow$ ", we may assume that $A = \text{diag}(\lambda_1, \ldots, \lambda_n)$ by spectral decomposition, where λ_i's are nonnegative. Then $AB + BA = 0$ says $\lambda_i b_{ij} + \lambda_j b_{ij} = 0$ for all i, j. Since $\lambda_i + \lambda_j \geq 0$, we have either $\lambda_i + \lambda_j = 0$ or $b_{ij} = 0$, i.e., $\lambda_i = \lambda_j = 0$ or $b_{ij} = 0$. It follows that $\lambda_i b_{ij} = 0 = \lambda_j b_{ij}$ for all i, j. Thus, $AB = 0 = BA$. (One may also use the Lyapunov equation.)

Alternatively, $AB + BA = 0 \Rightarrow B^*AB + B^*BA = 0 \Rightarrow B^*AB = -B^*BA$. Since the eigenvalues of B^*AB are nonnegative and the eigenvalues of $-B^*BA$ (as $-BAB^*$) are nonpositive, all the eigenvalues of B^*AB are zero. Thus, $B^*AB = 0$, implying $B^*A^{\frac{1}{2}}A^{\frac{1}{2}}B = 0 \Rightarrow A^{\frac{1}{2}}B = 0 \Rightarrow AB = 0$. Consequently, $B^*BA = 0$ and $BA = 0$.

4.36 (a) The eigenvalues of H are all real and satisfy $x^5 + x^3 + 2x = 4$ which has only one real solution $x = 1$. So all eigenvalues of H are equal to 1 and H is the identity matrix.

(b) $x^3 + 3x^2 + x = 0$ has no positive roots.

4.37 By spectral decomposition, let $A = U^* \operatorname{diag}(\lambda_1, \dots, \lambda_n)U$, where U is unitary and λ_i's are the (real) eigenvalues of A.

(a) Note that $x^2 + 1 \geq kx$, where $k \in [-2, 2]$, for all real x.

(b) $A - \lambda_{\min}(A)I_n \geq 0$, $\lambda_{\max}(A)I_n - A \geq 0$, and they commute.

4.38 (a) If $A = A^*$, then $A^2 = A^*A \geq 0$. Take $B = \left(\begin{smallmatrix} 1 & 2 & -1 \\ 0 & -1 & 1 \\ 0 & 0 & 1 \end{smallmatrix}\right)$, $B^2 = I$.

(b) If $A^* = -A$, then $-A^2 = (-A)A = A^*A \geq 0$.

(c) It is obvious that the main diagonal entries of an upper- (or lower-) triangular matrix are the eigenvalues of the matrix.

(d) This is readily seen from the spectral decomposition.

(e) If A is Hermitian and all entries on the main diagonal are eigenvalues of A, then A has to be diagonal. A proof goes as follows. Let $\lambda_1, \dots, \lambda_n$ be the (necessarily real) eigenvalues of $A = (a_{ij})$. Suppose that $\{\lambda_1, \dots, \lambda_n\} = \{a_{11}, \dots, a_{nn}\}$. Since A is Hermitian, $A^2 = A^*A$. Taking trace for both sides, we have

$$\sum_{i=1}^{n} \lambda_i^2 = \sum_{i,j=1}^{n} |a_{ij}|^2 = \sum_{i=1}^{n} \lambda_i^2 + \sum_{i \neq j} |a_{ij}|^2.$$

It follows that $a_{ij} = 0$ for all $i \neq j$, that is, A is diagonal.

(f) If the eigenvalues of a matrix A are exactly the singular values of A, then A need be positive semidefinite. Proof: Let $A = U^*TU$, where $T = (t_{ij})$ is upper-triangular, with $t_{ii} = \lambda_i$, $i = 1, 2, \dots, n$, being the eigenvalues of A. We show that T is nonnegative diagonal. Let the singular values of A be $\sigma_1, \sigma_2, \dots, \sigma_n$. Compute

$$\sum_{i=1}^{n} \sigma_i^2 = \operatorname{tr}(A^*A) = \operatorname{tr}(T^*T) = \sum_{i=1}^{n} |\lambda_i|^2 + \sum_{i<j} |t_{ij}|^2.$$

If $\{\lambda_1, \dots, \lambda_n\} = \{\sigma_1, \sigma_2, \dots, \sigma_n\}$, then $t_{ij} = 0$ for all $i < j$. So T is nonnegative diagonal, and A is positive semidefinite. See also Problem 4.144.

4.39   It must be shown that

$$(tA + \tilde{t}B)^*(tA + \tilde{t}B) \leq tA^*A + \tilde{t}B^*B$$

or

$$t^2A^*A + t\tilde{t}(B^*A + A^*B) + \tilde{t}^2B^*B \leq tA^*A + \tilde{t}B^*B,$$

which is

$$0 \leq t\tilde{t}(A^*A + B^*B - B^*A - A^*B) = t\tilde{t}(A^* - B^*)(A - B).$$

This is always true.

4.40 $tA^2 + (1 - t)B^2 - (tA + (1 - t)B)^2 = t(1 - t)(A - B)^2 \geq 0$. This yields the 1st inequality. Note that $(a - b)I \leq A - B \leq (b - a)I$ gives $(A - B)^2 \leq (b - a)^2I$. Since $t(1 - t) \leq \frac{1}{4}$, the 2nd inequality follows.

4.41   Since $A = (A^{\frac{1}{2}})^2$, we have $\lambda_i(A) = \lambda_i((A^{\frac{1}{2}})^2) = (\lambda_i(A^{\frac{1}{2}}))^2$. The eigenvalues of $I - A$ are one minus the eigenvalues of A. The eigenvalues of A^{-1} are the reciprocals of the eigenvalues of A. By taking $X = -I_2$, we see that the identity $\sigma_i(I - X) = 1 - \sigma_{n-i+1}(X)$ for singular values does not hold in general. However, it is true that $\sigma_i(X^{-1}) = \frac{1}{\sigma_{n-i+1}(X)}$ when X is nonsingular.

4.42   (a)   If $n = 1$, we have nothing to show. Let $n > 1$. Use the fact (Example 4.9) that if $A, B \geq 0$, then there exists an invertible matrix P such that P^*AP and P^*BP are both diagonal. If $A > 0$, then P can be chosen so that $P^*AP = I$.

      (b)   Use the fact in (a) and the Hölder inequality (which can be proved by induction): for nonnegative numbers a's and b's,

$$(a_1 \cdots a_n)^{\frac{1}{n}} + (b_1 \cdots b_n)^{\frac{1}{n}} \leq \left((a_1 + b_1) \cdots (a_n + b_n)\right)^{\frac{1}{n}}.$$

      (c)   Use the fact that $a^t b^{\tilde{t}} \leq ta + \tilde{t}b$ for $a, b \geq 0$ and $t \in [0, 1]$. For the special case, take $t = \frac{1}{2}$. Note that $2^k \leq 2^n$.

      (d)   Since $\sqrt{ab} \leq \frac{1}{2}(a + b)$ when $a, b \geq 0$ and by (a).

4.43   (a) $x^*(C^*AC - C^*BC)x = (Cx)^*(A - B)(Cx) \geq 0$.

      (b) $(A - B) + (C - D) \geq 0$.

      (c) $A - B \geq 0 \Rightarrow \text{tr}(A - B) \geq 0 \Rightarrow \text{tr}\, A \geq \text{tr}\, B$.

(d) $A - B \geq 0 \Rightarrow x^*(A - B)x \geq 0$, or $x^*Ax \geq x^*Bx$. Thus

$$\lambda_{\max}(A) = \max_{\|x\|=1} x^*Ax \geq \max_{\|x\|=1} x^*Bx = \lambda_{\max}(B).$$

(e) Note that $\det A = \det((A - B) + B) \geq \det B$.
Alternatively, if $\det B = 0$, it is trivial. Let $\det B \neq 0$. Then

$$\begin{aligned}
A \geq B \quad &\Rightarrow \quad B^{-\frac{1}{2}}AB^{-\frac{1}{2}} \geq I \\
&\Rightarrow \quad \det(B^{-\frac{1}{2}}AB^{-\frac{1}{2}}) \geq 1 \\
&\Rightarrow \quad \det A \geq \det B.
\end{aligned}$$

(f) Use Example 4.9(3).

(g) This can be proved in different ways. A direct proof: If $B = I$,

$$A \geq I \quad \Rightarrow \quad I \geq A^{-\frac{1}{2}}IA^{-\frac{1}{2}} = A^{-1}.$$

In general,

$$A \geq B \Rightarrow B^{-\frac{1}{2}}AB^{-\frac{1}{2}} \geq I \Rightarrow I \geq B^{\frac{1}{2}}A^{-1}B^{\frac{1}{2}} \Rightarrow B^{-1} \geq A^{-1}.$$

(h) First note that $A^{\frac{1}{2}} - B^{\frac{1}{2}}$ is Hermitian. It must be shown that
the eigenvalues of $A^{\frac{1}{2}} - B^{\frac{1}{2}}$ are nonnegative. Let

$$(A^{\frac{1}{2}} - B^{\frac{1}{2}})x = \lambda x, \quad x \neq 0.$$

Then

$$B^{\frac{1}{2}}x = A^{\frac{1}{2}}x - \lambda x.$$

Notice that (the Cauchy–Schwartz inequality)

$$|x^*y| \leq (x^*x)^{\frac{1}{2}} (y^*y)^{\frac{1}{2}}, \quad x, y \in \mathbb{C}^n.$$

Since $A \geq B$, we have $(x^*Ax)^{\frac{1}{2}} \geq (x^*Bx)^{\frac{1}{2}}$ and

$$\begin{aligned}
x^*Ax &= (x^*Ax)^{\frac{1}{2}}(x^*Ax)^{\frac{1}{2}} \geq (x^*Ax)^{\frac{1}{2}}(x^*Bx)^{\frac{1}{2}} \\
&\geq |(x^*A^{\frac{1}{2}})(B^{\frac{1}{2}}x)| = |x^*A^{\frac{1}{2}}(A^{\frac{1}{2}}x - \lambda x)| \\
&= |x^*Ax - \lambda x^*A^{\frac{1}{2}}x|,
\end{aligned}$$

which yields $\lambda x^*A^{\frac{1}{2}}x \geq 0$ or $\lambda x^*A^{\frac{1}{2}}x \geq x^*Ax \geq 0$. Thus $\lambda \geq 0$.
$A \geq B$ does not imply $A^2 \geq B^2$. Take $A = \left(\begin{smallmatrix} 2 & 1 \\ 1 & 1 \end{smallmatrix}\right)$ and $B = \left(\begin{smallmatrix} 1 & 1 \\ 1 & 1 \end{smallmatrix}\right)$.

4.44 (a) $AB = A^{\frac{1}{2}}A^{\frac{1}{2}}B$ and $A^{\frac{1}{2}}BA^{\frac{1}{2}}$ have the same eigenvalues, while the
latter is Hermitian, thus all eigenvalues are real; same for $A^{-1}B$.

(b) $A + B \geq 0 \iff I + A^{-\frac{1}{2}}BA^{-\frac{1}{2}} \geq 0 \iff$ all the eigenvalues of $A^{-\frac{1}{2}}BA^{-\frac{1}{2}}$ are greater than or equal to -1. Now it is sufficient to notice that $A^{-\frac{1}{2}}BA^{-\frac{1}{2}}$ and $A^{-1}B$ have the same eigenvalues.

(c) $r(AB) = r(B) = r(A^{\frac{1}{2}}BA^{\frac{1}{2}})$. The latter equals the number of nonzero eigenvalues of $A^{\frac{1}{2}}BA^{\frac{1}{2}}$, as it is Hermitian. Note that $A^{\frac{1}{2}}BA^{\frac{1}{2}}$ and AB have the same number of nonzero eigenvalues. If $A \geq 0$, then it is not true. Take $A = \begin{pmatrix} 1 & 0 \\ 0 & 0 \end{pmatrix}$ and $B = \begin{pmatrix} 0 & 1 \\ 1 & 0 \end{pmatrix}$. Then $r(AB) = 1$, while AB has no nonzero eigenvalues.

4.45   (a) $A = \begin{pmatrix} 0 & 1 \\ 1 & 0 \end{pmatrix}$, $B = \begin{pmatrix} 1 & 0 \\ 0 & -1 \end{pmatrix}$. The eigenvalues of AB are $\pm i$.

(b) Let $A \geq 0$. The AB has the same eigenvalues as the Hermitian matrix $A^{\frac{1}{2}}BA^{\frac{1}{2}}$. The eigenvalues of the latter are all real.

(c) Let $A > 0$. Then AB is similar to $A^{-\frac{1}{2}}(AB)A^{\frac{1}{2}} = A^{\frac{1}{2}}BA^{\frac{1}{2}}$, which is Hermitian, hence diagonalizable.

(d) $A = \begin{pmatrix} 1 & 1 \\ 1 & 1 \end{pmatrix}$, $B = \begin{pmatrix} 1 & 0 \\ 0 & -1 \end{pmatrix}$. $AB = \begin{pmatrix} 1 & -1 \\ 1 & -1 \end{pmatrix}$ is not diagonalizable.

4.46   Let $c = x^*Ax$, where $\|x\| = 1$. By the Cauchy–Schwartz inequality

$$\left| x^* \left(\frac{A + A^*}{2} \right) x \right| = \left| \frac{c + \bar{c}}{2} \right| \leq |c| = |x^*Ax| \leq (x^*A^*Ax)^{\frac{1}{2}}.$$

(Note that if $Bu = \mu u$, $u^*u = 1$, then $|\mu| = |u^*Bu|$.) We compute

$$\begin{aligned}
\left| \lambda \left(\frac{A + A^*}{2} \right) \right| &\leq \max_{x^*x=1} \left| x^* \left(\frac{A + A^*}{2} \right) x \right| \\
&\leq \max_{x^*x=1} (x^*A^*Ax)^{\frac{1}{2}} \\
&= \left(\max_{x^*x=1} x^*A^*Ax \right)^{\frac{1}{2}} \\
&= \sigma_{\max}(A).
\end{aligned}$$

For the trace inequality, noting that $A - A^*$ is skew-Hermitian, we have $\operatorname{tr}(A - A^*)^2 \leq 0$, which implies, by expanding and taking trace,

$$\operatorname{tr} A^2 + \operatorname{tr}(A^*)^2 \leq 2\operatorname{tr} A^*A, \quad \text{or} \quad \operatorname{tr}(A + A^*)^2 \leq 4\operatorname{tr} A^*A,$$

revealing,

$$\operatorname{tr} \left(\frac{A + A^*}{2} \right)^2 \leq \operatorname{tr}(A^*A).$$

4.47   (a) Since $(A^{\frac{1}{2}})^* = A^{\frac{1}{2}}$, $A^{\frac{1}{2}}BA^{\frac{1}{2}} = (A^{\frac{1}{2}})^*BA^{\frac{1}{2}} \geq 0$.

(b) $AB = A^{\frac{1}{2}}(A^{\frac{1}{2}}B)$ has the same eigenvalues as $A^{\frac{1}{2}}BA^{\frac{1}{2}} \geq 0$.

(c) AB is not positive semidefinite for $A = \left(\begin{smallmatrix} 1 & 0 \\ 0 & 0 \end{smallmatrix}\right)$, $B = \left(\begin{smallmatrix} 1 & 1 \\ 1 & 1 \end{smallmatrix}\right)$.

(d) If A and B commute, then AB is Hermitian, because

$$(AB)^* = B^*A^* = BA = AB.$$

As AB and $A^{\frac{1}{2}}BA^{\frac{1}{2}}$ have the same eigenvalues, $AB \geq 0$ by (a). Conversely, if $AB \geq 0$, then AB is Hermitian and it follows that

$$AB = (AB)^* = B^*A^* = BA.$$

(e) Use the trace cyclic property $\operatorname{tr}(XY) = \operatorname{tr}(YX)$.

(f) Let $\lambda_1(X), \ldots, \lambda_n(X)$ denote the eigenvalues of X. Since $AB^2A = (AB)(BA)$ and $BA^2B = (BA)(AB)$ have the same eigenvalues,

$$
\begin{aligned}
\operatorname{tr}(AB^2A)^{\frac{1}{2}} &= \sum_{i=1}^{n} \lambda_i\big((AB^2A)^{\frac{1}{2}}\big) = \sum_{i=1}^{n} \sqrt{\lambda_i(AB^2A)} \\
&= \sum_{i=1}^{n} \sqrt{\lambda_i(BA^2B)} = \operatorname{tr}(BA^2B)^{\frac{1}{2}}.
\end{aligned}
$$

(g) It may be assumed that $A = \operatorname{diag}(\lambda_1, \ldots, \lambda_n)$, all $\lambda_i \geq 0$. Suppose that $b_{11}, \ldots, b_{nn}$ are the main diagonal entries of B. Then

$$
\begin{aligned}
\operatorname{tr}(AB) &= \lambda_1 b_{11} + \cdots + \lambda_n b_{nn} \\
&\leq (\lambda_1 + \cdots + \lambda_n)(b_{11} + \cdots + b_{nn}) \\
&= \operatorname{tr} A \operatorname{tr} B \leq \tfrac{1}{2}\big((\operatorname{tr} A)^2 + (\operatorname{tr} B)^2\big).
\end{aligned}
$$

(h) Assume that $A = \operatorname{diag}(\lambda_1, \ldots, \lambda_n)$, all $\lambda_i \geq 0$. Then

$$
\begin{aligned}
\operatorname{tr}(AB) &= \lambda_1 b_{11} + \cdots + \lambda_n b_{nn} \\
&\leq \lambda_{\max}(A)(b_{11} + \cdots + b_{nn}) \\
&= \lambda_{\max}(A) \operatorname{tr} B.
\end{aligned}
$$

(i) Use the first inequality of (g) or show directly as follows:

$$
\begin{aligned}
\operatorname{tr}(AB) &= \lambda_1 b_{11} + \cdots + \lambda_n b_{nn} \\
&= \tfrac{1}{4}\big((2\lambda_1 b_{11} + \cdots + 2\lambda_n b_{nn}) \\
&\quad + (2\lambda_1 b_{11} + \cdots + 2\lambda_n b_{nn})\big) \\
&\leq \tfrac{1}{4}\big((\lambda_1^2 + b_{11}^2 + \cdots + \lambda_n^2 + b_{nn}^2) \\
&\quad + (2\lambda_1 b_{11} + \cdots + 2\lambda_n b_{nn})\big) \\
&\leq \tfrac{1}{4}(\lambda_1 + \cdots + \lambda_n + b_{11} + \cdots + b_{nn})^2 \\
&= \tfrac{1}{4}(\operatorname{tr} A + \operatorname{tr} B)^2.
\end{aligned}
$$

(j) Note that $A^2 + B^2 - AB - BA = (A - B)^2 \geq 0$. Take trace.

(k) The first inequality is the arithmetic and geometric mean inequality. The second one is because $\operatorname{tr} A^2 \leq (\operatorname{tr} A)^2$, $\operatorname{tr} B^2 \leq (\operatorname{tr} B)^2$.

4.48 (a) $(AB + BA)^* = B^*A^* + A^*B^* = BA + AB = AB + BA$.

(b) No in general. Take $A = \begin{pmatrix} 2 & 0 \\ 0 & 1 \end{pmatrix}$, $B = \begin{pmatrix} 1 & 1 \\ 1 & 1 \end{pmatrix}$.

(c) No. AB need not be Hermitian.

(d) Yes. $A^2 - AB - BA + B^2 = (A - B)^2 \geq 0$.

(e) No. They are not comparable (as one can easily find examples).

(f) No. Take $A = \begin{pmatrix} \frac{3}{2} & 0 \\ 0 & \frac{3}{2} \end{pmatrix}$ and $B = \begin{pmatrix} 3 & 0 \\ 0 & 0 \end{pmatrix}$ for both cases.

(g) Note that $\operatorname{tr}(XY) \geq 0$ when $X, Y \geq 0$. It follows that

$$\begin{aligned} \operatorname{tr}(CD) - \operatorname{tr}(AB) &= \operatorname{tr}(CD - CB) + \operatorname{tr}(CB - AB) \\ &= \operatorname{tr}(C(D - B)) + \operatorname{tr}((C - A)B) \geq 0. \end{aligned}$$

(h) See Example 4.9. Or one can show directly as follows:

$$\begin{aligned} \lambda_{\max}(B)I - B \geq 0 \quad &\Rightarrow \quad A^{\frac{1}{2}}(\lambda_{\max}(B)I - B)A^{\frac{1}{2}} \geq 0 \\ &\Rightarrow \quad \lambda_{\max}(B)A \geq A^{\frac{1}{2}}BA^{\frac{1}{2}} \\ &\Rightarrow \quad \lambda_{\max}(A)\lambda_{\max}(B) \geq \\ &\qquad \lambda_{\max}(A^{\frac{1}{2}}BA^{\frac{1}{2}}) = \lambda_{\max}(AB). \end{aligned}$$

(i) No. For three positive semidefinite matrices, the eigenvalues of ABC can be imaginary numbers. For instance, the eigenvalues of ABC are 0 and $8 + i$, where $A = \begin{pmatrix} 1 & 1 \\ 1 & 1 \end{pmatrix}$, $B = \begin{pmatrix} 2 & 1 \\ 1 & 1 \end{pmatrix}$, $C = \begin{pmatrix} 2 & i \\ -i & 1 \end{pmatrix}$. See Problem 6.84 for $|\lambda_{\max}(ABC)| \leq \lambda_{\max}(A)\lambda_{\max}(B)\lambda_{\max}(C)$.

(j) Use the result that $\lambda_{\max}(H) = \max_{\|x\|=1} x^*Hx$ for Hermitian H.

4.49 (a) Let $H = \begin{pmatrix} 1 & 0 & 1 \\ 0 & 1 & 0 \\ 1 & 0 & 0 \end{pmatrix}$, $K = \begin{pmatrix} 1 & 0 \\ 0 & 1 \end{pmatrix}$. Then K has 1 as an eigenvalue with multiplicity 2, while H has eigenvalues $1, 1.618, -0.618$.

(b) Take $A = \begin{pmatrix} 1 & 0 \\ 5 & 1 \end{pmatrix}$, $B = \begin{pmatrix} 1 & -5 \\ 0 & 1 \end{pmatrix}$. AB has two negative eigenvalues.

(c) No, because $\operatorname{tr}(A + B) = \operatorname{tr} A + \operatorname{tr} B > 0$.

(d) Take $A = \begin{pmatrix} 5 & 2 \\ 2 & 1 \end{pmatrix}$, $B = \begin{pmatrix} 3 & -1 \\ -1 & 1 \end{pmatrix}$, $C = \begin{pmatrix} 1 & 0 \\ 0 & 30 \end{pmatrix}$. Then A, B, and C are positive definite, the eigenvalues of ABC are $-5, -12$.

(e) No. Note that $\det(ABC) > 0$.

(f) Set $A = \left(\begin{smallmatrix} 1 & 0 \\ 0 & 0 \end{smallmatrix} \right)$, $B = \left(\begin{smallmatrix} 0 & 1 \\ 0 & 0 \end{smallmatrix} \right)$, $C = \left(\begin{smallmatrix} 0 & 0 \\ 0 & 1 \end{smallmatrix} \right)$. $\left(\begin{smallmatrix} A & B \\ B^* & C \end{smallmatrix} \right) \geq 0$, $\left(\begin{smallmatrix} A & B^* \\ B & C \end{smallmatrix} \right) \not\geq 0$.

4.50 (a) Let $A = S^*S$, where S is a nonsingular matrix. Since $S^{*-1}BS^{-1}$ is Hermitian, $T^*(S^{*-1}BS^{-1})T$ is diagonal for some unitary matrix T. Put $P = S^{-1}T$. Then $P^*AP = I$ and P^*BP is diagonal.

(b) If A and B are both positive semidefinite, then there exists an invertible matrix P such that P^*AP and P^*BP are both diagonal (see Example 4.9). If A is positive semidefinite and singular, and B is Hermitian, then such a P may not exist. Take $A = \left(\begin{smallmatrix} 1 & 0 \\ 0 & 0 \end{smallmatrix} \right)$ and $B = \left(\begin{smallmatrix} 0 & 1 \\ 1 & 0 \end{smallmatrix} \right)$. Then a simple computation shows the claim.

(c) The zeros of $\det(\lambda A - B)$ are all real. If $\lambda \leq 0$, then $-(\lambda A - B) = B - \lambda A > 0$ and $\det(\lambda A - B) = (-1)^n \det(B - \lambda A) \neq 0$. One may also use (a) to give a proof.

(d) By (a), let $P^*AP = I$ and $P^*BP = D$ be positive diagonal. As $\det(\lambda I - D) = 0$ has only solution $\lambda = 1$, we get $D = I$, $A = B$.

(e) Since B is positive definite, by (a), there exists a nonsingular matrix P such that $P^*BP = I$. Let $C = P^*(A - B)P$. Then $C \geq 0$. Since $A - \lambda B = (A - B) - (\lambda - 1)B$, we have $\det P^* \det(A - \lambda B) \det P = \det \big(C - (\lambda - 1)I \big)$. Thus $\lambda - 1 \geq 0$ and $\lambda \geq 1$.

4.51 Because $A \geq 0$ and $(A^{\frac{1}{2}})^2 = A$, it is sufficient to show that B commutes with A^2 if and only if B commutes with A.

Since A is positive semidefinite, let $A = U^*DU$, where U is unitary and $D = \text{diag}(\lambda_1, \ldots, \lambda_n)$, all $\lambda_i \geq 0$. Then $A^2B = BA^2$ if and only if $U^*D^2UB = BU^*D^2U$ if and only if $D^2(UBU^*) = (UBU^*)D^2$. Let $C = UBU^*$. Then $D^2C = CD^2$. We show that $DC = CD$. $D^2C = CD^2 \Rightarrow \lambda_i^2 c_{ij} = c_{ij}\lambda_j^2$ for all i and j. If $c_{ij} \neq 0$, then $\lambda_i = \lambda_j$. Thus $c_{ij}\lambda_i = c_{ij}\lambda_j$ for all i, j, and $DC = CD$. It follows that $D(UBU^*) = (UBU^*)D$ or $AB = BA$. The converse is obvious.

For the second part, repeatedly use the first part with $A, B \geq 0$.

4.52 We show that C commutes with $A + B$ first. For this, we show that C commutes with $(A + B)^2$ and then C commutes with $A + B$ by Problem 4.51. Since C is Hermitian and commutes with AB, $(AB)C = C(AB)$ implies $C^*(AB)^* = (AB)^*C^*$, that is, $C(BA) = (BA)C$. In other words, C commutes with BA. Now compute $C(A + B)^2$ and $(A + B)^2C$. Since C commutes with $A - B$, we have $C(A - B)^2 = (A - B)^2C$. Along with $CAB = ABC$ and $CBA = BAC$, we get $C(A + B)^2 = (A + B)^2C$. Thus $C(A + B) = (A + B)C$. It follows that C commutes with $2A = (A + B) + (A - B)$, thus C commutes with A. In a similar way, C commutes with B.

4.53 Let $Bx = \lambda x$, $x \neq 0$. Pre- and post-multiplying $A > B^*AB$ by x^* and x, respectively, we have $x^*Ax > |\lambda|^2 x^*Ax$. Thus $|\lambda| < 1$.

It does not hold for singular values. Take $A = \begin{pmatrix} 1 & 0 \\ 0 & 5 \end{pmatrix}$ and $B = \begin{pmatrix} 0 & 2 \\ 0 & 0 \end{pmatrix}$. Then $A - B^*AB = I_2 > 0$. But the largest singular value of B is 2.

4.54 (a) This is immediate from (b) because $r(X) = \dim \operatorname{Im} X$.

(b) By Example 4.11, $B = A^{\frac{1}{2}}Y$ for some Y. So $\operatorname{Im} B \subseteq \operatorname{Im} A^{\frac{1}{2}}$. Now it is sufficient to observe that $\operatorname{Im} A = \operatorname{Im} A^{\frac{1}{2}}$ because $\operatorname{Im} A \subseteq \operatorname{Im} A^{\frac{1}{2}}$ as $A = (A^{\frac{1}{2}})^2$, and A and $A^{\frac{1}{2}}$ have the same rank.

(c) By Example 4.11, $\begin{pmatrix} \operatorname{tr} A & \operatorname{tr} B \\ \operatorname{tr} B^* & \operatorname{tr} C \end{pmatrix} \geq 0$. Taking the determinant gives $\det \begin{pmatrix} \operatorname{tr} A & \operatorname{tr} B \\ \operatorname{tr} B^* & \operatorname{tr} C \end{pmatrix} \geq 0$. It follows that $|\operatorname{tr} B|^2 \leq \operatorname{tr} A \operatorname{tr} C$.

(d) By Example 4.11, $\begin{pmatrix} \det A & \det B \\ \det B^* & \det C \end{pmatrix} \geq 0$. Take the determinant.

(e) $(I, \pm I) \begin{pmatrix} A & B \\ B^* & C \end{pmatrix} \begin{pmatrix} I \\ \pm I \end{pmatrix} = A \pm (B + B^*) + C \geq 0$.

(f) $M = (m_{ij}) \geq 0 \Rightarrow e^t M e = \sum_{i,j} m_{ij} = \Sigma(M) \geq 0$, where e is the all-ones column vector of $2n$ components.

(g) $\begin{pmatrix} f^t & 0 \\ 0 & f^t \end{pmatrix} \begin{pmatrix} A & B \\ B^* & C \end{pmatrix} \begin{pmatrix} f & 0 \\ 0 & f \end{pmatrix} = \begin{pmatrix} \Sigma(A) & \Sigma(B) \\ \Sigma(B^*) & \Sigma(C) \end{pmatrix} \geq 0$, where f is the all-ones column vector of n components.

(h) $\begin{pmatrix} A & B \\ B^* & C \end{pmatrix} \circ \begin{pmatrix} \bar{A} & \overline{B} \\ B^t & \overline{C} \end{pmatrix} = \begin{pmatrix} (|a_{ij}|^2) & (|b_{ij}|^2) \\ (|b_{ij}|^2) & (|c_{ij}|^2) \end{pmatrix} \geq 0$. An application of

(g) implies $\begin{pmatrix} \sum_{i,j} |a_{ij}|^2 & \sum_{i,j} |b_{ij}|^2 \\ \sum_{i,j} |b_{ij}|^2 & \sum_{i,j} |c_{ij}|^2 \end{pmatrix} = \begin{pmatrix} \Sigma_2(A) & \Sigma_2(B) \\ \Sigma_2(B^*) & \Sigma_2(C) \end{pmatrix} \geq 0$.

Note: This is equivalent to the norm inequality $\|B\|_F^2 \leq \|A\|_F \|C\|_F$.

4.55 It is sufficient to note that
$$P^* \begin{pmatrix} A & B \\ B & A \end{pmatrix} P = \begin{pmatrix} A - B & 0 \\ 0 & A + B \end{pmatrix}, \quad P = \tfrac{1}{\sqrt{2}} \begin{pmatrix} I & I \\ -I & I \end{pmatrix}.$$

4.56 Let $A, B \in M_n(\mathbb{R})$. $A + iB \geq 0 \Leftrightarrow x^*(A + iB)x \geq 0$ for all $x \in \mathbb{C}^n$. Taking conjugate gives $y^*(A - iB)y \geq 0$, $y = \bar{x}$. It follows that $A + iB \geq 0$ if and only if $A - iB \geq 0$. Let $M = \begin{pmatrix} A & -B \\ B & A \end{pmatrix}$. Note that
$$M = S^* \begin{pmatrix} A + iB & 0 \\ 0 & A - iB \end{pmatrix} S, \quad S = \tfrac{1}{\sqrt{2}} \begin{pmatrix} I & iI \\ -iI & -I \end{pmatrix}.$$

4.57 $\begin{pmatrix} A & B \\ B^* & C \end{pmatrix} \geq 0 \Rightarrow \begin{pmatrix} C & B^* \\ B & A \end{pmatrix} \geq 0 \Rightarrow \begin{pmatrix} A & B \\ B^* & C \end{pmatrix} + \begin{pmatrix} C & B^* \\ B & A \end{pmatrix} = \begin{pmatrix} A+C & B+B^* \\ B+B^* & A+C \end{pmatrix} \geq 0.$

$$\begin{pmatrix} A+C & 0 \\ 0 & A+C \end{pmatrix} - \begin{pmatrix} A & B^* \\ B & C \end{pmatrix} = \begin{pmatrix} C & -B^* \\ -B & A \end{pmatrix} \geq 0$$

because

$$\begin{pmatrix} C & -B^* \\ -B & A \end{pmatrix} = \begin{pmatrix} 0 & -I \\ I & 0 \end{pmatrix} \begin{pmatrix} A & B \\ B^* & C \end{pmatrix} \begin{pmatrix} 0 & I \\ -I & 0 \end{pmatrix} \geq 0.$$

4.58 $\begin{pmatrix} A & B \\ B^* & C \end{pmatrix} \geq 0 \Rightarrow \begin{pmatrix} C & B^* \\ B & A \end{pmatrix} \geq 0.$ Taking sum and Hadamard product preserves the positivity. Take determinant of each block.

4.59 (a) Use the fact that $\det \begin{pmatrix} A & B \\ C & D \end{pmatrix} = \det(AD - CB)$ when $AC = CA$ (see Problem 2.56) to compute

$$\det M = \det \begin{pmatrix} 0 & A^* \\ A & 0 \end{pmatrix} = \det(-AA^*) = (-1)^n |\det A|^2.$$

(b) $\operatorname{tr} M = 0$. If $M \geq 0$, then $M = 0$, a contradiction to $A \neq 0$.

(c) Since $|\lambda I - M| = |\lambda^2 I - AA^*|$, the eigenvalues of M are the square roots of the eigenvalues of AA^* (the singular values of A).

(d) The eigenvalues of $N = I + M$ are $1 \pm \sigma_i$, $i = 1, \ldots, n$, and the singular values are $1 + \sigma_i, |1 - \sigma_i|$, $i = 1, \ldots, n$.

(e) The eigenvalues of $K = kI + M$ are $k \pm \sigma_i$, $i = 1, \ldots, n$, and the singular values are $|k \pm \sigma_i|$, $i = 1, \ldots, n$.

4.60 $|\lambda I_{2n} - M| = \lambda^n |\lambda I_n - (A + A^{-1})|$. So, the eigenvalues of M are $\lambda_i + \frac{1}{\lambda_i}$, $i = 1, \ldots, n$, plus n zeros, where λ_i's are the eigenvalues of A.

4.61 By computing A^*A, the singular values of A are $\sqrt{n}, \sqrt{n}, 1, \ldots, 1$. $r(A) = n$. Since A is real symmetric, the singular values are the absolute values of the eigenvalues of A. Notice that $\operatorname{tr} A = -(n-2)$. Since A is congruent to $\begin{pmatrix} 1 & 0 \\ 0 & -I_{n-1} - e^t e \end{pmatrix}$, where $e = (1, \ldots, 1)$ is a row vector of $n - 1$ components, A has $n - 1$ negative eigenvalues. Thus, the eigenvalues of A are $\sqrt{n}, -\sqrt{n}, -1, \ldots, -1$. One may also apply the eigenvalue interlacing theorem to get the same conclusion.

4.62 Since $\lambda_{\max}(PQ) \leq \lambda_{\max}(P)\lambda_{\max}(Q)$, where $P, Q \geq 0$ (Example 4.9),

$$\begin{aligned} \sigma_{\max}^2(AB) &= \lambda_{\max}(B^*A^*AB) = \lambda_{\max}(BB^*A^*A) \\ &\leq \lambda_{\max}(BB^*)\lambda_{\max}(A^*A) = \sigma_{\max}^2(A)\sigma_{\max}^2(B). \end{aligned}$$

To show the inequality for sum, let $\tilde{A} = \begin{pmatrix} 0 & A \\ A^* & 0 \end{pmatrix}$ and $\tilde{B} = \begin{pmatrix} 0 & B \\ B^* & 0 \end{pmatrix}$.

Then $\tilde{A}$ and $\tilde{B}$ are Hermitian matrices. The largest eigenvalue of $\tilde{A} + \tilde{B}$ is $\sigma_{\max}(A + B)$, the largest eigenvalue of $\tilde{A}$ is $\sigma_{\max}(A)$, and that of $\tilde{B}$ is $\sigma_{\max}(B)$. Since $\lambda_{\max}(\tilde{A} + \tilde{B}) \leq \lambda_{\max}(\tilde{A}) + \lambda_{\max}(\tilde{B})$, the desired inequality follows (see Rayleigh–Ritz min-max theorem).

For the last inequality, it is sufficient to notice that

$$A^2 - B^2 = \tfrac{1}{2}(A + B)(A - B) + \tfrac{1}{2}(A - B)(A + B).$$

4.63   We use the fact that $\det(P + Q) \geq \det P$ for $P > 0$ and $Q \geq 0$ with equality if and only if $Q = 0$. If B is nonsingular, then

$$\begin{aligned} \det A &= \det \begin{pmatrix} B & C \\ C^* & D \end{pmatrix} = \det \begin{pmatrix} B & 0 \\ 0 & D - C^*B^{-1}C \end{pmatrix} \\ &= \det B \det(D - C^*B^{-1}C) \leq \det B \, \det D \end{aligned}$$

If B is singular, use $B + \varepsilon I$ ($\varepsilon > 0$) in place of B in the above inequality, then let $\varepsilon \to 0^+$. Thus, it is always true that $\det A \leq \det B \det D$.

Now we consider the equality case. If $C = 0$, then equality obviously holds. If B or D is singular, then the above inequality shows that A is singular. So, $0 = \det A = \det B \det D$. Conversely, suppose equality holds and both B and D are nonsingular. Then $\det(D - C^*B^{-1}C) = \det D$ which yields $C^*B^{-1}C = 0$. It follows that $C = 0$.

4.64   For (a), let $A = E^*E \geq 0$. Hadamard's inequality (which is a special case of Fischer's inequality, Problem 4.63) says $\det A \leq \prod_{i=1}^{n} a_{ii}$.

For (b), let P be a permutation matrix such that $FP = (G, H)$. Use the Fischer inequality to $(FP)^*(FP) = (G, H)^*(G, H)$ to get $\det(F^*F) = \det(FP)^*(FP) = \det(G, H)^*(G, H) \leq \det G^*G \det H^*H$.

4.65   Note that $\begin{pmatrix} I & X^* \\ Y & I \end{pmatrix}^* \begin{pmatrix} I & X^* \\ Y & I \end{pmatrix} = \begin{pmatrix} I + Y^*Y & X^* + Y^* \\ X + Y & I + XX^* \end{pmatrix} \geq 0$. For the first inequality, take determinants of both sides. For the second inequality, take determinants of the blocks in the matrix on the right.

4.66   (a) $H = H^* = A^t - iB^t$ implies $A = A^t$ and $B = -B^t$. Thus, A is real symmetric and B is real skew-symmetric. To show $A \geq 0$, we show that the eigenvalues of A are all nonnegative.

Let U be a real orthogonal matrix such that $U^t A U$ is a real diagonal matrix with the eigenvalues of A on the diagonal. Since

$$(U^t B U)^t = U^t B^t U = -U^t B U,$$

the diagonal entries of $U^t BU$ are all equal to zero, and so are the diagonal entries of $iU^t BU$. However,

$$U^t HU = U^t AU + iU^t BU \geq 0,$$

thus the diagonal entries of $U^t HU$, which are the eigenvalues $\lambda_1(A), \ldots, \lambda_n(A)$ of A, are all nonnegative.

(b) Consider the 2×2 principal minor of H formed by the entries on the s-th and t-th columns and rows.

(c) Following the proof of (a) and using the Hadamard inequality

$$|H| = |U^t HU| \leq \lambda_1(A) \cdots \lambda_n(A) = |A|.$$

Equality holds if and only if $B = 0$ or A has a zero eigenvalue.

(d) By (c). The converse is not true. Take $H = \begin{pmatrix} 1 & 0 \\ 0 & 1 \end{pmatrix} + i \begin{pmatrix} 0 & 1 \\ -1 & 0 \end{pmatrix}$.

4.67 (a) Let $A = UDV$, where D has (i,i)-entries $\sigma_i(A)$, the singular values of A. Then $AA^* = UDD^t U^*$ and $A = UDV = AA^* S$, where $S = URV$, R is the matrix with (i,i)-entries $1/\sigma_i(A)$ if $\sigma_i(A) > 0$, and 0 otherwise, for which $DD^t R = D$.

(b) $S = (A^*)^{-1}$.

(c) $A = (AA^*)^{\frac{1}{2}} T$, where $T = (AA^*)^{\frac{1}{2}} S$, S is given in (a).

(d) Since S is unique, T is unique. A computation gives $T^* T = I$.

4.68 (I) Let A, B, C, D be the given matrices, respectively. Then

$$\ell(A) = \begin{pmatrix} 0 & 0 \\ 0 & 1 \end{pmatrix}, \quad \ell(A^*) = \begin{pmatrix} 1 & 0 \\ 0 & 0 \end{pmatrix};$$

$$\ell(B) = \ell(B^*) = B; \quad \ell(C) = \ell(C^*) = I_2;$$

$$\ell(D) = \frac{1}{\sqrt{2}} \begin{pmatrix} 1 & 1 \\ 1 & 1 \end{pmatrix}, \quad \ell(D^*) = \begin{pmatrix} \sqrt{2} & 0 \\ 0 & 0 \end{pmatrix}.$$

(II)

(a) $\left(\det \ell(A) \right)^2 = \det(A^* A) = |\det A|^2$.

(b) If $A \geq 0$, then $A^* = A$, and $A^* A = A^2$.

(c) Because $\ell(A) \geq 0$.

(d) $A^* A = V^* D^2 V$, so $\ell(A) = V^* DV$. Similarly $\ell(A^*) = UDU^*$.

(e) It is immediate from (d).

(f) $\ell(A) = \ell(A^*)$ if and only if $(A^*A)^{\frac{1}{2}} = (AA^*)^{\frac{1}{2}}$ if and only if $A^*A = AA^*$ since the positive semidefinite square root is unique.

(g) (Note: A can be an $m \times n$ matrix.) Upon computation

$$\begin{pmatrix} \ell(A) & A^* \\ A & \ell(A^*) \end{pmatrix} = \begin{pmatrix} V^* & 0 \\ 0 & U \end{pmatrix} \begin{pmatrix} D & D \\ D & D \end{pmatrix} \begin{pmatrix} V & 0 \\ 0 & U^* \end{pmatrix} \geq 0.$$

(h) $A\ell(A) = UDVV^*DV = UDU^*UDV = \ell(A^*)A$. Take $A = \begin{pmatrix} 0 & 1 \\ 2 & 0 \end{pmatrix}$. Then $\ell(A) = \begin{pmatrix} 2 & 0 \\ 0 & 1 \end{pmatrix}$, $A\ell(A) = \begin{pmatrix} 0 & 1 \\ 4 & 0 \end{pmatrix}$, $\ell(A)A = \begin{pmatrix} 0 & 2 \\ 2 & 0 \end{pmatrix}$.

(i) By Problem 4.51, we only show that H commutes with A^*A. $A^*AH = A^*HA = (HA)^*A = (AH)^*A = HA^*A$. It follows that H commutes with the square root of A^*A, that is, $\ell(A)$.

(j) By (d), $A = UDV = UVV^*DV = UDU^*UV$. Set $W = UV$.

(k) $\ell(A)A^{-1} = V^*DVV^{-1}D^{-1}U^{-1} = V^*U^{-1}$ is unitary. $A(\ell(A))^{-1}A^* = UDVV^{-1}D^{-1}(V^*)^{-1}V^*DU^* = UDU^* = \ell(A^*)$.

(l) We use $X \sim^e Y$ to mean that X and Y have the same eigenvalues. $(\ell(AB))^2 = B^*A^*AB \sim^e (BB^*)(A^*A) \sim^e (BB^*)^{\frac{1}{2}}(A^*A)(BB^*)^{\frac{1}{2}} = \ell(B^*)(\ell(A))^2\ell(B^*) = \big(\ell(\ell(A)\ell(B^*))\big)^2$. Then take the square roots. Note: The B^* on the right cannot be changed to B.

4.69 BB^* is invertible since $r(BB^*) = r(B)$. Use Schur complement to

$$\begin{pmatrix} A \\ B \end{pmatrix} (A^*, B^*) = \begin{pmatrix} AA^* & AB^* \\ BA^* & BB^* \end{pmatrix} \geq 0.$$

4.70   Note that $(A, B)^*(A, B) \geq 0$. For the second part, use Example 4.11(5). A direct proof goes as follows. If $r(A) < n$, then

$$|A^*A| = |A^*B| = |B^*A| = 0.$$

If $r(A) = n$, $B^*B \geq B^*A(A^*A)^{-1}A^*B$. Taking the determinants,

$$\begin{aligned} |B^*B| &\geq |B^*A(A^*A)^{-1}A^*B| \\ &= |B^*A||(A^*A)^{-1}||A^*B| \\ &= |B^*A||A^*A|^{-1}|A^*B|. \end{aligned}$$

$\begin{pmatrix} A^*A & B^*A \\ A^*B & B^*B \end{pmatrix} \not\geq 0$ in general. Take $A = (1, 0)$ and $B = (0, 1)$.

However, $\begin{pmatrix} |A^*A| & |B^*A| \\ |A^*B| & |B^*B| \end{pmatrix} = \begin{pmatrix} |A^*A| & |A^*B| \\ |B^*A| & |B^*B| \end{pmatrix}^t \geq 0$ as $P \geq 0 \Leftrightarrow P^t \geq 0$.

4.71   If $A \geq 0$, it is obvious that $\frac{A+A^*}{2} = (AA^*)^{\frac{1}{2}}$. For the converse, we
show that $A = A^*$, which will imply $A = \frac{A+A^*}{2} = (AA^*)^{\frac{1}{2}} \geq 0$.

$$\frac{A+A^*}{2} = (AA^*)^{\frac{1}{2}} \Rightarrow \frac{(A+A^*)^2}{4} = AA^*$$
$$\Rightarrow A^2 + AA^* + A^*A + (A^*)^2 = 4AA^*$$
$$\Rightarrow \mathrm{tr}(A^2 + (A^*)^2) = 2\,\mathrm{tr}(AA^*)$$
$$\Rightarrow \mathrm{tr}(A - A^*)^2 = 0 \quad \text{(Use skew-Hermicity)}$$
$$\Rightarrow A - A^* = 0, \text{ i.e., } A = A^*.$$

4.72 $\Leftarrow$: Let A be an $n \times n$ normal matrix with eigenvalues $\lambda_1, \ldots, \lambda_n$. By
spectral decomposition, let $A = U^*DU \in M_n(\mathbb{C})$, where U is unitary
and $D = \mathrm{diag}(\lambda_1, \ldots, \lambda_n)$. Write $\lambda_k = x_k + y_k i$, where $x_k, y_k \in \mathbb{R}$,
$k = 1, \ldots, n$. Since the eigenvalues λ_k of A are located on the circle
$|z - \frac{1}{2}| = \frac{1}{2}$, we have $(x_k - \frac{1}{2})^2 + y_k^2 = \frac{1}{4}$, implying $x_k = x_k^2 + y_k^2$, or
$\frac{\lambda_k + \overline{\lambda_k}}{2} = \lambda_k \overline{\lambda_k}$. So, $\frac{D+D^*}{2} = DD^*$ and $\frac{A+A^*}{2} = AA^*$.

$\Rightarrow$: We first show that $I - 2A$ is a unitary matrix.

$$(I-2A)(I-2A)^* = I - 2A^* - 2A + 4AA^* = I - 2(A+A^*) + 4\frac{A+A^*}{2} = I.$$

Since $I - 2A$ is unitary, the eigenvalues of $I - 2A$ have moduli one.
Thus, the eigenvalues of A lie on the circle $|z - \frac{1}{2}| = \frac{1}{2}$. Moreover,
$(I - 2A)(I - 2A)^* = (I - 2A)^*(I - 2A)$, implying $AA^* = A^*A$.

4.73   (a) $A \circ I = \mathrm{diag}(a_{11}, \ldots, a_{nn})$.

    (b) $A \circ J = A$.

    (c)   Note that $A^p \geq 0$ and $B^q \geq 0$. Apply Example 4.9(5).

    (d)   Use Example 4.9(5) to A (k times) to get $A \circ \cdots \circ A = (a_{ij}^k) \geq 0$.
Since $\bar{A} \geq 0$, $A \circ \bar{A} = (|a_{ij}|^2) \geq 0$. Use Example 4.9(5) again.

    (e)   Note that $\lambda_{\max}(A)I - A \geq 0$. Thus

$$\big(\lambda_{\max}(A)I - A\big) \circ B \geq 0 \quad \text{or} \quad \lambda_{\max}(A)(I \circ B) \geq A \circ B.$$

    The conclusion then follows since $\max_i\{b_{ii}\} \leq \lambda_{\max}(B)$.

    (f)   No. Take $A = \left(\begin{smallmatrix} 1 & 1 \\ 1 & 1 \end{smallmatrix}\right)$, $B = \left(\begin{smallmatrix} 1 & 0 \\ 0 & 1 \end{smallmatrix}\right)$.

    (g)   Note that $a_{ii}b_{ii} \leq \frac{1}{2}(a_{ii}^2 + b_{ii}^2)$ for all i. Take sum over i.

4.74   Compute the corresponding entries on both sides to get the identity.
The inequalities are immediate because the second (summation) term
of the right-hand side is positive semidefinite. See also Problem 6.129.

4.75   Let $M = \begin{pmatrix} A & A^{\frac{1}{2}} \\ A^{\frac{1}{2}} & I \end{pmatrix}$, $N = \begin{pmatrix} I & B^{\frac{1}{2}} \\ B^{\frac{1}{2}} & B \end{pmatrix}$. Then $M, N \geq 0$ and

$$M \circ N = \begin{pmatrix} A \circ I & A^{\frac{1}{2}} \circ B^{\frac{1}{2}} \\ A^{\frac{1}{2}} \circ B^{\frac{1}{2}} & B \circ I \end{pmatrix} = \begin{pmatrix} I & A^{\frac{1}{2}} \circ B^{\frac{1}{2}} \\ A^{\frac{1}{2}} \circ B^{\frac{1}{2}} & I \end{pmatrix} \geq 0.$$

4.76   (a)   Let $P = \begin{pmatrix} I & -B^{-1}C \\ 0 & I \end{pmatrix}$. One may verify that

$$P^* \begin{pmatrix} B & C \\ C^* & D \end{pmatrix} P = \begin{pmatrix} B & 0 \\ 0 & D - C^*B^{-1}C \end{pmatrix}.$$

Thus

$$\begin{pmatrix} B & C \\ C^* & D \end{pmatrix}^{-1} = P \begin{pmatrix} B^{-1} & 0 \\ 0 & (D - C^*B^{-1}C)^{-1} \end{pmatrix} P^*.$$

Direct computations reveal

$$U = B^{-1} + B^{-1}C(D - C^*B^{-1}C)^{-1}C^*B^{-1}$$

and

$$W = (D - C^*B^{-1}C)^{-1}.$$

Similarly, with D in the role of B,

$$W = D^{-1} + D^{-1}C^*(B - CD^{-1}C^*)^{-1}CD^{-1}$$

and

$$U = (B - CD^{-1}C^*)^{-1}.$$

(b)   By (a), U is invertible and $U^{-1} = B - CD^{-1}C^*$. It follows that

$$\begin{pmatrix} B - U^{-1} & C \\ C^* & D \end{pmatrix} = \begin{pmatrix} CD^{-1}C^* & C \\ C^* & D \end{pmatrix} =$$

$$\begin{pmatrix} C & 0 \\ 0 & I \end{pmatrix} \begin{pmatrix} D^{-1} & I \\ I & D \end{pmatrix} \begin{pmatrix} C^* & 0 \\ 0 & I \end{pmatrix} \geq 0.$$

4.77   (a)   2, 2, 0, 0; 2 and 0, n copies of each.

(b)   Notice that

$$\begin{pmatrix} A & A \\ A & A \end{pmatrix} \circ \begin{pmatrix} I_n & I_n \\ I_n & I_n \end{pmatrix} = \begin{pmatrix} \mathrm{diag}(A) & \mathrm{diag}(A) \\ \mathrm{diag}(A) & \mathrm{diag}(A) \end{pmatrix} \geq 0.$$

4.78   (a)   Note that

$$A^2 - 2A + I = (A - I)^2 = (A - I)^*(A - I) \geq 0.$$

Thus, $A^2 + I \geq 2A$. Pre- and post-multiplying by $A^{-\frac{1}{2}}$ yields

$$A + A^{-1} \geq 2I.$$

It can also be proved by writing A as $A = U^* \operatorname{diag}(\lambda_1, \ldots, \lambda_n)U$.

(b)   It is sufficient to observe that

$$\begin{pmatrix} A & I \\ I & A^{-1} \end{pmatrix} \circ \begin{pmatrix} A^{-1} & I \\ I & A \end{pmatrix} = \begin{pmatrix} A \circ A^{-1} & I \\ I & A \circ A^{-1} \end{pmatrix} \geq 0.$$

Another proof: Partition A and A^{-1} as

$$A = \begin{pmatrix} a & \alpha \\ \alpha^* & A_1 \end{pmatrix}, \quad A^{-1} = \begin{pmatrix} b & \beta \\ \beta^* & B_1 \end{pmatrix}.$$

Then by Problem 4.76

$$\left(A - \begin{pmatrix} \frac{1}{b} & 0 \\ 0 & 0 \end{pmatrix} \right) \circ \left(A^{-1} - \begin{pmatrix} 0 & 0 \\ 0 & A_1^{-1} \end{pmatrix} \right) \geq 0,$$

which yields

$$A \circ A^{-1} \geq \begin{pmatrix} 1 & 0 \\ 0 & A_1 \circ A_1^{-1} \end{pmatrix}.$$

The desired result follows by an induction on A_1.

4.79   (a)   Let $x = (x_1, \ldots, x_n)^t$ be an eigenvector of λ, and let

$$|x_i| = \max\{|x_1|, \ldots, |x_n|\} > 0.$$

Consider the i-th component of both sides of $Ax = \lambda x$.

(b) $Ae = e$, where $e = (1, \ldots, 1)^t$.

(c) $Ae = e$ results in $A^{-1}e = e$ if A^{-1} exists.

(d) $A = (|u_{ij}|^2)$. Since $U = (u_{ij})$ is a unitary matrix, each row (column) of U is a unit vector. Thus, (i) each row (column) sum of A is 1. For (ii) and (iii), note that $Ae = e$ and $A^t e = e$. 1 is a common eigenvalue and e is a common eigenvector.

4.80 (a) If $A > 0$, then $|A| > 0$ and $A^{-1} > 0$. For $x \neq 0$, $x^* A^{-1} x > 0$,

$$\begin{vmatrix} A & x \\ x^* & 0 \end{vmatrix} = \begin{vmatrix} A & 0 \\ 0 & -x^* A^{-1} x \end{vmatrix} = -|A|(x^* A^{-1} x) < 0.$$

If A is singular, then use $A + \epsilon I$, $\epsilon > 0$, for the above A. Then

$$\begin{vmatrix} A + \epsilon I & x \\ x^* & 0 \end{vmatrix} < 0.$$

Letting $\epsilon \to 0^+$ yields the nonnegativity of the determinant.

(b) Let $\delta = x^* A^{-1} x$. Then $\delta > 0$ for $A > 0$ and $x \neq 0$. We have

$$\begin{pmatrix} A & x \\ x^* & 0 \end{pmatrix}^{-1} = \begin{pmatrix} A^{-1} - \delta^{-1} A^{-1} x x^* A^{-1} & \delta^{-1} A^{-1} x \\ \delta^{-1} x^* A^{-1} & -\delta^{-1} \end{pmatrix}.$$

4.81 Since A is a real orthogonal matrix, that is, $A^t A = I$, we see $|\lambda| = 1$. So $a^2 + b^2 = 1$. $A(x + yi) = (a + bi)(x + yi)$ implies $Ax = ax - by$ and $Ay = ay + bx$. Thus, $x^t A^t = ax^t - by^t$. Since $A^t A = I$, we have $x^t x = x^t A^t A x = (ax^t - by^t)(ax - by) = a^2 x^t x + b^2 y^t y - 2ab x^t y$. Because $a^2 + b^2 = 1$, we obtain $b^2 x^t x = b^2 y^t y - 2ab x^t y$, which implies $2a x^t y = -b x^t x + b y^t y$, as $b \neq 0$. With this in mind, compute $x^t y$:

$$\begin{aligned}
x^t y = x^t A^t A y &= (ax^t - by^t)(ay + bx) \\
&= a^2 x^t y + ab x^t x - ab y^t y - b^2 y^t x \\
&= a^2 x^t y + ab x^t x - ab y^t y - b^2 x^t y \\
&= (a^2 - b^2) x^t y - a(-b x^t x + b y^t y) \\
&= (a^2 - b^2) x^t y - 2a^2 x^t y \\
&= -(a^2 + b^2) x^t y = -x^t y.
\end{aligned}$$

Thus, $x^t y = 0$, $0 = 2a x^t y = -b x^t x + b y^t y$, and $x^t x = y^t y$ as $b \neq 0$.

4.82 (a), (b), and (c) are easy.

(d) Let $Ux = \lambda x$, $x \neq 0$. Then $|\lambda| = 1$ because

$$|\lambda|^2 x^* x = (\lambda x)^* (\lambda x) = (Ux)^* (Ux) = x^* U^* U x = x^* x.$$

(e) $Ux = \lambda x$ implies $\frac{1}{\lambda} x = U^* x$.

(f) Let $U = V^* \operatorname{diag}(\lambda_1, \ldots, \lambda_n)V$, where V is a unitary matrix and the λ's are the eigenvalues of U, each of which equals 1 in absolute value. Let $y = Vx = (y_1, \ldots, y_n)^t$. Then y is a unit vector and

$$\begin{aligned} |x^*Ux| &= \left| \lambda_1 |y_1|^2 + \cdots + \lambda_n |y_n|^2 \right| \\ &\le |\lambda_1||y_1|^2 + \cdots + |\lambda_n||y_n|^2 \\ &= |y_1|^2 + \cdots + |y_n|^2 = 1. \end{aligned}$$

(g) $\|Ux\| = \sqrt{x^*U^*Ux} = \sqrt{x^*x} = 1.$

(h) Each column or row vector of U is a unit vector.

(i) Let $Ux = \lambda_1 x$, $Uy = \lambda_2 y$, $\lambda_1 \ne \lambda_2$. Then

$$(\overline{\lambda_1}\lambda_2)(x^*y) = (\lambda_1 x)^*(\lambda_2 y) = (Ux)^*(Uy) = x^*y,$$

which, with $\lambda_1 \ne \lambda_2$ and $|\lambda_1| = |\lambda_2| = 1$, implies $x^*y = 0$.

(j) The column vectors form a basis since U is nonsingular. They form an orthonormal basis since $u_j^* u_i = 1$ if $i = j$ and 0 otherwise.

(k) Note that any k rows of U are linearly independent. Thus, the rank of the submatrix of these rows is k. So, there is a $k \times k$ submatrix whose determinant is nonzero.

(l) It may be assumed that A is diagonal. Note that each $|u_{ii}| \le 1$.

Each of (a), (b), (c), (g), and (j) implies that U is unitary.

4.83 Verify $U^*U = I$. Note that $A^*(I - AA^*)^{\frac{1}{2}} = (I - A^*A)^{\frac{1}{2}}A^*$.

4.84 Let $u_1, u_2, \ldots, u_n$ be the columns of U. Consider the matrix U^*U whose (i, j)-entry is $u_i^* u_j$ and use the Hadamard inequality.

4.85 Use induction on n. Suppose that A is upper-triangular (by Schur triangularization). It can be seen by taking $x = (0, \ldots, 0, 1)^t$ that every entry except the last component in the last column of A is 0.

Alternatively, one may use the equality case of the Hadamard inequality to show that $A^*A = I$. Note that $\det(A^*A) = |\det A|^2 = 1$ and the diagonal entries of A^*A are $e_i^t(A^*A)e_i = \|Ae_i\|^2 \le 1$, where $e_i \in \mathbb{R}^n$ is the column vector with i-th component 1 and 0 elsewhere.

4.86 Verify that $U^*U = I$. Note that for any positive integer k,

$$1 + \omega^k + \omega^{2k} + \cdots + \omega^{(n-1)k} = 0.$$

4.87 Note that $\sum_{i,j} |x_{ij}|^2 = \operatorname{tr}(X^*X) = \sigma_1(X) + \cdots + \sigma_n(X)$. Compute

$$\operatorname{tr}\left((U \circ A)^*(U \circ A)\right) = \sum_{i,\,j=1}^{n} |u_{ij}|^2 |a_{ij}|^2,$$

$$\sigma_{\max}(U \circ A) \leq (\|U \circ A\|_F =) \left(\sum_{i,\,j=1}^{n} |u_{ij}|^2 |a_{ij}|^2 \right)^{\frac{1}{2}} (\leq \|A\|_F).$$

Take U to be the unitary matrix in Problem 4.86. See Problem 6.88.

4.88 By direct verifications. Note that $\pm i$ cannot be the eigenvalues of A.

4.89 If A and A^* commute, then their inverses commute. Thus

$$(A^{-1}A^*)^*(A^{-1}A^*) = A(A^*)^{-1}A^{-1}A^* = AA^{-1}(A^*)^{-1}A^* = I.$$

So $A^{-1}A^*$ is unitary. Conversely, if $A^{-1}A^*$ is unitary, $(A^{-1}A^*)^* = (A^{-1}A^*)^{-1}$. Thus $A(A^*)^{-1} = (A^*)^{-1}A$, yielding $AA^* = A^*A$.

4.90 $\begin{pmatrix} \cos\theta & \sin\theta \\ -\sin\theta & \cos\theta \end{pmatrix}$, $\begin{pmatrix} \cos\theta & \sin\theta \\ \sin\theta & -\cos\theta \end{pmatrix}$, $\theta \in \mathbb{R}$.

4.91 If $A^t = \operatorname{adj}(A)$, then $C_{ij} = a_{ij}$, where C_{ij} is the cofactor of a_{ij}, and $AA^t = A\operatorname{adj}(A) = |A|I_3$. It follows that $|A|^2 = |A|^3$. So $|A| = 0$ or $|A| = 1$. $A \neq 0$, say $a_{12} \neq 0$. Then $|A| = a_{11}C_{11} + a_{12}C_{12} + a_{13}C_{13} = a_{11}^2 + a_{12}^2 + a_{13}^2 > 0$. So $|A| = 1$ and A is real orthogonal.

4.92 $\mu u^*v = u^*(\mu v) = u^* Av = (Au)^*v = \lambda u^*v$. Since $\lambda \neq \mu$, $u^*v = 0$. For the non-Hermitian case, take $A = \begin{pmatrix} 1 & 1 \\ 0 & 0 \end{pmatrix}$. Then $u = (1,0)^t$ and $v = (1,-1)^t$ are eigenvectors corresponding to 1 and 0, $u^*v \neq 0$.

4.93 If $A^2 - B^2 = AB$ for real orthogonal A and B, we multiply by A^t from the left and by B^t from the right to get $AB^t - A^tB = I$. Taking trace yields $0 = n$, a contradiction. The statement also holds true for unitary matrices, that is, there do not exist unitary matrices A and B such that $A^2 - B^2 = AB$. Suppose otherwise, pre- and post-multiplying the equation by A^* and B^* reveals $AB^* - A^*B = I_n$. Taking the trace for both sides gives $\operatorname{tr}(AB^*) - \operatorname{tr}(A^*B) = n$. However, $\operatorname{tr}(AB^*) - \operatorname{tr}(A^*B) = \overline{\operatorname{tr}(BA^*)} - \operatorname{tr}(A^*B) = \overline{\operatorname{tr}(A^*B)} - \operatorname{tr}(A^*B)$, which is a pure imaginary number; it cannot be equal to n.

4.94 For simplicity, we use $|\cdot|$ for $\det(\cdot)$, the determinant. First note that

$$|A + B| = |A^t + B^t|.$$

Multiply both sides by $|A|$ and $|B|$, respectively, to get

$$|A||A + B| = |A||A^t + B^t| = |I + AB^t|,$$

$$|B||A + B| = |B|A^t + B^t| = |I + BA^t| = |(I + BA^t)^t| = |I + AB^t|.$$

Thus

$$(|A| - |B|)|A + B| = 0, \quad \text{and} \quad |A| = |B| \quad \text{if} \quad |A + B| \neq 0,$$

which, with $|A| + |B| = 0$, implies $|A| = |B| = 0$. Thus A and B are singular. This contradicts the orthogonality of A and B.

It is false for unitary matrices. Take $A = I_2$ and $B = iI_2$.

4.95 Since $A > 0$, there exists a real invertible matrix P such that $P^t A P = I$. Since $(P^t B P)^t = -P^t B P$, $P^t B P$ is real skew-symmetric and thus its eigenvalues are 0 or pure imaginary. Since $P^t B P$ is real, the non-real eigenvalues appear in conjugate pairs. Let T be a real invertible matrix such that $T^{-1}(P^t B P)T = \operatorname{diag}(\lambda_1, \lambda_2, \ldots, \lambda_n)$, where the λ_i are either 0 or nonreal complex numbers in conjugate pairs. It follows that $T^{-1}P^t(A + B)PT = \operatorname{diag}(1 + \lambda_1, 1 + \lambda_2, \ldots, 1 + \lambda_n)$. By taking determinants, we see that $|A + B| > 0$. Not true for complex matrices.

4.96 By a direct verification.

4.97 Use Problem 2.64 with a singular value decomposition of $A + iB$.

4.98 $A = R_1 + iR_2 = H_1 + iH_2$, where $R_1 = \frac{A + \bar{A}}{2}$ and $R_2 = \frac{A - \bar{A}}{2i}$ are real; $H_1 = \frac{A + A^*}{2}$ and $H_2 = \frac{A - A^*}{2i}$ are Hermitian. If $A \geq 0$, then $A^* = A$, implying $H_1 = A \geq 0$, $H_2 = 0$, $R_1^t = R_1$, and $R_2^t = -R_2$. $x^* A x = x^* R_1 x + (x^* R_2 x)i = x^* H_1 x + (x^* H_2 x)i \geq 0$ reveals that $x^* R_1 x \geq 0, x^* R_2 x = 0$ for all $x \in \mathbb{C}^n$, that is, $R_1 \geq 0$.

4.99 (a) No. (b) Yes. (c) No.

4.100 (a) is by direct verification. (b) and (c) are by Example 4.13(3),(4).

4.101 Use Example 4.9(3) and the fact that $f(x) = \frac{1}{x}$ is convex on $(0, \infty)$.

4.102 (a) $\Leftrightarrow$ (b) by a direct verification. We show that (a), (c), (d), and (e) are all equivalent to each other. To see (a) implies (c), use induction. If $n = 1$, there is nothing to prove. Suppose it is true for $n - 1$. For the case of n, let u_1 be a unit eigenvector belonging to the eigenvalue

λ_1 of A. Let U_1 be a unitary matrix with u_1 as its first column. Then $U_1^* A U_1$ is in the form

$$\begin{pmatrix} \lambda_1 & \alpha \\ 0 & A_1 \end{pmatrix}.$$

The normality of A yields $\alpha = 0$. (c) follows by induction.

It is obvious that (c) implies (a).

(c) $\Leftrightarrow$ (d): Note that $U^* A U = D$, where $D = \text{diag}(\lambda_1, \ldots, \lambda_n)$, if and only if $AU = UD$, or $Au_i = \lambda_i u_i$, where u_i is the i-th column of U.

(c) $\Leftrightarrow$ (e): If $Av = \lambda v$, $v \neq 0$, assume that v is a unit vector. Let $U = (v, U_1)$ be a unitary matrix. Since A is normal,

$$U^* A U = \begin{pmatrix} \lambda & 0 \\ 0 & A_1 \end{pmatrix}.$$

By taking conjugate transpose, we see that v is an eigenvector of A^* corresponding to $\bar{\lambda}$. For the other direction, let A be upper-triangular. Take $e_1 = (1, 0, \ldots, 0)^t$. Then e_1 is an eigenvector of A. If e_1 is an eigenvector of A^*, then the first column of A^* consists of zeros except the first component. Use induction hypothesis on n.

(f) $\Leftrightarrow$ (c): If $A^* = AU$, then $A^* A = A^* (A^*)^* = (AU)(AU)^* = AA^*$ and A is normal, hence (c) holds. To see the converse, let

$$A = S^* \text{diag}(\lambda_1, \lambda_2, \ldots, \lambda_n) S,$$

where S is unitary. Take $U = S^* \text{diag}(l_1, l_2, \ldots, l_n) S$, where $l_i = \frac{\bar{\lambda_i}}{\lambda_i}$ if $\lambda_i \neq 0$, and $l_i = 1$ otherwise, $i = 1, 2, \ldots, n$.

Similarly (g) is equivalent to (c).

(c) $\Rightarrow$ (h) is obvious. To see the converse, assume that A is an upper-triangular matrix and consider the trace of $A^* A$.

(i) $\Rightarrow$ (c): Let A be upper-triangular. Consider the diagonal of $A^* A$.

(j) $\Rightarrow$ (a): Note that $\text{tr}(XY) = \text{tr}(YX)$. By computation, we have

$$\text{tr}(A^* A - AA^*)^* (A^* A - AA^*) = \text{tr}(A^* A - AA^*)^2 =$$

$$\text{tr}(A^* A)^2 - \text{tr}\left((A^*)^2 A^2\right) - \text{tr}\left(A^2 (A^*)^2\right) + \text{tr}(AA^*)^2 = 0.$$

On the other hand, $\text{tr}(X^* X) = 0 \Leftrightarrow X = 0$, thus, $A^* A - AA^* = 0$.

(k) $\Rightarrow$ (a): $\|Ax\| = \|A^*x\| \Rightarrow x^*A^*Ax = x^*AA^*x \Rightarrow x^*(A^*A - AA^*)x = 0$ for all $x \in \mathbb{C}^n$. So $A^*A - AA^* = 0$, and A is normal.

(l) $\Rightarrow$ (a): By a direct verification.

(m) $\Rightarrow$ (a): Note that $\operatorname{tr}(A^*A - AA^*) = 0$.

(n) $\Rightarrow$ (j): If $AA^*A = A^*A^2$, then by multiplying A^* from the left

$$A^*AA^*A = (A^*)^2A^2.$$

Thus (j) is immediate by taking trace.

(o) $\Rightarrow$ (a): We show that (o) $\Rightarrow$ (j). Since A and $AA^* - A^*A$ commute,

$$A^2A^* + A^*A^2 = 2AA^*A.$$

Multiply both sides by A^* from the left to get

$$A^*A^2A^* + (A^*)^2A^2 = 2A^*AA^*A.$$

(j) follows by taking trace for both sides.

4.103   (a) Take $B = \frac{A+A^*}{2}$ and $C = \frac{A-A^*}{2i}$. Then $A = B + iC$. If $AA^* = A^*A$, a computation shows $BC = CB$. Conversely, if $BC = CB$, compute to verify that $AA^* = A^*A$.

(b) If A is normal, then $AA^* = A^*A$. Polar decomposition ensures $A = HP = PH$, where $H \geq 0$ and P is unitary. Alternatively, let $A = U\operatorname{diag}(\lambda_1, \ldots, \lambda_n)U^*$ and write $\lambda_k = |\lambda_k|e^{i\theta_k}$ for each λ_k (set $\theta_k = 0$ if $\lambda_k = 0$). Then $A = HP = PH$, where $H = U\operatorname{diag}(|\lambda_1|, \ldots, |\lambda_n|)U^*$, $P = U\operatorname{diag}(e^{i\theta_1}, \ldots, e^{i\theta_n})U^*$. Conversely, $A^*A = (PH)^*(PH) = H^2 = (HP)(HP)^* = AA^*$.

4.104   (a) We show that $\operatorname{Ker} A^* \subseteq \operatorname{Ker} A$. The other way is similar. Let

$$x \in \operatorname{Ker} A^* \quad \text{or} \quad A^*x = 0,$$

then $AA^*x = 0$ and $A^*Ax = 0$ since A is normal. Thus $x^*A^*Ax = (Ax)^*(Ax) = 0$ and $Ax = 0$, that is, $x \in \operatorname{Ker} A$.

(b) Let $x \in \operatorname{Im} A^*$ and $x = A^*y$. Since A is normal, by Problem 4.102(f), assume $A^* = AU$ for some unitary matrix U, then

$$x = A^*y = AUy \in \operatorname{Im} A.$$

Thus $\operatorname{Im} A^* \subseteq \operatorname{Im} A$. The other way around is similar.

(c) Since $n = \dim(\operatorname{Im} A) + \dim(\operatorname{Ker} A)$ and $\operatorname{Im} A^* = \operatorname{Im} A$, we show

$$\operatorname{Im} A^* \cap \operatorname{Ker} A = \{0\}.$$

Let $x = A^*y$ and $Ax = 0$. Then

$$0 = y^*Ax = y^*AA^*y = (A^*y)^*(A^*y) \;\Rightarrow\; x = A^*y = 0.$$

4.105 B and D are square matrices. By computation, $BB^* + CC^* = B^*B$. Taking trace results in $\operatorname{tr}(CC^*) = 0$. So, $C = 0$. B and D are normal.

4.106   First consider the case where A is a diagonal matrix. Let $A = \operatorname{diag}(\lambda_1, \ldots, \lambda_n)$. Then $AB = BA$ yields $\lambda_i b_{ij} = \lambda_j b_{ij}$. Thus $b_{ij} = 0$ or $\lambda_i = \lambda_j$, so $\overline{\lambda_i}\, b_{ij} = \overline{\lambda_j}\, b_{ij}$, which implies $A^*B = BA^*$. For the general case, let $A = U^* \operatorname{diag}(\lambda_1, \ldots, \lambda_n)U$ for some unitary matrix U, then use the above argument with UBU^* for B.

Alternatively, use the fact that A^* is a polynomial of A (Problem 6.91).

4.107   We show that $A\bar{A} = 0 \Leftrightarrow A^tA = 0$. The remaining parts are similar.

$\Rightarrow$: $A\bar{A} = 0 \Rightarrow A^*A\bar{A} = 0 \Rightarrow AA^*\bar{A} = 0 \Rightarrow \bar{A}A^tA = 0$.
Thus, $(A^tA)^*(A^tA) = A^*\bar{A}A^tA = 0$. So, $A^tA = 0$.

$\Leftarrow$: $(A\bar{A})^*(A\bar{A}) = A^tA^*A\bar{A} = A^tAA^*\bar{A} = 0$. So, $A\bar{A} = 0$.

4.108   (a) Let λ_1 be an eigenvalue of A and let $V_{\lambda_1} = \{\, x \mid Ax = \lambda_1 x \,\}$ be the eigenspace of λ_1. Since A and B commute, for $x \in V_{\lambda_1}$,

$$A(Bx) = B(Ax) = B(\lambda_1 x) = \lambda_1(Bx),$$

V_{λ_1} is an invariant subspace of $\mathbb{C}^n$ under B, regarded as a linear transformation on $\mathbb{C}^n$. As a linear transformation on V_{λ_1} over $\mathbb{C}$, B has an eigenvalue $\mu_1 \in \mathbb{C}$ and a unit eigenvector $u_1 \in V_{\lambda_1}$. Let U_1 be a unitary matrix whose first column is u_1. Then

$$U_1^*AU_1 = \begin{pmatrix} \lambda_1 & \alpha \\ 0 & A_1 \end{pmatrix} \quad \text{and} \quad U_1^*BU_1 = \begin{pmatrix} \mu_1 & \beta \\ 0 & B_1 \end{pmatrix}.$$

The normality of A and B implies that $\alpha = \beta = 0$, and that A_1 and B_1 are normal. Now apply induction hypotheses to A_1 and B_1 to get the desired result. (See Example 3.14.)

(b) It is immediate from (a). If the condition $AB = BA$ is dropped, the conclusion does not necessarily follow. Take

$$A = \begin{pmatrix} 1 & 0 \\ 0 & 0 \end{pmatrix}, \quad B = \begin{pmatrix} 1 & 1 \\ 1 & 1 \end{pmatrix}, \quad C = \begin{pmatrix} 0 & 1 \\ -1 & 0 \end{pmatrix}.$$

A, B, and C are normal, but AB and $B + C$ are not normal.

(c) If $AB^* = B^*A$, then $BA^* = A^*B$. It follows that

$$
\begin{aligned}
(AB)(AB)^* &= A(BB^*)A^* \\
&= (AB^*)(BA^*) \quad (B \text{ is normal}) \\
&= (B^*A)(A^*B) \\
&= B^*(AA^*)B \\
&= (B^*A^*)(AB) \quad (A \text{ is normal}) \\
&= (AB)^*(AB).
\end{aligned}
$$

Hence, AB is normal. Similarly, BA is normal.

(d) By a direct computation.

4.109 A is a diagonalizable matrix of eigenvalues $1, 1, -1$ or $1, -1, -1$.

4.110 If λ is an eigenvalue of A, then $\lambda^3 + \lambda = 0$, and λ is 0, i, and/or $-i$. The complex roots of real-coefficient polynomial appear in conjugate pairs, so do the eigenvalues of a real matrix. Thus, $\operatorname{tr} A = 0$.

4.111 If λ is an eigenvalue of A, then $\lambda^k = 1$. Thus, $\frac{1}{\lambda} = \bar{\lambda}$. Note that the eigenvalues of A^{-1} are exactly the reciprocals of the eigenvalues of A. Hence, $\operatorname{tr}(A^{-1}) = \overline{\operatorname{tr} A}$. Since $x^k = 1$ has at most k roots in $\mathbb{C}$, some λ_i's are the same when $k < n$, i.e., A has repeated eigenvalues.

4.112 (a) Use Jordan form of A. Let J be a Jordan block of A. Then $J^k = I$. J has to be 1×1. Thus A is diagonalizable. (b) For B, the characteristic polynomial is $\lambda^2 + 1$, which has no real solution. So B is not diagonalizable over $\mathbb{R}$. $B^4 = I_2$, B is diagonalizable over $\mathbb{C}$.

4.113 Consider Jordan canonical forms with J as an arbitrary Jordan block. (a) $A^k = 0$ for some k if and only if $J^k = 0$ if and only if all the eigenvalues of A are zero. (b) $A^2 = A$ if and only if $J^2 = J$ if and only if J is 1 or 0. (c) $A^2 = I$ if and only if $J^2 = I$ if and only if J is 1 or -1. For (d) and (e), one direction is easy by (b) and (c), respectively. For the converses, examine the cases of Jordan blocks.

See also Problem 2.96 using vector spaces to show (d). For (e), take $W_1 = \operatorname{Im}(I + A)$ and $W_2 = \operatorname{Im}(I - A)$. Then use dimension identity.

4.114 Let $A^m = 0$ and $B^n = 0$ for some positive integers n and m. Then $AB = BA$ implies $(AB)^k = A^k B^k = 0$, where $k \geq \min\{m, n\}$. For $A + B$, expanding $(A + B)^{m+n}$, since $AB = BA$, we see that every term in the expansion contains $A^p B^q$, where $p + q = m + n$. So either $p \geq m$ or $q \geq n$. Thus, $A^p B^q = 0$ and $(A + B)^{m+n} = 0$.

If $AB \neq BA$, take $A = \begin{pmatrix} 0 & 1 \\ 0 & 0 \end{pmatrix}$ and $B = \begin{pmatrix} 0 & 0 \\ 1 & 0 \end{pmatrix}$ as a counterexample.

4.115   Note that for every positive integer k,

$$AB = -BA \quad \Rightarrow \quad AB^{2k-1} = -B^{2k-1}A.$$

Thus if $B = AX + XA$, then

$$B^{2k} = (AX + XA)B^{2k-1} = A(XB^{2k-1}) - (XB^{2k-1})A,$$

implying $\operatorname{tr} B^{2k} = 0$ for all positive integers k (by the trace cyclic property). Consequently, B^2 is nilpotent, and so is B, a contradiction.

4.116   (a)   If $A^m = 0$, then $I = I - A^m = (I - A)(I + A + A^2 + \cdots + A^{m-1})$.
Thus $I - A$ is invertible and $(I - A)^{-1} = I + A + A^2 + \cdots + A^{m-1}$.

       (b)   Because all eigenvalues of A are 0. Or replace A by $-A$ in (a).

       (c)   Because all eigenvalues of A are 0.

       (d)   If A is diagonalizable, then $A = 0$ as all eigenvalues of A are 0.

4.117   Since $A^2 = A$ and $B^2 = B$, we have

$$(A + B)^2 = A^2 + AB + BA + B^2 = A + AB + BA + B.$$

If $(A + B)^2 = A + B$, then $AB + BA = 0$, that is, $AB = -BA$. Also,

$$AB = A^2B = A(AB) = A(-BA) = -(AB)A = BA^2 = BA.$$

It follows that $BA = -BA$ and $BA = 0$. So $AB = 0$.
If $AB = BA = 0$, then obviously $(A + B)^2 = A + B$.

4.118   When A and B are Hermitian, $(AB)^* = B^*A^* = BA$. Thus

$$(BA)^2 = \left((AB)^*\right)^2 = \left((AB)^2\right)^* = (AB)^* = B^*A^* = BA.$$

4.119   If $A^2 = A$, then there exists an invertible matrix T such that

$$A = T^{-1} \begin{pmatrix} I_r & 0 \\ 0 & 0 \end{pmatrix} T = T^{-1} \begin{pmatrix} I_r \\ 0 \end{pmatrix} (I_r, 0)T = BC,$$

where $B = T^{-1} \begin{pmatrix} I_r \\ 0 \end{pmatrix}$, $C = (I_r, 0)T$, both have rank r and $CB = I_r$.
The converse is easy. Use $A = T^{-1} \begin{pmatrix} I_r & 0 \\ 0 & 0 \end{pmatrix} T$ for the determinants.

4.120 Let T be an invertible matrix such that $T^{-1}AT = \text{diag}(I_r, 0_{n-r})$, where $1 \leq r \leq n-1$. Let $B = T\,\text{diag}(0, J)T^{-1}$, where J is the $k \times k$ Jordan block with main diagonal entries 0, $k \leq n-r$. Then one may verify that $AB = BA = 0$ and $(A+B)^{k+1} = (A+B)^k \neq (A+B)^{k-1}$.

4.121 Let r be the rank of A. Since $A^2 = A$, the eigenvalues of A are either 1 or 0. Using the Jordan form of A, by $A^2 = A$, we see that every Jordan block of A must be 1×1. Thus A is diagonalizable. It follows that (a) $\det(A + I) = 2^r$ and $\det(A - I) = 0$; (b) $r(A)$ equals the number of nonzero eigenvalues of A, which are 1's, and also equals $\text{tr}(A)$; (c) $\dim \text{Im}\, A = r(A) = \text{tr}(A)$.

4.122 (a) If $x \in \text{Im}\, A$, let $x = Ay$. Then $Ax = A^2y = Ay = x$. Conversely, if $x = Ax$, then obviously $x \in \text{Im}\, A$. (A need not be Hermitian.)

(b) For Hermitian A and B, we show $\text{Im}\, A = \text{Im}\, B$ implies $A = B$. For vector v, let $Bv = Au$. Then $ABv = A^2u = Au = Bv$. Similarly, $BAv = Av$. For any v, we compute $(A - B)^2v = A^2v + B^2v - ABv - BAv = Av + Bv - A(Bv) - B(Av) = 0$. Thus $(A - B)^2 = 0$. Since $A - B$ is Hermitian, $A = B$.

4.123 Note that the idempotent Hermitian matrices are the positive semidefinite matrices with eigenvalues 1's and 0's, and that if P_1 is a principal submatrix of a positive semidefinite matrix P, then

$$\lambda_{\min}(P) \leq \lambda_{\min}(P_1) \leq \lambda_{\max}(P_1) \leq \lambda_{\max}(P).$$

Since $B \leq A \Leftrightarrow AB = B$ is equivalent to

$$U^*BU \leq U^*AU \quad \Leftrightarrow \quad (U^*AU)(U^*BU) = U^*BU,$$

where U is unitary, we may only consider the case in which $B = \begin{pmatrix} I_r & 0 \\ 0 & 0 \end{pmatrix}$, where $r = r(B)$ is the rank of B. Partition A conformably as

$$A = \begin{pmatrix} M_1 & M_2 \\ M_2^* & M_3 \end{pmatrix}.$$

To show $B \leq A \Rightarrow AB = B$, note that $A - B \geq 0$ implies $M_1 - I_r \geq 0$ and that the eigenvalues of M_1 are all equal to 1. It follows that $M_1 = I_r$. Thus $M_2 = 0$ and $A = \begin{pmatrix} I_r & 0 \\ 0 & M_3 \end{pmatrix}$, where $M_3 \geq 0$. It is immediate that $AB = B$. Conversely, $AB = B$ results in $M_2 = 0, M_1 = I_r$.

4.124 (a) $\Leftrightarrow$ (b): $(P+Q)^2 = P^2 + PQ + QP + Q^2 = P + PQ + QP + Q$. It follows that $(P+Q)^2 = P + Q \Leftrightarrow PQ + QP = 0$.

(b) $\Rightarrow$ (c): $PQ + QP = 0 \Rightarrow PQ = -QP \Rightarrow PQ = PQ^2 = -QPQ$. Since QPQ is Hermitian, PQ is Hermitian. Thus, $PQ = (PQ)^* = Q^*P^* = QP$. Since $PQ + QP = 0$, we arrive at $PQ = 0$.

(c) $\Rightarrow$ (b): Sine $PQ = 0$, $0 = (PQ)^* = Q^*P^* = QP$.

(c) $\Leftrightarrow$ (d): This is because $(PQ)^* = QP$.

(d) $\Leftrightarrow$ (e): This is because $QP = 0 \Leftrightarrow Q(Px) = 0$ for all $x \in \mathbb{C}^n$.

(c) $\Leftrightarrow$ (f): This is because $PQ = 0 \Leftrightarrow P(Qx) = 0$ for all $x \in \mathbb{C}^n$.

4.125 All implications are straightforward. For instance, (d) $\Rightarrow$ (a): $A^2 = I$ and $A^*A = I$ imply $A^* = A$.

4.126 (a) By a direct verification.

(b) $A^2 - I = 0$ gives $(A+I)(A-I) = 0$. So the columns of $A - I$ are contained in the null space of $A+I$. Thus, $r(A+I)+r(A-I) \leq n$. On the other hand, $n = r(A) = r\big((A+I) + (A-I)\big) \leq r(A+I) + r(A-I)$. See Problems 2.96 and 4.113 for different proofs.

(c) If $Ax = \lambda x$, for some $x \neq 0$, then $\lambda = \pm 1$ because
$$x = Ix = A^2 x = A(Ax) = \lambda Ax = \lambda^2 x \Rightarrow \lambda^2 = 1.$$

(d) For any $v \in \mathbb{C}^n$, $\frac{1}{2}(v + Av) \in V_1$, $\frac{1}{2}(v - Av) \in V_{-1}$, so
$$v = \tfrac{1}{2}(v + Av) + \tfrac{1}{2}(v - Av) \in V_1 + V_{-1}.$$

Thus $\mathbb{C}^n = V_1 + V_{-1}$. It is a direct sum as $V_1 \cap V_{-1} = \{0\}$.

(e) Since $(A+I)(A-I) = 0$, $(A+I)\big((A-I)x\big) = 0$ for all x.

(a), (d), and (e) each imply $A^2 = I$. For (d), show $A^2 x = x$, $x \in \mathbb{C}^n$.

4.127 $ABA = B$ and $BAB = A$ imply $AB = BA^{-1}$ and $AB = B^{-1}A$. So $BA^{-1} = B^{-1}A$ and $A^2 = B^2$. $A = BAB = (ABA)(AB) = B^{-1}A^3B$. Thus, $BA = A^3B$. Note that $ABA = B$ gives $BA = A^{-1}B$. We have $A^{-1}B = A^3B$. Hence, $A^{-1} = A^3$, that is, $A^4 = I$. Similarly, $B^4 = I$.

4.128 Check $(A - B)(A + B) = (A + B)(B - A) = AB - BA$. Thus, $\mathrm{Im}(AB - BA)$ is contained in both $\mathrm{Im}(A - B)$ and $\mathrm{Im}(A + B)$, that is, $\mathrm{Im}(AB - BA) \subseteq \mathrm{Im}(A - B) \cap \mathrm{Im}(A + B)$. For the reverse inclusion, let $u \in \mathrm{Im}(A - B) \cap \mathrm{Im}(A + B)$ and write $u = (A + B)x = (A - B)y$.

Then $(A - B)u = (AB - BA)x$ and $(A + B)u = (BA - AB)y$. So, $2Au = (AB - BA)(x - y)$. As $A(AB - BA) = -(AB - BA)A$, we get

$$u = A^2 u = \tfrac{1}{2}A(AB - BA)(x - y) = \tfrac{1}{2}(AB - BA)A(y - x),$$

which is contained in $\mathrm{Im}(AB - BA)$.

4.129 If $A^2 = A$, then $\tfrac{1}{4}(B^2 + 2B + I) = \tfrac{1}{2}(B + I)$, which implies $B^2 = I$. If $B^2 = I$, then $A^2 = \tfrac{1}{4}(B^2 + 2B + I) = \tfrac{1}{4}(2I + 2B) = \tfrac{1}{2}(B + I) = A$.

4.130 (a)-(d) are by direct verifications. Use SVD for (e).

4.131 (a) $x^*(A + cI)x = x^*Ax + c$. (Note: $S + c = \{s + c \mid s \in S\}$.)

(b) $x^*(cA)x = c(x^*Ax)$. (Note: $cS = \{cs \mid s \in S\}$.)

(c) Let e_i be the column vector with the i-th component 1 and 0 elsewhere. Then $e_i^*Ae_i = a_{ii} \in F(A)$.

(d) Let $Av = \lambda v$, where λ is an eigenvalue and v is a unit eigenvector corresponding to λ. Then, $v^*Av = v^*(\lambda v) = \lambda \in F(A)$.

(e) No, in general. Take $A = \left(\begin{smallmatrix} 0 & 1 \\ 0 & 0 \end{smallmatrix}\right)$. Then $\sigma_1(A) = 1 \notin F(A)$.

(f) For any unitary U, $x^*U^*AUx = y^*Ay$, where $y = Ux$ is unit.

(g) If A is singular, then A has a zero eigenvalue. So $0 \in F(A)$. The converse is not true. Take $A = \left(\begin{smallmatrix} 0 & 1 \\ 1 & 0 \end{smallmatrix}\right)$, $y = \left(\begin{smallmatrix} 1 \\ 0 \end{smallmatrix}\right)$. Then $y^*Ay = 0$ and $0 \in F(A)$, but A is nonsingular (see also Problem 4.3). So, it is false to say that A is nonsingular if and only if $0 \notin F(A)$.

(h) A is nonsingular if and only if A^{-1} is nonsingular. Let y be a unit vector. Then $A^{-1}y \neq 0$. Let $z = \frac{1}{\|A^{-1}y\|}A^{-1}y$. Then $\|z\| = 1$, and $z^*Az = 0$ if and only if $y^*A^{-1}y = 0$ because

$$
\begin{aligned}
z^*Az &= \frac{1}{\|A^{-1}y\|^2}y^*(A^*)^{-1}A(A^{-1}y) \\
&= \frac{1}{\|A^{-1}y\|^2}y^*(A^*)^{-1}y \\
&= \frac{1}{\|A^{-1}y\|^2}(y^*A^{-1}y)^*.
\end{aligned}
$$

Thus, $0 \notin F(A)$ if and only if $0 \notin F(A^{-1})$.

(i) A closed interval on the x-axis. It is $[\lambda_{\min}, \lambda_{\max}]$.

(j) A closed interval on the nonnegative side of the x-axis.

(k) The fields of values are, respectively, $[-1,1]$ (Note that this A is invertible and $0 \in F(A)$); $[0,1]$; the closed disc centered at the origin with radius $\frac{1}{2}$; the closed elliptical disc with foci at 0 and 1, minor axis 1 and major axis $\sqrt{2}$; the closed line segment joining $(1,0)$ and $(1,1)$; and the triangle (including interior) with the vertices $(0,0), (1,0)$, and $(1,1)$.

4.132   By spectral decomposition, let $A = U^*DU$, where D is the diagonal matrix $\mathrm{diag}(\lambda_1, \lambda_2, \ldots, \lambda_n)$, and U is unitary. Then $F(A) = F(D) = \{\lambda_1|x_1|^2 + \lambda_2|x_2|^2 + \cdots + \lambda_n|x_n|^2 \mid x = (x_1, x_2, \ldots, x_n)^t \in \mathbb{C}^n, \|x\| = 1\}$ is the convex hull of $\lambda_1, \lambda_2, \ldots, \lambda_n$, that is, $F(A)$ is the polygon in the complex plane with vertices $\lambda_1, \lambda_2, \ldots, \lambda_n$.

The converse is not true. Let $A = \begin{pmatrix} B & 0 \\ 0 & C \end{pmatrix}$, where $B = \mathrm{diag}(1, -1, i, -i)$ and $C = \begin{pmatrix} 0 & \delta \\ 0 & 0 \end{pmatrix}$, $\delta > 0$. When δ is small enough, $F(C) \subsetneq F(B)$ and $F(A) = F(B)$ is the convex hull of the eigenvalues of A, but A is not normal because C is not normal.

4.133   Let x and y be unit vectors such that $x^*Ax = 0$ and $y^*Ay = 1$. Apparently, x and y are linearly independent. Let $A = H + Ki$ be the Cartesian decomposition of A, where H and K are Hermitian.

A simple computation gives $x^*Kx = y^*Ky = 0$. We may assume that the real part of x^*Ky is zero (otherwise, use $e^{i\theta}x$ in place of x so that $e^{-i\theta}x^*Ky$ is pure imaginary). So, $\mathrm{Re}(x^*Ky) = x^*Kx = y^*Ky = 0$.

Let $z(t) = \frac{1}{\|(1-t)x+ty\|}((1-t)x + ty)$, $t \in [0,1]$. Then $z = z(t)$ is a unit vector. By computation, we have $z^*Kz = 0$ and $z^*Az = z^*Hz + z^*Kzi = z^*Hz$ is real. Let $f(t) = z^*(t)Az(t)$. $f(t)$ is a real-valued continuous function on $t \in [0,1]$. As $f(0) = 0$ and $f(1) = 1$, by the intermediate value theorem, $[0,1] \subseteq \{z^*(t)Az(t) \mid t \in [0,1]\} \subseteq F(A)$.

Note: This result states that a numerical range is convex.

4.134   (a) We claim that $(x^*Ax)(x^*A^{-1}x) \geq 1$ for all unit $x \in \mathbb{C}^n$. Note that $\begin{pmatrix} A & I \\ I & A^{-1} \end{pmatrix} \geq 0$. For any unit x, we have $\begin{pmatrix} x^* & 0 \\ 0 & x^* \end{pmatrix} \begin{pmatrix} A & I \\ I & A^{-1} \end{pmatrix} \begin{pmatrix} x & 0 \\ 0 & x \end{pmatrix} = \begin{pmatrix} x^*Ax & 1 \\ 1 & x^*A^{-1}x \end{pmatrix} \geq 0$. Taking determinants yields $(x^*Ax)(x^*A^{-1}x) \geq 1$. Thus, it is impossible that both $x^*Ax < 1$ and $x^*A^{-1}x < 1$ hold.

(b) Take $A = \begin{pmatrix} 2 & 0 \\ 0 & \frac{1}{2} \end{pmatrix}$, $x = \frac{1}{\sqrt{2}}\begin{pmatrix} 1 \\ 1 \end{pmatrix}$. Then $x^*Ax = x^*A^{-1}x = \frac{5}{4} > 1$.

4.135   (a) $n!$.

(b) By a direct verification. Sum is not a permutation matrix.

(c) $PP^t = P^t P = I$.

(d) Permutation matrices are normal, thus diagonalizable.

(e) If P is a permutation matrix, then $PP^t = I$. So $P^{-1} = P^t$.

(f) If $R = t_1 P + t_2 Q$, then $r_{ij} = t_1 p_{ij} + t_2 q_{ij}$. As $0 < t_1, t_2 < 1$, $r_{ij} = 1$ if and only if $p_{ij} = q_{ij} = 1$. It follows that $R = P = Q$.

(g) Symmetric permutation matrices as $P^2 = I \Rightarrow P = P^{-1} = P^t$.

(h) Since every row sum of a permutation matrix is 1, 1 is an eigenvalue with $e = (1, \ldots, 1)^t$ as a corresponding eigenvector.

(i) $\begin{pmatrix} 0&1&0&0 \\ 1&0&0&0 \\ 0&0&0&1 \\ 0&0&1&0 \end{pmatrix}$ and $\begin{pmatrix} 0&0&0&1 \\ 0&0&1&0 \\ 0&1&0&0 \\ 1&0&0&0 \end{pmatrix}$ have two positive eigenvalues 1, 1.

(j) If P is a symmetric permutation matrix, then $P^2 = I$. If λ is an eigenvalue of such a P, then $\lambda^2 = 1$. So $\lambda = \pm 1$. If a 5×5 symmetric permutation matrix has two positive eigenvalues 1, 1, then the rest are three -1's, which results in a negative trace. But the trace of any permutation matrix cannot be negative.

(k) The eigenvalues of a symmetric permutation matrix are either 1 and/or -1. If two symmetric permutation matrices have the same number of positive eigenvalues, then they have exactly the same eigenvalues. Thus, they are similar (by diagonalization).

(l) If $B = 0$, then A contains n 1's, consequently, C contains no 1's, that is, $C = 0$. Similarly, $A = 0$ if and only if $D = 0$.

4.136 (a) By direct computations. Note that for matrix A, AP is the matrix obtained from A by moving the last column of A to the front as the first column of the resulting matrix AP.

(b) P^k is symmetric if and only if n is even and $k = \frac{n}{2}$.

(c) The eigenvalues of P are the primitive n-th roots of unity, that is, $1, \omega, \omega^2, \ldots, \omega^{n-1}$, where $\omega^n = 1$, $\omega^m \neq 1$ for any positive integer $m < n$. The eigenvalues of P^k are $1, \omega^k, \omega^{2k}, \ldots, \omega^{(n-1)k}$.

(d) Let
$$k_1 P + k_2 P^2 + \cdots + k_n P^n = 0.$$
Since the k's are in different positions of the matrix on the left-hand side, we see that all the k's must be equal to zero.

(e) Verify by definition. (In fact, any polynomial of P is normal.)

(f) $P^i + P^j$ is symmetric for $1 \le i, j < n$ if and only if $i + j = n$.
    If $i+j = n$, then $P^i + P^j = P^i + P^{n-i} = P^i + (P^i)^t$ is symmetric.
    Conversely, if $(P^i + P^j)^t = P^i + P^j$, then $(P^{-1})^i + (P^{-1})^j = P^i + P^j$. Multiplying both sides by P^{i+j}, we get $P^i + P^j = P^{i+j}(P^i + P^j)$. It follows that $(P^i + P^j)(I - P^{i+j}) = 0$. Since P is diagonalizable with eigenvalues $1, \omega, \omega^2, \ldots, \omega^{(n-1)}$, we have

$$\left(\omega^{ti} + \omega^{tj}\right)\left(1 - \omega^{t(i+j)}\right) = 0, \quad t = 0, 1, \ldots, n - 1.$$

In particular,

$$\left(\omega^i + \omega^j\right)\left(1 - \omega^{i+j}\right) = 0, \quad \left(\omega^{2i} + \omega^{2j}\right)\left(1 - \omega^{2(i+j)}\right) = 0.$$

We show $1 - \omega^{i+j} = 0$ and $n = i+j$. (Keep in mind $1 \le i, j < n$.)
If $1 - \omega^{i+j} \neq 0$, then $\omega^i + \omega^j = 0$. Let $j > i$. Then $\omega^{j-i} = -1$ and $\omega^{2(j-i)} = 1$. So, $n = 2(j - i)$ as $j < n$, and also $\omega^{2j} = \omega^{2i}$ which by the above display implies $1 - \omega^{2(i+j)} = 0$. Thus, n divides $2(i + j)$. So $2(i + j) = n, 2n$, or $3n$ (as $1 \le i, j < n$). Since $n = 2(j - i)$, we can rule out $2(i+j) = n$ and $2(i+j) = 3n$. Thus, $2(i + j) = 2n$ and $i + j = n$, contradicting $1 - \omega^{i+j} \neq 0$.

(g) P is a real orthogonal matrix, so it is diagonalizable over $\mathbb{C}$. Note that the characteristic polynomial of P is $\lambda^n - 1$. The eigenvalues of P are the primitive roots of $\lambda^n - 1 = 0$. When $n \ge 3$, P has nonreal eigenvalues. So it cannot be diagonalizable over the reals.

(h) Since $(i, n) = 1$, there are integers s and t such that $is + nt = 1$. Thus, $P = P^{is+nt} = (P^i)^s$. Because the eigenvalues of P are all distinct, the eigenvalues of P^i are all distinct. So, P and P^i have the same eigenvalues (the primitive n-th roots of unity). Since P and P^i are both diagonalizable over $\mathbb{C}$, they are similar.

Note: Every permutation matrix is a real orthogonal matrix, thus diagonalizable (over $\mathbb{C}$). So, two permutation matrices are similar if and only if they have the same eigenvalues. In fact, two permutation matrices are similar if and only if they are similar via a permutation matrix. This needs more work such as irreducibility, cyclic group theory, or combinatorics.

(i) For $n = 4$ and $i = 2$, P^2 has eigenvalues $1, 1, -1, -1$. P^2 cannot be similar to P, because the eigenvalues of P are $1, -1, i, -i$.

4.137 Since A is invertible, at least one summand in the expansion of the determinant $\det A$ is nonzero. Because A is a matrix of nonnegative integer entries, we may write A as $A = P + B$, where P is a permutation matrix and B is a nonnegative matrix. Notice that $P^m = I$. Thus, in the expansion of $A^n = (P + B)^n$, there is an arbitrary large number of terms $P^i B P^j$ that are identical to B if n is large enough. If B had a nonzero (thus positive) entry, then $\Sigma(A^n)$ would be unbounded as $n \to \infty$. Therefore, $B = 0$ and $A = P$ is a permutation matrix. For the case of union of entries, because the powers of A collectively have only finitely many entries, we must have $A^k = A^n$ for some k and n. So $A^p = I$ for some p. Then expand $I = A^p = (P + B)^p = P^p + P^{p-1}B + \cdots$. The sum of the entries on the left-hand side is m, and then so is on the right. Thus $B = 0$.

4.138   Since every row and every column of A have one and only one nonzero entry, which is either 1 or -1, and all other entries are equal to 0, it is easy to see that the powers A, A^2, A^3, ..., have the same property, that is, for all positive integers m, every row and every column of A^m have one and only one nonzero entry, which is either 1 or -1, and all other entries are 0. Since there are a finite number of matrices that have the property, we have $A^p = A^q$ for some positive integers p and q, $p < q$. Since A is nonsingular, we have $A^{q-p} = I$.

4.139 A is an $(n-1) \times n$ matrix. Let B be the submatrix of A by deleting the last column. Since each of the row sums of A is equal to zero, we may write $A = (B, BR)$, where $R = (-1, \cdots, -1)^t \in \mathbb{R}^{n-1}$. Then $\operatorname{tr}(RR^t) = \operatorname{tr}(R^t R) = n - 1$ and $\det(I_{n-1} + RR^t) = n$. Thus,

$$
\begin{aligned}
\det(AA^t) &= \det\left((B, BR)(B, BR)^t\right) = \det(BB^t + BRR^t B^t) \\
&= \det B \det(I_{n-1} + RR^t) \det B^t = n\left(\det B\right)^2.
\end{aligned}
$$

4.140   Denote the rows of $A = (a_{ij})$ by $r_1, r_2, \ldots, r_n$. Let S and T be the row and column indices of the $s \times t$ submatrix whose entries are all 1. Set $v = \sum_{i \in S} r_i$. Since the rows are mutually orthogonal, we have

$$
\|v\|^2 = \left(\sum_{i \in S} r_i\right)^t \left(\sum_{i \in S} r_i\right) = \sum_{i \in S} r_i^t r_i = sn.
$$

However, the j-th component of v is $\sum_{i \in S} a_{ij}$,

$$
\|v\|^2 = \sum_{j=1}^n \left(\sum_{i \in S} a_{ij}\right)^2 \geq \sum_{j \in T} \left(\sum_{i \in S} a_{ij}\right)^2 \geq s^2 t.
$$

It follows that $sn \geq s^2 t$, implying $st \leq n$.

4.141   Let r_i be the sum of the entries of row i of A. Then $S(A) = r_1 + r_2 + \cdots + r_n$. Since $a_{ik} = a_{ki}$ and by interchanging the summands, we get

$$
\begin{aligned}
S(A^2) &= \sum_{i=1}^{n}\sum_{j=1}^{n}\sum_{k=1}^{n} a_{ik}a_{kj} = \sum_{k=1}^{n}\left(\sum_{i=1}^{n}\sum_{j=1}^{n} a_{ki}a_{kj}\right) \\
&= \sum_{k=1}^{n}\left(\left(\sum_{i=1}^{n} a_{ki}\right)\left(\sum_{j=1}^{n} a_{kj}\right)\right) \\
&= \sum_{k=1}^{n} r_k^2 \geq \frac{1}{n}\left(\sum_{k=1}^{n} r_k\right)^2 = \frac{1}{n}(S(A))^2.
\end{aligned}
$$

4.142 $(A + tB)^3 = (t^3 - 2)I_3$, $B^3 = I_3$, $A^3 = -2I_3$.

4.143 $X^4 = (2pq)^2 I_2$, $Y^4 = (p^2 - q^2)^2 I_2$, $Z^4 = (p^2 + q^2)^2 I_2$.

4.144   (a) The singular values $\sigma_i(A)$ of a normal matrix $A = (a_{ij})$ are the absolute values of its eigenvalues. If the eigenvalues are exactly the diagonal entries, computing $\mathrm{tr}(A^*A)$, we have $\sum_{i,j} |a_{ij}|^2 = \sum_i \sigma_i^2(A) = \sum_i |a_{ii}|^2$. Thus, $a_{ij} = 0$ if $i \neq j$, i.e., A is diagonal.

     (b) Let $A = \begin{pmatrix} 1 & a & b \\ c & 2 & d \\ e & f & 3 \end{pmatrix}$. Compute the characteristic polynomial $p(\lambda)$ of A, then set $p(\lambda) = 0$ for $\lambda = 1, 2, 3$. Choose values of a, b, c, d, e, f. $A = \begin{pmatrix} 1 & 1 & -1 \\ 0.5 & 2 & 0.5 \\ 1 & 1 & 3 \end{pmatrix}$ is such a matrix having eigenvalues 1, 2, and 3.

     (c) $\mathrm{tr}\, A = a + d$. If a is an eigenvalue of A, then d is the other eigenvalue of A. Since A is normal, the singular values of A are $|a|$ and $|d|$. So, $\mathrm{tr}(A^*A) = |a|^2 + |b|^2 + |c|^2 + |d|^2 = |a|^2 + |d|^2$, implying $b = c = 0$. Thus A is (2×2) diagonal.

     (d) Consider $kI + \begin{pmatrix} 0 & x \\ x^t & 0 \end{pmatrix}$ to construct an example with a large k. Let $A = \begin{pmatrix} 3 & 1 & 1 \\ 1 & 3 & 0 \\ 1 & 0 & 3 \end{pmatrix}$. Then A is positive definite with 3 as an eigenvalue, and A is not diagonal. (Note: $B = \begin{pmatrix} 0 & 1 & 1 \\ 1 & 1 & 1 \\ 1 & 1 & 1 \end{pmatrix}$ is real symmetric, and 0 is an eigenvalue of B because the rank of B is 2.)

     (e) Positive semidefinite matrices are precisely the matrices whose singular values are exactly the same as eigenvalues.
     See also Problem 4.38.

4.145   Let U be a unitary matrix such that U^*AU is upper-triangular (by the Schur triangularization). Then the diagonal entries of U^*HU are $\mathrm{Re}\,\lambda_1, \mathrm{Re}\,\lambda_2, \ldots, \mathrm{Re}\,\lambda_n$. Now apply Example 4.14.

Hints and Answers for Chapter 5

5.1 $\mathbb{C}^2$ is a vector space over $\mathbb{R}$ of dimension 4 with respect to the usual addition $u + v$ and scalar multiplication ru, where $u, v \in \mathbb{C}^2$, $r \in \mathbb{R}$. However, $\mathbb{C}^2$ is not an inner product space over $\mathbb{R}$ with respect to the inner product $\langle u, v \rangle = v^* u$ because $v^* u$ need not be a real number. For example, $u = (i, 0)^t, v = (1, 0)^t$. Then $v^* u = i \notin \mathbb{R}$. One might attempt to identify $\mathbb{C}^2$ with $\mathbb{R}^4$ then use the standard inner product for $\mathbb{R}^4$ to define an inner product for $\mathbb{C}^2$. But this is not $\langle u, v \rangle = v^* u$.

5.2 $\left\| \frac{1}{\|u\|} u \right\| = \frac{1}{\|u\|} \|u\| = 1$. So $\frac{1}{\|u\|} u$ is a unit vector. For $v, w \in V$,

$$\langle v, \langle v, w \rangle w \rangle = \overline{\langle v, w \rangle} \langle v, w \rangle = |\langle v, w \rangle|^2.$$

$\langle \langle v, w \rangle v, w \rangle = \langle v, w \rangle^2 \neq |\langle v, w \rangle|^2$ over $\mathbb{C}$ (but it is true over $\mathbb{R}$).

5.3 The first identity is easy. For the second one, use $\langle u, \lambda v \rangle = \bar{\lambda} \langle u, v \rangle$.

5.4 (a) False. Take $u = v = (3, 0)$.

(b) True. By the Cauchy–Schwartz inequality.

(c) False. Take $u = v = (1, 1)$, $w = (1, -1)$.

(d) False. Take $u = v \neq 0$, $w = -u$.

(e) True. Use the triangle inequality to $u + v = (u + w) + (v - w)$.

5.5 (a) and (d) are true. (b) is false (it is true if $\lambda \geq 0$). (c) is false. (e) is false: take $x = (1, 0)$ and $y = (1, 1)$.

5.6 It is routine to check the positivity, homogeneity, and symmetry conditions. We show that the additivity condition is satisfied:

$$
\begin{aligned}
\langle u, v + w \rangle_3 &= \langle u, v + w \rangle_1 + \langle u, v + w \rangle_2 \\
&= \langle u, v \rangle_1 + \langle u, w \rangle_1 + \langle u, v \rangle_2 + \langle u, w \rangle_2 \\
&= \langle u, v \rangle_1 + \langle u, v \rangle_2 + \langle u, w \rangle_1 + \langle u, w \rangle_2 \\
&= \langle u, v \rangle_3 + \langle u, w \rangle_3.
\end{aligned}
$$

5.7 By the definition of a norm. If 5 is replaced by -5, it is not a norm.

5.8 (a) S_1 is the rhombus (including the interior points) with vertices $(1, 0), (0, 1), (-1, 0)$, and $(0, -1)$ in the xy-plane.

(b) S_2 is the unit disk centered at the origin.

(c) S_3 is the square (including the interior points) with vertices $(1,1)$, $(-1,1), (-1,-1)$, and $(1,-1)$.

5.9 Direct computations show that η meets all conditions for a norm. The set $\{(x,y) \in \mathbb{R}^2 \mid \eta(x,y) \le 1\}$ is the elliptical region $x^2 + 4y^2 \le 1$.

5.10 The first two properties of a norm are routine to verify. For the triangle inequality, compute $\left(\eta(x_1+x_2, y_1+y_2)\right)^2$ and use the Cauchy–Schwartz inequality to $u = (x_1 - y_1, \sqrt{k}y_1)$ and $v = (x_2 - y_2, \sqrt{k}y_2)$.

5.11 Use the Cauchy–Schwartz inequality $|\langle u, v\rangle|^2 = (v^t u)^2 \le \langle u, u\rangle \langle v, v\rangle$.

(a) $u = (1, \ldots, 1), v = (\sqrt{p_1}, \ldots, \sqrt{p_n})$.

(b) $u = (\sqrt{t_1}, \ldots, \sqrt{t_n}), v = (\sqrt{t_1}p_1, \ldots, \sqrt{t_n}p_n)$.

(c) $u = (\frac{1}{\sqrt{p_1}}, \ldots, \frac{1}{\sqrt{p_n}}), v = (\sqrt{p_1}, \ldots, \sqrt{p_n})$.

Equality holds in (a), (b), or (c) if and only if $p_1 = \cdots = p_n$.

5.12 We show by induction on n. By spectral decomposition, we may assume that $A = \operatorname{diag}(\lambda_1, \ldots, \lambda_n)$, where all $\lambda_i \ge 0$. Let $x = (x_1, \ldots, x_n)^t \in \mathbb{C}^n$ and $t_i = |x_i|^2$, $t_1 + \cdots + t_n = 1$. Then $(x^*Ax)^k \le x^*A^k x$ is essentially $(\lambda_1 t_1 + \cdots + \lambda_n t_n)^k \le \lambda_1^k t_1 + \cdots + \lambda_n^k t_n$ for $k \ge 1$. We first show the case of $n = 2$, i.e., $(\lambda_1 t_1 + \lambda_2 t_2)^k \le \lambda_1^k t_1 + \lambda_2^k t_2$, because of the convex graph of $f(x) = x^k$ on $[0, \infty)$ for $k \ge 1$.

Suppose it is true for $n - 1$. We reduce the case of n to 2 and $n - 1$. Let $0 < t_1 < 1$ and let $t_i' = \frac{t_i}{1 - t_1}$, $i \ge 2$. Then $t_2' + \cdots + t_n' = 1$ and

$$
\begin{aligned}
(\lambda_1 t_1 + \cdots + \lambda_n t_n)^k &= \left(\lambda_1 t_1 + (1 - t_1)\sum_{i=2}^{n}\lambda_i t_i'\right)^k \\
&\le \lambda_1^k t_1 + (1 - t_1)\left(\sum_{i=2}^{n}\lambda_i t_i'\right)^k \\
&\le \lambda_1^k t_1 + (1 - t_1)\sum_{i=2}^{n}\lambda_i^k t_i' \quad \text{(hypothesis)} \\
&= \lambda_1^k t_1 + \lambda_2^k t_2 + \cdots + \lambda_n^k t_n.
\end{aligned}
$$

Note: This is a so-called Jensen inequality.

The case of $k = 2$ can be directly shown as follows: for any unit x, $xx^* \le I_n \Rightarrow (Ax)^*(xx^*)(Ax) \le (Ax)^*(Ax)$, i.e., $(x^*Ax)^2 \le x^*A^2 x$.

5.13   Positivity: For $x = (x_1, x_2, x_3)^t$, $\lfloor x, x \rfloor = 2|x_2|^2 + |x_1|^2 + \overline{x_1}x_3 + x_1\overline{x_3} + 2|x_3|^2$. Since $|x_1|^2 + \overline{x_1}x_3 + x_1\overline{x_3} + |x_3|^2 \geq 0$, we see $\lfloor x, x \rfloor \geq 0$. Equality holds if and only if $x = 0$. Homogeneity: $\lfloor \lambda x, y \rfloor = y^*A(\lambda x) = \lambda y^*Ax = \lambda \lfloor x, y \rfloor$. Additivity: $\lfloor x, y + z \rfloor = (y + z)^*Ax = (y^* + z^*)Ax = y^*Ax + z^*Ax = \lfloor x, y \rfloor + \lfloor x, z \rfloor$. Hermitian Symmetry: Note that $A^* = A$. $\lfloor x, y \rfloor = y^*Ax = (x^*A^*y)^* = \overline{x^*Ay} = \overline{\lfloor y, x \rfloor}$.

If the $(2, 2)$-entry is changed to -2, then the positivity does not hold as $x^tAx = -2$ for $x = (0, 1, 0)^t$. So it is not an inner product (in our definition). If it is defined by $\lfloor x, y \rfloor = x^*Ay$, it is an inner product over $\mathbb{R}$ as $x^tAy = (x^tAy)^t = y^tAx$, but not over $\mathbb{C}$ since $\lfloor \lambda x, y \rfloor \neq \lambda \lfloor x, y \rfloor$.

See Problem 5.30 for the more general case.

5.14   By computations, $A = I_2$, having eigenvalues 1, 1, and

$$B = \begin{pmatrix} 1 & 0 & 0 & 1 \\ 0 & 0 & 0 & 0 \\ 0 & 0 & 0 & 0 \\ 1 & 0 & 0 & 1 \end{pmatrix}, \quad C = \begin{pmatrix} 1 & 0 & 0 & 0 \\ 0 & 0 & 1 & 0 \\ 0 & 1 & 0 & 0 \\ 0 & 0 & 0 & 1 \end{pmatrix}.$$

The eigenvalues of B and C are respectively 2, 0, 0, 0, and $1, 1, 1, -1$.

5.15   Expand $0 \leq (ux^* - vy^*)(ux^* - vy^*)^* = (ux^* - vy^*)(xu^* - yv^*)$.

5.16   Let $F = \begin{pmatrix} 1 & p \\ 2 & q \end{pmatrix}$ and $G = \begin{pmatrix} 1 & 1 + \frac{p}{2} \\ 1 + \frac{p}{2} & q \end{pmatrix}$. Then $f(x, y) = y^tFx$.

(a) $f(x, x) \geq 0$ implies $(p + 2)^2 \leq 4q$. (For example, $p = 0, q = 1$.) $f(x, x)$ is a quadratic form in x and $f(x, x) = x^tGx$.

(b) $(p - 2)y_1x_2 = (p - 2)x_1y_2$ for all x and y. Thus, $p = 2$.

(c) It holds for all $x, y \in \mathbb{R}^2$, $\lambda, p, q \in \mathbb{R}$.

(d) It holds for all $x, y, z \in \mathbb{R}^2$, $p, q \in \mathbb{R}$. In fact, f is bilinear:

$$f(\lambda x + y, z) = z^tF(\lambda x + y) = \lambda z^tFx + z^tFy = \lambda f(x, z) + f(y, z),$$

$$f(x, \lambda y + z) = (\lambda y + z)^tFx = \lambda y^tFx + z^tFx = \lambda f(x, y) + f(x, z).$$

(e) f defines an inner product, $f(x, y) = \langle x, y \rangle$, on $\mathbb{R}^2$ if $p = 2$, $q \geq 4$.

Note: Let $g(x, y) = y^tGx$. If $(p + 2)^2 < 4q$, then G is positive definite. In this case, g defines an inner product on $\mathbb{R}^2$. $f(x, y)$ and $g(x, y)$ agree when $x = y$, i.e., $f(x, x) = g(x, x) = x_1^2 + (p + 2)x_1x_2 + qx_2^2$.

If we take $p = 4, q = 10$, then $g(x, y) = g(y, x)$ for all $x, y \in \mathbb{R}^2$. However, $f(x, y) = 4 \neq 2 = f(y, x)$ for $x = (0, 1)^t$ and $y = (1, 0)^t$.

5.17   It is routine to show by definition that V is a subspace of $\mathbb{R}^4$. To find a basis for V, note that x_3 and x_4 are free variables. Setting $x_3 = 1$, $x_4 = 0$ and $x_3 = 0$, $x_4 = 1$, respectively, we have a basis for V: $\{(1,1,1,0),(1,-1,0,1)\}$. To find a basis for $V^\perp$, let $y = (y_1, y_2, y_3, y_4) \in V^\perp$. Then y is orthogonal to the basis vectors of V. So $y_1+y_2+y_3 = 0$ and $y_1-y_2+y_4 = 0$. This reveals $2y_1 = -y_3-y_4$ and $2y_2 = y_4 - y_3$. It is immediate that $(-1,0,1,1)$ and $(-1,1,0,2)$ form a basis for $V^\perp$ and $V^\perp = \{a(-1,0,1,1) + b(-1,1,0,2) \mid a,b \in \mathbb{R}\}$.

5.18 $\alpha_3 = (0, \frac{1}{\sqrt{2}}, -\frac{1}{\sqrt{2}}, 0)$, $\alpha_4 = (0, -\frac{1}{2}, -\frac{1}{2}, \frac{1}{\sqrt{2}})$.

5.19 $S^\perp = \text{Span}\{(-2,1,1,0),(2,-3,0,1)\}$.

5.20 $\langle u, v \rangle = -1$, $\|u\| = \sqrt{2}$, $\|v\| = \sqrt{5}$. The angle between vectors u and v is $\arccos \frac{-1}{\sqrt{10}} \approx 108.43°$. The angle between the one dimensional spaces $\text{Span}\{u\}$ and $\text{Span}\{v\}$ is $\arccos \frac{1}{\sqrt{10}} \approx 71.57°$.

5.21 $\mathbb{R}^\perp = 1^\perp = \{a + bx + cx^2 + dx^3 \in \mathbb{P}_4[x] \mid a + \frac{1}{2}b + \frac{1}{3}c + \frac{1}{4}d = 0\}$, and $f_1(x) = 1 - 2x$, $f_2(x) = 1 - 3x^2$, $f_3(x) = 1 - 4x^3$ form a basis for $\mathbb{R}^\perp$. $x^\perp = \{a + bx + cx^2 + dx^3 \in \mathbb{P}_4[x] \mid \frac{1}{2}a + \frac{1}{3}b + \frac{1}{4}c + \frac{1}{5}d = 0\}$.

5.22   (a) $\|1\| = \sqrt{2}$. $u(x) = 1$ is not a unit vector. $\pm\frac{1}{\sqrt{2}}$.

(b) $a + bx - 3ax^2 + cx^3$, $a,b,c, \in \mathbb{R}$.

(c) $\sqrt{\frac{3}{2}}x$, $\sqrt{\frac{5}{2}}x^2$. If $k \in \mathbb{R}$ is orthogonal to x^2, $\int_{-1}^{1} kx^2 dx = 0$, so $k = 0$.

(d) $\sqrt{\frac{3}{2}}x$, $\sqrt{\frac{5}{2}}x^2$, $\frac{1}{2\sqrt{2}}(3 - 5x^2)$, $\frac{3\sqrt{7}}{2\sqrt{2}}(x - \frac{5}{3}x^3)$.

(e) It suffices to show that the basis vectors are orthogonal. For instance, by integration, $\langle x, x^3 - \frac{3}{5}x \rangle = \int_{-1}^{1}(x^4 - \frac{3}{5}x^2)dx = 0$.

(f) Note that 1, x, $x^2 - \frac{1}{3}$, and $x^3 - \frac{3}{5}x$ are linearly independent.

5.23   (a) For $f(x) = 1$, $\langle f, f \rangle_1 = 2$, f is not unit. $\langle f, f \rangle_2 = 1$, f is unit.

(b) Take $1 - 3x^2$.

(c) Take $v(x) = 1$ and $w(x) = x$.

(d) $p(x) = 0$.

(e) No. $\langle u, u \rangle_1 \neq \langle u, u \rangle_2$ if $u \neq 0$.

(f) 1, x, $1 - 3x^2$, $x - \frac{5}{3}x^3$.

(g) $1 - 2x$, $1 - 3x^2$, $1 - 4x^3$.

(h) $W = \{ax^2 \mid a \in \mathbb{R}\}$.

5.24   Let V be the space of real-valued continuous functions on $[a, b]$. V is
an inner product space with respect to the inner product $\langle f_i, f_j \rangle = \int_a^b f_i(x) f_j(x)\,dx$. Let $A = (a_{ij})$ be the $n \times n$ matrix with

$$a_{ij} = \int_a^b f_i(x) f_j(x) = \langle f_i, f_j \rangle.$$

Then A is a Gram matrix. $\det A = 0$ if and only if A is singular, if
and only if $f_1, \ldots, f_n$ as vectors in V are linearly dependent.

5.25   (a) $\sqrt{2\pi}$.

(b) 1.

(c) $\sqrt{\pi}$.

(d) If $s(x)$ is an odd function, then $\langle 1, s \rangle = \int_{-\pi}^{\pi} s(x)\,dx = 0$.

(e) $\sin mx \cos nx$ is an odd function, the integration on $[-\pi, \pi]$ is 0.

5.26   The first three matrices are positive semidefinite. The last one is not
as the 4×4 matrix $\begin{pmatrix} u_1 u_1^* & u_2 u_1^* \\ u_1 u_2^* & u_2 u_2^* \end{pmatrix} \not\geq 0$ for $u_1 = (1, 0)^t$ and $u_2 = (0, 1)^t$.

5.27   For G, use the fact that if $A = (a_{ij}) \geq 0$ then the Hadamard product
$A \circ A = (a_{ij}^2) \geq 0$. The Gram matrix $(\langle u_j, u_i \rangle) \geq 0$. So $G = (\langle u_j, u_i \rangle^2) \geq 0$. For H, use the fact that if $A = (a_{ij}) \geq 0$ then
$\bar{A} = (\overline{a_{ij}}) \geq 0$, so $A \circ \bar{A} = (|a_{ij}|^2) \geq 0$. Thus, $H = (|\langle u_j, u_i \rangle|^2) \geq 0$.

5.28   Since $\|u\| < 1$ and $\|v\| < 1$, by the Cauchy–Schwartz inequality, we
have $|\langle x, y \rangle| < 1$, where x and y are any choices of u and v. Let θ be
the angle between u and v, let $a = \|u\|$ and $b = \|v\|$. It is sufficient
to show that $(1 - a^2)(1 - b^2) \leq (1 - ab\cos\theta)^2$ (in the real case),
which is equivalent to $ab(ab\sin^2\theta + 2\cos\theta) \leq a^2 + b^2$. Noting that
$ab\sin^2\theta + 2\cos\theta \leq \sin^2\theta + 2\cos\theta = 2 - (1 - \cos\theta)^2 \leq 2$, we
are done.

Alternative proof (this approach works for higher dimension). Recall
the power series expansion $\frac{1}{1-r} = 1 + r + r^2 + r^3 + \cdots$ when $|r| < 1$.

$$\begin{pmatrix} \frac{1}{1-\langle u,u \rangle} & \frac{1}{1-\langle u,v \rangle} \\ \frac{1}{1-\langle v,u \rangle} & \frac{1}{1-\langle v,v \rangle} \end{pmatrix} = \begin{pmatrix} \sum_k \langle u, u \rangle^k & \sum_k \langle u, v \rangle^k \\ \sum_k \langle v, u \rangle^k & \sum_k \langle v, v \rangle^k \end{pmatrix}$$

$$= \sum_k \begin{pmatrix} \langle u, u \rangle^k & \langle u, v \rangle^k \\ \langle v, u \rangle^k & \langle v, v \rangle^k \end{pmatrix}.$$

The positivity follows because the sum of positive semidefinite matrices is again positive semidefinite (with a little work on convergence)
and because $\begin{pmatrix} a & b \\ \bar{b} & c \end{pmatrix} \geq 0 \Rightarrow \begin{pmatrix} a^k & b^k \\ \bar{b}^k & c^k \end{pmatrix} \geq 0$ for any positive integer k.

5.29 If there exist n vectors $u_1, u_2, \ldots, u_n \in \mathbb{C}^n$ such that $A = (a_{ij})$, where $a_{ij} = \langle u_j, u_i \rangle$, then for any vector x, where $x = (x_1, x_2, \ldots, x_n)^t \in \mathbb{C}^n$,

$$x^* A x = \sum_{i,j} \overline{x_i} x_j a_{ij} = \sum_{i,j} \overline{x_i} x_j \langle u_j, u_i \rangle = \left\langle \sum_j x_j u_j, \sum_i x_i u_i \right\rangle \geq 0.$$

So $A \geq 0$. Conversely, if $A \geq 0$, we can write $A = B^* B$ for some n-square matrix B (see Chapter 4, p. 140). Then $a_{ij} = b_i^* b_j = \langle b_j, b_i \rangle$, where b_i is the i-th column of B. So A is a Gram matrix.

5.30 If $A > 0$, then one checks that $\langle x, y \rangle = y^* A x$ is an inner product. Conversely, for the given inner product $\langle \cdot, \cdot \rangle$, let $\langle e_i, e_j \rangle = a_{ji}$, where $\{e_1, \ldots, e_n\}$ is the standard basis of $\mathbb{C}^n$. Let $A = (a_{ji})$. Then $\langle x, y \rangle = y^* A x$. Since $\langle x, x \rangle = x^* A x > 0$ for all $x \neq 0$, A is positive definite.

5.31 (a) Take $A = \left(\begin{smallmatrix} 1 & 0 \\ 0 & 0 \end{smallmatrix} \right)$.

(b) Compute $\langle\langle u + v, u + v \rangle\rangle - \langle\langle u - v, u - v \rangle\rangle$ directly.

(c) $u = (1, 1)^t$ is such a vector that $\langle\langle u, u \rangle\rangle = 0$. Let $v = (x, y)^t$, where x, y are real. Then $v^t A v = x^2 - y^2$. Thus, K_0 consists of all points in the xy-plane such that $|x| = |y|$; K_+ contains all points with $|x| > |y|$; and K_- has all points with $|x| < |y|$.

(d) If $u, v \in W$, then $\langle\langle ku + v, w \rangle\rangle = k\langle\langle u, w \rangle\rangle + \langle\langle v, w \rangle\rangle = 0$, where k is any scalar. Thus, W is a subspace of V. For the last part, if $w \in K_0$, then $\langle\langle w, w \rangle\rangle = 0$. So $w \in W$. The converse is obvious.

5.32 Let $t_1 u_1 + \cdots + t_n u_n = 0$. Taking inner product with u_1 reveals $t_1 \langle u_1, u_1 \rangle = 0$. So $t_1 = 0$. Repeat this with u_2 to get $t_2 = 0$. Inductively, we obtain $t_1 = \cdots = t_n = 0$. The converse is false.

5.33 By the Cauchy–Schwartz inequality, if x is a unit vector, then

$$\langle y, x \rangle \langle x, y \rangle \leq \langle y, y \rangle, \quad \text{for any } y \in V.$$

Replacing y by $\mathcal{A}(x)$, we obtain

$$\langle \mathcal{A}(x), x \rangle \langle x, \mathcal{A}(x) \rangle \leq \langle \mathcal{A}(x), \mathcal{A}(x) \rangle.$$

5.34 Let $\alpha = \{\alpha_1, \ldots, \alpha_n\}$ be an orthonormal basis of V and let A be the matrix representation of $\mathcal{A}$ with respect to the basis α, that is, $\mathcal{A}(\alpha) = \alpha A$ (see Chapter 3, p. 95). Then $\langle \mathcal{A}x, x \rangle = y^* A y$, where y is the coordinate of x under α. If $\mathcal{A}$ is a positive map, then $y^* A y \geq 0$ for all $y \in \mathbb{C}^n$. It follows that A is positive semidefinite. The inequality is the same as Problem 5.12 because $\langle \mathcal{A}^k x, x \rangle = y^* A^k y$.

5.35 (a) If $x \in \operatorname{Ker} A$, then $Ax = 0$. Thus, $(I - A)x = x$ and $(I + A)x = x$. Since $I + A$ is nonsingular, $x = (I + A)^{-1}x$. It follows that $(I - A)(I + A)^{-1}x = (I - A)x = x$. Conversely, suppose $(I - A)(I + A)^{-1}x = x$. Since $I - A$ and $(I + A)^{-1}$ commute (because $I - A$ and $I + A$ commute), we have $(I + A)^{-1}(I - A)x = x$ or $(I - A)x = (I + A)x$, which implies $Ax = 0$, that is, $x \in \operatorname{Ker} A$.

   (b) Let $A = U^*DU$ be a spectral decomposition of A, where U is unitary and $D = \operatorname{diag}(\lambda_1, \ldots, \lambda_n)$, in which λ_i's are the eigenvalues of A. Let $y = Ux$. Then $\|y\| = \|x\|$, and $\|(I - A)(I + A)^{-1}x\| \leq \|x\|$ is the same as $\|(I - D)(I + D)^{-1}y\| \leq \|y\|$, or equivalently,

$$\sum_{i=1}^{n} \left(\frac{1 - \lambda_i}{1 + \lambda_i}\right)^2 |y_i|^2 \leq \sum_{i=1}^{n} |y_i|^2.$$

This is obvious because $A \geq 0$ and all λ_i's are nonnegative. Equality occurs if and only if $\lambda_i = 0$ or $y_i = 0$ for $i = 1, \ldots, n$, which is the same as $(I - D)(I + D)^{-1}y = y$, or equivalently, $(I - A)(I + A)^{-1}x = x$, namely, $x \in \operatorname{Ker} A$, by (a).

5.36 (a) Suppose otherwise that v_1, v_2, and v_3 are linearly dependent. It may be assumed that v_1, v_2, and v_3 are unit vectors and that

$$v_3 = \lambda_1 v_1 + \lambda_2 v_2.$$

Then

$$\langle v_1, v_3 \rangle = \lambda_1 + \lambda_2 \langle v_1, v_2 \rangle < 0 \quad \Rightarrow \quad \lambda_1 \langle v_1, v_2 \rangle + \lambda_2 \langle v_1, v_2 \rangle^2 > 0$$

and

$$\langle v_2, v_3 \rangle = \lambda_1 \langle v_1, v_2 \rangle + \lambda_2 < 0.$$

By subtraction,

$$\lambda_2(\langle v_1, v_2 \rangle^2 - 1) > 0 \quad \Rightarrow \quad \lambda_2 < 0 \quad \Rightarrow \quad \lambda_1 < 0.$$

Now compute to get $\langle v_4, v_3 \rangle > 0$, contradicting $\langle v_4, v_3 \rangle < 0$:

$$\langle v_4, v_3 \rangle = \lambda_1 \langle v_4, v_1 \rangle + \lambda_2 \langle v_4, v_3 \rangle > 0.$$

   (b) No, since the dimension of the xy-plane is 2. However, it is possible for three vectors in the xy-plane to have pairwise negative products. Take $v_1 = (1, 0)$, $v_2 = (-\frac{1}{2}, \frac{\sqrt{3}}{2})$, $v_3 = (-\frac{1}{2}, -\frac{\sqrt{3}}{2})$.

(c) v_1, v_2, v_3, v_4 need not be linearly independent (or dependent):

$$(1,0,0), \quad (-1,1,0), \quad (-1,-2,1), \quad (-1,-2,-6).$$

(d) $3, -\frac{3}{2}$. The maximum is attained when $u = v = w$, and the minimum is attained when the angles between any two of the unit vectors u, v, and w are equal to $\frac{2\pi}{3}$.

5.37 Let
$$x = l_1 v_1 + l_2 v_2 + \cdots + l_n v_n.$$
Taking inner product of both sides with v_i results in
$$l_i = \langle x, v_i \rangle, \quad i = 1, 2, \ldots, n.$$
By a direct computation,
$$\langle x, x \rangle = |l_1|^2 + |l_2|^2 + \cdots + |l_n|^2 \geq |l_1|^2 + |l_2|^2 + \cdots + |l_k|^2, \ k \leq n.$$
Equality holds if and only if $x \in \text{Span}\{v_1, \ldots, v_k\}$.

5.38 Since $\{v_1, v_2, \ldots, v_n\}$ is an orthonormal basis for V, we can write $x = \sum_{i=1}^n \langle x, v_i \rangle v_i$ and $y = \sum_{i=1}^n \langle y, v_i \rangle v_i$. A direct computation gives
$$\langle x, y \rangle = \sum_{i=1}^n \overline{\langle y, v_i \rangle} \langle x, v_i \rangle = \sum_{i=1}^n \langle x, v_i \rangle \langle v_i, y \rangle.$$

5.39 Check $\text{tr}(B^t A) = e^t (A \circ B) e$. One may also show directly that $e^t (A \circ B) e$ defines an inner product by observing that $A \circ A = (a_{ij}^2)$, $(\lambda A \circ B) = \lambda(A \circ B)$, $A \circ (B + C) = A \circ B + A \circ C$, and $A \circ B = B \circ A$.

5.40 (a) By verification.

(b) Note that $\text{tr}(A^* A) = \langle A, A \rangle$.

(c) Use the Cauchy–Schwartz inequality to A and B^*.

(d) Note that $\text{tr}(ABB^* A^*) = \text{tr}(A^* ABB^*)$. Use the fact that
$$\text{tr}(XY) \leq \text{tr} X \, \text{tr} Y, \quad X \geq 0, \ Y \geq 0,$$
which can be shown by first assuming that X is diagonal.

(e) Since $\|A\|^2 = \langle A, A \rangle = \text{tr}(A^* A)$ and $A^* A - AA^*$ is Hermitian,
$$\begin{aligned}
\|A^* A - AA^*\|^2 &= \text{tr}(A^* A - AA^*)^2 \\
&= 2\,\text{tr}(A^* AA^* A) - 2\,\text{tr}(A^* AAA^*) \\
&= 2\|A^* A\|^2 - 2\|A^2\|^2 \\
&\leq 2\|A^* A\|^2 \\
&\leq 2\|A\|^4.
\end{aligned}$$

(f) Take $X = A^*$, then use (b). Or set $X = E_{ij}$, where E_{ij} has 1 in the (i, j) position and 0 elsewhere. $\text{tr}(AX) = a_{ji} = 0$ for all i, j.

(g) If $X, Y \in W$, then $\text{tr}(kX + Y) = k\,\text{tr}\,X + \text{tr}\,Y = 0$ for any scalar k. So W is a subspace. The dimension of $M_n(\mathbb{C})$ is n^2. Since $\text{tr}\,X = x_{11} + x_{22} + \cdots + x_{nn} = 0$, all x_{ij} are free except one entry, say x_{nn}, so $\dim W = n^2 - 1$. Thus $\dim W^\perp = 1$. The identity matrix I is a basis for $W^\perp$ and $W^\perp$ consists of scalar matrices.

(h) Take $X = xx^*$. $\langle A, xx^* \rangle = \text{tr}(xx^*A) = \text{tr}(x^*Ax) = x^*Ax \geq 0$.

5.41   (a) No (not over $\mathbb{C}$; yes over $\mathbb{R}$.) (b) Yes. (c) No. (Yes over $\mathbb{R}$.) (d) No. (e) No. (f) No. (Yes over $\mathbb{R}$.) (g) Yes. (h) No in general. (i) Yes.

5.42   The two 2×2 matrices have the same Frobenius norm.

5.43 $\|A \circ B\|_F^2 = \sum_{i,j} |a_{ij}b_{ij}|^2 \leq \left(\sum_{i,j} |a_{ij}|^2 \right)\left(\sum_{i,j} |b_{ij}|^2 \right) = \|A\|_F^2 \|B\|_F^2.$

For the spectral norm, by Problem 4.74, we have $(A \circ B)^*(A \circ B) \leq (A^*A) \circ (B^*B)$. Using Problem 4.73e, we derive

$$
\begin{aligned}
\|A \circ B\|_{\text{sp}}^2 &= \sigma_{\max}^2(A \circ B) = \lambda_{\max}\big((A \circ B)^*(A \circ B)\big) \\
&\leq \lambda_{\max}\big((A^*A) \circ (B^*B)\big) = \|(A^*A) \circ (B^*B)\|_{\text{sp}} \\
&\leq \lambda_{\max}(A^*A)\lambda_{\max}(B^*B) \\
&= \sigma_{\max}^2(A)\sigma_{\max}^2(B) = \|A\|_{\text{sp}}^2 \|B\|_{\text{sp}}^2.
\end{aligned}
$$

5.44   It is a standard exercise to verify that each $\| \cdot \|$ in (a)-(d) is a norm on $M_n(\mathbb{C})$. We assume that U and V are unitary in the following.

(a) $\|UAV\|_F = \sqrt{\text{tr}(V^*A^*U^*UAV)} = \sqrt{\text{tr}(A^*A)} = \|A\|_F$. So the Frobenius norm is unitarily invariant. To show submultiplicativity, use the fact that if P, Q are positive semidefinite matrices, then $\text{tr}(PQ) \leq \text{tr}\,P\,\text{tr}\,Q$ (which is easily proven by assuming that P is diagonal). Now $\text{tr}(B^*A^*AB) = \text{tr}(BB^*A^*A) \leq \text{tr}(BB^*)\,\text{tr}(A^*A)$. It is immediate that $\|AB\|_F \leq \|A\|_F \|B\|_F$.

(b) $\|A\|_{\text{sp}} = \sigma_{\max}(A)$, while UAV and A have the same singular values. So the spectral norm is unitarily invariant. Since $\|ABx\|_2 = \|A(Bx)\|_2 \leq \|A\|_{\text{sp}} \|Bx\|_2 \leq \|A\|_{\text{sp}}\|B\|_{\text{sp}}\|x\|_2$, we see by taking "max" that the spectral norm is submultiplicative.

(c) $\|A\|_1 = \sum_{i,j} |a_{ij}|$ is not unitarily invariant. For $M_2(\mathbb{C})$, take $A = \left(\begin{smallmatrix} 1 & 0 \\ 0 & 0 \end{smallmatrix} \right)$ and $U = \frac{1}{\sqrt{2}} \left(\begin{smallmatrix} 1 & 1 \\ 1 & -1 \end{smallmatrix} \right)$. Then $\|A\|_1 = 1$, $\|AU\|_1 = \sqrt{2}$.

This norm is submultiplicative:

$$\|AB\|_1 = \sum_{i,j}\left|\sum_p a_{ip}b_{pj}\right| \le \sum_{i,j}\sum_p |a_{ip}b_{pj}|$$

$$\le \sum_{i,p,q,j}|a_{ip}b_{qj}| \le \sum_{i,p}|a_{ip}|\sum_{q,j}|b_{qj}| = \|A\|_1\|B\|_1.$$

(d) $\|A\|_\infty = \max_{i,j}|a_{ij}|$ is not unitarily invariant by considering $A = \left(\begin{smallmatrix}1&0\\0&0\end{smallmatrix}\right)$ and $U = \frac{1}{\sqrt{2}}\left(\begin{smallmatrix}1&1\\1&-1\end{smallmatrix}\right)$. It is not submultiplicative because
$1 = \|I_2\|_\infty = \|UU\|_\infty > \|U\|_\infty\|U\|_\infty = \frac{1}{\sqrt{2}}\frac{1}{\sqrt{2}} = \frac{1}{2}.$

Note that unitarily invariant property implies multiplicativity. This needs some more diligent and technical work from matrix analysis.

5.45   (a)   Let $Au = \lambda u$, $u \ne 0$. We may assume that u is a unit vector. Then $u^*Au = \lambda u^*u = \lambda$, implying $\rho \le \omega$. The Cauchy–Schwartz inequality yields $|x^*Ax|^2 \le x^*A^*Ax$ for unit x, revealing $\omega \le \sigma$.

(b) ρ is not a norm as $\rho(A) = 0 \not\Rightarrow A = 0$. Note that $x^*Ax = 0$ for all $x \in \mathbb{C}^n$ yields $A = 0$. ω and σ are (vector) norms on $M_n(\mathbb{C})$.

(c) σ is submultiplicative. ρ and ω are not. Take $A = \left(\begin{smallmatrix}0&1\\0&0\end{smallmatrix}\right)$, $B = \left(\begin{smallmatrix}0&0\\1&0\end{smallmatrix}\right)$, and $A = \left(\begin{smallmatrix}0&I_3\\0&0\end{smallmatrix}\right)_{4\times4}$, $B = \left(\begin{smallmatrix}0_2&I_2\\0_2&0_2\end{smallmatrix}\right)_{4\times4}$ as counterexamples.

(d) σ is unitarily invariant.

5.46 $\left(\begin{smallmatrix}a&b\\-b&a\end{smallmatrix}\right)$, $a, b \in \mathbb{C}$.

5.47   Let $P = (u_1, \ldots, u_n)$. Since $u_1, \ldots, u_n$ are linearly independent, P is invertible. Let $A = PDP^{-1}$, where $D = \mathrm{diag}(\lambda_1, \ldots, \lambda_n)$. Then $AP = PD$ reveals $Au_1 = \lambda_1 u_1$, ..., $Au_n = \lambda_n u_n$.

If B is also a matrix such that $Bu_1 = \lambda_1 u_1$, ..., $Bu_n = \lambda_n u_n$. Then $BP = PD$, and $B = PDP^{-1} = A$. Note: A is unique up to permutation similarity (i.e., the ordering of $\lambda_1, \ldots, \lambda_n$).

If $u_1, \ldots, u_n$ are (nonzero) orthogonal, then $v_1 = \frac{1}{\|u_1\|}u_1, \ldots, v_n = \frac{1}{\|u_n\|}u_n$ are orthonormal. Let $Q = (v_1, \ldots, v_n)$. Then Q is real orthogonal, that is, $Q^{-1} = Q^t$. Thus, QDQ^t is real symmetric, and

$$QDQ^t = QDQ^{-1} = (v_1, \ldots, v_n)D(v_1, \ldots, v_n)^{-1}$$
$$= (u_1, \ldots, u_n)D(u_1, \ldots, u_n)^{-1} = PDP^{-1} = A.$$

See also Problem 3.91 and Problem 6.29

5.48   (a) $u^\perp$ is the y-axis, $v^\perp$ is the line $y = x$.

(b) $u^\perp \cap v^\perp = \{0\}$.

(c) $\{u, v\}^\perp = \{0\}$.

(d) $(\mathrm{Span}\{u, v\})^\perp = \{0\}$.

(e) $\mathrm{Span}\{u^\perp, v^\perp\} = \mathbb{R}^2$.

5.49   (a) Use the Cauchy–Schwartz inequality. Note that $\mathrm{Re}\langle u, v \rangle \leq |\langle u, v \rangle|$.

(b) $\|u - v\|^2 = \langle u, u \rangle + \langle v, v \rangle - 2\,\mathrm{Re}\langle u, v \rangle \geq (\|u\| - \|v\|)^2$.

(c) $\|u + v\|^2 \|u - v\|^2 = (\|u\|^2 + \|v\|^2)^2 - 4(\mathrm{Re}\langle u, v \rangle)^2$.

(d) Compute $\|u + v\|^2 - \|u - v\|^2$ directly to get $4\langle u, v \rangle$.

(e) $\|u + v\|^2 - \|u - v\|^2 = 4\,\mathrm{Re}\langle u, v \rangle$. Replace v with iv, then add.

5.50   (a) By a direct computation.

(b) Suppose $\|x + y\| = \|x\| + \|y\|$. On the one hand, for $s, t \geq 0$,

$$\|sx + ty\| \leq s\|x\| + t\|y\|.$$

On the other hand, assuming $t \geq s$ (similarly for $s > t$),

$$\begin{aligned}
\|sx + ty\| &= \|t(x + y) - (t - s)x\| \\
&\geq \big|\, t\|x + y\| - (t - s)\|x\| \,\big| \\
&= s\|x\| + t\|y\|.
\end{aligned}$$

Thus, $\|sx + ty\| = s\|x\| + t\|y\|$. The other direction is obvious by setting $s = t = 1$. This result can also be shown by examining the proof of the triangle inequality for inner product spaces.

(c) By a direct computation. x and y can be considered as two sides of a rhombus, and $x + y$ and $x - y$ are the diagonals of the rhombus. The diagonals of a rhombus are perpendicular.

(d) If x and y are orthogonal, a simple computation gives

$$\|x + y\|^2 = \|x\|^2 + \|y\|^2 + \langle x, y \rangle + \langle y, x \rangle = \|x\|^2 + \|y\|^2.$$

(e) The converse of (d) is true over $\mathbb{R}$; false over $\mathbb{C}$: $x = 1$, $y = i$.

(f) Take $x = (1, 0)$ and $y = (0, 1)$ for both cases. This says that ℓ_1-norm and ℓ_∞-norm are not derived from an inner product. However, ℓ_2-norm is derived from the standard inner product.

5.51   (a) $\|x\|_2 = \sqrt{\langle x, x \rangle} = (|x_1|^2 + \cdots + |x_n|^2)^{\frac{1}{2}}$.

(b) It is easy to show that $\|x\|_1 = \sum_{i=1}^{n} |x_i|$ is a norm on $\mathbb{R}^n$. To show it is not induced by any inner product, we use Problem 5.50(a). Take $x = (1, 1, 0, \ldots, 0)$ and $y = (-1, 1, 0, \ldots, 0)$ in $\mathbb{R}^n$. Then $\|x\|_1 = \|y\|_1 = \|x + y\|_1 = \|x - y\|_1 = 2$. However, $\|x + y\|_1^2 + \|x - y\|_1^2 = 8 \neq 2\|x\|_1^2 + 2\|y\|_1^2 = 16$. So it is not an induced norm.

(c) It is obvious that $\|x\|_2^2 \leq \|x\|_1^2$, so $\|x\|_2 \leq \|x\|_1$. By the Cauchy–Schwartz inequality, $\|x\|_1 = \sum_{i=1}^{n} 1 \cdot |x_i| \leq \sqrt{n}\|x\|_2$. It follows that $\frac{1}{\sqrt{n}}\|x\|_1 \leq \|x\|_2 \leq \|x\|_1$.

(d) $\|x\|_1^2 = \|x\|_2^2 + 2\sum_{i \neq j} |x_i x_j|$. $\|x\|_1 = \|x\|_2$ if and only if $|x_i x_j| = 0$ for all $i \neq j$ if and only if x has at most one nonzero component.

5.52 (a) Note that $\frac{x}{1+x} = \frac{1}{1+\frac{1}{x}}$ is strictly increasing on $(0, \infty)$. Compute

$$
\begin{aligned}
d(x, y) &= \frac{|x - y|}{1 + |x - y|} = \frac{1}{1 + \frac{1}{|x-y|}} \quad (x \neq y) \\
&\leq \frac{1}{1 + \frac{1}{|x-z|+|z-y|}} \quad (x \neq z \text{ or } y \neq z) \\
&= \frac{|x - z| + |z - y|}{1 + |x - z| + |z - y|} \\
&= \frac{|x - z|}{1 + |x - z| + |z - y|} + \frac{|z - y|}{1 + |x - z| + |z - y|} \\
&\leq \frac{|x - z|}{1 + |x - z|} + \frac{|z - y|}{1 + |z - y|} \\
&= d(x, z) + d(z, y).
\end{aligned}
$$

(b) $g(x)$ is not a norm on $\mathbb{R}$ because $g(\lambda x) \neq |\lambda| g(x)$ in general.

5.53 Note that $(\|x\| + \|y\|)^2 = \|x\|^2 + \|y\|^2 + 2\|x\|\|y\| \leq 2(\|x\|^2 + \|y\|^2)$. By the triangle inequality, $\|x \pm y\| \leq \|x\| + \|y\|$. It follows that

$$
\frac{\|x + y\|^2 + \|x - y\|^2}{\|x\|^2 + \|y\|^2} \leq \frac{4(\|x\|^2 + \|y\|^2)}{\|x\|^2 + \|y\|^2} = 4.
$$

Now setting $x = u + v$ and $y = u - v$, we obtain

$$
\frac{4\|u\|^2 + 4\|v\|^2}{\|u + v\|^2 + \|u - v\|^2} \leq 4,
$$

which is the same as

$$
\frac{\|u + v\|^2 + \|u - v\|^2}{\|u\|^2 + \|v\|^2} \geq 1.
$$

5.54   Notice that for all $x, y \in \mathbb{C}^n$ (in columns)

$$\begin{pmatrix} x^* & 0 \\ 0 & y^* \end{pmatrix} \begin{pmatrix} A & B^* \\ B & C \end{pmatrix} \begin{pmatrix} x & 0 \\ 0 & y \end{pmatrix} = \begin{pmatrix} x^*Ax & x^*B^*y \\ y^*Bx & y^*Cy \end{pmatrix} \geq 0.$$

Then take the determinant. For the special cases, observe that

$$\begin{pmatrix} A & A \\ A & A \end{pmatrix} \geq 0, \quad \begin{pmatrix} A & I \\ I & A^{-1} \end{pmatrix} \geq 0, \quad \begin{pmatrix} \ell(M) & M^* \\ M & \ell(M^*) \end{pmatrix} \geq 0.$$

5.55   (a) Let $\dim V = n$. First $W \cap W^\perp = \{0\}$ because if x is contained in W and $W^\perp$, then $\langle x, x \rangle = 0$, yielding $x = 0$. Thus $W + W^\perp$ is a direct sum. To show $V = W \oplus W^\perp$, let $\dim W = s$ and $\dim W^\perp = t$. We show $s + t = n$. On the one hand, $n \geq \dim(W \oplus W^\perp) = \dim W + \dim W^\perp = s + t$. On the other hand, if $s + t < n$, then $(W \oplus W^\perp)^\perp$ contains nonzero vectors, say v. This means v is orthogonal to all vectors in W (so $v \in W^\perp$) and $W^\perp$. Thus, $v = 0$, a contradiction.

The dimension identity, decomposition of v, and (b) are immediate.

5.56   Necessity: Let $v = w_0 + w'$, where $w_0 \in W, w' \in W^\perp$. Then $v - w_0 \in W^\perp$. Thus, for any $w \in W$, $w_0 - w \in W$, $\langle v - w_0, w_0 - w \rangle = 0$. Thus

$$\|v - w\|^2 = \|(v - w_0) + (w_0 - w)\|^2 = \|v - w_0\|^2 + \|w_0 - w\|^2$$

and

$$\|v - w_0\| \leq \|v - w\|.$$

Sufficiency: Let w_0 be the orthogonal projection of v onto W. Assume that $v_0 \in W$, $v_0 \neq w_0$, and $\|v - v_0\| \leq \|v - w\|$ for all $w \in W$. Then $v_0 - w_0$ lies in W and is orthogonal to $v - w_0$ (which is in $W^\perp$). Thus

$$\begin{aligned} \|v - v_0\|^2 &= \|(v - w_0) + (w_0 - v_0)\|^2 \\ &= \|(v - w_0)\|^2 + \|(w_0 - v_0)\|^2 \\ &> \|v - w_0\|^2, \end{aligned}$$

contradicting $\|v - v_0\| \leq \|v - w\|$ for all $w \in W$.

5.57   (a) False. (b) False. (c) True. (d) False. Let $V = \{(x_1, x_2, x_3, \dots)$ $| \; x_n \in \mathbb{R}, \sum_{n=1}^\infty |x_n|^2 < \infty\}$ with the usual inner product. Let W be the subspace of V consisting of the sequences that all but finitely many components are zero. Then $W^\perp = \{0\}$. There does not exist a subspace W' of V such that $W' + W = V$ and $W' \perp W$. (e) False for infinite dimensional inner product spaces in general (use the example in (d)). It is true for finite dimensional inner product spaces V.

5.58 P consists of (x, y, z) such that $\langle (x, y, z), (1, -2, 1) \rangle = 0$. $(1, -2, 1)$ is a basis for $P^\perp$ (a line). Find the projection of $(1, 0, 1)$ onto $P^\perp$ to be $p = \frac{1}{3}(1, -2, 1)$. The distance from $(1, 0, 1)$ to P is $\|p\| = \sqrt{\frac{2}{3}}$.

5.59 $\frac{\sqrt{6}}{4}$.

5.60   We show (a). (b) is similar. Since $W_1 \subseteq W_1 + W_2$, $(W_1 + W_2)^\perp \subseteq W_1^\perp$. Likewise, $(W_1 + W_2)^\perp \subseteq W_2^\perp$. So $(W_1 + W_2)^\perp \subseteq W_1^\perp \cap W_2^\perp$. Now suppose $u \in W_1^\perp \cap W_2^\perp$. Then $\langle u, w_1 \rangle = 0$ for all $w_1 \in W_1$ and $\langle u, w_2 \rangle = 0$ for all $w_2 \in W_2$. Thus $\langle u, w \rangle = 0$ for all $w \in W_1 + W_2$.

5.61   We show $(\operatorname{Im} A)^\perp = \operatorname{Ker} A^t$. Replacing A by A^t gives the 2nd identity. If $x \in \operatorname{Ker} A^t$, then $A^t x = 0$. For any y, $(Ay)^t x = y^t A^t x = 0$. Thus $x \in (\operatorname{Im} A)^\perp$. Conversely, if $x \in (\operatorname{Im} A)^\perp$, then $x \perp Ay$ for any y. So $(Ay)^t x = y^t(A^t x) = 0$ for all y. Therefore, $A^t x = 0$, and $x \in \operatorname{Ker} A^t$.

5.62   Since $p + \dim S^\perp = \dim V = n$, $\dim S^\perp = n - p < q$. The q vectors $v_1, v_2, \ldots, v_q$ are all contained in $S^\perp$, they must be linearly dependent.

5.63

(a)   Obvious.

(b)   Obvious.

(c)   If $v \in S \cap S^\perp$, then $\langle v, v \rangle = 0$, so $v = 0$. If $0 \notin S$, then $S \cap S^\perp = \emptyset$.

(d) $0 \in S^\perp \cap \operatorname{Span} S$. If $v \in S^\perp \cap \operatorname{Span} S$, then $\langle v, v \rangle = 0$, so $v = 0$.

(e)   If $u, v \in S^\perp$, then $\langle u + kv, s \rangle = \langle u, s \rangle + k \langle v, s \rangle = 0$, k is a scalar.

(f)   If $S^\perp \neq \{0\}$, then there is a nonzero $v \in S^\perp$, $v \notin \operatorname{Span} S$, so $\operatorname{Span} S \neq V$. Conversely, if $\operatorname{Span} S \neq V$, then there is a nonzero w in $(\operatorname{Span} S)^\perp$, w is contained in $S^\perp$. So $S^\perp \neq \{0\}$.

(g)   For any $s \in S$, $\langle s, x \rangle = 0$ for all $x \in S^\perp$, so $s \in (S^\perp)^\perp$.

(h)   If $x \in T^\perp$, then x is orthogonal to any element in T, in particular, to any element in S because $S \subseteq T$. Thus, $x \in S^\perp$.

(i)   Since $S \subseteq \operatorname{Span} S$, $(\operatorname{Span} S)^\perp \subseteq S^\perp$. Conversely, if x is orthogonal to all elements of S, then x is orthogonal to all elements in $\operatorname{Span} S$. So $S^\perp \subseteq (\operatorname{Span} S)^\perp$. Thus, $S^\perp = (\operatorname{Span} S)^\perp$.

(j)   If W is a subspace of V (of finite dimension), then $(W^\perp)^\perp = W$. Note: This need not be true for infinite dimensional V.

(k)   This is because S is a subspace if and only if $S = \operatorname{Span} S$.

(l)   Note that $S^\perp$ is a subspace.

(m)   This is because $S^\perp = (\operatorname{Span} S)^\perp$.

(n) Note that $S^\perp \cap (S^\perp)^\perp = \{0\}$. $S^\perp + (S^\perp)^\perp$ is a direct sum.

5.64 (a) Use the Gram–Schmidt process. To have positive diagonal entries for R, multiply R from on the left by a diagonal unitary matrix if necessary. For uniqueness, let $A = Q_1 R_1 = Q_2 R_2$. We show $Q_1 = Q_2$ and $R_1 = R_2$. Q_1 and Q_2 are unitary. $A^* A = R_1^* Q_1^* Q_1 R_1 = R_2^* Q_2^* Q_2 R_2 = R_1^* R_1 = R_2^* R_2$. Thus $R_1 R_2^{-1} = (R_1^*)^{-1} R_2^*$. Since the left-hand side is upper-triangular and the right-hand side is lower-triangular, we have $R_1 R_2^{-1} = (R_1^*)^{-1} R_2^* = D$ is a positive diagonal matrix. So $D^* = D$ and $D^2 = DD^* = I$. Thus $D = I$, implying $R_1 = R_2$ and $Q_1 = Q_2$.

(b) By the QR factorization, $A^{\frac{1}{2}} = QR$. So, $A = R^* Q^* Q R = R^* R$.

Note: The factorizations can be extended to singular matrices, but Q, R, and C need not be unique, R may have zero diagonal entries.

5.65 We want $AT = TD$, where D is diagonal. First, find the eigenvalues of A and the corresponding eigenvectors (which are orthogonal for distinct eigenvalues as A is Hermitian), then find an orthonormal basis for $\mathbb{R}^3$ from the eigenvectors by the Gram–Schmidt process.

$$T = \begin{pmatrix} \frac{1}{\sqrt{2}} & -\frac{1}{\sqrt{6}} & \frac{1}{\sqrt{3}} \\ -\frac{1}{\sqrt{2}} & -\frac{1}{\sqrt{6}} & \frac{1}{\sqrt{3}} \\ 0 & \frac{2}{\sqrt{6}} & \frac{1}{\sqrt{3}} \end{pmatrix}, \quad T^t AT = \operatorname{diag}(2,2,8).$$

5.66 The null space is a subspace of $\mathbb{R}^3$ with dimension 2. $\alpha_1 = (3,0,-1)$ and $\alpha_2 = (-2,1,0)$ are two linearly independent solutions to the equation. By the Gram–Schmidt process, $\gamma_1 = \frac{1}{\sqrt{10}}(3,0,-1)$, $\gamma_2 = \frac{1}{\sqrt{35}}(-1,5,-3)$ are orthonormal and form a basis for the null space.

5.67 $S = \operatorname{Span}\{u_1, u_2\}$, where $u_1 = (-3,0,1,0)$, $u_2 = (2,-1,0,1)$, and $S^\perp = \operatorname{Span}\{v_1, v_2\}$, where $v_1 = (1,2,3,0)$, $v_2 = (0,1,0,1)$.

5.68 First find a basis for the solution space:

$$\alpha_1 = (1,0,0,-5,-1), \quad \alpha_2 = (0,1,0,-4,-1), \quad \alpha_3 = (0,0,1,4,1).$$

Now use the Gram–Schmidt process to obtain an orthonormal basis:

$$\begin{aligned} \gamma_1 &= \tfrac{1}{3\sqrt{3}}(1,0,0,-5,-1), \\ \gamma_2 &= \tfrac{1}{3\sqrt{15}}(-7,9,0,-1,-2), \\ \gamma_3 &= \tfrac{1}{3\sqrt{35}}(7,6,15,1,2). \end{aligned}$$

5.69   Note that $\alpha_3 = \alpha_2 - \alpha_1 - \alpha_4$ and $\alpha_1, \alpha_2, \alpha_4$ are linearly independent. Thus, $\alpha_1, \alpha_2, \alpha_4$ comprise a basis for $\mathrm{Span}\{\alpha_1, \alpha_2, \alpha_3, \alpha_4\}$. Using the Gram–Schmidt process, we find an orthonormal basis:

$$\gamma_1 = \frac{1}{\sqrt{3}}\begin{pmatrix} 1 \\ -1 \\ 0 \\ 1 \end{pmatrix}, \quad \gamma_2 = \frac{1}{\sqrt{15}}\begin{pmatrix} 3 \\ 2 \\ 1 \\ -1 \end{pmatrix}, \quad \gamma_3 = \frac{1}{\sqrt{3}}\begin{pmatrix} 0 \\ 1 \\ -1 \\ 1 \end{pmatrix}.$$

5.70   By row operations on A, one gets a basis for the row space. Use the Gram–Schmidt process to find an orthonormal basis (not unique):

$$\begin{aligned} \gamma_1 &= \tfrac{1}{\sqrt{6}}(1, 0, 2, 1), \\ \gamma_2 &= (0, 1, 0, 0), \\ \gamma_3 &= \tfrac{1}{\sqrt{30}}(-5, 0, 2, 1). \end{aligned}$$

5.71   Let $\dim V = n$, $\dim V_1 = s$, $\dim V_2 = t$, $s < t$. Then $\dim V_1^{\perp} = n - s \geq 1$. Let $V_3 = V_2 \cap V_1^{\perp}$. By the dimension identity, we have

$$\dim V_3 = \dim V_2 + \dim V_1^{\perp} - \dim(V_2 + V_1^{\perp}) \geq t + (n-s) - n = t - s > 0.$$

This implies that $V_3 = V_2 \cap V_1^{\perp} \neq \{0\}$. It follows that for some $u \in V_2$, $u \neq 0$, $\langle x, u \rangle = 0$ for all $x \in V_1$.

5.72 $[\cdot, \cdot]$ is an inner product if and only if $\mathcal{A}$ is invertible.

5.73   Linearity of $\mathcal{L}$ is easy to show. For the positivity of $\mathcal{L}$, we compute $\langle \mathcal{L}(v), v \rangle = \langle \sum_{i=1}^{n} \langle v, u_i \rangle u_i, v \rangle = \sum_{i=1}^{n} |\langle v, u_i \rangle|^2 \geq 0$. So, $\mathcal{L}$ is positive. To show that $\mathcal{L}$ is nonsingular, let $\mathcal{L}(v) = 0$. Then $\sum_{i=1}^{n} \langle v, u_i \rangle u_i = 0$. Because $\{u_1, \ldots, u_n\}$ is a basis, $\langle v, u_i \rangle = 0$, $i = 1, \ldots, n$. Thus, $v = 0$.

5.74   Let $(\mathcal{A}(u_1), \ldots, \mathcal{A}(u_n)) = (u_1, \ldots, u_n)A$, where A is the matrix representation of $\mathcal{A}$ with respect to the basis (see Chapter 3, p. 95). It follows that $\mathcal{A}(u_j) = a_{1j}u_1 + a_{2j}u_2 + \cdots + a_{nj}u_n$. Since $\{u_1, u_2, \ldots, u_n\}$ is orthonormal, taking the inner product with u_i gives

$$\langle \mathcal{A}(u_j), u_i \rangle = \langle a_{1j}u_1 + a_{2j}u_2 + \cdots + a_{nj}u_n, u_i \rangle = a_{ij}.$$

5.75   (a) $\langle \mathcal{A}(v), w \rangle - \langle \mathcal{B}(v), w \rangle = \langle \mathcal{A}(v) - \mathcal{B}(v), w \rangle = 0$ for all $w \in W$. Setting $w = \mathcal{A}(v) - \mathcal{B}(v)$ reveals $\mathcal{A}(v) - \mathcal{B}(v) = 0$, i.e., $\mathcal{A} = \mathcal{B}$.

(b) Assuming that $\mathcal{A}$ is linear, we show that $\mathcal{C}$ is linear as follows:

$$
\begin{aligned}
\langle v, \mathcal{C}(w_1 + kw_2) \rangle &= \langle \mathcal{A}(v), w_1 + kw_2 \rangle \\
&= \langle \mathcal{A}(v), w_1 \rangle + \bar{k} \langle \mathcal{A}(v), w_2 \rangle \\
&= \langle v, \mathcal{C}(w_1) \rangle + \bar{k} \langle v, \mathcal{C}(w_2) \rangle \\
&= \langle v, \mathcal{C}(w_1) \rangle + \langle v, k\mathcal{C}(w_2) \rangle \\
&= \langle v, \mathcal{C}(w_1) + k\mathcal{C}(w_2) \rangle,
\end{aligned}
$$

that is, $\mathcal{C}(w_1 + kw_2) = \mathcal{C}(w_1) + k\mathcal{C}(w_2)$, and $\mathcal{C}$ is linear.

5.76 (a) Existence: Let $\{e_1, \ldots, e_n\}$ be an orthonormal basis of V. Define $\mathcal{B}(e_j) = \sum_{i=1}^{n} \overline{\langle \mathcal{A}(e_i), e_j \rangle}\, e_i$, $j = 1, \ldots, n$. Then $\mathcal{B}$ is a linear operator on V. It is easy to check that $\langle \mathcal{A}(e_i), e_j \rangle = \langle e_i, \mathcal{B}(e_j) \rangle$ for all i, j. It follows that $\langle \mathcal{A}(x), y \rangle = \langle x, \mathcal{B}(y) \rangle$ for all x and y. Such a $\mathcal{B}$ is an adjoint of $\mathcal{A}$, namely, $\mathcal{A}^*$ exists. Uniqueness: If $\langle x, \mathcal{C}(y) \rangle = \langle \mathcal{A}(x), y \rangle = \langle x, \mathcal{A}^*(y) \rangle$ for all x and y, then $\mathcal{C} = \mathcal{A}^*$.

(b) It is sufficient to notice that

$$
\begin{aligned}
\langle \mathcal{A}(x), y \rangle &= \langle x, \mathcal{A}^*(y) \rangle = \overline{\langle \mathcal{A}^*(y), x \rangle} \\
&= \overline{\langle y, (\mathcal{A}^*)^*(x) \rangle} = \langle (\mathcal{A}^*)^*(x), y \rangle.
\end{aligned}
$$

(c) If $x \in \operatorname{Ker} \mathcal{A}^*$, then $\mathcal{A}^*(x) = 0$. For any $y \in V$, $\mathcal{A}(y) \in \operatorname{Im} \mathcal{A}$,

$$
\langle x, \mathcal{A}(y) \rangle = \langle \mathcal{A}^*(x), y \rangle = 0.
$$

Hence $\operatorname{Ker} \mathcal{A}^* \subseteq (\operatorname{Im} \mathcal{A})^{\perp}$. The other way is similarly shown.

(d) Similar to (c).

(e) $\dim \operatorname{Im} \mathcal{A}^* = \dim(\operatorname{Ker} \mathcal{A})^{\perp} = n - \dim(\operatorname{Ker} \mathcal{A}) = \dim \operatorname{Im} \mathcal{A}$.

(f) By (c), (d), and $V = W \oplus W^{\perp}$, where W is a subspace of V.

(g) This is because $\langle \mathcal{A}(e_i), e_j \rangle = \langle e_i, \mathcal{A}^*(e_j) \rangle = \overline{\langle \mathcal{A}^*(e_j), e_i \rangle}$. More explicitly, let $A = (a_{ij})$ be the matrix of $\mathcal{A}$ under the basis, i.e., $(\mathcal{A}(e_1), \ldots, \mathcal{A}(e_n)) = (e_1, \ldots, e_n)A$. Then $\mathcal{A}(e_i) = \sum_{j=1}^{n} a_{ji} e_j$ and $\mathcal{A}^*(e_i) = \sum_{j=1}^{n} \overline{a_{ij}} e_j$ for each i. So the matrix of $\mathcal{A}^*$ is A^*.

(h) $\langle (\mathcal{AB})(x), y \rangle = \langle x, (\mathcal{AB})^*(y) \rangle$. On the other hand, $\langle (\mathcal{AB})(x), y \rangle = \langle \mathcal{B}(x), \mathcal{A}^*(y) \rangle = \langle x, \mathcal{B}^* \mathcal{A}^*(y) \rangle$ for all x, y. Thus, $(\mathcal{AB})^* = \mathcal{B}^* \mathcal{A}^*$. Similarly, it is straightforward to show $(\mathcal{A} + \mathcal{B})^* = \mathcal{A}^* + \mathcal{B}^*$.

5.77 This is because $\mathcal{A}(x) = Ax$ and $\mathcal{A}^*(y) = A^* y$.

5.78 The adjoint operator $\mathcal{A}^*$ of $\mathcal{A}$ is given by $\mathcal{A}^*(Y) = A^*Y$ because

$$\langle \mathcal{A}(X), Y \rangle = \langle AX, Y \rangle = \text{tr}(Y^*AX) = \text{tr}\left((A^*Y)^*X\right) = \langle X, A^*Y \rangle.$$

5.79 (a) Let $\mathcal{A}(u) = \lambda u$, u is a unit vector. Then $\lambda = \langle \lambda u, u \rangle = \langle \mathcal{A}(u), u \rangle = \langle u, \mathcal{A}^*(u) \rangle = \langle u, \mathcal{A}(u) \rangle = \langle u, \lambda u \rangle = \bar{\lambda}$. So λ is real.

(b) Let $u_1, \ldots, u_n$ be an orthonormal basis. If $\mathcal{A}^* = \mathcal{A}$, by Problem 5.74, the matrix $A = (a_{ij})$ of $\mathcal{A}$ with respect to the basis is given by $a_{ij} = \langle \mathcal{A}(u_j), u_i \rangle = \langle u_j, \mathcal{A}^*(u_i) \rangle = \langle u_j, \mathcal{A}(u_i) \rangle = \overline{\langle \mathcal{A}(u_i), u_j \rangle} = \overline{a_{ji}}$, so $A^* = A$, i.e., A is Hermitian. Conversely, $a_{ij} = \overline{a_{ji}} \Rightarrow \langle u_j, \mathcal{A}^*(u_i) \rangle = \langle u_j, \mathcal{A}(u_i) \rangle$ for all i, j. So $\mathcal{A} = \mathcal{A}^*$.

(c) Let λ be an (real) eigenvalue of $\mathcal{A}$ and let V_λ be the eigenspace of λ over $\mathbb{F}$ ($= \mathbb{C}$ or $\mathbb{R}$). V_λ is invariant under $\mathcal{A}$ because if $u \in V_\lambda$ then $\mathcal{A}(u) \in V_\lambda$ as $\mathcal{A}(\mathcal{A}(u)) = \mathcal{A}(\lambda u) = \lambda(\mathcal{A}(u))$. There exists a set of orthonormal vectors that form a basis of V_λ and every basis vector is an eigenvector of $\mathcal{A}$ corresponding to λ. Write $V = V_\lambda \oplus V_\lambda^\perp$. We show that $V_\lambda^\perp$ is invariant under $\mathcal{A}$. Let $x \in V_\lambda^\perp$. Then $\langle x, y \rangle = 0$ for all $y \in V_\lambda$. Thus $\langle \mathcal{A}(x), y \rangle = \langle x, \mathcal{A}(y) \rangle = \langle x, \lambda y \rangle = \bar{\lambda}\langle x, y \rangle = 0$, that is, $\mathcal{A}(x) \in V_\lambda^\perp$. Now use induction on the dimension of the vector space for $V_\lambda^\perp$.

5.80 Let $\lambda_1, \lambda_2, \ldots, \lambda_n$ be the eigenvalues of $\mathcal{A}$, and let $v_1, v_2, \ldots, v_n$ be corresponding eigenvectors that form an orthonormal basis for V. Suppose further that the first m eigenvalues are positive and the rest are not. If $m < k$, then there exists a nonzero vector w such that

$$w \in W \cap \text{Span}\{v_{m+1}, \ldots, v_n\}.$$

Write

$$w = c_{m+1}v_{m+1} + \cdots + c_n v_n.$$

Since $\{v_1, v_2, \ldots, v_n\}$ is an orthonormal set,

$$\langle \mathcal{A}(w), w \rangle = \left\langle \sum_{i=m+1}^n c_i \lambda_i v_i, \sum_{i=m+1}^n c_i v_i \right\rangle = \sum_{i=m+1}^n |c_i|^2 \lambda_i \leq 0.$$

This contradicts $\langle \mathcal{A}(x), x \rangle > 0$ for all nonzero $x \in W$. So, $m \geq k$.

5.81 Compute $\langle \mathcal{A}(ku_1 + u_2), v \rangle = -\langle ku_1 + u_2, \mathcal{A}(v) \rangle = -k\langle u_1, \mathcal{A}(v) \rangle - \langle u_2, \mathcal{A}(v) \rangle = k\langle \mathcal{A}(u_1), v \rangle + \langle \mathcal{A}(u_2), v \rangle = \langle k\mathcal{A}(u_1) + \mathcal{A}(u_2), v \rangle$ for all $v \in V$. Thus, $\mathcal{A}(ku_1 + u_2) = k\mathcal{A}(u_1) + \mathcal{A}(u_2)$, and $\mathcal{A}$ is linear.

Let x be an eigenvector of $\mathcal{A}$ corresponding to eigenvalue λ. Then $\langle \mathcal{A}(x), x \rangle = -\langle x, \mathcal{A}(x) \rangle$ implies that $\langle \lambda x, x \rangle = -\langle x, \lambda x \rangle$. Thus $\lambda \langle x, x \rangle = -\bar{\lambda} \langle x, x \rangle$. It follows that $\lambda + \bar{\lambda} = 0$ and $\lambda = 0$ if λ is real.

$\langle \mathcal{A}^2(u), v \rangle = -\langle \mathcal{A}(u), \mathcal{A}(v) \rangle = \langle u, \mathcal{A}^2(v) \rangle$. Thus, $\mathcal{A}^2$ is self-adjoint, so it is diagonalizable. (The matrix of $\mathcal{A}^2$ is Hermitian.)

Note: Such a map $\mathcal{A}$ exists, for example, $\mathcal{A} : \left(\begin{smallmatrix} x \\ y \end{smallmatrix} \right) \mapsto \left(\begin{smallmatrix} y \\ -x \end{smallmatrix} \right)$ for $\mathbb{R}^2$.

5.82   Both $\mathcal{A}$ and $\mathcal{B}$ have to be invertible. Suppose otherwise, say $\mathcal{A}(w) = 0$ for some nonzero $w \in V$. Then $0 = \langle \mathcal{A}(w), \mathcal{B}(w) \rangle = \langle w, w \rangle$, a contradiction to the positivity condition. Since

$$\langle \mathcal{A}(u), \mathcal{B}(v) \rangle = \langle \mathcal{B}^*(\mathcal{A}(u)), v \rangle = \langle u, v \rangle, \quad \text{for all } u, v \in V,$$

we conclude that $\mathcal{B}^* \mathcal{A}$ is identity. So $\mathcal{A}^{-1} = \mathcal{B}^*$. Similarly, $\mathcal{B}^{-1} = \mathcal{A}^*$.

5.83   By computation, $\langle \mathcal{A}(x+y) - \mathcal{A}(x) - \mathcal{A}(y), \mathcal{A}(x+y) - \mathcal{A}(x) - \mathcal{A}(y) \rangle = 0$ (after expanding the left-hand side, all $\mathcal{A}$'s on the left can be dropped by the condition $\langle \mathcal{A}(x), \mathcal{A}(y) \rangle = \langle x, y \rangle$). Thus, $\mathcal{A}(x + y) = \mathcal{A}(x) + \mathcal{A}(y)$. Similarly, one can prove $\langle \mathcal{A}(kx) - k\mathcal{A}(x), \mathcal{A}(kx) - k\mathcal{A}(x) \rangle = 0$. By the previous problem, $\mathcal{A}^{-1} = \mathcal{A}^*$. This gives $\mathcal{A}^*\mathcal{A} = \mathcal{A}\mathcal{A}^* = \mathcal{I}_V$.

5.84   (a) For any two vectors x and y in V, by computation, we have

$$0 = \langle \mathcal{A}(x + y), x + y \rangle = \langle \mathcal{A}(x), y \rangle + \langle \mathcal{A}(y), x \rangle$$

$$0 = i\langle \mathcal{A}(x + iy), x + iy \rangle = \langle \mathcal{A}(x), y \rangle - \langle \mathcal{A}(y), x \rangle.$$

So, $\langle \mathcal{A}(x), y \rangle = 0$ for all $x, y \in V$. Setting $y = \mathcal{A}(x)$ reveals $\langle \mathcal{A}(x), \mathcal{A}(x) \rangle = 0$. Thus, $\mathcal{A}(x) = 0$ for all x in V, i.e., $\mathcal{A} = 0$.

(b) $\mathcal{A}$ need not be zero if $\mathbb{C}$ is changed to $\mathbb{R}$. Define $\mathcal{A}$ by $\mathcal{A} \left(\begin{smallmatrix} a \\ b \end{smallmatrix} \right) = \left(\begin{smallmatrix} -b \\ a \end{smallmatrix} \right)$ for $\mathbb{R}^2$. Then $\langle \mathcal{A}(v), v \rangle = 0$ for all $v = \left(\begin{smallmatrix} a \\ b \end{smallmatrix} \right) \in \mathbb{R}^2$.

(c) $A = \left(\begin{smallmatrix} 1 & 0 \\ 0 & -1 \end{smallmatrix} \right)$, $v = \left(\begin{smallmatrix} 1 \\ 1 \end{smallmatrix} \right)$. Then $v^* A v = 0$, but $v^* A^2 v \neq 0$.

(d) This is nearly trivial in view of Hermitian matrices.

5.85   Take $V = \mathbb{R}^2$ and define $\mathcal{L}(x, y) = \left(\sqrt{\frac{1}{2}(x^2 + y^2)}, \sqrt{\frac{1}{2}(x^2 + y^2)} \right)$ and $\mathcal{D}(x, y) = (x + 1, y + 1)$ from $\mathbb{R}^2$ to $\mathbb{R}^2$. Then $\mathcal{L}$ preserves length and $\mathcal{D}$ preserves distance, but neither $\mathcal{L}$ nor $\mathcal{D}$ is linear. Note that a map that preserves inner product is necessarily linear. See Problem 5.83.

5.86   If $\mathcal{L}^* \mathcal{L} = \mathcal{I}$, then $\mathcal{L}$ is an isometry (preserving vector's length) as

$$\|\mathcal{L}(v)\|^2 = \langle \mathcal{L}(v), \mathcal{L}(v) \rangle = \langle \mathcal{L}^* \mathcal{L}(v), v \rangle = \langle v, v \rangle = \|v\|^2.$$

Conversely, if $\|\mathcal{L}(v)\| = \|v\|$, then $\langle \mathcal{L}^*\mathcal{L}(v), v \rangle = \langle \mathcal{L}(v), \mathcal{L}(v) \rangle = \langle v, v \rangle$. So $\langle (\mathcal{L}^*\mathcal{L} - \mathcal{I})(v), v \rangle = 0$ for all $v \in V$. By Problem 5.84, $\mathcal{L}^*\mathcal{L} - \mathcal{I} = 0$, i.e., $\mathcal{L}^*\mathcal{L} = \mathcal{I}$. One may also use Example 5.6: $\|\mathcal{L}(v)\| = \|v\|$ for all $v \in V$ implies $\langle \mathcal{L}(u), \mathcal{L}(w) \rangle = \langle u, w \rangle$ for all $u, w \in V$. Thus, $\langle \mathcal{L}^*\mathcal{L}(u), w \rangle = \langle u, w \rangle$ for all $u, w \in V$. It follows that $\mathcal{L}^*\mathcal{L} = \mathcal{I}$.

5.87 (a) Let $u, v \in V$ and let $u = \sum_{i=1}^{n} x_i \alpha_i$, $v = \sum_{j=1}^{n} y_j \alpha_j$. Then

$$
\begin{aligned}
\langle \mathcal{A}(u), \mathcal{A}(v) \rangle &= \Big\langle \sum_{i=1}^{n} x_i \mathcal{A}(\alpha_i), \sum_{j=1}^{n} y_j \mathcal{A}(\alpha_j) \Big\rangle \\
&= \sum_{i,j=1}^{n} x_i \overline{y_j} \langle \mathcal{A}(\alpha_i), \mathcal{A}(\alpha_j) \rangle \\
&= \sum_{i,j=1}^{n} x_i \overline{y_j} \langle \alpha_i, \alpha_j \rangle \\
&= \Big\langle \sum_{i=1}^{n} x_i \alpha_i, \sum_{j=1}^{n} y_j \alpha_j \Big\rangle = \langle u, v \rangle.
\end{aligned}
$$

(b) $\mathcal{A}$ need not be orthogonal. Take $\mathbb{R}^2$ with $e_1 = (1, 0)$, $e_2 = (0, 1)$ and define $\mathcal{A}(a, b) = (a + b, 0)$ for $(a, b) \in \mathbb{R}^2$. Then $\mathcal{A}(e_1) = e_1$ and $\mathcal{A}(e_2) = e_1$. $\mathcal{A}$ is a linear transformation satisfying $\langle \mathcal{A}(e_i), \mathcal{A}(e_i) \rangle = \langle e_1, e_1 \rangle = 1$, $i = 1, 2$. But $\mathcal{A}$ is not an orthogonal transformation: $\langle e_1, e_2 \rangle = 0$, $\langle \mathcal{A}e_1, \mathcal{A}e_2 \rangle = \langle e_1, e_1 \rangle = 1 \neq 0$. Neither $\langle \mathcal{A}(u), \mathcal{A}(v) \rangle = \langle u, v \rangle$ nor $\|\mathcal{A}(u)\| = \|u\|$ holds. For the latter, take $u = (1, 2)$. Then $\|u\| = \sqrt{5}$ and $\|\mathcal{A}(u)\| = \|e_1 + 2e_1\| = \|(3, 0)\| = 3$.

5.88 As $\langle \mathcal{A}(x), y \rangle + \langle x, \mathcal{A}(y) \rangle = \langle x, \mathcal{A}^*(y) \rangle + \langle x, \mathcal{A}(y) \rangle = \langle x, (\mathcal{A}^* + \mathcal{A})y \rangle = 0$ for all x, y, $\mathcal{A}^* + \mathcal{A} = 0$. So $\mathcal{A}$ is skew-adjoint and a matrix representation of $\mathcal{A}$ under a basis is a skew-symmetric real matrix. If n is odd, the determinant of such a matrix is zero. So the matrix, thus $\mathcal{A}$, is singular. Note that if n is even, then $\mathcal{A}$ need not be singular. Take $A = \left(\begin{smallmatrix} 0 & 1 \\ -1 & 0 \end{smallmatrix} \right)$. Over the complex number field $\mathbb{C}$, take $A = iI_3$.

5.89 (a) Answer to the question is negative in general; it is affirmative if $\alpha_1, \ldots, \alpha_n$ are linearly independent. (b) The sufficiency is immediate from Example 5.6(1). For the necessity, without loss of generality, let $\{\alpha_1, \ldots, \alpha_t\}$ be a basis for $\mathrm{Span}\{\alpha_1, \ldots, \alpha_n\}$. It can be shown that $\{\beta_1, \ldots, \beta_t\}$ is a basis for $\mathrm{Span}\{\beta_1, \ldots, \beta_n\}$. Now let $\{u_1, \ldots, u_{n-t}\}$ and $\{v_1, \ldots, v_{n-t}\}$ be orthonormal bases, respectively, for

$$ (\mathrm{Span}\{\alpha_1, \ldots, \alpha_t\})^{\perp} \quad \text{and} \quad (\mathrm{Span}\{\beta_1, \ldots, \beta_t\})^{\perp}. $$

Then

$$\{\alpha_1, \ldots, \alpha_t, u_1, \ldots, u_{n-t}\} \quad \text{and} \quad \{\beta_1, \ldots, \beta_t, v_1, \ldots, v_{n-t}\}$$

are two bases for V. Now we define $\mathcal{A}$ as follows: for each i and j,

$$\mathcal{A}(\alpha_i) = \beta_i, \ i \le t; \ \mathcal{A}(u_j) = v_j, \ j \le n - t$$

and for $w \in V$, if

$$w = \sum_{i=1}^{t} x_i \alpha_i + \sum_{i=1}^{n-t} y_i u_i,$$

then

$$\mathcal{A}(w) = \sum_{i=1}^{t} x_i \beta_i + \sum_{i=1}^{n-t} y_i v_i.$$

Then $\mathcal{A}$ is linear. $\mathcal{A}$ is orthogonal because $\langle \mathcal{A}(\omega_i), \mathcal{A}(\omega_j) \rangle = \langle \omega_i, \omega_j \rangle$ for all i, j, where $\{\omega_1, \ldots, \omega_n\} = \{\alpha_1, \ldots, \alpha_t, u_1, \ldots, u_{n-t}\}$.

5.90   Sufficiency is obvious. Necessity: Expand both sides of $\langle \mathcal{A}(x+y), \mathcal{A}(x+y) \rangle = \langle \mathcal{B}(x+y), \mathcal{B}(x+y) \rangle$ to get $\langle \mathcal{A}(x), \mathcal{A}(y) \rangle = \langle \mathcal{B}(x), \mathcal{B}(y) \rangle$. Take a basis $\alpha_1, \ldots, \alpha_n$ for V. For $\{\mathcal{A}(\alpha_1), \ldots, \mathcal{A}(\alpha_n)\}$ and $\{\mathcal{B}(\alpha_1), \ldots, \mathcal{B}(\alpha_n)\}$, by Problem 5.89, there exists an orthogonal transformation $\mathcal{C}$ such that $\mathcal{C}(\mathcal{B}(\alpha_i)) = \mathcal{A}(\alpha_i)$. It follows that $\mathcal{A} = \mathcal{C}\mathcal{B}$.

5.91   It is straightforward to verify that each map from V to $\mathbb{F}$ is linear.

5.92   First show $W_1 \cap W_2 = \{0\}$. If $\alpha \in W_1 \cap W_2$, then $\alpha = \mathcal{T}(\alpha)$ and $\alpha = \beta - \mathcal{T}(\beta)$ for some $\beta \in V$. Compute $\langle \alpha, \alpha \rangle = \langle \alpha, \beta - \mathcal{T}(\beta) \rangle = \langle \alpha, \beta \rangle - \langle \alpha, \mathcal{T}(\beta) \rangle = \langle \alpha, \beta \rangle - \langle \mathcal{T}(\alpha), \mathcal{T}(\beta) \rangle = \langle \alpha, \beta \rangle - \langle \alpha, \beta \rangle = 0$. So $\alpha = 0$. Since $x = \mathcal{T}(x) + (x - \mathcal{T}(x))$ for all $x \in V$, $V = W_1 \oplus W_2$.

5.93   (a) By definition

$$
\begin{aligned}
\mathcal{L}_v(ax + by) &= \langle ax + by, v \rangle = a\langle x, v \rangle + b\langle y, v \rangle \\
&= a\mathcal{L}_v(x) + b\mathcal{L}_v(y).
\end{aligned}
$$

(b) Notice that $\mathcal{L}(ax + by) = \mathcal{L}_{ax+by}$ and that (over the real $\mathbb{R}$)

$$
\begin{aligned}
\mathcal{L}_{ax+by}(u) &= \langle u, ax + by \rangle = a\langle u, x \rangle + b\langle u, y \rangle \\
&= a\mathcal{L}_x(u) + b\mathcal{L}_y(u) = (a\mathcal{L}_x + b\mathcal{L}_y)(u) \\
&= \big(a\mathcal{L}(x) + b\mathcal{L}(y)\big)(u).
\end{aligned}
$$

Thus, $\mathcal{L}(ax + by) = a\mathcal{L}(x) + b\mathcal{L}(y)$, and $\mathcal{L}$ is linear.

(c) If $\mathcal{L}(v_1) = \mathcal{L}(v_2)$ for some v_1 and v_2, then $\mathcal{L}_{v_1} = \mathcal{L}_{v_2}$ and

$$\langle u, v_1 \rangle = \langle u, v_2 \rangle, \quad \text{for every } u \in V.$$

Thus, $\langle u, v_1 - v_2 \rangle = 0$ for all u. So, $v_1 = v_2$ and $\mathcal{L}$ is one-to-one. To show that $\mathcal{L}$ is onto from V to V^*, suppose $f \in V^*$. We need to show that there exists $v \in V$ such that $\mathcal{L}(v) = f$. Let $\{e_1, e_2, \ldots, e_n\}$ be an orthonormal basis for V and let

$$v = f(e_1)e_1 + \cdots + f(e_n)e_n.$$

Then $\mathcal{L}_v(e_i) = \langle e_i, v \rangle = f(e_i)$ for every i. It follows that $\mathcal{L}_v = f$.

(d) For orthonormal basis $\{e_1, e_2, \ldots, e_n\}$ of V, define $f_i \in V^*$ by

$$f_i(e_j) = \begin{cases} 1 & \text{if } i = j, \\ 0 & \text{if } i \neq j. \end{cases}$$

Then $\{f_1, f_2, \ldots, f_n\}$ is a basis for V^*.

(e) $f_1(x, y) = x$ and $f_2(x, y) = y$ form a basis for $(\mathbb{R}^2)^*$.

(f) Let V be a vector space of dimension n over a field $\mathbb{F}$. The set of all linear functionals from V to $\mathbb{F}$ is a vector space over $\mathbb{F}$ with respect to the usual (pointwise) addition and scalar multiplication for functions. Such a vector space associated with V is called the *dual space* of V and is denoted by V^*.

Given a basis $\alpha_1, \alpha_2, \ldots, \alpha_n$ of V, define f_i by $f_i(\alpha_j) = 1$ if $i = j$, or 0 if $i \neq j$. Each f_i is a well-defined linear functional. We claim that $f_1, f_2, \ldots, f_n$ are linearly independent. Suppose $c_1 f_1 + c_2 f_2 + \cdots + c_n f_n = 0$. Then $(c_1 f_1 + c_2 f_2 + \cdots + c_n f_n)(\alpha_i) = 0(\alpha_i)$ for each i, i.e., $c_i f_i(\alpha_i) = c_i = 0$. On the other hand, every linear functional f is a linear combination of $f_1, f_2, \ldots, f_n$ as

$$f = f(\alpha_1)f_1 + f(\alpha_2)f_2 + \cdots + f(\alpha_n)f_n.$$

$\{f_1, f_2, \ldots, f_n\}$ is said to be a *dual basis* of the dual space V^* relative to the basis $\{\alpha_1, \ldots, \alpha_n\}$ of V. Note that every $v \in V$ can be uniquely written as $v = f_1(v)\alpha_1 + \cdots + f_n(v)\alpha_n$.

5.94 (a) Extend u to an orthonormal basis $\{u, u_2, \ldots, u_n\}$ of V. Then

$$\mathcal{A}(u) = -u, \quad \mathcal{A}(u_i) = u_i, \quad i = 2, \ldots, n.$$

$\mathcal{A}$ maps an orthonormal basis to an orthonormal basis.

(b) We compute (over $\mathbb{R}$)

$$
\begin{aligned}
\langle \mathcal{A}(x), y \rangle &= \langle x - 2\langle x, u \rangle u, y \rangle \\
&= \langle x, y \rangle - 2\langle x, u \rangle \langle u, y \rangle \\
&= \langle x, y \rangle - \langle x, 2\langle y, u \rangle u \rangle \\
&= \langle x, y - 2\langle y, u \rangle u \rangle.
\end{aligned}
$$

Thus, $\mathcal{A}^*(y) = y - 2\langle y, u \rangle u = \mathcal{A}(y)$, that is, $\mathcal{A}^* = \mathcal{A}$.

(c) Note that u is a unit vector. Direct computations reveal

$$
\langle \mathcal{A}^2(x), y \rangle = \langle \mathcal{A}(x), \mathcal{A}^*(y) \rangle = \langle x - 2\langle x, u \rangle u, y - 2\langle y, u \rangle u \rangle = \langle x, y \rangle.
$$

(d) The matrix of $\mathcal{A}$ under the basis in (a) is $A = \begin{pmatrix} -1 & 0 \\ 0 & I_{n-1} \end{pmatrix}$. Thus $\det A = -1$. A matrix representation of $\mathcal{A}$ under a different basis is similar to A; it has the same determinant as A.

(e) For any given orthonormal basis, suppose the coordinates of x and $\mathcal{A}(x)$ with respect to the orthonormal basis are y_0 and z_0, respectively. Let the coordinate of u under the basis be v. Then

$$
z_0 = y_0 - 2(v^t y_0)v = I y_0 - 2(vv^t)y_0 = (I - 2vv^t)y_0.
$$

Thus the matrix representation of $\mathcal{A}$ has the form $I - 2vv^t$.

Alternatively, rewrite $A = \begin{pmatrix} -1 & 0 \\ 0 & I_{n-1} \end{pmatrix} = I_n - \begin{pmatrix} 2 & 0 \\ 0 & 0 \end{pmatrix} = I_n - 2e_1 e_1^t$, where $e_1 = (1, 0, \ldots, 0)^t$. The matrices of $\mathcal{A}$ with respect to other orthonormal bases are orthogonally similar to A. It follows that $TAT^t = T^t T - 2Te_1 e_1^t T^t = I_n - 2vv^t$, where $v = Te_1$ and T is an orthogonal matrix.

(f) Apply $\mathcal{A}$ to $x = ku + y$.

(g) Denote by V_1 the eigenspace of $\mathcal{B}$ corresponding to 1 and suppose that $\{u_1, u_2, \ldots, u_{n-1}\}$ is a basis for V_1. Let $\lambda \neq 0$ be an eigenvalue of $\mathcal{B}$ having unit eigenvector $u_n \in V_1^{\perp}$. Then $u_1, u_2, \ldots, u_n$ form a basis for V. Since $\mathcal{B}$ is orthogonal and 1 is an eigenvalue of $\mathcal{B}$ with multiplicity $n - 1$, we see that λ has to be -1. Thus

$$
\mathcal{B}(u_i) = u_i, \quad i = 1, 2, \ldots, n-1, \quad \mathcal{B}(u_n) = -u_n.
$$

Take $w = u_n$. Then $\mathcal{B}(x) = x - 2\langle x, w \rangle w$.

See also Problem 4.??.

5.95   Verify $(\mathcal{L}_v - \mathcal{I})(\mathcal{L}_w - \mathcal{I}) = 0$ with $\mathcal{L}_v^* = \mathcal{L}_v$. Or alternatively, compute

$$
\begin{aligned}
\mathcal{L}_v\mathcal{L}_w(x) &= \mathcal{L}_v\big(\mathcal{L}_w(x)\big) = \mathcal{L}_w(x) - 2\langle \mathcal{L}_w(x), v\rangle v\\
&= x - 2\langle x, w\rangle w - 2\langle (x - 2\langle x, w\rangle w), v\rangle v\\
&= x - 2\langle x, w\rangle w - 2\langle x, v\rangle v\\
&= -x + \big(x - 2\langle x, w\rangle w\big) + \big(x - 2\langle x, v\rangle v\big)\\
&= -\mathcal{I}(x) + \mathcal{L}_v(x) + \mathcal{L}_w(x).
\end{aligned}
$$

5.96   (a) By a direct computation, $A^2 = A$.

(b) $\operatorname{Im} A$ and $\operatorname{Ker} A$ are invariant subspaces for any A (as a map).

(c) $\operatorname{Im} A = \{\left(\begin{smallmatrix} a \\ 0 \end{smallmatrix}\right)\}$, $\operatorname{Ker} A = \{\left(\begin{smallmatrix} b \\ -b \end{smallmatrix}\right)\}$, $a, b \in \mathbb{R}$. $\operatorname{Im} A \cap \operatorname{Ker} A = \{0\}$.

(d) No. Take $u = (2, 0)^t \in \operatorname{Im} A$ and $v = (1, -1)^t \in \operatorname{Ker} A$.

(e) A has two distinct eigenvalues, so A is diagonalizable. This means that under some basis of $\mathbb{R}^2$, the matrix A when considered as a linear map on $\mathbb{R}^2$ has a diagonal matrix representation.

(f) If U is a unitary matrix such that $U^*AU = D$, where $D = \left(\begin{smallmatrix} 1 & 0 \\ 0 & 0 \end{smallmatrix}\right)$. Then $AU = UD$. A simple computation shows U takes the form $\left(\begin{smallmatrix} x & y \\ 0 & -y \end{smallmatrix}\right)$. Apparently, such a unitary matrix U does not exist. This means that there does not exist an orthonormal basis for $\mathbb{R}^2$ under which the matrix A when considered as a linear transformation on $\mathbb{R}^2$ has a diagonal matrix representation.

5.97   (a) For $v \in V$, write (uniquely) $v = w + w'$, where $w \in W$ and $w' \in W^\perp$. Define $\mathcal{P}(v) = w$. Then $\mathcal{P}(w) = w$ and $\mathcal{P}(w') = 0$. Choose a basis for W and a basis for $W^\perp$ to form a basis for V. It is straightforward to show that $\mathcal{P}$ is a linear map.

(b) The decomposition $v = w + w'$ is uniquely determined by W.

(c) $\mathcal{P}^2(v) = \mathcal{P}(w) = w = \mathcal{P}(v)$. To see $\mathcal{P}^* = \mathcal{P}$, let $v, u \in V$. Write $v = w_1 + w_1'$, $u = w_2 + w_2'$, $w_1, w_2 \in W$ and $w_1', w_2' \in W^\perp$. Then

$$\langle \mathcal{P}(v), u\rangle = \langle w_1, w_2 + w_2'\rangle = \langle w_1, w_2\rangle = \langle w_1 + w_1', w_2\rangle = \langle v, \mathcal{P}(u)\rangle.$$

So $\mathcal{P}(u) = \mathcal{P}^*(u)$ for all u, that is, $\mathcal{P}$ is self-adjoint.

For the converse, if $\mathcal{P}^2 = \mathcal{P}$ and $\mathcal{P}^* = \mathcal{P}$, we show that $\mathcal{P}$ is an orthogonal projection onto $W = \operatorname{Im}\mathcal{P}$. We need to show $\mathcal{P}(w) = w$ for every $w \in \operatorname{Im}\mathcal{P}$ and $\mathcal{P}(w') = 0$ if $w' \in (\operatorname{Im}\mathcal{P})^\perp$. Let $w = \mathcal{P}(z)$. Then $\mathcal{P}(w) = \mathcal{P}(\mathcal{P}(z)) = \mathcal{P}^2(z) = \mathcal{P}(z) = w$. Let $w' \in (\operatorname{Im}\mathcal{P})^\perp$. Then $\langle w', \mathcal{P}(v)\rangle = 0$ for all $v \in V$. This means $\langle \mathcal{P}^*(w'), v\rangle = \langle \mathcal{P}(w'), v\rangle = 0$ for all $v \in V$. So $\mathcal{P}(w') = 0$.

(d) This is because $V = W \oplus W^{\perp}$ and $W = \operatorname{Im} \mathcal{P}$.

(e) Since W is a proper subspace, there is a subspace $W^{\dagger}$ such that

$$V = W \oplus W^{\dagger}, \quad W^{\dagger} \nsubseteq W^{\perp}, \quad W^{\perp} \nsubseteq W^{\dagger}.$$

For each $v \in V$, we can uniquely write $v = w_1 + w_1' = w_2 + w_2'$, where $w_1, w_2 \in W$ and $w_1' \in W^{\perp}$, $w_2' \in W^{\dagger}$. For $w \in W$, $w = w_1 = w_2$. Define $\mathcal{T}(v) = w_2$. Then $\mathcal{T}$ is a linear map on V. $\mathcal{T}(w) = \mathcal{P}(w)$ for all $w \in W$. However, $\mathcal{T} \neq \mathcal{P}$ on V. This is seen as follows: Take a nonzero $w_1' \in W^{\perp} \setminus W^{\dagger}$ and let $w_1' = u_2 + u_2'$, where $0 \neq u_2 \in W, u_2' \in W^{\dagger}$. Then $\mathcal{P}(w_1') = 0$, $\mathcal{T}(w_1') = u_2 \neq 0$.

(f) $\langle \mathcal{P}(v), v \rangle = \langle w, w + w' \rangle = \langle w, w \rangle \geq 0$.

(g) Since $w \perp w'$, $\|\mathcal{P}(v)\| = \|w\| \leq \|w + w'\| = \|v\|$.

(h) $(\mathcal{I} - \mathcal{P})(w') = w' - \mathcal{P}(w') = w'$ and $(\mathcal{I} - \mathcal{P})(w) = w - \mathcal{P}(w) = 0$.

(i) Note that $v = \mathcal{P}(v) + (\mathcal{I} - \mathcal{P})(v)$ and $\langle \mathcal{P}(v), (\mathcal{I} - \mathcal{P})(v) \rangle = 0$.

(j) An orthogonal transformation maps an orthonormal basis to an orthonormal basis, while an orthogonal projection maps some vectors in an orthonormal basis to themselves and maps the rest vectors in the basis to zero. Identity is the only orthogonal transformation that is an orthogonal projection (extreme case).

5.98 (a) $\{\beta_1, \beta_2, \ldots, \beta_n\}$ is linearly independent.

(b) $\{\alpha_1, \alpha_2, \ldots, \alpha_n\}$ and $\{\beta_1, \beta_2, \ldots, \beta_n\}$ are orthonormal bases.

(c) Each $\beta_i = 0$ or appears once in orthonormal $\{\alpha_1, \alpha_2, \ldots, \alpha_n\}$.

5.99 (a) $\operatorname{Im} \mathcal{A}$ and $\operatorname{Ker} \mathcal{A}$ are $\mathcal{A}$-invariant subspaces. Sine $x = \mathcal{A}x + (x - \mathcal{A}x)$ for every $x \in V$ and $\operatorname{Im} \mathcal{A} \perp \operatorname{Ker} \mathcal{A}$, we have $V = \operatorname{Im} \mathcal{A} \oplus \operatorname{Ker} \mathcal{A}$, that is, V is the direct sum of $\mathcal{A}$-invariant subspaces that are mutually orthogonal. So an orthonormal basis of $\operatorname{Im} \mathcal{A}$ and an orthonormal basis of $\operatorname{Ker} \mathcal{A}$ comprise an orthonormal basis of V. Because $\mathcal{A}^2 = \mathcal{A}$, the restriction of $\mathcal{A}$ on $\operatorname{Im} \mathcal{A}$ is nonsingular, thus it is the identity; the restriction of $\mathcal{A}$ on $\operatorname{Ker} \mathcal{A}$ is zero. Thus, $\mathcal{A}$ is self-adjoint.

(b) We show that $\|\mathcal{A}(x)\| \leq \|x\|$ implies $\operatorname{Im} \mathcal{A} \perp \operatorname{Ker} \mathcal{A}$. Suppose $\operatorname{Im} \mathcal{A} \not\perp \operatorname{Ker} \mathcal{A}$. Then $\langle x_0, x_1 \rangle \neq 0$ for some $x_0 \in \operatorname{Ker} \mathcal{A}$ and $x_1 \in \operatorname{Im} \mathcal{A}$. We may assume that $\langle x_0, x_1 \rangle$ is real negative (or use $k x_0$ in place of x_0 for some scalar k). Put $x = x_0 + t x_1$, $t > 0$. We compute

$$\|x\|^2 = \|x_0\|^2 + t^2 \|x_1\|^2 + 2t \langle x_0, x_1 \rangle,$$

$$\|\mathcal{A}(x)\|^2 = \|t\mathcal{A}(x_1)\|^2 = t^2\|\mathcal{A}(x_1)\|^2 = t^2\|x_1\|^2,$$

as $\mathcal{A}(x_1) = x_1$ for $x_1 \in \operatorname{Im}\mathcal{A}$. If t is large enough, then $\|x\| < \|\mathcal{A}x\|$, contradicting the assumption that $\|\mathcal{A}(x)\| \le \|x\|$ for all $x \in V$.

It follows that $V = \operatorname{Im}\mathcal{A} \oplus \operatorname{Ker}\mathcal{A}$ is the direct sum of two $\mathcal{A}$-invariant orthogonal subspaces. $\mathcal{A}$ on $\operatorname{Im}\mathcal{A}$ is the identity map and on $\operatorname{Ker}\mathcal{A}$ is the zero map. Therefore, $\mathcal{A}$ is self-adjoint.

Note: This is a restatement of Problem 3.110. In both cases, $\mathcal{A}$ is an orthogonal projection, $\operatorname{Im}\mathcal{A}$ is the eigenspace of the eigenvalue 1, and $\operatorname{Ker}\mathcal{A}$ is the eigenspace of the eigenvalue 0. See Problem 3.150.

5.100   No. For $V = \mathbb{R}^2$, let $\mathcal{A}\begin{pmatrix} x \\ y \end{pmatrix} = \begin{pmatrix} 2 & 0 \\ 0 & 0 \end{pmatrix}\begin{pmatrix} x \\ y \end{pmatrix} = \begin{pmatrix} 2x \\ 0 \end{pmatrix}$. Then $\mathcal{A}^* = \mathcal{A}$, $\operatorname{Im}\mathcal{A} = \{\begin{pmatrix} a \\ 0 \end{pmatrix} \in \mathbb{R}^2 \mid a \in \mathbb{R}\}$, and $\operatorname{Ker}\mathcal{A} = \{\begin{pmatrix} 0 \\ b \end{pmatrix} \in \mathbb{R}^2 \mid b \in \mathbb{R}\}$. Obviously, $\mathbb{R}^2 = \operatorname{Im}\mathcal{A} \oplus \operatorname{Ker}\mathcal{A}$, but $\mathcal{A}^2 \ne \mathcal{A}$.

5.101   (a) $\Leftrightarrow$ (b): This is Problem 5.97(c).

(a) $\Rightarrow$ (c): Let $\{\alpha_1, \dots, \alpha_n\}$ be an orthonormal basis of V. By Problem 5.74, the matrix representation of $\mathcal{L}$ with respect to the basis is $L = (l_{ij})$, where $l_{ij} = \langle \mathcal{L}(\alpha_j), \alpha_i \rangle$. Since $\mathcal{L}^* = \mathcal{L}$, we obtain

$$l_{ji} = \langle \mathcal{L}(\alpha_i), \alpha_j \rangle = \overline{\langle \alpha_j, \mathcal{L}(\alpha_i) \rangle} = \overline{\langle \mathcal{L}^*(\alpha_j), \alpha_i \rangle} = \overline{\langle \mathcal{L}(\alpha_j), \alpha_i \rangle} = \overline{l_{ij}}.$$

That is, L is a Hermitian matrix. We show that $A^2 = A$ as follows:

The (i, j)-entry of the matrix representation of $\mathcal{L}^2$ is $\langle \mathcal{L}^2(\alpha_j), \alpha_i \rangle$. Since $\mathcal{L}^2(\alpha_i) = \mathcal{L}(\alpha_i)$, $\langle \mathcal{L}^2(\alpha_j), \alpha_i \rangle = \langle \mathcal{L}(\alpha_j), \alpha_i \rangle = l_{ij}$. So $L^2 = L$.

(c) $\Rightarrow$ (a): $L^* = L$ says $l_{ji} = \overline{l_{ij}}$, that is, $\langle \mathcal{L}^*(\alpha_j), \alpha_i \rangle = \langle \mathcal{L}(\alpha_j), \alpha_i \rangle$. Since $\{\alpha_1, \dots, \alpha_n\}$ is an orthonormal basis, $\langle \mathcal{L}^*(u), v \rangle = \langle \mathcal{L}(u), v \rangle$ for all $u, v \in V$, so $\mathcal{L}^* = \mathcal{L}$. Similarly, $L^2 = L$ results in $\langle \mathcal{L}^2(\alpha_j), \alpha_i \rangle = \langle \mathcal{L}(\alpha_j), \alpha_i \rangle$ and $\langle \mathcal{L}^2(u), v \rangle = \langle \mathcal{L}(u), v \rangle$ for all $u, v \in V$, so $\mathcal{L}^2 = \mathcal{L}$.

(c) $\Leftrightarrow$ (d): By spectral decomposition for Hermitian matrices.

(a) $\Rightarrow$ (e): $V = \operatorname{Im}\mathcal{L} \oplus (\operatorname{Im}\mathcal{L})^{\perp}$. Choose an orthonormal basis of $\operatorname{Im}\mathcal{L}$ and an orthonormal basis of $(\operatorname{Im}\mathcal{L})^{\perp}$. Together they form a basis for V and the matrix representation of $\mathcal{L}$ under the basis is $\begin{pmatrix} I_r & 0 \\ 0 & 0 \end{pmatrix}$.

(e) $\Rightarrow$ (a): If $\{\alpha_1, \dots, \alpha_n\}$ is an orthonormal basis of V and the matrix of $\mathcal{L}$ under the basis is $\begin{pmatrix} I_r & 0 \\ 0 & 0 \end{pmatrix}$, then $\mathcal{L}(\alpha_i) = \alpha_i$ for $i = 1, \dots, r$, $\mathcal{L}(\alpha_i) = 0$ for $i > r$. $\mathcal{L}$ is an orthogonal projection onto $\operatorname{Im}\mathcal{L}$.

5.102   We show the statements with orthogonal projection matrices P and Q.

(a) If $P + Q$ is an orthogonal projection, then $(P + Q)^2 = P + Q$ which yields $PQ + QP = 0$. Thus, $PQ = -QP$ and $PQ = P^2 Q = P(PQ) = -PQP$. Since PQP is Hermitian, PQ is Hermitian. So $PQ = (PQ)^* = Q^* P^* = QP$. $PQ + QP = 0$ reveals $PQ = 0$. Conversely, if $PQ = 0$, then $0 = (PQ)^* = QP$. Thus, $(P+Q)^2 = P^2 + PQ + QP + Q^2 = P + Q$ and $(P+Q)^* = P^* + Q^* = P + Q$. It follows that $P + Q$ is an orthogonal projection (onto its range).

(b) If PQ is an orthogonal projection (onto its range $\mathrm{Im}(PQ)$), then PQ is Hermitian. So $PQ = (PQ)^* = Q^* P^* = QP$. Conversely, if $PQ = QP$, then $(PQ)^2 = PQPQ = P^2 Q^2 = PQ$ and $(PQ)^* = Q^* P^* = QP = PQ$. This means PQ is an orthogonal projection.

See also Problem 4.124. Note that in (b), $\mathrm{Im}(PQ) = \mathrm{Im}\, P \cap \mathrm{Im}\, Q$.

5.103 For subspaces W_1 and W_2 of finite dimension, $(W_1 \cap W_2)^\perp = W_1^\perp + W_2^\perp$. For an orthogonal projection $\mathcal{P}$, $(\mathrm{Im}\, \mathcal{P})^\perp = \mathrm{Ker}\, \mathcal{P}$. Thus

$$(\mathrm{Im}\, \mathcal{P} \cap \mathrm{Im}\, \mathcal{Q})^\perp = (\mathrm{Im}\, \mathcal{P})^\perp + (\mathrm{Im}\, \mathcal{Q})^\perp = \mathrm{Ker}\, \mathcal{P} + \mathrm{Ker}\, \mathcal{Q}.$$

5.104 (a) $\mathcal{L}_1 : (x, y, z) \mapsto (x, 0, 0)$ is an orthogonal projection onto W.

(b) $\mathcal{L}_2 : (x, y, z) \mapsto (x, y, 0)$ is an orthogonal projection onto the xy-plane in $\mathbb{R}^3$; it is not an orthogonal projection onto W.

(c) $\mathcal{L}_3 : (x, y, z) \mapsto (x, y + z, 0)$. $\mathcal{L}_3^*(x, y, z) \mapsto (x, y, y)$.

5.105 We show if $\mathcal{L}$ is normal, i.e., $\mathcal{L}^* \mathcal{L} = \mathcal{L} \mathcal{L}^*$, then $\|\mathcal{L}(v)\| = \|\mathcal{L}^*(v)\|$. The reverse direction is similar. Upon computation, we have

$$\begin{aligned} \|\mathcal{L}(v)\|^2 &= \langle \mathcal{L}(v), \mathcal{L}(v) \rangle = \langle v, \mathcal{L}^* \mathcal{L}(v) \rangle = \langle v, \mathcal{L} \mathcal{L}^*(v) \rangle \\ &= \langle \mathcal{L}^*(v), \mathcal{L}^*(v) \rangle = \|\mathcal{L}^*(v)\|^2. \end{aligned}$$

5.106 (a) For $v \in V$,

$$v = \mathcal{I}(v) = \mathcal{P}_1(v) + \cdots + \mathcal{P}_m(v).$$

Thus

$$V = \mathrm{Im}\, \mathcal{P}_1 + \cdots + \mathrm{Im}\, \mathcal{P}_m.$$

For direct sum, it is sufficient to show that

$$\dim V = \dim(\mathrm{Im}\, \mathcal{P}_1) + \cdots + \dim(\mathrm{Im}\, \mathcal{P}_m).$$

Take a basis for V and suppose that the matrix representation for $\mathcal{P}_i$ is P_i, $i = 1, \ldots, m$. By Problem 3.150,

$$\dim(\mathrm{Im}\, \mathcal{P}_i) = \mathrm{tr}\, P_i = r(P_i), \quad i = 1, \ldots, m.$$

It follows that

$$\begin{aligned} \dim V &= n = \operatorname{tr} I_n = \operatorname{tr} P_1 + \cdots + \operatorname{tr} P_m \\ &= \dim(\operatorname{Im} \mathcal{P}_1) + \cdots + \dim(\operatorname{Im} \mathcal{P}_m). \end{aligned}$$

To see $\mathcal{P}_i \mathcal{P}_j = 0$ for $i \neq j$, let $v \in V$. Then

$$\mathcal{P}_j(v) = \Big(\sum_{i=1}^{n} \mathcal{P}_i \Big) \mathcal{P}_j(v) = \sum_{i=1}^{n} (\mathcal{P}_i \mathcal{P}_j)(v).$$

Note that

$$\mathcal{P}_j(v) \in \operatorname{Im} \mathcal{P}_j, \quad \mathcal{P}_i \mathcal{P}_j(v) \in \operatorname{Im} \mathcal{P}_i$$

and that V is a direct sum of $\operatorname{Im} \mathcal{P}_j$'s. It follows that

$$(\mathcal{P}_i \mathcal{P}_j)(v) = 0, \quad i \neq j.$$

(b) Choose a basis for each $\operatorname{Im} \mathcal{P}_i$, $i = 1, \ldots, m$, and put all these vectors together to form an orthonormal basis for V by defining the inner product of any two distinct vectors in the basis to be 0, and the inner product of any vector with itself to be 1.

(c) Obviously

$$\operatorname{Im} \mathcal{P}_1 + \cdots + \operatorname{Im} \mathcal{P}_m + \cap_{i=1}^{m} \operatorname{Ker} \mathcal{P}_i \subseteq V.$$

Let

$$\mathcal{T} = \mathcal{I} - \mathcal{P}_1 - \cdots - \mathcal{P}_m.$$

Then for $x \in V$,

$$\mathcal{T}(x) = x - \mathcal{P}_1(x) - \cdots - \mathcal{P}_m(x).$$

Thus for each i,

$$\begin{aligned} \mathcal{P}_i(\mathcal{T}(x)) &= \mathcal{P}_i \big(x - \mathcal{P}_1(x) - \cdots - \mathcal{P}_m(x) \big) \\ &= \mathcal{P}_i(x) - \mathcal{P}_i^2(x) = 0. \end{aligned}$$

So

$$\mathcal{T}(x) \in \cap_{i=1}^{m} \operatorname{Ker} \mathcal{P}_i.$$

Since

$$x = \mathcal{P}_1(x) + \cdots + \mathcal{P}_m(x) + \mathcal{T}(x),$$

we have

$$V = \operatorname{Im} \mathcal{P}_1 + \cdots + \operatorname{Im} \mathcal{P}_m + \cap_{i=1}^{m} \operatorname{Ker} \mathcal{P}_i.$$

To show it is a direct sum, let

$$x_1 + \cdots + x_m + y = 0,$$

where

$$x_i \in \operatorname{Im} P_i, \quad y \in \cap_{i=1}^{m} \operatorname{Ker} \mathcal{P}_i.$$

Notice that $\mathcal{P}_i(x_i) = x_i$ and that $\mathcal{P}_i \mathcal{P}_j = 0$. Applying $\mathcal{P}_i$ to both sides of the above identity yields $x_i = 0$, $i = 1, \ldots, m$, $y = 0$.

5.107   (a)   If $\mathcal{T} = a\mathcal{I}$ (or $b\mathcal{I}$), then a (or b) is the only (repeated) eigenvalue of $\mathcal{T}$. If $\mathcal{T} \neq a\mathcal{I}, b\mathcal{I}$, we show that a and b are the eigenvalues of $\mathcal{T}$. Suppose $b\mathcal{I} - \mathcal{T} \neq 0$, and let $u = (b\mathcal{I} - \mathcal{T})(v) \neq 0$ for some v. Then $(a\mathcal{I} - \mathcal{T})(u) = 0$, so $\mathcal{T}(u) = au$, and a is an eigenvalue of $\mathcal{T}$. Similarly, b is an eigenvalue because $(b\mathcal{I} - \mathcal{T})(a\mathcal{I} - \mathcal{T}) = 0$. Let λ be an eigenvalue of $\mathcal{T}$ and $\mathcal{T}(w) = \lambda w$ for some $w \neq 0$. As

$$(a\mathcal{I} - \mathcal{T})(b\mathcal{I} - \mathcal{T})(w) = (a\mathcal{I} - \mathcal{T})(bw - \lambda w)$$
$$= (b - \lambda)(a\mathcal{I} - \mathcal{T})(w) = (b - \lambda)(a - \lambda)w = 0,$$

we see $\lambda = a$ or $\lambda = b$. $\mathcal{T}$ has no other eigenvalues than a and b.

(b)   For $v \in V$, write $v = v_1 + v_2$, where

$$v_1 = \frac{1}{b - a}(b\mathcal{I} - \mathcal{T})(v), \quad v_2 = \frac{1}{a - b}(a\mathcal{I} - \mathcal{T})(v).$$

One checks that $v_1 \in \operatorname{Ker}(a\mathcal{I} - \mathcal{T})$ and $v_2 \in \operatorname{Ker}(b\mathcal{I} - \mathcal{T})$. Thus

$$V = \operatorname{Ker}(a\mathcal{I} - \mathcal{T}) + \operatorname{Ker}(b\mathcal{I} - \mathcal{T}).$$

To show it is a direct sum, let $v \in \operatorname{Ker}(a\mathcal{I} - \mathcal{T}) \cap \operatorname{Ker}(b\mathcal{I} - \mathcal{T})$. Then $\mathcal{T}(v) = av$ and $\mathcal{T}(v) = bv$. Since $a \neq b$, $v = 0$.

(c)   Note that $\operatorname{Ker}(a\mathcal{I} - \mathcal{T})$ and $\operatorname{Ker}(b\mathcal{I} - \mathcal{T})$ are invariant subspaces (eigenspaces V_a and V_b) under $\mathcal{T}$. Since $V = \operatorname{Ker}(a\mathcal{I} - \mathcal{T}) + \operatorname{Ker}(b\mathcal{I} - \mathcal{T})$ is a direct sum, choose a basis for $\operatorname{Ker}(a\mathcal{I} - \mathcal{T})$ and a basis for $\operatorname{Ker}(b\mathcal{I} - \mathcal{T})$ to form a basis for V. The restriction of $\mathcal{T}$ on $\operatorname{Ker}(a\mathcal{I} - \mathcal{T})$ is $a\mathcal{I}$ and the restriction of $\mathcal{T}$ on $\operatorname{Ker}(b\mathcal{I} - \mathcal{T})$ is $b\mathcal{I}$. Thus, the matrix representation of $\mathcal{T}$ with respect to the basis is $\operatorname{diag}(aI_k, bI_{n-k})$, where k is some positive integer.

(d) Let $\mathcal{T}^* = \mathcal{T}$. Let $u \in \text{Ker}(a\mathcal{I} - \mathcal{T})$ and $v \in (b\mathcal{I} - \mathcal{T})$. Then $\mathcal{T}(u) = au$ and $\mathcal{T}(v) = bv$. $\langle u, \mathcal{T}(v) \rangle = \langle \mathcal{T}^*(u), v \rangle = \langle \mathcal{T}(u), v \rangle$, implying $b\langle u, v \rangle = a\langle u, v \rangle$. Since $a \neq b$, $\langle u, v \rangle = 0$.

5.108   (a) Let $f, g \in W^\circ$, i.e., $f(x) = g(x) = 0$ for all $x \in W$. Then $(f + kg)(x) = f(x) + kg(x) = 0$ for all $x \in W$ and scalars $k \in \mathbb{F}$. It follows that W° is a subspace of V^*.

(b) For the dimension identity, let the dimension of V be n and let the dimension of W be s. Let $w_1, \ldots, w_s$ be an orthonormal basis of W and extend it to an orthonormal basis of V: $w_1, \ldots, w_s, w_{s+1}, \ldots, w_n$. Consider the dual basis $f_i(w_j) = \delta_{ij}$ of V^*. Since $f_i(w_j) = 0$ for each $j \leq s$ and $i > s$, we see $f_{s+1}, \ldots, f_n \in W^\circ$. We claim that $f_{s+1}, \ldots, f_n$ are linearly independent. Suppose $x_{s+1}f_{s+1} + \cdots + x_n f_n = 0$. Then $(x_{s+1}f_{s+1} + \cdots + x_n f_n)(w_{s+1}) = (x_{s+1}f_{s+1})(w_{s+1}) = 0$, yielding $x_{s+1} = 0$. Inductively, all x_i's are 0. It follows that $f_{s+1}, \ldots, f_n$ comprise a basis for W°. Thus, $\dim W^\circ = n - s = \dim V - \dim W$.

(c) If $f \in W_1^\circ \cap W_2^\circ$, then $f(u) = f(v) = 0$ for all $u \in W_1$ and $v \in W_2$. So $f(u + v) = f(u) + f(v) = 0$ for all $u + v \in W_1 + W_2$, i.e., $f \in (W_1 + W_2)^\circ$. Conversely, if $f \in (W_1 + W_2)^\circ$, then $0 = f(u+0) = f(u) \Rightarrow f \in W_1^\circ$. Similarly, $f \in W_2^\circ$. So $f \in W_1^\circ \cap W_2^\circ$.

5.109   (1) and (2) involve no inner product nor linear transformation. (3) and (5) are for inner product spaces. Answers to (a), (b), and (c):

(1) $W_1 + W_2 = \{w_1 + w_2 \mid w_1 \in W_1, w_2 \in W_2\} = \text{Span}(W_1 \cup W_2)$ is the sum of W_1 and W_2; it is the smallest subspace of V that contains both W_1 and W_2.

(2) $W_1 \oplus W_2$ is the direct sum of W_1 and W_2, where $W_1 \cap W_2 = \{0\}$. Each element $w \in W_1 \oplus W_2$ has a unique decomposition as a sum of elements in W_1 and W_2.

    In the xy-plane, take W_1 and W_2 to be two lines passing through the origin. Unless the lines coincide, the sum is direct.

(3) $W_1 \oplus_\perp W_2$, the orthogonal sum, is the direct sum of W_1 and W_2, where W_1 and W_2 are mutually orthogonal. Note that an orthogonal sum is necessarily a direct sum.

    In the xy-plane, if two lines passing through the origin are perpendicular to each other, then the sum is orthogonal.

(4) $W_1 \oplus_\mathcal{A} W_2$, $\mathcal{A}$-invariant direct sum, is the direct sum of W_1 and W_2, where W_1 and W_2 are $\mathcal{A}$-invariant subspaces.

In the xy-plane, take W_1 to be the x-axis and W_2 to be the line $y = -x$, and let $\mathcal{A}$ be the linear transformation $\mathcal{A} : (x, y) \mapsto (x + y, 0)$. Then $W_1 + W_2$ is an $\mathcal{A}$-invariant direct sum.

(5) $W_1 \oplus_{\mathcal{A}, \perp} W_2$, $\mathcal{A}$-invariant orthogonal direct sum, is the direct sum of W_1 and W_2 which are $\mathcal{A}$-invariant and mutually orthogonal.

In the xy-plane, take W_1 and W_2 to be the x- and y-axes, respectively, and let $\mathcal{A}$ be the linear transformation $\mathcal{A} : (x, y) \mapsto (2x, -y)$. ($\mathcal{A}$ extends the vectors on the x-axis and reflects the ones on the y-axis.) Then W_1 and W_2 are $\mathcal{A}$-invariant and mutually orthogonal. Note that $\mathcal{A}$ is not an orthogonal projection.

(d) A matrix of $\mathcal{A}$ in (4) under some basis has the form $\begin{pmatrix} A_1 & 0 \\ 0 & A_2 \end{pmatrix}$.

(e) $\mathcal{A}$ need not be an orthogonal transformation nor an orthogonal projection. See the example in (5). What we can say is that a matrix of $\mathcal{A}$ with respect to some orthonormal basis has the form $\begin{pmatrix} A_1 & 0 \\ 0 & A_2 \end{pmatrix}$.

Hints and Answers for Chapter 6

6.1   (a) $(1,2) \boxplus (3,4) = (-5,10)$.

      (b) $(0,0) \boxplus (x,y) = (0,0)$.

      (c) $(x,y) \boxplus (y,x) = (0, x^2 + y^2)$.

      (d) $(1,2) \boxplus (x,y) = (x - 2y, 2x + y) \neq (0,0)$ unless $x = y = 0$.

      (e) $(1,0) \boxplus (x,y) = (x,y)$ for all $(x,y) \in V$.

      (f)   By a direct computation.

      (g)   If $(a,b) \boxplus (x,y) = (x,y)$, then $ax - by = x$ and $bx + ay = y$, which hold for all x and y if and only if $a = 1, b = 0$.

      (h)   No. By (e), $(1,0)$ is the zero element for addition $\boxplus$. By (b), $(0,0) \boxplus (x,y) = (0,0) \neq (1,0)$. $-v$ does not exist for $v = (0,0)$.

6.2   We repeatedly make use of the dimension identity and the inequality: $\dim(U \cap W) = \dim U + \dim W - \dim(U + W) \geq \dim U + \dim W - n$.

$$\dim(V_1 \cap \cdots \cap V_m)$$
$$= \dim V_1 + \dim(V_2 \cap \cdots \cap V_m) - \dim(V_1 + V_2 \cap \cdots \cap V_m)$$
$$\geq \dim V_1 + \dim(V_2 \cap \cdots \cap V_m) - n$$
$$\geq \dim V_1 + \dim V_2 + \dim(V_3 \cap \cdots \cap V_m) - 2n$$
$$\vdots$$
$$\geq \dim V_1 + \dim V_2 + \cdots + \dim V_m - (m-1)n > 0.$$

6.3   Let V be a vector space. If $V = \{0\}$, then V has no proper subspace and the statement is trivial. Suppose that V contains more than one element and let $V_1, \ldots, V_m$ be m (finite number) proper subspaces of V such that $V = \cup_{i=1}^{m} V_i$. We may further assume that no V_i is contained in the union of the remaining subspaces V_j's.

Let $v_1 \in V_1 \setminus \cup_{i \neq 1}^{m} V_i$ and $v_2 \in V_2 \setminus \cup_{i \neq 2}^{m} V_i$. For any nonzero scalar x, $u = v_1 + xv_2$ does not lie in $V_1 \cup V_2$ (or it would contradict the assumption that $v_1 \notin V_2$ and $v_2 \notin V_1$). Therefore, $u = v_1 + xv_2 \in \cup_{i>2} V_i$ for any nonzero x. For a nonzero distinct scalar sequence $x_1, x_2, \ldots$, all vectors $u_1 = v_1 + x_1 v_2, u_2 = v_1 + x_2 v_2, \ldots$, will lie in $\cup_{i>2} V_i$. Some two vectors u_p and u_q must lie in the same set, say

V_r, $r > 2$. It follows that $u_p - u_q = (x_p - x_q)v_2 \in V_r$, which implies $v_2 \in V_r$, $r > 2$. This contradicts the selection of $v_2 \in V_2 \setminus \cup_{i \neq 2}^m V_i$.

Note: The statement is false over a finite field. Take $V = \mathbb{Z}_2 \times \mathbb{Z}_2$ over $\mathbb{Z}_2$. Let $W_1 = \{(0,0), (1,0)\}$, $W_2 = \{(0,0), (0,1)\}$, $W_3 = \{(0,0), (1,1)\}$. Then V is the union of the three proper subspaces.

6.4   Assume the vectors of $\mathbb{R}^3$ are in columns. We need a nonsingular matrix T with no zero entries such that $(\beta_1, \beta_2, \beta_3) = (\alpha_1, \alpha_2, \alpha_3)T$. Then

$$v = (\beta_1, \beta_2, \beta_3)\begin{pmatrix} 1 \\ 1 \\ 1 \end{pmatrix} = (\alpha_1, \alpha_2, \alpha_3)T\begin{pmatrix} 1 \\ 1 \\ 1 \end{pmatrix} = (\alpha_1, \alpha_2, \alpha_3)\begin{pmatrix} 1 \\ 2 \\ 3 \end{pmatrix}.$$

Thus, $T\begin{pmatrix} 1 \\ 1 \\ 1 \end{pmatrix} = \begin{pmatrix} 1 \\ 2 \\ 3 \end{pmatrix}$. There are many choices of such a T. We can construct invertible matrices P and Q such that $TQ = P$, where the first column of P is $(1, 2, 3)^t$ and the first column of Q is $(1, 1, 1)^t$. Then $T = PQ^{-1}$. Below is an example of P, Q, and T, respectively,

$$\begin{pmatrix} 1 & 4 & 5 \\ 2 & 6 & -7 \\ 3 & -5 & 8 \end{pmatrix}, \quad \begin{pmatrix} 1 & -2 & 2 \\ 1 & 2 & -1 \\ 1 & 3 & 4 \end{pmatrix}, \quad \frac{1}{23}\begin{pmatrix} -4 & -3 & 30 \\ -15 & 75 & -14 \\ 66 & -8 & 11 \end{pmatrix}.$$

Since $(\beta_1, \beta_2, \beta_3) = (\alpha_1, \alpha_2, \alpha_3)T$ and no entry of T is zero, none of $\beta_1, \beta_2, \beta_3$ is a linear combination of any two vectors of $\alpha_1, \alpha_2, \alpha_3$.

6.5   Let E_{ij} denote the $n \times n$ matrix with the (i, j)-entry 1 and 0 elsewhere.

(a)   No. It is not closed with respect to the scalar multiplication. Take $A = I_n$ and $k = i$. Then kA is not Hermitian.

(b)   Yes. All $E_{st} + E_{ts}$, $s < t$, E_{ss}, and F_{pq}, $p < q$, form a basis, where $F_{pq} = i(E_{pq} - E_{qp})$ is the $n \times n$ matrix with (p, q)-entry i, (q, p)-entry $-i$, and 0 elsewhere. The dimension is n^2.

Take H_2 as an example. Let $a, c, x, y \in \mathbb{R}$, $b = x + iy$. Then

$$\begin{pmatrix} a & b \\ \bar{b} & c \end{pmatrix} = \begin{pmatrix} a & x+yi \\ x-yi & c \end{pmatrix} = a\begin{pmatrix} 1 & 0 \\ 0 & 0 \end{pmatrix} + x\begin{pmatrix} 0 & 1 \\ 1 & 0 \end{pmatrix} + y\begin{pmatrix} 0 & i \\ -i & 0 \end{pmatrix} + c\begin{pmatrix} 0 & 0 \\ 0 & 1 \end{pmatrix}.$$

(c)   Yes. All $E_{st} + E_{ts}$, $s < t$, E_{ss} form a basis. $\dim S_n = \frac{n(n+1)}{2}$.

(d)   Yes. All $E_{st} + E_{ts}$, $s < t$, E_{ss}, iE_{ss}, and G_{pq}, $p < q$, form a basis, where $G_{pq} = i(E_{pq} + E_{qp})$. The dimension is $n(n+1)$.

(e)   No. $H_n \cap S_n$ is the set of all $n \times n$ real symmetric matrices. It is not closed under the scalar multiplication over $\mathbb{C}$.

(f) Yes. $H_n \cap S_n$ is the set of all $n \times n$ real symmetric matrices. $E_{st} + E_{ts}$, $s < t$, E_{ss}, form a basis. The dimension is $\frac{n(n+1)}{2}$.

6.6 (a) Let $\operatorname{Im} R = \operatorname{Im} S$. Since $(PRQ)x = P(R(Qx)) = P(Sy) = (PSQ)(Q^{-1}y)$ for some y, we see $\operatorname{Im}(PRQ) \subseteq \operatorname{Im}(PSQ)$. Similarly, we have $\operatorname{Im}(PSQ) \subseteq \operatorname{Im}(PRQ)$. Thus equality holds.

(b) Let $\operatorname{Im}(PRQ) = \operatorname{Im}(PSQ)$ for some invertible matrices P and Q. Since $Rx = P^{-1}(PRQ)(Q^{-1}x) = P^{-1}(PSQ)z = SQz = S(Qz)$ for some z, we have $\operatorname{Im} R \subseteq \operatorname{Im} S$. Likewise, $\operatorname{Im} S \subseteq \operatorname{Im} R$. It follows that $\operatorname{Im} R = \operatorname{Im} S$.

(c) $\operatorname{Im} R \neq \operatorname{Im}(PRQ)$ when $R = \left(\begin{smallmatrix} 1 & 0 \\ 0 & 0 \end{smallmatrix}\right)$ and $P = Q = \left(\begin{smallmatrix} 0 & 1 \\ 1 & 0 \end{smallmatrix}\right)$.

(d) True. $(RQ)x = R(Qx) \Rightarrow \operatorname{Im}(RQ) \subseteq \operatorname{Im} R$, and also $Ry = (RQ)(Q^{-1}y) \Rightarrow \operatorname{Im} R \subseteq \operatorname{Im}(RQ)$. Thus, $\operatorname{Im} R = \operatorname{Im}(RQ)$.

(e) No. Take $R = \left(\begin{smallmatrix} 1 & 0 \\ 0 & 0 \end{smallmatrix}\right)$ and $P = \left(\begin{smallmatrix} 0 & 1 \\ 1 & 0 \end{smallmatrix}\right)$. Then $\operatorname{Im} R \neq \operatorname{Im}(PR)$.

6.7 We say that A has Property-1 if $|\det X| \leq 1$ for every proper square submatrix X of A. By permutating the columns of A, we can list all 1's and -1's in the first row of A from left to right without changing $|\det A|$ and Property-1. If we multiply the columns that have -1's in the first row by -1, $|\det A|$ will remain the same and Property-1 will hold. Thus, we may assume that A has first row $(1,\ldots,1,0,\ldots 0)$. Now look at the first column, and move the rows with 1's and -1's up, then change the signs of the rows with -1's in the first column so that the first column is $(1,\ldots,1,0,\ldots 0)^t$. Thus, $A = \left(\begin{smallmatrix} 1 & 1 & \cdots \\ 1 & \times & \cdots \\ \cdots & \cdots & \end{smallmatrix}\right)$, where $\times$ is 0, 1, or -1. The $\times$ cannot be -1 unless $(1 \times \ldots)$ is the last row, otherwise there is a proper submatrix X with $|\det X| = 2$, contradicting Property-1. For the same reason, subtracting row 1 from other rows that have 1's in the first column does not result in a -2. Therefore, we may assume that $A = \left(\begin{smallmatrix} 1 & e \\ 0 & A_1 \end{smallmatrix}\right)$, where $e = (1,\ldots,1,0,\ldots 0)$ and A_1 is a square matrix with entries 0, 1, and (or) -1's. Repeat this process until the submatrix $\left(\begin{smallmatrix} 1 & 1 \\ 1 & -1 \end{smallmatrix}\right)$ is obtained at end. It follows that $|\det A| = 2$.

6.8 If A is singular, i.e., $\det A = 0$, then we have nothing to show. If a row or a column of A contains just one nonzero entry with modulus less than or equal to 1, by deleting the row and column of the entry, we go down to lower dimension with our discussion. Note that A contains at most $2n$ nonzero entries. If a column contains three or more nonzero entries, then some column contains at most one nonzero entry which reduces the discussion to the previous case.

So, we assume that $\det A \neq 0$ and each row and column has exactly two nonzero entries (one positive and one negative). We use a cross $\times$ for a nonzero entry (regardless of its sign at this step). Note that $|\det A|$ does not change if we permute rows or columns.

We claim that A can be brought to the following form (with absent entries being 0's) or a block-diagonal matrix with blocks in the form:

$$F = \begin{pmatrix} \times & \times & & \\ & \ddots & \ddots & \\ & & \ddots & \times \\ \times & & & \times \end{pmatrix}.$$

To begin with, there are two nonzero entries in the 1st row. Move the two columns that contain nonzero entries in the 1st row all the way to the left. Move the row that has the other nonzero entry in the 2nd column up to the 2nd row. Now if the $(2,1)$ position is a $\times$, then A contains a submatrix $\left(\begin{smallmatrix} \times & \times \\ \times & \times \end{smallmatrix} \right)$ in form F (or F_2, 2 for the size). Then our discussion reduces to a lower dimension. If the $(2,1)$ position is a 0, then move the column that contains the other nonzero entry in the 2nd row to the 3rd column. Note that the (new) 2nd column already contains two $\times$'s, and the $(3,2)$-entry has to be 0. Move the row that has the nonzero in the 3rd column up as the 3rd row. If $(3,1)$-entry is nonzero, then A contains a submatrix $\left(\begin{smallmatrix} \times & \times & 0 \\ 0 & \times & \times \\ \times & 0 & \times \end{smallmatrix} \right) = F_3$ in form F. Then it is the case of a lower dimension again. Otherwise, the entries to the left of the $(3,3)$-entry are zeros, one finds the other nonzero entry on the 3rd row and moves the column left as the 4th column. Repeat this process until a principal submatrix in form F is obtained (which reduces to lower dimensions) or to the end with the last two rows in the form $\left(\begin{smallmatrix} 0 & 0 & \cdots & \times & \times \\ \times & 0 & \cdots & 0 & \times \end{smallmatrix} \right)$ in which the $(n,1)$ position has to be a $\times$ because each row and column has two $\times$'s. Thus, A is F_n.

We have shown that either A itself is F_n or A is a block-diagonal matrix with blocks in F form (possibly of different sizes).

We now prove that if every row of A contains at most one positive entry and at most one negative entry, then $|\det A| \leq \prod_{i=1}^{n} \max_j |a_{ij}|$. It is sufficient to show that the inequality holds true for $A = F_n$.

Since $|\det A|$ does not change if we change the sign of a row or a column (i.e., times -1), we can multiply a row with a negative diagonal entry by -1 to make it positive, then the other nonzero entry in the row is automatically negative. So we may assume that $A = F_n$ is

$$A = \begin{pmatrix} a_1 & -b_1 & & \\ & \ddots & \ddots & \\ & & \ddots & -b_{n-1} \\ -b_n & & & a_n \end{pmatrix}, \quad a_1, \ldots, a_n, b_1, \ldots, b_n > 0.$$

We complete the proof by computing

$$\begin{aligned} |\det A| &= |a_1 \cdots a_n - b_1 \cdots b_n| \\ &\leq \max\{a_1 \cdots a_n, b_1 \cdots b_n\} \\ &\leq \max\{a_1, b_1\} \cdots \max\{a_n, b_n\} \leq 1. \end{aligned}$$

6.9   We first assume that all a_i's and b_i's are nonzero.

M is skew-symmetric. If n is odd, $\det M = 0$ by the fact that the determinant of an odd-ordered real skew-symmetric matrix is zero. Suppose that n is even (one may start with the case $n = 4$). Write

$$\begin{aligned} M = A_1 &= \begin{pmatrix} 0 & a_1b_2 & a_1b_3 & a_1b_4 & \cdots & a_1b_n \\ -a_1b_2 & 0 & a_2b_3 & a_2b_4 & \cdots & a_2b_n \\ -a_1b_3 & -a_2b_3 & 0 & a_3b_4 & \cdots & a_3b_n \\ -a_1b_4 & -a_2b_4 & -a_3b_4 & 0 & \cdots & a_4b_n \\ \vdots & \vdots & \vdots & \vdots & \ddots & \vdots \\ -a_1b_n & -a_2b_n & -a_3b_n & -a_4b_n & \cdots & 0 \end{pmatrix} \\ &= D_1 \begin{pmatrix} 0 & b_2 & \alpha \\ -b_2 & 0 & a_2\alpha \\ -\alpha^t & -a_2\alpha^t & A_2 \end{pmatrix} D_1. \end{aligned}$$

where $D_1 = \mathrm{diag}(a_1, 1, \ldots, 1)$ and $\alpha = (b_3, b_4, \ldots, b_n)$.

Working on the matrix in the middle, we multiply the first two rows from the left by $(\alpha^t, a_2\alpha^t) \left(\begin{smallmatrix} 0 & b_2 \\ -b_2 & 0 \end{smallmatrix} \right)^{-1}$, then add it to the third block row to get $\left(\begin{smallmatrix} 0 & b_2 & \alpha \\ -b_2 & 0 & a_2\alpha \\ 0 & 0 & \tilde{A}_2 \end{smallmatrix} \right)$, where $\tilde{A}_2 = A_2 - (\alpha^t, a_2\alpha^t) \left(\begin{smallmatrix} 0 & b_2 \\ -b_2 & 0 \end{smallmatrix} \right)^{-1} \left(\begin{smallmatrix} \alpha \\ a_2\alpha \end{smallmatrix} \right)$.

A simple computation reveals $(\alpha^t, a_2\alpha^t) \left(\begin{smallmatrix} 0 & b_2 \\ -b_2 & 0 \end{smallmatrix} \right)^{-1} \left(\begin{smallmatrix} \alpha \\ a_2\alpha \end{smallmatrix} \right) = 0$. Thus, $\det M = \det A_1 = a_1^2 b_2^2 \det A_2$. Notice that A_2 has the same pattern as A_1. The desired identity follows. For $n = 4$, $\det M = (a_1 a_3 b_2 b_4)^2$.

If some a_i's or b_i's are zero, use a continuity argument, in other words, replace these zero a_i's and b_i's with a small positive variable ε. Let $\varepsilon \to 0^+$. Note that the determinant is a continuous function of ε.

6.10 We show by induction. If $n = 2$, it is easy to check. Suppose it is true for $n - 1$, where $n \geq 3$. Let $\det A \neq 0$. A contains no zero row. If every row of A contains two 1's, then one 1 is in P and the other 1 is in Q. By adding all columns of P and all columns of Q times -1 would result in a zero column, and A would be singular. Thus, at least one row of A contains precisely one 1. By expanding $\det A$ along this row, we get $\det A = \pm \det B$, where B is a submatrix of size $n - 1$ obtained from A. By the induction hypothesis, $\det B = \pm 1$.

6.11 Recall the Hadamard inequality, if $P = (p_{ij})$ is an $n \times n$ positive semidefinite matrix, then $\det P \leq \prod_{i=1}^{n} p_{ii}$. Set $P = AA^*$. Then

$$
|\det A|^2 \leq \prod_{i=1}^{n} \left(\sum_{j=1}^{n} |a_{ij}|^2 \right) \leq \prod_{i=1}^{n} (k \max_{j} |a_{ij}|^2) \leq k^n \prod_{i=1}^{n} (\max_{j} |a_{ij}|^2).
$$

6.12 $A \operatorname{adj}(A) = \det A \, I_n$ and $\operatorname{adj}(B) B = \det B \, I_n$. Since $\det A$ and $\det B$ are relatively prime integers, there exist integers p and q such that $p \det A + q \det B = 1$. Take $X = p \operatorname{adj}(A)$ and $Y = q \operatorname{adj}(B)$.

6.13 Let $e_j \in \mathbb{R}^n$ be the column vector with j-th component 1 and 0 elsewhere. Then $M I_n = (Me_1, \ldots, Me_n)$. Thus, Me_j is the j-th column of M. Likewise, $f_i^t M$ is the i-th row of M, where $f_i \in \mathbb{R}^m$ is the column vector with i-th component 1 and 0 elsewhere. Then the submatrix of M lying in the rows $i_1, i_2, \ldots, i_p$ and columns $j_1, j_2, \ldots, j_q$ is XMY, where $X = (f_{i_1}, \ldots, f_{i_p})^t$ and $Y = (e_{j_1}, \ldots, e_{j_q})$.

6.14 Let $r(X - Y) = 1$. $Y^2 \in \{XY, YX\}$ is equivalent to $Y^3 = YXY$ because $Y^3 = YXY \Leftrightarrow Y(X - Y)Y = 0 \Leftrightarrow Y(X - Y) = 0$ or $(X - Y)Y = 0$ (by the inequality $r(ABC) \geq r(AB) + r(BC) - r(B)$).

Note: It is possible that $Y^2 = XY$ but $\neq YX$. Take $X = \begin{pmatrix} 0 & 0 & 0 \\ -1 & 0 & 0 \\ 0 & 1 & 0 \end{pmatrix}$ and $Y = \begin{pmatrix} 0 & 0 & 0 \\ 1 & 0 & 0 \\ 0 & 1 & 0 \end{pmatrix}$. $r(X - Y) = 1$, $Y^3 = YXY$, $Y^2 = XY$, $Y^2 \neq YX$.

6.15 Compute:
$(A - B)^2 = A^2 + B^2 - (AB + BA) = A + B - (AB + BA)$,
$(A + B)^2 = A^2 + B^2 + (AB + BA) = A + B + (AB + BA)$.

Adding and then moving $(A + B)^2$ to the right-hand side, we obtain

$$
(A - B)^2 = 2(A + B) - (A + B)^2 = (A + B)(2I - A - B).
$$

Since $A - B$ is invertible, $A + B$ is invertible. One may check

$$
AB - BA = (I - A - B)(A - B), \quad AB + BA = -(I - A - B)(A + B).
$$

It follows that

$$r(AB - BA) = r\big((I - A - B)(A - B)\big) = r(I - A - B) = r(AB + BA).$$

Note that for $A = B = I$, $\det(A + B) \neq 0$ but $\det(A - B) = 0$.

6.16 $(A + iB)(A - iB) = A^2 + B^2 - i(AB - BA) = (1 - i)(AB - BA)$. Thus, $|\det(A + iB)|^2 = (1 - i)^n \det(AB - BA)$. If $AB - BA = A^2 + B^2$ is nonsingular, then $(1 - i)^n$ need to be real. A simple computation shows that n is a multiple of 4.

6.17 (a) $\begin{pmatrix} A & \beta \\ \alpha & \gamma \end{pmatrix} = \begin{pmatrix} A & 0 \\ 0 & 0 \end{pmatrix} + \begin{pmatrix} 0 & \beta \\ \alpha & \gamma \end{pmatrix}$, and $r\begin{pmatrix} 0 & \beta \\ \alpha & \gamma \end{pmatrix} \leq 2$. So $k = 2$.

(b) If $r(A) = 1$ and M is invertible, then $r(M) = n + 1 \leq 1 + 2$. Thus, $n \leq 2$ and M is 2×2 or 3×3.

(c) $r(A) \geq r(M) - 2 = n + 1 - 2 = n - 1$.

Note that $\begin{pmatrix} 0 & 1 \\ 1 & 1 \end{pmatrix}$ has rank 2, and $\begin{pmatrix} 1 & 1 & 0 \\ 1 & 1 & 1 \\ 0 & 1 & 1 \end{pmatrix}$ has rank 3.

6.18 (a) The system $Ax = b$ has a solution if and only if $r(A) = r(A, b)$. In case there is a solution, say x_0, if $r(A) < n - 1$, then $\dim \mathrm{Ker}\, A = n - r(A) > 1$. Let u and v be two different solutions for $Ax = 0$, i.e., $Au = Av = 0$, $u \neq v$. Then $A(u + x_0) = b$ and $A(v + x_0) = b$. Because $u + x_0 \neq v + x_0$, the solution to $Ax = b$ is not unique. That is, in order that the solution be unique, $r(A) = r(A, b) = n - 1$.

Conversely, if $r(A) = r(A, b) = n - 1$, then $Ax = b$ is solvable. Suppose u and v are both solutions to $Ax = b$ in $(n - 1)$ unknowns. Then $Au = Av = b$, thus $A(u - v) = 0$. Because $r(A) = n - 1$, $\dim \mathrm{Ker}\, A = (n - 1) - (n - 1) = 0$. So $u - v = 0$, i.e., $u = v$. In fact, since the columns of A are linearly independent, $b = Ax$ as a linear combination of the columns of A is unique. Alternatively, there is an $(n - 1) \times n$ matrix P such that $PA = I_{n-1}$, so $Ax = b$ gives $x = Pb$.

(b) If $r(B) \neq r(B, c)$, then $By = c$ has no solution. Suppose $r(B) = r(B, c)$, then $By = c$ has a solution (for n unknowns), say y_0. As $r(B) \leq n - 1$, $\dim \mathrm{Ker}\, B = n - r(B) \geq n - (n - 1) = 1$. Let y_1 and y_2 be two different solutions of $Bz = 0$. Then $B(y_1 + y_0) = B(y_2 + y_0) = c$. Because $y_1 + y_0 \neq y_2 + y_0$, $By = c$ has different solutions. In fact, the solution set is $\{y_0 + z \mid Bz = 0\} = y_0 + \mathrm{Ker}\, B$ (affine space).

6.19 $\mathfrak{R}(XAX^*) = \frac{1}{2}(XAX^* + XA^*X^*) = \frac{1}{2}X(A + A^*)X^* = X\mathfrak{R}(A)X^*$.

If $X = \begin{pmatrix} I_k & 0 \\ 0 & 0 \end{pmatrix}$, then $XAX^* = \begin{pmatrix} A_k & 0 \\ 0 & 0 \end{pmatrix}$, where A_k is the leading $k \times k$ principal submatrix of A. It follows that $\mathfrak{R}(A_k) = (\mathfrak{R}(A))_k$.

6.20   (a) and (b) are by direct verifications. For (c), let $A = H_1 + iK_1 = H_2 + iK_2$, where H_1, H_2, K_1, K_2 are Hermitian matrices. Then $H_1 - H_2 = i(K_2 - K_1)$. The left-hand side is Hermitian; the right-hand side is skew-Hermitian. So $H_1 - H_2 = K_2 - K_1 = 0$, that is, $H_1 = H_2, K_1 = K_2$. Thus, the Cartesian decomposition of A is unique. For (d), by $\sigma_{\max}(X + Y) \leq \sigma_{\max}(X) + \sigma_{\max}(Y)$ (Problem 4.62), we have

$$\begin{aligned} \sigma_{\max}(A) &= \sigma_{\max}(H + iK) \leq \sigma_{\max}(H) + \sigma_{\max}(iK) \\ &= \max_j |\lambda_j(H)| + \max_j |\lambda_j(K)|. \end{aligned}$$

6.21   (a) By a direct computation: $B^*B = BB^*$.

   (b) $B = \frac{1}{2} \begin{pmatrix} H+iK & H-iK \\ H-iK & H+iK \end{pmatrix}$. If B is invertible, let $B^{-1} = \begin{pmatrix} X & P \\ Y & Q \end{pmatrix}$. Then

$$HX + iKX + HY - iKY = 2I, \quad HX - iKX + HY + iKY = 0,$$

   implying $H(X + Y) = I$ and $iK(X - Y) = I$. So H and K are invertible. Conversely, suppose that H^{-1} and K^{-1} exist. One verifies directly that $B^{-1} = \frac{1}{2} \begin{pmatrix} H^{-1}-iK^{-1} & H^{-1}+iK^{-1} \\ H^{-1}+iK^{-1} & H^{-1}-iK^{-1} \end{pmatrix}$.

   (c) Compute B^*B in terms of H and K, then compare to $\begin{pmatrix} I & 0 \\ 0 & I \end{pmatrix}$.

   (d) Obvious. Note that $A \geq 0 \Leftrightarrow \begin{pmatrix} A & A \\ A & A \end{pmatrix} \geq 0$.

6.22   (a) Let $\alpha = (a_1, a_2, a_3)$, $\beta = (b_1, b_2, b_3)$, $\gamma = (c_1, c_2, c_3)$, and let $p_1 = a_1 b_1 c_1$, $p_2 = a_2 b_2 c_2$, $p_3 = a_3 b_3 c_3$. The orthogonality of α, β, and γ implies $a_1 b_1 + a_2 b_2 + a_3 b_3 = 0$, $a_1 c_1 + a_2 c_2 + a_3 c_3 = 0$, $c_1 b_1 + c_2 b_2 + c_3 b_3 = 0$. Rewriting this in matrix form, we have

$$\begin{pmatrix} \frac{1}{a_1} & \frac{1}{a_2} & \frac{1}{a_3} \\ \frac{1}{b_1} & \frac{1}{b_2} & \frac{1}{b_3} \\ \frac{1}{c_1} & \frac{1}{c_2} & \frac{1}{c_3} \end{pmatrix} \begin{pmatrix} p_1 \\ p_2 \\ p_3 \end{pmatrix} = 0.$$

   The rank of the coefficient matrix is at most 2. It follows that the rows $\tilde{\alpha}, \tilde{\beta}$, and $\tilde{\gamma}$ are linearly dependent.

   (b) This is immediate from (a).

   (c) For 2×2, take $U = \frac{1}{\sqrt{2}} \begin{pmatrix} 1 & 1 \\ -1 & 1 \end{pmatrix}$. $\tilde{U} = \sqrt{2} \begin{pmatrix} 1 & 1 \\ -1 & 1 \end{pmatrix}$ is nonsingular.

6.23   The minimal polynomial of the matrix is $\lambda^3 - z\lambda^2 - y\lambda - x$.

6.24   Let $p_A(t) = \det(tI - A)$ be the characteristic polynomial of A. Let $\lambda_1, \ldots, \lambda_k$ be distinct eigenvalues of A with algebraic multiplicities $p_1, \ldots, p_k$, respectively, $p_1 + \cdots + p_k = n$. Then $p_A(t) = \prod_{i=1}^{k}(t - \lambda_i)^{p_i}$.

Let $m_A(t) = \prod_{i=1}^{k}(t - \lambda_i)^{t_i}$ be the minimal polynomial of A.

Let J be a Jordan canonical form of A with Jordan blocks $J_{i_j}(\lambda_i)$ associated with the eigenvalue λ_i, where $J_{i_j}(\lambda_i)$ has size $i_j \times i_j$, $j = 1, \ldots, q_i$, $i_1 + \cdots + i_{q_i} = p_i$, $i = 1, \ldots, k$.

Every factor $(t - \lambda_i)$ of $p_A(t)$ is a factor of $m_A(t)$ and every factor $(t - \lambda_i)$ of $m_A(t)$ appears in $p_A(t)$. The power t_i of $(t - \lambda_i)$ in $m_A(t)$ is the order of the largest Jordan block associated to λ_i. So each $t_i \leq p_i$.

For example, if A is a 6×6 matrix, and if the Jordan blocks of A are $(2), \left(\begin{smallmatrix} 2 & 1 \\ 0 & 2 \end{smallmatrix}\right), \left(\begin{smallmatrix} 2 & 1 & 0 \\ 0 & 2 & 1 \\ 0 & 0 & 2 \end{smallmatrix}\right)$, then $p_A(t) = (t-2)^6$, $m_A(t) = (t-2)^3$. A matrix B with Jordan blocks $\left(\begin{smallmatrix} 2 & 1 & 0 \\ 0 & 2 & 1 \\ 0 & 0 & 2 \end{smallmatrix}\right)$ and $\left(\begin{smallmatrix} 2 & 1 & 0 \\ 0 & 2 & 1 \\ 0 & 0 & 2 \end{smallmatrix}\right)$ has the same characteristic and minimal polynomials as the matrix A.

Note: The characteristic and minimal polynomials are insufficient to determine the similarity of matrices. In contrast, the Jordan canonical form carries much more information about the original matrix. Jordan canonical form best characterizes the matrix up to similarity. It is easy to find the characteristic and minimal polynomials if the Jordan blocks are given. However, determining the Jordan blocks of a given matrix need some effort and diligent work. General theory on Jordan canonical (normal) forms requires to introduce generalized eigenvectors or Smith rational forms, or more advanced tools.

6.25   The minimal polynomial of A is the monic polynomial that annihilates A with the smallest degree. So, the degree of the minimal polynomial of A is equal to the dimension of $\mathrm{Span}\{I, A, A^2, \ldots\}$.

Let $m_A(t) = \prod_{i=1}^{k}(t - \lambda_i)^{t_i}$ be the minimal polynomial of A, where λ_i's are distinct eigenvalues of A. The power t_i of $(t - \lambda_i)$ in $m_A(t)$ is the order of the largest Jordan block associated to λ_i. Thus, the degree of the minimal polynomial of A is $t_1 + \cdots + t_k$ and it is the sum of the orders of the Jordan blocks each of which is the largest in size for the corresponding eigenvalue. For example, if 1 is an eigenvalue of A and if there are two Jordan blocks corresponding to the eigenvalue 1 having sizes, say, 2×2, and 3×3, then the order of the largest Jordan block associated to 1 is 3; $(t - 1)^3$ appears in $m_A(t)$.

6.26

$$\begin{pmatrix} \lambda & 1 & & 0 \\ & \lambda & \ddots & \\ & & \ddots & 1 \\ 0 & & & \lambda \end{pmatrix} = \begin{pmatrix} 0 & & & 1 \\ & & \ddots & \\ 1 & & & 0 \end{pmatrix}\begin{pmatrix} 0 & & & \lambda \\ & & \ddots & 1 \\ & \ddots & & \\ \lambda & 1 & & 0 \end{pmatrix}.$$

6.27 Write $J = \lambda I + N$ and expand $J^k = (\lambda I + N)^k$. Note that for each $i < n$, N^i is obtained from N^{i-1} by pushing the 1's to the right for one line (losing one 1 each time); $N^n = 0$. Therefore, the (i,j)-entry of J^k is given by $\binom{k}{j-i}\lambda^{k-(j-i)}$, with $\lambda^0 = 1$, $\binom{k}{0} = 1$, $\binom{k}{j-i} = 0$ if $j - i < 0$ or $j - i > k$. Alternative proof: use induction on k and compute the (i,j)-entry of $J^k = J^{k-1}J$ by $(J^k)_{ij} = \sum_{t=1}^{n}(J^{k-1})_{it}(J)_{tj}$.

6.28 (a) $\left(\begin{smallmatrix} 1 & 2 \\ 4 & 3 \end{smallmatrix}\right)$ has eigenvalues $5, -1$, and $\left(\begin{smallmatrix} 1 & 3 \\ 4 & 2 \end{smallmatrix}\right)$ has eigenvalues $5, -2$. So the eigenvalues of A and B are $5, 5, -1, -2$.

   (b) Since the eigenvalues of each 2×2 submatrix in (a) are distinct, the two 2×2 matrices are diagonalizable. So A is diagonalizable. If B were diagonalizable, then it would be similar to $\mathrm{diag}(5, 5, -1, -2)$, and the rank of $5I - B$ would be 2. However, $r(5I - B) = 3$. Thus, B cannot be diagonalizable.

6.29 Given linearly independent (eigen-) vectors α and β (for eigenvalue 1), we find a vector γ orthogonal to α and β. Apply the Hilbert–Schmidt process on α, β, γ to get orthonormal vectors α', β', γ'. Let $P = (\alpha', \beta', \gamma')$. Then P is an orthonormal matrix. Set $A = PDP^t$, where $D = \mathrm{diag}(1, 1, -1)$. Then A is real symmetric. $AP = PD$ implies $A\alpha' = \alpha', A\beta' = \beta', A\gamma' = -\gamma'$. Since $V_1 = \mathrm{Span}\{\alpha', \beta'\} = \mathrm{Span}\{\alpha, \beta\}$ and α' and β' are eigenvectors of A associated to 1, we see that β is also an eigenvector of A associated to 1.

For the uniqueness of A (up to the ordering of 1, 1, -1), we show that A is independent of the choices of the bases of $\mathrm{Span}\{\alpha, \beta\}$ and the eigenvectors $w = k\gamma$ belonging to -1, where $k \neq 0$. Let $u, v \in \mathrm{Span}\{\alpha, \beta\}$, where u and v are linearly independent. Let $(u, v) = (\alpha, \beta)T_2$ for a 2×2 invertible matrix T_2. Then

$$\begin{aligned} &(u, v, w)D(u, v, w)^{-1} \\ &= ((\alpha, \beta)T_2, k\gamma)D((\alpha, \beta)T_2, k\gamma)^{-1} \\ &= (\alpha, \beta, \gamma)\left(\begin{smallmatrix} T_2 & 0 \\ 0 & k \end{smallmatrix}\right)\left(\begin{smallmatrix} I_2 & 0 \\ 0 & -1 \end{smallmatrix}\right)\left(\begin{smallmatrix} T_2 & 0 \\ 0 & k \end{smallmatrix}\right)^{-1}(\alpha, \beta, \gamma)^{-1} \\ &= (\alpha, \beta, \gamma)D(\alpha, \beta, \gamma)^{-1}. \end{aligned}$$

Choosing orthonormal $\{u, v, w\}$ (such as $\{\alpha', \beta', \gamma'\}$) ensures the symmetry of A. In fact, A does not depend on the choices of bases.

We can also construct a real symmetric A without using the Hilbert–Schmidt process. $\alpha = (1, 0, 1)^t$ and $\beta = (0, 1, 1)^t$ are not orthogonal. We need to find two orthogonal vectors in the space spanned by α, β. Let $x(1, 0, 1)^t + y(0, 1, 1)^t = (x, y, x + y)^t$. Choose $(x, y, x + y)^t$ so that it is orthogonal to $\alpha = (1, 0, 1)^t$ by setting $x + x + y = 0$ with $x = -1, y = 2$, that is, $\tilde{\beta} = (-1, 2, 1)^t$ is orthogonal to α.

Now choose γ so that it is orthogonal to α and $\tilde{\beta}$ by solving $\gamma^t \alpha = 0$ and $\gamma^t \tilde{\beta} = 0$. It is easy to find such a vector, say $\gamma = (1, 1, -1)^t$. Now normalize α, $\tilde{\beta}$, and γ (by dividing by their lengths, respectively) to obtain orthonormal vectors α', β', and γ'. Let $W = (\alpha', \beta', \gamma')$. Then W is an orthogonal matrix, and $A = WDW^t$ is real symmetric:

$$A = \frac{1}{3} \begin{pmatrix} 1 & -2 & 2 \\ -2 & 1 & 2 \\ 2 & 2 & 1 \end{pmatrix}, \quad W = (\alpha', \beta', \gamma') = \begin{pmatrix} \frac{1}{\sqrt{2}} & \frac{-1}{\sqrt{6}} & \frac{1}{\sqrt{3}} \\ 0 & \frac{2}{\sqrt{6}} & \frac{1}{\sqrt{3}} \\ \frac{1}{\sqrt{2}} & \frac{1}{\sqrt{6}} & \frac{-1}{\sqrt{3}} \end{pmatrix}.$$

One may also check that $A = UDU^{-1} = VDV^{-1} = WDW^t$, where

$$U = (\alpha, \beta, \gamma) = \begin{pmatrix} 1 & 0 & 1 \\ 0 & 1 & 1 \\ 1 & 1 & -1 \end{pmatrix}, \quad V = (\alpha, \tilde{\beta}, \gamma) = \begin{pmatrix} 1 & -1 & 1 \\ 0 & 2 & 1 \\ 1 & 1 & -1 \end{pmatrix}.$$

Note: $V^t V$ is diagonal, but VV^t is not.

See also Problem 3.91 and Problem 5.47.

6.30 Use singular value decomposition. Both equal the spectral norm.

6.31 (i) $|\det A| = |\lambda_1 \cdots \lambda_n| = \sigma_1 \cdots \sigma_n$. So $\sigma_n = 0 \Leftrightarrow \lambda_n = 0$, that is, $\sigma_n = 0$ if and only if A is singular if and only if 0 is an eigenvalue.
(ii) If $Au_1 = \lambda_1 u_1$, where u_1 is a unit vector, then $u_1^* A^* A u_1 = |\lambda_1|^2 \leq \max_{\|x\|=1} x^* A^* A x = \sigma_1^2$. So $\sigma_1 \geq |\lambda_1|$, that is, the spectral norm is greater than or equal to the spectral radius.
(iii) If A is singular, then $\sigma_n = 0$. If A is nonsingular, then $\frac{1}{\sigma_n(A)} = \sigma_1(A^{-1}) \geq |\lambda_1(A^{-1})| = \frac{1}{|\lambda_n(A)|}$. So $\sigma_n \leq |\lambda_n|$. See also Problem 3.77.

6.32 A is real symmetric. So A has three real eigenvalues. A contains a 2×2 principal submatrix $\left(\begin{smallmatrix} a & 1 \\ 1 & a \end{smallmatrix} \right)$ which has eigenvalues $a + 1$ and $a - 1$. By the Cauchy eigenvalue interlacing theorem, $\lambda_{\max}(A) \geq a + 1$ and $\lambda_{\min}(A) \leq a - 1$. 0 (as a 1×1 submatrix) is on the main diagonal of A, so $\lambda_{\max}(A) \geq 0 \geq \lambda_{\min}(A)$. Thus, the claimed inequalities follow.

$\det A = 2xy - a(x^2 + y^2)$. If $a = 1$, A is singular if and only if $y = x$.

6.33 By computation, we have

$$A\hat{+}(B\hat{+}C) = \tfrac{1}{4}(ABC + ACB + BCA + CBA);$$

$$(A\hat{+}B)\hat{+}C = \tfrac{1}{4}(ABC + BAC + CAB + CBA).$$

So $A\hat{+}(B\hat{+}C) \neq (A\hat{+}B)\hat{+}C$ in general. For the trace identity, it is sufficient to observe that $\operatorname{tr}(BAC + CAB) = \operatorname{tr}(ACB + BCA)$.

6.34 $y = -x$. The eigenvalues are $0, 1, 2$, independent of the values of x.

6.35 A has eigenvalues $5, -1$, C has eigenvalues $5, -2$. So A and C are diagonalizable over $\mathbb{Q}$ (rationals). B has eigenvalues $\frac{5\pm\sqrt{33}}{2}$. Since $\sqrt{33}$ is irrational, B is diagonalizable over $\mathbb{R}$ (not over $\mathbb{Q}$). D has eigenvalues $\frac{5\pm\sqrt{15}i}{2}$, so D is diagonalizable over $\mathbb{C}$, not over $\mathbb{R}$ or $\mathbb{Q}$.

6.36 $A(t)$ is normal for any $t \in \mathbb{R}$. So it is unitarily diagonalizable by spectral decomposition. Since $\det(\lambda I - A(t)) = \lambda^2 - 2\lambda \cos t + 1$, $A(t)$ has eigenvalues $\cos t \pm \sqrt{\cos^2 t - 1}$. For $A(t)$ to have real eigenvalues, $\cos t = \pm 1$. Thus, $t = k\pi$ for all integers k, and $A = \pm I_2$.

$A(t)$ is real, it is diagonalizable over $\mathbb{C}$ but not over $\mathbb{R}$ unless $t = k\pi$.

6.37 Case 1: Let A have two real eigenvalues, say, a and b. If $a \neq b$, then A is similar to $\left(\begin{smallmatrix} a & 0 \\ 0 & b \end{smallmatrix}\right)$ over $\mathbb{R}$. If $a = b$, $A = aI_2$ or A is not diagonalizable. In the latter case, A is similar to the Jordan block $\left(\begin{smallmatrix} a & 1 \\ 0 & a \end{smallmatrix}\right)$ over $\mathbb{R}$.

Case 2: Let A have nonreal eigenvalues. Since A is real and 2×2, the two eigenvalues of A are a conjugate pair. Let $\lambda_1 = a + bi$ and $\lambda_2 = a - bi$ be the eigenvalues of A, where $a, b \in \mathbb{R}$, $b \neq 0$. Then A is similar to $\left(\begin{smallmatrix} a+bi & 0 \\ 0 & a-bi \end{smallmatrix}\right)$ over $\mathbb{C}$. Note that $\left(\begin{smallmatrix} a+bi & 0 \\ 0 & a-bi \end{smallmatrix}\right) = B^{-1}\left(\begin{smallmatrix} a & -b \\ b & a \end{smallmatrix}\right)B$, where $B = \frac{1}{\sqrt{2}}\left(\begin{smallmatrix} 1 & 1 \\ -i & i \end{smallmatrix}\right)$, and $\left(\begin{smallmatrix} a & -b \\ b & a \end{smallmatrix}\right) = r\left(\begin{smallmatrix} \cos\theta & -\sin\theta \\ \sin\theta & \cos\theta \end{smallmatrix}\right)$, $r = a^2 + b^2$, $\cos\theta = \frac{a}{a^2+b^2}$, $\sin\theta = \frac{b}{a^2+b^2}$. Now use the fact that if two real matrices are similar over $\mathbb{C}$, then they are similar over $\mathbb{R}$. See Problem 3.69.

6.38 (a) Since B is nonsingular, $B^{-1}AB = \lambda A$. Taking determinants of both sides gives $\lambda^n \det A = \det A$. As $\det A \neq 0$, $\lambda^n = 1$.

(b) $B^{-1}AB = \lambda A$ reveals that A and λA have the same eigenvalues. If all eigenvalues of A are zero, then $A^n = 0$, i.e., A is nilpotent. If A has a nonzero eigenvalue, say a, then $\lambda \neq 0$, and $\lambda a, \lambda^2 a, \ldots$ are all eigenvalues of A. So $\lambda^p = \lambda^q$ for some p, q. Thus, $|\lambda| = 1$.

(c) Since $B^{-1}AB = -A$, we have $\det A = (-1)^n \det A$. If n is odd, then $\det A = 0$ and 0 is an eigenvalue. If n is even, A need not be singular. Take $A = \begin{pmatrix} 0 & 1 \\ -1 & 0 \end{pmatrix}$ and $B = \begin{pmatrix} 0 & 1 \\ 1 & 0 \end{pmatrix}$.

(d) $A = \begin{pmatrix} 1 & 0 & 0 \\ 0 & 0 & 0 \\ 0 & 0 & -1 \end{pmatrix}$, $B = \begin{pmatrix} 0 & 0 & 1 \\ 0 & 1 & 0 \\ 1 & 0 & 0 \end{pmatrix}$, $\lambda = -1$.

6.39 If $A = 0$, we have nothing to prove. Let $A \neq 0$. A cannot be a scalar matrix as $\operatorname{tr} A = 0$. Let v be a nonzero vector so that v and Av are linearly independent. Extend $\{v, Av\}$ to a basis and let P be the matrix with the basis vectors as columns. Then $AP = A(v, Av, \dots) = (Av, A^2 v, \dots) = (v, Av, \dots) \begin{pmatrix} 0 & * & * \\ 1 & * & * \\ 0 & * & * \end{pmatrix} = P \begin{pmatrix} 0 & * \\ * & B \end{pmatrix}$, where B is a matrix of size $n - 1$. Thus, $P^{-1}AP = \begin{pmatrix} 0 & * \\ * & B \end{pmatrix}$, i.e., A is similar to $\begin{pmatrix} 0 & * \\ * & B \end{pmatrix}$. Inductively, A is similar to a matrix with 0's on the main diagonal.

6.40 Let A have a repeated eigenvalue λ with algebraic multiplicity $k > 1$. Consider the Jordan blocks of A associated with λ. If all these Jordan blocks are 1×1, then λI_k commutes with a $k \times k$ nilpotent matrix with $(1, 2)$-entry 1 and all others 0. If a Jordan block of λ has size $p \times p$, $p > 1$, write it as $\lambda I_p + N_p$. Then N_p is nilpotent. With N_p, one can easily construct a nilpotent matrix commuting with A.

Conversely, suppose $AN = NA$, where N is nonzero nilpotent. If A has no repeated eigenvalues, then A is diagonalizable. We may assume that $A = \operatorname{diag}(\lambda_1, \dots, \lambda_n)$. $AN = NA$ implies $\lambda_i n_{ij} = \lambda_j n_{ij}$ for all i, j. Because N is nonzero nilpotent, some off-diagonal entry is nonzero, say, $n_{st} \neq 0$, where $s \neq t$. Then $\lambda_s = \lambda_t$, a contradiction.

6.41 If J is a Jordan block such that $J^2 = 0$, then $J = 0$ or $J = \begin{pmatrix} 0 & 1 \\ 0 & 0 \end{pmatrix}$. By Jordan canonical form of A, we see that the rank of A is equal to the number of Jordan blocks $\begin{pmatrix} 0 & 1 \\ 0 & 0 \end{pmatrix}$. Thus, $r(A) \leq \frac{n}{2}$ if n is even and $r(A) \leq \frac{n-1}{2} < \frac{n}{2}$ if n is odd. (Or $r(A) \leq \lfloor \frac{n}{2} \rfloor$, the floor function.)

6.42 Let λ be an eigenvalue of X. If $X^2 + X + I = 0$, then $\lambda^2 + \lambda + 1 = 0$ which has solutions $\frac{1}{2}(-1 \pm \sqrt{3}i)$. The characteristic polynomial of real X is real. As the pure imaginary roots to a real polynomial occur in conjugate pairs, the characteristic polynomial of X has to have an even degree, i.e., n is even. For $n = 2$, $X = -\frac{1}{2} \begin{pmatrix} 1 & -\sqrt{3} \\ \sqrt{3} & 1 \end{pmatrix}$ satisfies the equation. If $n = 2k$, then use k copies of such blocks on diagonal.

6.43 Note the fact that if $r(AB) = 0$, i.e., $AB = 0$, where $A, B \in M_n(\mathbb{C})$, then the columns of B are contained in $\operatorname{Ker} A$, so $r(A) + r(B) \leq n$.

If $0 = A^5 + A^3 + 2A^2 + 2I = (A^2 + I)(A^3 + 2I)$, then $r(A^2 + I) + r(A^3 + 2I) \leq n$. Now we show the reverse inequality. Since 0 and 2 are not roots of $x^5 + x^3 + 2x^2 + 2$, 0 and 2 are not eigenvalues of A. Thus, A and $2I - A$ are invertible. Using this, we have

$$
\begin{aligned}
r(A^2 + I) + r(A^3 + 2I) &= r\big((-A)(A^2 + I)\big) + r(A^3 + 2I) \\
&= r(-A^3 - A) + r(A^3 + 2I) \\
&\geq r(2I - A) = n.
\end{aligned}
$$

6.44 (a) $Ae = \lambda_1 e$, where $e = (1, 1, \ldots, 1)^t$.

(b) The rank of $\lambda_1 I - A$ is $n - 1$ because the $(n - 1) \times (n - 1)$ submatrix located in the upper-right corner is nonsingular (as its determinant is $(-1)^{n-1}\lambda_n(\lambda_1 + \lambda_2)\cdots(\lambda_1 + \lambda_{n-1}) \neq 0$). Thus, the eigenspace V_{λ_1} associated to λ_1 consists of all multiples of e.

6.45 The characteristic polynomial of a real square matrix is a real-coefficient polynomial, its nonreal complex roots (if any) appear in conjugate pairs including powers. The determinant of a matrix is the product of its eigenvalues, and the eigenvalues of an orthogonal matrix are all 1 in absolute value. So if the order of the matrix is odd, excluding the nonreal eigenvalues (if any), -1 (if any) as an eigenvalue appears even number of times, and there is at least one 1 as eigenvalue.

6.46 A 2×2 matrix M is diagonalizable over $\mathbb{C}$ if and only if $M = kI$ for some scalar $k \in \mathbb{C}$, or M has two distinct eigenvalues.

Let $\Delta = b^2 - 4ac$ be the discriminant of $ax^2 + bx + c = 0$.

A: The characteristic polynomial of A is $p_A(\lambda) = \lambda^2 - 2\lambda + 1 - xy$, with $\Delta = 4xy$, which has two distinct roots (over $\mathbb{C}$) if and only if $xy \neq 0$. Thus, A is diagonalizable if and only if $x = y = 0$ or $xy \neq 0$.

B: $p_B(\lambda) = \lambda^2 - (1 + y)\lambda - x + y$, $\Delta = 4x + (1 - y)^2$. $p_B(\lambda)$ has two distinct roots if and only if $x \neq -\frac{1}{4}(1 - y)^2$. As a result, any matrix in the form $\begin{pmatrix} 1 & 1 \\ -\frac{1}{4}(1-y)^2 & y \end{pmatrix}$, like $\begin{pmatrix} 1 & 1 \\ -1 & -1 \end{pmatrix}$, is not diagonalizable.

C: C is the transpose of B; it is diagonalizable if and only if B is diagonalizable. So, C is diagonalizable if and only if $x \neq -\frac{1}{4}(1 - y)^2$.

D: $p_D(\lambda) = \lambda^2 - (x + y)\lambda + xy - 1$, $\Delta = (x - y)^2 + 4$, $\Delta = 0$ if and only if $y = x \pm 2i$. So D is diagonalizable if and only if $y \neq x \pm 2i$.

If x and y are real, then D is always diagonalizable over $\mathbb{R}$. In order that the four matrices are all diagonalizable (not necessarily through the same invertible matrix) if and only if (x, y) is $(0, 0)$, or the point (x, y) does not lie on the x-axis, y-axis, or the curve (parabola) $4x + (1 - y)^2 = 0$ in the xy-plane. When x and y are both positive, then (x, y) is not on the axes or the parabola. So they are all diagonalizable.

6.47   (a)   Let A have eigenvalues λ_1 and λ_2. If $\lambda_1 \neq \lambda_2$, then A is diagonalizable over $\mathbb{C}$, and it has a square root. If $\lambda_1 = \lambda_2 = \lambda \neq 0$, then A is similar to $\left(\begin{smallmatrix} \lambda & 0 \\ 0 & \lambda \end{smallmatrix}\right)$ or $\left(\begin{smallmatrix} \lambda & 1 \\ 0 & \lambda \end{smallmatrix}\right)$. In the first case, A is a scalar matrix and it has a square root. For the second case, let $\sqrt{\lambda}$ be any (fixed) square root of $\lambda \neq 0$, $x = \frac{1}{2\sqrt{\lambda}}$, and set $X = \left(\begin{smallmatrix} \sqrt{\lambda} & x \\ 0 & \sqrt{\lambda} \end{smallmatrix}\right)$. Then $X^2 = \left(\begin{smallmatrix} \lambda & 1 \\ 0 & \lambda \end{smallmatrix}\right)$. It follows that A has a square root.

If $A \neq 0$ has repeated eigenvalue 0, then A is similar to $\left(\begin{smallmatrix} 0 & 1 \\ 0 & 0 \end{smallmatrix}\right)$. It is easy to check that there does not exist a matrix X such that $X^2 = \left(\begin{smallmatrix} 0 & 1 \\ 0 & 0 \end{smallmatrix}\right)$. Thus, in this case, A has no square root.

     (b)   Let $A = \left(\begin{smallmatrix} a & b \\ c & d \end{smallmatrix}\right)$. We compute $\det(\lambda I - A) = \det\left(\begin{smallmatrix} \lambda - a & -b \\ -c & \lambda - d \end{smallmatrix}\right) = \lambda^2 - (a+d)\lambda + ad - bc$. The eigenvalues of A are $\frac{a+d \pm \sqrt{(a-d)^2 + 4bc}}{2}$. If both eigenvalues of A are 0, then $a + d = 0$ and $ad - bc = 0$. Thus, if $a + d \neq 0$ or $a^2 + bc \neq 0$, then A has a nonzero eigenvalue. So, by (a), A has a square root.

     (c)   Since $\operatorname{tr} A \neq \det A$, A has nonzero eigenvalue(s). Then use (a).

     (d)   Since A is 2×2, we have $A^2 - (\operatorname{tr} A)A + (\det A)I = 0$. Thus $A^2 + (\det A)I = (\operatorname{tr} A)A$. Let $\sqrt{y}$ denote any (fixed) square root of y. If $(\operatorname{tr} A)^2 \neq 4\det A$, then $\operatorname{tr} A + 2\sqrt{\det A} \neq 0$. Verify directly that $X^2 = A$, where $X = \frac{1}{\sqrt{\operatorname{tr} A + 2\sqrt{\det A}}}(A + \sqrt{\det A}\, I)$.

6.48   (a)   The eigenvalues of $A(z)$ are $\pm\sqrt{1 + z^2}$.

     (b)   If $z \neq \pm i$, then $A(z)$ has two distinct eigenvalues, thus $A(z)$ is diagonalizable, in particular, for all $z \in \mathbb{R}$.

     (c)   If $z \neq \pm i$, see (b). If $z = \pm i$, $A(z)$ has a repeated eigenvalue 0, it is not similar to a diagonal matrix. So it is similar to $\left(\begin{smallmatrix} 0 & 1 \\ 0 & 0 \end{smallmatrix}\right)$.

     (d)   For $z = i$, take $u = (2, 2i)^t$ and $v = (1, -i)^t$. Then $A(i)u = 0$ and $A(i)v = u$. So the matrix representation of $A(i)$ as a linear transformation on $\mathbb{C}^2$ under the basis $\{u, v\}$ is $\left(\begin{smallmatrix} 0 & 1 \\ 0 & 0 \end{smallmatrix}\right)$.

6.49   (a)   The eigenvalues of A are $\pm\sqrt{ab}$; those of $B = I + A$ are $1 \pm \sqrt{ab}$.

     (b)   Let $P = \left(\begin{smallmatrix} \sqrt{a} & \sqrt{a} \\ -\sqrt{b} & \sqrt{b} \end{smallmatrix}\right)$. Then P is invertible. One may check that
$$P^{-1}AP = \left(\begin{smallmatrix} -\sqrt{ab} & 0 \\ 0 & \sqrt{ab} \end{smallmatrix}\right) \text{ and } P^{-1}BP = \left(\begin{smallmatrix} 1 - \sqrt{ab} & 0 \\ 0 & 1 + \sqrt{ab} \end{smallmatrix}\right).$$

(c) $A^2 = abI$. $A^{2k} = (ab)^k I$; $A^{2k+1} = (ab)^k A = \begin{pmatrix} 0 & a(ab)^k \\ b(ab)^k & 0 \end{pmatrix}$.

$$B^n = P \begin{pmatrix} 1-\sqrt{ab} & 0 \\ 0 & 1+\sqrt{ab} \end{pmatrix}^n P^{-1} = \frac{1}{2} \begin{pmatrix} \alpha^n+\beta^n & \sqrt{\frac{a}{b}}(\alpha^n-\beta^n) \\ \sqrt{\frac{b}{a}}(\alpha^n-\beta^n) & \alpha^n+\beta^n \end{pmatrix},$$

where $\alpha = 1 + \sqrt{ab}$, $\beta = 1 - \sqrt{ab}$.

6.50 (a) Let λ and μ be the eigenvalues of A. Since A is integral, $\lambda\mu = \det A$ and $\lambda + \mu = \operatorname{tr} A$ are both integral. If λ and μ lie in the open unit disk, then $\lambda = \mu = 0$.

(b) Let λ and μ be the integral eigenvalues of A. Since they are different, we have $QAQ^{-1} = \begin{pmatrix} \lambda & 0 \\ 0 & \mu \end{pmatrix}$, equivalently, $QA = \begin{pmatrix} \lambda & 0 \\ 0 & \mu \end{pmatrix} Q$, for some rational matrix Q. Let m be the common denominator of all entries in Q, then replace Q with $Z = mQ$.

6.51 (a) A is a non-diagonalizable Jordan block over $\mathbb{C}$.

(b) The eigenvalues of A are $1 + 3z$ and $1 - z$. They are different if $z \neq 0$. Thus over the rational field $\mathbb{Q}$, A is diagonalizable. Scaling by the common denominator gives the desired result.

(c) $\begin{pmatrix} 1 & 1 \\ -1 & 1 \end{pmatrix} \begin{pmatrix} 1 & 1 \\ 1 & 1 \end{pmatrix} \begin{pmatrix} 1 & 1 \\ -1 & 1 \end{pmatrix}^{-1} = \begin{pmatrix} 2 & 0 \\ 0 & 0 \end{pmatrix}$.

(d) Let D be an integral matrix. Then the eigenvalues of D are necessarily integers. This is because of the Rational Root Theorem which states that a polynomial of integer coefficients has rational roots only in the form p/q, where p and q are (relative prime) integers, and q divides the leading coefficient. In this case, the leading coefficient of $\det(\lambda I - D)$ is 1. So $q = 1$.

If D is similar to a diagonal rational matrix T, then T (of the eigenvalues of D) is integral. Let $P^{-1}DP = T$, where P is rational, then $DP = PT$. By scaling, we can choose P to be a nonsingular integral matrix. (Note: P^{-1} need not be integral.) The converse is trivial.

(e) E has two real, non-rational eigenvalues $\pm\sqrt{2}$.

(f) F has two complex, non-real eigenvalues $1 \pm i$.

6.52 (a) A has two eigenvalues i and $-i$. So $A = P^{-1} \begin{pmatrix} i & 0 \\ 0 & -i \end{pmatrix} P$ for some complex P. Since $\begin{pmatrix} i & 0 \\ 0 & -i \end{pmatrix} = S^{-1} \begin{pmatrix} 0 & 1 \\ -1 & 0 \end{pmatrix} S$, where $S = \begin{pmatrix} 1 & i \\ i & 1 \end{pmatrix}$, we have $A = Q^{-1} \begin{pmatrix} 0 & 1 \\ -1 & 0 \end{pmatrix} Q$, where $Q = SP$ is a complex matrix. Such a Q can be chosen to be real. See Problem 3.69.

(b) Since $A^2 + I_4$ is singular, i is an eigenvalue of A. Because A is 4×4 and real, we can assume that the eigenvalues of A are

$i, -i, a, b$. tr $A \neq 0$ yields $\{i, -i\} \cap \{a, b\} = \emptyset$ (the empty set). Let $A = P^{-1} \left(\begin{smallmatrix} A_1 & 0 \\ 0 & A_2 \end{smallmatrix} \right) P$, where P is an invertible real matrix, $A_1 = \left(\begin{smallmatrix} 0 & 1 \\ -1 & 0 \end{smallmatrix} \right)$, and A_2 is a real 2×2 matrix. Take $B = P^{-1} \left(\begin{smallmatrix} B_1 & 0 \\ 0 & 0 \end{smallmatrix} \right) P$, where $B = \left(\begin{smallmatrix} 0 & 1 \\ 1 & 0 \end{smallmatrix} \right)$. Then $A_1 B_1 + B_1 A_1 = 0$, and $AB + BA = 0$.

(c) Since i is an eigenvalue of A, $-i$ is also an eigenvalue of A; they are both simple (i.e., their algebraic multiplicities are one). Now use a similar argument to the second part of (b).

6.53 Let $P^{-1} A P = J$, where P is an invertible complex matrix. Since all eigenvalues of A are real, J is real. Let $P = R_1 + i R_2$, where R_1 and R_2 are real. Then $AP = PJ$ implies $AR_1 = R_1 J$ and $AR_2 = R_2 J$. If $R_2 = 0$, then $P = R_1$ is real. Let $R_2 \neq 0$. Consider $T = R_1 + t R_2$, where $t \in \mathbb{R}$. Because $\det(R_1 + t R_2) = 0$ has at most n roots t, we can choose a real t such that $T = R_1 + t R_2$ is nonsingular. Then $AT = AR_1 + t AR_2 = R_1 J + t R_2 J = TJ$. See also Problem 3.69.

6.54 $Au = iu \Rightarrow Au_1 = -u_2, Au_2 = u_1$. Obviously, neither u_1 nor u_2 is zero. u_1 and u_2 are linearly independent over $\mathbb{R}$. Suppose otherwise that $u_1 = r u_2$. Then $u = (r+i) u_2$, implying $Au_2 = i u_2$, contradicting Au_2 being real. Thus $\{u_1, u_2\}$ is a basis of $\mathbb{R}^2$. It is routine to show that v and w are linearly independent, and they also form a basis for $\mathbb{R}^2$. $A(u_1, u_2) = (u_1, u_2) \left(\begin{smallmatrix} 0 & 1 \\ -1 & 0 \end{smallmatrix} \right)$ and $A(v, w) = (v, w) \left(\begin{smallmatrix} 0 & -1 \\ 1 & 0 \end{smallmatrix} \right)$.

6.55 Let A be an idempotent matrix, i.e., $A^2 = A$. Then the eigenvalues of A are 0's and 1's. Consider the Jordan canonical form of A. If J is a Jordan block, then $J^2 = J$. One can easily check that J is 1×1. It follows that A is similar to a matrix in the form $\mathrm{diag}(1, \ldots, 1, 0, \ldots, 0)$, where the number of 1's is the rank of the matrix. Thus, any two $n \times n$ idempotent matrices of the same rank are similar.

6.56 Suppose $X^2 = A$. Since $A^n = 0$, all the eigenvalues of A are zeros, so the eigenvalues of X are all zeros. Let J be the Jordan canonical form of X. Then the diagonal of J contains only zeros. Since $X^2 = A$, A is similar to J^2. But $(J^2)^{n-1} = 0$, contradicting $A^{n-1} \neq 0$.

6.57 $A^* A = \left(\begin{smallmatrix} 1 & x \\ x^* & I_n + x^* x \end{smallmatrix} \right) = I_{n+1} + \left(\begin{smallmatrix} 0 & x \\ x^* & x^* x \end{smallmatrix} \right)$. The rank of $\left(\begin{smallmatrix} 0 & x \\ x^* & x^* x \end{smallmatrix} \right)$ is 0 if $x = 0$ and 2 if $x \neq 0$. If $x = 0$, then $A^* A = I_{n+1}$, the conclusion is trivially true. Let $x \neq 0$. $\left(\begin{smallmatrix} 0 & x \\ x^* & x^* x \end{smallmatrix} \right)$ is Hermitian and has $n - 2$ zero eigenvalues and two nonzero eigenvalues, one positive (say p) and one negative (say $-q$). So the eigenvalues of $A^* A$ are $1+p, 1, \ldots, 1, 1-q$. Since $\det(A^* A) = |\det A|^2 = 1$, $\lambda_{\max} \cdot \lambda_{\min} = (1+p)(1-q) = 1$.

To compute the eigenvalues of A^*A, we consider AA^* which has the same set of eigenvalues as A^*A. (Note: xx^* as a number is easier to handle than x^*x as a matrix.) $AA^* = I_{n+1} + \begin{pmatrix} xx^* & x \\ x^* & 0 \end{pmatrix}$. Let $a = xx^*$.

$$\det \begin{pmatrix} \lambda - a & -x \\ -x^* & \lambda I_n \end{pmatrix} = \det \begin{pmatrix} \lambda - a - \frac{1}{\lambda}a & -x \\ 0 & \lambda I_n \end{pmatrix}, \quad \lambda \neq 0.$$

So the nonzero eigenvalues of $\begin{pmatrix} xx^* & x \\ x^* & 0 \end{pmatrix}$ satisfy $\lambda - a - \frac{1}{\lambda}a = 0$, i.e., $\lambda^2 - \lambda a - a = 0$. It follows that the eigenvalues of A^*A are

$$\lambda_{\max} = 1 + \frac{a + \sqrt{a^2 + 4a}}{2}, \quad 1, \ldots, 1, \quad \lambda_{\min} = 1 + \frac{a - \sqrt{a^2 + 4a}}{2}.$$

By the above formula, a simple computation also gives $\lambda_{\max} \cdot \lambda_{\min} = 1$. Note: The problem can be stated in terms of the singular values of A.

6.58 (a) Since $Y = I - X$, $XY = X - X^2 = (I - X)X = YX$.

(b) By (a), $(I + X)^{-1}(I + Y)^{-1} = (I + Y)^{-1}(I + X)^{-1}$, equivalently, $(I + Y)(I + X) = (I + X)(I + Y)$, which implies $XY = YX$.

(c) By (b), $XY = YX$, and since X and Y are both positive definite, they are simultaneously unitarily diagonalizable, i.e., $X = U^*D_1U$ and $Y = U^*D_2U$, where U is unitary, D_1 and D_2 are positive diagonal. Note that for positive scalars x and y, the identity $(1 + x)^{-1} + (1 + y)^{-1} = 1$ yields $x = y^{-1}$. Thus, $X = Y^{-1}$.

6.59 Note that for permutation matrices P and Q, the singular value identity $\sigma_i(PBQ) = \sqrt{\lambda_i(Q^*B^*P^*PBQ)} = \sqrt{\lambda_i(B^*B)} = \sigma_i(B)$ holds. Through permutations, we can write $PBQ = \begin{pmatrix} A & 0 \\ 0 & 0 \end{pmatrix}$. Then the singular values of B are the singular values of A plus some zeros.

The answer to the second part is negative. Take $A = \begin{pmatrix} 1 & 1 \\ 1 & 1 \end{pmatrix}$, and insert some zeros "randomly" to get, say, $B = \begin{pmatrix} 1 & 0 & 1 \\ 1 & 1 & 0 \end{pmatrix}$. Then A and B have different singular values: $\sigma(A) = \{2, 0\}$, $\sigma(B) = \{\sqrt{3}, 1, 0\}$.

6.60 (a) Every square matrix is similar to its transpose (Problem 3.68). $\begin{pmatrix} 0 & A \\ 0 & 0 \end{pmatrix}$ is similar to $\begin{pmatrix} 0 & 0 \\ A^t & 0 \end{pmatrix}$, which is permutation-similar to $\begin{pmatrix} 0 & A^t \\ 0 & 0 \end{pmatrix}$.

(b) By Jordan canonical form, B and C have the same Jordan blocks.

(c) If the eigenvalues of A are all real, then the Jordan canonical form J of A is a real matrix. Let $P^{-1}AP = J$. Then $P^*A^*(P^*)^{-1} = J^* = J^t$. Since J and J^t are similar, A and A^* are similar.

Note: A need not be similar to A^* in general. Take $A = iI_2$.

(d) Let $A = \begin{pmatrix} i & 1 & 0 & 0 \\ 0 & i & 0 & 0 \\ 0 & 0 & -i & 0 \\ 0 & 0 & 0 & -i \end{pmatrix}$. Then A and A^* have the same eigenvalues, but they are not similar as they have different Jordan blocks.

(e) Use singular value decomposition.

(f) No. Take $A = \begin{pmatrix} 1 & i \\ 1 & i \end{pmatrix}$. Then $AA^t = 0$, $A^t A \neq 0$.

6.61 (a) No. Take $A = (1)$, $B = (0)$.

(b) No. Take $A = \begin{pmatrix} 1 & 0 \\ 0 & 0 \end{pmatrix}$ and $B = \begin{pmatrix} 0 & 1 \\ 0 & 0 \end{pmatrix}$.

(c) Yes. Permutation-similar: $\begin{pmatrix} 0 & I \\ I & 0 \end{pmatrix} \begin{pmatrix} A & B \\ 0 & 0 \end{pmatrix} \begin{pmatrix} 0 & I \\ I & 0 \end{pmatrix} = \begin{pmatrix} 0 & 0 \\ B & A \end{pmatrix}$.

(d) No. Take $A = (1), B = (1)$.

(e) Yes. A matrix is always similar to its transpose.

(f) No. Take $A = (i), B = (0)$. Then consider the eigenvalues.

If A is nonsingular, in addition to (c) and (e), the block matrix in (b) is also similar to $\begin{pmatrix} A & B \\ 0 & 0 \end{pmatrix}$ because $\begin{pmatrix} A & B \\ 0 & 0 \end{pmatrix}$ is similar to $\begin{pmatrix} A & 0 \\ 0 & 0 \end{pmatrix}$ as

$$\begin{pmatrix} I & A^{-1}B \\ 0 & I \end{pmatrix} \begin{pmatrix} A & B \\ 0 & 0 \end{pmatrix} \begin{pmatrix} I & -A^{-1}B \\ 0 & I \end{pmatrix} = \begin{pmatrix} A & 0 \\ 0 & 0 \end{pmatrix}.$$

Similarly, $\begin{pmatrix} A & 0 \\ B & 0 \end{pmatrix}$ is similar to $\begin{pmatrix} A & 0 \\ 0 & 0 \end{pmatrix}$. So $\begin{pmatrix} A & 0 \\ B & 0 \end{pmatrix}$ and $\begin{pmatrix} A & B \\ 0 & 0 \end{pmatrix}$ are similar.

6.62 (a) $P^{-1}MP = N$, where $P = \begin{pmatrix} 0 & I \\ I & 0 \end{pmatrix}$.

(b) Similar to (a).

(c) $M = \begin{pmatrix} A & 0 \\ 0 & B \end{pmatrix} \begin{pmatrix} I & I \\ I & I \end{pmatrix} = \begin{pmatrix} A & 0 \\ 0 & B \end{pmatrix} \begin{pmatrix} I \\ I \end{pmatrix} (I, I)$ has the same nonzero eigenvalues as $(I, I) \begin{pmatrix} A & 0 \\ 0 & B \end{pmatrix} \begin{pmatrix} I \\ I \end{pmatrix} = A + B$. The others are similar.

(d) $A = \begin{pmatrix} 0 & 1 \\ 0 & 0 \end{pmatrix}$, $B = \begin{pmatrix} 0 & -1 \\ 0 & 1 \end{pmatrix}$. Then $r(M) = 1$, $r(R) = 2$.

(e) $M = \begin{pmatrix} A & 0 \\ 0 & B \end{pmatrix} \begin{pmatrix} I & I \\ I & I \end{pmatrix}$. $R = \begin{pmatrix} I & I \\ I & I \end{pmatrix} \begin{pmatrix} A & 0 \\ 0 & B \end{pmatrix} = \begin{pmatrix} A & 0 \\ 0 & B \end{pmatrix}^{-1} M \begin{pmatrix} A & 0 \\ 0 & B \end{pmatrix}$.

(f) If $n = 1$, $\begin{pmatrix} a & a \\ b & b \end{pmatrix}$ and $\begin{pmatrix} a & b \\ a & b \end{pmatrix} = \begin{pmatrix} a & a \\ b & b \end{pmatrix}^t$ are similar for all scalars a, b.

(g) $M^t = \begin{pmatrix} A^t & B^t \\ A^t & B^t \end{pmatrix} = \begin{pmatrix} A & B \\ A & B \end{pmatrix} = R$. By the fact that a square matrix is similar to its transpose, M is similar to $M^t = R$.

6.63 Observe that if X is nonsingular, then $\begin{pmatrix} X & Y \\ 0 & 0 \end{pmatrix}$ is similar to $\begin{pmatrix} X & 0 \\ 0 & 0 \end{pmatrix}$ because $\begin{pmatrix} I & XY^{-1} \\ 0 & I \end{pmatrix} \begin{pmatrix} X & Y \\ 0 & 0 \end{pmatrix} \begin{pmatrix} I & -X^{-1}Y \\ 0 & I \end{pmatrix} = \begin{pmatrix} X & 0 \\ 0 & 0 \end{pmatrix}$. Likewise, $\begin{pmatrix} X & 0 \\ Y & 0 \end{pmatrix}$ is similar to $\begin{pmatrix} X & 0 \\ 0 & 0 \end{pmatrix}$. Thus, $\begin{pmatrix} X & Y \\ 0 & 0 \end{pmatrix}$ and $\begin{pmatrix} X & 0 \\ Y & 0 \end{pmatrix}$ are similar.

$AB = \begin{pmatrix} PT & 0 \\ RT & 0 \end{pmatrix}$ is similar to $\begin{pmatrix} PT & 0 \\ 0 & 0 \end{pmatrix}$, $BA = \begin{pmatrix} TP & TQ \\ 0 & 0 \end{pmatrix}$ is similar to $\begin{pmatrix} TP & 0 \\ 0 & 0 \end{pmatrix}$. So AB and BA are similar as $PT = P(TP)P^{-1}$.

6.64 Let $A = UDV$ be a singular value decomposition of A. Then $A = UDV = (UDV)(V^*DU^*)(U\tilde{D}V) = AA^*B$, where $B = U\tilde{D}V$ and $\tilde{D}$ is such a diagonal matrix (by taking the reciprocals of the nonzero diagonal entries of D) that $DD\tilde{D} = D$. Alternatively, use the fact that $\text{Im}(AA^*) = \text{Im}\,A$. (See Problem 2.102 and Problem 3.103.)

By the above fact, there exists a matrix E such that $A^* = A^*AE$. Thus $A^*C = A^*AEC$. $\begin{pmatrix} A^*A & A^*C \\ 0 & 0 \end{pmatrix}$ and $\begin{pmatrix} A^*A & 0 \\ 0 & 0 \end{pmatrix}$ are similar because

$$\begin{pmatrix} A^*A & 0 \\ 0 & 0 \end{pmatrix} = \begin{pmatrix} I & EC \\ 0 & I \end{pmatrix}\begin{pmatrix} A^*A & A^*C \\ 0 & 0 \end{pmatrix}\begin{pmatrix} I & -EC \\ 0 & I \end{pmatrix}.$$

6.65 (a) A is similar to $B = A^t$, but AA^t is not similar to A^tA in general. Take $A = \begin{pmatrix} 1 & i \\ 0 & 0 \end{pmatrix}$. Then $AA^t = 0$ but $A^tA \neq 0$.

(b) If AB and BA are similar, then obviously $r(AB)^k = r(BA)^k$, $k = 1, 2, \ldots, n$. For the converse, we show that AB and BA have the same (upper-triangular) Jordan canonical form (up to permutation of the Jordan blocks). It is known that AB and BA have the same eigenvalues. The condition that $r(AB)^k = r(BA)^k$, $k = 1, 2, \ldots, n$, implies that AB and BA have the same Jordan blocks corresponding to the zero eigenvalues (if any).

We now show that AB and BA have the same Jordan blocks corresponding to the nonzero eigenvalues (if any). Note that $\begin{pmatrix} AB & A \\ 0 & 0 \end{pmatrix} = P^{-1}\begin{pmatrix} 0 & A \\ 0 & BA \end{pmatrix}P$, where $P = \begin{pmatrix} I & 0 \\ B & I \end{pmatrix}$. Thus, $\begin{pmatrix} \lambda I - AB & -A \\ 0 & \lambda I \end{pmatrix}$ is similar to $\begin{pmatrix} \lambda I & -A \\ 0 & \lambda I - BA \end{pmatrix}$. One can verify the fact that $r\begin{pmatrix} X & A \\ 0 & Y \end{pmatrix}^k = n + r(Y^k)$, where X is an $n \times n$ invertible matrix. It follows that for any $\lambda \neq 0$ and $k = 1, 2, \ldots, n$

$$\begin{aligned} n + r(\lambda I - AB)^k &= r\begin{pmatrix} \lambda I - AB & -A \\ 0 & \lambda I \end{pmatrix}^k \\ &= r\begin{pmatrix} \lambda I & -A \\ 0 & \lambda I - BA \end{pmatrix}^k \\ &= n + r(\lambda I - BA)^k. \end{aligned}$$

Consequently, $r(\lambda I - AB)^k = r(\lambda I - BA)^k$. Setting λ to be each of the eigenvalues of AB (or BA), we see that AB and BA have the same Jordan blocks for nonzero eigenvalues.

(c) (i) $A\bar{A}$ is similar to $\bar{A}A$ because $r(A\bar{A})^k = r\big(\overline{(A\bar{A})^k}\big) = r(\bar{A}A)^k$.
(ii) AA^* is similar to A^*A as $r(AA^*)^k = r(A^*A)^k$. (iii) $\bar{A}A^t = \bar{A}(\bar{A})^*$ is similar to $A^t\bar{A} = (\bar{A})^*\bar{A}$. (iv) No. See (a).

(d) For Hermitian A and B, $r(AB)^k = r\big((AB)^k\big)^* = r(B^*A^*)^k = r(BA)^k$. By part (a), AB and BA are similar. See Problem 4.17.

6.66   Let $A = UDV^*$ be a singular value decomposition of A, where U and V are unitary, and D is nonnegative diagonal. Then $B = VDV^*$ and $C = UDU^*$. Let x_i be the i-th row of matrix X, $i = 1, 2, \ldots, n$. Then

$$a_{ij} = u_i Dv_j^*, \quad b_{ij} = v_i Dv_j^*, \quad c_{ij} = u_i Du_j^*.$$

By the Cauchy–Schwartz inequality,

$$
\begin{aligned}
|a_{ij}|^2 &= |u_i D^{\frac{1}{2}} D^{\frac{1}{2}} v_j^*|^2 \\
&= |(u_i D^{\frac{1}{2}})(v_j D^{\frac{1}{2}})^*|^2 \\
&\leq (u_i Du_i^*)(v_j Dv_j^*) = c_{ii}b_{jj}.
\end{aligned}
$$

The answer for the last question is negative by taking $A = \left(\begin{smallmatrix} 0 & 1 \\ 0 & 0 \end{smallmatrix}\right)$.

6.67   (a) Let $p(x)$ be the characteristic polynomial of A. By the Cayley–Hamilton theorem, $p(A) = 0$, so $I, A, \ldots, A^n$ are linearly dependent and A^n is a linear combination of $I, A, \ldots, A^{n-1}$. Thus, for $k \geq n$, $A^k \in \operatorname{Span}\{I, A, \ldots, A^{n-1}\} = \mathbb{P}_n[A]$ and $\mathbb{P}_k[A] = \mathbb{P}_n[A]$. It follows that $\mathbb{P}_1[A] \subseteq \mathbb{P}_2[A] \subseteq \cdots \subseteq \mathbb{P}_n[A] = \mathbb{P}_{n+1}[A] = \cdots$

(b) The smallest number k for which $\mathbb{P}_k[A] = \mathbb{P}_n[A]$ is the degree of the minimal polynomial of A. If k_0 denotes the degree, then

$$\mathbb{P}_1[A] \subset \cdots \subset \mathbb{P}_{k_0}[A] = \mathbb{P}_{k_0+1}[A] = \cdots = \mathbb{P}_n[A] = \mathbb{P}_{n+1}[A] = \cdots$$

(c) The minimal polynomial (which is the same as the characteristic polynomial in this case) of A is $p(x) = (x - \lambda_1) \cdots (x - \lambda_n)$. So $\dim \mathbb{P}_n[A] = n$ and $I, A, \ldots, A^{n-1}$ form a basis of $\mathbb{P}_n[A]$.

(d) The minimal polynomials of A and B are $x^2(x-2)$ and $(x-2)^2$, respectively. So, $\mathbb{P}_3[A] = \operatorname{Span}\{I, A, A^2\}$, $\mathbb{P}_3[B] = \operatorname{Span}\{I, B\}$.

6.68   Since $(f_1(x), f_2(x)) = 1$, there exist $g_1(x), g_2(x) \in \mathbb{P}[x]$ such that $g_1(x)f_1(x) + g_2(x)f_2(x) = 1$. So $g_1(A)f_1(A) + g_2(A)f_2(A) = I$. For $v \in V$, let $v_1 = g_2(A)f_2(A)X$ and $v_2 = g_1(A)f_1(A)X$. Then

$$v = Iv = g_1(A)f_1(A)v + g_2(A)f_2(A)v = v_2 + v_1.$$

Since $v \in V = \mathrm{Ker} f(A)$, $f(A)v = 0$, $f_1(A)v_1 = g_2(A)f(A)v = 0$, implying $v_1 \in V_1$. For the same reason, $v_2 \in V_2$. Thus, $V = V_1 + V_2$. If $v \in V_1 \cap V_2$, then v_1 and v_2 are both equal to 0. Thus, $V_1 \cap V_2 = \{0\}$. It follows that $V = V_1 \oplus V_2$.

If $w \in V_1$, then $f_1(A)Aw = Af_1(A)w = A0 = 0$, implying $Aw \in V_1$. So, V_1 is invariant under A. Similarly, V_2 is invariant under A.

Note: Consider the characteristic polynomial $p(x)$ of A and factor it as $p(x) = x^k q(x)$, where k is a nonnegative integer and $q(0) \neq 0$. Taking bases of $V_1 = \mathrm{Ker} A^k$ and $V_2 = \mathrm{Ker} q(A)$, we see that A is similar to a matrix in the form $\begin{pmatrix} E & 0 \\ 0 & N \end{pmatrix}$, where E is nilpotent and N is nonsingular. (Note: N and E can be absent.)

6.69 Since $f_1(x)$ and $f_2(x)$ are relatively prime, there exist polynomials $s(x)$ and $t(x)$ such that $s(x)f_1(x) + t(x)f_2(x) = 1$. This yields $s(A)f_1(A) + t(A)f_2(A) = I_n$. Let $M = \begin{pmatrix} f_1(A) & 0 \\ 0 & f_2(A) \end{pmatrix}$. Then $r(M) = r(f_1(A)) + r(f_2(A))$. On the other hand, by elementary operations,

$$\begin{pmatrix} f_1(A) & 0 \\ 0 & f_2(A) \end{pmatrix} \rightarrow \begin{pmatrix} f_1(A) & 0 \\ s(A)f_1(A) & f_2(A) \end{pmatrix}$$

$$\rightarrow \begin{pmatrix} f_1(A) & 0 \\ s(A)f_1(A) + t(A)f_2(A) & f_2(A) \end{pmatrix}$$

$$= \begin{pmatrix} f_1(A) & 0 \\ I_n & f_2(A) \end{pmatrix}$$

$$\rightarrow \begin{pmatrix} 0 & -f_1(A)f_2(A) \\ I_n & f_2(A) \end{pmatrix} = \begin{pmatrix} 0 & -f(A) \\ I_n & f_2(A) \end{pmatrix}.$$

The desired rank identity follows at once.

6.70 (a) $\mathcal{A}\begin{pmatrix} -2 \\ 3 \end{pmatrix} = \mathcal{A}\left(\begin{pmatrix} 1 \\ 0 \end{pmatrix} - 3\begin{pmatrix} 1 \\ -1 \end{pmatrix}\right) = \begin{pmatrix} 1 \\ 4 \end{pmatrix} - 3\begin{pmatrix} -1 \\ 1 \end{pmatrix} = \begin{pmatrix} 4 \\ 1 \end{pmatrix}$.

(b) $\begin{pmatrix} x \\ y \end{pmatrix} = (x+y)\begin{pmatrix} 1 \\ 0 \end{pmatrix} - y\begin{pmatrix} 1 \\ -1 \end{pmatrix}$. So $\mathcal{A}\begin{pmatrix} x \\ y \end{pmatrix} = \begin{pmatrix} x+2y \\ 4x+3y \end{pmatrix} = \begin{pmatrix} 1 & 2 \\ 4 & 3 \end{pmatrix}\begin{pmatrix} x \\ y \end{pmatrix}$.

(c) $A_e = \begin{pmatrix} 1 & 2 \\ 4 & 3 \end{pmatrix}$.

(d) $\mathcal{A}(\alpha_1) = 5\alpha_1 - 4\alpha_2$, $\mathcal{A}(\alpha_2) = 0\alpha_1 - \alpha_2$. So $A_\alpha = \begin{pmatrix} 5 & 0 \\ -4 & -1 \end{pmatrix}$.

(e) $P = \begin{pmatrix} 1 & 1 \\ 0 & -1 \end{pmatrix}$. $P^{-1}A_e P = A_\alpha$.

(f) $5, -1$.

6.71 Linearity of $\mathcal{A}$ is easily checked. The matrices of $\mathcal{A}$ under the bases are

$$\begin{pmatrix} a & 0 & b & 0 \\ 0 & a & 0 & b \\ c & 0 & d & 0 \\ 0 & c & 0 & d \end{pmatrix}, \quad \begin{pmatrix} a & b & 0 & 0 \\ c & d & 0 & 0 \\ 0 & 0 & a & b \\ 0 & 0 & c & d \end{pmatrix}.$$

6.72 $\mathcal{L}$ is linear: $\mathcal{L}(X + kY) = A(X + kY)B = AXB + kAYB = \mathcal{L}(X) + k\mathcal{L}(Y)$. Let $X = \begin{pmatrix} a & b \\ c & d \end{pmatrix}$. Then $\mathcal{L}(X) = AXB = \begin{pmatrix} 3a+4c & 3b+4d \\ a+2c & b+2d \end{pmatrix}$. Since

$$\begin{pmatrix} 3a+4c & 3b+4d \\ a+2c & b+2d \end{pmatrix} = (3a+4c)E_{11}+(a+2c)E_{21}+(3b+4d)E_{12}+(b+2d)E_{22},$$

we obtain the matrix representation of $\mathcal{L}$ with respect to the (ordered) basis $\{E_{11}, E_{21}, E_{12}, E_{22}\}$ of $M_2(\mathbb{R})$:

$$L = \begin{pmatrix} 3 & 4 & 0 & 0 \\ 1 & 2 & 0 & 0 \\ 0 & 0 & 3 & 4 \\ 0 & 0 & 1 & 2 \end{pmatrix}.$$

So $\operatorname{tr} L = 10$, $\det L = 4$, and the eigenvalues are $\frac{5 \pm \sqrt{17}}{2}$, $\frac{5 \pm \sqrt{17}}{2}$. Since the matrices of $\mathcal{L}$ under different bases are similar, the trace, determinant, and eigenvalues are independent of the choices of bases.

6.73   (a)   It is straightforward to show that $\mathcal{L}$ is a linear map.

(b) $k = 1$: $\operatorname{Im} \mathcal{L}$ is the set of all real symmetric matrices.

$k = -1$: $\operatorname{Im} \mathcal{L}$ is the set of all real skew-symmetric matrices.

$k = 2$: $\operatorname{Im} \mathcal{L} = M_n(\mathbb{R})$ which is proven directly by showing that $\mathcal{L}$ is onto. Let $A = \begin{pmatrix} a & b \\ c & d \end{pmatrix} \in M_2(\mathbb{R})$ and let $X = \begin{pmatrix} x_1 & x_2 \\ x_3 & x_4 \end{pmatrix}$ be such that $X + 2X^t = A$. Then $x_1 = \frac{1}{3}a, x_4 = \frac{1}{3}d$, $x_2 + 2x_3 = b$, and $2x_2 + x_3 = c$. X exists and is uniquely determined by A.

(c)   If $n = 1$, $\mathcal{L}(X) = (k+1)X$. $\mathcal{L}$ is invertible if and only if $k \neq -1$. Let $n \geq 2$. If $k = \pm 1$, it is easy to find a nonzero X such that $\mathcal{L}(X) = 0$, namely, $\mathcal{L}$ is not invertible. We claim that if $k \neq \pm 1$ then $\mathcal{L}$ is invertible. It suffices to show that $\mathcal{L}$ maps only 0 to 0. Let $\mathcal{L}(X) = X + kX^t = 0$. Then $X = -kX^t$ and $X^t = -kX$. Thus, $X = k^2 X$. Since $k^2 \neq 1$, we have $X = 0$.

(d) $n = 2, k = 1$. Take the standard basis $\{E_{11}, E_{12}, E_{21}, E_{22}\}$. Then

$$\mathcal{L}(E_{11}) = E_{11} + E_{11}^t = \begin{pmatrix} 2 & 0 \\ 0 & 0 \end{pmatrix} = 2E_{11} + 0E_{12} + 0E_{21} + 0E_{22}$$
$$\mathcal{L}(E_{12}) = E_{12} + E_{12}^t = \begin{pmatrix} 0 & 1 \\ 1 & 0 \end{pmatrix} = 0E_{11} + E_{12} + E_{21} + 0E_{22}$$
$$\mathcal{L}(E_{21}) = E_{21} + E_{21}^t = \begin{pmatrix} 0 & 1 \\ 1 & 0 \end{pmatrix} = 0E_{11} + E_{12} + E_{21} + 0E_{22}$$
$$\mathcal{L}(E_{22}) = E_{22} + E_{22}^t = \begin{pmatrix} 0 & 0 \\ 0 & 2 \end{pmatrix} = 0E_{11} + 0E_{12} + 0E_{21} + 2E_{22}.$$

So the matrix of $\mathcal{L}$ with respect to the standard basis is the following L_1. Similarly, one finds matrix L_2 when $n = 2, k = 2$.

$$L_1 = \begin{pmatrix} 2 & 0 & 0 & 0 \\ 0 & 1 & 1 & 0 \\ 0 & 1 & 1 & 0 \\ 0 & 0 & 0 & 2 \end{pmatrix}, \quad L_2 = \begin{pmatrix} 3 & 0 & 0 & 0 \\ 0 & 1 & 2 & 0 \\ 0 & 2 & 1 & 0 \\ 0 & 0 & 0 & 3 \end{pmatrix}.$$

Note that L_1 is singular, while L_2 is nonsingular. Moreover, one may also identify a matrix in $M_2(\mathbb{R})$ with a vector in $\mathbb{R}^4$ and convert the problem to a problem in $\mathbb{R}^4$. For instance, $E_{12} \sim (0,1,0,0)^t$, $E_{21} \sim (0,0,1,0)^t$, and $\left(\begin{smallmatrix} 0 & 0 \\ 0 & 2 \end{smallmatrix}\right) \sim (0,0,0,2)^t$, etc.

(e) $\mathcal{L}$ is self-adjoint because, for the standard inner product of $M_n(\mathbb{R})$,

$$
\begin{aligned}
\langle \mathcal{L}(X), Y \rangle &= \operatorname{tr}(Y^t \mathcal{L}(X)) = \operatorname{tr}(Y^t(X + kX^t)) \\
&= \operatorname{tr}(Y^t X) + k \operatorname{tr}(Y^t X^t) = \operatorname{tr}(Y^t X) + k \operatorname{tr}(XY) \\
&= \operatorname{tr}(Y^t X) + k \operatorname{tr}(YX) = \operatorname{tr}((Y^t + kY)X) \\
&= \langle X, Y + kY^t \rangle = \langle X, \mathcal{L}(Y) \rangle.
\end{aligned}
$$

6.74 $\dim \operatorname{Im} \mathcal{A} + \dim \operatorname{Ker} \mathcal{A} = 3^2 = 9$. $\operatorname{Ker} \mathcal{A}$ consists of the matrices $X \in M_3(\mathbb{R})$ such that $AX = XA$; X is of the form $\left(\begin{smallmatrix} * & * & 0 \\ * & * & 0 \\ 0 & 0 & * \end{smallmatrix}\right)$, with 5 independent variables. So $\dim \operatorname{Ker} \mathcal{A} = 5$. Thus, the dimension of the kernel of $\mathcal{A}$ is 5 and the dimension of the range of $\mathcal{A}$ is 4.

6.75 The eigenvalues of L are 4, 2, 2. $\left(\begin{smallmatrix} L & 0 & 0 \\ 0 & L & 0 \\ 0 & 0 & L \end{smallmatrix}\right)$ is the matrix representation of $\mathcal{L}$ under the basis $\{E_{11}, E_{21}, E_{31}, E_{12}, E_{22}, E_{32}, E_{13}, E_{23}, E_{33}\}$. It follows that the trace of $\mathcal{L}$ is $3 \times \operatorname{tr} L = 24$.

6.76 The matrix representation of $\mathcal{A}$ with respect to the standard basis is

$$
A = \begin{pmatrix} \frac{1}{2} & 0 & \frac{\sqrt{3}}{2} \\ 0 & 1 & 0 \\ -\frac{\sqrt{3}}{2} & 0 & \frac{1}{2} \end{pmatrix}.
$$

The characteristic polynomial of A is $(\lambda - 1)(\lambda^2 - \lambda + 1)$. So A has three distinct eigenvalues $1, \frac{1 \pm \sqrt{3}i}{2}$, two of which are nonreal. Thus A is diagonalizable over $\mathbb{C}$ but not over $\mathbb{R}$.

The basis with respect to which $\mathcal{A}$ is diagonalizable over $\mathbb{C}$ consists of the eigenvectors of A associated to the eigenvalues. By computation, we have $\lambda_1 = 1, u_1 = (0,1,0)$; $\lambda_2 = \frac{1+\sqrt{3}i}{2}, u_2 = (-i,0,1)$; and $\lambda_3 = \frac{1-\sqrt{3}i}{2}, u_3 = (i,0,1)$. The matrix of $\mathcal{A}$ under $\{u_1, u_2, u_3\}$ is diagonal.

6.77 (a) Since $\mathcal{A}^2 = \mathcal{B}^2 = \mathcal{I}$, the eigenvalues of $\mathcal{A}$ and $\mathcal{B}$ are ± 1. Because $\mathcal{AB} + \mathcal{BA} = 0$, neither $\mathcal{A}$ nor $\mathcal{B}$ is $\pm \mathcal{I}$. So $\mathcal{A}$ has eigenvalues 1 and -1; so does $\mathcal{B}$. The eigenspace of $\mathcal{A}$ for the eigenvalue 1 is $V_1(\mathcal{A}) = \{u \in \mathbb{R}^2 \mid \mathcal{A}(u) = u\}$, and for -1, $V_{-1}(\mathcal{A}) = \{v \in \mathbb{R}^2 \mid \mathcal{A}(v) = -v\}$. Moreover, $\dim V_1(\mathcal{A}) = \dim V_{-1}(\mathcal{A}) = 1$.

For $u \in V_1(\mathcal{A})$, $\mathcal{B}(u) = \mathcal{B}(\mathcal{A}(u)) = -\mathcal{A}(\mathcal{B}(u))$, so $\mathcal{B}(u) \in V_{-1}(\mathcal{A})$, that is, $\mathcal{B}$ maps $V_1(\mathcal{A})$ to $V_{-1}(\mathcal{A})$. Similarly, $\mathcal{B}$ maps $V_{-1}(\mathcal{A})$ to $V_1(\mathcal{A})$. Note that $\mathcal{A}$ and $\mathcal{B}$ are nonsingular.

(b) Take any nonzero vectors $u \in V_1(\mathcal{A})$ and $v \in V_{-1}(\mathcal{A})$. Then the matrix of $\mathcal{A}$ under the basis $\{u, v\}$ is $\left(\begin{smallmatrix} 1 & 0 \\ 0 & -1 \end{smallmatrix}\right)$. Since $\mathcal{B}(u) \in V_{-1}(\mathcal{A})$, $\mathcal{B}(v) \in V_1(\mathcal{A})$, let $\mathcal{B}(u) = \lambda v$ and $\mathcal{B}(v) = \mu u$. Then the matrix of $\mathcal{B}$ under $\{u, v\}$ is $\left(\begin{smallmatrix} 0 & \mu \\ \lambda & 0 \end{smallmatrix}\right)$, $\lambda\mu = 1$.

Set $u' = \frac{1}{\lambda}u$ and use $\{u', v\}$ as a basis. The matrix of $\mathcal{A}$ with respective to $\{u', v\}$ remains the same, i.e., $\left(\begin{smallmatrix} 1 & 0 \\ 0 & -1 \end{smallmatrix}\right)$, the matrix of $\mathcal{B}$ with respect to $\{u', v\}$ is $\left(\begin{smallmatrix} 0 & 1 \\ 1 & 0 \end{smallmatrix}\right)$, as desired.

(c) Take any two nonzero vectors $u \in V_1(\mathcal{A})$ and $v \in V_{-1}(\mathcal{A})$ as basis I. The matrix of $\mathcal{A}$ under this basis is $\left(\begin{smallmatrix} 1 & 0 \\ 0 & -1 \end{smallmatrix}\right)$. Take any two nonzero vectors $x \in V_1(\mathcal{B})$ and $y \in V_{-1}(\mathcal{B})$ as basis II. The matrix of $\mathcal{B}$ under this basis is $\left(\begin{smallmatrix} 1 & 0 \\ 0 & -1 \end{smallmatrix}\right)$.

It is impossible to have one basis under which both $\mathcal{A}$ and $\mathcal{B}$ have the same matrix, i.e., $\left(\begin{smallmatrix} 1 & 0 \\ 0 & -1 \end{smallmatrix}\right)$, in this case. Otherwise $\mathcal{A} = \mathcal{B}$, $\mathcal{AB} + \mathcal{BA} = 2\mathcal{A}^2 = 0$, contradicting $\mathcal{A}^2 = \mathcal{I}$.

6.78 A is the matrix of $\mathcal{A}$ under α, so $\mathcal{A}(\alpha) = \alpha A$, where $\alpha = (\alpha_1, \ldots, \alpha_n)$. Let $v = x_1\alpha_1 + \cdots + x_n\alpha_n = \alpha x \in V$, where $x = (x_1, \ldots, x_n)^t$. Then

$$\mathcal{A}(v) = x_1\mathcal{A}(\alpha_1) + \cdots + x_n\mathcal{A}(\alpha_n) = (\mathcal{A}(\alpha_1), \ldots, \mathcal{A}(\alpha_n))x = \alpha(Ax).$$

Thus, $\operatorname{Im}\mathcal{A} = \{\alpha y \mid y = Ax, x \in \mathbb{F}^n\}$, where $\mathbb{F}$ is the underlying field.

6.79   Note that the matrix representation of $\mathcal{L}$ is a permutation matrix; it is always diagonalizable over $\mathbb{C}$. Whether $\mathcal{L}$ is diagonalizable over $\mathbb{R}$ depends on if the permutation matrix has only real eigenvalues or not. (Note: The matrix is reducible to smaller blocks over $\mathbb{R}$ if $\dim V > 2$.)

6.80   If $\mathcal{L}$ has a real eigenvalue, say λ, there is a nonzero vector v such that $\mathcal{L}(v) = \lambda v$. Take $W = \{kv \mid k \in \mathbb{R}\}$. Then W is an invariant subspace under $\mathcal{L}$ and the dimension of W is one.

If $\mathcal{L}$ has no real eigenvalues, then n is even and the roots of the characteristic polynomial of $\mathcal{L}$ appear in conjugate pairs. Let $\lambda = a + bi$ be an eigenvalue of $\mathcal{L}$, $a, b \in \mathbb{R}$, $b \neq 0$. Choose a basis α of V and let A be the (real) matrix representation of $\mathcal{L}$ relative to α. Let $Au = \lambda u$ (over $\mathbb{C}$) with $u = u_1 + iu_2$, where $u_1, u_2 \in \mathbb{R}^n$, not both equal to zero. Then $Au = \lambda u$ implies $Au_1 = au_1 - bu_2$ and $Au_2 = au_2 + bu_1$. Let w_1 and w_2 be the vectors in V which have the coordinates u_1 and u_2, respectively, with respect to the basis α. (Or explicitly, $w_1 = \alpha u_1$ and $w_2 = \alpha u_2$.) Let $W = \text{Span}\{w_1, w_2\}$. Then W has dimension one or two and it is invariant under $\mathcal{L}$.

For $n = 2$, take $V = \mathbb{R}^2$ and let $\mathcal{L}\left(\begin{smallmatrix} x \\ y \end{smallmatrix}\right) = \left(\begin{smallmatrix} -y \\ x \end{smallmatrix}\right)$. Suppose W is an invariant subspace of dimension one. Then any two vectors in W are linearly dependent. If $0 \neq w = \left(\begin{smallmatrix} a \\ b \end{smallmatrix}\right) \in W$. Then $\mathcal{L}(w) = \left(\begin{smallmatrix} -b \\ a \end{smallmatrix}\right) = k\left(\begin{smallmatrix} a \\ b \end{smallmatrix}\right)$ for some $k \in \mathbb{R}$. A computation gives $k^2 = -1$, a contradiction.

6.81 It is not hard to show that $B = 2(I+A)^{-1} - I \Leftrightarrow A = 2(I+B)^{-1} - I$. Moreover, one verifies that $2(I + X)^{-1} - I = (I + X)^{-1}(I - X)$.

$\Rightarrow$: Let $A^* = -A$. Use the above formula to show $B^*B = I$ directly.

$\Leftarrow$: Let $B^*B = I$. Then $B^* = B^{-1}$.

$$
\begin{aligned}
A^* + A &= 2(I + B^*)^{-1} - I + 2(I + B)^{-1} - I \\
&= 2(B^*B + B^*)^{-1} + 2(I + B)^{-1} - 2I \\
&= 2(I + B)^{-1}B + 2(I + B)^{-1} - 2I = 0.
\end{aligned}
$$

6.82 We only show the first identity. The identity is the same as

$$(I - A^*)^{-1}(I - A^*A)(I - A)^{-1} = (I - A)^{-1} + (I - A^*)^{-1} - I$$

which is verified by pre- and post-multiplying by $I - A^*$ and $I - A$.

6.83 $I = (I - A) + A$, all singular values of I are equal to 1. Use the fact that $\sigma_{\min}(X) + \sigma_{\min}(Y) \leq \sigma_i(X + Y) \leq \sigma_{\max}(X) + \sigma_{\max}(Y)$ for all singular values $\sigma_i(X + Y)$ of $X + Y$ (see Problem 4.62).

6.84 Take $A = \left(\begin{smallmatrix} 5 & 2 \\ 2 & 1 \end{smallmatrix}\right)$, $B = \left(\begin{smallmatrix} 2 & -1 \\ -1 & 1 \end{smallmatrix}\right)$, $C = \left(\begin{smallmatrix} 1 & 0 \\ 0 & 5 \end{smallmatrix}\right)$, all positive semidefinite. Then $ABC = \left(\begin{smallmatrix} 8 & -15 \\ 3 & -5 \end{smallmatrix}\right)$. The eigenvalues of ABC are $\frac{1}{2}(3 \pm \sqrt{11}i)$.

$|\lambda(ABC)| \leq \sigma_{\max}(ABC) \leq \sigma_{\max}(A)\sigma_{\max}(B)\sigma_{\max}(C)$ (Problem 4.62). Note that $\sigma_{\max}(P) = \lambda_{\max}(P)$ if P is positive semidefinite.

6.85 $C - B^*A^{-1}B = \left(\begin{smallmatrix} \frac{1}{2} & \frac{1}{2} \\ \frac{1}{2} & x - \frac{3}{2} \end{smallmatrix}\right)$ is positive semidefinite if $x \geq 2$.

$C - BA^{-1}B^* = \left(\begin{smallmatrix} -\frac{1}{2} & 0 \\ 0 & x-1 \end{smallmatrix}\right)$ is not positive semidefinite for any x.

6.86 If $A \geq 0$, then $\det A \geq 0$. Thus $abc - a|y|^2 - c|x|^2 \geq 0$. Dividing by ac gives $b \geq |x|^2 a^{-1} + |y|^2 c^{-1}$. Conversely, $b \geq |x|^2 a^{-1} + |y|^2 c^{-1}$ implies $abc \geq a|y|^2 + c|x|^2$, $ab \geq |x|^2$, $bc \geq |y|^2$, that is, all principal minors of A are nonnegative, consequently, A is positive semidefinite.

Note that a, b, c need be positive as a precondition; otherwise it may not be true. For example, $a = b = c = -1$, $x = y = 1$.

6.87 For the following matrix A, $\det A = 0$ but $\det A^{[3]} = 48$.

$$A = \begin{pmatrix} 1 & 1 & 1 \\ 1 & 2 & 0 \\ 1 & 0 & 2 \end{pmatrix}.$$

6.88 Since $\|X\|_F^2 = \operatorname{tr}(X^*X) = \sum_{i,j} |x_{ij}|^2$, where $X = (x_{ij})$, we have

$$
\begin{aligned}
\|U \circ A\|_F &= \left(\sum_{i,j=1}^{n} |u_{ij}|^2 |a_{ij}|^2 \right)^{\frac{1}{2}} \\
&\leq \left(\sum_{i,j=1}^{n} |u_{ij}|^4 \right)^{\frac{1}{4}} \left(\sum_{i,j=1}^{n} |a_{ij}|^4 \right)^{\frac{1}{4}} \quad \text{(Cauchy–Schwarz)} \\
&\leq \left(\sum_{i,j=1}^{n} |u_{ij}|^2 \right)^{\frac{1}{4}} \left(\sum_{i,j=1}^{n} |a_{ij}|^4 \right)^{\frac{1}{4}} \quad \text{(U is unitary)} \\
&\leq \sqrt[4]{n} \left(\sum_{i,j=1}^{n} |a_{ij}|^4 \right)^{\frac{1}{4}}.
\end{aligned}
$$

6.89 A is positive semidefinite, B is not because $\det B < 0$. This example shows that matrix $(a_{ij}) \geq 0$ does not imply matrix $(|a_{ij}|) \geq 0$.

6.90 $UA = BV$ implies $UAA^*U^* = BVV^*B^* = BB^*$, i.e., $UA^2U^* = B^2$. Uniqueness of positive semidefinite square root reveals $UAU^* = B$. Thus, $UA = BU$, and $UA = BU = BV$. So $U = V$ as B is invertible.

If $UA = VB$, then $V^*UA = BI$. Thus, $V^*U = I$, and $U = V$.

This yields uniqueness of polar decomposition of the nonsingular case.

6.91 This follows from the *Lagrange Interpolation Theorem* which states: If $\{a_1, \ldots, a_n\}$ and $\{b_1, \ldots, b_n\}$ two sets of numbers in field $\mathbb{F}$ ($= \mathbb{R}$ or $\mathbb{C}$), and if $a_1, \ldots, a_n$ are distinct, then there exists a polynomial p of degree at most $n - 1$ over $\mathbb{F}$ such that $p(a_i) = b_i$, $i = 1, \ldots, n$. $p(x)$ is given by $p(x) = b_1 f_1(x) + \cdots + b_n f(x)$, where $f_i(x) = \prod_{1 \leq k \leq n, \, k \neq i} \frac{x - a_k}{a_i - a_k}$. (Alternatively, by Vandermonde matrix.) Note that if A is normal, then $A = U^*DU$, where $D = \operatorname{diag}(\lambda_1, \ldots, \lambda_n)$ is a diagonal matrix of the eigenvalues of A. Take $a_i = \lambda_i$, $b_i = \bar{\lambda}_i$, $i = 1, \ldots, n$. (In this case, the a_i's need not be distinct.)

6.92 Let $a_1, \ldots, a_n$ represent the eigenvalues of A. When A is positive semidefinite (or Hermitian), there is a unitary matrix U such that $A = U^* \operatorname{diag}(a_1, \ldots, a_n) U$, where a_i's are nonnegative (or real).

    (a) Lagrange interpolation enures a real-coefficient polynomial $p(x)$ such that $p(a_i) = \sqrt{a_i}$ when all $a_i \geq 0$. Then $p(A) = A^{\frac{1}{2}}$.

    (b) Similarly, there exists a real-coefficient polynomial $f(x)$ such that $f(b_i) = |b_i|$ for each i. Then $(B^*B)^{\frac{1}{2}} = f(B)$ (for Hermitian B).

    (c) Apply (a) to $\left(\begin{smallmatrix} A & 0 \\ 0 & B \end{smallmatrix}\right)$. Note that p and f have degrees $n-1$ or less. g has degree $2n-1$ or less.

6.93   Consider the matrix $A = U^*DU$. $\det(U_k^*DU_k) \geq 0$ for all U_k ensures that all principal minors of A are nonnegative. So A is positive semidefinite. Note that for each k, there are $\binom{n}{k}$ U_k's.

6.94   (a) Take $A = \left(\begin{smallmatrix} x & 0 \\ 0 & 1 \end{smallmatrix}\right)$ and $B = \left(\begin{smallmatrix} 2 & 1 \\ 1 & 1 \end{smallmatrix}\right)$. Then $AB + BA = \left(\begin{smallmatrix} 4x & x+1 \\ x+1 & 2 \end{smallmatrix}\right)$. $\det(AB + BA) = -x^2 + 6x - 1$ is negative if $x > 6$, say.

    (b) $(A \pm B)^*(A \pm B) \geq 0 \Rightarrow \pm(AB + BA) \leq A^2 + B^2$.

    (c) Note that $(A, B)^*(A, B) = \left(\begin{smallmatrix} A^2 & AB \\ BA & B^2 \end{smallmatrix}\right) \geq 0$ and $\left(\begin{smallmatrix} B^2 & BA \\ AB & A^2 \end{smallmatrix}\right) \geq 0$. Taking the sum, we have $\left(\begin{smallmatrix} A^2+B^2 & AB+BA \\ AB+BA & A^2+B^2 \end{smallmatrix}\right) \geq 0$. Taking the determinants of the blocks gives $\left(\begin{smallmatrix} \det(A^2+B^2) & \det(AB+BA) \\ \det(AB+BA) & \det(A^2+B^2) \end{smallmatrix}\right) \geq 0$. Taking the determinant of 2×2 matrix reveals the inequality.

    (d) $\|AB+BA\|_{\mathrm{sp}} = \sigma_{\max}(AB+BA) \leq \sigma_{\max}(A^2+B^2) = \|A^2+B^2\|_{\mathrm{sp}}$.

6.95   (a) $0 \leq A \leq I \Rightarrow 0 \leq A^2 \leq A \leq I$. Likewise, $0 \leq B^2 \leq B \leq I$. So, $0 \leq (A + B - \frac{1}{2}I)^2 = (A^2 - A) + (B^2 - B) + AB + BA + \frac{1}{4}I \Rightarrow AB + BA + \frac{1}{4}I \geq 0 \Rightarrow -\frac{1}{4}I \leq AB + BA$. $(A - B)^2 = A^2 + B^2 - AB - BA \geq 0 \Rightarrow AB + BA \leq A^2 + B^2 \leq A + B \leq 2I$. A similar argument yields $-(AB + BA) \leq A + B$.

    (b) Take $A = \frac{1}{4}\left(\begin{smallmatrix} 1 & \sqrt{3} \\ \sqrt{3} & 3 \end{smallmatrix}\right) \geq 0$ and $B = \frac{1}{4}\left(\begin{smallmatrix} 1 & -\sqrt{3} \\ -\sqrt{3} & 3 \end{smallmatrix}\right) \geq 0$. Then $0 \leq A, B \leq I_2$, and $AB + BA = \frac{1}{4}\left(\begin{smallmatrix} -1 & 0 \\ 0 & 3 \end{smallmatrix}\right)$ has $-\frac{1}{4}$ as an eigenvalue.

    (c) It is impossible to find some A and B such that $AB+BA = -\frac{1}{4}I_n$ because $\mathrm{tr}(AB + BA) = 2\,\mathrm{tr}(AB) \geq 0$, while $\mathrm{tr}(-\frac{1}{4}I_n) < 0$.

    (d) $A(I - B) + B(I - A) \geq 0 \Leftrightarrow AB + BA \leq A + B$; see part (a).

    (e) We first show the inequality on the right. Let $C = X + iY$, where X and Y are Hermitian matrices. Compute

$$0 \leq CC^* = (X + iY)(X - iY) = X^2 + Y^2 - i(XY - YX).$$

Thus

$$i(XY - YX) \leq X^2 + Y^2.$$

Setting $X = A - \frac{1}{2}I$ and $Y = B - \frac{1}{2}I$, we obtain

$$i(AB - BA) \leq (A^2 - A) + (B^2 - B) + \tfrac{1}{2}I \leq \tfrac{1}{2}I.$$

The inequality on the left follows immediately as we switch A and B to get $i(BA - AB) \leq \frac{1}{2}I$, implying $-\frac{1}{2}I \leq i(AB - BA)$. See also Problems 4.23 and 4.25.

6.96   Let x be a unit row vector. Then x^*x is an $n \times n$ positive semidefinite matrix of rank 1. Note that $\begin{pmatrix} xx^* & x \\ x^* & 0 \end{pmatrix} = \begin{pmatrix} 1 & x \\ x^* & 0 \end{pmatrix} = \begin{pmatrix} 1 & 0 \\ x^* & I \end{pmatrix} \begin{pmatrix} 1 & 0 \\ 0 & -x^*x \end{pmatrix} \begin{pmatrix} 1 & x \\ 0 & I \end{pmatrix}$ has one positive eigenvalue and one negative eigenvalue, plus $n - 1$ zeros. The inertia of $\begin{pmatrix} xx^* & x \\ x^* & 0 \end{pmatrix}$ is the triple $(1, 1, n-1)$. So is the inertia of $\begin{pmatrix} x^*x & x^* \\ x & 0 \end{pmatrix}$ as $\begin{pmatrix} I & 0 \\ -x & 1 \end{pmatrix} \begin{pmatrix} x^*x & x^* \\ x & 0 \end{pmatrix} \begin{pmatrix} I & -x^* \\ 0 & 1 \end{pmatrix} = \begin{pmatrix} x^*x & 0 \\ 0 & -1 \end{pmatrix}$.

To find the nonzero eigenvalues of $\begin{pmatrix} xx^* & x \\ x^* & 0 \end{pmatrix}$, compute $\det \begin{pmatrix} \lambda - 1 & -x \\ -x^* & \lambda I_n \end{pmatrix} = \lambda^n(\lambda - 1 - \frac{1}{\lambda})$. Thus, the nonzero eigenvalues of $\begin{pmatrix} xx^* & x \\ x^* & 0 \end{pmatrix}$ are $\frac{1 \pm \sqrt{5}}{2}$.

For the nonzero eigenvalues of $\begin{pmatrix} x^*x & x^* \\ x & 0 \end{pmatrix}$, compute $\det \begin{pmatrix} \lambda I - x^*x & -x^* \\ -x & \lambda \end{pmatrix} = \lambda \det \left(\lambda I - (1 + \frac{1}{\lambda})x^*x \right) = \lambda \cdot (\lambda - 1 - \frac{1}{\lambda}) \cdot \lambda^{n-1}$. So, the nonzero eigenvalues of $\begin{pmatrix} x^*x & x^* \\ x & 0 \end{pmatrix}$ are also $\frac{1 \pm \sqrt{5}}{2}$. (The two matrices are similar.)

6.97   (a) $AA^* = 14$ and $A^*A = \begin{pmatrix} 4 & -2 & -6i \\ -2 & 1 & 3i \\ 6i & -3i & 9 \end{pmatrix}$.

(b)   Let $A = UDV$ be a singular value decomposition of A, where U is an $m \times m$ unitary matrix, V is an $n \times n$ unitary matrix, and D is the $m \times n$ nonnegative matrix with the nonzero singular values $s_1, \ldots, s_r$ of A on the diagonal $(1, 1), \ldots, (r, r)$. Then

$$M_0 = \begin{pmatrix} A^*A & A^* \\ A & 0 \end{pmatrix} = \begin{pmatrix} V^* & 0 \\ 0 & U \end{pmatrix} \begin{pmatrix} D^tD & D^t \\ D & 0 \end{pmatrix} \begin{pmatrix} V & 0 \\ 0 & U^* \end{pmatrix}.$$

As $\begin{pmatrix} V^* & 0 \\ 0 & U \end{pmatrix}$ is unitary, M_0 is unitarily similar to $\begin{pmatrix} D^tD & D^t \\ D & 0 \end{pmatrix}$ which contains $s_1^2, s_2^2, \ldots, s_r^2$ respectively in $(1, 1), (2, 2), \ldots, (r, r)$ positions, $s_1, s_2, \ldots, s_r$ in $(1, n+1), (2, n+2), \ldots, (r, n+r)$ positions, $s_1, s_2, \ldots, s_r$ in $(n+1, 1), (n+2, 2), \ldots, (n+r, r)$ positions, and

0's elsewhere, or we may write (with many 0's absent)

$$
\begin{pmatrix} D^t D & D^t \\ D & 0 \end{pmatrix} =
\begin{pmatrix}
s_1^2 & & & & s_1 & & \\
& s_2^2 & & & & s_2 & \\
& & \ddots & & & & \ddots \\
& & & s_r^2 & & & & s_r \\
s_1 & & & & 0 & & \\
& s_2 & & & & 0 & \\
& & \ddots & & & & \ddots \\
& & & s_r & & & & 0
\end{pmatrix}_{(m+n)\times(m+n)}
$$

which is permutation-similar to the 2×2 block diagonal matrix

$$
\begin{pmatrix}
\begin{smallmatrix} s_1^2 & s_1 \\ s_1 & 0 \end{smallmatrix} & & & & \\
& \begin{smallmatrix} s_2^2 & s_2 \\ s_2 & 0 \end{smallmatrix} & & & \\
& & \ddots & & \\
& & & \begin{smallmatrix} s_r^2 & s_r \\ s_r & 0 \end{smallmatrix} & \\
& & & & 0
\end{pmatrix}.
$$

Note that the eigenvalues of 2×2 matrix $\begin{pmatrix} s^2 & s \\ s & 0 \end{pmatrix}$ are $\frac{s^2 \pm |s|\sqrt{s^2+4}}{2}$.
It follows that the eigenvalues of M_0 are

$$
\frac{s_1^2 \pm s_1\sqrt{s_1^2+4}}{2}, \ \ldots, \ \frac{s_r^2 \pm s_r\sqrt{s_r^2+4}}{2}, \ 0, \ldots, 0,
$$

If the rank of A is r, then the rank of M_0 is $2r$. So is $r(N_0)$ by
switching A and A^* (of the same nonzero singular values).

(c) The eigenvalues of the $(m+n) \times (m+n)$ matrix M_t are

$$
\overbrace{0,\ldots,0}^{n-r}, \ \overbrace{t,\ldots,t}^{m-r}, \ \tfrac{1}{2}\left((s_i^2+t) \pm \sqrt{(s_i^2-t)^2+4s_i^2}\right), \ i=1,\ldots,r
$$

and the eigenvalues of the $(m+n) \times (m+n)$ matrix N_t are

$$
\overbrace{0,\ldots,0}^{m-r}, \ \overbrace{t,\ldots,t}^{n-r}, \ \tfrac{1}{2}\left((s_i^2+t) \pm \sqrt{(s_i^2-t)^2+4s_i^2}\right), \ i=1,\ldots,r.
$$

If $A^*A = I_n$, then $s_1 = \cdots = s_r = 1$ in the above displays.

(d) If $m = n$, then M_t and N_t have the same eigenvalues. Since they
are both Hermitian, they are unitarily similar. In particular,
$\begin{pmatrix} AA^* & A \\ A^* & 0 \end{pmatrix}$ is unitarily similar to $\begin{pmatrix} A^*A & A^* \\ A & 0 \end{pmatrix}$, which is also seen via

$$
I_{2n} + \begin{pmatrix} AA^* & A \\ A^* & 0 \end{pmatrix} = \begin{pmatrix} A & I \\ I & 0 \end{pmatrix}\begin{pmatrix} A^* & I \\ I & 0 \end{pmatrix}, \quad I_{2n} + \begin{pmatrix} A^*A & A^* \\ A & 0 \end{pmatrix} = \begin{pmatrix} A^* & I \\ I & 0 \end{pmatrix}\begin{pmatrix} A & I \\ I & 0 \end{pmatrix}
$$

as XY and YX have the same nonzero eigenvalues. However, this approach gives no information about the eigenvalues of the individual block matrices, nor do the following identities:

$$\begin{pmatrix} AA^* & A \\ A^* & 0 \end{pmatrix} = \begin{pmatrix} A & 0 \\ 0 & I_n \end{pmatrix} \begin{pmatrix} I_n & I_n \\ I_n & 0_n \end{pmatrix} \begin{pmatrix} A^* & 0 \\ 0 & I_n \end{pmatrix}.$$

6.98 $MP = PM \Rightarrow QPQ^{-1}P = PQPQ^{-1}$. Multiplying by $Q^{-\frac{1}{2}}$ from the left-hand side and by $Q^{\frac{1}{2}}$ from the right-hand side, we obtain

$$Q^{\frac{1}{2}}PQ^{-\frac{1}{2}} \cdot Q^{-\frac{1}{2}}PQ^{\frac{1}{2}} = Q^{-\frac{1}{2}}PQ^{\frac{1}{2}} \cdot Q^{\frac{1}{2}}PQ^{-\frac{1}{2}}.$$

Denote $N = Q^{\frac{1}{2}}PQ^{-\frac{1}{2}}$. Then $NN^* = N^*N$, that is, N is normal. On the other hand, N has the same eigenvalues as P that are all positive. Thus N is Hermitian, i.e., $N^* = N$. Hence $Q^{-\frac{1}{2}}PQ^{\frac{1}{2}} = Q^{\frac{1}{2}}PQ^{-\frac{1}{2}}$. It follows that $QP = PQ$. Consequently, $M = P$.

6.99   The determinant is the product of all eigenvalues. Since A is real, the characteristic polynomial of A has real coefficients, the eigenvalues of A are either real or complex (non-real) numbers that occur in conjugate pairs. We claim that the real ones are positive.

Let λ be an eigenvalue of A and let x be a corresponding (complex) unit eigenvector. Then $x^*(A + A^t)x = \lambda + \bar{\lambda} > 0$. Thus the real part of λ is positive. In particular, if λ is real, then $\lambda > 0$.

6.100   Reduce the positivity of the inverse block matrix to an obvious case:

$$\begin{pmatrix} (A+A)^{-1} & (A+B)^{-1} \\ (B+A)^{-1} & (B+B)^{-1} \end{pmatrix} \geq 0$$

$$\Leftrightarrow \quad (2B)^{-1} \geq (A+B)^{-1}(2A)(A+B)^{-1}$$

$$\Leftrightarrow \quad 2^{-1}(A+B)B^{-1}(A+B) \geq 2A$$

$$\Leftrightarrow \quad AB^{-1}A + B \geq 2A$$

$$\Leftrightarrow \quad (B-A)B^{-1}(B-A) \geq 0.$$

6.101   Consider the space of improperly integrable functions on $[0, \infty)$ with the inner product $\langle f, g \rangle = \int_0^\infty f(x)g(x)dx$. Let

$$u_1(x) = e^{-a_1 x}, \ u_2(x) = e^{-a_2 x}, \ \ldots, \ u_n(x) = e^{-a_n x}.$$

Then

$$\frac{1}{a_i + a_j} = \int_0^\infty e^{-(a_i + a_j)x}dx = \int_0^\infty e^{-a_i x}e^{-a_j x}dx = \langle u_i, u_j \rangle.$$

So $((a_i+a_j)^{-1})$ (a Cauchy matrix) is Gramian and is positive semidefinite. It is immediate that $G = ((i+j-1)^{-1})$ is positive semidefinite. Another proof: Show that the principal minors are nonnegative.

6.102 (a) $\begin{pmatrix} I & I \\ -I & I \end{pmatrix}\begin{pmatrix} H & K \\ K & H \end{pmatrix}\begin{pmatrix} I & -I \\ I & I \end{pmatrix} = 2\begin{pmatrix} H+K & 0 \\ 0 & H-K \end{pmatrix}$. (See Problem 4.55.)

(b) Let $K = U^*DU$, where $D = \mathrm{diag}(k_1, \ldots, k_n)$ and U is unitary. Then $\ell(K) = U^*\tilde{D}U$, where $\tilde{D} = \mathrm{diag}(|k_1|, \ldots, |k_n|)$. Apparently, $\ell(K) \geq \pm K$. So $H \geq \ell(K) \geq \pm K$.

(c) $H \geq \pm K$ does not imply $H \geq \ell(K)$ in general. Let $2 < x < 2\sqrt{2}$. Set $H = \begin{pmatrix} x & 1-i \\ 1+i & x \end{pmatrix}$ and $K = \begin{pmatrix} 0 & -1-i \\ -1+i & 0 \end{pmatrix}$. Then $H > 0$ and K is Hermitian, $H + K = \begin{pmatrix} x & -2i \\ 2i & x \end{pmatrix} > 0$, and $H - K = \begin{pmatrix} x & 2 \\ 2 & x \end{pmatrix} > 0$. However, $\ell(K) = \begin{pmatrix} \sqrt{2} & 0 \\ 0 & \sqrt{2} \end{pmatrix}$, $H - \ell(K) = \begin{pmatrix} x-\sqrt{2} & 1-i \\ 1+i & x-\sqrt{2} \end{pmatrix} \not\geq 0$.

6.103 Let A be of size $n \times n$. If $r = 0$ or n, we have nothing to show. Let $0 < r < n$ and let $i_-(\cdot)$ denote the number of negative eigenvalues.

Let $A = UDV$ be a singular value decomposition of A, where U, V are unitary, and $D = \begin{pmatrix} D_r & 0 \\ 0 & 0 \end{pmatrix}$, D_r is an $r \times r$ positive diagonal matrix. Then $H = UDV + V^*DU^*$. We may further assume $U = I$. Then $H = DV + V^*D = \begin{pmatrix} A_r & X \\ X^* & 0 \end{pmatrix}$, where A_r is an $r \times r$ Hermitian matrix. If A_r is invertible, $i_+(H) = i_+(A_r) + i_-(X^*A_r^{-1}X) \leq i_+(A_r) + i_-(A_r) = r$. If A_r is singular, let $E = \begin{pmatrix} aI_r & 0 \\ 0 & 0 \end{pmatrix}$, where a is a positive number such that $A_r + aI_r$ is nonsingular. Then $i_+(H) \leq i_+(H + E) \leq r$.

6.104 Let $\sigma_1(A)$ (or $\|A\|_{\mathrm{sp}}$) be the spectral norm of A, i.e., the largest singular value of A. (a) means $\sigma_1(A) \leq 1$. (b) is the same as $AA^* \leq I_m$, which is equivalent to $\sigma_1^2(A) \leq 1$. So (a) $\Leftrightarrow$ (b). Similarly, (a) $\Leftrightarrow$ (c). (d) $\Leftrightarrow$ (b) is by taking the Schur complement, that is, $\begin{pmatrix} I_n & A^* \\ A & I_m \end{pmatrix} \geq 0 \Leftrightarrow \begin{pmatrix} I_n & 0 \\ -A & I_m \end{pmatrix}\begin{pmatrix} I_n & A^* \\ A & I_m \end{pmatrix}\begin{pmatrix} I_n & -A^* \\ 0 & I_m \end{pmatrix} = \begin{pmatrix} I_n & 0 \\ 0 & I_m-AA^* \end{pmatrix} \geq 0 \Leftrightarrow I_m - AA^* \geq 0$. Similarly, (e) $\Leftrightarrow$ (c). Moreover, $\begin{pmatrix} I_n & A^* \\ A & I_m \end{pmatrix}$ and $\begin{pmatrix} I_m & A \\ A^* & I_n \end{pmatrix}$ are permutation-similar: $\begin{pmatrix} 0 & I_m \\ I_n & 0 \end{pmatrix}\begin{pmatrix} I_n & A^* \\ A & I_m \end{pmatrix}\begin{pmatrix} 0 & I_n \\ I_m & 0 \end{pmatrix} = \begin{pmatrix} I_m & A \\ A^* & I_n \end{pmatrix}$.

6.105 (a) If P and Q are contractions, then $\sigma_1(PQ) \leq \sigma_1(P)\sigma_1(Q) \leq 1$. So PQ is a contraction. Note: $P + Q$ need not be a contraction.

(b) Since C is a contraction, we assume that $C = UDV$ is a singular value decomposition of C, where D is a diagonal matrix with diagonal entries less than or equal to 1. Then, $\mathrm{tr}(C^*AC) = \mathrm{tr}(V^*DU^*AUDV) = \mathrm{tr}(DU^*AUD) \leq \mathrm{tr}(U^*AU) = \mathrm{tr}\, A$. However, $C^*AC \not\leq A$ in general. Take $A = \begin{pmatrix} 1 & 0 \\ 0 & 0 \end{pmatrix}$ and $C = \begin{pmatrix} 0 & 1 \\ 1 & 0 \end{pmatrix}$.

(c) Let $A = \left(\begin{smallmatrix} 1 & 0 \\ 0 & 5 \end{smallmatrix}\right), B = \left(\begin{smallmatrix} 0 & 2 \\ 0 & 0 \end{smallmatrix}\right)$. Then $B^*B \leq A$ but $BB^* \not\leq A$.

(d) First let A be invertible. Then $DAD \leq A \Leftrightarrow A^{-\frac{1}{2}}DADA^{-\frac{1}{2}} \leq I \Leftrightarrow \lambda_{\max}(A^{-\frac{1}{2}}DADA^{-\frac{1}{2}}) \leq 1 \Leftrightarrow \lambda_{\max}(A^{-1}DAD) \leq 1$. Since $0 \leq D \leq I$, we get $\lambda_{\max}(A^{-1}DAD) \leq \lambda_{\max}(A^{-1}DA)\lambda_{\max}(D) = \lambda_{\max}^2(D) \leq 1$. So $DAD \leq A$. If A is singular, use $A + \epsilon I$, where $\epsilon > 0$. Then $D(A + \epsilon I)D \leq A + \epsilon I$. Thus, $DAD \leq A + \epsilon(I - D^2)$ for all $\epsilon > 0$. Letting $\epsilon \to 0^+$ yields $DAD \leq A$.

6.106

$$\begin{pmatrix} I_n & X & 0 \\ X^* & I_n & X \\ 0 & X^* & I_n \end{pmatrix} \geq 0 \quad \Leftrightarrow \quad \begin{pmatrix} I_n & 0 & 0 \\ -X^* & I_n & 0 \\ 0 & 0 & I_n \end{pmatrix} \begin{pmatrix} I_n & X & 0 \\ X^* & I_n & X \\ 0 & X^* & I_n \end{pmatrix} \begin{pmatrix} I_n & -X & 0 \\ 0 & I_n & 0 \\ 0 & 0 & I_n \end{pmatrix}$$

$$= \begin{pmatrix} I_n & 0 & 0 \\ 0 & I_n - X^*X & X \\ 0 & X^* & I_n \end{pmatrix} \geq 0$$

$$\Leftrightarrow \begin{pmatrix} I_n & 0 & 0 \\ 0 & I_n - X^*X - XX^* & 0 \\ 0 & 0 & I_n \end{pmatrix} \geq 0$$

$$\Leftrightarrow I_n - X^*X - XX^* \geq 0.$$

6.107 Applying simultaneous row and column (*-congruent) operations gives

$$\begin{pmatrix} A & X & 0 \\ X^* & B & Y \\ 0 & Y^* & C \end{pmatrix} \to \begin{pmatrix} A & 0 & 0 \\ 0 & B - X^*A^{-1}X & Y \\ 0 & Y^* & C \end{pmatrix}$$

$$\to \begin{pmatrix} A & 0 & 0 \\ 0 & B - X^*A^{-1}X - YC^{-1}Y^* & 0 \\ 0 & 0 & C \end{pmatrix}.$$

Y and Y^* in the inequality cannot be switched, that is, the positive semidefiniteness of M does not imply $B \geq X^*A^{-1}X + Y^*C^{-1}Y$. Take, for example, $A = C = I_2, X = 0, B = \left(\begin{smallmatrix} 1 & 0 \\ 0 & 5 \end{smallmatrix}\right), Y = \left(\begin{smallmatrix} 0 & 0 \\ 2 & 0 \end{smallmatrix}\right)$.

6.108 If $n = 1$, it is trivial. We assume that $n \geq 2$.

Consider the function $f(x_1, \ldots, x_n) = \prod_{i=1}^n x_i - \sum_{i=1}^n x_i$, where $x_i \in [0, 1]$, $i = 1, \ldots, n$. The extreme values of f are $1 - n$ (when all $x_i = 1$) and 0 (when all $x_i = 0$), respectively. So $1 - n + \sum_{i=1}^n x_i \leq \prod_{i=1}^n x_i$.

Let $\sigma_1, \ldots, \sigma_n$ be the singular values of C. Because $\sum_{i,j=1}^n |c_{ij}|^2 = \operatorname{tr}(CC^*) = \sum_{i=1}^n \sigma_i^2$ and $|\det C|^2 = \det(CC^*) = \prod_{i=1}^n \sigma_i^2$, with $x_i = \sigma_i^2$, $i = 1, \ldots, n$, the above discussion leads to the first inequality.

The second inequality is the Hadamard determinantal inequality for the matrix CC^*, while the last inequality is due to the fact that $\sum_{i=1}^n |c_{ij}|^2 \leq \sigma_{\max}(C) \leq 1$ for each j.

If all three equalities occur simultaneously, then C is unitary.

6.109   Write $M = \frac{1}{n}ee^t$, where e is the column vector whose n components are all equal to 1. Note that $Ae = e$ since A is doubly stochastic. So 1 is an eigenvalue of A. Moreover, $(A - M)e = Ae - Me = 0$.

Let $1, \lambda_2, \ldots, \lambda_n$ be all the eigenvalues of A (counting multiplicities) over $\mathbb{C}$. We show that the eigenvalues of $A - M$ are $0, \lambda_2, \ldots, \lambda_n$.

Let $e, e_2, \ldots, e_n$ be an orthogonal set of eigenvectors of A belonging to the eigenvalues $1, \lambda_2, \ldots, \lambda_n$, respectively. (This is possible for normal matrices over $\mathbb{C}$.) Then $e^t e_i = 0$ and $Me_i = \frac{1}{n}ee^t e_i = 0$. Thus

$$(A - M)e_i = Ae_i - Me_i = Ae_i = \lambda_i e_i, \quad i = 2, \ldots, n.$$

If A is positive semidefinite, then all $\lambda_i \geq 0$. So $A - M \geq 0$.

Note: Consequently, $x^* A x \geq x^* M x = \frac{1}{n}|\sum_{i=1}^{n} x_i|^2$ for all $x \in \mathbb{C}^n$.

6.110   For any matrix X, $r(X) = r(X^*X) = r(XX^*)$. For normal A and B, $r(AB) = r(B^*A^*AB) = r(B^*AA^*B) = r(A^*BB^*A) = r(A^*B^*BA)$ $= r(BA)$. For $A = \begin{pmatrix} 1 & 1 \\ 1 & 1 \end{pmatrix}, B = \begin{pmatrix} 1 & 1 \\ -1 & -1 \end{pmatrix}, AB = 0, BA \neq 0$.

6.111   If AB and BA are normal, then they are diagonalizable. On the other hand, AB and BA have the same eigenvalues. Thus, AB and BA are similar. By a direct verification, C and D are normal. Computations reveal $CD = \begin{pmatrix} 0 & 0 & 0 & 1 \\ 0 & 0 & 0 & 0 \\ 0 & 1 & 0 & 0 \\ 0 & 0 & 0 & 0 \end{pmatrix}$ and $DC = \begin{pmatrix} 0 & 0 & 0 & 0 \\ 1 & 0 & 0 & 0 \\ 0 & 0 & 0 & 0 \\ 0 & 1 & 0 & 0 \end{pmatrix}$. One can check that $(CD)^2 = 0$ and $(DC)^2 \neq 0$. Thus, CD and DC cannot be similar.

Alternatively, use Jordan canonical forms. The Jordan form of CD is $J_2(0) \oplus J_2(0)$, while the Jordan form of DC is $J_3(0) \oplus J_1(0)$. (Here $J_k(\lambda)$ means the $k \times k$ Jordan block with λ's on the main diagonal.)

6.112   Let $A = D - E$. We show $x^* A x > 0$ for all nonzero $x \in \mathbb{C}^n$. Note that $x^* A x = \sum_i \frac{1}{\lambda_i}|x_i|^2 - |\sum_i x_i|^2$. By the Cauchy–Schwartz inequality,

$$\left(\sum_i |x_i|\right)^2 = \left(\sum_i \sqrt{\lambda_i} \cdot \frac{1}{\sqrt{\lambda_i}}|x_i|\right)^2 \leq \sum_i \lambda_i \cdot \sum_i \frac{1}{\lambda_i}|x_i|^2 < \sum_i \frac{1}{\lambda_i}|x_i|^2.$$

6.113   (a) $\Leftrightarrow$ (b): By Problem 6.60, $\bar{A}$ is similar to $\bar{A}^t = A^*$.

(a) $\Rightarrow$ (c): If A is similar to $\bar{A}$, then A and $\bar{A}$ have exactly the same Jordan blocks (including the number of times they appear). We only consider the Jordan blocks associated with non-real eigenvalues $\lambda = x + iy$. If $J(\lambda)$ is a Jordan block of A, so is $J(\bar{\lambda})$. Thus, A is similar to $\begin{pmatrix} J(\lambda) & & \\ & J(\bar{\lambda}) & \\ & & \ddots \end{pmatrix}$. Look at the 2×2 case $J(\lambda) = \begin{pmatrix} \lambda & 1 \\ 0 & \lambda \end{pmatrix}$. (General case can be done through the same process but requires more work.)

One checks that $\begin{pmatrix} J(\lambda) & 0 \\ 0 & J(\bar{\lambda}) \end{pmatrix}$ is similar to $\begin{pmatrix} \lambda & 0 & 1 & 0 \\ 0 & \bar{\lambda} & 0 & 1 \\ 0 & 0 & \lambda & 0 \\ 0 & 0 & 0 & \bar{\lambda} \end{pmatrix}$ via permutations.

As $\begin{pmatrix} \lambda & 0 \\ 0 & \bar{\lambda} \end{pmatrix}$ is similar to the real matrix $\begin{pmatrix} x & y \\ -y & x \end{pmatrix}$ via $P = \frac{1}{\sqrt{2}} \begin{pmatrix} 1 & 1 \\ i & -i \end{pmatrix}$ (Problem 1.31 or Problem 2.63), we see that $\begin{pmatrix} J(\lambda) & 0 \\ 0 & J(\bar{\lambda}) \end{pmatrix}$ is similar to a real matrix, concluding that A is similar to a real matrix.

(c) $\Rightarrow$ (a): If A is similar to a real matrix R, say, $A = Q^{-1}RQ$. Then $\bar{A} = \bar{Q}^{-1} R \bar{Q}$ is also similar to R. So A and $\bar{A}$ are similar.

(a) $\Rightarrow$ (d): Following the proof of (a) $\Rightarrow$ (c), since $J(\bar{\lambda})$ is similar to $(J(\bar{\lambda}))^t = (J(\lambda))^*$, $\begin{pmatrix} J(\lambda) & 0 \\ 0 & J(\bar{\lambda}) \end{pmatrix}$ is similar to $\begin{pmatrix} J(\lambda) & 0 \\ 0 & (J(\lambda))^* \end{pmatrix}$ which equals $\begin{pmatrix} 0 & J(\lambda) \\ (J(\lambda))^* & 0 \end{pmatrix} \begin{pmatrix} 0 & I \\ I & 0 \end{pmatrix}$, the product of two Hermitian matrices (say HK for the general case). So $A = V^{-1}(HK)V = V^{-1}H(V^{-1})^*V^*KV$.

(d) $\Rightarrow$ (b): If $A = HK$, where H and K are Hermitian, then $A^* = (HK)^* = K^*H^* = KH$ is similar to $HK = A$ by Problem 4.17.

6.114 (a) $A^{-\frac{1}{2}}(AB)A^{\frac{1}{2}} = A^{\frac{1}{2}}BA^{\frac{1}{2}}$ is Hermitian and diagonalizable.

(b) Since U^*AUB is similar to $AUBU^*$, by spectral decomposition, we may assume $A = \begin{pmatrix} D_r & 0 \\ 0 & 0 \end{pmatrix}$, where D_r is a positive diagonal matrix and r is the rank of A. Because B is positive semidefinite, we can write $B = \begin{pmatrix} S^* \\ T^* \end{pmatrix} (S, T) = \begin{pmatrix} S^*S & S^*T \\ T^*S & T^*T \end{pmatrix}$, where S is $n \times r$. Then $AB = \begin{pmatrix} D_r S^*S & D_r S^*T \\ 0 & 0 \end{pmatrix}$. $D_r S^*S$ is diagonalizable. Note that $\text{Im}(D_r S^*T) \subseteq \text{Im}(D_r S^*) = \text{Im}(D_r S^*S)$ as $r(S^*) = r(S^*S)$ and $\text{Im}(S^*S) \subseteq \text{Im}(S^*)$. The diagonalization of AB is by the fact that $\begin{pmatrix} C & D \\ 0 & 0 \end{pmatrix}$ is similar to $\begin{pmatrix} C & 0 \\ 0 & 0 \end{pmatrix}$ if $\text{Im} D \subseteq \text{Im} C$, i.e., $D = CE$ for some E, as $\begin{pmatrix} I & E \\ 0 & I \end{pmatrix}\begin{pmatrix} C & D \\ 0 & 0 \end{pmatrix}\begin{pmatrix} I & -E \\ 0 & I \end{pmatrix} = \begin{pmatrix} C & 0 \\ 0 & 0 \end{pmatrix}$. (See Problem 3.103.)

(c) $A = \begin{pmatrix} 1 & 1 \\ 1 & 1 \end{pmatrix}$, $B = \begin{pmatrix} 1 & 0 \\ 0 & -1 \end{pmatrix}$. $AB = \begin{pmatrix} 1 & -1 \\ 1 & -1 \end{pmatrix}$ is not diagonalizable.

(d) By Problem 6.113 (c) and (d).

6.115 We assume that none of A, B, and C is zero. Hermicity is inherent in the definition of positive semidefiniteness (over $\mathbb{C}$), that is, a positive semidefinite matrix is necessarily Hermitian. So when AB and ABC are positive semidefinite, AB and ABC are Hermitian. To show the converses, we use the fact that the eigenvalues of the product of two positive semidefinite matrices are nonnegative because $AB = A^{\frac{1}{2}}A^{\frac{1}{2}}B$ has the same eigenvalues as $A^{\frac{1}{2}}BA^{\frac{1}{2}}$, which is positive semidefinite.

(a) This is trivial as Hermitian AB has nonnegative eigenvalues.

(b) Let ABC be Hermitian. If A is invertible, then $A^{-\frac{1}{2}}(ABC)A^{-\frac{1}{2}} = A^{\frac{1}{2}}(BC)A^{-\frac{1}{2}}$, implying ABC has only nonnegative eigenvalues as BC does. Thus, ABC is positive semidefinite if A is nonsingular.

Now consider the case where A is singular. If A and C contain a zero diagonal entry in the same position, say $a_{11} = c_{11} = 0$, a simple computation shows the problem reduces to the matrices of size $n-1$. Then we can repeat this process (or use induction).

Let $a_{ii} + c_{ii} > 0$ for $i = 1, \ldots, n$. Then $a_{ii} + tc_{ii} > 0$ for all i and $t > 0$. Let $A_t = (A + tC)$, $t > 0$. Then $\det A_0 = \det A = 0$, and $\det(A_t)$ (not a constant) is a polynomial in t having at most n roots. There is $\delta > 0$ such that A_t is invertible for all $t \in (0, \delta)$.

Since $A_t BC = (A + tC)BC = ABC + tCBC = CBA + tCBC = CB(A + tC) = CBA_t$, $A_t BC$ is Hermitian. It follows that $A_t BC$ is positive semidefinite. As is known, a Hermitian matrix P is positive semidefinite if and only if all principal minors of P are nonnegative. So all minors of $P_t = A_t BC$ are nonnegative. Because every minor of P_t is a continuous function in t, letting $t \to 0^+$, we see that all principal minors of ABC are nonnegative. Thus ABC is positive semidefinite. (Note: One may use continuity of eigenvalues, but be aware of the difference between topological and functional continuities. The statement does not hold for four positive definite matrices.)

6.116 If $P = P_1 P_2$, where $P_1, P_2 \geq 0$, then P has the same number of nonnegative eigenvalues as $P_1^{\frac{1}{2}} P_2 P_1^{\frac{1}{2}}$ which is positive semidefinite. Thus, all eigenvalues of P are nonnegative real numbers. By Problem 6.114(b), P is diagonalizable. So P is similar to a nonnegative diagonal matrix. Note: $P = P_1 P_2$ need not be positive semidefinite.

Conversely, if P is similar to a nonnegative diagonal matrix, say $A = V^{-1}DV$, then $A = \left(V^{-1}D^{\frac{1}{2}}(V^{-1})^*\right)\left(V^*D^{\frac{1}{2}}V\right)$ is the product of two positive semidefinite matrices.

6.117 Let c be such a complex number that $|c| = 1$ and $-c^2$ is not an eigenvalue of A. Let $B = cI + \bar{c}A$. Then B is nonsingular, $A\bar{B} = B$.

6.118 The inequality is unitarily invariant, that is, it holds if and only if it holds when $A = (a_{ij})$, $B = (b_{ij})$, and $C = (c_{ij})$ are replaced with U^*AU, U^*BU, and U^*CU for any $n \times n$ unitary matrix U. So we may assume that A is a nonnegative diagonal matrix.

The claimed inequality is equivalent to

$$\sum_i a_{ii}c_{ii} - \sum_{i,j} |b_{ij}|^2 \le \sum_i a_{ii} \sum_j c_{jj} - \sum_i \overline{b_{ii}} \sum_i b_{ii},$$

which is the same as

$$\sum_{i \neq j} \overline{b_{ii}} b_{jj} - \sum_{i \neq j} |b_{ij}|^2 \le \sum_{i \neq j} a_{ii} c_{jj}.$$

In fact, the following stronger inequality holds:

$$\sum_{i \neq j} \overline{b_{ii}} b_{jj} \le \sum_{i \neq j} a_{ii} c_{jj}.$$

Note that $\begin{pmatrix} A & B \\ B^* & C \end{pmatrix} \ge 0 \Rightarrow \begin{pmatrix} C & B^* \\ B & A \end{pmatrix} \ge 0$. Extracting the 2×2 principal submatrices in the (i,i) and (j,j) positions respectively from the block matrices, we have $\begin{pmatrix} a_{ii} & b_{ii} \\ \overline{b_{ii}} & c_{ii} \end{pmatrix} \ge 0$ and $\begin{pmatrix} c_{jj} & \overline{b_{jj}} \\ b_{jj} & a_{jj} \end{pmatrix} \ge 0$. So, by taking the entrywise product, $\begin{pmatrix} a_{ii}c_{jj} & b_{ii}\overline{b_{jj}} \\ \overline{b_{ii}}b_{jj} & c_{ii}a_{jj} \end{pmatrix} \ge 0$. The desired inequality is immediate from the fact that $\begin{pmatrix} x & y \\ \overline{y} & z \end{pmatrix} \ge 0$ gives $y + \overline{y} \le x + z$.

6.119   Let $f(t) = at^3 + bt^2 + ct + d$. Then $f \in \{t, t^2\}^{\perp}$. By computation,

$$\frac{a}{5} + \frac{b}{4} + \frac{c}{3} + \frac{d}{2} = 0, \quad \frac{a}{6} + \frac{b}{5} + \frac{c}{4} + \frac{d}{3} = 0.$$

Then $a = \frac{5}{2}c + 10d$ and $b = -\frac{10}{3}c - 10d$. $d = f(0) = 1$. Thus

$$f(t) = (\tfrac{5}{2}c + 10)t^3 - (\tfrac{10}{3}c + 10)t^2 + ct + 1, \quad c \in \mathbb{R}.$$

6.120   (a)   For $f, g \in \mathbb{P}_3[x]$, $k \in \mathbb{R}$, $\mathcal{L}(f + kg) = f(x) + kg(x) - (f(-x) + kg(-x)) = f(x) - f(-x) + k(g(x) - g(-x)) = \mathcal{L}(f) + k\mathcal{L}(g)$.

(b) $\operatorname{Im} \mathcal{L} = \{bx \mid b \in \mathbb{R}\}$.

(c)   If $\mathcal{L}(f) = -f$, then $a = b = c = 0$, and $f = 0$.

(d)   Start with the basis $\{1, x, x^2\}$ to get the orthonormal basis

$$1, \quad \sqrt{3}\,x, \quad \tfrac{\sqrt{5}}{2}(-1 + 3x^2).$$

(e)   For any $r \in \mathbb{R}$ and $f(x) = bx \in \operatorname{Im} \mathcal{L}$, $\langle r, f \rangle = \frac{1}{2} \int_{-1}^{1} rbx\,dx = 0$.

(f) $W = \{a + cx^2 \mid a, c \in \mathbb{R}\}$.

(g) For any $p(x) = bx \in \operatorname{Im}\mathcal{L}$, $\frac{1}{2}\mathcal{L}(p) = \frac{1}{2}(bx + bx) = bx = p(x)$. So $\frac{1}{2}\mathcal{L}$ is the identity on $\operatorname{Im}\mathcal{L}$. For any $q(x) \in W$, $\frac{1}{2}\mathcal{L}(q) = 0$.

6.121 It is routine to verify that $\|\cdot\|$ is a norm. Take $f(x) = 1$ and $g(x) = x$. Then $\|f + g\|^2 + \|f - g\|^2 = 5 \neq 4 = 2\|f\|^2 + 2\|g\|^2$, which does not obey the parallelogram identity for induced norms (Problem 5.50).

6.122 Define $f_i(\alpha_j) = \delta_{ij}$, $1 \leq i, j \leq 3$. Then for any $x \in \mathbb{R}^3$, let $x = (x_1, x_2, x_3) = a\alpha_1 + b\alpha_2 + c\alpha_3$. We solve for a, b, and c to get $a = \frac{1}{2}(x_1 - x_2 + x_3)$, $b = x_2$, and $c = \frac{1}{2}(-x_1 - x_2 + x_3)$. It follows that

$$f_1(x) = a = \tfrac{1}{2}(x_1 - x_2 + x_3)$$
$$f_2(x) = b = x_2$$
$$f_3(x) = c = \tfrac{1}{2}(-x_1 - x_2 + x_3).$$

6.123 (a) Since $M^*M = MM^*$, we have $A^*A + C^*C = AA^* + BB^*$. Taking trace reveals $\operatorname{tr}(A^*A) + \operatorname{tr}(C^*C) = \operatorname{tr}(AA^*) + \operatorname{tr}(BB^*)$. As $\operatorname{tr}(A^*A) = \operatorname{tr}(AA^*)$, $\operatorname{tr}(B^*B) = \operatorname{tr}(C^*C)$. For any matrix X $\operatorname{tr}(X^*X) = \|X\|_F^2$, $\|B\|_F = \|C\|_F$. $\|A\|_F \neq \|D\|_F$ for $\left(\begin{smallmatrix}1 & 0\\0 & 0\end{smallmatrix}\right)$.

(b) Example 4.1 shows $|\det A| = |\det D|$. Since $\left(\begin{smallmatrix}0 & I\\I & 0\end{smallmatrix}\right)\left(\begin{smallmatrix}A & B\\C & D\end{smallmatrix}\right) = \left(\begin{smallmatrix}C & D\\A & B\end{smallmatrix}\right)$ is unitary, $|\det B| = |\det C|$. Below is a direct proof. If M is unitary, then $A^*A + C^*C = AA^* + BB^* = I$. Thus, $C^*C = I - A^*A$, $BB^* = I - AA^*$. So B and C have the same singular values. We write $\sigma_j(B) = \sigma_j(C) = \sqrt{1 - \sigma_j^2(A)}$. Then

$$
\begin{aligned}
|\det B|^2 &= \prod_j \sigma_j^2(B) \\
&= \prod_j \sigma_j^2(C) = |\det C|^2 \\
\Rightarrow\quad & |\det B| = |\det C|.
\end{aligned}
$$

6.124 (a) Let $U = (u_1, \ldots, u_n)$. Then U is unitary, and $U^*HU = D$, where $D = \operatorname{diag}(\lambda_1, \ldots, \lambda_n)$. Obviously, $U^*H^2U = D^2$. Computing the (k, k)-entries of both sides reveals the identity for λ_k^2.

(b) For the identity of $\det(H_k)$, $H = UDU^*$ implies $H^{-1} = UD^{-1}U^*$ and $\det(H)H^{-1} = \det(H)UD^{-1}U^*$. The left-hand side is the adjugate of H. Thus, $\operatorname{adj}(H) = \det(H)U\operatorname{diag}(\frac{1}{\lambda_1}, \ldots, \frac{1}{\lambda_n})U^*$. Comparing the (k, k)-entries on both sides yields the identity.

(c) We use the idea in the above discussion.

$$\lambda I - H = U(\lambda I - D)U^*$$
$$\Rightarrow (\lambda I - H)^{-1} = U(\lambda I - D)^{-1}U^*$$
$$\Rightarrow \mathrm{adj}(\lambda I - H) = \det(\lambda I - H)U(\lambda I - D)^{-1}U^*$$
$$\Rightarrow \det(\lambda I - H)_i = \det(\lambda I - H)\sum_{j=1}^{n}|u_{ij}|^2(\lambda - \lambda_j)^{-1}$$
$$\Rightarrow (\lambda - \mu_{i1})\cdots(\lambda - \mu_{i(n-1)}) = \sum_{j=1}^{n}|u_{ij}|^2 d_j(\lambda),$$

where $d_j(\lambda)$ denotes $(\lambda - \lambda_1)\cdots(\lambda - \lambda_n)$ with $(\lambda - \lambda_j)$ absent. Setting $\lambda = \lambda_j$ and diving both sides by $d_j(\lambda_j)$ gives the identity.

6.125 (a) It is routine to show that the conditions of a vector space are met.

(b) f is linear with respect to the first component x because

$$f(ax_1 + bx_2, y) \;=\; (ax_1 + bx_2)^t y = (ax_1^t + bx_2^t)y$$
$$=\; a(x_1^t y) + b(x_2^t y) = af(x_1, y) + bf(x_2, y).$$

Similarly, one can show that f is linear with respect to y. f is not a linear transformation from V to $\mathbb{R}$ because

$$f\big(k(x,y)\big) = f(kx, ky) = k^2 x^t y = k^2 f(x,y) \neq k f(x,y).$$

Likewise, g is linear with respect to x and y individually, but it is not linear overall with respect to both x and y.

(c) $g(x,y) = xy^t$ is a matrix of rank 0 or 1. So $\mathrm{Im}\,g$ is the set of all 2×2 real matrices of rank 0 or 1. It is not a subspace of $M_2(\mathbb{R})$. Take $u = (e_1, e_1) \in V$ and $v = (e_2, e_2) \in V$, where $e_1 = (1,0)^t$, $e_2 = (0,1)^t$. Then $g(u) = \left(\begin{smallmatrix}1&0\\0&0\end{smallmatrix}\right)$ and $g(v) = \left(\begin{smallmatrix}0&0\\0&1\end{smallmatrix}\right)$. $g(u) + g(v) = \left(\begin{smallmatrix}1&0\\0&1\end{smallmatrix}\right) \notin \mathrm{Im}\,g$. This also shows that $\mathrm{Im}\,g$ is not a subspace. Recall that if $\mathcal{L}$ is a linear map from a vector space V to a vector space W, then $\mathrm{Im}\,\mathcal{L}$ is a subspace of W.

Note: f and g are examples of multi-linear transformations.

6.126 Through permutations we may assume that the principal submatrix is located in the upper-left corner: $A = \left(\begin{smallmatrix}A_{11} & A_{12}\\A_{21} & A_{22}\end{smallmatrix}\right)$, where $A_{11} = A[\alpha]$.

(a) $B = \left(\begin{smallmatrix}B_{11} & B_{12}\\B_{21} & B_{22}\end{smallmatrix}\right)$. Then $(B^*B)[\alpha] = B_{11}^*B_{11} + B_{21}^*B_{21} \geq B_{11}^*B_{11}$.

(b) Set $B = P$ in (a).

(c) Set $B = P^{\frac{1}{2}}$ in (a), then take square roots.

(d) Since P is invertible, $P[\alpha] = P_{11}$ is invertible. We compute

$$\begin{pmatrix} I & 0 \\ -P_{21}P_{11}^{-1} & I \end{pmatrix} P \begin{pmatrix} I & -P_{11}^{-1}P_{12} \\ 0 & I \end{pmatrix} = \begin{pmatrix} P_{11} & 0 \\ 0 & \widetilde{P_{11}} \end{pmatrix},$$

where $\widetilde{P_{11}} = P_{22} - P_{21}P_{11}^{-1}P_{12}$ (which is positive definite). So

$$P^{-1} = \begin{pmatrix} I & -P_{11}^{-1}P_{12} \\ 0 & I \end{pmatrix} \begin{pmatrix} P_{11}^{-1} & 0 \\ 0 & (\widetilde{P_{11}})^{-1} \end{pmatrix} \begin{pmatrix} I & 0 \\ -P_{21}P_{11}^{-1} & I \end{pmatrix}.$$

Computing the (1,1) block on the right-hand side reveals

$$(P^{-1})[\alpha] = P_{11}^{-1} + P_{11}^{-1}P_{12}(\widetilde{P_{11}})^{-1}P_{21}P_{11}^{-1} \geq P_{11}^{-1} = (P[\alpha])^{-1}.$$

See also Example 4.12 and Problem 4.76.

6.127 Proof 1: $A \circ B$ is a principal submatrix of $A \otimes B > 0$.

Proof 2: By Example 4.9, $A, B \geq 0 \Rightarrow A \circ B \geq 0$. Now for $A, B > 0$, the diagonal entries as well as the eigenvalues of A and B are all positive. Let λ_n be the smallest eigenvalue of B. Then $B - \lambda_n I \geq 0$. Thus $A \circ (B - \lambda_n I) \geq 0$. This implies that $u^*(A \circ (B - \lambda_n I))u \geq 0$ for any nonzero $u = (u_1, \ldots, u_n)^t \in \mathbb{C}^n$. On the other hand,

$$\begin{aligned} u^*(A \circ (B - \lambda_n I))u &= u^*(A \circ B)u - \lambda_n u^*(A \circ I)u \\ &= u^*(A \circ B)u - \lambda_n \sum_{i=1}^{n} a_{ii}|u_i|^2. \end{aligned}$$

It follows that $u^*(A \circ B)u \geq \lambda_n \sum_{i=1}^{n} a_{ii}|u_i|^2 > 0$, that is, $A \circ B > 0$.

6.128 By Problem 3.112, $A \circ B$ is a principal submatrix of $A \otimes B$. Write $A \circ B = (A \otimes B)[\alpha]$. Using Problem 6.126, we obtain $(A \circ B)^{-1} = ((A \otimes B)[\alpha])^{-1} \leq ((A \otimes B)^{-1})[\alpha] = (A^{-1} \otimes B^{-1})[\alpha] = A^{-1} \circ B^{-1}$. Setting $B = A^{-1}$ reveals an interesting inequality: $A \circ A^{-1} \geq I$.

6.129 Note that $\begin{pmatrix} X^* \\ Y^* \end{pmatrix}(X, Y) = \begin{pmatrix} X^*X & X^*Y \\ Y^*X & Y^*Y \end{pmatrix} \geq 0$ and $\begin{pmatrix} Y^*Y & Y^*X \\ X^*Y & X^*X \end{pmatrix} \geq 0$.

Taking Hadamard product gives $\begin{pmatrix} (X^*X) \circ (Y^*Y) & (X^*Y) \circ (Y^*X) \\ (Y^*X) \circ (X^*Y) & (Y^*Y) \circ (X^*X) \end{pmatrix} \geq 0$.

Then add all the blocks in the matrix. See Problems 4.24 and 4.74:

$$(X^*X + Y^*Y) \pm (X^*Y + Y^*X) \geq 0,$$

$$(XX^*) \circ (YY^*) - (X \circ Y)(X^* \circ Y^*) \geq 0.$$

6.130 $\begin{pmatrix} A+B & X^*+Y^* \\ X+Y & XA^{-1}X^*+YB^{-1}Y^* \end{pmatrix} = \begin{pmatrix} A & X^* \\ X & XA^{-1}X^* \end{pmatrix} + \begin{pmatrix} B & Y^* \\ Y & YB^{-1}Y^* \end{pmatrix} \geq 0.$

Taking the Schur complement of $A+B$ gives the desired inequality.

6.131 Let the eigenvalues of a Hermitian matrix be ordered decreasingly.

(a) By the Cauchy eigenvalue interlacing theorem, $\lambda_i(H_1) \leq \lambda_i(H)$. If $\lambda_i(H_1) > 0$, then $\lambda_i(H) > 0$. So $i_+(H_1) \leq i_+(H)$. For negative eigenvalues, $i_-(H_1) = i_+(-H_1) \leq i_+(-H) = i_-(H)$.

(b) If $K \leq H$, then $\lambda_i(K) \leq \lambda_i(H)$. If $\lambda_i(K) > 0$, then $\lambda_i(H) > 0$. Thus, $i_+(K) \leq i_+(H)$. But $i_-(K) \leq i_-(H)$ need not be true. For example, if $H = I$, $K = -I$, then $i_-(K) = n$, $i_-(H) = 0$.

(c) Because $i_\pm(X^*HX) \leq r(X^*HX) \leq \min\{r(H), r(X)\}$.

(d) Assume that $H = \begin{pmatrix} I_p & 0 & 0 \\ 0 & -I_q & 0 \\ 0 & 0 & 0 \end{pmatrix} = \begin{pmatrix} I_p & 0 & 0 \\ 0 & 0 & 0 \\ 0 & 0 & 0 \end{pmatrix} - \begin{pmatrix} 0 & 0 & 0 \\ 0 & I_q & 0 \\ 0 & 0 & 0 \end{pmatrix} = P - Q,$ where $p = i_+(H)$, $q = i_-(H)$, $p+q = r(H)$, P and Q are positive semidefinite. Then $X^*HX \leq X^*HX + X^*QX = X^*PX$. Thus, $i_+(X^*HX) \leq i_+(X^*PX) \leq p = i_+(H)$. For negative eigenvalues, $i_-(X^*HX) = i_+(-X^*HX) \leq i_+(-H) = i_-(H)$.

(e) $i_0(H) = n - r(H) \leq n - r(X^*HX) \leq m - r(X^*HX) = i_0(X^*HX)$. Note that if $m < n$, then $i_0(H) \leq i_0(X^*HX)$ need not be true. Take $H = \begin{pmatrix} 1 & 0 \\ 0 & 0 \end{pmatrix}$ with $X = \begin{pmatrix} 1 \\ 1 \end{pmatrix}$. Then $X^*HX = (1)$. Thus, $i_0(X^*HX) = 0$, while $i_0(H) = 1$.

6.132 Using Problem 4.6 with $x = e$, $y = \sqrt{n}\,e_1$, and $w = \frac{x-y}{\|x-y\|}$, we get

$$P = I_n - 2ww^t$$

$$= I_n - \frac{1}{n-\sqrt{n}} \begin{pmatrix} 1-\sqrt{n} \\ 1 \\ \vdots \\ 1 \end{pmatrix} (1-\sqrt{n}, 1, \ldots, 1).$$

This matrix P is real (symmetric) and orthogonal. A simple computation reveals (a) $Pe = \sqrt{n}\,e_1$. With this, we have (b) $(Pe)(Pe)^t = Pee^tP^t = P^tEP = \begin{pmatrix} n & 0 \\ 0 & 0 \end{pmatrix}$. For (c), let $Q = \begin{pmatrix} 1 & 0 \\ 0 & H \end{pmatrix}$, where H is the previous P of order $n-1$. Then Q is a real (symmetric) orthogonal,

One computes to get $Q^tFQ = \begin{pmatrix} 1 & \sqrt{n-1} & 0 & \ldots & 0 \\ \sqrt{n-1} & 0 & 0 & \ldots & 0 \\ 0 & 0 & 0 & \ldots & 0 \\ \vdots & \vdots & \vdots & \ldots & \vdots \\ 0 & 0 & 0 & \ldots & 0 \end{pmatrix} = \begin{pmatrix} N & 0 \\ 0 & 0 \end{pmatrix}.$

Notes: We can also use the following matrix S in place of P:

$$S = I_n - \frac{1}{n+\sqrt{n}} \begin{pmatrix} 1+\sqrt{n} \\ 1 \\ \vdots \\ 1 \end{pmatrix} (1+\sqrt{n}, 1, \ldots, 1).$$

Apparently S is real symmetric. Write $u = (1+\sqrt{n}, 1, \ldots, 1)^t$. Then uu^t has rank one with nonzero eigenvalue $\operatorname{tr}(uu^t) = u^t u = 2(n+\sqrt{n})$. It follows that the eigenvalues of S are $-1, 1, \ldots, 1$. Thus, S is orthogonal. A simple computation shows $Se = \sqrt{n}\,e_1$. Moreover, since F is real symmetric, it is also unitarily (real orthogonally) diagonalizable. The diagonalization of $\left(\begin{smallmatrix} 1 & \sqrt{n} \\ \sqrt{n} & O \end{smallmatrix} \right)$ requires some computations.

See related Problems 3.51, 4.6, and 5.94. See Problem 3.16 for a matrix in the form $\left(\begin{smallmatrix} 1 & a \\ a & 0 \end{smallmatrix} \right)$. To find a 2×2 real orthogonal matrix that diagonalizes $\left(\begin{smallmatrix} 1 & a \\ a & 0 \end{smallmatrix} \right)$, normalize the orthogonal eigenvectors.

6.133 Let $f(x) = \sum_k a_k x^k$, where $a_k \geq 0$ for all k. Then $a_k A^k \geq 0$. Since any (finite) sum of positive semidefinite matrices is positive semidefinite, we obtain $f(A) \geq 0$. For the positivity of $\bigl(f(a_{ij})\bigr)$, recall that the entrywise (Hadamard) product of two positive semidefinite matrices is positive semidefinite (Example 4.9). We have

$$\bigl(f(a_{ij})\bigr) = \Bigl(\sum_k a_k a_{ij}^k\Bigr) = \sum_k a_k \bigl(a_{ij}^k\bigr) = \sum_k a_k A^{[k]} \geq 0,$$

where $A^{[k]} = \bigl(a_{ij}^k\bigr) = A \circ \cdots \circ A$ is the k-th Hadamard power of A.

6.134 Case I: $A = I$. Use the fact $\det(P+Q) \geq \det P + \det Q$ if $P, Q \geq 0$.

$$\begin{aligned}
\det M &= \det \left(\begin{pmatrix} I & 0 \\ -B^*A^{-1} & I \end{pmatrix} M \right) = \det \begin{pmatrix} A & B \\ 0 & C-B^*A^{-1}B \end{pmatrix} \\
&= \det A \, \det(C - B^*A^{-1}B) = \det(C - B^*B) \\
&\leq \det C - \det(B^*B) = \det A \det C - |\det B|^2.
\end{aligned}$$

In the above, $P = C - B^*B$, $Q = B^*B$. (See Problem 4.42.)

Case II: A is nonsingular. Convert it to Case I. Set $R = A^{-\frac{1}{2}}$. Then

$$\begin{aligned}
(\det A)^{-1} \det M &= \det \left(\begin{pmatrix} R & 0 \\ 0 & I \end{pmatrix} M \begin{pmatrix} R & 0 \\ 0 & I \end{pmatrix} \right) \\
&= \det \begin{pmatrix} I & RB \\ B^*R & C \end{pmatrix} \leq \det C - \det(RBB^*R) \\
&= \det C - (\det A)^{-1}|\det B|^2,
\end{aligned}$$

implying $\det M \leq \det A \det C - |\det B|^2$. (Note: One may combine cases I and II in one step. However, the case $A = I$ is more transparent and easy to deal with, and this is usually the way to get started.)

Case III: A is singular. Let $A_\epsilon = A + \epsilon I$, $\epsilon > 0$. By Case II, we have

$$\det \begin{pmatrix} A_\epsilon & B \\ B^* & C \end{pmatrix} \leq \det A_\epsilon \det C - |\det B|^2.$$

Letting $\epsilon \to 0^+$ reveals the desired determinantal inequality.

6.135   Let $f(A, B) = (\operatorname{tr} B)A + (\operatorname{tr} A)B - (AB + BA)$. Note that f is linear with respect to A and B individually, that is, f is bilinear.

By spectral decomposition (see (4.13), Chapter 4, p. 149), we write

$$A = \sum_{i=1}^n \lambda_i u_i u_i^*, \quad B = \sum_{j=1}^n \mu_j v_j v_j^*,$$

where u_i's and v_j's are orthonormal eigenvectors associated to the nonnegative eigenvalues λ_i's of A and μ_j's of B, respectively. Then

$$f(A, B) = f\left(\sum_i \lambda_i u_i u_i^*, \sum_j \mu_j v_j v_j^*\right) = \sum_{i,j} \lambda_i \mu_j f(u_i u_i^*, v_j v_j^*).$$

It is sufficient to show that $f(uu^*, vv^*) \geq 0$ for any vectors $u, v \in \mathbb{C}^n$. We may assume that u^*v is real; otherwise, replace v by $\tilde{v} = e^{i\theta}v$ such that $e^{i\theta}u^*v$ is real. Note that $f(uu^*, vv^*) = f(uu^*, \tilde{v}\tilde{v}^*)$. We derive

$$
\begin{aligned}
f(uu^*, vv^*) &= \operatorname{tr}(vv^*)uu^* + \operatorname{tr}(uu^*)vv^* - (uu^*vv^* + vv^*uu^*) \\
&= (v^*v)uu^* + (u^*u)vv^* - (u^*v)uv^* - (v^*u)vu^* \\
&= (v^*v)uu^* + (u^*u)vv^* - (v^*u)uv^* - (u^*v)vu^* \\
&= (uv^* - vu^*)(uv^* - vu^*)^* \geq 0.
\end{aligned}
$$

Note: Assuming that u^*v is real is an often used trick. Without the assumption, since $(u^*v)^t = v^t\bar{u}$, one can show the inequality directly via $(uv^t - vu^t)(uv^t - vu^t)^* \geq 0$. Alternatively, use Problem 5.15 in which appropriate x and y are chosen. Or derive the inequality from $(I, -v^*uI)\begin{pmatrix} uu^* & uv^* \\ vu^* & vv^* \end{pmatrix}\begin{pmatrix} I \\ -u^*vI \end{pmatrix} \geq 0$ for unit vectors u and v in $\mathbb{C}^n$.

Notation

$\mathbb{R}$	real numbers
$\mathbb{C}$	complex numbers
$\mathbb{Q}$	rational numbers
$\mathbb{F}$	a field, usually $\mathbb{F} = \mathbb{C}$ or $\mathbb{R}$ in this book
$\mathbb{R}^n$	(column) vectors with n real components
$\mathbb{C}^n$	(column) vectors with n complex components
$\mathbb{P}_n[x]$	real polynomials with degree less than n
$\mathbb{P}[x]$	real polynomials with any finite degree
i	$i = \sqrt{-1}$ if it is a complex number
	i like $j, k...$ etc is also used as an index
$\lvert c \rvert$	absolute value of complex number c
$\bar{c}$	conjugate of complex number c
$\operatorname{Re} c$	real part of complex number c
$\mathcal{C}[a, b]$	real-valued continuous functions on $[a, b]$
$\mathcal{C}(\mathbb{R})$	real-valued continuous functions on $\mathbb{R}$
$\mathcal{C}_\infty(\mathbb{R})$	real-valued functions of derivatives of all orders
(a) $\Rightarrow$ (b)	If (a) then (b)
(a) $\Leftrightarrow$ (b)	(a) if and only if (b)
$S \setminus T$	set minus, i.e., it contains all elements in S not in T
$e_1, \ldots, e_n$	standard basis for $\mathbb{R}^n$ or $\mathbb{C}^n$
$M_{m \times n}(\mathbb{F})$	$m \times n$ matrices with entries in $\mathbb{F}$
$M_n(\mathbb{F})$	$n \times n$ matrices with entries in $\mathbb{F}$
$\frac{df}{dt}, '$	derivative of f with respect to t
y''	second derivative of y
I_n, I	identity matrix of size $n \times n$ (or inferred size from context)
$\dim V$	dimension of vector space V
$W_1 + W_2$	sum of subspaces W_1 and W_2
$W_1 \oplus W_2$	direct sum of subspaces W_1 and W_2
$\operatorname{Span} S$	vector space spanned by the elements in S
$A, B, \ldots$	matrices
$\mathcal{A}, \mathcal{B}, \ldots$	linear transformations (in calligraphy)
E_{ij}	matrix with the (i, j)-entry 1 and 0 elsewhere
$A = (a_{ij})$	matrix A with (i, j)-entry a_{ij}
$\lvert A \rvert$	determinant of matrix A (mostly for block matrices)
$\det A$	determinant of matrix A
$r(A)$	rank of matrix A
$\operatorname{tr} A$	trace of matrix A
A^t, v^t	transpose of matrix A, vector v
$\bar{A}$	conjugate of matrix A

452

A^*, v^*	conjugate transpose of matrix A, vector v
$A^{-1}, \mathcal{A}^{-1}$	inverse of matrix A, linear map $\mathcal{A}$
$\mathrm{adj}(A)$	adjugate or cofactor matrix of matrix A
$A[\alpha]$	principal submatrix of A with rows (columns) indexed by α
$\mathfrak{R}(A)$	$\frac{A+A^*}{2}$
$\mathfrak{I}(A)$	$\frac{A-A^*}{2i}$. Cartesian decomposition: $A = \mathfrak{R}(A) + i\mathfrak{I}(A)$
$\mathrm{diag}(\lambda_1, \ldots, \lambda_n)$	diagonal matrix with $\lambda_1, \ldots, \lambda_n$ on the main diagonal
$\mathrm{Ker}\,A$	kernel or null space of matrix A, i.e., $\mathrm{Ker}\,A = \{x \mid Ax = 0\}$
$\mathrm{Ker}\,\mathcal{A}$	kernel of transformation $\mathcal{A}$, i.e., $\mathrm{Ker}\,\mathcal{A} = \{x \mid \mathcal{A}(x) = 0\}$
$\mathrm{Im}\,A$	image, range, or column space of matrix A, i.e., $\mathrm{Im}\,A = \{Ax\}$
$\mathrm{Im}\,\mathcal{A}$	image or range of transformation $\mathcal{A}$, i.e., $\mathrm{Im}\,\mathcal{A} = \{\mathcal{A}(x)\}$
$F(A)$	numerical range or field of values of A, i.e., $\{x^*Ax \mid \|x\| = 1\}$
$|\lambda I - A|$	characteristic polynomial of A
$\lambda_i(A)$	eigenvalues of A (usually decreasingly ordered if all real)
$\sigma_i(A)$	singular values of A (usually decreasingly ordered)
$\lambda_{\max}(A), \lambda_{\min}(A)$	largest, smallest eigenvalues of matrix A
$\sigma_{\max}(A), \sigma_{\min}(A)$	largest, smallest singular values of matrix A
$\|x\|$	length of $x \in \mathbb{C}^n$, i.e., $\|x\| = \sqrt{x^*x}$ (unless otherwise stated)
$\|v\|$	(induced) norm of vector v, i.e., $\|v\| = \sqrt{\langle x, x \rangle}$
$\|A\|_F$	Frobenius norm of matrix A, i.e., $\|A\|_F = (\sum_{i,j} |a_{ij}|^2)^{\frac{1}{2}}$
$\|A\|_{\mathrm{sp}}$	spectral norm of matrix A, i.e., $\|A\|_{\mathrm{sp}} = \sigma_{\max}(A)$
$\|A\|_{\mathrm{op}}$	operator norm of matrix A, i.e., $\|A\|_{\mathrm{op}} = \max_{\|x\|=1} \|Ax\|$
$\|A\|_2$	2-norm of matrix A (note: $\|A\|_2 = \|A\|_{\mathrm{op}} = \|A\|_{\mathrm{sp}} = \sigma_{\max}(A)$)
$A \geq 0$	A is a positive semidefinite (Hermitian) matrix
$A \geq B$	$A - B \geq 0$, where A and B are Hermitian
$A > 0$	A is a positive definite matrix
$A^{\frac{1}{2}}$	positive semidefinite square root of matrix $A \geq 0$
$\ell(A)$	the modulus of A, i.e., $\ell(A) = (A^*A)^{\frac{1}{2}}$
$[A, B]$	commutator $AB - BA$
$A \circ B$	Hadamard (Schur) product of A and B, i.e., $A \circ B = (a_{ij}b_{ij})$
$A \otimes B$	Kronecker product of A and B, i.e., $A \otimes B = (a_{ij}B)$
$A \oplus B$	direct sum of matrices A and B, i.e., $A \oplus B = \left(\begin{smallmatrix} A & 0 \\ 0 & B \end{smallmatrix}\right)$
δ_{ij}	Kronecker delta
$\mathcal{A}^*, \mathcal{B}^*, \ldots$	adjoints of linear transformations $\mathcal{A}, \mathcal{B}, \ldots$
$\mathcal{A}(W)$	$\{\mathcal{A}(w) \mid w \in W\}$
$\langle \cdot, \cdot \rangle$	inner product
$u^\perp$	space of the vectors that are orthogonal to vector u
$W^\perp$	space of the vectors that are orthogonal to all vectors in W
$W \perp V$	W and V are mutually orthogonal
V^*	dual space of the vector space V
$\mathcal{P}$	orthogonal projection
$\mathrm{Proj}_v(u)$	projection of u onto v
$\left|\begin{smallmatrix} A & B \\ C & D \end{smallmatrix}\right|$	determinant of the block matrix $\left(\begin{smallmatrix} A & B \\ C & D \end{smallmatrix}\right)$

Main References

- Barreira L., and C. Valls, *Exercises in Linear Algebra*. World Scientific, 2016.

- Bhatia R. and C. Davis. *A Better Bound on the Variance*. The American Mathematical Monthly, Vol. 107, No. 4 (April, 2000), pp. 353–357.

- Carlson D., C. Johnson, D. Lay, and A. Porter. *Linear Algebra Gems*. Mathematical Association of America, 2002.

- de Souza P. N., and J.-N. Silva, *Berkeley problems in mathematics*. Springer, 2004 (3rd ed). https://math.berkeley.edu/programs/graduate/prelim-exams.

- Denton P., S. Parke, T. Tao, and X. Zhang. *Eigenvectors from Eigenvalues: a survey of a basic identity in linear algebra*. DOI: https://doi.org/10.1090/bull/1722, Bull. Amer. Math. Soc., 2021.

- Garcia S. R., and R. A. Horn. *Second Course in Linear Algebra*. Cambridge University Press, 2017.

- Horn R. A., and C. R. Johnson. *Matrix Analysis*. Cambridge University Press, 2013 (2nd ed).

- Li T.-T. (editor). *Problems and Solutions in Mathematics*. World Scientific, 2011 (2nd ed).

- Li Z.-L. (editor). *Beijing Normal University Graduate Entrance Exams in Mathematics (1978–2007)*. Beijing Normal University Press, 2007.

- Marcus M., and H. Minc. *A Survey of Matrix Theory and Matrix Inequalities*. Dover, 1992.

- Marshall A. W., I. Olkin, and B. C. Arnold. *Inequalities: Theory of Majorization and Its Applications*. Springer, 2011 (2nd ed).

- Ou W.-Y., C.-X. Li, and P. Zhang. *Graduate Entrance Exams in Math with Solutions*. Jilin University Press, 1998 (in Chinese).

- Qian J.-L. *Selected Problems in Higher Algebra*. Beijing Central University of Nationalities Press, 2002 (in Chinese).

- Shi M.-R. *600 Linear Algebra Problems with Solutions*. Beijing Press of Science and Technology, 1985 (in Chinese).

- Wang B.-Y. *Foundations of Majorization Inequalities*. Beijing Normal University Press, 1990 (in Chinese).

- Wigner, E.P. *On Weakly Positive Matrices*. Canadian Journal of Mathematics, Volume 15, 1963, pp. 313–317.

- Yanovsky I. *Linear Algebra: Graduate Level Problems and Solutions* (2005). https://www.math.ucla.edu/ yanovsky/handbooks/linear-algebra.pdf.

- Zhang F. *Matrix Theory: Basic Results and Techniques*. Springer, 2011 (2nd ed).

Postface – A Few Open Problems

Traditional linear algebra is composed of basic theory on vector spaces of finite dimensions and matrix algebra. The problems collected in the book are in this area; some problems are easy, most are at moderate difficulty.

There are many unsolved problems concerning matrices. These problems are neither linear nor algebraic (in the traditional sense); and they have great interests and applications in a variety of mathematics and scientific areas. New ideas and machinery may be needed to solve the problems. We present a few that are easily stated and understood to feed the reader.

Nonnegative Inverse Eigenvalue Problem (NIEP)

Given a set of n real numbers $\lambda_1, \ldots, \lambda_n$, does there exist an $n \times n$ matrix with nonnegative entries whose eigenvalues are exactly $\lambda_1, \ldots, \lambda_n$? In other words, find necessary and sufficient conditions for the existence of an $n \times n$ nonnegative matrix whose spectrum is the prescribed set.

This is a hard, long-standing open problem. (See R. Loewy, *Some additional notes on the spectra of non-negative symmetric 5×5 matrices*, Electronic J. of Linear Algebra, Volume 37 (January 2021) 1–13.)

Chollet Conjecture

Recall that the determinant of an $n \times n$ matrix $A = (a_{ij})$ is defined by $\det(A) = \sum_{p \in S_n} \text{sgn}(p) \prod_{t=1}^{n} a_{tp(t)}$, where S_n is the symmetric group on $\{1, \ldots, n\}$ and $\text{sgn}(p) = 1$ if p is an even permutation or -1 if p is an odd permutation. By removing $\text{sgn}(p)$, we get the so-called *permanent* of A:

$$\text{per}(A) = \sum_{p \in S_n} \prod_{t=1}^{n} a_{tp(t)}.$$

In view of the classical result (of Oppenheim) $\det(A \circ B) \geq \det(A) \det(B)$, where $A = (a_{ij})$ and $B = (b_{ij})$ are $n \times n$ positive semidefinite matrices, and $A \circ B = (a_{ij} b_{ij})$ is the entrywise (also known as Hadamard or Schur) product of A and B, Chollet (J. Chollet, *Is there a permanental analogue to Oppenheim's inequality?* Amer. Math. Monthly 89, No. 1 (1982) 57–58.) proposed a question about the permanent. He conjectured that

$$\text{per}(A \circ B) \leq \text{per}(A) \, \text{per}(B).$$

Lieb Permanent Dominance Conjecture

With a little knowledge of groups, a problem on permanent follows.

Let H be a subgroup of S_n and let χ be a character of degree m of H. It is conjectured that for any $n \times n$ positive semidefinite matrix A,

$$\frac{1}{m} \sum_{p \in H} \chi(p) \prod_{i=1}^{n} a_{ip(i)} \leq \operatorname{per}(A).$$

(E. Lieb, *Proofs of some conjectures on permanents*, J. Math. and Mech. Vol. 16, No. 2 (1966) 127–134. For more open questions, see F. Zhang, *An update on a few permanent conjectures*, Spec. Matrices 4 (2016) 305–316.)

Hadamard Matrix Conjecture

Probably the most well-known and extremely difficult open problem (in combinatorics) concerning matrices is the Hadamard matrix conjecture. A *Hadamard matrix* of order n is an $n \times n$ matrix, say H, whose entries are either $+1$ or -1 and whose rows are mutually orthogonal, i.e., $HH^t = nI$. It is known that n must be 1, 2 or a multiple of 4 for such a matrix to exist.

Example: $H_4 = \begin{pmatrix} 1 & 1 & 1 & 1 \\ 1 & -1 & 1 & -1 \\ 1 & 1 & -1 & -1 \\ 1 & -1 & -1 & 1 \end{pmatrix}$ is a Hadamard matrix of order 4.

Question: If n is a positive integer divisible by 4, is there a square matrix H of order n, with entries $+1$ or -1, such that $HH^t = nI$? That is, does a Hadamard matrix of order $4k$ exist for every positive integer k?

One can find online an enormous list of references on the topic.

The readers who are interested in the study of matrices may find the following comprehensive monographs helpful: Horn and Johnson's *Matrix Analysis* and *Topics in Matrix Analysis* (Cambridge University Press).

Fuzhen Zhang
(张福振)
zhang@nova.edu
zfznova@yahoo.com
Fort Lauderdale, Florida

Index

"Mathematics is not a spectator sport. Actively working on well-chosen problems is the best way to understand and master any area of mathematics. Students at all levels will benefit from putting pencil to paper and solving Professor Zhang's linear algebra & matrix problems."

– Roger Horn, University of Utah

"This is a large collection of good problems of varied levels of difficulty. They offer readers many opportunities for thinking of clever ways to apply classic theorems."

– Jane Day, San Jose State University

"I had thoroughly enjoyed the first edition of Professor Zhang's delightful volume of problems when I was a student. When the second edition appeared, I used some of its problems for my Linear Algebra class at Berkeley. This volume is clearly a labor of love, enriched by Professor Zhang's years of editorial service on the problem sections of MAA's *American Mathematical Monthly* and ILAS's *IMAGE*. I cannot wait to see the third edition."

– Lek-Heng Lim, University of Chicago

"When I was a student, I suffered from the lack of sufficiently interesting exercises on linear algebra: the problems in my textbooks were sometimes boring, with a lot of calculation but little need of imagination. This book offers a remedy for this situation and is therefore very welcome."

– Fabio Mainardi, MAA-reviews for the 2nd edition

Fuzhen Zhang is a professor of Mathematics at Nova Southeastern University, Fort Lauderdale, Florida, USA. He received his Ph.D. in Mathematics from the University of California at Santa Barbara (UCSB), M.S. from Beijing Normal University, and B.Sci. from Shenyang Normal University (China). He is the author of *Matrix Theory: Basic Results and Techniques* and the editor of *The Schur Complement and Its Applications*.

Printed in the United States
by Baker & Taylor Publisher Services